D1219159

# Plant-Cast Precast and Prestressed Concrete

# Plant-Cast Precast and Prestressed Concrete

David A. Sheppard

William R. Phillips

**McGraw-Hill Publishing Company**

New York St. Louis San Francisco Auckland Bogotá
Caracas Hamburg Lisbon London Madrid Mexico
Milan Montreal New Delhi Oklahoma City
Paris San Juan São Paulo Singapore
Sydney Tokyo Toronto

Library of Congress Cataloging-in-Publication Data

Sheppard, David A.
Plant-cast precast and prestressed concrete.

Phillips's name appears first in earlier editions.
Includes index.
1. Precast concrete. 2. Prestressed concrete.
I. Phillips, William R. II. Title.
TP885.A2S444 1989 624.1′8341
ISBN 0-07-056760-3 89-12388



1234567890 DOC/DOC 895432109

ISBN 0-07-056760-3

*The editors for this book were Joel Stein and Beatrice E. Eckes, the designer was Naomi Auerbach, and the production supervisor was Richard Ausburn. It was set in Century Schoolbook by Frank Scheuer.*

*Printed and bound by R.R. Donnelley & Sons Company.*

# Contents

# Preface

This design guide on plant-cast precast and prestressed concrete is the result of a project conceived and funded by the Prestressed Concrete Manufacturers Association of California (PCMAC).

The first edition was developed by California Polytechnic State University at San Luis Obispo, in collaboration with certain key industry personnel representing PCMAC, and was published in 1977. Since it was desired to have the text serve equally for practicing design professionals and for students, extensive technical editing of the first edition was required. The resulting second edition, published in 1980, was used mainly in support of educational programs for students, architects, and engineers conducted by the Prestressed Concrete Manufacturers Association of California. The current third edition is a result of a combination of updating to current Uniform Building Code provisions, addition of new material, and a major "face lifting" by McGraw-Hill. The purpose of the text, however, remains as originally conceived in 1976: to introduce the student, architect, or engineer to precast concrete and its potential in the constant search for efficiency and form and to eliminate some of the current mystery and misconceptions surrounding its design and use.

The result is a practical and thorough text covering the items one must know in order to design plant-cast precast and prestressed concrete in a "how to do it" fashion. In Parts 1 and 2, a good general discussion of the mechanics and advantages of precast concrete production and use is given, followed by presentation of related design aspects of the material, as well as detailed design procedures and the present state of the art of overall structural design with precast and prestressed concrete as permitted by the Uniform Building Code. Part 3 covers the design of components, with an emphasis on production and erection considerations that must be understood for efficient design. Finally, in Part 4, a discussion of the importance of good specifications is presented, as well as new material on glass-fiber reinforced concrete (GFRC), metric design, additional material on seismic design, and computer programming applications.

Overall, the text tells the reader what he or she really needs to know to design prestressed and precast concrete: what is known prior to beginning the design; what preliminary design aids are required to eliminate wasted time with unnecessary trial designs; what stresses are critical and why (and which ones can usually be ignored); and, finally, what are inherent pitfalls that we should warn the reader about in the design of a particular component (camber, deflection, connections, creep effects, loss of prestress, effects of extreme seismic disturbance, temporary handling stresses, erection considerations, tolerances, etc.).

This is the real significance of the text. The various existing design manuals and design texts presuppose a knowledge of the essential practical aspects of plant-cast precasting which for the most part is lacking at both the student level and the level of the practicing professional. This book bridges the gap between theory and practice.

As a prerequisite to using this text, it is mandatory that the student or designer be familiar with the principles of basic reinforced-concrete design. For this reason, the elements of ultimate strength theory related to flexure and shear design are not covered. However, they are treated in example form in the component design sections, as well as in the *PCI Design Handbook*.

In closing, we would like to dedicate this book to Lloyd Compton, president of L.A. Compton Group in Visalia, California: Owing to his vision, prescience, and direction, the original project to develop the text was conceived, funded, and implemented.

William R. Phillips
*Los Osos, California*

David A. Sheppard
*Sonora, California*

Part

# 1

# Introduction

**Figure 1.1** Beneficial Standard Life Insurance Company Building, Los Angeles. Sierra white granite aggregate was revealed by deep sandblasting in the plant. These handsome exterior T-shaped units also support vertical floor loads for this building, constructed in 1966. *(Photograph by Eugene Phillips.)*

Chapter

# 1

# Precast Concrete: A Material and a Method

Precast concrete: what is it? Basically, precast concrete is defined as concrete which is cast in some location other than its final position in the finished structure. Precast concrete elements are reinforced either with mild steel reinforcing bars or with prestressing strands. When prestressing is employed for the production of precast concrete members, the method generally used is pretensioning, in which the strands are tensioned prior to pouring the concrete in long lines in the precasting operation. Prestressing is discussed in greater detail in Chap. 16, "Precast Prestressed Concrete."

Precast concrete is both a construction method and a construction material. Precast concrete is produced under rigid quality control conditions in a precasting plant. The concrete strengths used range from 4000 to 6000 lb/in$^2$, with the higher strengths being preferred to ensure durability and high cycle production rates in the plant. The forms used are of better quality than those normally used for cast-in-place concrete; hence, truer shapes and better finishes are obtained. Cast-in-place concrete requires more formwork, and forms can be reused only up to 10 times. For precast concrete, finished wood and fiberglass forms may be used up to 50 times with minor rework; concrete and metal forms have practically unlimited service lives. With precast concrete, the architect is offered greater freedom in design with a lower total form and product cost, given sufficient project size and repetition of units.

By placing the forms on a vibrating table or by placing external vibrators on the forms, a higher degree of consolidation is obtained. Along with a lower water-cement ratio and higher cement contents, this results in higher strengths and truer surfaces and corners.

Repetitious casting of elements also lends itself to pretensioning, saving reinforcing and ensuring crack control during handling and under service loads.

In using precasting, there is greater control of surface texture, allowing the achievement of many finishes which are not easily obtainable with cast-in-place concrete.

As a material, precast concrete may be used either as cladding or structurally.

As a method of construction, precast concrete can greatly reduce total project construction time since the units or components are cast and stockpiled while other phases of the building process are performed. By using steam or controlled curing methods, precast concrete can be cast in the afternoon and cured at elevated ambient temperatures for 1 to 10 hours during the night, permitting the units to be removed from the forms the next morning and the forms to be readied for reuse the same day. With steam or controlled curing the concrete can reach two-thirds of its ultimate strength in 10 to 14 hours, whereas cast-in-place concrete forms usually cannot be removed in less than 7 days without reshoring.

In using precast concrete, continuous uninterrupted erection of components is possible, quickly forming the structural frame and enclosing the building. Precast concrete also lends itself to fast-track construction methods. This total saving in time equates to lower interest paid on the construction loan, earlier occupancy, and quicker return on the owner's investment.

Precast concrete is a unit system of construction wherein a number of identical or similar components are assembled to produce the total building or structure. The precast components may be prestressed piling, single-story or multistory precast columns, prestressed concrete girders, single-T, double-T, hollow-core plank, or solid-slab floor or roof members, and wall panels consisting of solid units, single or double-T members, or insulated sandwich units. The wall units may be load-bearing or non-load-bearing. Other types of precast components are curtain wall cladding panels, sunshades, complete cell units such as hotel rooms or apartment units of one or more rooms, and architectural precast concrete units serving as fire-protective covering for steel columns and spandrel beams or as forms for poured-in-place concrete. The selection of other types of precast units is limited only by the imagination of the architect and the mold maker's skill.

At this point, it should be stated that a project should be of a certain minimum size in order to spread out certain fixed costs such as plant setup, mold costs, and erection mobilization costs over a sufficiently broad base to make the use of precast concrete economically viable. Minimum project sizes are those that would generate 10,000 $ft^2$ of architectural precast concrete panels, 15,000 $ft^2$ of prestressed concrete deck members, or 1000 $ft^2$ of standard prestressed or precast concrete components such as girders, columns, or piling.

The structural system is usually a shear wall lateral-load-resisting system; however, a ductile moment-resisting frame is made possible by using precast shell units and poured-in-place concrete. The actual system used depends on the functional requirements of a particular project.

The total structural system may consist of precast concrete or of a combination of precast concrete and either structural steel or poured-in-place concrete. The components may be structural prestressed, architectural precast, or a combination of the two. The potential of architectural expression is unlimited.

The feasibility of the systems building concept is due to a large number of identical parts, leading to industrialized production, combined with a guaranteed long-term market for the system. The economic advantage is fully realized only if the planning rules of modular coordination with many like units are observed. The component size limit is usually controlled by transportation and erection considerations.

The coarse aggregates used may be either lightweight or normal-weight, subject to local availability and cost. Where an exposed aggregate surface is desired, the same aggregate can be used throughout the section. Alternatively, the section can be composed of two mixes, with the more expensive special aggregate used in the portion to be exposed and a standard structural concrete mix used for the backup.

Both 40- and 60-ksi yield strength reinforcing bars are used; 40-ksi yield steel is usually employed for no. 3 and no. 4 bars, and 60-ksi yield steel is used for no. 5 bars and larger. Prestressing strand of 270-ksi specified ultimate strength is normally used. Strand is available in $^3/_8$-, $^7/_{16}$-, and $^1/_2$-in diameters. Where a smaller strand is used, say, $^1/_4$- or $^5/_{16}$-in, then grade 250-ksi is used. Welded wire fabric is commonly used to reinforce both structural and architectural precast concrete elements. Cement may be natural gray, white, or colored with an additive. Concrete may also be colored by adding predetermined amounts of color additives when the concrete is batched.

A number of sketches and project applications will be used in the initial chapters to illustrate applications and practical aspects of precast and prestressed concrete manufacture and installation. Following these are chapters that discuss design considerations which help in understanding the uniqueness of precast concrete technology.

Chapter

# 2

# Typical Framing with Standard Components

Certain standard precast and prestressed concrete structural shapes are made by precast concrete manufacturers. These components are used together to form structures which fall into two general categories:

1. Frame structures
2. Shear wall structures

Various combinations of these systems form subcategories that allow the architect to optimize the function of the building and at the same time to minimize the architectural constraints imposed by the structural lateral-force-resisting system of the building. The Uniform Building Code currently permits the use of precast concrete elements to resist seismic lateral forces, provided the design and detailing used satisfy the code requirements for cast-in-place concrete. To date, most structures have used the precast elements as "pin-ended" non-lateral-load-resisting elements, with the lateral forces being resisted by other methods of construction, such as masonry or poured-in-place concrete shear walls. The use of the vertical precast or prestressed concrete elements to resist lateral forces has been limited to situations in which the load transfer is direct from the precast concrete element to the supporting foundation or the resisting soil medium, such as for double-T wall panels in industrial building construction and for bearing or sheet piling.

In this chapter, we shall first see how the basic building blocks of the precast concrete industry are put together in buildings, and then we shall briefly survey seismic design of precast prestressed concrete buildings as it is related to current design concepts used for cast-in-place concrete and indicated areas of required research.

In Fig. 2.1 are shown the basic standard structural shapes to build structures in precast concrete. Whenever possible, the designer should attempt to use the standard shapes and sizes available in the region of the jobsite. The resulting savings in mold costs will reduce product cost. Just about any structural configuration may be made by using these shapes in combination with one another. In addition to normal deck systems consisting of T's or slabs, "spread" deck systems are often used. These feature channels or T's spread apart with hollow-core plank or solid slabs spanning transversely between the stemmed members. Alternative cross sections should be permitted in bidding so that particular manufacturers can bid their most competitive sections.

| | Approximate size ranges | | |
|---|---|---|---|
| | Width | Depth | Span |
| Double T 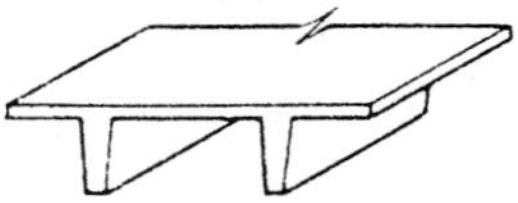 | 4–12 ft | 10–41 in | 30–90 ft |
| Single T 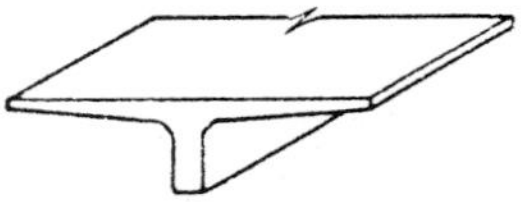 | 6–12 ft | 16–48 in | 30–110 ft |
| Channel slab 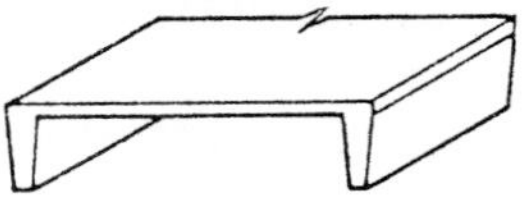 | 6–12 ft | 24–42 in | 40–90 ft |
| Flat slab 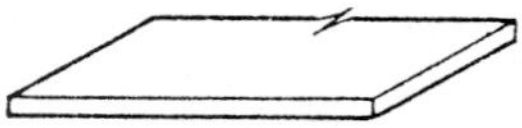 | 8–12 ft | 3–6 in | 14–22 ft (35 ft with shoring) |
| Hollow-core plank 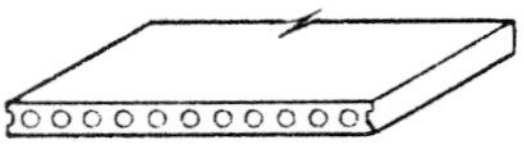 | 3 ft 4 in–8 ft | 6–12 in | 16–42 ft |
| Rectangular girder 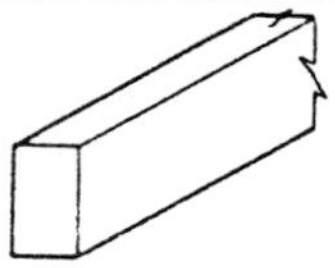 | 12–36 in | 18–48 in | 24–70 ft |

**Figure 2.1** Standard structural shapes.

| | Approximate size ranges | | |
|---|---|---|---|
| | Width | Depth | Span |
| Inverted-T girder | 12–24 in | 18–48 in | 24–48 ft |
| Ledger beam | 12–30 in | 18–48 in | 24–48 ft |
| Column | 10–24 in | 12–24 in | |
| Bearing pile | 12–24 in | 12–24 in | |
| Sheet pile | 4–8 ft | 10–16 in | |

**Figure 2.1** Standard structural shapes.*(Continued)*

In Chap. 7, the selection of a module is discussed. The hypothetical building in Fig. 2.2 illustrates the concept of using standard components on an 8-ft module. For the two-story structure, only every other interior column is taken up to the roof, since the girders supporting the relatively light roof loads can span a much greater distance than the girders supporting the heavier floor loads.

The exterior wall panels are 8 ft wide in both directions, making it mandatory that the 8-ft module be held in the direction of deck member span as well as in the modular width direction of the deck member. When the exterior elements enclosing the building in the deck member span direction are nonmodular, the module in the deck member span is 1 ft. If the wall panels on the ends of the building are designed as non-load-bearing cladding panels, ledger beams are required to support the ends of the double T's and carry the loads to the columns. This situation also allows the panels to be nonmodular in width if required. Load-bearing panels would normally be designed in a modular width, with corbels on the inside face to support the double-T stems.

On the T floor and roof, a structural topping is generally used to form the horizontal diaphragm to transmit wind and seismic lateral forces to vertical or lateral load-resisting elements in the structure, which in this case are the exterior wall panels. The prestressed concrete deck elements may also be used without a topping, in which case the diaphragm may be formed by interconnecting units together with flange weld plates or by tying the units together with poured-in-place concrete closures and shear friction reinforcing. The interior girders may be either inverted-T girders, with the T's resting on the ledges of the girder, or rectangular with the T's sitting on top. The columns are usually made in one piece, with corbels to support girders framing in at intermediate levels. The wall panels are supported on grade beams spanning between spread footings or on continuous-strip footings. Where prestressed concrete piling is used, the footing is called a pile cap. The wall panels are interconnected by flange weld plates and are also connected to the supporting footing by welded or grouted connections, as well as being connected to the poured-in-place floor slab. Single or double T's may also be used as wall panels, with openings for fenestration being placed between the T stems in the flanges.

If spread deck systems are used, the module is dependent upon the standard width of the infill product spanning between the spread-apart supporting channels, single T's, or double T's. Among several types of supporting elements used with spread channels are double columns (one at each channel stem), tree columns (tree arms support channels), and solid masonry or precast wall columns, designed to take lateral forces as well as the floor vertical loads.

As stated at the beginning of this chapter, the Uniform Building Code requires that precast concrete elements or systems may provide lateral force resistance, provided that the resulting construction and confinement details conform to those required for cast-in-place concrete. In addition, the connections must be monolithic near points of maximum stress, capable of development to cause the building to act as a single, integral unit, and detailed to eliminate brittle-type failures. Precast concrete construction has the potential to provide part or all of the building structure by employing three basic design philosophies:

1. The precast and prestressed concrete elements function primarily as forms and/or "pin-ended" structural elements and reduce or eliminate the need for temporary shoring and forming. The precast concrete units remain as an integral part of the final construction and contain reinforcing to carry vertical loads (or perpendicular wind or seismic forces on panel elements) to independent lateral-load-resisting systems through diaphragms or properly designed connectors. The precast and prestressed concrete elements do not form part of the lateral-load-resisting system in the building.

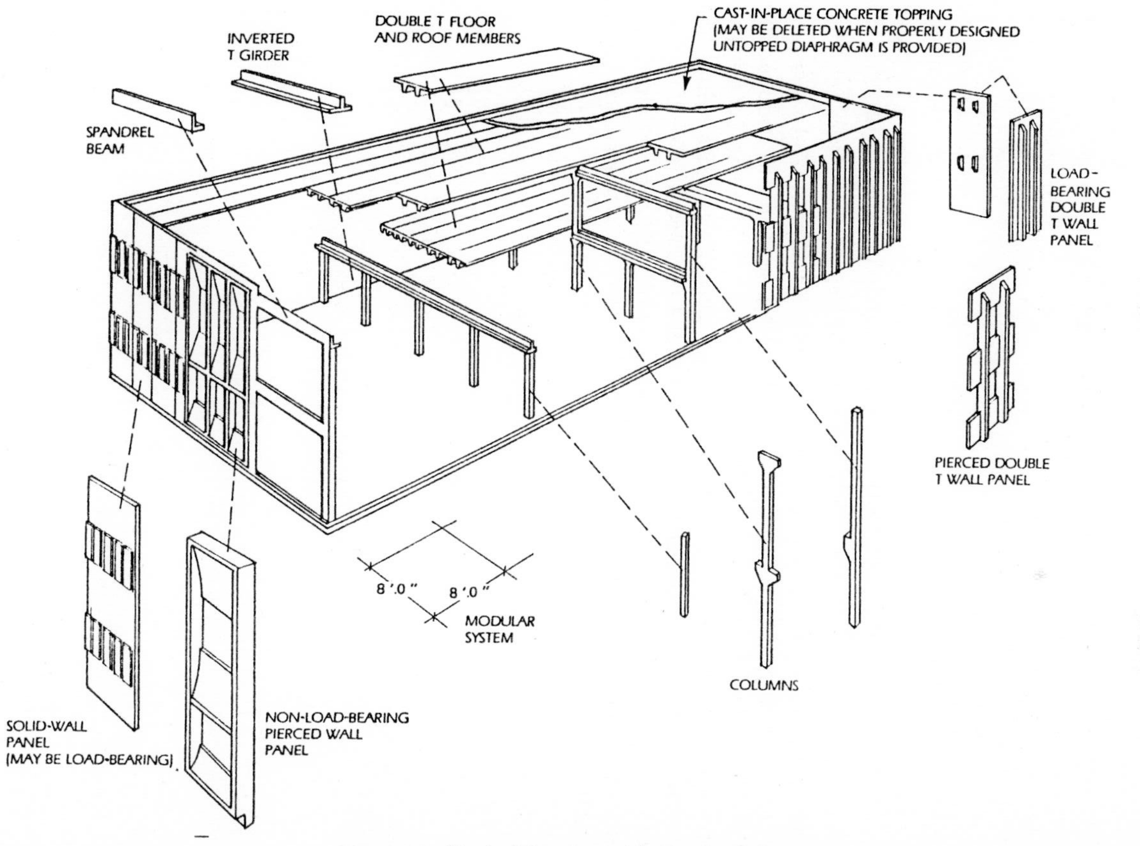


**Figure 2.2** Typical framing with standard shapes.

2. The precast and prestressed concrete elements form a structural system unto itself by supporting vertical loads as well as by serving to form the lateral-load-resisting system in transmitting these forces to the foundation. The individual precast concrete pieces are integrated into a total structure by various types of mechanical or grouted connection devices. Cast-in-place concrete is not normally a part of the system except when it is used to simplify certain connections or ties between elements.

3. Precast and prestressed concrete elements are used in various combinations of the two previous concepts by taking advantage of the positive aspects of both systems. For example, precast concrete shear walls with mechanical connectors may be combined with cast- in-place concrete ductile moment-resisting frames to provide the required total damping and energy dissipation to accommodate cyclic seismic loading.

Unfortunately, there are very few test data to substantiate the performance of connections under cyclic loading such as is induced by earthquake ground motions. In Figs. 2.3 through 2.7 we shall comment upon the seismic design of the indicated structure types as related to current design concepts used for cast-in-place concrete. We shall also indicate potential areas of application for prestressed concrete in using design philosophies 2 and 3 and will allude to indicated required research.

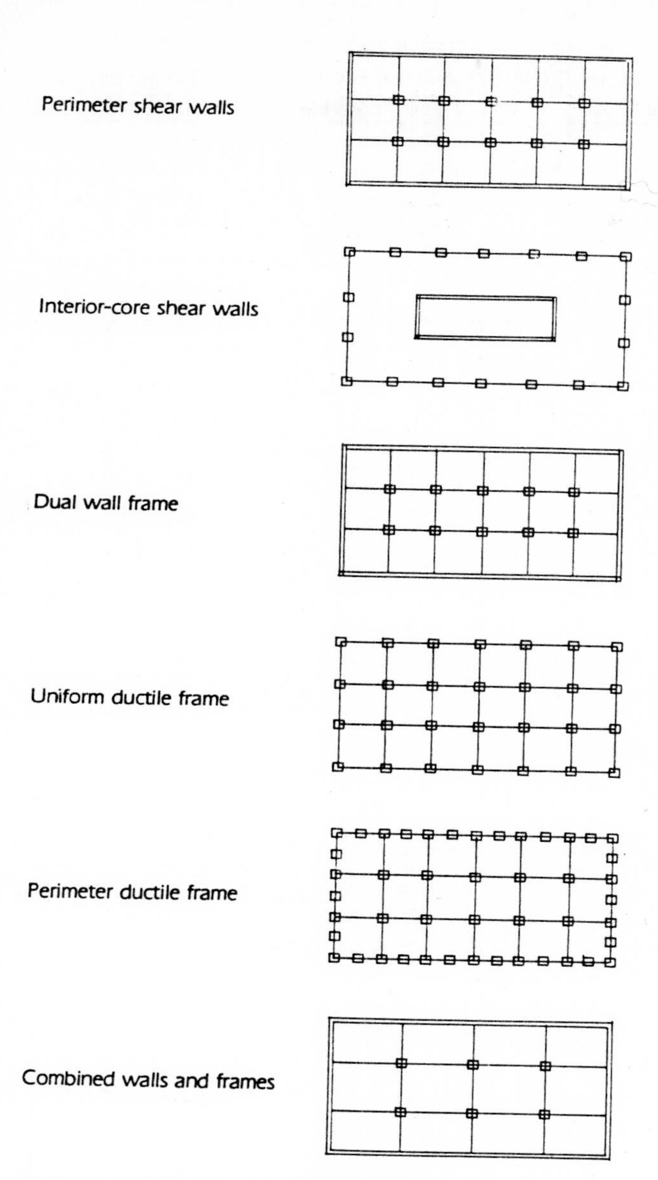


**Figure 2.3** Basic structural configurations.

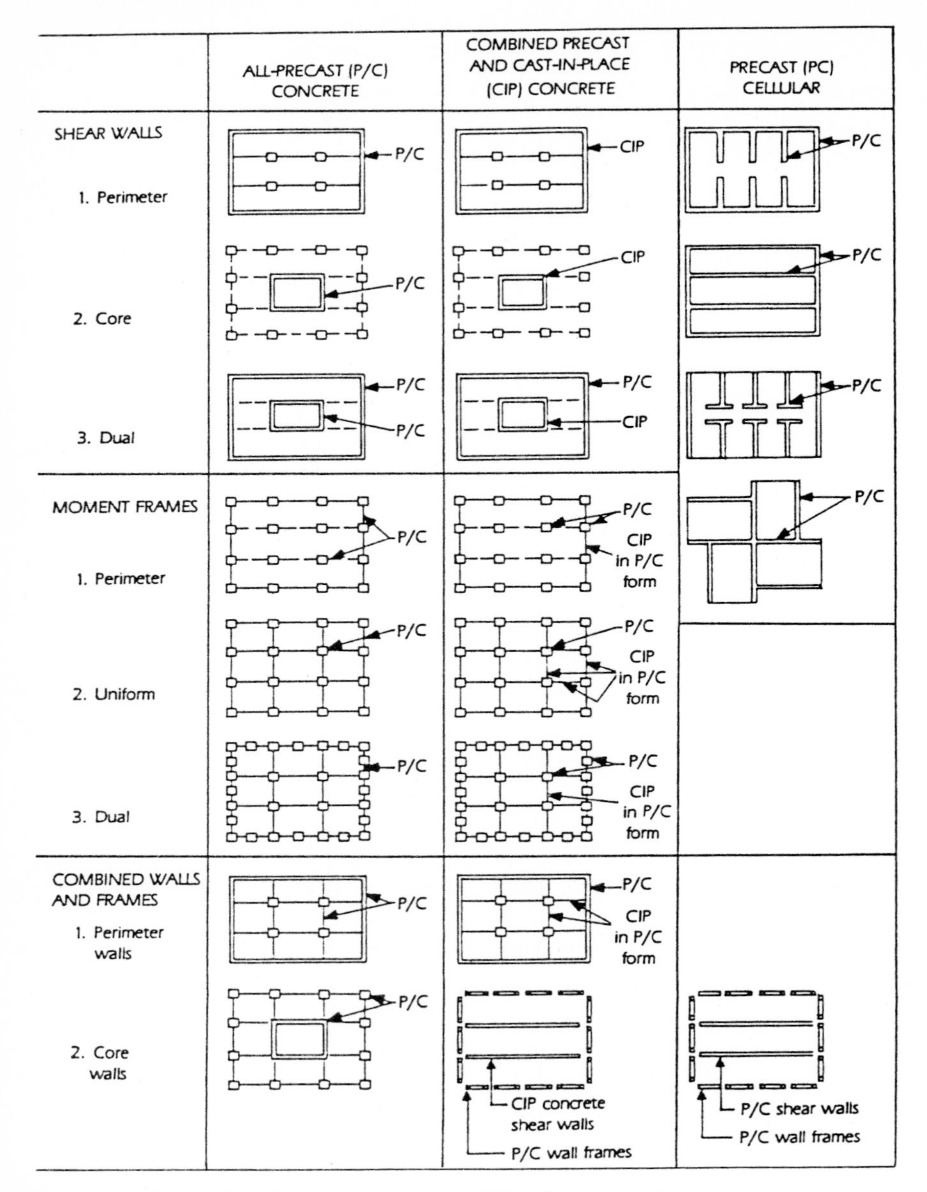


**Figure 2.4** Categories of precast concrete building framing systems.

## 1. PERIMETER LOCATION

All precast

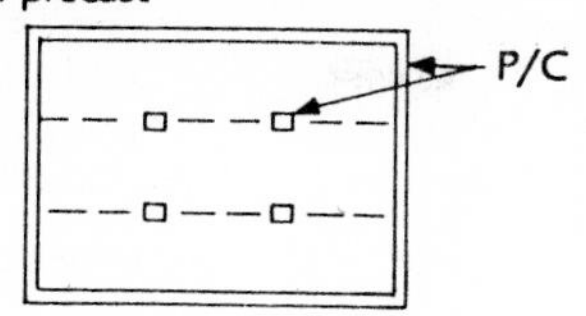


Combined precast and cast-in-place

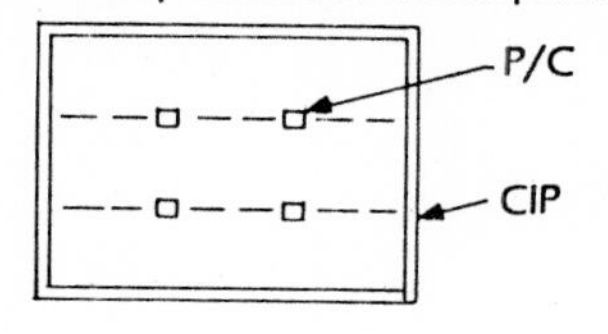


Precast cellular

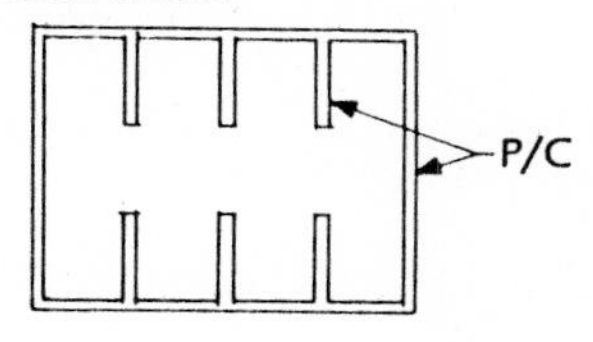


### SEISMIC PERFORMANCE COMMENTARY

**Permitted under Uniform Building Code**

- Horizontal and vertical connections are provided by cast-in-place concrete closures. Horizontal joints, where provided, may use a combination of grouting, dry packing, and/or cast-in-place concrete to provide a monolithic connection. Vertical joints may require shear castellations, horizontal loop steel, and field-placed reinforcing.

- Vertical connections, such as weld plates, may be used to transfer shear provided they are designed to perform elastically under design earthquake loading.

- Boundary elements are required in walls with aspect ratio $h/d > 4$.

- Openings in walls should be carefully located to avoid local frame action in panels.

- A continuous horizontal strut or diaphragm is required to transmit shear; a positive connection is required between wall and supporting foundation.

**Areas Requiring Further Research**

- Coupled-wall concept, utilizing intentionally designed yield hinges in coupling links to provide system ductility.

## 2. CORE LOCATION

All precast

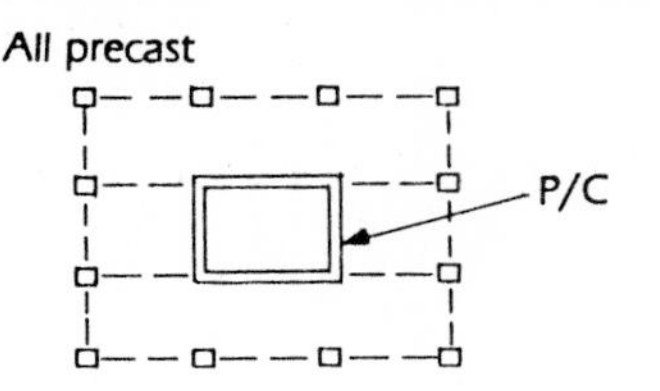


**Figure 2.5** Shear wall systems.

### SEISMIC PERFORMANCE COMMENTARY

**Permitted under Uniform Building Code**

- Same as for perimeter location.

**Areas Requiring Further Research**

- Same as for perimeter location.

- Precast beam and column elements support vertical loads only.

- Interior-core walls should be coupled and analyzed for torsion. Caution should be exercised in designing taller structures with only core walls providing lateral rigidity.

Combined precast and cast-in-place

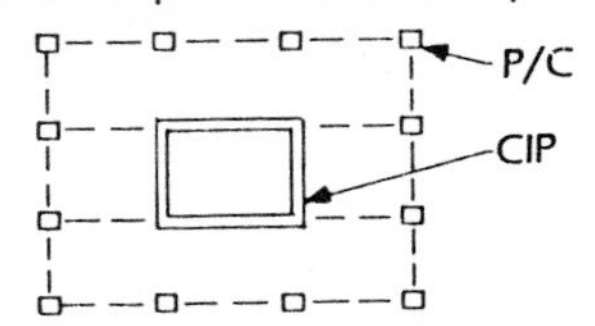


Precast cellular

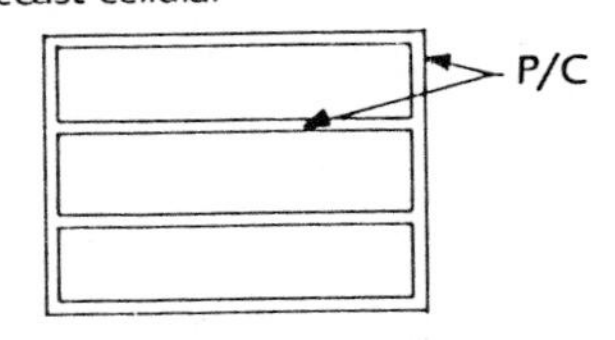


3. DUAL LOCATION

All precast

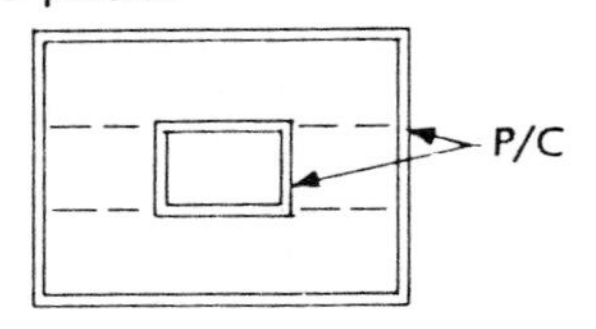


Combined precast and cast-in-place

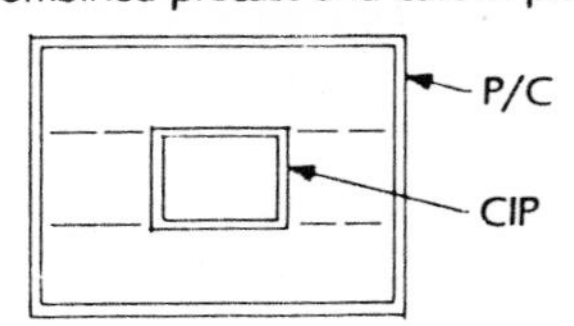


Precast cellular

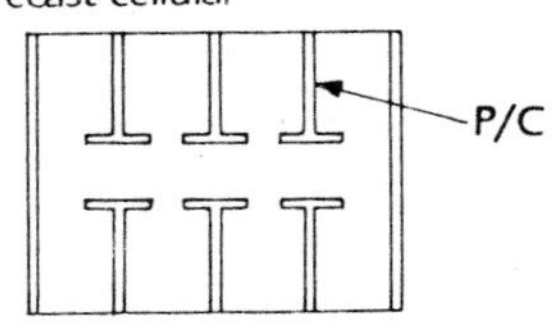


**Figure 2.5** Shear wall systems. *(Continued)*

**SEISMIC PERFORMANCE COMMENTARY**

- Same comments as for preceding shear wall systems.

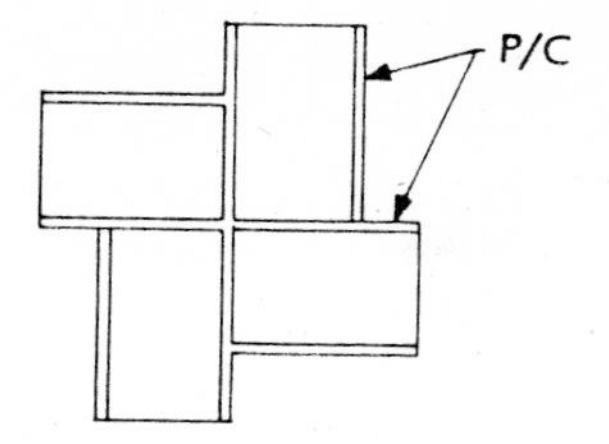


**Figure 2.5** Shear wall systems. *(Continued)*

1. PERIMETER LOCATION

All Precast

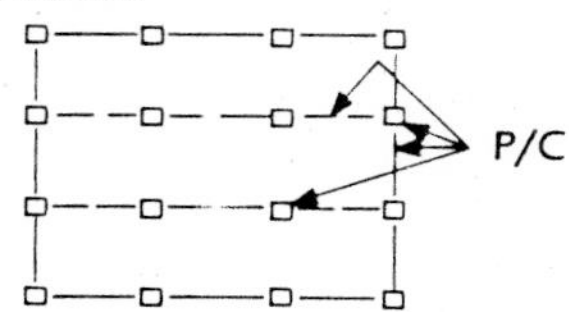


Combined precast and cast-in-place

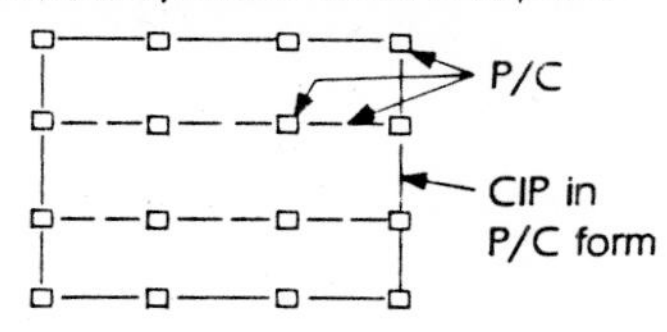


2. UNIFORM LOCATION

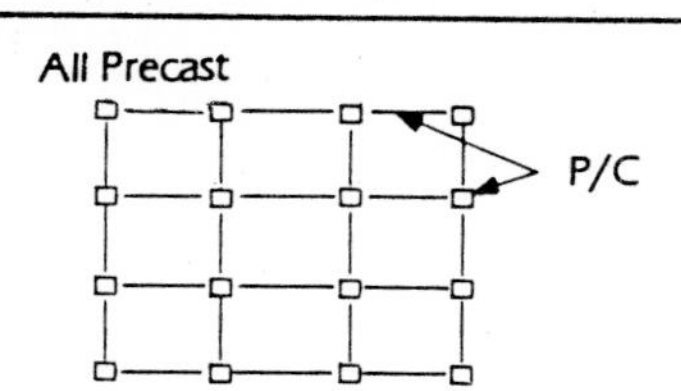


Combined precast and cast-in-place

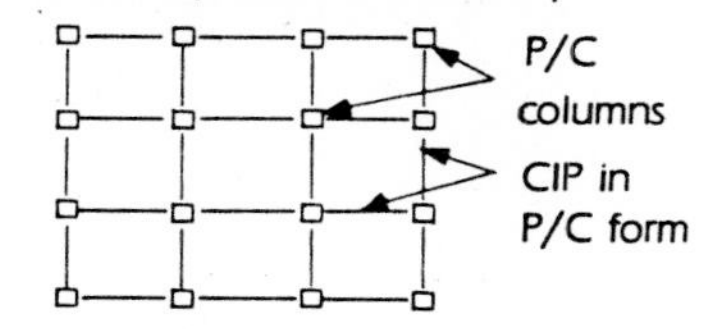


**Figure 2.6** Moment frame systems.

## SEISMIC PERFORMANCE COMMENTARY

### Permitted under Uniform Building Code

• All precast or combined precast concrete systems use detailing conforming to the ductile moment-resisting frame provisions of the code. The most frequently used method employs single-unit or tree columns with extended reinforcing spliced to field-placed longitudinal bars at noncritical locations. U-shaped spandrels may incorporate required confinement reinforcing as well as containing closure pours, forming a complete monolithic structure.

• For perimeter frames interior beams and columns support vertical loads only.

• For uniform location, all elements provide lateral rigidity.

### Areas Requiring Further Research

• Great potential exists for forming ductile moment-resisting frames by the posttensioning of precast concrete beam and column segments. This type of construction can achieve great energy dissipation with the prestressed concrete system remaining in the elastic cracked range.

Much of this research has been performed by Prof. Robert Park and others at the University of Canterbury in New Zealand. Detailed connection and specific system testing remain to be performed.

3. FRAMED TUBE

All precast

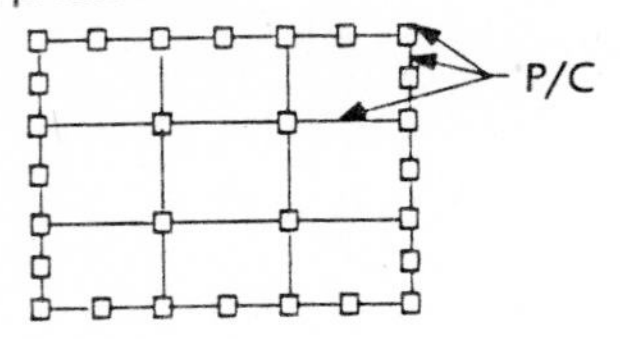


Combined precast and cast-in-place

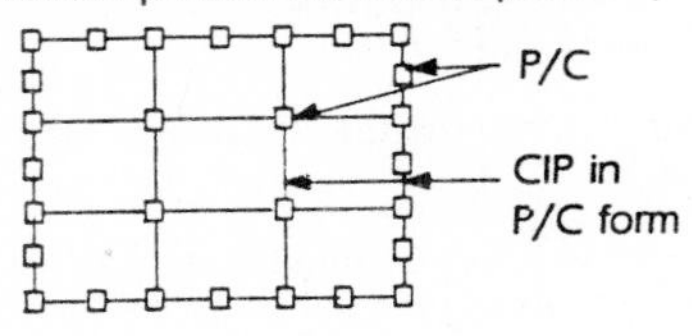


4. TUBE-IN-TUBE

All precast

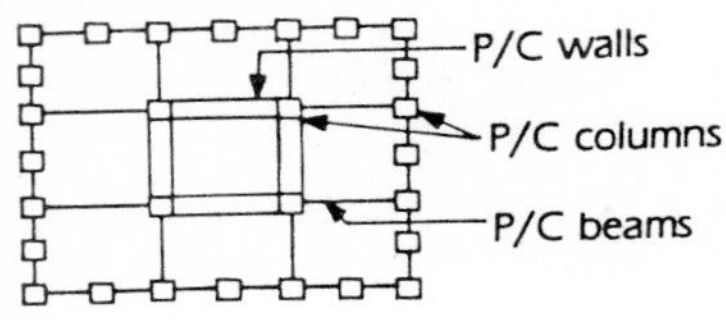


Combined precast and cast-in-place

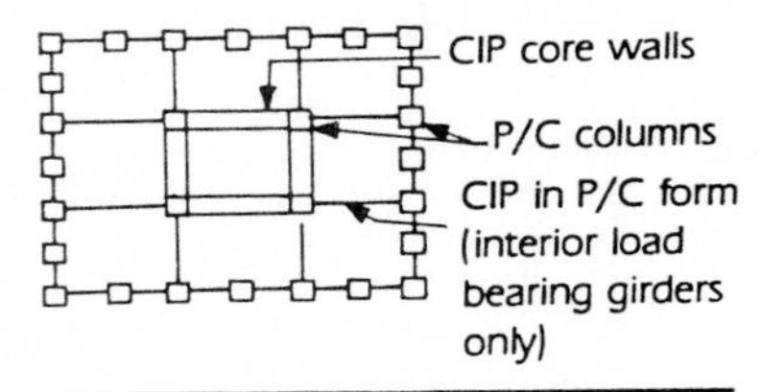


5. MULTIPLE TUBE

All precast

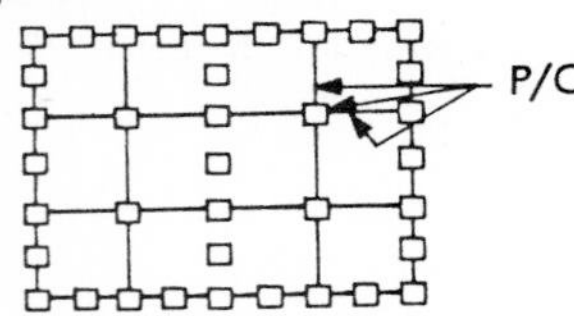


**Figure 2.6** Moment frame systems. *(Continued)*

Combined precast and cast-in-place

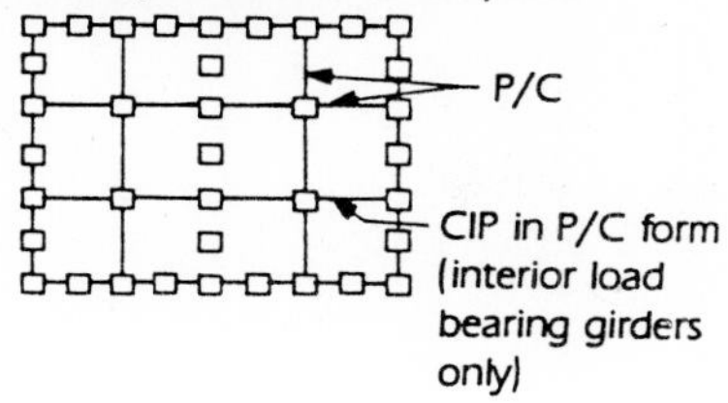


**Figure 2.6** Moment frame systems. *(Continued)*

1. PERIMETER WALLS AND INTERIOR FRAMES

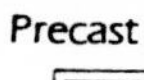


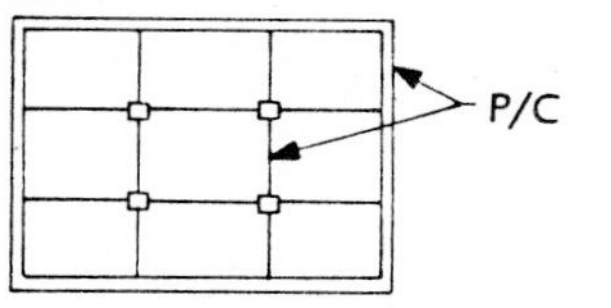


Combined precast and cast-in-place

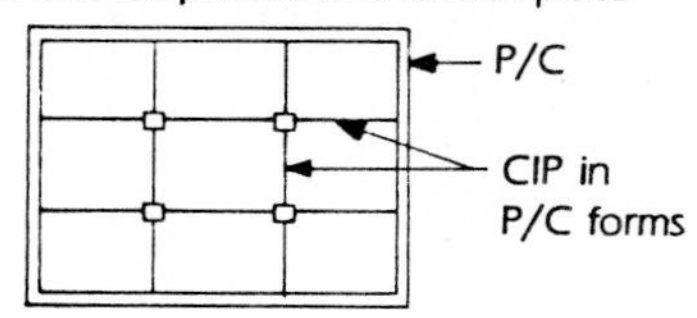


2. CORE WALLS AND EXTERIOR FRAMES

Precast

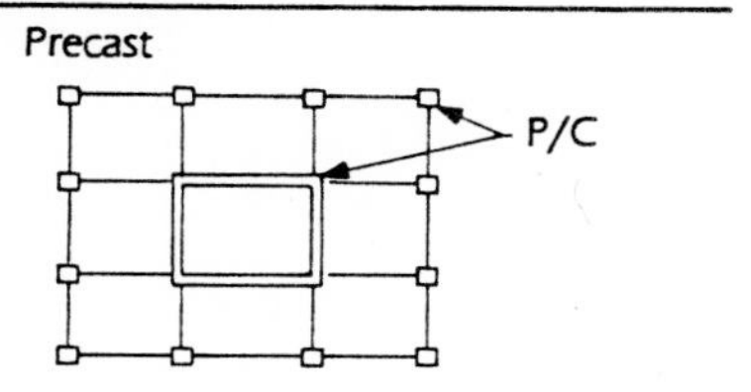


**Figure 2.7** Combined wall and frame systems.

## SEISMIC PERFORMANCE COMMENTARY

• Comments made regarding walls and frames in Figs. 2.5 and 2.6 apply here.

• Precast concrete wall frames are usually proportioned so that the resultant beam stiffness is less than the frame column stiffness. If the beam element is a deep stiff section, such as a spandrel, then the column element should be designed to perform elastically under the design earthquake loading.

Combined precast and cast-in-place

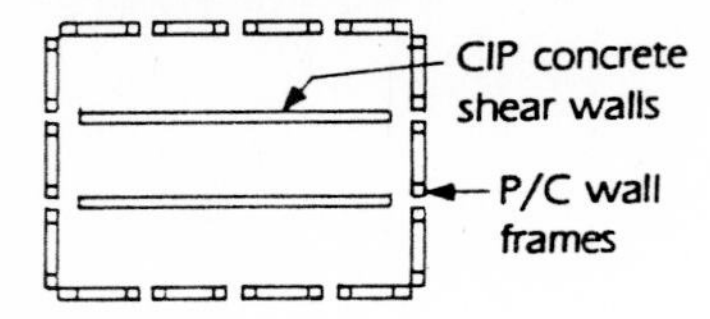


Precast cellular

P/C shear walls

P/C wall frames

**Figure 2.7** Combined wall and frame systems. *(Continued)*

**Figure 2.8 Union Bank--Oceangate Building, Long Beach, California.** A perimeter moment frame serves to provide lateral stability for this 16-story building, shown under construction in 1975. The precast concrete columns were spliced vertically at certain intervals and contained reinforcing extending out into U-shaped precast spandrels at each floor. Field reinforcing bar splices were made, and additional confinement reinforcing was placed prior to making cast-in-place concrete closures. The completed construction conformed to 1976 Uniform Building Code requirements for cast-in-place concrete ductile moment-resisting frames. The floor system consists of 12-ft-wide by 36-in-deep single T's with 12-in-thick stems, which are framed into the exterior columns and an intermediate vertical-load-carrying frame. The space between the spread-apart single T's is spanned with 8-in-deep hollow-core plank. The single T's were shored at the flange tips as well as below the stems until the cast-in-place concrete topping and closures were poured and cured. (See Chap. 15 for a more extensive pictorial coverage of various types of total precast concrete buildings that can be constructed by using current Uniform Building Code provisions.)

Chapter

# 3

# Precast Plant Operation

Precast and prestressed concrete plants are organized generally along the lines shown in Fig. 3.1. Very large organizations would have more specialized positions within the various categories shown, but basically an efficient flow of work can be effected with this organization.

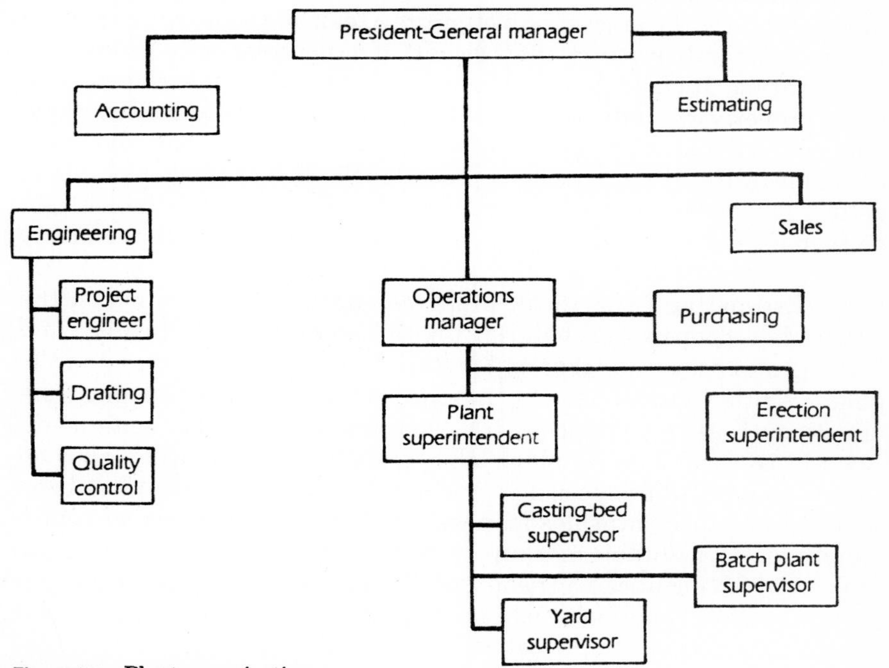


**Figure 3.1** Plant organization.

The president or general manager, in addition to overseeing the entire operation, maintains close fiscal control over the company. Daily feedback on estimated versus actual costs is given to accounting from the plant and field. Any significant variance from programmed costs warrants the accounting manager's immediate attention. Similar rigid controls are maintained over the estimating function. All estimates and quotations over a certain dollar amount require the president's review and approval. Sales has limited authority in reviewing and approving quotations. Pricing decisions should be made only by the president.

The Sales Department keeps current on work that is coming up for bid. Sales also maintains a projections file so that longer-range forecasts may be made. This information is a result of data gained from contacts with architects, engineers, and general contractors. Sales furnishes budget prices on work being analyzed by designers in preliminary planning stages. There are two principal ways in which sales obtains contracts for the precasting firm. In the first the sales manager prepares a subcontract proposal for a list of general contractors bidding on plans and specifications for projects designed in the traditional architect-engineer approach. The second method, which is gaining in popularity especially in parking structures, housing, and other types of structures where the bulk of the project is precast, is the design-construct contract with an owner, developer, or construction manager. The design-construct approach requires the precaster to organize a team of people with required design skills and construction services that do not exist in the precaster's organization. It places heavy reliance on sales and engineering personnel to properly evaluate the project for design variables and risks which may not be apparent from the outline drawings and information made available by the owner. In either process, sales has the responsibility of defining the extent of the work provided by the precaster in the contract agreement and coordinating with the other departments in the company as required.

The Estimating Department prepares total cost information on the project based upon the material takeoff, construction schedule, and existing and projected labor and material costs in the plant and in the field. Written quotations are obtained for major items bought from others, such as miscellaneous iron, hauling, and erection equipment. Actual historical data on plant labor are used in estimating labor on future projects where possible. When all the project costs have been totaled, the president, sales manager, and estimating manager confer and decide on a suitable markup, or gross margin, to be assigned to the project, which includes overhead, general and administrative costs, and profit. The total plant and field *costs*, added to the *gross margin*, equal the *selling price* for the project.

If the precaster is the successful bidder, a written subcontract or contract agreement is signed with the contractor or owner as applicable. Then the Sales Department releases information to accounting, engineering, and operations, instructing them to proceed with the project. Engineering will perform its product design and proceed to prepare drawings based upon approval of the product design by the architect and the structural engineer. For architectural precast concrete projects, additional approvals are required: one for shape, which permits mold fabrication to proceed; and one for design, which allows reinforcing and embedded items to be ordered. Finish approval is usually handled by the use of a mock-up, which is provided for in the specifications. Operations prepares casting schedules while engineering approvals are being obtained. Project coordination in a small to medium-sized company is usually handled by operations; in large precast concrete companies, a special project coordinator or project manager position is usually established to make sure that the project proceeds in a timely manner.

The Engineering Department usually has the responsibility for quality control. Each plant should have its own inspection and testing facilities as a part of an established quality control program. This program is conducted as outlined in the Prestressed Concrete Institute (PCI) manuals for quality control (PCI MNL 116 and 117). These in-plant quality control programs are periodically certified by either the PCI Plant Certification Program or by the International Conference of Building Officials (ICBO) Plant Certification Program as a part of the ICBO fabricator approval registration system.

The purchasing agent, in addition to buying materials required for fabricating the precast elements, is also responsible for contracting for outside services, such as hauling and erection, when these capabilities do not exist within the precaster's organization.

The production of the units proceeds with the fabrication of molds where required, fabrication of reinforcing cages, placing and stressing of prestressing strands, placing embedded items and inserts, placing and vibrating concrete, curing concrete (usually overnight), stripping the precast concrete units, performing any required finishing and patching, and placing units into storage. Throughout the production process, ongoing quality control (QC) checks are performed at certain critical stages, such as after bed layout, during stressing, prior to casting (cylinder making, slump, and air content checks), and prior to release of prestress. Quality control also performs an after-casting check to assure dimensional correctness, satisfactory placement of inserts and block-outs, etc. At that time, the QC seal of approval is placed on the unit, along with the piece mark and date cast. Units are then taken to storage, where they are stored by shipping load if and when possible. This is done when the erection sequence is known prior to shipping.

The erection superintendent has a very important position. He or she must ascertain that the site is indeed ready to receive precast concrete elements. The superintendent coordinates with the general contractor to assure proper access and availability of lines and grades; checks interfacing structures and supporting bearings provided by others to assure they are satisfactory; checks locations of anchor bolts, inserts, and embedded items required to erect the precast units to assure that the assumed tolerances on the drawings are realistic; and continues to maintain close control over the erection to make sure that required coordination is given to allow quick erection of the units. The erection superintendent also makes sure that required erection bracing is installed as shown on the bracing drawings to assure the safety of jobsite personnel. Finally, the superintendent sees that all required cleanup is performed and punch list items taken care of to fulfill requirements for payment of any retainage owed to the precaster by the general contractor.

**Figure 4.1** Meridian Hill Park, Washington, D.C. John J. Earley pioneered the development of exposed aggregate concrete as an architectural finish, the forerunner of plant finishing techniques still used to this day. This railing element and the walls in the foreground were fabricated in situ in about 1920. The control over mix proportions with a low water-cement ratio and the retardation of surface matrix (using sugar syrup coating on the forms) resulted in exposed natural aggregates to achieve the desired color, excellent weathering characteristics, and durability. Inspection of this 60-year-old project in December 1979 revealed the concrete in an as-new condition, supporting the statement that properly proportioned concrete in a properly designed structure has an infinite life even in severe extremes of temperature and weathering such as are experienced in the northeastern United States. *(Photograph by David A. Sheppard.)*

Chapter

# 4

# History of Precast Prestressed Concrete

The history of concrete dates back to the Roman Empire. Cement consisting of slaked lime and pozzolana (volcanic ash containing silica) was mixed with gravel, broken tile, and brick to form a concrete. This material was placed between or on forms to create domes, vaults, and walls. The majority of the concrete was mixed and cast in place. As the art was perfected, concrete was used to create sculptures, fountains, and decorative cast stone. Marble or cut stone was the preferred material for facing walls and for sculptured surfaces; however, in some areas concrete was used. (Slaked lime had been discovered prior to the Roman Empire.)

The weakness of masonry and concrete in tension was recognized at an early age. In Egypt as early as 2000 B.C., metal was used to help tie stone together. Because of the problems involved in producing the metal, it was not used to a great extent. During the Roman period, iron chains were used with concrete to resist the thrust in domes. Iron straps were used in some architectural-type structures.

Between the fall of the Roman Empire and the 1700s, the production of concrete was not widespread. This was due to the lack of availability of pozzolana. In England in 1756, John Smeaton rediscovered the art of making hydraulic cement by using natural cement rock. In the early nineteenth century Canvass White discovered natural cement rock in

Madison County, New York, which he used in the construction of the Erie Canal. Portland cement was invented in 1824 in England and was first produced in the United States in 1871. In 1850 Joseph Monier of France developed reinforced concrete, the art of combining metal with concrete. He used this new material to cast garden pots, tubs, tanks, and sculptures. This was probably when precast reinforced concrete was first used.

The date of the earliest development of the concept of prestressed concrete is unknown. In 1886 P. H. Jackson, a San Francisco structural engineer, obtained a patent on a system of tightening steel rods through voided precast concrete blocks to form slabs. In 1888 C. E. W. Doehring of Germany secured a patent for reinforced concrete with metal tensioned prior to loading. Neither of these methods was successful because of the loss of prestressing through concrete creep and shrinkage and the relaxation of the low-strength steels used. In 1908 C. R. Steiner of the United States developed a method of retightening the reinforcing bar after shrinkage and creep had occurred. In 1925 R. E. Dill of the United States used high-strength steel bars, coated to prevent the concrete from bonding to the steel, which were posttensioned. Owing to the high cost of the high-strength steel, this method was not economically feasible.

The greatest impetus to the development of a practical system of prestressing came from the work of E. Freyssinet of France, who started using high-strength steel wires for prestressing in 1928. With a yield point of approximately 180,000 lb/in$^2$ the wires were tensioned to around 150,000 lb/in$^2$, leaving an effective prestressing force of over 125,000 lb/in$^2$ after losses. Freyssinet worked with both pretensioned and posttensioned systems, but a German, E. Hoyer, is generally credited with developing the first practical system of pretensioning. In the late 1930s and early 1940s several practical end-anchorage systems for posttensioned work were perfected, the most widely accepted being those of Freyssinet and Gustave Magnel, a Belgian professor.

Principally on the basis of the work of these two men, France and Belgium led in the development of prestressed concrete as a structural material following World War II, but England, Germany, the Netherlands, and Switzerland quickly followed suit and made important contributions.

The development of prestressing in the United States was along somewhat different lines. Instead of linear prestressing, circular prestressing took the lead, primarily in the work of the Preload Company, which developed special wire-winding machines. Between 1935 and 1953, over 700 circular tanks were constructed by this system in the United States and other countries. The first major linear prestressed structure in the United States was the Walnut Lane Bridge in Philadelphia. Begun in 1949, it is a 160-ft posttensioned highway span utilizing the Magnel system of end anchorage. The bridge was designed by Magnel using I girders which were cast on a falsework and posttensioned. The first pretensioned beams in the United States were cast in 1951 for a 24-ft-high bridge near Hershey, Pennsylvania. The first prestressed railway bridge in the United States was constructed on the Burlington line in 1954.

The Arroyo Seco footbridge near Pasadena was the first prestressed bridge to be built in California. Built in 1950, it used a button-ended rod and anchorage system. By 1951 the concept of prestressing was well accepted. In 1954 the Prestressed Concrete Institute was founded in the United States. The first widely used textbook on prestressed concrete was written by T. Y. Lin during his Fulbright fellowship while studying under Gustave Magnel. The book was published in 1954. By 1957 there were over 60 prestressed bridges contracted for in California, and as a result the Bridge Division of Public Roads published *Criteria for Prestressed Concrete Bridges*. In 1956 the 24-mi Lake Pontchartrain Bridge near New Orleans was completed. It used pretensioned and precast elements.

In the United States, after the success of the Walnut Lane Bridge in 1949, several small precast concrete plants began experimenting and producing pretensioned prestressed concrete components for building structures. These were hollow slabs, flat slabs, channel slabs, beams, single T's, double T's, and piling. The first high-rise prestressed concrete building, the Diamond Head apartment house, was built in 1956-1957 in Hawaii. Designed by Alfred Yee, this building was 14 stories tall and consisted of prestressed I beams and cast-in-place slabs. In Germany, Ulrich Finsterwalder designed many creative bridges using prestressed concrete. A good example is the Mangfall Bridge at Dorching. It is a two-level cantilever bridge built in 1957.

The more rapid growth of precasting in Europe can be attributed to the major rebuilding task facing most of the European countries following the devastation of World War II, coupled with critical material shortages. Structural steel in particular was scarce and expensive, while labor costs were still quite low.

In the United States the development of precast technology was slower, but after 1950 it developed more rapidly as the physical capabilities were proved in Europe. Rising steel costs, material shortages during the Korean conflict, the expanded highway construction program, and the development of mass production methods to minimize labor costs have all been factors leading to the expansion of the use of plant-cast prestressed and precast concrete.

Throughout the United States from 1901 through 1940 there was much experimenting with precast concrete. Most of these experiments were on building facades and consisted of wall panels cast on tilt beds at the jobsite or in a plant and trucked to the jobsite. A good example exists in several storefronts constructed between 1906 and 1912 in Des Moines. The panels were cast on a tilt platform at the jobsite. In several portions of the panels, hollow walls were produced by first casting a 2-in layer of concrete and then a 2-in layer of sand followed by a second 2-in layer of concrete. The two concrete layers were tied together with reinforcing. As the wall was tilted up, the sand was washed out. In 1906 a railroad bridge was also precast and lifted by a railroad crane to its final position. During this period, plant-cast precast concrete was also used to produce decorative elements for buildings, such as capitals for columns, railings, etc. With the advent of prestressing and high-strength concrete, the technology developed as we know it today: a fabrication method, using quality control and efficiency in design, production, and erection, in which concrete strengths of 6000 $lb/in^2$ are common as compared with lower strengths achieved prior to 1950.

For a more detailed review of early history of the precast and prestressed concrete industry in North America, the interested reader can refer to the excellent series of articles entitled "Reflections on the Beginning of Prestressed Concrete in America" (Parts 1 through 9, May–June 1978 through May–June 1980, *PCI Journal*) and available in one text from the Prestressed Concrete Institute in Chicago.

Chapter

# 5

# The Future of Precast Concrete

We are on the threshold of a new era in modern construction methodology: the era of prefabrication. After several false starts, precast concrete will finally come into its own, principally owing to the increasing cost of energy and the role that this will play in the resultant gradual restructuring of our society. The beginnings of this new era can already be seen in Europe and parts of South America. Two basic trends will highlight the advanced development of plant-cast precast concrete technology:

1. Standard precast concrete shapes, similar in concept to the format in the structural steel industry, will be developed. This standardization of both structural and architectural forms in precast concrete will facilitate modular design and will be a natural complement to the establishment of the design-build concept as the predominant building method in constructing total precast and prestressed concrete structures.
2. With the increasing scarcity of precious fossil fuels, the automobile will gradually fade from the scene as the predominant mode of transportation. This will force a drastic restructuring of our entire way of living. Precast and prestressed concrete will be the prevailing structural material used in building the new society and living environment of the twenty-first century; housing and living groups will re-form in vertical

clusters containing essential services for day-to-day living, surrounded by open areas and parks, and interconnected by monorails and people movers. Feeder lines serviced by electric-powered vehicles will move people individually and in groups to the mass transit systems. Existing freeways will serve as the rights-of-way for the mass transportation systems. Rail transit will reassert itself as the predominant mode of long-distance conveyance for goods and people. The use of fossil fuels will be restricted to major industrial production, essential air transportation, and national defense. At the same time, this will cause traditional on-site construction methods to be no longer economically feasible. Plant-cast precast concrete systems building will be used to provide energy-efficient permanent structures, with components being installed with electric hoisting equipment. Solar energy systems will interface with precast concrete sandwich panel construction both above and below grade. Prestressed concrete, with its efficient ratio of supported load to weight of steel reinforcing, will be utilized to the maximum. Hybrid materials will be used to replace mild steel reinforced concrete and structural steel as building materials owing to the increasing cost and scarcity of fossil fuels for steel production. Glass-fiber reinforced concrete, polymer concrete, and other composites exhibiting high tensile strength will be required. Other manufacturing processes not currently employed in the precast industry will be used in production of these materials, such as extrusion molding, sintering, ceramic processes, and automation of all sorts for conveying, vibrating, removal of excess moisture to facilitate early stripping, and computerized cutting, storing, and handling methods.

The beginnings of the coming change are already apparent. Some of these are evident in the United States today, such as:

Monorail structures

Prestressed concrete railroad ties

Mass transportation structures

People movers

Solar housing

Systems building

Glass-fiber reinforced concrete cladding panels for buildings

Earth-sheltered housing

In the Soviet Union all prestressed precast concrete shapes are standardized in a national products manual used by architects and engineers throughout the country. Polymer-impregnated concrete is used extensively there for power poles; studies in the United States on glass–flyash composites show that such power poles can be produced at one-fourth of the cost of comparable wood poles. Cities in Europe are already rearranging living modes to adapt to an automobileless society and will not suffer the travail that will accompany adjustment in the United States, which to date has been delayed for political and other reasons.

The architect of the future will face an ever-increasing challenge to design for the greatest structural and construction efficiencies. Economics point toward the increasing use of precast concrete with all its advantages. In many ways, new opportunities are created for designers in devising different architectural forms as well as totally precast seismic-resistant structures, combining the advantages of precasting and pretensioning with posttensioning.

Chapter

# 6

# Advantages of Plant-Cast Precast Concrete

High-quality precast and prestressed concrete components are plant-manufactured under ideal quality control conditions while foundation and site work proceed at the same time, allowing delivery and erection from truck to structure on precise and predetermined construction schedules. On larger projects, this "telescoping" of critical-path functions results in significantly reduced construction time, decreasing high on-site labor costs and interim financing charges and allowing earlier initial occupancy and use of the completed structure. Other advantages of plant-cast precast and prestressed concrete are outlined below:

1. *Shallow construction depth.* Prestressed concrete deck members achieve long spans in a minimum of construction depth, thereby reducing the overall building height and the space requiring cooling and heating.

2. *High load capacity.* Prestressed concrete possesses the structural strength and rigidity essential to accommodate heavy loads, such as those resulting from heavy manufacturing equipment.

3. *Durability.* Precast concrete is exceptionally resistant to weathering, abrasion, impact, corrosion, and the general ravages of time.

4. *Long spans.* Fewer supporting columns or walls result in more

usable floor space, allowing greater flexibility in interior design, efficiency, and economy.

5. *Flexibility for expansion.* Precast concrete components can easily be designed to facilitate future horizontal or vertical expansion. Thus, with the design of precast concrete buildings being on a modular system, necessary additions can be made with low removal cost.

6. *Long economic life.* Precast concrete buildings give added years of service with a minimum of repairs and maintenance.

7. *Low maintenance.* Dense precast and prestressed concrete components factory-cast in steel, concrete, or fiberglass molds exhibit smooth surfaces which resist moisture penetration, fungus, and corrosion. High-density concrete reduces the size and quality of surface voids, thereby resisting the accumulation of dirt and dust. Prestressing controls crack propagation, thus assuring product integrity. The clean, maintenance-free properties of precast concrete produce operating-cost savings for the owner and have advantages for specific occupancies, such as food-processing plants and electronics-manufacturing facilities.

8. *Moldability into desired shapes and forms.* Precast concrete, being a plastic material when poured, may be cast in complex shapes and have intricate surface textures. With optimum reuse of forms, three-dimensional sculptural designs are possible which cannot be achieved with vertically formed site-cast concrete or "tilt-up" site-cast precast concrete.

9. *Attractive appearance.* With architectual treatment of surfaces, the pattern, texture, and color variations of precast and prestressed concrete are practically unlimited. Its inherent beauty enhances the image of the occupant by suggesting good taste, stability, and permanence. This increases salability and rentability, keeping occupancy high.

10. *Ready availability.* Daily cyclic casting of standard structural shapes assures that mass-produced structural components can be furnished to satisfy fast-track construction schedules.

11. *Economy.* In addition to low-first-cost advantages, precast concrete construction results in other cost savings. Expensive on-site labor costs are reduced. Higher-strength, lighter prestressed concrete elements result in savings in foundation costs. By quickly enclosing the building during favorable weather conditions, the interior finish work of other trades may proceed during inclement weather or in winter. Faster construction with factory-cast precast concrete elements results in reduced general contractor overhead.

12. *Quality control.* Plant-cast precast concrete components are fabricated under optimum conditions of forming, fabrication, and placement of reinforcing cages and embedded items, vibration of low-slump

concrete, and curing not achievable at the jobsite. Other special manufacturing techniques are made possible in the factory. In-plant quality control programs monitor these high standards of manufacturing.

13. *Fire resistance.* Precast concrete, being noncombustible, is an excellent material to prevent the spread of fire within a building or between buildings. This inherent benefit in concrete construction assures occupant safety and results in low fire insurance premiums. (See Chap. 12 for an expanded discussion of fire resistance.)

14. *Low noise transmission.* One method of reducing sound transmission is to increase the mass of the barrier or intervening elements such as walls and floors. The density of precast concrete provides excellent sound reduction properties. Precast concrete is excellent for occupancies such as multistory apartments, condominiums, hotels, motels, music studios, auditoriums, and schools. Good sound control results in lower tenant turnover. (See Chap. 11 for an expanded discussion of sound transmission.)

15. *Energy conservation.* Precast concrete, in addition to being able to incorporate thermal insulation cast in during its manufacture, also contributes additional savings owing to the inherent thermal lag it develops in modifying the rate at which heat or cold moves through the material. (See Chap. 10 for an expanded discussion of energy conservation.)

16. *Control of creep and shrinkage.* Since precast and prestressed concrete elements are usually kept in storage after casting for 30 to 60 days prior to delivery to the jobsite, a significant portion (50 percent or more) of the long-term creep and shrinkage movements usually occurs before the components are incorporated into the structure. This reduces the amount of long-term movement that must be recognized in the building design when compared with cast-in-place and posttensioned concrete construction. Plant-cast precast concrete elements use high-strength concrete with a lower water-cement ratio which minimizes the amount of potential creep movement.

17. *Elimination of formwork.* Precast and prestressed concrete elements may be erected without falsework, maintaining traffic over traveled ways. Precast elements are quickly erected in buildings, providing an instant working platform for other trades. Precast concrete elements may also serve as forms for cast-in-place concrete, such as in column covers or spandrel units, which also comprise the finished exterior for the building.

18. *Speed of construction.* Precast concrete construction reduces construction time by reducing on-site forming and casting for cast-in-place concrete structures and by eliminating the time-consuming infill construction required for steel-framed buildings with metal decking, exte-

rior enclosures, hung ceilings, and fireproofing operations. Other trades also have instant access to the erected prestressed concrete deck members, which form a stable, secure floor. This saving in time afforded by plant-cast precast construction results in significant interest savings in the builder's construction financing, as well as reducing the general contractor's overhead. Interim financing charges and loan fees for construction loans are normally 3 to 6 percent over the prime rate. Earlier occupancy also brings earlier returns on the initial investment.

Part

# 2

# Design Considerations

**Figure 7.1** Administration Building, Civic Center, Inglewood, California. Note how the modular regularity of the exterior precast panels coincides with the structural module selected within. *(Photograph by Eugene Phillips.)*

# Chapter 7

# Selection of a Module

To achieve the full advantages of plant precasting, a large number of like units must be used in the building design. This requires that a module be established. The module selected must be convenient for the precast manufacturing operation as related to product and casting-bed widths, in-plant handling, transportation to the jobsite, and erection. The module must also work with the intended use of the building. In that regard, the architect may want to consult with local precasters at an early stage in the design process.

The module selected is subject to limitations depending upon whether the direction being considered is the dimension of the deck member span or the dimension across the width of the individual deck member units. A fairly wide latitude in the direction of member span is available, subject to the load-carrying capacity of the deck member being considered. Shipping and handling may limit the optimum length to less than 80 ft. Therefore, unless there are wall panel modules which would dictate this dimension, the module in the direction of floor span may vary in increments of 1 ft.

In the direction of deck member widths, a larger module is normally chosen. Hollow-core planks are made in 4-ft, 8-ft, and 40-in widths. (Three 40-in planks equal 10 ft.) Solid slabs are made in widths of 8 or

12 ft. Double T's and single T's are made in widths of 8, 10, or 12 ft. Member widths may be made up to 12 ft and still be shipped without requiring escorts. Considering all these variables, the most common module selected is 8 ft, which leads to optimum bay widths of from 24 to 80 ft, varying in multiples of 8 ft. Of course, depth limitations on supporting girder sizes will usually limit the bay width to between 24 and 40 ft. In any event, the module selected in the direction across the width of the member should *always* be a multiple of 2 ft. The floor-to-floor height need not be modular.

In summary, the following are suggested modules for use in project planning and layout:

- In the direction of member span (unless dictated by wall panel dimensions): 1 ft.
- In the direction transverse to member span: 8 ft.

Chapter

# 8

# Transportation

Precast concrete manufacturers are usually responsible not only for the production but also for the shipping and erection of the units. Some manufacturers have their own equipment for transporting units to the jobsite. Others subcontract transportation. Units are normally shipped by truck. Consequently, the routing selected as well as trucking regulations has a definite effect upon unit size and weight.

Precast concrete units are usually supported on two points to avoid undue strain caused by the flexibility of the truck bed en route to the jobsite. As such, the units must also be designed for two-point support (see Fig. 8.1). Often the same two locations are used in stripping and plant handling. For erection, a two-point system is used if the unit is a double T, inverted T, L beam, rectangular beam, or hollow-core slab. However, if the unit is a wall panel which must be rotated, then one or more of these two points and a top insert can be used for loading. Where the size of the precast unit is such that more than two supports are required, then a rocker system is used (Fig. 8.2). (See also Figs. 8.3 and 8.4.)

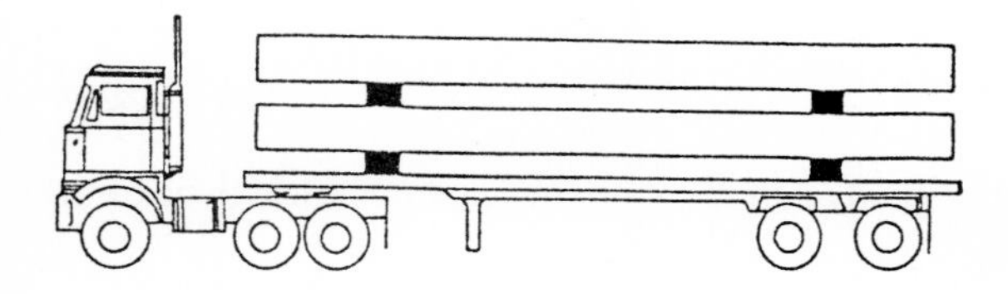

**Figure 8.1**
Typical two-point support.

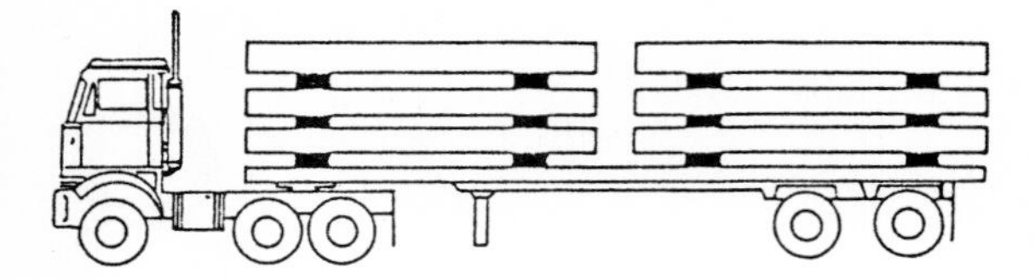

**Figure 8.2**
Rocker system.

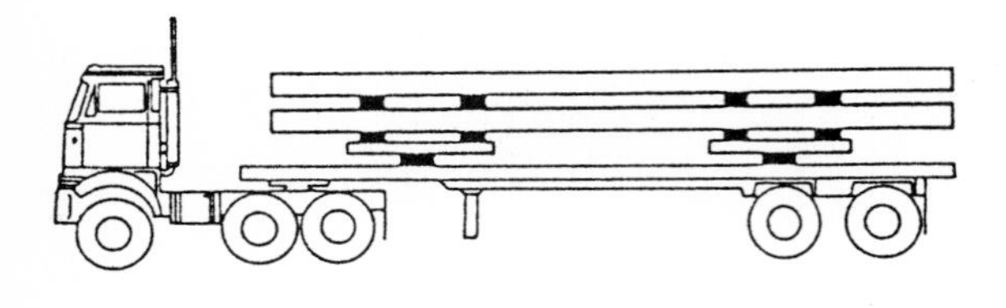

**Figure 8.3**
Wall panels laid flat.

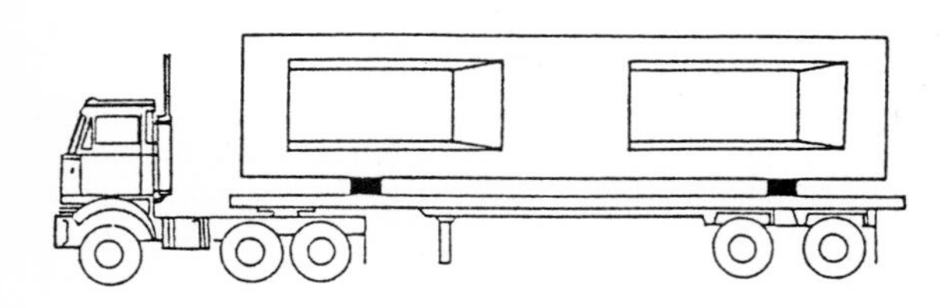

**Figure 8.4**
Wall panels supported edgewise.

The interstate highway system and most major highways and bridges are designed for HS20-44 loading. However, secondary road systems may restrict the maximum weight and height owing to bridge limitations. Thus it is important to check out the route if heavy or tall loads are to be transported. The California vehicle code pertaining to weight allowable without special permissions is complex, but the following information will serve as a guide. Excerpts from the *Vehicle Code Size and Weight Law Summary* are given here. Unless special permission is obtained, the following limitations are mandatory.

*Width.* Maximum width of a vehicle or load is 96 in.

*Vehicle height.* No vehicle shall exceed a height of 13 ft 6 in measured from the surface on which the vehicle stands except that a mast or boom of a forklift may extend up to 14 ft.

*Load height.* No load on a vehicle may exceed a height of 14 ft measured from the surface on which the vehicle stands.

*Vehicle length.*

(1) No vehicle shall exceed a length of 40 ft except a vehicle in combination.

ation.

(2) Length of combination of vehicles. No combination of vehicles shall exceed 65 ft when coupled together except that the following combinations may not exceed 60 ft:

(*a*) Truck-tractor and semitrailer.
(*b*) Truck-tractor, auxiliary dolly, and semitrailer.

(3) Extension devices. Any extension or device including any adjustable axle added to the front or rear of a vehicle used to increase the carrying capacity of a vehicle shall be included in measuring the length of the vehicle. A drawbar shall be included in measuring the overall length of a combination of vehicles.

*Load length*

(*a*) *Front projections*. No load may extend more than 3 ft in front of the front bumper or foremost part of the front tires if there is no front bumper.

(*b*) *Rear projections*. The load upon any motor vehicle alone or independent load upon a trailer or semitrailer shall not extend to the rear beyond the last point of support for a distance greater than that equal to two-thirds of the vehicle carrying the load, except that the wheelbase of a trailer shall be considered as the distance between the rearmost axles of the semitrailer.

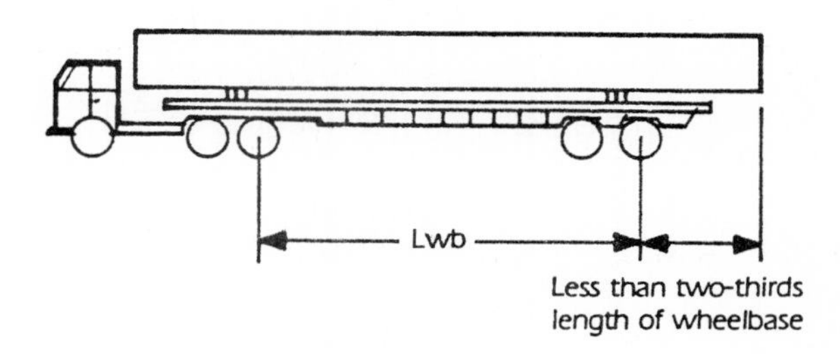


(*c*) *Combinations of vehicles*. The load upon any combination of vehicles shall not exceed 75 ft measured from the front extremity of the vehicle or load to the rear extremity of the last vehicle or load.

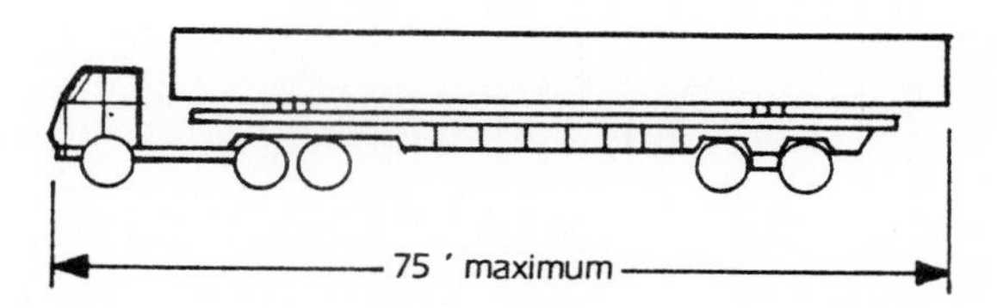

| Transport vehicle | | Component | | | Vehicle | | | Max. total weight tons, † | Special provisions |
|---|---|---|---|---|---|---|---|---|---|
| | | ℓ | w | h | ℓ | w | h | | |
| 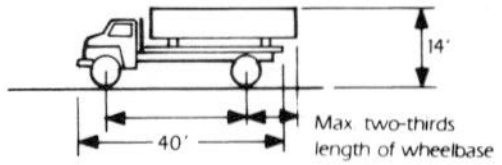 | Truck | * | 8' 4" | 14' | 40' | 8' | 13' 6" | 23.25 | None |
| | Over width or length | | | | | | | | Permission |
| 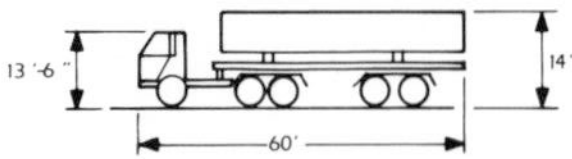 | Semitrailer; no overhang | * | 8' 4" | 14' | 60' | 8' | 13' 6" | 40 | None |
| | Over width | | | | | | | | Permission |
| 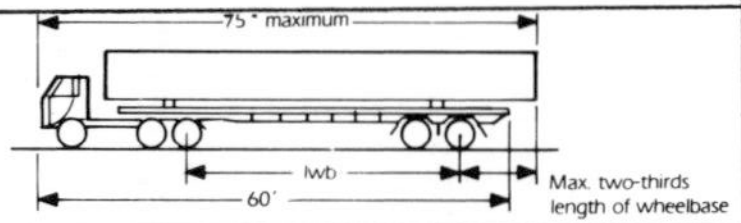 | Semitrailer with overhang | * | 8' 4" | 14' | 60' | 8' | 13' 6" | 40 | None |
| | Over width | | | | | | | | Permission |
| 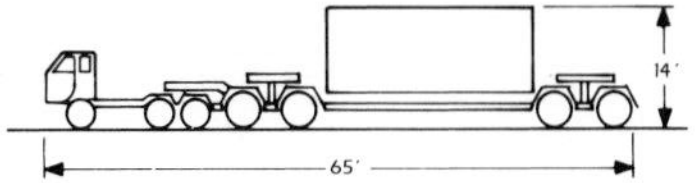 | Low loader | * | 8' 4" | 14' | 65' | 8' | 13' 6" | 40 | None |
| | Over width | | | | | | | | Permission |
| 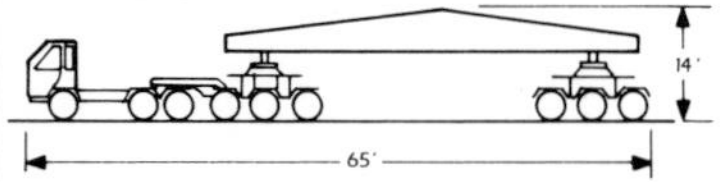 | With drawbar | * | 8' 4" | 14' | 65' | 8' | 13' 6" | 40 | None |
| | Special transports | . . . | . . . | 14' | . . . | 13' 6" | | | Permission needed |

* Length varies subject to the vehicle and its wheelbase.
† Verify weight with table and department of motor vehicles.

**Figure 8.5** Size and weight provisions of the California vehicle code.

*Axle gross weight*

(*a*) The gross weight imposed upon the highway by the wheels on any one axle of a vehicle shall not exceed 20,000 lb, and the gross weight upon anyone wheel, or wheels, supporting one end of an axle, and resting upon the roadway, shall not exceed 10,500 lb, except that the gross weight imposed upon the highway by the wheels on any front steering axle of a motor vehicle shall not exceed 12,500 lb.

(*b*) The following vehicles are exempt from the front axle weight limits specified in this section:

(1) Trucks transporting vehicles.

(2) Dump trucks.

(3) Cranes.

(4) Transit-mix concrete or cement trucks and trucks that mix concrete or cement at or adjacent to a jobsite.

(5) Motor vehicles that are not commercial vehicles.

(6) Truck or truck-tractor with a front axle at least 4 ft to the rear of the foremost part of the truck or truck-tractor, not including the front bumper.

(7) Trucks equipped with a fifth wheel when towing a semitrailer.

**TABLE 8.1 Gross Weight to Length**

| Distance between extremes of any group of two or more consecutive axles, ft | Gross weight, lb | | | | |
|---|---|---|---|---|---|
| | 2 axles | 3 axles | 4 axles | 5 axles | 6 axles |
| 4 | 34,000 | 34,000 | 34,000 | 34,000 | 34,000 |
| 5 | 34,000 | 34,000 | 34,000 | 34,000 | 34,000 |
| 6 | 34,000 | 34,000 | 34,000 | 34,000 | 34,000 |
| 7 | 34,000 | 34,000 | 34,000 | 34,000 | 34,000 |
| 8 | 34,000 | 34,000 | 34,000 | 34,000 | 34,000 |
| 9 | 39,000 | 42,500 | 42,500 | 42,500 | 42,500 |
| 10 | 40,000 | 43,500 | 43,500 | 43,500 | 43,500 |
| 11 | 40,000 | 44,000 | 44,000 | 44,000 | 44,000 |
| 12 | 40,000 | 45,000 | 50,000 | 50,000 | 50,000 |
| 13 | 40,000 | 45,500 | 50,500 | 50,500 | 50,500 |
| 14 | 40,000 | 46,500 | 51,500 | 51,500 | 51,500 |
| 15 | 40,000 | 47,000 | 52,000 | 52,000 | 52,000 |
| 16 | 40,000 | 48,000 | 52,500 | 52,500 | 52,500 |
| 17 | 40,000 | 48,500 | 53,500 | 53,500 | 53,500 |
| 18 | 40,000 | 49,500 | 54,000 | 54,000 | 54,000 |
| 19 | 40,000 | 50,000 | 54,500 | 54,500 | 54,500 |
| 20 | 40,000 | 51,000 | 55,500 | 55,500 | 55,500 |

**TABLE 8.1 Gross Weight to Length (*Continued*)**

| Distance between extremes of any group of two or more consecutive axles, ft | Gross weight, lb | | | | |
|---|---|---|---|---|---|
| | 2 axles | 3 axles | 4 axles | 5 axles | 6 axles |
| 21 | 40,000 | 51,500 | 56,000 | 56,000 | 56,000 |
| 22 | 40,000 | 52,500 | 56,500 | 56,500 | 56,500 |
| 23 | 40,000 | 53,000 | 57,500 | 57,500 | 57,500 |
| 24 | 40,000 | 54,000 | 58,000 | 58,000 | 58,000 |
| 25 | 40,000 | 54,500 | 58,500 | 58,500 | 58,500 |
| 26 | 40,000 | 55,500 | 59,500 | 59,500 | 59,500 |
| 27 | 40,000 | 56,000 | 60,000 | 60,000 | 60,000 |
| 28 | 40,000 | 57,000 | 60,500 | 60,500 | 60,500 |
| 29 | 40,000 | 57,500 | 61,500 | 61,500 | 61,500 |
| 30 | 40,000 | 58,500 | 62,000 | 62,000 | 62,000 |
| 31 | 40,000 | 59,000 | 62,500 | 62,500 | 62,500 |
| 32 | 40,000 | 60,000 | 63,500 | 63,500 | 63,500 |
| 33 | 40,000 | 60,000 | 64,000 | 64,000 | 64,000 |
| 34 | 40,000 | 60,000 | 64,500 | 64,500 | 64,500 |
| 35 | 40,000 | 60,000 | 65,500 | 65,500 | 65,500 |
| 36 | 40,000 | 60,000 | 66,000 | 66,000 | 66,000 |
| 37 | 40,000 | 60,000 | 66,500 | 66,500 | 66,500 |
| 38 | 40,000 | 60,000 | 67,500 | 67,500 | 67,500 |
| 39 | 40,000 | 60,000 | 68,000 | 68,000 | 68,000 |
| 40 | 40,000 | 60,000 | 68,500 | 70,000 | 70,000 |
| 41 | 40,000 | 60,000 | 69,500 | 72,000 | 72,000 |
| 42 | 40,000 | 60,000 | 70,000 | 73,280 | 73,280 |
| 43 | 40,000 | 60,000 | 70,500 | 73,280 | 73,280 |
| 44 | 40,000 | 60,000 | 71,500 | 73,280 | 73,280 |
| 45 | 40,000 | 60,000 | 72,000 | 76,000 | 80,000 |
| 46 | 40,000 | 60,000 | 72,500 | 76,500 | 80,000 |
| 47 | 40,000 | 60,000 | 73,500 | 77,500 | 80,000 |
| 48 | 40,000 | 60,000 | 74,000 | 78,000 | 80,000 |
| 49 | 40,000 | 60,000 | 74,500 | 78,500 | 80,000 |
| 50 | 40,000 | 60,000 | 75,500 | 79,000 | 80,000 |
| 51 | 40,000 | 60,000 | 76,000 | 80,000 | 80,000 |
| 52 | 40,000 | 60,000 | 76,500 | 80,000 | 80,000 |
| 53 | 40,000 | 60,000 | 77,500 | 80,000 | 80,000 |
| 54 | 40,000 | 60,000 | 78,000 | 80,000 | 80,000 |
| 55 | 40,000 | 60,000 | 78,500 | 80,000 | 80,000 |
| 56 | 40,000 | 60,000 | 79,500 | 80,000 | 80,000 |
| 57 | 40,000 | 60,000 | 80,000 | 80,000 | 80,000 |
| 58 | 40,000 | 60,000 | 80,000 | 80,000 | 80,000 |
| 59 | 40,000 | 60,000 | 80,000 | 80,000 | 80,000 |
| 60 | 40,000 | 60,000 | 80,000 | 80,000 | 80,000 |

*Ratio of weight to length.* The total gross weight in pounds imposed on the highway by any group of two or more consecutive axles shall not exceed that given for the respective distance in Table 8.1.

In addition to the weights specified above, two consecutive sets of tandem axles may carry a gross weight of 34,000 lb each if the overall distance between the first and last axles of such consecutive sets of tandem axles is 36 ft or more. The gross weight of each set of tandem axles shall not exceed 34,000 lb and the gross weight of the two consecutive sets of tandem axles shall not exceed 68,000 lb.

The above weights are gross loads. To determine the net payload, the weight of the transporting unit must be deducted. Once the maximum payload has been determined, the number of units to be included within the payload can then be determined. If the weight exceeds that in the table, then a special permit is required. If the shipping route includes minor roads with bridges, then the highway department should be consulted as to the load capacity of the bridges.

**Overweight and oversize permits.** Permits for oversize and overweight vehicles and loads may be issued under certain conditions by the California Department of Transportation with respect to highways under its jurisdiction or by local authorities with respect to highways under their jurisdiction. Some of the increases allowed with permits are as follows:

1. *On state roads (CAL-TRANS)*
   - Increased load width from 8 to 12 ft — no escort required.
   - Load width from 12 to 14 ft — front escort required. (Absolute maximum width: 14 ft 0 in.)
   - Overweight load — no escort required.
   - Over 100 ft in length (load plus truck) — front escort required.

2. *On city and county roads, local jurisdiction may grant permits subject to the following limitations:*
   - Increased load width from 8 to 14 ft — no escort required.
   - Load width from 14 to 15 ft — front and rear escorts required.
   - Load width from 15 to 20 ft — front and rear escorts required; loads may only be moved at night between 1 and 6 A.M.
   - Overweight load — no escort required.

Permit loads for overweight loads are limited by bridge classification. CAL-TRANS uses a color code system to rate the capacity of bridges. The lighter classification is green, and the heavier classification is purple. Some of the axle loads for various transportation configurations are given in Table 8.2. Most other state trucking regulations are similar to California's.

Table 8.2 CAL-TRANS Weight Chart

| | Color class | Tractor axle width, ft | Maximum gross weight, lb | Axles 2 and 3, lb | Axles 4 and 5, lb | Axles 4, 5, and 6, lb | Axles 6 and 7, lb |
|---|---|---|---|---|---|---|---|
| 1. Three-axle tractor and two-axle semitrailer | Green<br>Purple | 8 | NA<br>NA | 40,200<br>46,300 | 40,200<br>46,300 | | |
| 2. Three-axle tractor and two-axle semitrailer (16 tires on 4 and 5) | Green<br>Purple | 8 | NA<br>NA | 40,200<br>46,300 | 46,000<br>53,100 | | |
| 3. Three-axle tractor and two-axle semitrailer (16 tires on 4 and 5) | Green<br>Purple | Minimum, 10 | ...<br>NA | 40,200<br>46,300 | 49,700<br>57,400 | | |
| 4. Three-axle tractor and three-axle semitrailer | Green<br>Purple | 8 | ......<br>...... | 40,200<br>46,300 | ......<br>...... | 43,200<br>49,900 | |
| 5. Three-axle tractor and two-axle auxiliary trailer and two-axle semitrailer (16 tires on 4 and 5, 6 and 7) | Green<br>Purple | 8 | 124,000<br>142,000 | 40,200<br>46,300 | 46,000<br>53,100 | ......<br>...... | 46,000<br>53,100 |
| 6. Three-axle tractor and two-axle auxiliary trailer and two-axle semitrailer (16 tires on 4 and 5, 6 and 7) | Green<br>Purple | Minimum, 10 | 130,000<br>152,000 | 40,200<br>46,300 | 49,700<br>57,400 | ......<br>...... | 49,700<br>57,400 |

| Component loadings | Green | Purple |
|---|---|---|
| Single axle | 24,000 | 27,000 |
| Tandem axles | 40,200 | 46,300 |
| Three-axle group | 43,200 | 49,900 |
| Trunnion tandem axles (8 ft or over but less than 10 ft out-to-out width on tires) | 46,000 | 53,100 |
| Trunnion tandem axles (10 ft minimum out-to-out width on tires) | 49,700 | 57,400 |

NOTE: Five- or seven-axle trunnion tandem units may be allowed bonus weights in excess of those listed here if certain conditions are met. The California Division of Highways Bridge Department will rate a certain unit upon presentation of a drawing containing the following information:

*a.* Axle spacing, hub to hub, between each axle.
*b.* Tires per axle.
*c.* Out-to-out width of tires on axle.
*d.* Tare weight for steering, drive, and trailer axles (connected in train).
*e.* Kingpin location referenced to an axle.
*f.* License and equipment number of tractor, auxiliary trailer, and/or semitrailer.

# Chapter 9

# Erection

If any one element is the key to a successful plant-cast precast project, it is erection. Enormous sums of money can be lost by inefficient installation; on the other hand, understanding the problems and solutions involved with putting the precast pieces together in the field can give astute precasters a competitive edge and dramatize the most significant aspect of plant-cast precast concrete construction: speed. Proper planning through all phases of the building process is essential to ensure efficient erection. Preliminary design decisions by the architect and the structural engineer, the attention to detail given by the precast concrete manufacturer at bid time, the manufacturer's product design prior to fabrication, and the coordination between the erector and the general contractor are all vital to the success of the erection phase. Everyone involved with the project should understand the importance of erection and how it is affected by so many factors.

## Preliminary Design Decisions by the Architect-Engineer

Designers usually make decisions as to sizes and weights of precast concrete elements early in the project. As a general statement, component weights should be limited to 11 tons or less for both architectural and structural precast concrete projects unless precast concrete erection personnel are consulted. Heavier weights of up to 22 tons may be handled provided good access is available immediately adjacent to the erection point. Designers should be aware of the importance of proper access, especially in tight city sites between existing structures or on stepped, spread-out buildings where heavy panels coupled with long reaches may require heavy and expensive erection equipment not available locally. Realistic tolerances between precast concrete elements and other materials are required. For example, construction tolerances for cast-in-place concrete or structural steel supporting elements must usually be added to fabrication tolerances for the precast concrete elements to determine realistic precast concrete panel clearances, ledge dimensions for cast-in-place concrete corbels, etc. Other materials such as window walls, partitions, and glazing should be designed and specified by recognizing the existence of tolerances required for proper precast concrete installation. Most important, the structural engineer should define connection requirements early in the project for architectural precast concrete cladding panels.

## Precaster Decisions at Bid Time

Precast concrete manufacturers usually seal their fate in the decisions they make in estimating at bid time. This is especially true of decisions affecting erection. Care should be taken that the designed connections work and allow for required erection tolerances. If redesign is required at bid time, make sure that the general contractor is aware of the effect of changes on other trades; otherwise these extras will become back charges to the precast concrete manufacturer's account. Also, precast concrete manufacturers should determine the construction schedule for the erection of the precast concrete elements and tie this into the subcontract agreement, with escalation provisions for increases in field labor, equipment rental rates, and fuel for delays in the project. Delays of 1 year or more are not uncommon on large projects, and these delays can have disastrous effects on erection costs owing to inflation.

## Product Design

Specific installation and erection considerations for individual precast and prestressed concrete components are covered in Part 3.

The shape of the member influences the type of hoisting equipment required for erection, the number of workers required to set or guide the member in place, the sequence of placement, the time and difficulty of attaining the required vertical and horizontal alignments, the time and difficulty of completing temporary and permanent connections, and the time and difficulty of releasing the member from the lifting device (crane).

Inasmuch as each shape poses different erection problems, the shape of the member influences the design of the structural connections, which should facilitate installation as well as in-place stability. Vertical member shapes (load-bearing wall panels and columns) tend to require the most care during erection because these shapes often act as benchmarks for the members which they support. Erection procedures must be developed to make sure that load-carrying panels and columns are plumb. Connections for these members should be designed to facilitate temporary alignment as well as permanent alignment and stability.

The wide flange and narrow stem of the single T make that shape a relatively difficult member to erect, in that a single T may fall over on its side unless it is braced or otherwise stabilized in some manner. Stability is often accomplished by a combination of detailing and sequencing. For example, the first single T in the erection sequence is welded by flange weld plates to another member or members in the structural system, the second T is connected to the first, and subsequent T's are erected in a similar manner, each being connected to the previously erected T with these erection connections, which may also form part of the final connection system for the precast elements.

The remaining structural shapes (double T's, beams, and slabs) tend to be relatively easy to erect, in that they do not pose any special stability problems and often require no temporary connections during the erection sequence. The erection crew simply places these shapes into position as quickly as possible (without erection connections) until all units are erected; then the crew goes back and completes the permanent connections at a later time.

Precast concrete cladding panels are probably the most critical from the standpoint of the interrelation of product design and erection. The connections should be designed to obtain immediate stability and alignment, maximizing erection productivity. These connections should accommodate building tolerances as well as product tolerances. The selection of bolted, welded, or grouted connections should always be made with erection kept in mind. Auxiliary bolted connections are often selected to provide alignment and temporary erection bracing even though the final connection may be made at some future time.

Certain drawings are essential for both installation and field inspection personnel:

1. *Erection drawings.* Building elevations and/or plans showing member layout with piece marks. It is strongly recommended that the marking procedure used for structural steel also be used for precast elements; i.e., the piece mark on the top left of the piece corresponds to

the same location on erection drawings (right side for opposite-hand members).

2. *Connection detail sheets.* Cross-referenced to erection drawings and bills of material.

3. *Erection hardware bills.* Loose erection materials should be clearly marked by piece mark or, better yet, color-coded and cross-referenced to connection details. This material should be segregated into individual kegs or cans at the jobsite marked with corresponding piece marks and color code.

4. *Bracing plan.* Erection bracing and shoring required for precast concrete elements that are not stable or built into the structure prior to release from the erection equipment. This plan is prepared by a registered civil engineer in the employ of the precast concrete manufacturer and furnished to the general contractor and field installation and inspection personnel prior to erection.

5. *Anchor bolt–insert plate location plan.* Prepared by the precaster for other trades setting hardware cast in and interfacing with precast concrete elements. Many times the precaster will also provide setting templates for critical cast-in-place embedments.

6. *Field worksheet.* Showing alterations required to be performed to existing structures prior to erecting precast concrete.

Product handling, as it relates to removing elements from the truck and rotating into the vertical position as required ("tripping up"), should be shown on the erection drawings. (See Chap. 19.)

## Coordination with the General Contractor Prior to Beginning Erection

Before beginning erection of precast or prestressed concrete elements, certain coordination is required. First, it should be determined that there is sufficient access to, around, and inside the site, as applicable, on well-compacted roads. Other trades should not interfere with the uninterrupted installation of the precast concrete elements. Good baselines and control points for line and grade should be provided by the general contractor. The precast erector lays out the project to make sure that embedded items are cast in correctly and that the building frame, where supporting precast concrete elements, is installed within allowable tolerances and provides sufficient marks so that the precast concrete elements may be installed without a lot of juggling and repositioning to correctly space out the precast concrete units. The sequence of erection is reconfirmed with the general contractor in order to establish the casting priority for production of the precast concrete elements. On the basis of this sequence of erection, the erector makes up load lists for the plant so that the pieces are shipped as they are required for installation on the building. The load list will also indicate the location of individual piece marks on the truck so that time is not lost in the field in double-handling pieces.

Additional coordination is sometimes required when a crane supplied by the owner or general contractor is available for the erection of precast concrete units. Subcontract agreements in this case should read that the crane is "fully manned and operated for the exclusive use of the precaster during precast concrete erection."

## Erection Equipment

Erection equipment used for installing precast concrete components usually varies with the height of the building.

1. Tall buildings (greater than 16 stories) (architectural precast concrete cladding panels)
   *a.* Fixed tower crane
   *b.* Monorail system with Chicago boom
   *c.* Guy derrick; stiff-leg crane
2. Medium-sized buildings (5 to 16 stories) (architectural and structural precast)
   *a.* Portable tower crane or fixed tower crane
   *b.* Crawler crane (140 to 200 tons)
   *c.* Rubber-tired truck crane (125 to 140 tons)*
3. Low-rise buildings (up to 4 stories) (architectural and structural precast)
   *a.* Rubber-tired truck crane (50 to 140 tons)
   *b.* Hydro to 50 tons (GFRC; light precast)

The vast majority of precast concrete falls into the low-rise category erected with truck cranes. Erection capacities of some typical truck cranes are listed in Table 9.1

*Note availbility of 225-ton rubber-tired cranes in some areas.

**TABLE 9.1 Typical Rubber-Tired Truck Crane Capabilities**

| 50-ton P & H, 100-ft boom (16,000-lb counterweight) | |
|---|---|
| 30-ft radius | 22 tons |
| 50-ft radius | 10 tons |
| 70-ft radius | 6 tons |

**TABLE 9.1** (***Continued***)

| Crane / radius | Capacity |
|---|---|
| 80-ton American, 110-ft boom (full counterweight) | |
| 40-ft radius | 20 tons |
| 60-ft radius | 11 tons |
| 80-ft radius | 7 tons |
| 100-ton P&H, 120-ft boom (25,000-lb counterweight) | |
| 40-ft radius | 21 tons |
| 50-ft radius | 16 tons |
| 70-ft radius | 10 tons |
| 90-ft radius | 7 tons |
| 125-ton American, 140-ft boom (34,000-lb counterweight) | |
| 50-ft radius | 21 tons |
| 60-ft radius | 16 tons |
| 80-ft radius | 10 tons |
| 110-ft radius | 6 tons |
| 140-ton Manitowoc, 140-ft boom (74,000-lb counterweight) | |
| 65-ft radius | 21 tons |
| 80-ft radius | 16 tons |
| 100-ft radius | 11 tons |
| 120-ft radius | 8 tons |

NOTE: All capacities shown are with outriggers extended and set. Radius is from center pin (center of rotation) to boom tip. All loads are lifted over the side or rear of the crane.

Often it is necessary to store precast concrete elements temporarily on the ground because of occasional problems with loads out of sequence, etc. Jobsite storage should be performed in the same manner as plant storage, using proper supporting dunnage in the proper locations. Care should be taken to protect the precast components from staining by mud, etc.

Erection crews vary in composition, depending upon the extent and variety of work performed at the jobsite. In general, the basic erection crew for either architectural or structural precast consists of a crane operator and oiler, an ironworker supervisor and four ironworkers, and workers with additional skills, depending upon other work required as a part of the erection, such as welding (welders), grouting (masons), patching (cement finishers), and shoring (laborers). Sometimes the ironworker also serves as the overall erection supervisor, but it is wiser for the precast concrete manufacturer to have its own employee on the job to provide overall supervision and coordination with the plant and the general contractor.

Certain major decisions affecting erection are usually made early in project planning. Their impact is seen much later, when it can be said that hindsight is better than foresight. One of these decisions involves the interrelation of cast-in-place concrete work that supports precast concrete elements. In general, cast-in-place concrete walls, shafts, frames, or other supporting elements for deck members should be designed so that they can all be cast before precast erection begins, allowing the precast concrete to be installed in one continuous operation without expensive "on and off" charges for cranes and resultant loss in productivity as compared with continuous operation. The erection mode which is most efficient for total precast concrete buildings is by bay, erecting a stack of precast members required to complete a vertical portion of the structure without moving the crane and then moving to the next crane setup point, where the operation is repeated. Long vertical elements such as wall panels or columns are tripped up into the vertical position by using two lines. Handling inserts are used also as final connection points to the buildings where possible. "Choking" precast elements (lifting with long cable slings completely around an element without using fixed handling points) is employed only as a last resort for reasons of safety and potential damage to the precast surfaces. If it is used, choking should be done with protective dunnage and pads or with load-rated nylon slings to preclude scarring.

In closing, personnel safety is the most important concern for the precast erector. Totally precast concrete structures are veritable houses of cards until closures are made connecting the elements together and to the lateral-force-resisting system. Adequate bracing to resist wind loads or seismic forces, as shown on bracing plans prepared by a registered civil engineer, is mandatory for all projects of this nature. Cladding panels should be tied into the structure immediately before releasing from the erection equipment. Written procedures should be furnished to erection personnel for this type of erection.

In conclusion, prior planning of every crane move is recommended in order to minimize lost time in the field. In Fig. 9.1 are shown schematic representations for crane sequencing for light and heavy erection loads.

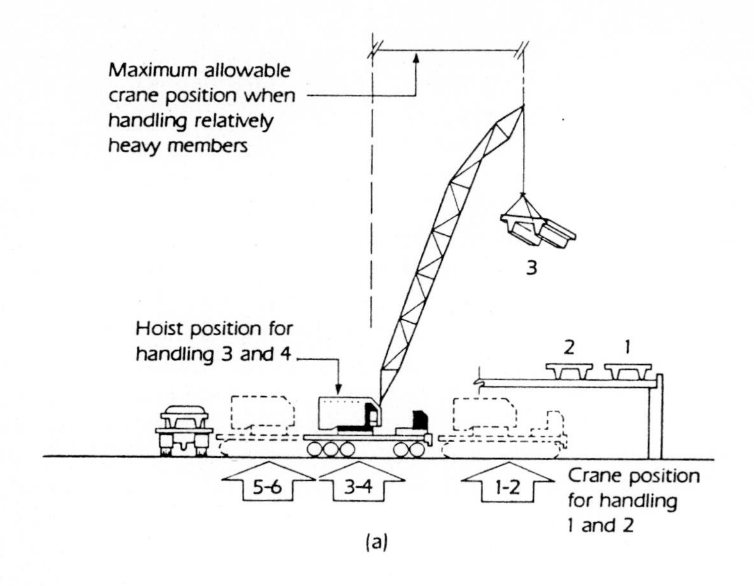


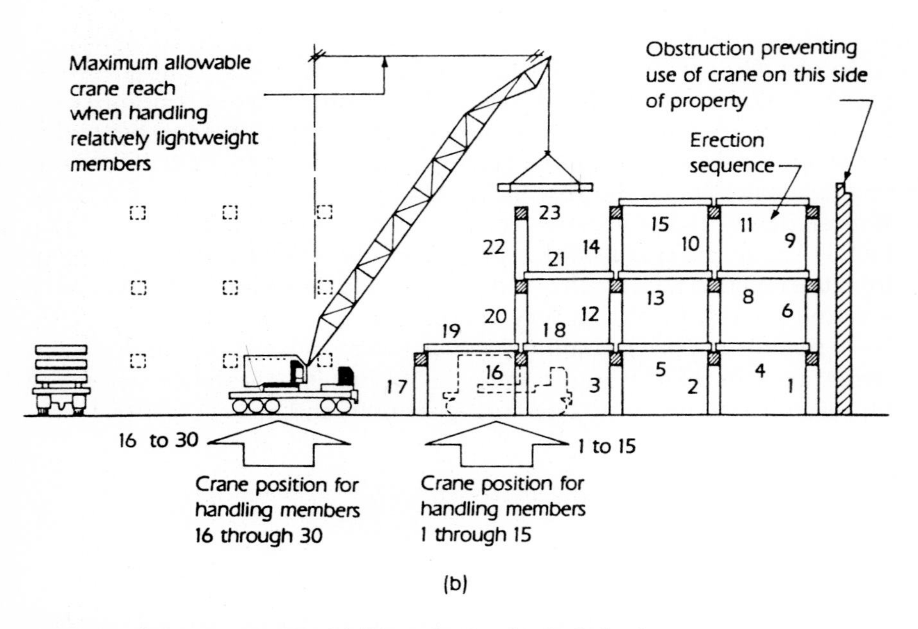


**Figure 9.1** Crane sequencing. (*a*) Heavy loads. (*b*) Light loads.

Chapter

# 10

# Energy Conservation

In California the State Energy Resources Conservation and Development Commission has adopted energy conservation regulations for buildings. The regulations for nonresidential buildings became effective on February 6, 1977. The regulations for residential buildings went into effect on February 22, 1975. The primary purpose of the regulations is to conserve depleting energy sources such as natural gas, fuel oil, and electrical energy. These regulations were adopted in the California administrative code.

The requirements are in terms of $U$ values. The $U$ value is defined as the transmission of heat flow through a square foot of wall, floor, or roof surface, air to air, per degree of temperature difference in British thermal units per hour. The British thermal unit is that quantity of heat necessary to raise the temperature of one pound of water one degree Fahrenheit. $U$ is equal to the reciprocal of the sum of the resistance to heat flow: $U = 1/R$.

$$U = \frac{1}{R_o + R_a + R_1 + R_2 + R_3 + R_4 + R_i}$$

where $R_i = \dfrac{1}{f_i}$ = resistance of inside air film

$R_o = \dfrac{1}{f_i}$ = resistance of outside air film

$R_a = \dfrac{1}{C_a}$ = resistance of core air space

$$R_1 = \frac{x_1}{k_1} = \frac{1}{C_1} = \text{resistance of first material (concrete shell)}$$

$$R_2 = \frac{x_2}{k_2} = \frac{1}{C_2} = \text{resistance of second material (core insulation)}$$

$$R_3 = x_3 = \frac{1}{C_3} = \text{resistance of third material (concrete wall material)}$$

$$R_4 = \frac{x_4}{k_4} = \frac{1}{C_4} = \text{resistance of fourth material (finished wall material)}$$

$$U_{ow} = \frac{U_{\text{wall}} A_{\text{wall}} \, \text{MCF} + U_{\text{window}} A_{\text{window}} + U_{\text{door}} A_{\text{door}} + UA_{\text{other elements}}}{A_{ow}}$$

where $U_{ow}$ = average thermal transmission of gross wall area
$A_{ow}$ = gross wall area
MCF = mass correction factor used for walls or roofs

The mass correction factor takes into account the thermal inertia properties of concrete. Dense concrete walls and roofs store large quantities of heat or cold and respond slowly to temperature changes. This is also referred to as the "flywheel effect." Concrete construction possesses an inherent advantage due to its mass inertia, which affects the thermal dynamic response of the building. For example, lightweight wood, metal, or glass exteriors cannot store large quantities of heat whereas heavier, more massive materials in walls, partitions, floors, and roofs store greater amounts of heat and subsequently release it as demand is increased by a change in the weather and by solar effects. Portland Cement Association Publication EB089.01B, *Simplified Thermal Design of Building Envelopes for Use with ASHRAE Standard 90-75*, contains charts which allow one to determine the MCF to be used to adjust theoretical $U$ values for thermal lag which results in as much as a 35 percent reduction in the $U$ value. This significantly reduces energy consumption in heating and cooling.

If a wall panel is made up of both solid and hollow portions or insulated portions, then the following averaging method is used:

$$U_{\text{panel}} = \frac{U_{\text{solid}} A_{\text{solid}} + U_{\text{hollow}} + U_{\text{insulated}} A_{\text{insulated}}}{A_{\text{panel}}}$$

Note that percentages can be used in lieu of areas in this equation.

## Residential Buildings

The requirements for residential buildings in terms of maximum $U$ values are as follows:

| | |
|---|---|
| Walls $U$ = 0.08 | Walls and spandrels with insulation between framing and the effects of the framing not considered |
| Walls $U$ = 0.095 | Walls and spandrels with insulation not affecting the framing of the framing system considered |
| Walls $U$ = 0.12 | Walls weighing between 26 and 40 lb/ft$^2$ and in areas of 3500 or fewer degree-days |
| Walls $U$ = 0.16 | Walls weighing over 40 lb/ft$^2$ and in areas of 3500 or fewer degree-days |
| Ceiling or roofs $U$ = 0.05 | With insulation between framing members and the effects of the framing not considered |
| Ceiling or roofs $U$ = 0.06 | With insulation not penetrated by the framing or the effects of the framing considered |
| Floors $U$ = 0.10 | In areas with from 3001 to 4500 degree-days |
| Floors $U$ = 0.08 | In areas with over 4500 degree-days |

Glazed portions of walls should be a maximum of 20 percent of the gross floor area for low-rise buildings and 40 percent of the exterior wall area for high-rise buildings. A high-rise building is defined as containing four or more stories, excluding basements, parking, and nonhabitable areas. The Uniform Building Code requires a minimum of 10 percent of the floor area of habitable rooms to be in glazing with one-half of that amount openable for ventilation unless mechanical ventilation is provided (Uniform Building Code, Sec. 1205).

The regulations also state that the $U$ value of any component of the walls, ceilings, roof system, or floor including the glazing may be increased and the $U$ value of other components decreased until the overall heat gain or loss of the total building does not exceed that total resulting from conformance with the stated $U$ values. The building official may approve any alternative design including designs utilizing nondepleting energy systems such as solar or wind, provided the official finds that the total system does not use more depleting energy than that given by the stated $U$ values.

The preceding paragraph allows the architect to increase the glazing area above the maximum if he or she uses insulating glass (double glazing) subject to the degree-days for the area. Alternatively, the architect may decrease the $U$ value for the walls, roof, or floors. The reverse is also true; that is, the architect may increase the $U$ value of the walls, roof, or floor system if he or she uses insulating glass without increasing the window area or decreases the window area. Thus it now becomes possible for the architect to balance the area and $U$ values of the various materials so as not to exceed the maximum total allowable heat loss or gain.

The zone related to the annual heating degree-days would determine whether or not floor and foundation insulation would be necessary as well as whether or not insulating glass would be required.

*Zone 1.* 2500 or fewer heating degree-days will not require insulation of foundation walls or walls of heated basements or crawl spaces.

*Zone 2.* For 2500 to 3000 heating degree-days, foundation walls of heated spaces must have a $U$ value of 0.15 but floors need not be insulated.

*Zone 3.* For 3000 to 4500 heating degree-days, floors over unheated basements or crawl spaces must have a $U$ value of 0.10.

*Zone 4.* For 4500 or more heating degree-days, the floor must have a $U$ value of 0.8 and all windows in the building must have insulating or special glazing.

## Nonresidential Buildings

For nonresidential buildings using the standard design concept, minimum requirements for the thermal design of the exterior envelope are as follows: for a building which is both heated and cooled, the more stringent of either the heating or the cooling requirements shall govern. The $U$ value of any component such as the walls, roof, ceiling, or floor may be increased and the $U$ value of other components decreased until the overall heat gain or loss for the entire building envelope does not exceed the total resulting from the stated $U$ values. The overall $U$ value of the wall, roof, or floor portion of the envelope is governed by the degree-days of the building location for heating and by the latitude for cooling. The maximum $U_{ow}$ ($U$ overall for walls), the maximum $U_{or}$ ($U$ overall for roofs), and the $U_f$ ($U$ overall for floors) related to heating degree-days are indicated in Figs. 10.1, 10.2, and 10.3. Values that may be used for the mass correction factor (MCF) are given in Table 10.1.

**Table 10.1** MCF Values

| Weight of wall construction, lb/ft² | MCF * |
|---|---|
| 0–25 | 1.00 |
| 26–40 | 0.85 |
| 41–80 | 0.75 |
| 81 and above | 0.65 |

*For areas with more than 3500 degree-days, MCF = 1.00.

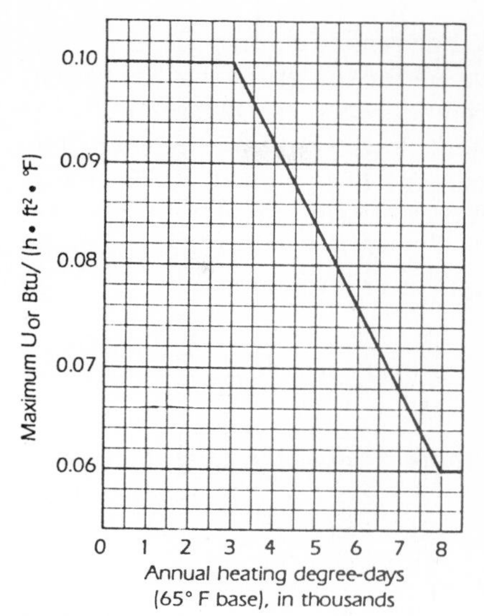


**Figure 10.1**

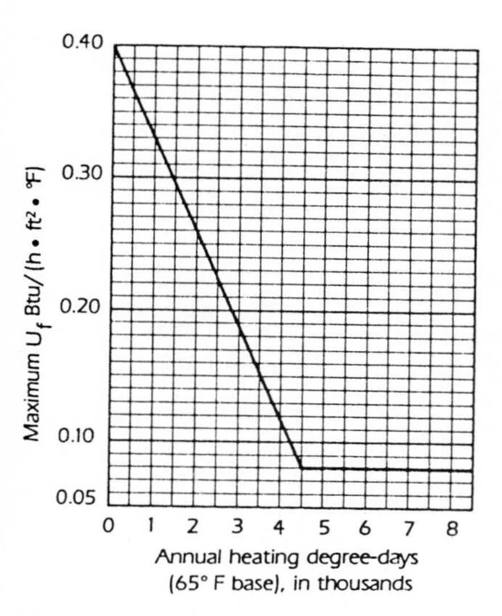


**Figure 10.2**

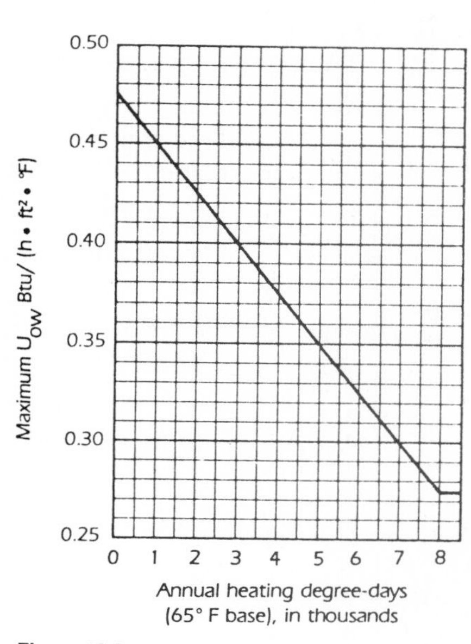


**Figure 10.3**

For roof or ceiling areas, $U_{or}$ is the combined thermal transmittance value for the gross roof area including skylights. The roof assembly is considered to be those components of the roof envelope through which heat flows. For buildings which are heated only, skylight areas up to 5 percent of the gross roof area are exempt from the $U_{or}$ calculations. Skylight area in excess of 5 percent must be included in the calculations. The $U_f$ value of the floor system is for heated spaces over unheated spaces or outdoors.

$$U_{ow} = \frac{U_{wall}\, A_{wall}\, \text{MCF} + U_{window}\, A_{window} + U_{door}\, A_{door}}{A_{ow}}$$

where $U_{ow}$ = average thermal transmittance of the gross wall area, Btu/(h·ft²·°F)

$A_{ow}$ = external exposed above-grade gross wall area of the building that faces heated spaces, ft²

$U_{wall}$ = thermal transmittance of all elements of the opaque wall area, adjusted for the effect of framing in the insulated building section, Btu/(h·ft²·°F)

$A_{wall}$ = opaque wall area, ft²

MCF = mass correction factor, value given in Table 10.1. (This mass correction value corrects for thermal storage in the wall system and is related to the mass of the wall and the annual degree-days for the area.)

$U_{window}$ = thermal transmittance of window area, Btu/(h· ft²· °F).

$A_{window}$ = window area including sash, ft²

$U_{door}$ = thermal transmittance of the door, considered as an assembly, including the frame, Btu/(h·ft²·°F).

$A_{door}$ = door area including frame, ft²

Note that where more than one type of wall, window, and/or door is used, the term or terms for the exposure shall be expanded into its subelements, as $U_{wall_1} A_{wall_1} \text{MCF}_1 + U_{wall_2} + A_{wall_2} \text{MCF}_2$,

$$U_{or} = \frac{U_{roof}\, A_{roof} + U_{skylight}\, A_{skylight}}{A_{or}}$$

where $U_{or}$ = average thermal transmittance of the gross wall roof or ceiling area, Btu/(h·ft²·°F)

$A_{or}$ = external exposed gross roof or ceiling area of the building over heated spaces, ft²

$U_{roof}$ = thermal transmittance of all elements of the opaque roof or ceiling area, adjusted for the effect of framing in the insulated building section, Btu/(h·ft²·°F)

$A_{roof}$ = opaque roof or ceiling area, ft²

$U_{skylight}$ = thermal transmittance of the skylight area, Btu/(h·ft²·°F)

$A_{skylight}$ = skylight area, ft²

Note that where more than one type of roof or ceiling and/or skylight is used, the $U \times A$ term for that exposure shall be expanded into its subelements, as $U_{roof_1} A_{roof_1} + U_{roof_2} A_{roof_2}$, etc.

## Cooling

For cooling, the overall thermal transmittance value of the walls in British thermal units per hour per square foot for the gross exterior wall area including window areas that enclose interior cooled spaces shall not exceed the value shown in Figs. 10.4 and 10.5. Windows or portions of

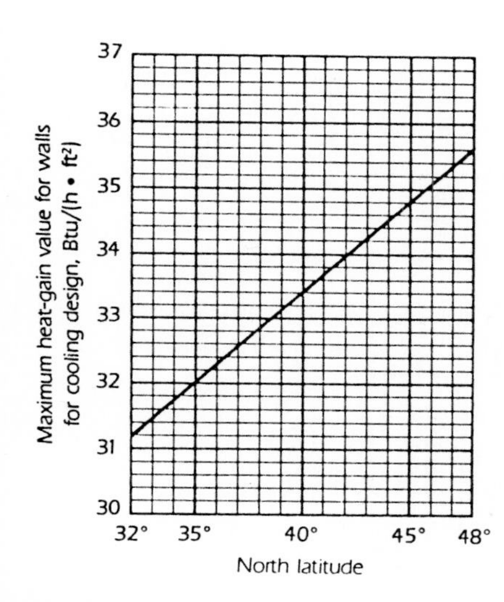


Figure 10.4

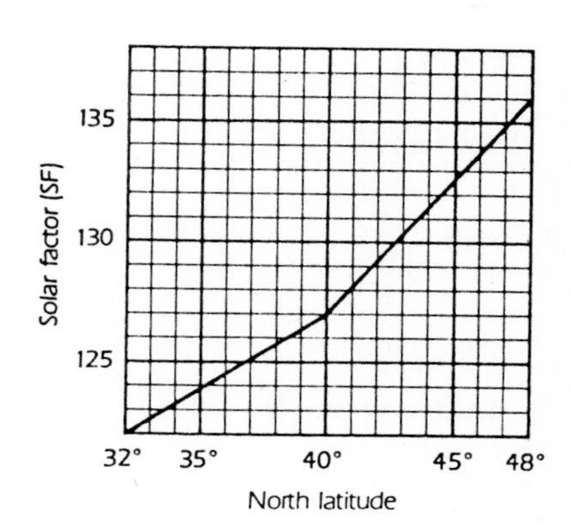


Figure 10.5

windows that, because of orientation or fixed exterior shading devices which are never exposed to direct sunlight between April 21 and October 21, shall be considered as having a solar gain factor (SF) of 30 Btu/h.

For cooling, the overall thermal transmittance value of the roof/ceiling in British thermal units per hour per square foot for the gross area of the exterior roof consisting of opaque roof areas and fenestration that enclose interior cooled spaces shall not exceed the value $41 \times U_{or}$ for heating. The following equation is used to determine acceptable combinations of opaque roof and fenestration areas:

$$\text{Btu/(h·ft}^2) = \frac{41U_r A_r \times A_c M_c + 118\,\text{SC}_s A_s + TU_s A_s}{A_{or}}$$

(overall thermal transfer value)

where
$U_r$ = thermal transmittance of opaque roof, Btu/(h·ft²·°F)
$A_r$ = area of opaque roof, ft²
$A_c$ = absorptance coefficient
$M_c$ = mass coefficient (This factor compensates for the storage effect of walls due to their mass.)
$\text{SC}_s$ = shading coefficient of skylights
$A_s$ = area of skylights, ft²
$T$ = temperature difference between exterior and interior design conditions, °F
$U_s$ = thermal transmittance of skylight, Btu/(h·ft²·°F)

**TABLE 10.2**

| Weight, lb/ft² | Class | $M_c$ |
|---|---|---|
| 0–15 | Light | 1.00 |
| 16–40 | Medium | 0.92 |
| 41 and above | Heavy | 0.84 |

**TABLE 10.3**

| Surface | Absorptance | $A_c$ |
|---|---|---|
| Asphalt, "dark roof" | 0.90 | 1.00 |
| Gravel | 0.70 | 0.79 |
| ASHRAE "light roof" | 0.45 | 0.52 |
| Intense white | 0.35 | 0.42 |

The following equation is used for walls:

Btu/(h·ft²)

$$= \frac{(U_w \times A_w \times \mathrm{TD_{eq}}) + (A_f \times \mathrm{SF} \times \mathrm{SC}) + (U_f \times A_f \times T)}{A_{ow}}$$

where $U_w$ = thermal transmittance of opaque walls, Btu/(h·ft²·°F)

$A_w$ = area of opaque wall, ft²

$\mathrm{TD_{eq}}$ = value given in Table 10.4

$A_f$ = area of fenestration, ft²

SF = solar factor, Btu/(h·ft²) (Values are given in Fig. 10.5.)

SC = shading coefficient of fenestration

$U_f$ = thermal transmittance of fenestration, Btu/(h·ft²·°F)

$T$ = temperature difference between exterior and interior design conditions, °F

$A_{ow}$ = total area of wall opposite cooled spaces, ft²

Note that where more than one type of wall and/or fenestration is used, the respective term or terms shall be expanded into subelements such as $(U_{w1} \times A_{w1} \times \mathrm{TD_{eq1}}) + (U_{w2} \times A_{w2} \times \mathrm{TD_{eq2}})$ ,etc.

**TABLE 10.4**

| Weight of wall construction, lb/ft² | $\mathrm{TD_{eq}}$ |
|---|---|
| 0--25 | 44 |
| 26--40 | 37 |
| 41--70 | 30 |
| 70 and above | 23 |

## Summary of the Energy Requirements for California

The following are the requirements for residential occupancies:

| | |
|---|---|
| Ceilings or roofs with framing not penetrating the insulation, maximum $U$ value | 0.06 |
| Floors in areas of 3001 to 4500 degree-days, maximum $U$ value | 0.10 |
| Walls weighing 40 lb/ft²,* maximum $U$ value | 0.16 |
| Walls weighing between 26 and 40 lb/ft², maximum $U$ value | 0.12 |

*Precast concrete wall panels usually weigh in excess of 40 lb/ft². Thus the standard value is 0.16 for residential occupancies.

Window area, maximum (single glazing)

| | |
|---|---|
| For low-rise buildings, three or fewer stories | 20 percent of floor area |
| For high-rise buildings | 40 percent of exterior |

If the glass area is reduced or insulative glass is used, the wall $U$ value may be adjusted so that the total heat loss based upon the standard is not increased.

Requirements for commercial buildings in various California cities are shown in Table 10.5 (the maximum $U$ value is based upon the annual degree-days).

**TABLE 10.5**

| City | Annual heating degree-days | Design temperature | | Heating | | | Latitude,° |
|---|---|---|---|---|---|---|---|
| | | Winter | Summer | $U_{ow}$ | $U_{or}$ | $U_{of}$ | |
| San Diego | 1439 | 42 | 83 | 0.44 | 0.10 | 0.30 | 33.0 |
| Los Angeles | 2061 | 42 | 86 | 0.43 | 0.10 | 0.25 | 34.0 |
| Santa Barbara | 2290 | 34 | 84 | 0.42 | 0.10 | 0.25 | 34.4 |
| Visalia | 2546 | 32 | 100 | 0.41 | 0.10 | 0.22 | 36.4 |
| San Jose | 2656 | 34 | 88 | 0.41 | 0.10 | 0.22 | 37.2 |
| San Francisco | 3080 | 42 | 80 | 0.40 | 0.10 | 0.18 | 37.7 |
| Sacramento | 2782 | 29 | 97 | 0.41 | 0.10 | 0.20 | 38.4 |

From the preceding data it can be seen that the roof system must be insulated to achieve a $U$ value of 0.10 or 0.06. If the floor system is over an open or unheated space, it must also be insulated to achieve a $U$ value between 0.18 and 0.30 or 0.10 for residential occupancies.

To achieve an average $U$ value between 0.40 and 0.44, the wall system must also be partially or totally insulated. The portion to be insulated and to what degree are determined by the quantity and type of glass and the quantity of uninsulated concrete in the precast pierced wall units.

The wall insulation is furnished by one of several methods:

1. Insulation applied to the inside wall surface with wallboard attached.

2. Full sandwich wall panels, i.e., two concrete wythes, one structural and one nonstructural.

3. A partial sandwich panel which is a total structural unit. However, the problems here may be differential temperature volume changes with induced shear stresses in the webs parallel to the flanges.

In using the $U_{ow}$ equation, the designer has the option of selecting the quantity and type of glass and varying the thickness and type of insulation in the sandwich panel portion to achieve the required $U$ value for the wall assembly.

## Examples of Precast Concrete Heat Transmission Values

### 1. Solid wall, no insulation

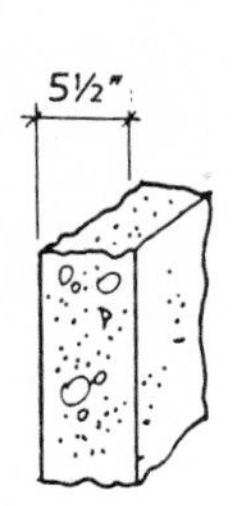


| | Normal-weight concrete, $R$ | Lightweight concrete, $R$ |
|---|---|---|
| Outside-surface air film | 0.17 | 0.17 |
| 5 ½-in concrete | 0.41 | 1.04 |
| Inside-surface air film | 0.68 | 0.68 |
| $R$, total | 1.26 | 1.89 |
| $U = 1/R$ | 0.79 | 0.528 |

NOTE: $R$ per inch of normal-weight concrete = 0.075. $R$ per inch of lightweight concrete, 100 lb/ft³ - 0.24. $R$ per inch of lightweight concrete, 110 lb/ft³ = 0.19. Lightweight structural concrete is usually 110 lb/ft³.

### 2. Solid wall with insulation on one side

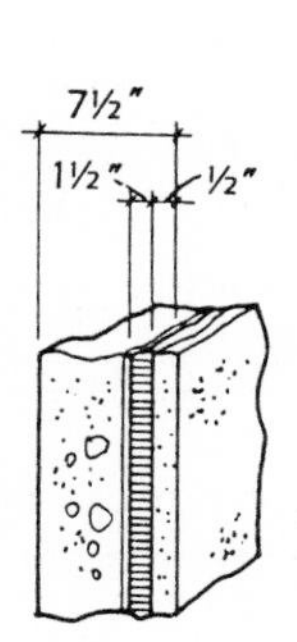


| | Normal-weight concrete, $R$ | Lightweight concrete, $R$ |
|---|---|---|
| Outside-surface air film | 0.17 | 0.17 |
| 5 ½-in concrete | 0.41 | 1.04 |
| 1 ½-in polystyrene | 6.25 | 6.25 |
| ½-in gypsum board | 0.45 | 0.45 |
| Inside-surface air film | 0.68 | 0.68 |
| $R$, total | 7.96 | 8.58 |
| $U = 1/R$ | 0.126 | 0.116 |

NOTE: for insulation other than that noted above correct the insulation $R$ value according to the following:

| Thickness, in | 1 | 1 ½ | 2 |
|---|---|---|---|
| Polyurethane | 6.25 | 9.38 | 12.50 |
| Polystyrene | 4.17 | 6.26 | 8.34 |
| Fiberglass board | 4.00 | 6.00 | 8.00 |
| Fiberglass blanket | 3.70 | 5.55 | 7.40 |

### 3. Sandwich wall panels

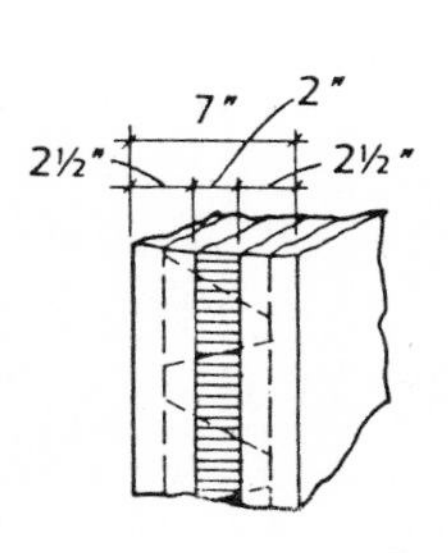


| | Normal-weight concrete, $R$ | Lightweight concrete, $R$ |
|---|---|---|
| Outside-surface air film | 0.17 | 0.17 |
| 2 ½-in concrete | 0.188 | 0.48 |
| 2-in polystyrene | 8.34 | 8.34 |
| 2 ½-in concrete | 0.188 | 0.48 |
| Inside-surface air film | 0.68 | 0.68 |
| $R$, total | 9.57 | 10.15 |
| $U = 1/R$ | 0.104 | 0.098 |

### 4. Marble-faced sandwich panel

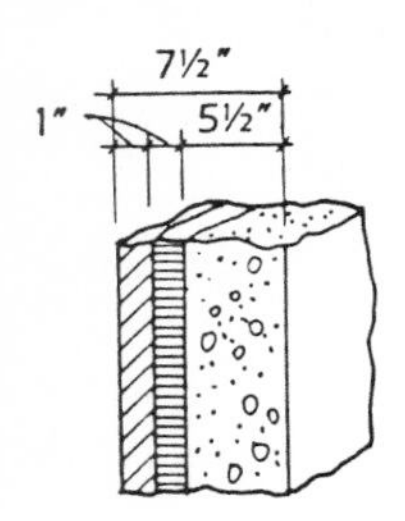


| | Normal-weight concrete, $R$ | Lightweight concrete, $R$ |
|---|---|---|
| Outside-surface air film | 0.17 | 0.17 |
| 1-in marble facing | 0.07 | 0.07 |
| 1-in polystyrene | 4.17 | 4.17 |
| 5 ½-in concrete | 0.41 | 1.04 |
| Inside-surface air film | 0.68 | 0.68 |
| $R$, total | 5.5 | 6.13 |
| $U = 1/R$ | 0.18 | 0.163 |

### 5. Ribbed sandwich panel

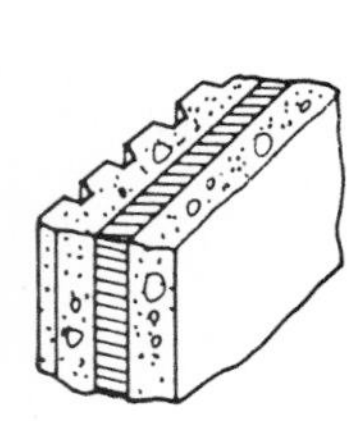

| | Normal-weight concrete, $R$ | Lightweight concrete, $R$ |
|---|---|---|
| Outside-surface air film | 0.17 | 0.17 |
| 3-in average concrete thickness | 0.225 | 0.57 |
| 1-in polystyrene | 6.25 | 6.25 |
| 2 ½-in concrete | 0.188 | 0.48 |
| Inside-surface air film | 0.68 | 0.68 |
| $R$, total | 7.51 | 8.15 |
| $U = 1/R$ | 0.133 | 0.123 |

### 6. Hollow-core wall panel

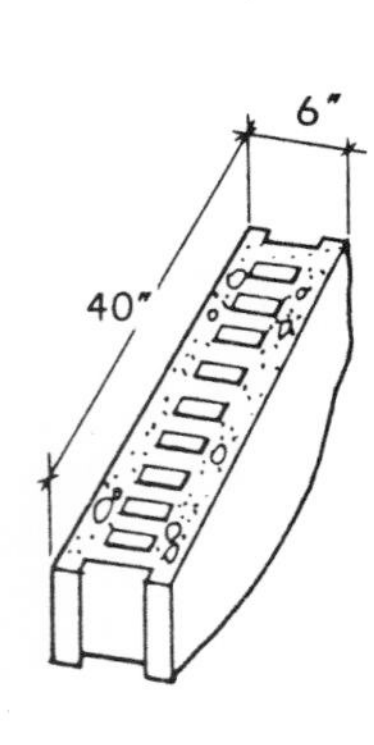


| Cell portion of panel = 59 percent of panel length | Normal-weight concrete, $R$ | Lightweight concrete, $R$ |
|---|---|---|
| Outside-surface air film | 0.17 | 0.17 |
| 1-in concrete | 0.08 | 0.19 |
| Air space ¾ in and larger | 0.97 | 0.97 |
| 1 ¼-in concrete | 0.09 | 0.24 |
| Inside-surface air film | 0.68 | 0.68 |
| $R$, total | 1.99 | 2.25 |
| $U = 1/R$ | 0.50 | 0.45 |

| Noncell portion of panel = 41 percent of panel length | Normal-weight concrete, $R$ | Lightweight concrete, $R$ |
|---|---|---|
| Outside-surface air film | 0.17 | 0.17 |
| 6-in concrete | 0.45 | 0.14 |
| Inside-surface air film | 0.68 | 0.68 |
| $R$, total | 1.3 | 1.99 |
| $U = 1/R$ | 0.77 | 0.50 |

$$U_{av} = \frac{U_1 A_1 + U_2 A_2}{A_{av}} = \frac{0.50 \times 0.59 + 0.77 \times 0.41}{1.00} = 0.61$$

$$= 0.45 \times 0.59 + 0.50 \times 0.41 = 0.47$$

NOTE: If the cells are filled with vermiculite, $R$ = 3-in average × 2.08/in = 6.24 in lieu of the 0.97 for the air space and $U_c$ = 0.137, 0.133, and $U_{av}$ = 0.396, 0.394.

## 7. Flat-slab roof system

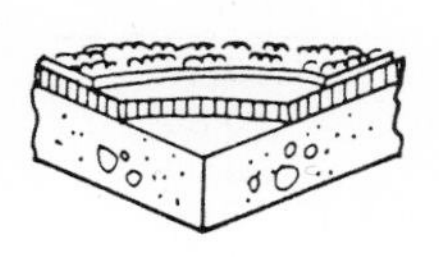

| | Normal-weight concrete, $R$ | Lightweight concrete, $R$ |
|---|---|---|
| Outside-surface air film | 0.17 | 0.17 |
| 3/8-in built-up roofing | 0.33 | 0.33 |
| 1 1/2-in polyurethane | 9.38 | 9.38 |
| 5-in concrete | 0.38 | 0.95 |
| Inside-surface air film | 0.61 | 0.61 |
| $R$, total | 10.87 | 11.44 |
| $U = 1/R$ | 0.092 | 0.0875 |

## 8. Double or single T's

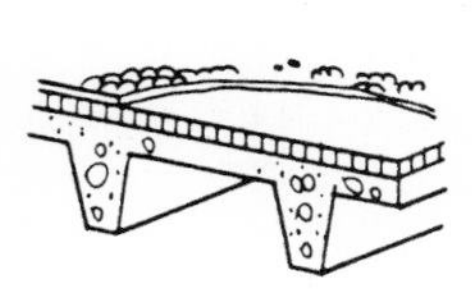

| | Normal-weight concrete, $R$ | Lightweight concrete, $R$ |
|---|---|---|
| Outside-surface air film | 0.17 | 0.17 |
| 3/8-in built-up roofing | 0.33 | 0.33 |
| 1 1/2-in polystyrene | 6.26 | 6.26 |
| 2 1/2-in concrete flange | 0.188 | 0.48 |
| Acoustical tile | 1.89 | 1.89 |
| Inside-surface air film | 0.61 | 0.61 |
| $R$, total | 9.448 | 9.74 |
| $U = 1/R$ | 0.1058 | 0.1027 |

## 9. Hollow-core slab, 8 in

| Cell portion of panel = 56 percent of panel width | Normal-weight concrete, $R$ | Lightweight concrete, $R$ |
|---|---|---|
| Outside-surface air film | 0.17 | 0.17 |
| 3/8-in built-up roofing | 0.33 | 0.33 |
| 3-in vermiculite insulating concrete topping | 3.00 | 3.00 |
| 1 1/4-in concrete | 0.09 | 0.24 |
| Air space 3/4 in and larger | 0.97 | 0.97 |
| 1 1/2-in concrete | 0.11 | 0.28 |
| Inside-surface air film | 0.61 | 0.61 |
| 1 1/2-in fiberglass roof insulation | 6.00 | 6.00 |
| $R$, total | 11.28 | 11.6 |
| $U = 1/R$ | 0.088 | 0.086 |

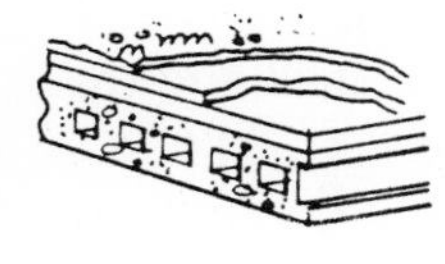

**9. Hollow-core slab, 8 in (*Continued*)**

| Noncell portion of panel = 44 percent of panel width | Normal-weight concrete, $R$ | Lightweight concrete, $R$ |
|---|---|---|
| Outside-surface air film | 0.17 | 0.17 |
| 3/8-in built-up roofing | 0.33 | 0.33 |
| 1 1/2-in fiberglass | 6.00 | 6.00 |
| 3-in vermiculite concrete topping | 3.00 | 3.00 |
| 8-in concrete | 0.60 | 1.52 |
| Inside-surface air film | 0.61 | 0.61 |
| $R$, total | 10.71 | 11.63 |
| $U = 1/R$ | 0.093 | 0.086 |

$$U_{av} = \frac{U_1 A_1 + U_2 A_2}{A_1 + A_2} = \frac{0.88 \times 0.57 + 0.093 \times 0.44}{1.0}$$

0.09 and 0.086

## Example of Nonresidential Building

Given the following office building to be constructed in San Jose, California,

- Heating degree-days for San Jose is 2650.
- The maximum value of $U_{ow}$ from the chart for 2650 degree-days is 0.41.
- If glass constitutes 40 percent of the wall area and single glazing is proposed, then $U_{glass} = 1.13 \times 40$ percent = 0.45, which is greater than 0.41, hence is not permitted.
- If insulating glass is used and the $U$ value is 0.67, then

$$U_{ow} = \frac{U_{wall} A_{wall}\, \text{MCF} + U_{glass} A_{glass}}{A_{ow}}$$

$$= \frac{0.41 \times 1.00 - 0.67 \times 0.40}{0.60 \times 0.75} = U_{wall}$$

$$U_{wall} = 0.237$$

$$\text{MCF} = 0.75 \text{ for walls between 41 and 80 lb/ft}^2$$

If the glass area were reduced to 30 percent of the wall area , then

$$\frac{0.41\times 1.00-1.13\times 0.30}{0.70\times 0.75}=U_{\text{wall}}=0.135$$

If insulating glass is used,

$$\frac{0.41\times 1.00-0.67\times 0.30}{0.70\times 0.75}=U_{\text{wall}}=0.398$$

If glass were reduced to 25 percent of the wall area, then

$$\frac{0.41\times 1.00-1.13\times 0.25}{0.70\times 0.75}=U_{\text{wall}}=0.227$$

The above combinations of glass area and type of glass will give the maximum $U$ value permitted for the walls unless the $U$ value of the roof is decreased. From the required wall $U$ value the type of wall panel insulative system can be selected. If the wall panel system is first selected, then the type and maximum area of glass can be determined.

The $U$ value for the roof system without skylights for the degree-day rating of San Jose is 0.10. For floors over unheated or outdoor areas the maximum $U$ value is 0.21.

## Nonresidential Building in Visalia, California

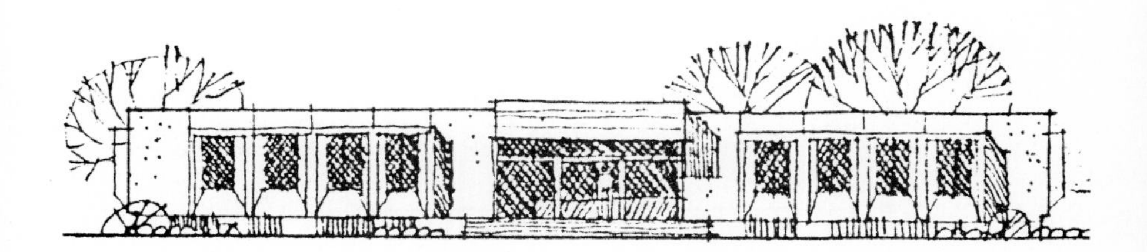

### Heating requirements

In Visalia the annual degree-days value is 2546. The design temperature for winter is 32°F, and that for summer is 100°F. $U_{ow}=0.41$, $U_{or}=0.10$, and $U_{of}=0.215$. The roof must be insulated. Therefore use double T's with 1 1/2-in polystyrene; $U=0.10$.

$A_f$ = area of glass = 6 ft×7 ft×32 = 1344 ft²

$A_{wp}$ = area of pierced panels less glass

= (8×13×32) - 1344 = 1984 ft²

| | | |
|---|---|---|
| $A_d$ | = area of glass entrance doors = (8×13×6) = | 624 ft² |
| $A_{ws}$ | = area of nonpierced panels = (8×13×22) = | 2288 ft² |
| $A_{ow}$ | = area, total wall = | 6240 ft² |
| $U_{wp}$ | = $U$ of 5 ½-in concrete, lightweight = 0.53 | |
| $U_f$ | = $U$ of insulating glass = 0.67 | |
| $U_d$ | = $U$ of glass for entrance doors = 1.13 | |
| MCF | = mass correction factor for walls = 0.95 | |

$$U_{ow} = \frac{U_w A_w \text{ MCF} + U_f A_f + U_d A_d}{A_{ow}}$$

$$U_{ws} \text{ maximum} = \frac{(U_{ow} + A_{ow}) - (U_{wp} A_{wp} \text{ MCF}) - (U_f A_f) - (U_d A_d)}{A_{ws} \text{ MCF}}$$

$$= \frac{(0.41 \times 6240) - (0.53 \times 1984 \times 0.95) - (0.69 \times 1344) - (624 \times 1.13)}{2288 \times 0.75}$$

$$= 0.0957$$

Solid panels must be insulated by (1) insulation applied to the inside surface or (2) sandwich precast panel.

1. 5 ½-in lightweight walls with 1 ½-in polyurethane insulation on the surface

$$U = \frac{1}{\Sigma R} = \frac{1}{R_{oa} + R_c + R_1 + R_{gb} + R_{1a}}$$

$$= \frac{1}{0.17 + 1.04 + 9.38 + 0.45 + 0.68} = 0.085$$

2. Sandwich panel of 2 ½-in lightweight concrete, 1 ½-in polyurethane, and 2-in lightweight concrete

$$U = \frac{1}{\Sigma R} = \frac{1}{R_{oa} + R_c + R_f + R_c + R_{1a}}$$

$$= \frac{1}{0.17 + 0.48 + 9.38 + 0.48 + 0.68} = 0.0894$$

**Cooling requirements**

The latitude of Visalia is 36.4°. Requirements are as follows:

Walls, maximum Btu/(h·ft$^2$) = 32.5
SF (solar factor) = 125; for north orientation or shade, 30
SC (shading factor) = 0.95 for doors; 0.90 for insulation glass
$TD_{eq}$ (temperature differential equivalent) = 30 (if wall weighs between 41 and 70 lb/ft$^2$)

$$\Delta T = 22°$$

$$\text{Btu/}(\text{h}\cdot\text{ft}^2) = \frac{(U_w A_w \text{TD}_{eq}) + (A_f S_f \text{SC}) + U_f A_f T}{A_{ow}}$$

If the building is placed north and south at 12 noon, the altitude angle is 66.3° and the azimuth angle is 0°. Hence the east and west walls are shaded. For a 2-ft projection at the top of the glass there is shading:

tan < × overhang = height of shaded area

tan 66.3° × 2 ft = 4.5 ft

| | |
|---|---|
| $U_w A_w \text{TD}_{eq}$ | |
| Solid panel = 0.089×2288×30 | 6,109 |
| Pierced panel less glass = 0.53×1984×30 = | 31,545 |
| $A_f S_f$ SC | |
| Insulating glass | |
| North, east, west = 6 ft×7 ft×24×30×0.90 = | 27,216 |
| South, shaded = 6 ft×4.5×8×30×0.9 = | 5,832 |
| South, unshaded = 6 ft×25×8.125×0.9 = | 13,500 |
| Noninsulating glass | |
| North = 24×13×30×0.95 = | 8,892 |
| South, shaded = 24×8×30×0.95 = | 5,492 |
| South, unshaded = 24×5×115×0.95 = | 14,250 |
| $U_f A_f \Delta T$ | |
| Insulating glass = 0.67×1344×22 = | 19,810 |
| Doors = 1.13×624×22 = | 15,512 |
| | 148,138 |

148,138/6240 = 23.7 Btu/ft$^2$. The actual 23.7 Btu/ft$^2$ is less than the maximum allowed, 32.5 Btu/ft$^2$; therefore the wall system is acceptable.

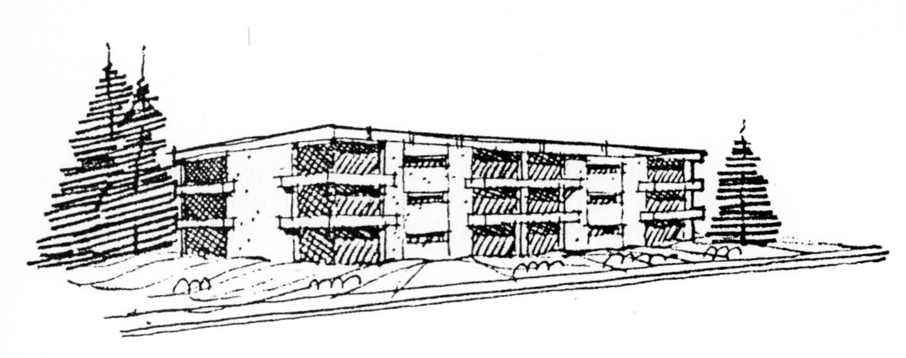

**Three-Story (Low-Rise) Apartment Building**

Floor area = 60 ft × 120 ft = 7200 ft$^2$

Total floor area = 3 × 7200 ft$^2$ = 21,600 ft$^2$

Perimeter 3 × 10 ft × 2( 6 + 120) = 10,800 ft$^2$

Area of glass = 3552 ft$^2$

Standard requirements for heating are as follows:

Standard glass at 20 percent of floor area

= 0.20 × 21,600 × 1.13 = 4881.6

Wall weight = 5/12 × 110 = 45.8 lb/ft$^2$; $U_w$ = 0.16

= 0.800 × 10,800 × 0.16 = 1382.4

Maximum envelope allowed ($U$ × area) 6264.0

**Design solution 1**

Use insulating glass, $U$ = 0.67: $A_f \times U_f$ = 3552 × 0.67 = 2379.8.

$$\text{Wall maximum } U \text{ value, } U_w = \frac{6264 - 2379.8}{(A_w)\ 10{,}900 - 3552} = 0.5359$$

$U$ value of 5 ½-in lightweight concrete wall is 0.529, and 0.529 is less than 0.5359. Therefore, wall insulation for exterior walls is not needed. Roof $U$ value maximum is 0.06, and therefore insulation must be used over the precast concrete roof system.

**Design solution 2**

Use standard glass, $U$ = 1.13: $A_f \times U_f$ = 3552 × 1.13 = 4013.7.

$$\text{Wall maximum } U \text{ value, } U_w = \frac{6264 - 4013.7}{(A_w)\ 10{,}800 - 3552} = 0.31$$

To achieve the required $U$ value, the wall must be insulated by (1) applying insulation to the inside surface or (2) using sandwich panels. For a sandwich panel with a minimum of 5 in of concrete,

$$R \text{ insulation minimum} = 1/U_w - (R_{oa} + R_{conc} + R_{in})$$
$$= 1/0.31 - (0.17 + 0.96 + 0.68)$$
$$= 1.41 \quad (1/2\text{-in polyurethane}) = 3.12$$
$$(1/2\text{-in polystyrene}) = 2.08$$

**Multistory (High-Rise) Apartment Building**

Perimeter = 8 stories × 10 ft × 2(56 + 120) = 27,200 ft$^2$

Glass area = 9000 ft$^2$ (9000/27,200 = 33 percent glass)

Maximum 40 percent of envelope = 0.40 × 27,200 = 10,880 ft$^2$

Standard requirements for heating are as follows:

| | |
|---|---:|
| Standard glass = 0.40 × 27,200 × 1.13 = | 12,294.4 |
| Walls = 0.60 × 27,200 × 0.16 = | 2,611.2 |
| Maximum envelope allowed ($U$ × area) | 14,905.6 |

**Design solution 1**

Standard glass $U$ = 0.67: $A_f \times U_f$ = 9000 × 0.67 = 6030

Wall maximum $U$ value

$$= \frac{U_{ow} - U_f}{A_{ow} - A_{glass}} = U_w$$

$$= \frac{14,905.6 - 6030}{27,200 - 9000} = 0.4876$$

Since 0.4876 is less than 0.529 for a lightweight 5½-in precast wall panel, insulation must be used. Use either (1) insulation applied to the inside surface or (2) precast sandwich panels.

### Design solution 2

If insulating glass and noninsulated wall panels of 5½-in lightweight precast concrete are used, the following maximum glass area is allowable:

$$XA_w U_f + (1 - X)A_w U_w = 14{,}905.6$$

$$X \times 27{,}200 \times 0.67 + (1 - X)(27{,}200 \times 0.53) = 14{,}905.6$$

$$X = 0.128, \text{ or } 12.8 \text{ percent glass}$$

$$\text{Maximum glass area} = 3497 \text{ ft}^2$$

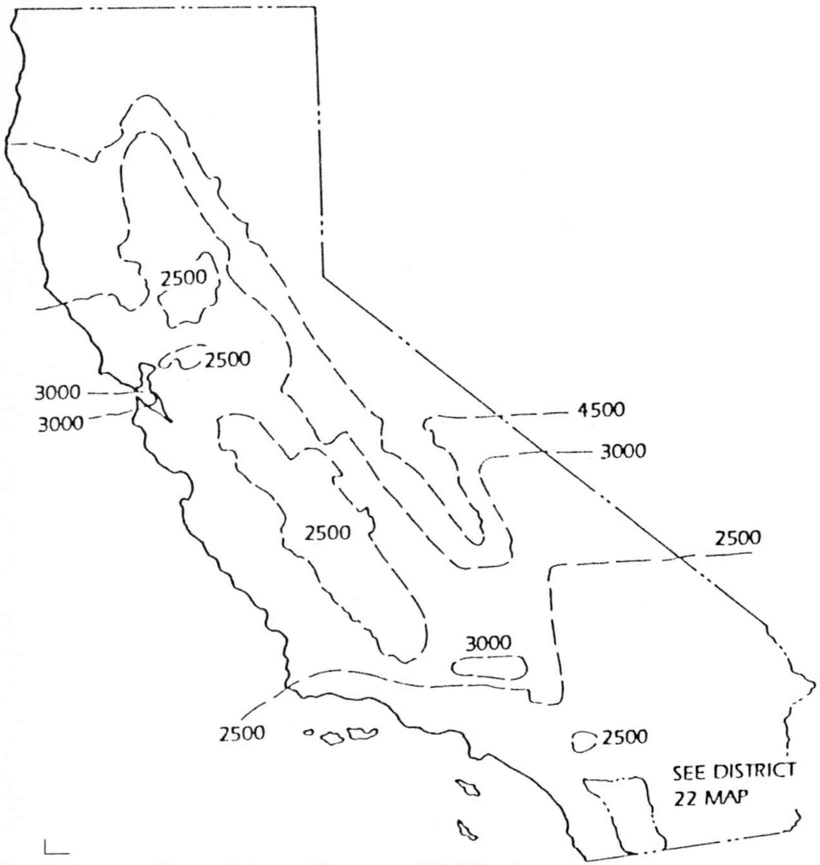


Figure 10.6 General degree-day map of California.

Figure 11.1 Sound barrier wall, Seal Beach, California. (*above*) 6000 ft of 18-ft-high barrier wall deflect noise from the San Diego Freeway over immediately adjacent residential areas. (*below*) Twelve-ft-wide prestressed concrete units are supported laterally by prestressed concrete joists anchored in pole-type drilled footings.

# Chapter 11

# Sound Transmission

For Group R occupancies such as hotels, motels, apartment houses, condominiums, and other non-single-family dwellings, regulations have been established for sound control by the state of California, the Department of Housing and Urban Development (HUD), and the Uniform Building Code, Chap. 35. According to the Uniform Building Code, airborne and impact sound insulation must be provided for walls and floor-ceiling assemblies that separate dwelling units from each other and from public spaces (i.e., interior corridors and service areas). The airborne sound insulation must not fall below a sound transmission class (STC) of 50 or, if field-tested, 45. All openings for pipes, ducts, electrical outlets, and fixtures are to be sealed and treated. Entrance doors from interior corridors together with their permanent seals must not have an STC rating of less than 26. All floor-ceiling assemblies between separate units or guest rooms shall provide sound impact insulation which will give an impact insulation class (IIC) of 50. Floor coverings may be included in the assembly to obtain the rating, but they must be retained as a permanent part of the assembly.

In addition, the state of California and HUD give the following requirements: For noise insulation from exterior sources, the interior community noise equivalent level (CNEL) shall not exceed 45 dB in any habitable room with all the doors and windows closed. For residential locations having an exterior CNEL value greater than 60 dB, an acoustical analysis is required to show that the structure has been designed to meet the interior CNEL value of 45. The exterior value shall be determined by the local jurisdiction in accordance with its general plan.

The unit of sound measurement is the decibel (dB), developed by Alexander Graham Bell. It is the log base of sound energy times 10; i.e.,

| For an energy of | 10 | 100 | 1000 |
|---|---|---|---|
| Log base | 1 | 2 | 3 |
| dB | 10 | 20 | 30 |

Sound is measured by special sound level instruments consisting of a microphone, an amplifier, and output instrumentation.

The following is a list of the approximate sound pressure levels of common everyday noises:

| | |
|---|---|
| Wind and leaves | 10 dB |
| Quiet conversation | 30 dB |
| Refrigerator | 30 dB |
| Quiet radio | 40 dB |
| Conversation | 50 dB |
| Loud radio | 70 dB |
| Speech | 70 dB |
| Car | 90 dB |
| Motorcycle | 110 dB |
| Rock band | 110 dB |
| Sonic boom | 120 dB |
| Jet plane taking off | 130 dB |

The changes of sound level are:

| | |
|---|---|
| 1–2 dB | Not noticeable |
| 3 dB | Just noticeable |
| 5 dB | Clearly noticeable |
| 10 dB | Twice as loud |
| 20 dB | Much louder |

The three basic terms used in sound transmission and sound control with which the designer should be familiar are:

1. *SPL (sound pressure level).* The intensity of sound present in a room, or outside environment, measured in decibels.

2. *STC (sound transmission class).* The amount of sound reduction given by a wall or floor-ceiling assembly. Basically, the SPL on one side of the barrier, less the STC of the barrier medium, equals the SPL on the other side of the medium. This is also the sound transmission loss (STL) of the medium.

3. *IIC (impact insulation class).* A designation given to a floor assembly to indicate its ability to *deaden* impact noises going through the floor-ceiling assembly. Insulating media or floating floor systems give good values of IIC.

Sound interferes with speech, privacy, concentration, sleep, and community stability. The degree of annoyance depends on the SPL, extent of pure tones, time of day, season, what you are doing and who hears it, and background levels. Annoyance is reduced by a volume reduction in the source, reduction or absorption along the sound's path, and masking (use of cover-up sound).

Sound is reduced by the following methods: (1) mass, each doubling of mass resulting in a reduction of 6 dB; (2) discontinuity, use of air spaces of a minimum of 2 in, with each doubling of the air space resulting in a reduction of 5 dB, and dissimilar materials (materials which vibrate at different frequencies); (3) distance, with a 6-dB reduction for doubling of distance from a point source and a 3-dB reduction for doubling the distance from a line source; and (4) use of barriers such as trees, earth, mounds, constructed walls, etc.

The way in which the STC of a wall or floor determines acoustical privacy is illustrated in Fig. 11.2. A good sound barrier should reduce the noise originating in one room to below the background noise level in an adjacent room. The higher the values of the STC and IIC, the more desirable the material or system.

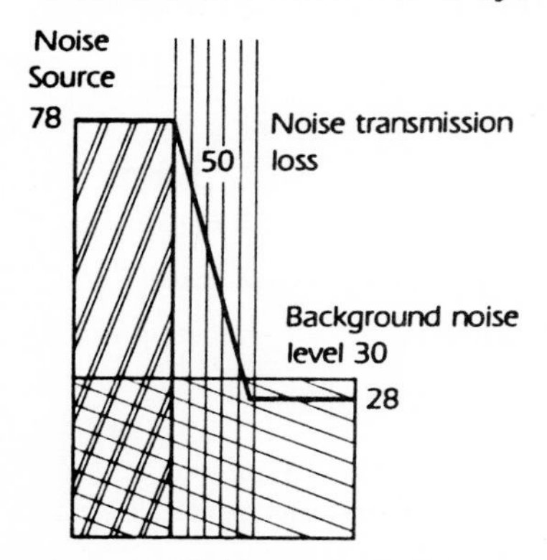


**Figure 11.2** Reduction in noise by a sound barrier.

Sound is made up of waves at various frequencies from 0 to 4000 Hz. Figure 11.3 presents an example of sound transmission loss through a 6-in block wall.

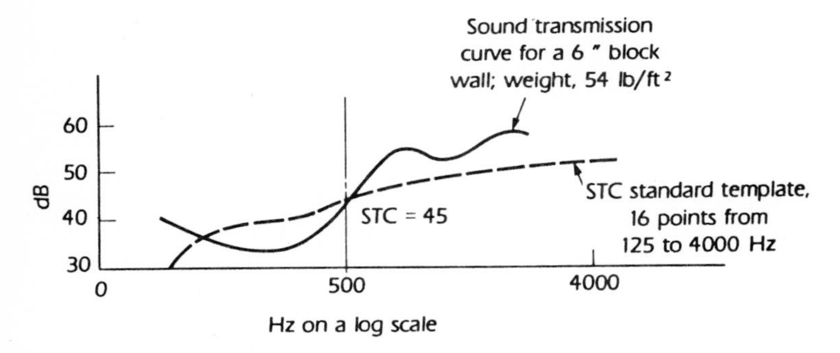


**Figure 11.3** The point where the standard sound transmission curve crosses the true material curve at 500 Hz is the sound transmission class (STC) rating for that material.

The third edition of the *PCI Design Handbook* (pages 9-25 through 9-33) has an excellent summary of acoustical properties of various precast concrete systems and other related materials used in building construction.

Sound leakage occurs from under doors, through louvers, wall openings, and ducts, and over hung-ceiling spaces, as shown in Figs. 11.4 and 11.5.

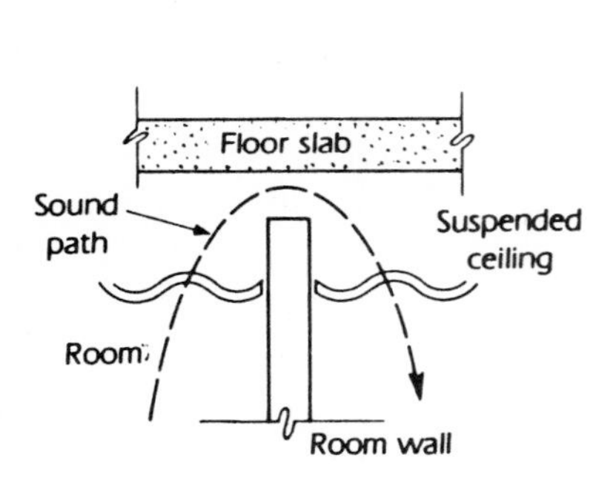


**Figure 11.4** Example of sound leakage.

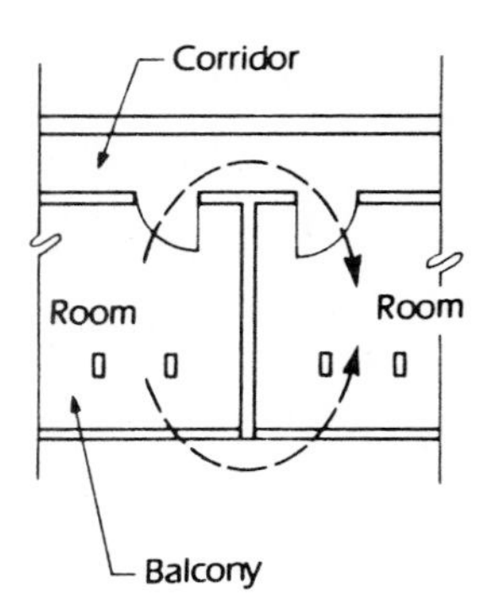


**Figure 11.5** Example of sound leakage.

Figure 11.6 shows various types of sound barriers used in design.

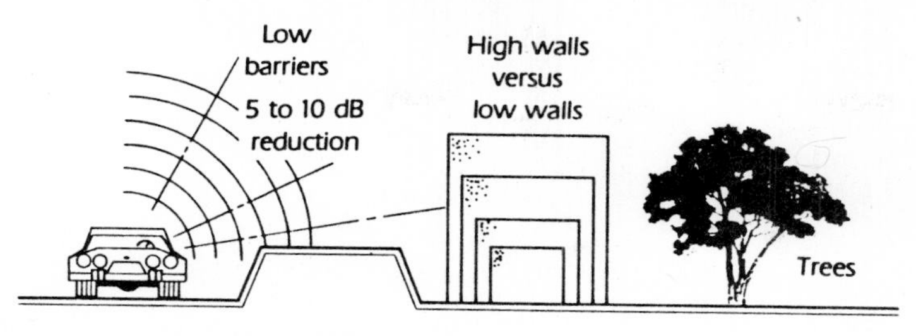


**Figure 11.6** Types of sound barriers.

In California precast and prestressed concrete units have been used to form barriers to deflect freeway noise over adjacent residential areas (see Figs. 11.1 and 11.7). The effectiveness of these barriers is dramatic, with complete silence and lack of audible traffic noise behind the wall. High walls (over 12 ft) are especially suited to the use of plant-cast precast prestressed concrete units.

**Figure 11.7** Sound barrier wall, Seal Branch, California. This sound barrier wall extends along the 605 Freeway on the north side between Valley View Street and Seal Beach Boulevard. The prestressed concrete solution was the low bid in a series of alternative systems and materials specified for design-construct proposals. Additional savings were realized from time saved by using precast concrete as compared with slow on-site construction methods. Sound is almost totally attenuated behind this wall.

**Figure 12.1** Los Cerros Intermediate School fire, Danville, California, June 2, 1979. Over 30 fires occurred in 1979 in California's nonfireproof schools, many of which were sprinklered. In this case, as in many others, arson was suspected. Not only did the San Ramon Valley School District have to pay the first $100,000 in cash for repairs, in accordance with the deductible provisions of its insurance policy, but it also spent over $200,000 in additional funds for temporary school quarters for the displaced students. As a result of another school fire (Neil Armstrong Elementary School, San Ramon), the school district in a lawsuit cited the architects for their failure to design the school to prevent the spread and extent of the fire.

Chapter

# 12

# Fire Resistance and Fireproofing

The Uniform Building Code requires that the designer consider fire resistance in the design of structures. The amount of fire protection required is determined by various provisions in the code, which can be somewhat confusing to one unfamiliar with the logic used. Therefore, the procedure for determining the required fire resistance of building components is given here.

1. *Determine occupancy type.* Section 501, Table 5-A, and Chaps. 6, 7, 8, 9, 10, 11, and 12 of the Uniform Building Code define the type of occupancy of the structure according to the type of use envisioned.
2. *Determine fire zone.* The intended location of the building will affect required fire resistance of certain portions of the structure.
3. *Determine proximity to property lines.* Increased fire-resistive requirements are imposed for exterior walls that are too close to property lines.
4. *Determine type of construction.* The type of construction (Type I, II, III, IV, or V) required for various components will be indicated by the occupancy type, the height of the building, and the area per tier. Types of construction are defined in Chaps. 17 through 22 of the Uniform Building Code. Tables 5-C and 5-D give types of construction required for the areas and heights of the structure being considered. (See also Sec.

505 for area requirements for multistory buildings.) Section 709 and Table 7-A give specific requirements for open parking structures.

5. *Determine required component fire rating.* On the basis of type of construction required, determine the fire rating required for the building components being designed. Table 17-A of the Uniform Building Code gives a breakdown of types of construction and required fire ratings in hours for the component. Fire ratings for components are determined by testing with a standard procedure as outlined in ASTM E-119. The component being tested is subjected to a specific fuel load in a standard furnace, which develops a specified temperature and time relationship calibrated to this fuel load. The fire rating of the component is the maximum time that the component can satisfy each of three basic "endpoint" criteria:

*a.* Support of the design applied load

*b.* Surface and interior integrity preventing the passage of flame through the member

*c.* For floors or roofs, limitation on the temperature of the unexposed surface of the deck

When any of these criteria is exceeded, the test is stopped and the time recorded is the fire rating of the component or system. Shown below is a part of Table 17-A.

| | Noncombustible | | | | Combustible | | | | |
|---|---|---|---|---|---|---|---|---|---|
| | Type I | Type II | | | Type III | | Type IV | Type V | |
| Building element | Fire resistance | Fire resistance | 1h | N | 1h | N | HT | 1h | N |
| Exterior bearing walls | 4 | 4 | 1 | 4 | 4 | 4 | 4 | 1 | N |
| Exterior nonbearing walls | 4 | 4 | 1 | 4 | 4 | 4 | 4 | 1 | N |
| Interior bearing walls | 3 | 2 | 1 | 1 | 1 | N | 1 | 1 | N |
| Interior nonbearing walls | 1 | 1 | 1 | 1 | 1 | N | 1 or HT | 1 | N |
| Structural frame | 3 | 2 | 1 | 1 | 1 | N | 1 or HT | 1 | N |
| Floors | 2 | 2 | 1 | 1 | 1 | N | HT | 1 | N |
| Roofs | 2 | 1 | 1 | 1 | 1 | N | HT | 1 | N |

NOTE: N = no fire rating required; HT = heavy timber

6. *Determine thickness and concrete cover requirements.* Chapter 43 of the Uniform Building Code lists fire-resistive standards for materials and components based upon the fire rating requirements determined above. Table 43-A gives required cover over reinforcing and strand for precast and prestressed concrete members. Table 43-B gives required thickness of walls, and Table 43-C gives required thickness of floors and roofs, to satisfy fire ratings required. Some of these requirements are listed in Table 12.1.

**TABLE 12.1 Thickness and Cover Requirements**

| Component | 4 h | 3 h | 2 h | 1 h |
|---|---|---|---|---|
| Solid concrete walls, thickness, in | 7 | 6 1/2 | 5 | 3 1/2 |
| Concrete floors and roofs (normal-weight concrete), thickness, in | . . . | 6 1/2 | 5 | 3 1/2 |
| Concrete floors and roofs (lightweight concrete), thickness, in | . . . | 5 | 4 | 3 |
| Concrete coverage, pretensioned members, Grade A concrete* | | | | |
| Beams and girders | 4 | 3 | 1 1/2 | 1 1/2 |
| Solid slabs | | 2 | 1 1/2 | 1 |
| Concrete coverage, reinforcing bars, Grade A concrete* | | | | |
| Columns and girders 12 in wide and larger | 1 1/2 | 1 1/2 | 1 1/2 | 1 1/2 |
| Concrete joists | 1 1/4 | 1 1/4 | 1 | 3/4 |
| Floor and roof slabs | 1 | 1 | 3/4 | 3/4 |

*For sand-lightweight concrete, cover requirements may be reduced by 25 percent but by not less than 1 1/2 in in girders and columns or 3/4 in in slabs.

The Uniform Building Code also allows the use of PCI MNL-124 (*Design for Fire Resistance of Precast Prestressed Concrete*), which is incorporated in UBC Standard 43-9. This publication gives analytical procedures which may be used to determine required cover and concrete thickness of prestressed and precast concrete components. Fire ratings of hollow-core planks are based upon tests performed by the particular licensing agent of the product under consideration. These tests have been performed in accordance with the requirements of ASTM E-119. Fire resistance ratings of precast and prestressed components and built-up assemblies can also be determined by referring to *Fire Resistance Ratings*, published by the American Insurance Association, New York, and the *Fire Resistance Index*, published by Underwriters Laboratories, Chicago.

## Architectural Precast Concrete Cladding Units as Fire Protection

Such units may be used to fireproof structural steel members, thus permitting steel-framed buildings to be classified as Type I and Type II construction. Architectural precast concrete units are used to enclose bracing systems, spandrel beams, girders, and columns. These precast concrete enclosures contribute elements of color, texture, and form to help express the design theme of the architect. The entire enclosure can give a sculptured effect to a column or frame. These cladding units are quickly erected and therefore reduce construction time when compared with on-site methods of providing exterior fireproofing. The joints between cladding units must be protected in order to provide the fire rating required for the individual cladding units (see PCI MNL-124).

## Fire Resistance and Sprinklers

The Uniform Building Code allows increases in maximum allowable heights and floor areas when an automatic sprinkler system is installed (Secs. 506 and 507). Section 508 allows elimination of fire rating requirements for all 1-hour construction when the building is sprinklered. Chapter 33 permits reducing the required number of exits and increasing the distance to exits when sprinklers are used. These reductions in life safety requirements in exchange for automatic sprinklers are known as "tradeoffs." The intent of these tradeoffs is to make it economically viable for owners to provide for sprinklers in the design of the building. This practice is not new and has been permitted for many years in United States building codes. Recently there has been an increased push by special-interest groups to effect further reductions in life safety requirements for buildings by using the tradeoff concept. The danger in this concept is that total reliance for fire safety and life safety is based upon 100 percent reliability of the sprinkler system. Statistics for past performance of sprinklers indicate that human and mechanical factors render this assumption invalid. The Prestressed Concrete Institute, in an industry position paper, indicates "strong objection to the concept of tradeoffs and endorses the following positive recommendations":

1. Building structures must be designed to minimize the possibility of failure when subject to fire.
2. Codes should be reoriented toward reducing fire hazards, smoke-generating materials , and combustible finishes.
3. Compartmentation is a proven method of providing life safety for building occupants and should be provided to contain a fire within a limited area.
4. Sprinklers should be required in hazardous areas, particularly where combustible contents exist. However, the structural integrity or life safety aspects of a building must not be impaired.

In summary, sprinklers are not totally reliable. Safe designs should take into account the possibility of sprinkler failure rather than the hope of sprinkler performance. Use of proven noncombustible materials, compartmentation, smoke detection, and selective use of sprinkler systems offer the best life safety system.

Compartmentation is the breaking up of the floor area into smaller subdivisions that contain the spread of fire, as is obtained with fire-resistive partitions between living units in apartments, etc. This aspect of construction has contained fires within relatively small areas until fire-fighting equipment and personnel have arrived. Chapter 38 of the Uniform Building Code requires sprinklers in practically all types of occupancies, independent of type of construction, and does not recognize the tremendous life safety aspects of compartmentation. The insurance industry, however, recognizes the advantage of fire-resistant construction; recent rate comparisons indicate a 4:1 ratio in favor of concrete for structure and contents. Extended coverage rates for a nonfireproofed steel structure are 9 times those for a precast concrete structure.

PCI MNL-124 provides an analytical method of evaluating the fire endurance of structures made of precast and prestressed concrete. This manual brings together information from many sources and permits the engineer to design for fire endurance without referring directly to standard tests. The manual is a positive contribution for designers, building officials, and insurance underwriters—for anyone concerned with fire safety of buildings.

**Figure 13.1** Oyster Point Marina breakwater, South San Francisco, California. Adequate cover, coupled with proper concrete mix proportioning and plant-casting quality control techniques, assures that these prestressed concrete sheet piling units will have longevity even when subjected to the heavy saline environment of San Francisco Bay and the great potential for corrosion in the splash zone above low water.

Chapter

# 13

# Minimum Concrete Coverage for Reinforcing

Minimum clear thicknesses of concrete cover are provided over reinforcing bars or prestressed strands in precast concrete members to provide sufficient protection from the elements and to satisfy the applicable requirements of fire protection. The greater of these two basic requirements governs the amount of cover required.

1. *Fire protection.* The Uniform Building Code uses minimum-cover criteria as outlined in Sec. 4303 and gives required cover dimensions in Table 43-A as one means to satisfy required fire rating requirements for precast and prestressed concrete components in structures.

2. *Protection from the elements.* In an alkaline concrete environment, reinforcing develops an isolating iron oxide film that prevents corrosion and makes additional protection unnecessary, such as by galvanizing or by providing cathodic protection. Minimum cement contents and low water-cement ratios, coupled with compliance with the cover requirements listed in the Uniform Building Code, will suffice to provide plant-cast precast concrete elements that will last indefinitely. Section 2607 gives the requirements for plant-cast precast elements and prestressed concrete members. For components that will be subjected to severe ocean exposure, minimum cover should be increased to $2^1/_2$ in over main reinforcing and prestress strands and to $1^1/_2$ in over secondary reinforcing (Prof. Ben C. Gerwick, Jr. in ACI Publication SP47-14). In this regard it should be reiterated that increased cover is just one of the durability requirements, which include proper water-cement ratio, minimum cement content, proper aggregates, proper tricalcium aluminate content in the cement along with maximum alkali content, and, in some cases, extended moist curing.

Chapter

# 14

# Design Loads

The design of precast concrete is similar to the design of buildings in steel, wood, or cast-in-place concrete. The exceptions are that precast units must also be designed for handling and that special consideration must be given to joinery. These loads are identified and their general nature discussed in this chapter. The manner in which the precast components are designed to resist these loads is covered more fully in Part 3. The design of total precast and prestressed concrete structures to resist seismic lateral forces, to the extent permitted by the Uniform Building Code, is covered in Chaps. 2 and 15. In general, prestressed and precast concrete components and/or structures are subjected to four major categories of loading :

1. Vertical
   *a*. Dead loads
   *b*. Live loads
2. Lateral
   *a*. Wind loads
   *b*. Seismic loads
3. Volumetric-change-induced loads
   *a*. Shrinkage
   *b*. Creep
   *c*. Temperature
4. Handling
   *a*. Stripping

*b*. Shipping
*c*. Erection

## Dead Loads

Dead loads consist of the weight of the structural system plus all permanently attached materials and equipment. In the design of a system, assumptions are often made for the dead load of the member or members being designed. If the assumption is in error by more than 10 percent, then a correction is necessary. The designer will often use the manufacturer's product information tables to determine member dead load prior to product design. Some typical dead loads of floors, ceilings, roofs, and walls are given in tabular form on page 11-2 of the third edition of the *PCI Design Handbook*.

## Live Loads

Live loads are the result of temporary, transitory weights of people, vehicles, snow, or moving equipment. Governing building codes recommend minimum live loads. The architect or engineer is responsible for defining the final design live loads to be used. These may or may not be the same as minimum code recommendations. The final values are based upon experience according to the use or occupancy of the building. Table 23-A of the Uniform Building Code gives some recommended minimum live loads to be used in designing buildings. If the tributary area supported by a member is large, a reduction is allowed because of the small probability that loading will occur over the entire area at any one time. Section 2306 of the Uniform Building Code bases this reduction on the amount of area supported over 150 $ft^2$. The amount of live-load reduction is limited to 40 percent for members receiving load from one level only and 60 percent for other members. Live loads exceeding 100 $lb/ft^2$ may not be reduced.

## Wind Loads

Wind produces dynamic forces with varying magnitudes due to gusts, altitude, direction, building shape, and shadow or funneling effects caused by surrounding structures. In the design of lower structures, wind loads are treated as static forces whose magnitude depends upon building location and component height above the ground. In the Uniform Building Code wind loads are determined by geographical location in wind pressure zones shown in Fig. 4, Chap. 23, and by the height above the ground and the severity of exposure. Section 2311 gives other requirements for wind design.

## Seismic Loads

Seismic ground motions occur in random patterns in both vertical and horizontal directions. They are very complex and nonpredictable. Some of the variables are focal-point depth, type of fault action, wave propagation through different rock and soil strata, fault location, magnitude, and probability. In any earthquake there are usually foreshocks, main shock, and aftershocks. No two earthquakes have ever been recorded to have occurred at the same focal point or of the same acceleration pattern. High-rise buildings are designed to withstand both known earthquake patterns and human-caused patterns. They are designed by the dynamic concept and compared with the static concept. Acceptable limits are set for story drift.

Low-rise buildings and precast concrete components used in buildings are designed according to the static concept, with story drift limits also considered. The static concept is referred to in the Uniform Building Code as the equivalent-lateral-force (ELF) concept. Section 2312 of the Uniform Building Code details the requirements of this procedure. Chapter 15 of this book goes into greater detail on the basic concepts of seismic design theory.

Seismic activity causes the application of another type of loading on the members and their connections to one another, diaphragm elements, or lateral stiffening elements: repetitive loading. Repetitive loads are cyclical loads which are applied, removed, and reapplied and, in the case of seismic cyclic loads, are characterized by reversals in direction of application. Many structural configurations using precast concrete which appear to offer promise in resisting cyclic seismic loads must have their performance verified under the influence of repetitive (cyclic) loading.

## Forces Induced from Volumetric Changes

Volume change movements resulting from shrinkage, creep, and temperature, if restrained, can induce significant forces in precast concrete component connections and interfacing materials in the structure. The significance of these three factors varies with the type of precast member being considered. In general, all three factors significantly affect precast prestressed members, while only temperature changes significantly affect nonprestressed members; and the effect of each factor increases as the length (or size) of the member increases.

### Shrinkage

The degree of precast concrete shrinkage (reduction of concrete volume) brought about by the evaporation of water used in the concrete mix is normally less than that of cast-in-place concrete of the same strength because low water-cement ratios are specified for precast concrete mixes. On the other hand, some precast units often tend to be relatively long; therefore, the total shrinkage (total change in dimension) of a single unit is sufficiently large to warrant close attention during the design process. Short nonprestressed wall panels (approximately 30 ft or less) constructed with low-slump concrete are usually not much affected by shrinkage. However, many wall panels are 40 ft or more in length and are usually prestressed and fabricated with higher-slump concrete. Such long prestressed wall panels are significantly affected by shrinkage.

All precast prestressed beams must be carefully designed to account for the effects of concrete volume change brought about by shrinkage. Shrinkage acts to shorten the member. As the member shortens, two important things happen: (1) the force in the prestressed tendons decreases, and (2) the member to which the beam is attached may be subjected to additional forces induced by the shortening movement. The effects of loss of prestress force are considered when selecting the initial prestress force to be applied to the member. The effects of possible transfer or redistribution of forces through differential movement of structural members are considered in design of the connection details.

Column sections are similar to wall panel sections in that the effects of shrinkage are highly dependent upon the factors of slump, length, presence or absence of prestressing, and magnitude of the prestressing force.

### Creep

Creep is the long-term deformation of a material under sustained load. Precast prestressed members have a greater tendency to creep than do members which are not prestressed. Creep tends to reduce the initial prestress forces placed on a member and to reduce its length at the centroid of the prestress force. The effects of creep must be carefully considered in order to prevent the development of excessive stresses and deflections due to loss of prestress within the prestressed member and to prevent the transference of forces caused by the reduction in length of the prestressed member (i.e., connection details must be designed so that no unconsidered force is developed by a change in length of the prestressed member).

The effect of creep on precast nonprestressed members is the same as that for cast-in-place concrete of equivalent strength and consistency. However, the high-strength and low-slump consistency of concrete used in precast operations is generally more resistant to the effects of creep than the concrete normally used in cast-in-place operations.

The effect of creep on wall panels, beams, and columns is similar to that of shrinkage except that (1) creep continues over a longer period of time, (2) lightweight concrete creeps more than normal concrete, and (3) creep on load-bearing wall panels and columns can be significant even when the panels or columns are not prestressed.

### Temperature

The effects of temperature on precast nonprestressed structural members are treated in the same manner as for cast-in-place members which use deformed bars as the primary reinforcing element.

When prestressing strands are used as the primary reinforcing elements, the designer must pay special attention to the effect of temperature changes because (1) such members tend to be relatively long and the total dimensional change induced by the temperature change, therefore, is represented as one relatively large change at one location (as contrasted with a series of relatively small changes spread out over many locations); and (2) changes in temperature often have a significant effect on the force developed by prestressing strands embedded in long members. (An increase in temperature causes a significant lessening of the prestress force on a strand which is not yet embedded in a member. However, once the member is cast and the strand is bonded to the concrete, an increase in temperature usually results in a negligible increase in strand tension.) Temperature changes must be carefully considered when designing connections for all structural members (walls, beams, and columns) to make sure that no unexpected or unaccounted-for force develops within the structural system. Of special significance is the effect of temperature-induced forces on long, thin spandrel elements and their connections, as outlined in Chap. 20.

The magnitudes of expected movements resulting from shrinkage, creep, and temperature strains for various precast and prestressed concrete components and structures can be estimated by using the design information given on pages 3-5 and 3-15 of the third edition of the *PCI Design Handbook*. Designers should consider these movements and forces in the preliminary design stages of projects, especially in composite structures consisting of precast prestressed concrete components with poured-in-place concrete closures or posttensioned slabs. The location and arrrangement of lateral stiffening elements are critical in light of long-term creep and shrinkage movements to preclude severe distress and cracking. Closer spacing for contraction joints coupled with temporary gaps in posttensioned slab construction to allow initial creep and shrinkage movements to occur is an essential practical consideration in this type of construction.

## Handling Loads

The term "handling" is included here as a loading classification to emphasize the fact that there exists a general set or collection of loads which are extremely important variables affecting the design of precast members but which are not even relevant to the design of most cast-in-place members.

Handling loads include those induced by the production operations of stripping, storing, transporting, and erecting. During each of these operations, the position of the member may vary; the combination of applied loads (dead, suction, mechanical bonding, impact) may vary, the location and characteristics of the member supports may vary, and the strength of the concrete may be significantly different at the time of each operation.

### Dead loads (member self-weight)

A discussion of dead loads is reintroduced to emphasize that such loads are a major production factor as well as an in-place factor to be considered. The larger the dead load of the member, the greater the work and time involved to move that member. Usually a precast member is moved in the following manner: it is first lifted from the casting bed and placed on a storage pad, then lifted from the storage pad and placed on a trailer for transportation to the site, and finally lifted from the trailer to its final resting place. Thus a minimum of three lifting operations normally is involved. When the dead load of the member is reduced, the size of the equipment needed to move the member is reduced and the time to move the member usually is also reduced. However, it should be noted that it is generally less costly to handle one large member than to handle two small members which are equivalent in weight to the large one.

### Suction

Suction forces are developed between concrete and the form in which it is placed. These forces are relatively insignificant for cast-in-place concrete, and the stresses developed by the form-stripping operation are usually ignored in the structural analysis of the member. In contrast, the suction forces on precast wall panels are significantly large and must be given special consideration during the form-stripping operation. Suction is insignificant in precast prestressed beam design since, when the tensile forces are transferred from the prestressed strands to the concrete in which the strands are embedded, the concrete beam shortens in the form bed; this shortening releases, or "breaks," the suction between form and concrete. Thus, there are no suctional forces to resist when removing precast pretensioned beams from their forms.

### Mechanical bonding

Mechanical bonding forces are often developed between concrete and the form in which it is placed. Forces developed by mechanical bonding during form stripping are similar to those created by suction in that they are relatively insignificant for cast-in-place concrete but often tend to be significant during the stripping operation for all types of precast units. Careful attention must be paid to the architectural detailing of the shape of wall panels to minimize the amount of mechanical bond that develops at panel edges, faces, and openings.

The sides of precast beams are never formed as parallel lines. The form sides are always canted slightly in order to assure that no mechanical bond will develop between the form and the concrete. (Recommended minimal slopes, or draft, are discussed more fully in Chap. 22.) If it were possible to maintain two perfectly parallel sides, it would be unnecessary to provide draft in the form sides. However, there is sometimes some slight bowing of the form sides. The sides are sloped in order to make sure that even after the bowing occurs, the concrete may be lifted straight up without encountering any mechanical keys formed by excessive deflections of the form sides.

The amount of draft depends on the design of the formwork itself and how resistant it is to deflecting under the weight of the concrete and the forces generated by the vibrating effort. It is important to give careful consideration to the effects of mechanical bonding regardless of whether the concrete is precast or cast in place. Poorly detailed forms for plant-manufactured precast concrete can result in a reduction in the number of times that a form can be reused, excessive stripping costs, and undesirable damage to the concrete finish.

### Impact

Impact loads are induced in all precast members when they are transported on trucks. The producer attempts to support the member at only two transverse points on the truck trailer (preferably above the trailer axles). Supporting the member at these points tends to minimize the amount of bouncing and thus minimizes the impact forces placed on the member.

A detailed summary of handling forces, impact loads, and design procedures is given in Chap.19.

**Figure 15.1** Gregory Bateson State Office Building, Sacramento, California. 8-ft-wide by 24-in-deep prestressed concrete double T's span between precast concrete frames that take wind and seismic lateral forces to the foundations as well as providing support for vertical loads. The lateral analysis for the building was performed by using the equivalent-lateral-force (ELF) design procedure outlined in the Uniform Building Code.

Chapter

# 15

# Seismic Design and Structural Systems

Three types of structural systems are used in buildings to resist lateral forces resulting from wind or seismic disturbances. These systems are:

1. Shear wall systems
2. Moment frame systems
3. Combined wall and frame systems

The present state of the art in precast concrete construction and the current level of acceptability in the Uniform Building Code and the *Recommended Lateral Force Requirements* of the Structural Engineers Association of California (referred to as the Blue Book) restrict the use of precast concrete in zones of high seismicity to shear wall systems and systems employing details conforming to the ductile moment-resisting frame design provisions of the Uniform Building Code for cast-in-place concrete. In this chapter we shall briefly discuss an empirical dynamic theory of the seismic response of buildings in resisting seismic forces and the current equivalent-lateral-force (ELF) design provisions of the Uniform Building Code and present a procedure which may be used for the rational analysis of shear wall buildings in resisting lateral forces. Finally, some design examples are given to demonstrate further the use of principles involved in shear wall design with precast concrete elements.

## Equivalent-Lateral-Force Design

The Uniform Building Code provisions attempt to reduce complex dynamic relationships resulting from seismic ground motion effects on buildings into a loading situation resulting from equivalent static lateral forces. In general, the actual induced lateral forces from a severe earthquake are much greater than the forces resulting from the use of the equivalent-lateral-force procedure. A frequently used method of determining seismic forces on buildings consists of performing a dynamic analysis based upon a response spectrum for a severe earthquake. The 1940 El Centro earthquake is commonly used as a model. This earthquake is the most severe for which we have accurate recorded data, and it is estimated that the likelihood of its occurrence would be once in 50 years. The maximum recorded ground acceleration for the El Centro quake was 0.33 *g*.

The graph of base shear versus building period (Fig. 15.2) shows the

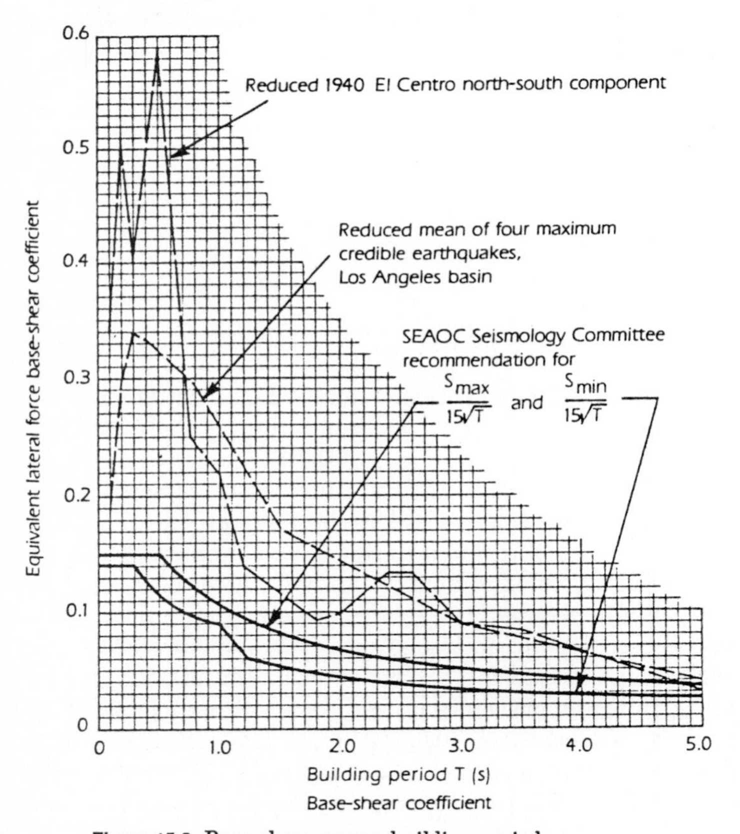


**Figure 15.2** Base shear versus building period.

large difference between actual seismic forces and those resulting from the ELF procedure. This difference in energy demand must be satisfied by inelastic deformation of the structure. In the design of stiff shear wall buildings, the Uniform Building Code assigns severe penalties in the form of a higher $K$ factor. In ductile frame structures where inelastic action can take place by the yielding of the structure or by the formation of plastic hinges to absorb energy, the ELF penalties are less severe. The required energy dissipation is provided by lateral translation of the structure. Ductility, or ductility demand, is defined as the ratio of the inelastic movement of the structure to the elastic movement. Owing to practical considerations, the code requires that these plastic hinges be made to form in the girder elements of building frames rather than in the columns. In other words, the stiffness of the columns in a ductile moment-resisting frame should be greater than the girder stiffness. Therefore, use of the code provisions requires that we provide for the required ductility by inelastic action in the girder elements while the columns respond elastically, with the result that the seismic disturbance will not cause the collapse of the building, provided drift limitations are also satisfied. This combination of elastic and inelastic deformation in the girders, in satisfying ductility demand or lateral ductility, is called elastoplastic response.

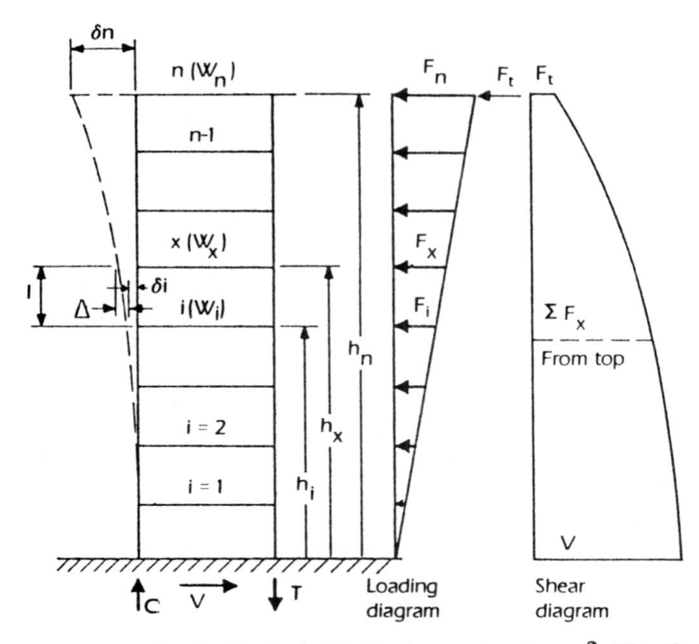


**Figure 15.3** Applied loads and deflections to structure. $\delta_i$ /H is the story drift and is limited to 0.005 × the $K$ factor.

**TABLE 15.1 SEAOC Minimum Earthquake Forces for Structures**

$$V = ZIKCSW$$

$$C = \frac{1}{15\sqrt{T}}$$

$$T = 2\pi\sqrt{\frac{\sum_{i=1}^{n} W_i d_i^2}{g\left[\left(\sum_{i=1}^{n-1} F_i d_i\right) + (F_t + F_n)\, d_n\right]}}$$

$$T = \frac{0.05h_n}{\sqrt{D}}$$

$$T = 0.10N$$

$$S = 1.0 + \frac{T}{T_s} - 0.5\left(\frac{T}{T_s}\right)^2; \quad \frac{T}{T_s} \leq 1$$

$$S = 1.2 + 0.6\frac{T}{T_s} - 0.3\left(\frac{T}{T_s}\right)^2; \quad \frac{T}{T_s} > 1$$

$$V = F_t + \sum_{i=1}^{n} F_i$$

$$F_t = 0.07TV$$

$$F_x = \frac{(V - F_t) W_x h_x}{\sum_{i=1}^{n} W_i h_x}$$

## Seismic-Resistant Design Concepts

The design of structures in the United States for lateral seismic loads generally follows the provisions of the Uniform Building Code. The basis for this code is the *Recommended Lateral Force Requirements and Commentary* of the Structural Engineers Association of California (SEAOC). These documents establish criteria and guidelines which have as their primary object that a minor earthquake should cause little or no damage and that a major earthquake should not cause collapse of the structure. As a corollary, the building should behave elastically under expected frequent earthquakes or normal wind loadings and should be capable of inelastic cyclic behavior under infrequent strong earthquakes or abnormal wind loadings. (See Table 15.1.)

Since code values for seismic forces are factored downward from expected or recorded earthquake motions, it is incumbent that the designer think not in terms of code values but of realistic forces which may be 3 to 4 times these code minima (see Fig. 15.2). Thus, for concrete structures, cracking of members can be expected, with an increase in building period and resulting increased drift and deformation. Rocking about the foundations can also increase building deformation. These conditions result in increased energy absorption and dictate that ductility must be preserved in the members and between members if failure is to be prevented.

The seismic loads that are applied to a building are influenced considerably by the $K$ coefficient in the lateral force formula system: a value of 0.67 for ductile-type frames to a value of 1.33 for bearing-wall buildings without frames and carrying both vertical and lateral loads. It is possible for the designer to incorporate shear walls into the structural system so that the $K$ value can be 0.67, 0.80, 1.0, or 1.33, the value depending on whether the framing is ductile, vertical load-carrying or bearing-wall. When designers are using $K = 1.0$, they must assure themselves that should the walls be heavily damaged during an earthquake, the framing will remain intact to carry the vertical loads.

The SEAOC *Recommended Lateral Force Requirements and Commentary* notes that "the minimum design forces prescribed by the SEAOC Recommendations are not to be implied as the actual forces to be expected during an earthquake. The actual motions generated by an earthquake may be expected to be significantly greater than the motions used to generate the prescribed minimum design forces. The justification for permitting lower values for design are manyfold and include: increased strength beyond working stress levels, damping contributed by all the building elements, an increase in ductility by the ability of members to yield beyond elastic limits, and other redundant contributions."

Most of the current seismic codes in use in the United States are based on these recommendations, including the Uniform Building Code, which is the reference for this chapter. On this basis, the ductility, damping, and drift considerations for a precast structure or for the individual elements and their connections must conform to the same requirements and limitations as for cast-in-place concrete. As an alternative, precast concrete construction could be considered as a separate type of structural system, with recognition given to its own particular response characteristics and behavior. To accomplish this will require extensive research and testing, which are already being carried out in many parts of the world.

The SEAOC *Commentary*, in discussing structures other than buildings which do not have significant damping and do not have elements which are capable of yielding or which may fail by jeopardizing the safety of the structure, suggests that a minimum $K$ value of 2.0 should be used.

It is therefore recommended that structures in areas of high seismicity have a degree of ductility to eliminate this high design load.

### Seismic Performance Characteristics

Since the seismic response of a structure is the action of resisting inertial forces generated by the mass of the elements within the structure as the ground beneath it accelerates and decelerates in a random pattern from earthquake motions, both horizontally and vertically, the structural design must incorporate a shear resistance capable of transferring these loads with smooth continuity from the top to the foundations. This capacity is developed by designing various vertical elements within the structure for continuity and stiffness, based on the concept that in order to achieve the best solution abrupt changes should be avoided. The basic vertical assemblies used for this purpose are rigid frames, braced frames, shear walls, and various combinations of these.

These inertial loads must be accumulated at each floor level and transferred to the vertical shear elements. In addition, for various reasons some of these vertical elements may be discontinuous or have minor changes in stiffness at some level. Since the participation of each element depends on its stiffness relative to total system stiffness and on its position within the structure, any changes in load participation must be redistributed through the floor system. Therefore, it becomes necessary to have a relatively rigid horizontal beam or diaphragm element to function as this transfer member to the vertical elements. The logical "members" are the floors and roofs, which must then be designed for these horizontal forces in addition to their normal vertical load functions.

A concept basic to designing for seismic forces is continuity and providing for a smooth transfer of forces without abrupt changes. This points to a critical part of precast construction, the connections. The cyclic nature of forces to be transferred is not always duplicated by the usual static test loading procedures, and thus many of the current design data for connections may not be valid for seismic design unless arbitrarily low allowable elastic design values are used, which in many cases will not appreciably affect the overall construction cost. Thus, many "non-ductile" precast concrete building systems in lower height ranges may be designed by using this philosophy.

### Terminology

Basic design concepts for earthquake-resistive design include consideration of the following terminology:

1. *Elastic design*. This is the basic design concept used to establish

the physical configuration of a structure or a detail. It is based on stresses factored below the yield point of the material.

2. *Ductility*. This is a measure of the inelastic strength of a material and is usually defined as the ratio of the maximum deformation prior to ultimate failure to the deformation at initial yield. Thus a brittle material which fractures at the yield point has a ductility factor of 1.0. Our concern for ductility is based on the recognition that we usually cannot realistically design elastically for 100 percent resistance to potential earthquake forces except in low-level structures. In the first place we don't know the actual maximum magnitude of forces which may be induced into a structure although an apparent range of forces is identifiable. Even if we did know with certainty, it usually is not economically feasible to design to remain within the elastic range of material stresses. Therefore, the present seismic design concept generally used is based on elastic design criteria for some "expected" level of earthquake excitation rather than on the maximum possible level. The reserve strength of the total structure is thus expected to withstand various degrees of short-term overstress or inelastic action which will prevent structural collapse under the so-called maximum credible earthquake. This does imply various degrees of structural damage, although some buildings in the San Fernando 1971 earthquake appeared to have experienced localized excursions of inelastic action with no visible signs of permanent damage.

3. *Damping*. This is a measure of the ability of a member or a structure to absorb the energy generated by the application of an external repetitive-type loading. This energy absorption reduces the oscillations resulting from the loading and is expressed as a percentage of critical damping, which is the minimum viscous damping that will allow a displaced system to return to its initial position without oscillation.

4. *Drift*. Drift is the unit lateral displacement of a structure. Our concern with drift is based on the comfort of building occupants, the minimizing of secondary stresses in a structure due to eccentricities resulting from the offsetting of vertical load-carrying members (the $P$-$\Delta$ effect), and the effects on nonstructural elements from lateral displacement between floors. These nonstructural elements include elevators, partitions, windows, exterior cladding, vertical pipes, ducts, and shafts.

Five references give a good overview on seismic design as related to construction with precast and prestressed concrete elements:

Clough, Douglas P.: *Design of Connections for Precast Prestressed Concrete Buildings for the Effects of Earthquake,* Prestressed Concrete Institute, Chicago, 1986.

*Design of Prefabricated Concrete Buildings for Earthquake Loads*, ATC-8, Applied

Technology Council, care of Structural Engineers Association of California, Sacramento, 1981.
Freeman, Sigmund: "Seismic Design Criteria for Multistory Precast Prestressed Buildings," *PCI Journal*, vol. 24, November–December 1979.
Hawkins, Neil M. : "State-of-the-Art Report on Seismic Resistance of Prestressed and Precast Concrete Structures," *PCI Journal*, November–December 1977 and January – February 1978.
Parme, Alfred L.: "American Practice in Seismic Design," *PCI Journal*, vol. 17, no. 4, July–August 1972.

## Seismic Forces

Seismic forces are illustrated in Fig. 15.4.

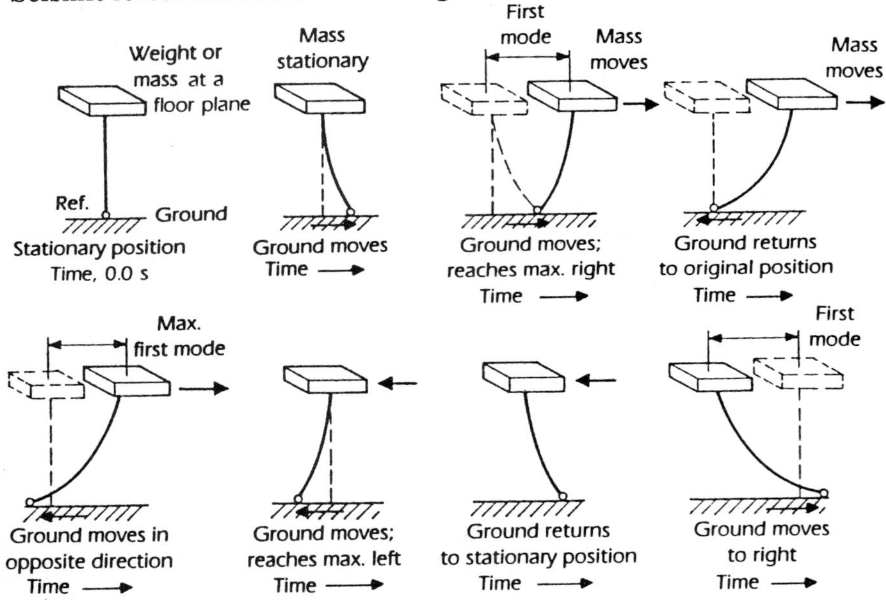


**Figure 15.4** How seismic forces affect buildings.

$F_{\text{total}} = ZKICSW_{DL}$

$V$ = base shear

$V = ZKCSW_{DL}$

Force = mass × acceleration = $ma$

Weight = mass × acceleration of gravity = $mg$

Force = $ma$ = weight $(a/g)$

Let $(a/g)$ = seismic factors

Seismic factors = $ZKICS$

$K$ = factor based upon building rigidity

= 1.33 for stiff shear wall systems

= 0.67 for ductile moment-resisting frames

= 0.80 for dual combination of frame and braced frame or shear wall

= 1.00 for braced-frame system

$C$ = factor relating to the building's period of vibration

$S$ = soil interaction factor

$Z$ = seismic zone factor related to probability and intensity

$I$ = importance of the building's remaining operative after the earthquake

The measurement of acceleration during an earthquake is shown in Fig. 15.5, the movement of typical mode shapes in Fig. 15.6, and the movement of symmetrical buildings in Fig. 15.7.

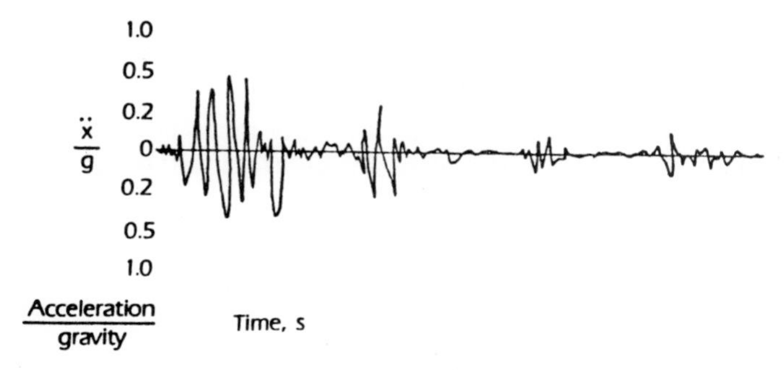


**Figure 15.5** Acceleration measurement during an earthquake.

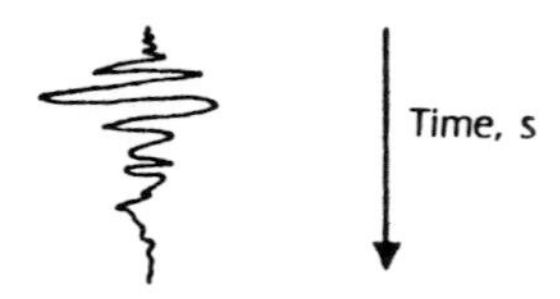

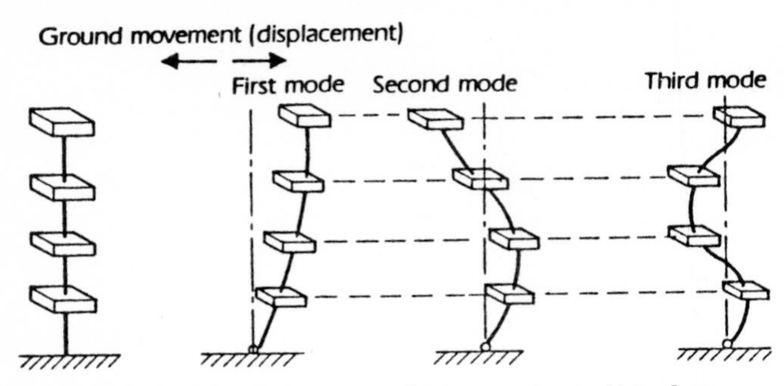


**Figure 15.6** Typical mode shapes during a seismic disturbance.

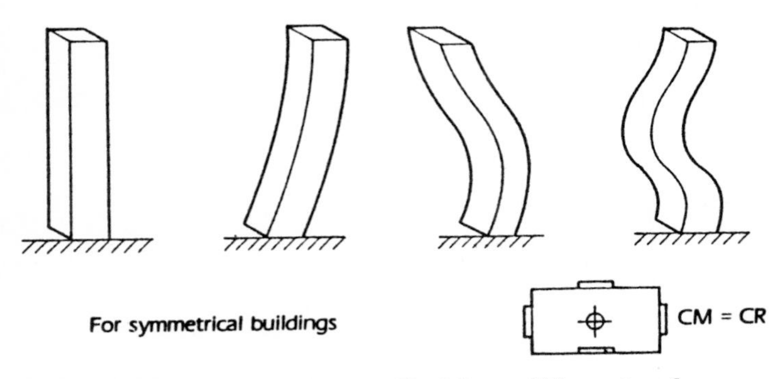


**Figure 15.7** Movement of symmetrical buildings. CM = center of mass of the building; CR = center of rigidity of all stiffening elements.

For nonsymmetrical buildings, where the center of mass is not in the same location as the center of rigidity, a torsional effect will occur (see Figs. 15.8, 15.9, and 15.10).

The torsional moment = $Fe$. The Uniform Building Code requires a minimum $e$ of $0.05\,\ell_{max}$ of the building.

$$f_{V1} = F_V \frac{r_1}{\Sigma r_1 + r_2}$$

$$f_{V2} = F_V \frac{r_2}{\Sigma r_1 + r_2}$$

$$f_{T1} = M_T \frac{r_1 d_1}{\Sigma_i^j rd^2}\ ; j = 4i = 1$$

$$f_{T2} = M_T \frac{r_2 d_2}{\Sigma_1^4 rd^2}$$

$$f_{T3} = M_T \frac{r_3 d_3}{\Sigma_1^4 rd^2}$$

$$f_{T4} = M_T \frac{r_4 d_4}{\Sigma_1^4 rd^2}$$

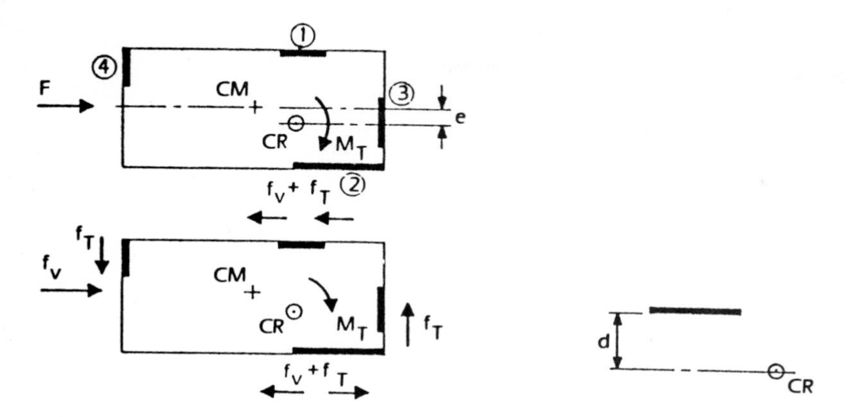


**Figure 15.8** Torsional effect in nonsymmetrical buildings.

$d$ is the distance measured from the centroid of the wall on a line normal to the wall passing through the centroid of the wall to a point on a line passing through the center of rigidity of the building.

For concrete floor systems, the floor and roof diaphragms are considered rigid. As such, concrete, precast concrete, or masonry walls are designed for forces proportional to their relative stiffnesses.

If the wall is cast integrally with the floor slab, then the wall is assumed to be fixed top and bottom. If the wall is precast concrete or if the floor-roof system is flexible, such as wood or metal decking without a concrete topping, then the wall is designed as a cantilever (see Fig. 15.11).

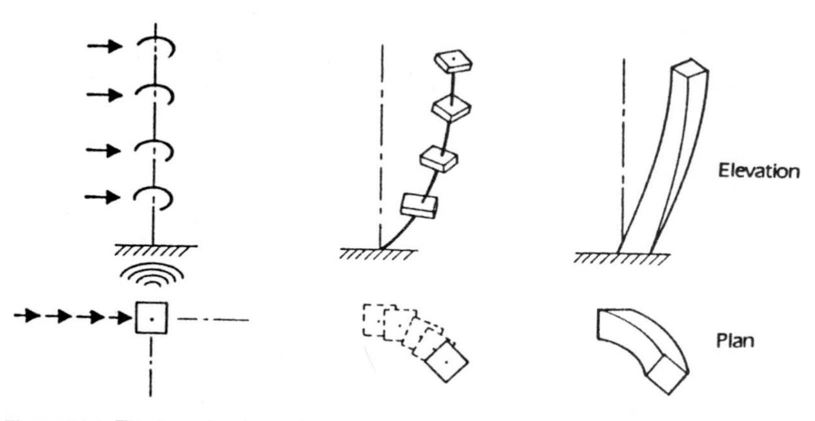


**Figure 15.9** First mode: lateral and torsion.

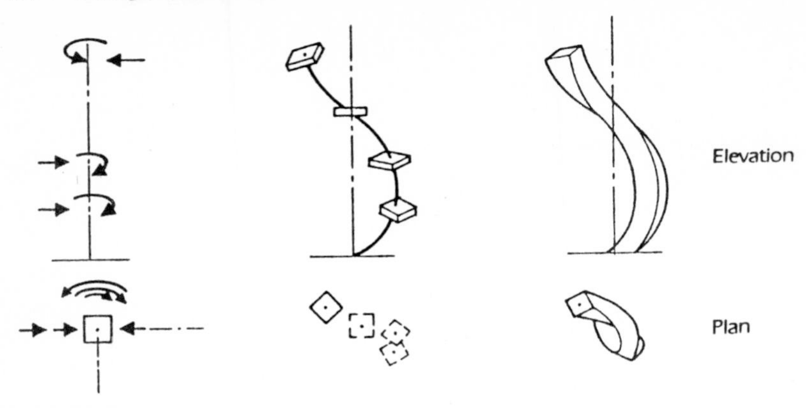


**Figure 15.10** Second mode: lateral and torsion.

$$r = \frac{1}{\Delta_T} = \frac{1}{\Delta_b + \Delta_s}$$

Fixed wall or panel

$$\Delta_T = \Delta_b + \Delta_s = \frac{Ph^3}{12EI} + \frac{1.2\,Ph}{AG}$$

$$I = \frac{tL^3}{12} \qquad A = tL$$

$$\text{Let } P = 1.0\,G$$

$$\Delta_T = \frac{1}{Et}\left(\frac{h}{L}\right)^3 + \frac{1.2}{Gt}\left(\frac{h}{L}\right)$$

Cantilever wall or panel

$$\Delta_T = \Delta_b + \Delta_s = \frac{Ph^3}{3EI} + \frac{1.2\,Ph}{AG}$$

$$\Delta_T = \frac{4}{Et}\left(\frac{h}{L}\right)^3 + \frac{1.2}{Gt}\left(\frac{h}{L}\right)$$

**Figure 15.11** Cantilever wall or panel.

For static analysis we assume the first mode first in one direction, then in a direction 90° to the first. We calculate the seismic force by using the equation $V = ZKICSW_{DL}$. For a one-story building, the force is applied at the roof plane. For a two-story building, the force is divided equally between the roof and second-story planes. For a multistory building, the force is proportional to the roof and each floor by the equations (see Fig. 15.12)

$$V = F_t + \sum_{i=1}^{n} F_t$$

$$F_t = 0.07TV$$

$$F_{T,\max} = 0.23V$$

$$F_T = 0 \text{ when } T \text{ is } 0.7 \text{ s or less}$$

$$F_x = (V - F_T)\frac{w_x h_x}{\sum_{i=1}^{n} w_i h_i}$$

See also Figs. 15.13 and 15.14.

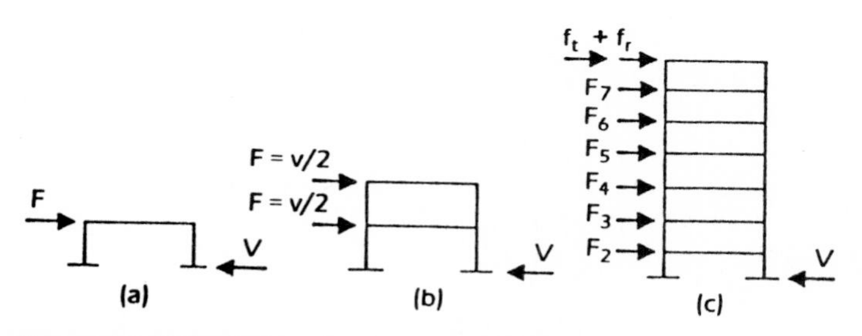


**Figure 15.12** Seismic force for (*a*) a one-story building, (*b*) a two-story building, and (*c*) a multistory building.

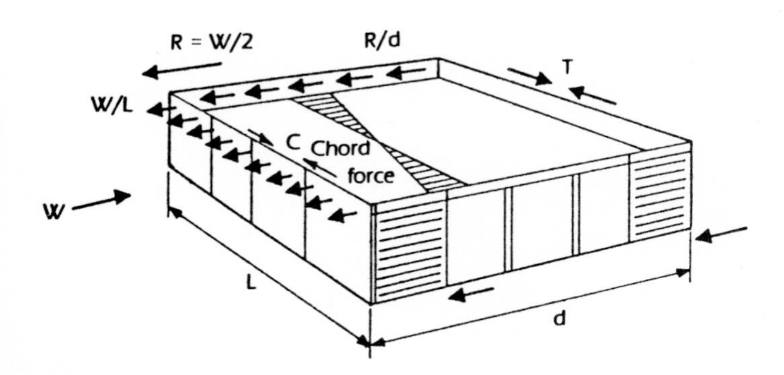


**Figure 15.13** Seismic force in a symmetrical building. $M_{ext} = 1/8\ W/L \times L^2 = WL/8$. $M_{int} = Td$. Therefore, $Td = WL/8d$.

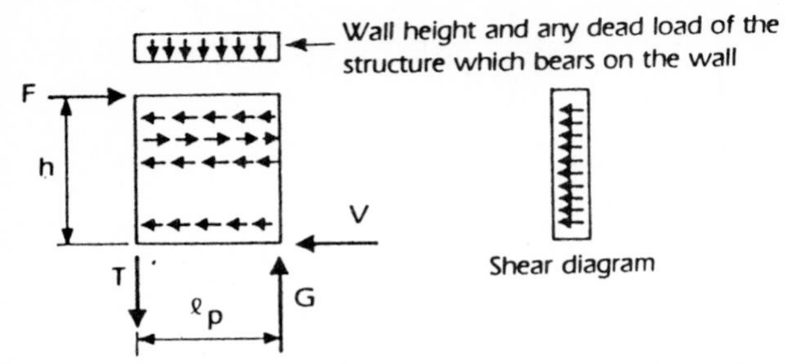


**Figure 15.14** Wall weight and dead loads bearing on the wall.

$$V = F \quad C = \frac{Fh}{L_p} + \frac{(W_w + W_s)\ L_p}{2} \qquad T = \frac{Fh}{L_p} - \frac{(W_w + W_s)L_p}{2}$$

where $W_w$ = wall weight, lb/ft; and $W_s$ = dead load of structure supported by the wall, lb/ft.

When the center of mass and the center of rigidity coincide, the effect is as shown in Fig. 15.15*a*.

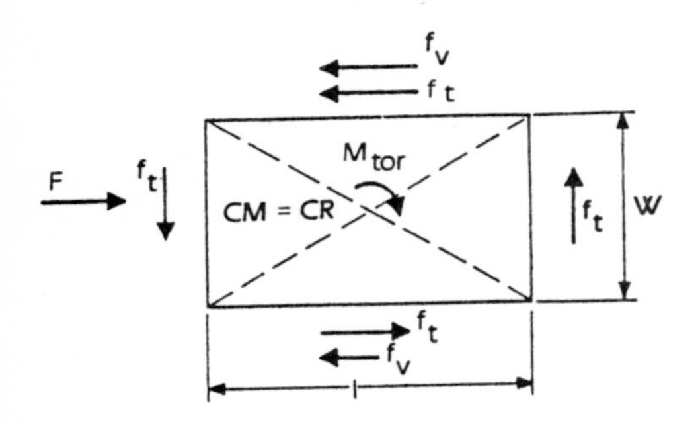


**Figure 15.15*a*** Building in which the centers of mass and rigidity coincide. $M_{tor}$ = the product of $F$ times - 0.05$L$ or 0.05$w$, whichever is greater.

When the center of mass and the center of rigidity differ, the effect is as shown in Fig. 15.15*b*.

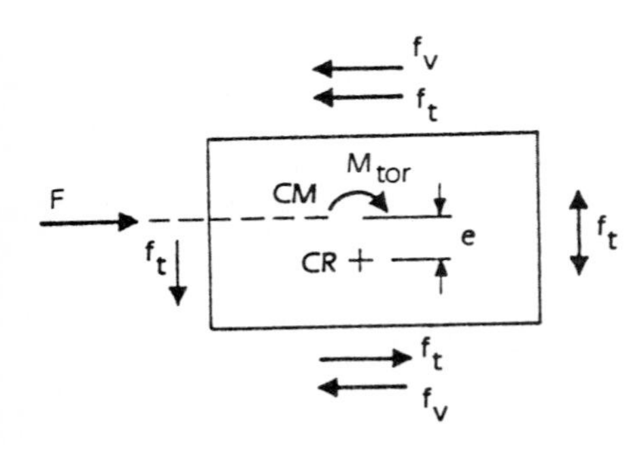


**Figure 15.15*b*** Building in which the centers of mass and rigidity differ. $M_{tor}$ = $Fe$. CM = center of mass; CR = center of rigidity.

Three cases with wall segments on two, three, and four sides are shown in Figs. 15.16*a*, 15.16*b*, and 15.16*c*.

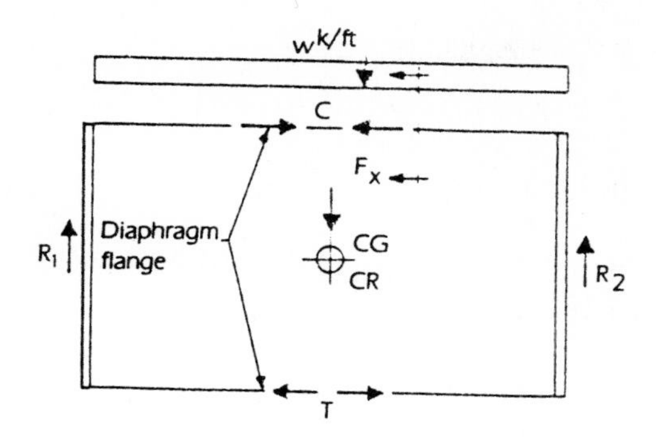


**Figure 15.16*a*** Case A: simple beam action with walls at two ends. $F_x = wd$. $R_1 = R_2 = F_x/2$. $T = C = F_x d/8h$.

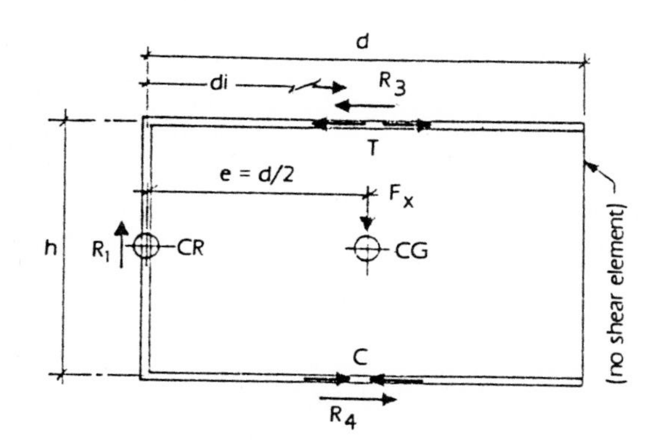


**Figure 15.16*b*** Case B: cantilever beam action with walls at three sides. $R_1 = F_x$. $Me = eF_x = d/2(F_x)$. $R_3 = R_4 = dF_x/2h$. $T_i =$

$$C_i = \frac{Me - \frac{wd_i^2}{2}}{h}.$$

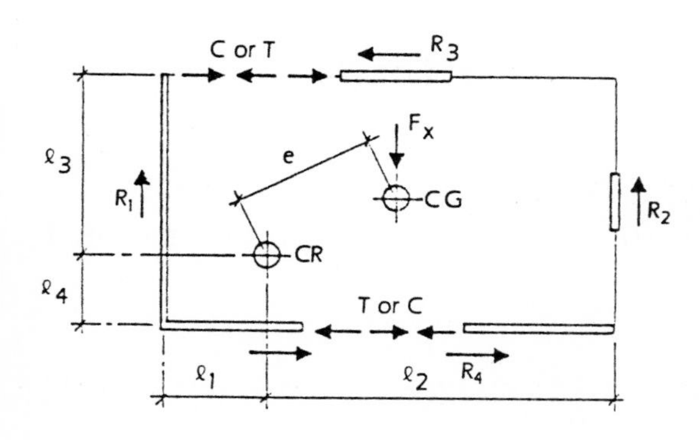


**Figure 15.16*c*** Case C: general case with wall segments at four sides: drag struts (typical). $Me_{cc} = eF_x$. $Rw = f(kw, Lw^2)$. $k$ = wall stiffness. $L$ = wall location from CR. $T_i = C_i$ = flexural + drag force components.

The following paragraphs show how lateral forces are proportional to shear walls.

$$R_{vx} = F \frac{rx}{\Sigma r} \quad F_{Tx} = M_{\text{tor}} \frac{r_x d_x}{\Sigma rd^2}$$

where $d_x$ is the distance from the center of rigidity to the resisting element.

The rigidity of the wall panel is determined from the concept illustrated in Figs. 15.17 and 15.18.

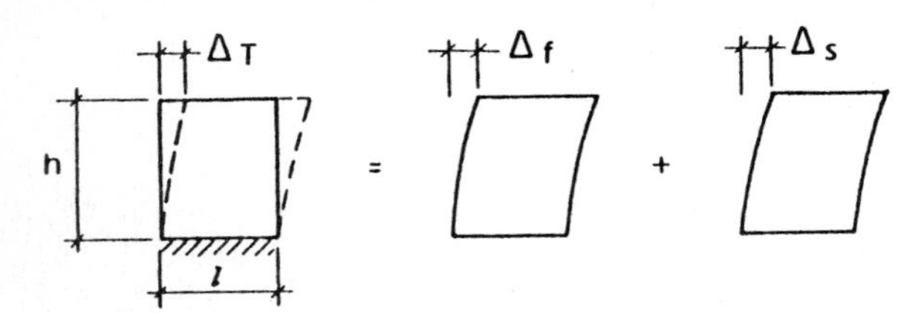


**Figure 15.17** Cantilever wall.

$$\Delta_T = \Delta_f + \Delta_s = \frac{Ph^3}{3E_c I} + \frac{1.2Ph}{AE_g}$$

where $E_c$ = modulus of elasticity $= 33w\ 1.5\sqrt{f_c'}$

$= (33)(144)^{1.5}\sqrt{f_c'}$

$E_g$ = shear modulus $0.4E_c$

$I$ = moment of inertia in direction of bending; for panels of uniform thickness $I = (tL^2)/12$

$$\frac{E_c}{E_g} = \frac{E_c}{0.4E_c} = 2.5$$

Assume $P = 1$, $t = 1$, and $E_g = 1$. Then,

$$\Delta = \frac{h^5}{3 \times 2.5 \times \frac{L^3}{12}} + 1.2\frac{h}{L} = 1.6\left(\frac{h}{L}\right)^3 + 1.2\frac{h}{L}$$

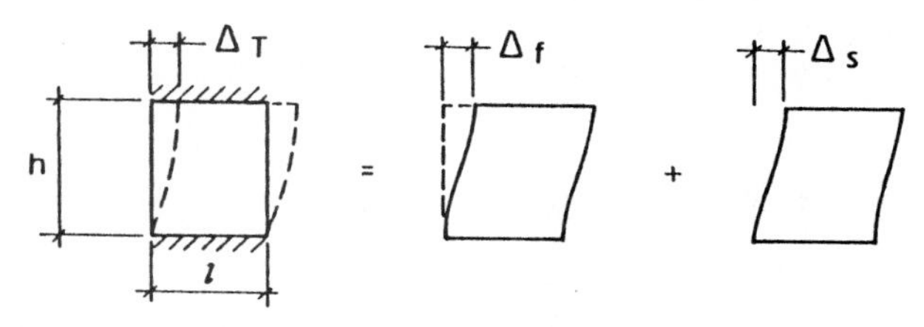


Figure 15.18 Fixed wall or pier.

$$r = \text{rigidity or stiffness of the wall panel} = \frac{1}{\Delta T}$$

$$\Delta_T = \Delta_f + \Delta_s$$

$$= \frac{Ph^3}{12E_cI} + \frac{1.2Ph}{AE_g} = \frac{h^3}{2.5L^3} + \frac{1.2h}{L} = 0.4\left(\frac{h}{L}\right)^3 + 1.2\frac{h}{L}$$

For a pierced wall panel the relationship shown in Fig. 15.19 is used.

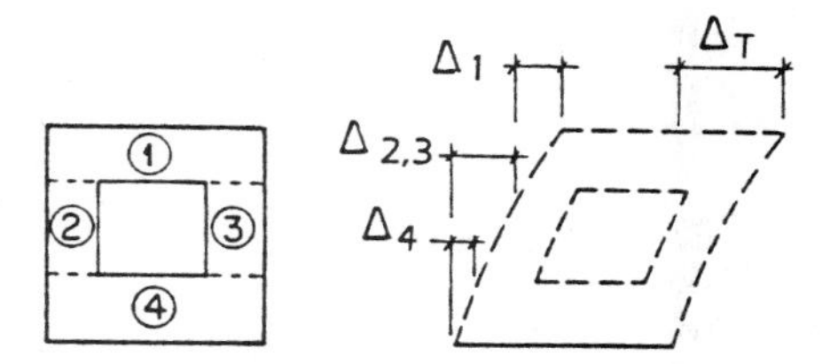


Figure 15.19 Pierced wall or panel.

$$\Delta_T = \Delta_1 + \Delta_4 + \frac{1}{\frac{1}{\Delta_2} + \frac{1}{\Delta_3}} \qquad r = \frac{1}{\Delta_T} = \frac{1}{\Delta_1 + \Delta_4 + \frac{1}{\frac{1}{\Delta_2} + \frac{1}{\Delta_3}}}$$

For walls with uniform-width piers the distribution of the shear force within such walls is shown in Fig 15.20.

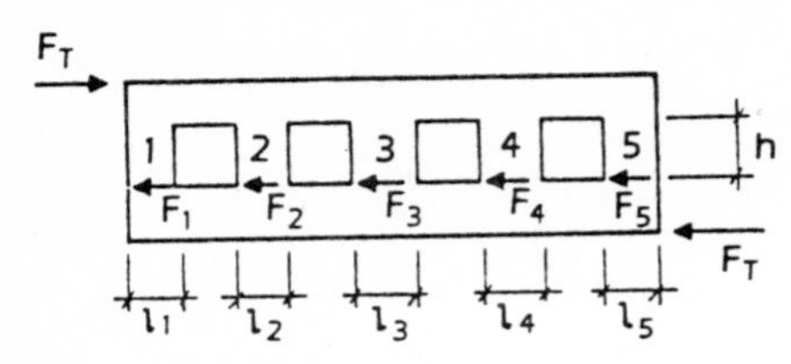


**Figure 15.20** Wall with uniform-width piers. $L_1 = L_2 = L_3 = L_4 = L_5$. $F_1 = F_2 = F_3 = F_4 = F_5$. $F_1 = F_T \times \dfrac{1}{\text{number of piers}}$ or $F_1 = F_T \times 1/5$.

For walls with variable-width piers the distribution within such walls is $F_1 = F_T\left(\dfrac{r_1}{\Sigma r}\right)$(see Fig. 15.21).

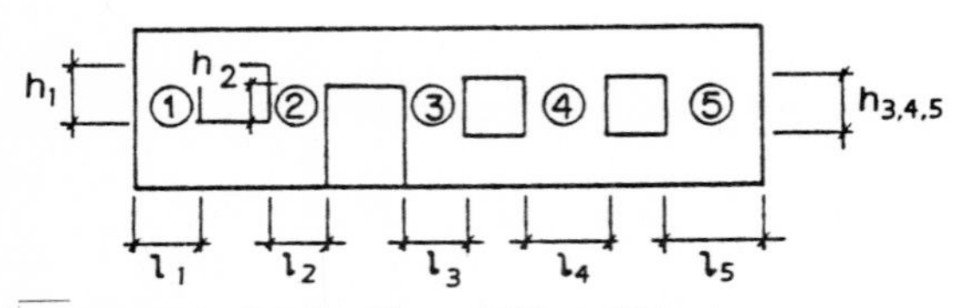


**Figure 15.21** Wall with variable-width piers.

To assist in determining the shear stress in the pier the following format is used:

Wall: ______ $F_T$ = ______

| Pier | $h$ | $L$ | $\dfrac{h}{L}$ | $0.4\dfrac{h}{L}$ | $+\ 1.2\dfrac{h}{L}$ | $= \Delta$ | $r = \dfrac{1}{\Delta}$ | $\dfrac{r}{\Sigma_r} \times F_T$ | $= F_P$ | $V_u = \dfrac{F_P U}{\phi t\,(L - 0.17)12}$ |
|---|---|---|---|---|---|---|---|---|---|---|
| | | | | | | | | | | |

To determine the center of mass the procedure illustrated in Fig. 15.22 and the accompanying tabular form is used.

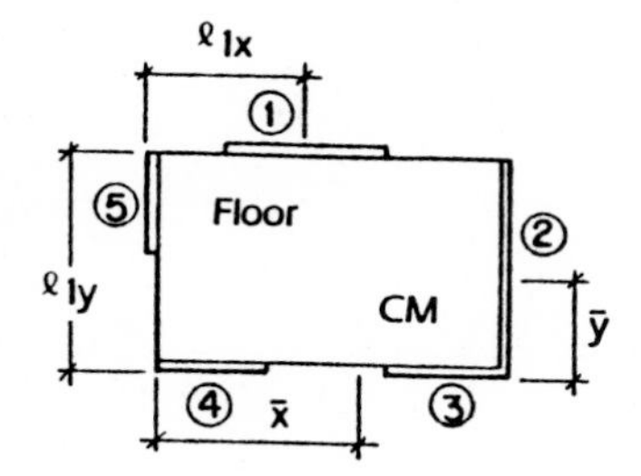


**Figure 15.22** Determining the center of mass. $x = \dfrac{\Sigma wx}{\Sigma w}$. $y = \dfrac{\Sigma wy}{\Sigma w}$.

| Element | Area | $w$/ft² | $w$ | Distance from zero reference | | $wx$ | $wy$ |
|---|---|---|---|---|---|---|---|
| | | | | $x$ | $y$ | | |
| Floor system | | | | | | | |
| Wall 1 | | | | | | | |
| 2 | | | | | | | |
| 3 | | | | | | | |
| 4 | | | | | | | |
| 5 | | | | | | | |
| | | Σ | | | Σ | | |

To determine the center of rigidity the procedure illustrated in Fig. 15.23 is used.

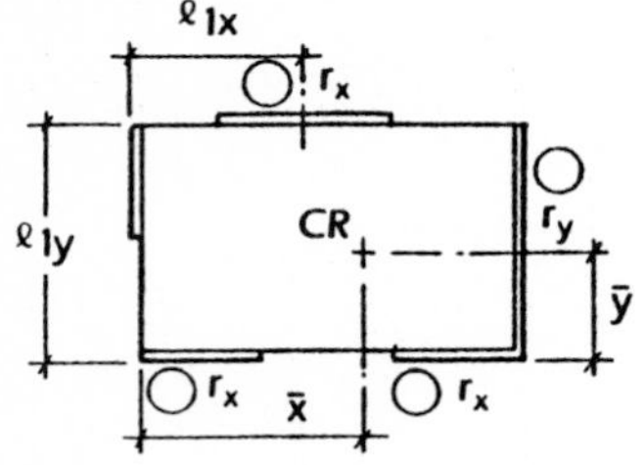


**Figure 15.23** Determining the center of rigidity. $\bar{x} = \dfrac{r_y x}{\Sigma r_y}$. $\bar{y} = \dfrac{r_x y}{\Sigma r_x}$.

| Wall | Distance from zero reference | | $r_x$ | $r_y$ | $r_x y$ | $r_y x$ |
|---|---|---|---|---|---|---|
| | $x$ | $y$ | | | | |
| 1 | | | | | | |
| 2 | | | | | | |
| 3 | | | | | | |
| 4 | | | | | | |
| 5 | | | | | | |
| | | Σ | | | | |

To determine the forces resisted by the walls, including torsion, the procedure illustrated in Fig. 15.24 is used.

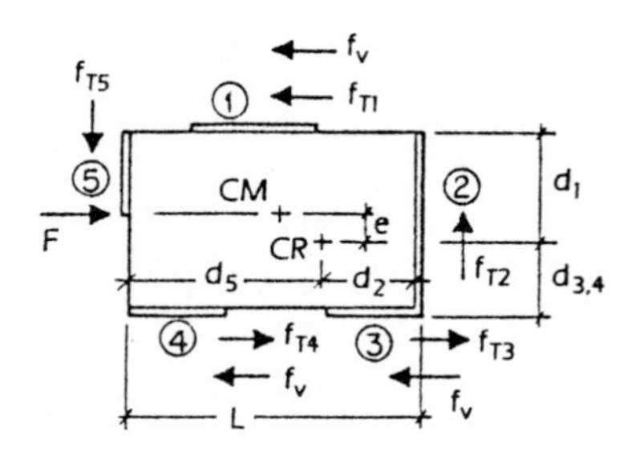


**Figure 15.24** Determining forces resisted by the walls. $M_{tor} = Fe$; $e_{min} = 0.05L$. $F_T = M_{tor}\,(rd)/\Sigma rd^2$

| Wall | $r_x$ | $r_y$ | $d$ | $rd$ | $rd^2$ | $F_v = \frac{r_x}{\Sigma rx} \times F$ | $F_{tor} = M_{tor} \frac{rd}{\Sigma rd^2}$ | $F_s = F_v + F_{tor}$ |
|---|---|---|---|---|---|---|---|---|
| 1 | | | | | | | | |
| 2 | | | | | | | | |
| 3 | | | | | | | | |
| 4 | | | | | | | | |
| 5 | | | | | | | | |

$$V_u = \frac{F_s U}{\phi t(L - 0.17\text{ft})12\text{in}/1\text{ft}} = \frac{2F_s}{0.85t(L - 0.17\text{ft})12\text{in}/1\text{ft}}$$

The following two example problems demonstrate the design principles that have been discussed for single-story buildings. The equivalent-lateral-force procedure of the Uniform Building Code is used to satisfy seismic design requirements.

## Design Example 1: Typical Office Building (Fig. 15.25)

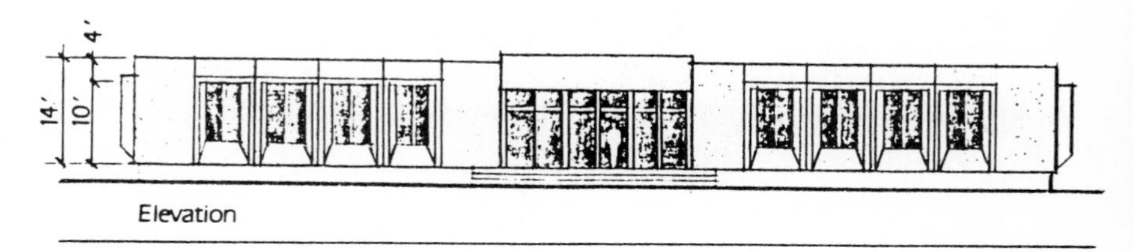


**Figure 15.25** Elevation of a typical office building.

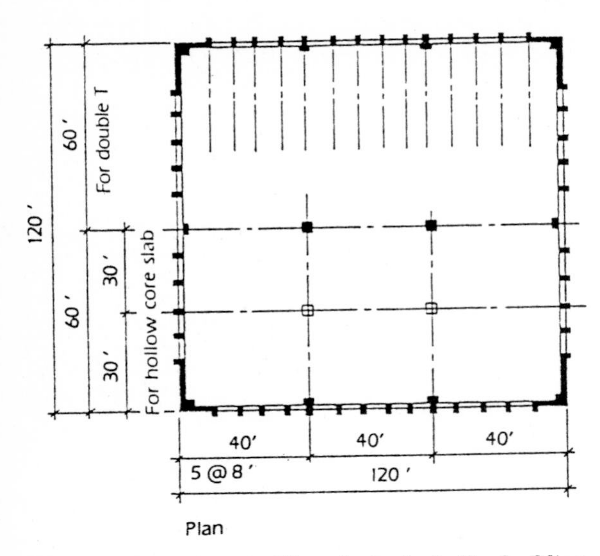


**Figure 15.25 (*continued*)** Plan of a typical office building.

## Analysis of example building for seismic forces

$V = ZIKCSW$

$V = 1 \times 1 \times 1.33 \times 0.14W$

$V = 0.1862W$

$Z$ for Zone 4 = 1

$I$ for an office building = 1

$K$ for a building with shear walls and without a vertical moment-resisting frame = 1.33

$C$ = factor based upon period of the building

$S$ = soil interaction factor

$CS$ = 0.14 maximum

$W$ = dead weight of the building

**Weight of the building.** The building weight is calculated as follows.

Roof system:

| | |
|---|---|
| Roofing | 6 lb/ft² |
| Insulation | 1 lb/ft² |
| Ceiling system | 5 lb/ft² |
| Sand-lightweight double T | 50 lb/ft² |
| | 62 lb/ft² x 120²/1000 = 892.8 kips |

| | |
|---|---|
| Inverted-T girder; assume 800 lb/ft × 120/1000 = | 96 kips |
| Ledger beams; two at 600 lb/ft × 120/1000 = | 144 kips |
| Wall panels; assume average 5-in thickness: | |

$$\frac{5\text{ ft}}{12\text{ in}} \times 144\text{ lb/ft}^2 \times \frac{14}{2} \times \frac{2\text{ walls} \times 120\text{ft}}{1000} = \frac{101\text{ kips}}{1234\text{ kips}}$$

$$V = 0.1862 \quad W = 0.1862 \times 1234 = 230\text{ kips}$$

Since the building is symmetrical, the center of mass and the center of rigidity both occur at the same point and in the center of the building. The Uniform Building Code requires the building to be designed for a minimum torsional moment based upon a minimum $e$ distance of 5 percent of the maximum building dimension at that level.

$$0.05 \times 120\text{ ft} = 6\text{ ft} = e \quad M_{\text{tor}} = 6 \times 229.7 = 1378.2\text{ kip}\cdot\text{ft}$$

**Walls.** See Fig. 15.26.

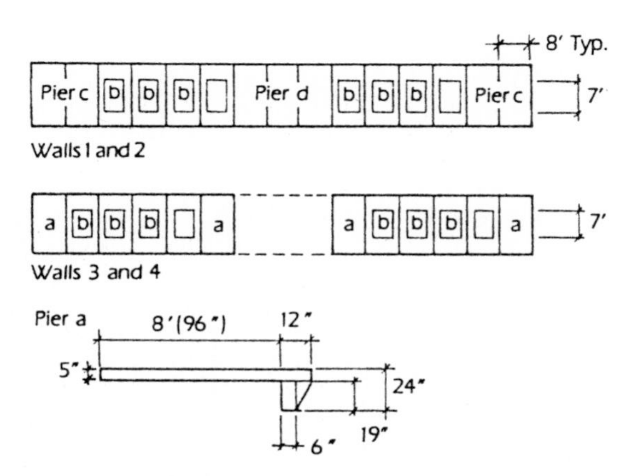


**Figure 15.26** Walls 1, 2, 3, and 4; pier $a$.

$$\text{CG} = \frac{\Sigma A_d}{\Sigma A} = \frac{\left[5(96+12)\left(\frac{46+12}{2}\right)\right] + (19\times 6\times 99) + \left(\frac{19+6}{2}\times 104\right)}{5\times 108 + 19\times 6 + \frac{18\times 6}{2}}$$

$$= 65.23\text{ in}$$

$$\text{Area} = \left(5\times 108 + 19\times 16 + \frac{19\times 6}{2}\right)\times\frac{1}{144} = 4.93\text{ ft}^2$$

$$I = \frac{bd^3}{3} + \frac{bd^3}{3} + A_{y^2} + A_{y^2} + \frac{bd^3}{12} + \frac{bd^3}{36}$$

$$I = \left( \frac{5 \times 65.22^3}{3} + \frac{5 \times 42.78^3}{3} + 6 \times 19 \times 33.76^2 + \frac{6 \times 19}{2} \times 38.78^2 \right.$$

$$\left. + \frac{19 \times 6^3}{12} + \frac{19 \times 6^3}{36} \right) \frac{1}{12^4}$$

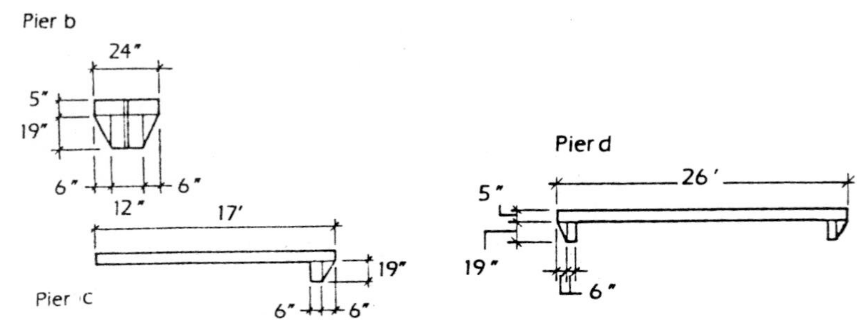


**Figure 15.26 (continued)** Piers *b*, *c*, *and d*.

$$\text{Area} = \left( 5 \times 24 + 19 \times 12 + \frac{2 \times 6 \times 19}{2} \right) \times \frac{1}{144} = 3.2\,\text{ft}^2$$

$$I = \left( \frac{5 \times 24^3}{12} + \frac{19 \times 12^3}{12} + \frac{2 \times 6 \times 19}{2} \times 8^2 + \frac{19 \times 6^3}{36} \right) \times \frac{1}{12^4}$$

$$= 0.8\ \text{ft}^4$$

$$\text{CG} = \frac{\Sigma A_d}{\Sigma A} = \frac{\left( \frac{5}{12} \times \frac{17^2}{2} \right) + \left( \frac{6 \times 19}{144} \times 16.5 \right) + \left( \frac{6 \times 19}{2 \times 144} \times 16.67 \right)}{\left( \frac{5}{12} \times 17 \right) + \left( \frac{6 \times 19}{144} \right) + \left( \frac{6 \times 19}{2 \times 144} \right)}$$

$$\text{CG} = 9.76\ \text{ft}$$

$$\text{Area} = \left( \frac{5}{12} \times 17 \right) + \left( \frac{6 \times 19}{144} \right) + \left( \frac{6 \times 19}{2 \times 144} \right) = 8.3\,\text{ft}^2$$

$$I = \left( \frac{5}{12} \times \frac{9.656^3}{3} \right) + \left( \frac{5}{12} \times \frac{7.344^3}{3} \right) + \left( \frac{6 \times 19}{144} \times 6.59^2 \right)$$

$$+ \left( \frac{6 \times 19}{2 \times 144} \times 7^2 \right) + \frac{19 \times 6^3}{12^4 \times 12} + \frac{19 \times 6^3}{12^4 \times 36} = 234$$

$$\text{Area} = \left(\frac{5}{12} \times 26\right) + \left(\frac{2 \times 19 \times 6}{144}\right) + \left(\frac{2 \times 19 \times 6}{2 \times 144}\right) = 13.20\text{ft}^2$$

$$I = \left(\frac{5 \times 26^3}{12 \times 12}\right) + \left(\frac{6 \times 19}{144} \times 12.25^2\right) + \left(\frac{16 \times 19}{144} \times 12.25^2\right)$$

$$+ \left(2 \times \frac{16 \times 19}{2 \times 144} \times 12.67^2\right) + \left(\frac{19 \times 6^3}{12} \times \frac{1}{12^4} \times 2\right) + \left(\frac{2 \times 19 \times 6^3}{36 \times 12^4}\right)$$

$$= 975 \text{ ft}^4$$

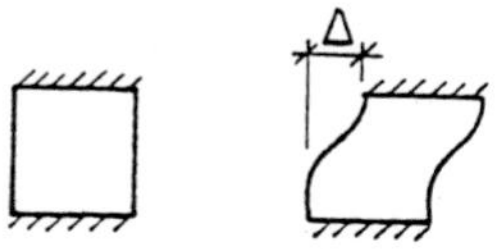

$$\Delta_T = \Delta_f + \Delta_s = \frac{Ph^3}{12E_cI} + \frac{1.2Ph}{AE_G}$$

Let $P = 1$, $E_c = 1$, $E_G = 0.4$, $E_c = 0.4$, and $h = 7$;
then $\Delta_T = h^3/12I + 3h/A$.

**Walls 1 and 2**

| Pier | $h$ | $h^3$ | $I$ | $A$ | $\frac{h^3}{12I}$ + | $\frac{3h}{A}$ = | $\Delta$ | $R = \frac{1}{\Delta}$ | $\frac{R}{\Sigma R}$ |
|---|---|---|---|---|---|---|---|---|---|
| $c$ | 7 | 343 | 233.8 | 8.27 | 0.122 | 2.54 | 2.66 | 0.375 | 0.249 |
| $b$ | 7 | 343 | 0.77 | 3.2 | 37.1 | 6.56 | 43.68 | 0.02289 | 0.0152 |
| $b$ | 7 | 343 | 0.77 | 3.2 | 37.1 | 6.56 | 43.68 | 0.02289 | 0.0152 |
| $b$ | 7 | 343 | 0.77 | 3.2 | 37.1 | 6.56 | 43.68 | 0.02289 | 0.0152 |
| $d$ | 7 | 343 | 975 | 13.2 | 0.0273 | 1.59 | 1.62 | 0.6172 | 0.4102 |
| $b$ | 7 | 343 | 0.77 | 3.2 | 37.1 | 6.56 | 43.68 | 0.02289 | 0.0152 |
| $b$ | 7 | 343 | 0.77 | 3.2 | 37.1 | 6.56 | 43.68 | 0.02289 | 0.0152 |
| $b$ | 7 | 343 | 0.77 | 3.2 | 37.1 | 6.56 | 43.68 | 0.02289 | 0.0152 |
| $a$ | 7 | 343 | 233.8 | 8.27 | 0.122 | 2.54 | 2.66 | 0.375 | 0.249 |
| | | | | | | | $\Sigma R$ = | 1.5045 | 1.00 |

**Walls 3 and 4**

| Pier | $h$ | $h^3$ | $I$ | $A$ | $\frac{h^3}{12I}$ + | $\frac{3h}{A}$ = | $\Delta$ | $R = \frac{1}{\Delta}$ | $\frac{R}{\Sigma R}$ |
|---|---|---|---|---|---|---|---|---|---|
| *a* | 7 | 343 | 39.0 | 4.94 | 0.733 | 4.25 | 4.89 | 0.20 | 0.213 |
| *b* | 7 | 343 | 0.77 | 3.2 | 37.1 | 6.56 | 43.68 | 0.023 | 0.0245 |
| *b* | 7 | 343 | 0.77 | 3.2 | 37.1 | 6.56 | 43.68 | 0.023 | 0.0245 |
| *b* | 7 | 343 | 0.77 | 3.2 | 37.1 | 6.56 | 43.68 | 0.023 | 0.0245 |
| *a* | 7 | 343 | 39.0 | 4.94 | 0.733 | 4.25 | 4.89 | 0.20 | 0.213 |
| *a* | 7 | 343 | 39.0 | 4.94 | 0.733 | 4.25 | 4.89 | 0.20 | 0.213 |
| *b* | 7 | 343 | 0.77 | 3.2 | 37.1 | 6.56 | 43.68 | 0.023 | 0.0245 |
| *b* | 7 | 343 | 0.77 | 3.2 | 37.1 | 6.56 | 43.68 | 0.023 | 0.0245 |
| *b* | 7 | 343 | 0.77 | 3.2 | 37.1 | 6.56 | 43.68 | 0.023 | 0.0245 |
| *a* | 7 | 343 | 39.0 | 4.94 | 0.733 | 4.25 | 4.89 | 0.20 | 0.213 |
| | | | | | | | $\Sigma R$ = | 1.5045 | 1.00 |

If the walls are cast in place or precast as one wall and the windows are of the same height, the rigidity of the wall is the sum of the rigidities of the piers.

If the precast panels are not connected along their vertical edges with weld plates or shear keys, then each panel is considered as an independent wall with its own rigidity.

If the panels are connected along their vertical edges with weld plates or shear keys, the condition would fall between the two in the preceding paragraphs, acting compositely near that of the first. The use of grouted shear keys is preferred owing to the effects of volume change under tension of the weld plates under concentrated forces.

This example will assume that shear keys are used and that the wall once assembled will act similarly to a cast-in-place wall. Since the windows are of the same height and the walls are of the same height, the rigidity of a wall will be based upon the sum of the rigidities of the piers. (See Figs. 15.27 and 15.28.)

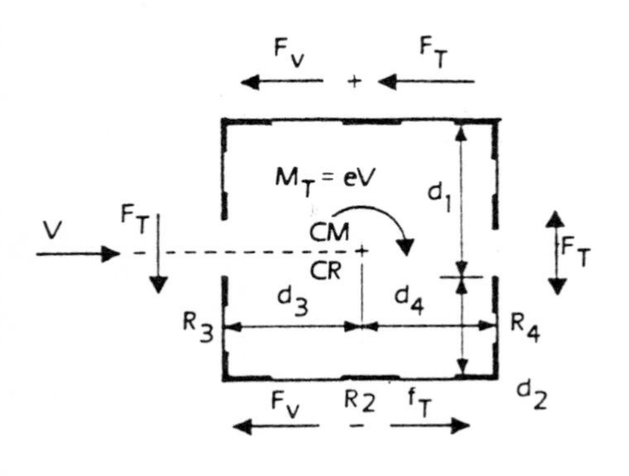


**Figure 15.27** Determining rigidity.

$$F_{U1} = V \times \frac{R_1}{\Sigma R_{\text{parallel}}}.$$

$$F_{T1} = M_T \times \frac{R_1 D_1}{\Sigma R d^2}.$$

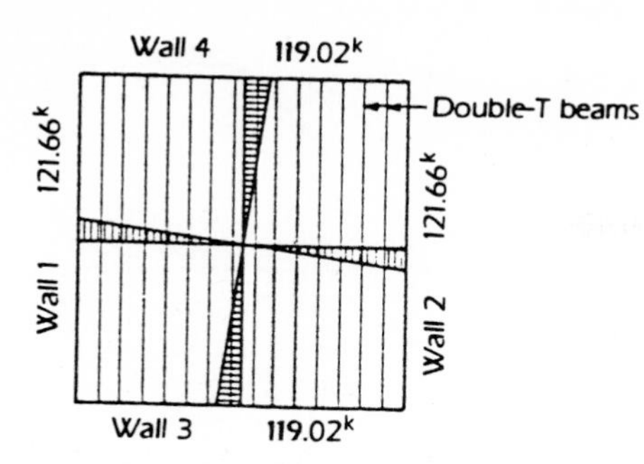


**Figure 15.28** Rigidity of walls.

From the following:

$$R_1 = R_2 = 1.5 \qquad R_3 = R_4 = 0.939 \qquad d = 60 \text{ ft}$$

$$V = 229.7^{k} \qquad M_T = 1378.2^{\text{ft k}}$$

For wall 1:

$$F_{v\,1} = V \frac{R_1}{\Sigma R_{\text{parallel}}} = 229.7 \times \frac{1.5}{2 \times 1.5} = \qquad 114.6^{k}$$

$$F_{T\,1} = \frac{R_1 d_1}{\Sigma R d^2}$$

$$\frac{R_1 d_1}{\Sigma R d^2} = 1378.2 \times \frac{1.5 \times 60}{(2 \times 1.5 \times 60^2) + (2 \times 0.939 \times 60^2)} = \frac{7.06^{k}}{121.66^{k}}$$

For wall 2:

$$F_{v\,1} = V \frac{R_1}{\Sigma R_{\text{parallel}}} = 229.7 \times \frac{0.939}{2 \times 0.939} = \qquad 114.6^{k}$$

$$F_T = M_T \frac{R_3 d_3}{\Sigma R d^2}$$

$$1378.2 \times \frac{0.939 \times 60}{(2 \times 1.5 \times 60^2) + (2 \times 0.939 \times 60^2)} = \frac{4.42^{k}}{119.02^{k}}$$

*Weld clips.* $U = 1.4$; $\phi = 0.85$; $T = C = \phi Asfy$.
Let $\theta = 45°$
For weld clips with no. 4 bar:

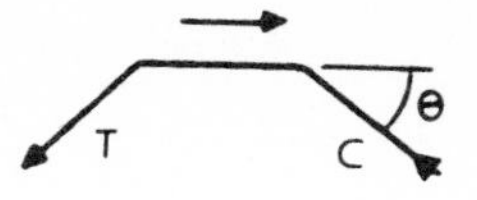


$$\frac{2\phi Asfy \cos\theta}{U} = \frac{2 \times 0.85 \times 0.2 \times 40 \times 0.707}{1.4} = 6.85 \text{ kips/clip allowable}$$

$$\frac{119.02}{6.85} = 18 \text{ clips} \therefore \text{ use two clips for each double T on end.}$$

$$\frac{121.66}{6.85} = 18 \text{ clips} \quad \frac{120\text{ft}}{18} = 6\text{ft on center along edge}$$

For weld clips with no. 5 bar:

$$\frac{2\phi Asfy \cos\theta}{U} = \frac{2 \times 0.85 \times 0.31 \times 40 \times 0.707}{1.4} = 10.6^{k}$$

Note that $\phi$, the capacity reduction factor, is not required for development length. Some designers, however, will use 0.9 or 0.85 (0.9 for tension and bending and 0.85 for shear).

$$\text{Development length for tension} = \frac{0.04 A_b f_y}{f'_c}$$

$$\text{Minimum} = 12 \text{ in or } 0.0004 d_b f_y$$

For all lightweight concrete multiply by 1.33. For sand-lightweight concrete multiply by 1.18.

$$\text{For no. 4 bar} = (0.04 \times 0.20 \times 40{,}000)/\sqrt{6000} \times 1.18 = 4.87\text{in}$$
$$= 0.0004 \times 0.5 \times 40{,}000 \times 1.18 = 9.44\text{in}$$

A minimum of 12 in governs for tension.

$$\text{Development length for compression} = \frac{0.02 d_b f_y}{f'_c}$$

$$\text{Minimum} = 8 \text{ in or } 0.0003 f_y d_b$$

$$\text{For no. 4 bar} = (0.02 \times 0.5 \times 40{,}000)/\sqrt{6000} = 5.16\text{in}$$
$$= 0.0003 \times 40{,}000 \times 0.5 = 6 \text{ in}$$

A minimum of 8 in governs for compression.

Since tension of 12 in is greater than 8 in for compression, a 12-in leg is required for the insert as a minimum.

**Weld capacity.** See Fig. 15.29.

$$T_w = \phi \times 25\text{ ksi } L_w t_w$$

where $t_w$ = thickness of weld at the throat; $\phi = 0.70$. Use E-70 XX electrode.

For no. 4 insert, 1/4-in weld rod to rod:

$$\frac{2A_s f_y \times 0.707}{\phi 25\text{ksi } t_w} = \text{length of weld (minimum)}$$

$$\frac{2 \times 0.20 \times 40 \times 0.707}{0.7 \times 25 \times 0.25} = 2.6 \text{ or } 3\text{ in}$$

Minimum length = 4 × weld size    4 × 1/4 = 1 in minimum

For $f_y$ = 60 ksi reinforcing,

$$\frac{2 \times 0.2 \times 60 \times 0.707}{0.7 \times 25 \times 0.25} = 3.88 \text{ or } 4\text{ in}$$

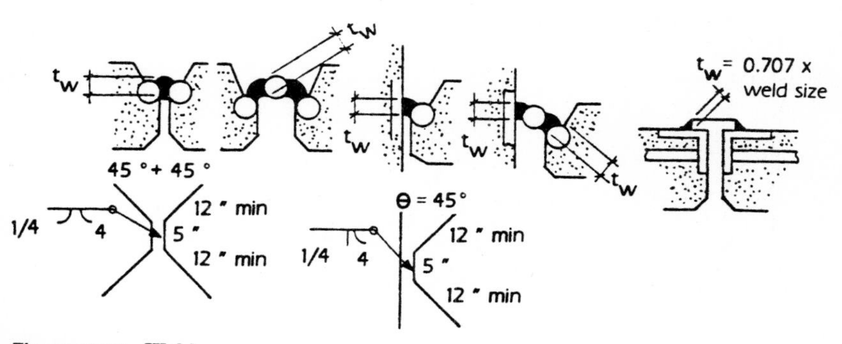


**Figure 15.29** Welds.

**Shear in concrete flange of double T.** For 2-in flange

$$V_u = \frac{(f_v + f_t)U}{} = \frac{121.66\text{ kip} \times 1.4}{120\text{ ft}} = 142^k/\text{ft}$$

$$v_u = \frac{V_u}{\phi b_w t} = \frac{1.42^k/\text{ft}}{0.85 \times 12\text{in/ft} \times 2\text{in}} = 0.069 \text{ ksi or } 69 \text{ lb/in}^2$$

$$< 131.66 \ (\text{OK})$$

$$v_c \text{ allowable without shear reinforcing} = 2\sqrt{f'_c} = 2\sqrt{6000}$$

$$= 154.9 \text{ lb/in}^2$$

For sand-lightweight concrete use $0.85 \times 2\sqrt{f'_c} = 131.66$.

*Chord stress.* See Fig. 15.30.

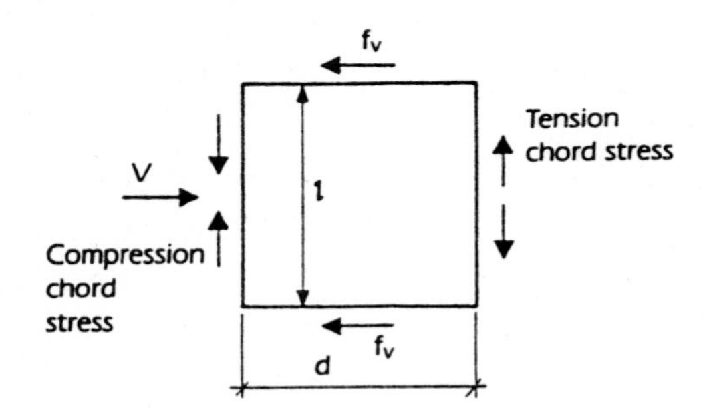


**Figure 15.30** Chord stress.

$$\text{Chord stress} = \frac{V\ell}{8d} = \frac{229.7^k \times 120\text{ft}}{8 \times 120\text{ft}}$$

$$T = 28.71 \text{ kips}$$

$$TU = \phi A_s f_y$$

$$A_s = \frac{TU}{\phi f_y} = \frac{28.71 \times 1.4}{0.9 \times 60} = 0.744$$

2 no. 6 bars = 2 × 0.44 = 0.88 > 0.74 (OK)

or 1 no. 8 bar = 0.79 > 0.744 (OK)

**Placement of chord reinforcing.** See Fig. 15.31.

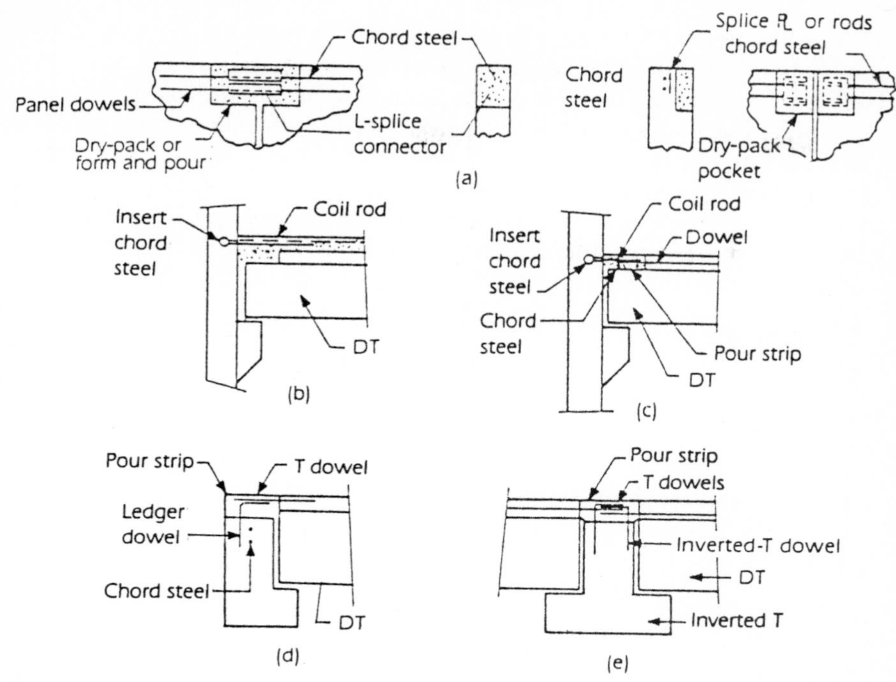


**Figure 15.31** Placement of chord reinforcing. (*a*) Precast panels tied together. (*b*) Pour strip with topping. (*c*) Pour strip without topping. (*d*) Ledger beam. (*e*) Connection between double T's without topping.

## Example problem 2

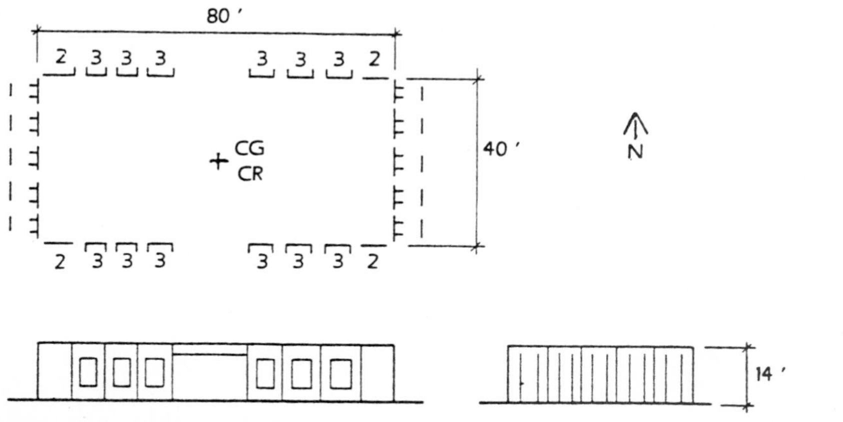


**Figure 15.32** Example problem 2.

The wall panels at the roof line are neither fully fixed nor pinned. If the panels are doweled into the cast-in-place topping, they are assumed to be fixed. If there is no cast-in-place topping, the connection consists of weld plates, and the panel is non-load-bearing, the condition is nearer that of a finned condition or cantilever. If the wall is load-bearing, it is usually considered as fixed. Some designers will always assume the top to be fixed. If the shear walls are always long without openings, as in an apartment building, some designers will base the rigidity only on the shear deflection, neglecting the flexural component. (See Fig. 15.32.)

For a fixed top, the equations are:

$$R = \frac{1}{\alpha} \qquad \alpha = \alpha_f + \alpha_s = \frac{h^3}{12EI} + \frac{1.2h}{AG}$$

For a pinned top, the equations are

$$R = \frac{1}{\alpha} \qquad \alpha = \alpha_f + \alpha_s = \frac{h^3}{3EI} + \frac{1.2h}{A_1G_1}$$

For concrete $G = 0.4E_c$

$V_b = ZIKCSW_D$

$Z$ for Zone 4 = 1

$I$ for an office building = 1

$K$ for a building with shear walls (box system) = 1.33

$CS$ = 0.14 (for a one-story building usually taken as 0.14)

$W_D$ for roof = 6 lb/ft$^2$ roofing, 1 lb/ft$^2$ insulation 1 lb/ft$^2$ miscellaneous, and 36 lb/ft$^2$ double T's = 42 lb/ft$^2$ [(40 × 80)/1000] = 134.4 kips

$W_D$ for walls = 14/2 × [2(80 + 40)/1000] × 70 lb/ft$^2$ assumed weight) = 117.6 kips

$$V_b = 1 \times 1 \times 1.33 \times 0.14\,(134.4 + 117.6) = 47 \text{ kips}$$

$$M_{tor} = 0.05 \times 80 \text{ ft} \times 47 = 188 \text{ ft} \cdot \text{kips}$$

**Wall panel 1.** See Fig 15.33.

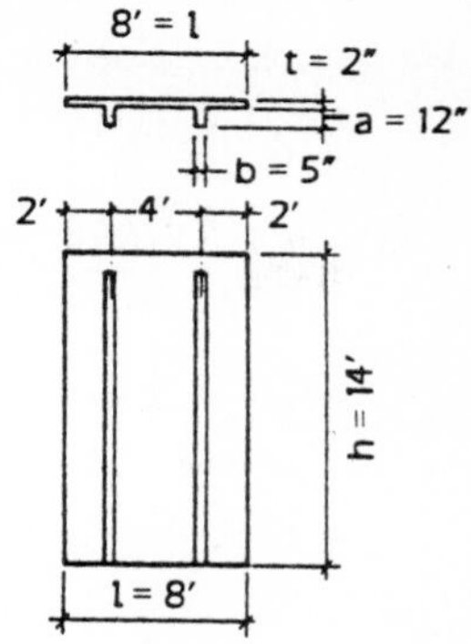

8′ = l
t = 2″
a = 12″
b = 5″
2′ 4′ 2′
h = 14′
l = 8′

Figure 15.33 Wall panel 1.

$$I = \frac{2 \times (8 \times 12)^3}{12} = \frac{t\ell^3}{12} = 147{,}456$$

$$2 \times 5 \times 12\,\text{in} \times 24^2 = 2Ay^2 = 69{,}120$$

$$\frac{2ab^3}{12} = \frac{2 \times 12 \times 5^3}{12} = \frac{250}{216{,}826\,\text{in}^4}$$

$$A = (2 \times 8) + (2 \times 5 \times 12) = 136\,\text{in}^2$$

Let $E_c = 1.$ $G = 0.4\,E_c = 0.4^4$ $h = 14 \times 12 = 168$

$$\alpha = \alpha_f + \alpha_s = \frac{h^3}{12eI} + \frac{1.2h}{A\,0.4\,E} = \frac{168}{12 \times 216{,}826} + \frac{1.2 \times 168}{136 \times 0.4} = 3.717$$

$$K = \frac{1}{\alpha} = 0.269$$

**Wall panel 2.** See Fig 15.34.

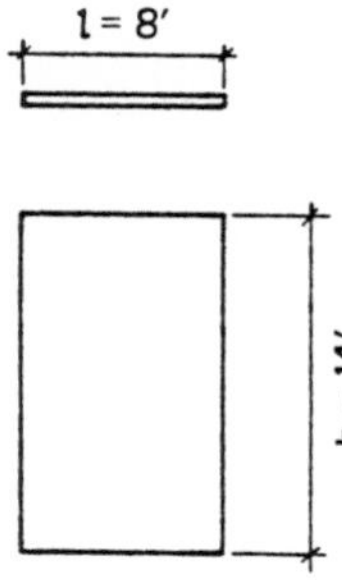

l = 8′
h = 14′

Figure 15.34 Wall panel 2.

$$I = \frac{t\ell^3}{12} = \frac{5 \times (8 \times 12)^3}{12} = 368{,}640$$

$$A = t\ell = 5 \times 8 \times 12 = 480$$

$$\alpha = \alpha_f + \alpha_s = \frac{168}{12 \times 368{,}640} + \frac{1.2 \times 168}{480 \times 0.4} = 1.05$$

$$K = \frac{1}{\alpha} = 0.95$$

**Wall panel 3.** See Fig. 15.35.

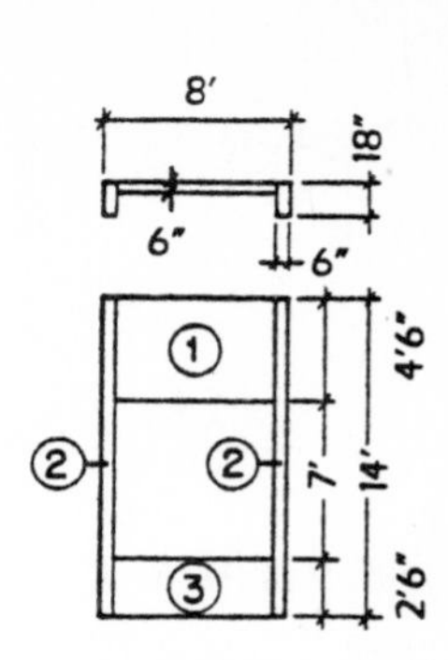

**Figure 15.35** Wall panel 3.

Part 1:

$$I = \frac{t\ell^3}{12} = \frac{6 \times 84^3}{12} = 296,352$$

$$+ 2Ay^2 = 2 \times 6 \times 18 \times 45^2 = \frac{437,400}{733,752}$$

$$A = (6 \times 84) + (2 \times 6 \times 18) = 720$$

$$h = 54 \text{ in}$$

$$\alpha = \alpha_f + \alpha_s = \frac{54}{12 \times 733,752} + \frac{1.2 \times 54}{720 \times 0.4} = 0.225$$

Part 2:

$$I = \frac{18 \times 6^3}{12} = 324$$

$$A = 6 \times 18 = 108$$

$$\alpha = \alpha_f + \alpha_s = \frac{84}{12 \times 324} + \frac{1.2 \times 84}{108 \times 0.4} = 2.35$$

Part 3:

$$I = 733,752$$

$$A = 720$$

$$\alpha = \alpha_f + \alpha_s = \frac{30}{12 \times 733,752} + \frac{1.2 \times 30}{720 \times 0.4} = 0.125$$

$$\alpha_p = \alpha_1 + \alpha_3 + \frac{1}{\frac{1}{\alpha_2} + \frac{1}{\alpha_2}} = 0.225 + 0.125 + \frac{1}{\frac{1}{2.35} + \frac{1}{2.35}}$$

$$\alpha_{p_3} = 1.525$$

$$K = \frac{1}{\alpha_{p_3}} = \frac{1}{1.525} = 0.655$$

East wall = west wall; 5 T panels, type 1 (see Fig. 15.36):

$$\alpha = \frac{1}{\frac{5}{\alpha_1}} = \frac{1}{\frac{5}{3.717}} = 0.7434 \quad K = \frac{1}{\alpha} = \frac{1}{0.7434} = 1.345$$

North wall = south wall:

$$\alpha = \frac{1}{\frac{2}{\alpha_f^{\ 2}} + \frac{6}{\alpha_f^{\ 2}}} = \frac{1}{\frac{2}{1.05} + \frac{6}{1.525}} = 0.171 \quad K = \frac{1}{\alpha} = \frac{1}{0.171} = 5.839$$

$$f_v = V_b \frac{R}{\Sigma R} = \frac{5.839}{2 \times 5.839} \times 47 = 23.5$$

$$F_T = M_T \frac{Rd}{\Sigma Rd^2}$$

23.5 ± 2.44

$47^k$

188

23.5 ± 2.44

**Figure 15.36** Design of walls.

| | $R$ | $d$, ft | $Rd$ | $Rd^2$ | $(Rd / \Sigma Rd^2) = M_T$ |
|---|---|---|---|---|---|
| North | 5.859 | 20 | 116.78 | 2335.6 | 2.44 |
| South | 5.859 | 20 | 116.78 | 2335.6 | 2.44 |
| East | 1.345 | 40 | 53.8 | 2152 | 1.12 |
| West | 1.345 | 40 | 53.8 | 2152 | 1.12 |
| | | | | 8975 | |

Design east and west walls for 23.5 + 1.12 kips = 24.62 kips; north and south walls, for 23.5 + 2.44 kips = 25.94 kips.

**East and west walls.** See Fig. 15.37.

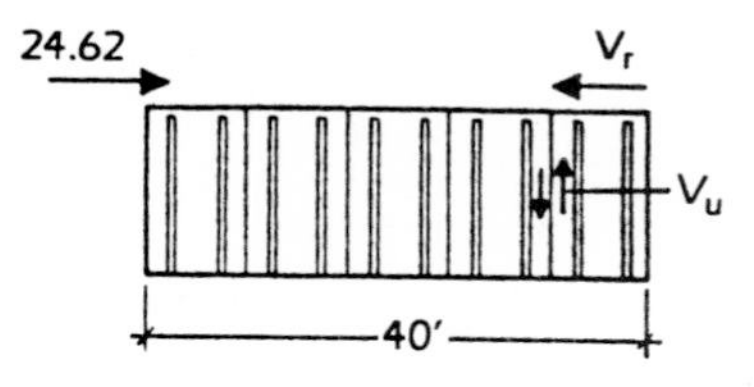


**Figure 15.37** East and west walls..

Horizontal shear stress in a panel per foot

$$= \frac{24.62}{40\,\text{ft}} = V_r = 615\,\text{lb/ft}$$

$$615 \times U = 615 \times 2.0 = 1230\,\text{lb/ft}$$

$$\text{Allowable} = \phi\,\lambda\,2\sqrt{f'_c} = 0.85 \times 0.85 \times 2 \times \sqrt{5000}$$

$$= 102\,\text{lb/in}^2 \text{ or } 102 \times 2 \times 12 = 2448\,\text{lb/ft}$$

Since 2448 is greater than 1230, this is OK.

$$v_v - v_r = 1230 \text{ lb/ft}$$

$$\text{Vertical shear for the panel} = \frac{1230 \text{ lb/ft} \times 14 \text{ ft}}{1000}$$

$$= 17.220 \text{ kips/ panel edge}$$

Use three weld connectors:

$$\frac{17.2}{3} = 5.73 \text{ kips/ connector}$$

Use no. 4.

$$V_u = \frac{2A_s \phi f y}{1.414} = \frac{2 \times 0.2 \times 0.9 \times 40}{1.414}$$

$$= 10.18 \text{ kips/connector}$$

$$5.73 \text{ required} < 10.18 \text{ allowable (OK)}$$

$$\text{Weld length} = \frac{10.18^k}{42 \times 1/3 \times 1/2} = 1.46$$

Use minimum of 2 in of weld.

**North and south walls.** See Fig. 15.38.

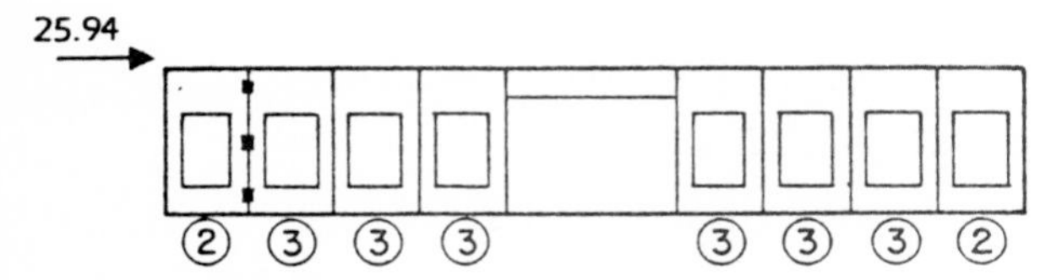


**Figure 15.38** North and south walls..

Shear force in panel 1:

$$F \times \frac{K_1}{\Sigma K} = 25.94 \times \frac{0.95}{5.839} = 4.22 \text{kips}$$

$$\text{Shear/ ft} = \frac{4.22}{8 \text{ ft}} = 0.527^k/\text{ft or } 527 \text{ lb/ ft}$$

$$\text{Allowable} = \frac{\phi \lambda 2\sqrt{f'_c} \times t \times 12 \text{ in}}{U} = \frac{0.85 \times 0.85 \times 2\sqrt{5000} \times 5 \times 12}{1.4}$$

$$= 4378 \text{ lb/ ft} > 527 \text{ (OK)}$$

$$V_u = V_r = 527 \times U = 527 \times 1.4\ 738 \text{ lb/ft}$$

For the total height,

$$738 \times 14 \text{ ft} = 10{,}332 \text{ ft}$$

$$\text{Minimum shear key area} = \frac{V_u}{\text{allowable shear stress}}$$

$$= \frac{10{,}332}{0.85 \times 0.5 \times 2\sqrt{5000}} = \frac{10{,}332}{102} = 102 \text{ in}^2$$

Use three shear key groups: 102/3 = 34 in/group.
Minimum shear key area based upon shear friction:

$$\text{Allowable per ACI} \begin{cases} 800 \text{ lb/in}^2 \times \lambda = 800 \times 0.85 = 680 \\ \lambda \times 0.2 f_c' = 0.85 \times 0.2 \times 5000 = 850 \end{cases}$$

$$\text{Allowable per PCI: } 1000 \times \lambda^2 = 1000 \times 0.85^2 = 722$$

$$\frac{V_u}{680} = \frac{10{,}332}{680} = 15.2 \text{ in}^2$$

Use three shear key groups: 15.2/3 = 5.1 in² (see Fig. 15.39).

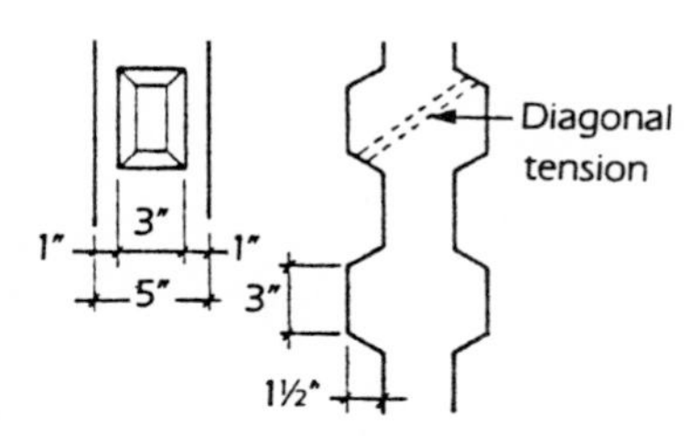


**Figure 15.39** Shear key groups.

$$3 \text{ in} \times 3 \text{ in} \times 2 \text{ keys} = 18 \text{ in}^2/\text{group} > 5.1 \text{ in}^2 \text{ group (OK)}$$

$$A_{vf} = \frac{V_u}{\phi f y \mu} = \frac{14.7}{0.9 \times 40 \times 1.4} = 0.29$$

Use minimum no. 4 bar top and bottom.

**Check overturning of north and south panels.** See Fig. 15.40.

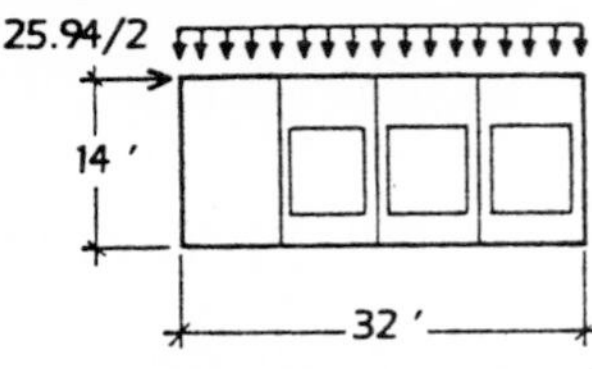


$$M_{cT} = \frac{25.94}{2} \times 14\text{ ft} = 181\text{ ft}\cdot\text{kips}$$

**Figure 15.40** Overturning of north-south panels.

$$M_{cT} = \frac{25.94}{2} \times 14\text{ ft} = 181\text{ ft}\cdot\text{kips}$$

$$\text{Weight of panel 2} = \frac{5}{12} \times \frac{115 \times 14}{1000} \times 8 = 5.36\text{ kips}$$

$$\text{Weight of panel 3} = \frac{6 \times 18}{144} \times 14 \times \frac{2 \times 115}{1000} = 2.41$$

$$+ \frac{6}{12} \times \frac{115 \times 7}{1000} \times (4.5 + 2.5) = \underline{2.81}$$

$$5.225\text{ kips}$$

$$M_R = (5.36 \times 28) + (5.22 \times 20) + (5.22 \times 12) + (5.22 \times 4)$$

$$= 338\text{ ft}\cdot\text{kips}$$

$$\frac{M_R}{M_{\text{total}}} = \frac{338}{181} = 1.86 > 1.5$$

Hence no hold-down is required. There is no tension in the panel.

**Shear in mullions.** Check panel 3 mullions for shear.

$$F\frac{K}{\Sigma K} = 25.94 \times \frac{0.655}{5.839} = 2.90\text{ kips}$$

There are two mullions; hence, 2.90/2 = 1.45 kips/mullion.

$$\text{Mullion area} = 6\text{ in} \times 18\text{ in} \qquad \frac{1.45}{6 \times 18} = 0.0134 \text{ or } 13.4\text{ lb/in}^2$$

$$\text{Allowable} = \frac{\lambda 2\sqrt{f_c'}}{U} = \frac{0.85 \times 2 \times \sqrt{5000}}{1.4} = 86\text{ lb/in}^2$$

Since 86 is greater than 13.4, this is OK.

**Moment.** Check for moment.

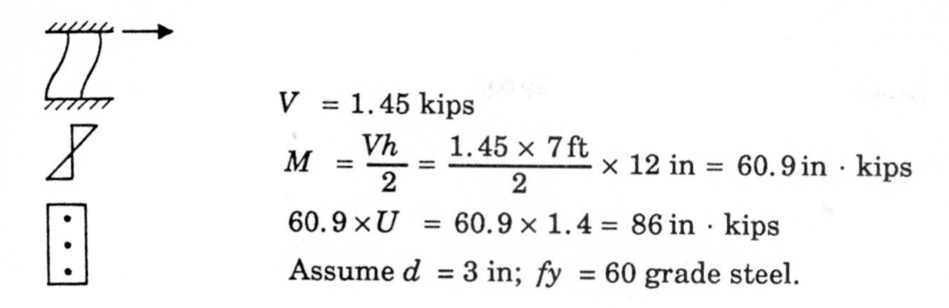

$$V = 1.45 \text{ kips}$$

$$M = \frac{Vh}{2} = \frac{1.45 \times 7\,\text{ft}}{2} \times 12 \text{ in} = 60.9 \text{ in} \cdot \text{kips}$$

$$60.9 \times U = 60.9 \times 1.4 = 86 \text{ in} \cdot \text{kips}$$

Assume $d = 3$ in; $fy = 60$ grade steel.

$$a = d - \sqrt{\frac{-2M}{\phi 0.85 f'_c\, b} + d^2} = 3 - \sqrt{\frac{-2 \times 86}{0.9 \times 0.85 \times 5 \times 18} + 3^2} = 0.51$$

$$A_s \text{ required} = \frac{0.85 f'_c ba}{fy} = \frac{0.85 \times 5 \times 18 \times 0.51}{60} = 0.65 \text{ in}^2$$

Use three no. 5 bars.

$$3 \times 0.31 = 0.93 > 0.65 \quad \text{(OK)}$$

## Buildings Comprising Precast and Prestressed Concrete Components

The buildings shown in Figs. 15.41, 15.42, 15.43, and 15.44 are all under construction, allowing us to see how the precast components fit together to form the structural frame or enclosure.

**Figure 15.41** Fairchild Semiconductor plant, San Jose, California. This three-level structure features an 18-in-deep by 12-ft-wide double T wall panel 65 ft high; 41-in-deep by 12-ft-wide double T's span form the exterior panels to the interior precast concrete beams and columns. Lateral support is furnished by the double-T wall panels, which are interconnected by flange weld plates to form a long, stiff wall diaphragm. Such structures can be designed to perform elastically to resist maximum design earthquake forces, thus negating the need to provide ductility. This plant was one of two identical industrial buildings erected here in 1976. (*Photograph courtesy of L. A. Compton.*)

Figure 15.42 Kaiser Hospital parking structure, Walnut Creek, California. Precast concrete columns and prestressed concrete beams form a vertical load-carrying frame for this garage structure, shown under construction in 1974. The deck in this type of structure is cast-in-place concrete, either mild-steel-reinforced or posttensioned. The prestressed concrete girders are shored at midspan or at third points concurrently with the floor-slab-forming operation. Lateral resistance to wind and seismic forces is provided by masonry or cast-in-place concrete shear walls. Contraction joints and shear walls are positioned by recognizing the effects of long-term creep and shrinkage occurring in cast-in-place posttensioned structures. (*Photograph courtesy of L. A. Compton.*)

**Figure 15.43** Syntex R-6 chemical research building, Palo Alto, California. Architecturally colored and exposed precast concrete spandrels and columns form a vertical load-carrying exterior frame supporting 8-ft-wide by 18-in-deep prestressed concrete double T's. Lateral rigidity is provided by interior cast-in-place concrete shear walls, which could have been furnished just as easily as precast concrete units, thus saving valuable construction time.

**Figure 15.44** Union Bank–Oceangate Building, Long Beach, California. Details conforming to Uniform Building Code provisions for cast-in-place concrete ductile moment-resisting frames were used in designing this 16-story structure. U-shaped precast concrete spandrels frame into the exterior precast concrete columns at each level. Reinforcing bars in the columns were spliced to field-placed reinforcing bars inside the spandrel zone. After placing additional confinement reinforcing, the closures were poured along with the topping over the spread single-T and 8-in hollow-core deck system. Here a 36-in-deep single T is being erected by a 200-ton Manitowoc crane. Use of a fixed tower crane in the core would probably have been more efficient and economical.

**Figure 15.45** Precast concrete column frames 12 ft wide by 50 ft high being erected at the Gregory Bateson State Office Building. Each 36-ton frame was erected directly over a matching set of prestressed concrete piling. *(Photograph courtesy of Ben Shook, Office of the State Architect, Sacramento, California.)*

Figures 15.46 through 15.54 all depict some of the construction stages of the Gregory Bateson State Office Building in Sacramento. Basic lateral resistance is furnished by ductile moment-resisting frames which are totally precast in one direction and combine with cast-in-place concrete girders in the other. Prestressed concrete piling deliver structure loads to the soil. Prestressed concrete double T's form the deck system for the floors and roof.

**Figure 15.46** Precast concrete frame being positioned over 12-in-square prestressed concrete piling. The frames were plant-sandblasted to reveal light buff aggregate and sand combined with buff cement. The tolerance for pile placement was ±2 in. A special bearing plate was grouted on the pile before erecting the column frame. Note the rebar extending from the bottom of the column that is to be embedded in a poured-in-place pile cap. Note also the notched surface and holes for reinforcing bars to complete the ductile moment-resisting frame in the direction transverse to the frame. *(Photograph courtesy of Ben Shook, Office of the State Architect, Sacramento, California.)*

**Figure 15.47** Precast concrete frame bents being positioned and braced while the pile cap is being formed. Note cable bracing connected to steel collars around the prestressed concrete piling units. Every frame had to be braced and guyed in four directions in this manner. Reinforcing-bar design and detailing provide fixity at the pile cap after it has been formed and poured. *(Photograph courtesy of Ben Shook, Office of the State Architect, Sacramento, California.)*

**Figure 15.48** Aerial view showing erected frame bents. Note on the right the cast-in-place concrete beams that have been formed and poured to provide the ductile moment-resisting frame in the direction transverse to the frame. Plant-cast prestressed concrete double T's will span the 40-ft gap between frames and will be supported on the cast-in-place concrete beams spanning between the frames. *(Photograph courtesy of Ben Shook, Office of the State Architect, Sacramento, California.)*

**Figure 15.49** A closer view of the precast concrete frames and cable bracing. In the background can be seen the forming operation for the cast-in-place concrete beam spanning between bents to provide frame action in that direction. The field-poured girders were made with the same mix design used in the precast concrete frame and subsequently were field-sandblasted, providing a close match with the precast finish. *(Photograph courtesy of Ben Shook, Office of the State Architect, Sacramento, California.)*

**Figure 15.50** Another view of the precast concrete frames. Four no. 11 bars ran the full height of each column without reinforcing-bar splices. This would have been difficult to achieve in the field if the bents had been cast-in-place concrete. The notched surface at the juncture of the cast-in-place concrete beam and the precast concrete frame could have been replaced by a sandblasted-retarded interface to achieve the same design result. *(Photograph courtesy of Ben Shook, Office of the State Architect, Sacramento, California.)*

**Figure 15.51** Placing reinforcing for cast-in-place concrete frame girders. Two no. 11 bars are threaded through each top hole, while one no. 11 bar extends through each bottom hole. These holes were then carefully grouted solid to assure an airtight closure without entrapped air bubbles, thus precluding the possibility of long-term corrosion of the bars in the column sleeve. Then the beam forming was completed, and the cast-in-place concrete was poured. The design was carried out in accordance with the equivalent-lateral-force provisions of the Uniform Building Code, with $K = 0.67$. Note the closed-loop stirrup reinforcement being placed by workers. *(Photograph courtesy of Ben Shook, Office of the State Architect, Sacramento, California.)*

**Figure 15.52** Plant-cast prestressed concrete double T's being erected. In the upper photograph, notice the extension of the cast-in-place girder beyond the outside face of the exterior frame in order to develop adequately the field-placed girder reinforcing.

**Figure 15.53** Double-T floor installation in progress.

**Figure 15.54** Two views of the partially completed structural frame with various precast and cast-in-place concrete elements in place.

**Figure 16.1** Strand stressing: double-T casting line. The bed supervisor (*left*) watches the pressure gauge indicating the force being imparted in the strand by the hydraulic jack. The quality control inspector measures the resulting elongation on the strand as a check against gauge pressure. This stressing operation is being conducted during the routine setup of the double-T bed for daily production of 8-ft-wide by 24-ft-deep double T's for the Gregory Bateson State Office Building in Sacramento, California.

Chapter

# 16

# Precast Prestressed Concrete

What is prestressed concrete? Prestressed concrete is concrete made by the introduction of an internal compressive force in a concrete member which counteracts the tensile forces produced by external loads imposed either during handling or in service. This internal force is introduced by pulling high-strength cables, or strands, as they are more correctly called, to within 70 percent or more of the ultimate strength of the strand and then releasing this force into the concrete. Prestressing may be accomplished in two ways.

1. *Pretensioning.* The strands are stressed before the concrete is poured and hardened around them. When the concrete has reached a sufficiently high compressive strength to withstand the force imposed by the strand force (release strength), the strands are cut or burned and the prestress is released into the member. This is the method used to produce plant-cast precast prestressed concrete members.
2. *Posttensioning.* The concrete element is cast with conduit or ducts sufficiently large to receive bundles of strand, wires, or high-strength rods, which are called tendons. The tendons are stressed *after* the concrete has reached sufficient strength to withstand the force imposed by the tendons. Posttensioning is usually done with cast-in-place concrete at the jobsite. Chapter 17 discusses how this method of prestressing interfaces with plant-cast precast pretensioned concrete.

So now we have factory-cast high-quality, high-strength concrete combined with high-strength steel strand, both materials being stressed to very high percentages of their respective ultimate strengths, yet remaining within their respective elastic renges. In most design situations, the concrete remains in the uncracked range, thereby more efficiently using the available depth and making possible shallower depths and longer spans for horizontal elements as well as for prestressed concrete wall panels. Precast concrete structural members are usually prestressed, whereas architectural precast concrete cladding units and load-bearing panels are not.

Precast pretensioned concrete elements are double-T floor and wall units, single T's, inverted-T girders, channel units, rectangular beams, hollow-core plank floor and wall units, piling, columns, and solid flat slabs.

Prestressing is used to introduce precompression in the area of the beam cross section where tension will be produced by superimposed loads. As the member is subjected to dead and live loads, the resulting tension is counteracted by the precompression induced by the force imparted to the concrete by the prestressing strands. Thus, one offsets the other, with the end product being a net compression stress throughout the beam or a small value of tension on the bottom of a deck member, usually less than the modulus of rupture (tensile strength of concrete).

It is common for precast concrete to have high compression strengths of up to 7000 lb/in$^2$ for hard-rock concrete, with 6000 lb/in$^2$ being routinely attainable. Lightweight aggregate concrete can be made to achieve 5000 lb/in$^2$ 28-day strengths. In tension, the strength of concrete, or modulus of rupture, is approximately $7.5\sqrt{f'_c}$ . The tensile strength for sand-lightweight aggregate concrete is 85 percent of this value. By prestressing, the inefficiency of the concrete in tension is eliminated and replaced with high-strength steel strand, which has an ultimate tensile strength of 270,000 lb/in$^2$.

The pretensioning force is introduced into the concrete by placing high-tensile steel strands in the mold before the concrete is placed and stressing the strands with hydraulic jacks. The concrete is placed, vibrated, and screeded; then it is cured by live steam or radiant heat. After 10 to 12 hours the concrete reaches a minimum compressive strength of approximately 3500 to 4000 lb/in$^2$. As the concrete hardens, it bonds to the tensioned steel strands. Test cylinders are broken to ascertain the strength. If the strength is equal to or above the required minimum for stress transfer (release strength), then the strands are cut, releasing the tension into the precast elements. The strands are held firmly by the bond developed by the concrete at the ends of the beams. The tension in the strands thus imparts a compressive force in the concrete (precompression).

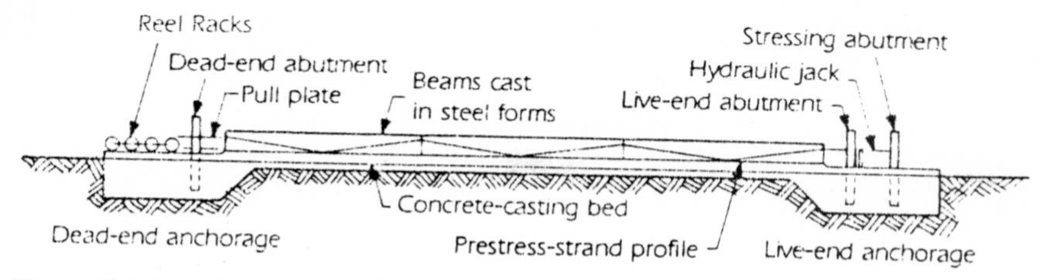


**Figure 16.2** A typical stressing bed.

Plant-cast prestressed concrete products are cast in long, continuous beds of up to 800 ft or more for some products such as piling and hollow-core planks (see Fig. 16.2). The length of the bed decreases with the increasing complexity of the unit and the amount of reinforcing required, with 400 ft being an average. Long lines enable the prestressed concrete manufacturer to cast multiple units end to end, with bulkhead separators between individual units of specified lengths for a particular project. The maximum bed length is the longest length which a standard casting crew can set up and pour on a daily cycle, optimizing plant labor.

Figures 16.3, 16.4, 16.5, 16.6, 16.7, and 16.8 show in diagrammatic form how prestressing works.

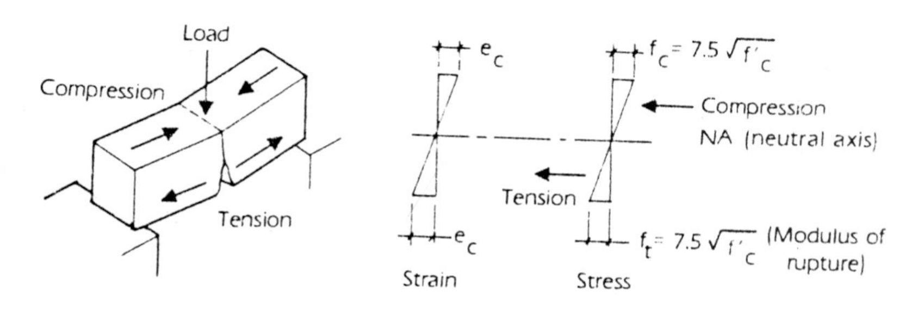


**Figure 16.3** Nonreinforced-concrete beam.

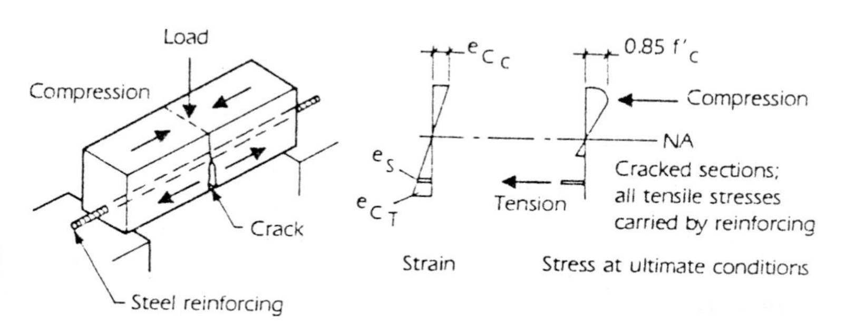


**Figure 16.4** Reinforced-concrete beam.

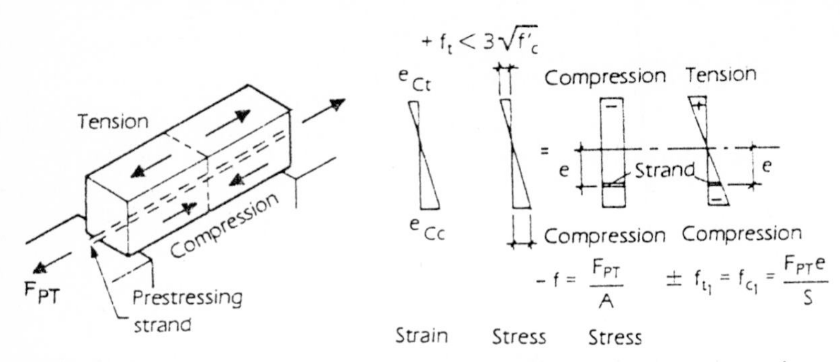


**Figure 16.5** A prestressed concrete beam at release of prestress; - = compression; + = tension.

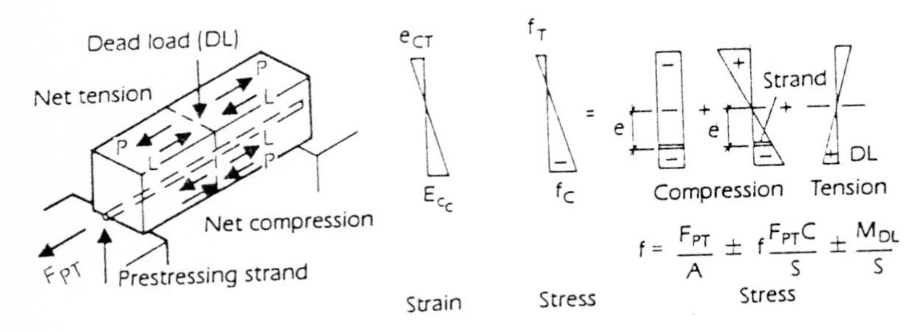


**Figure 16.6** A prestressed concrete beam with dead load applied. This condition must be evaluated for each of the following: (1) removal from the form and in-plant handling, including impact; (2) shipping and erection, including impact; (3) in final position without topping; and (4) in final position with topping.

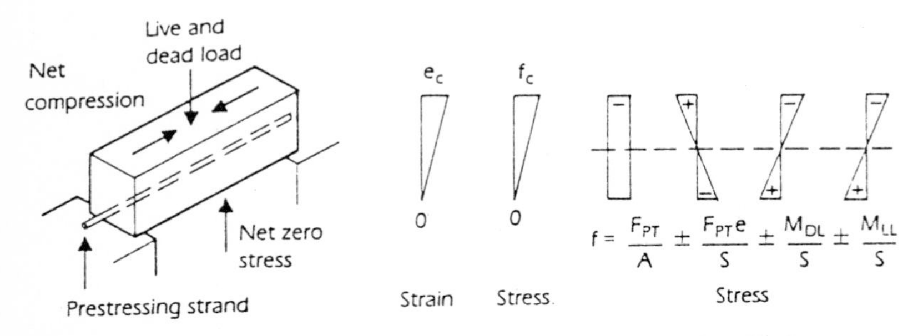


**Figure 16.7** A prestressed concrete beam with dead and live load applied; - = compression; + = tension.

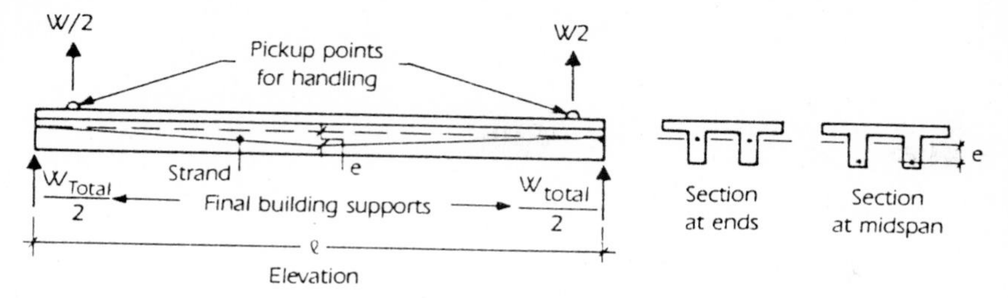


**Figure 16.8** Strand placement. For the most efficient design strands should be depressed.

Ideally, handling at the fifth points during lifting or storage produces equal negative and positive bending moments. However, most prestressed concrete products are lifted and supported near the ends since the member is designed to support service loads at these points (see Fig. 16.9). Also, prestressed concrete deck elements, other than those with especially designed cantilevers, should be stored at the ends so that excessive camber and deflection growth due to creep do not occur.

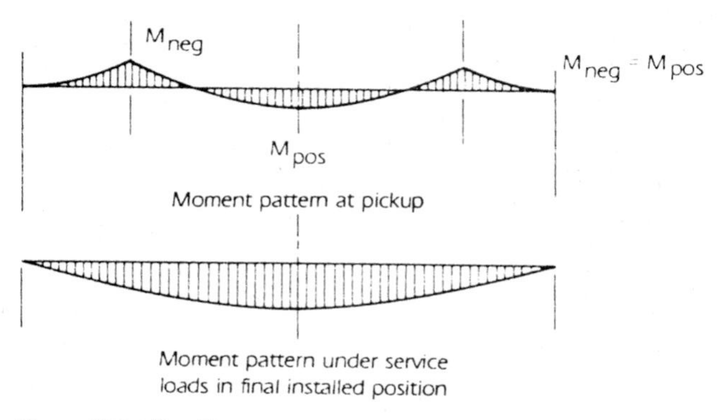


**Figure 16.9** Bending-moment patterns.

In closing, it is important to remember that other factors, such as shear, deflection, and ultimate-strength capacity, also affect the design of prestressed concrete members. Since these other design considerations are covered in prerequisite concrete design courses and in the *PCI Design Handbook*, they will not be repeated here. The component design sections show detailed analysis of these important design aspects relating to prestressed concrete members in example form. The intent in this chapter is to explain to the student and the designer the facets of prestressed concrete member analysis that are unique in comparison with ordinary reinforced concrete.

# Chapter 17

# Posttensioning and Its Interface with Precast Concrete

As opposed to pretensioning, in which the strands are tensioned before the concrete is poured and set around the strands, posttensioning is the method for prestressing in which the tendons are tensioned after the concrete is poured and cured. Posttensioning is used primarily for cast-in-place construction and precast segmental construction.

In the casting operation, ducts are placed in the proper position in the forms and concrete cast around them. The ducts are sufficiently large to receive a bundle of 12 to 15 strands, which are stressed when the concrete has reached sufficient strength to withstand the stresses imposed. After stressing, the remaining space in the duct is usually filled with grout to form what is called a "bonded tendon." (The condition of leaving the void ungrouted is referred to as an "unbonded tendon.") The term "tendon" refers to strands that are stressed together as a group in a posttensioning duct.

One advantage of posttensioning is that the prestress force can be applied in stages during construction as dead loads are applied. Posttensioning also lends itself to continuous or cantilevered flexural member design since connection limitations at column joints do not exist and the excess of prestress force required at cantilevered design conditions may be terminated at the first interior span.

In addition to $\frac{1}{2}$-in $\phi$-$270^k$ strand, tendons may be composed of other sizes of strands, button-ended wire, or high-strength bars. When unbonded tendons are used, the Uniform Building Code requires that a minimum area of bonded reinforcing be provided in addition to the required posttensioning.

Occasionally, precast concrete manufacturers become involved with posttensioning either as a part of their plant operations or as an interface condition to be recognized in the overall structural design of a building. The following categories outline the areas of potential involvement with posttensioning:

1. When production facilities or stressing abutments and anchorages cannot handle the magnitude of the prestressing forces involved, the elements are fabricated with mild steel reinforcing or are partially pretensioned and then posttensioned after stripping and when the concrete has reached sufficient strength to receive the posttensioning force.

2. When span or headroom requirements limit the available depth of the section so that an abnormally high value of release strength is required, then the section is posttensioned.

3. Segmental construction for bridges with plant-cast longitudinal segments, such as was provided for the Dumbarton Bridge project (see Fig. 17.1).

**Figure 17.1** Dumbarton Bridge, South San Francisco Bay. Spans 150 ft long were formed by posttensioning two 75-ft-long, 70-ton plant-cast segments together. The 7-ft 6-in-deep hollow girders were plant-cast in match-cast half segments and hauled to the jobsite. This marriage of plant precasting and jobsite posttensioning provided the most economical solution for this project.

4. Long-span segmental construction: spans of 180 to 800 ft with plant-cast transverse segments.
5. Large-diameter-cylinder piling plant-posttensioned to 750 to 800 lb/in$^2$.
6. Architectural precast concrete walls posttensioned vertically to form shear-resisting elements.
7. Plant-cast precast segments posttensioned vertically, horizontally, or both to build liquid storage tanks.
8. Parking structures, where the beams and columns are plant-cast precast prestressed concrete and the deck slab is cast-in-place posttensioned concrete.

## Design Considerations

The design of posttensioned concrete differs somewhat in several areas from the design of pretensioned concrete. These areas should be recognized by the designer:

1. The anchorage zone at the end of the member is subjected to large confined compressive forces that, if not reinforced properly, could cause longitudinal cracking or bursting of the concrete at the end of the member. As opposed to pretensioned concrete, in which the force buildup is gradual along the development length of the strand and in which sufficient space exists between strands to keep stress transfer uniformly distributed, all the force of the posttensioning tendon is delivered to a small area by the anchorage bearing plate at the end of the member, as required by Sec. 2618(h) of the Uniform Building Code.
2. Sufficient area must be provided to accommodate the physical dimensions of the bearing plate required with the multistrand wedge system that will be used.
3. Ducts should be of sufficient gauge to withstand the pressures imposed by vibrating equipment and not collapse. Temporary polyvinyl chloride pipe installed during casting and removed afterward will also serve to prevent the occurrence of collapsing posttensioning ducts.
4. Segment interface areas are match-cast (the mating segment is cast against the element with which it will subsequently be united in the field). Interface details that will prevent leakage during the grouting operation are selected.
5. The conduit may be positioned in a parabolic profile to use the prestress force more efficiently, when compared with shape of the externally applied moment diagram.
6. In addition to the stress losses considered in pretensioned designs (elastic shortening, strand relaxation, creep, and shrinkage), slippage at anchorage fixtures and conduit friction and wobble losses must also be considered.

7. Creep and shrinkage movements in cast-in-place concrete posttensioned structures will be considerably greater than those associated with buildings with plant-cast precast and prestressed components. Posttensioned deck slabs, for example, will exhibit creep and shrinkage movements more than double those associated with plant-cast decks. This should be taken into account in positioning shear walls and in selecting expansion joint locations. Posttensioning jacking points should be located at positions in the floor where temporary gaps can be left, to be poured after some initial percentage of creep and shrinkage has occurred. However, this will cause some delay in project completion.

## Posttensioning Procedures

Unbonded tendons are coated with a corrosion inhibitor and encased in a plastic sheath to prevent bond with the concrete. (Additional precautions to ensure satisfactory performance in both normal and corrosive environments are given in a specification presented in the March–April 1985 issue of the *PCI Journal.*)

Bonded tendons are placed in a flexible or rigid galvanized metal duct. The ducts are cast in the concrete, but the tendons usually are not placed in the duct until just before the posttensioning operation. If the prestressing steel is placed in the duct prior to casting the concrete, then it may require coating with a corrosion inhibitor, which will not prevent bond between the tendon and the grout.

If the concrete is to be steam-cured, then the tendons are not placed in the ducts until after the steam curing has been completed. Prior to placing the tendons the ducts are flushed with air under pressure or water with 0.1 lb of slaked lime or quicklime per gallon of water. Once the strand has been installed, if left to set more than 10 days it must be coated with a corrosion inhibitor. (If it is left less than 10 days, this is not necessary.) Grout is pumped in at 100 $lb/in^2$ pressure, with water content not exceeding 5 gal of water per sack of cement; admixtures may also be used.

A complete overview of posttensioning is presented in the *Post-Tensioning Manual* published by the Post-Tensioning Institute.

**Figure 18.1** Ambassador College Science and Fine Arts Building, Pasadena, California. Classically sculptured precast panels grace these twin buildings. The honeycombed units, supported at the second floor, are vertical load-bearing units. They feature smooth white concrete contrasted with recessed areas of exposed black granite and solar bronze glazing.

# Chapter 18

# Textures and Finishes for Architectural Precast Concrete

Precast concrete cast in steel, concrete, wood, or fiberglass molds possesses a smooth surface which may be acceptable, depending upon the architect's desired design effect. The concrete may be white (white cement and white sand; see Fig. 18.2), natural gray, or colored with a pigment. With smooth concrete, minute imperfections readily show up under bright light and in dark shadows. A smooth as-cast surface is very absorbent and is easily stained by oil, smog, soot, dust, and handling. To offset this, the surface is often textured or sculptured or the coarse aggregate is exposed. By exposing the aggregate, a large portion of the surface becomes coarse aggregate. Coarse aggregate has the greatest resistance to absorption, abrasion, and staining from adverse weathering. Depending upon the color and shape of the aggregate, many interesting effects are possible. The various surface textures used with precast concrete are discussed in the following sections.

## Marble or Granite Facing

Precast concrete panels can be faced with marble or granite during the casting operation (see Fig. 18.3). Though expensive, these materials do have the greatest resistance to absorption, abrasion, and staining.

**Figure 18.2** Mark Taper Forum Music Center, Los Angeles, California. It features bas-relief panels in white cement by using rubber form liners with nine different patterns that are repeated around the building facade.

Marble and granite are beautiful and resist adverse weathering. The stone sheets are drilled and metal anchors attached with epoxy glue or expansive inserts. If the panel is not insulated, a polyethylene bond breaker sheet is applied and the concrete backing is cast. The metal anchors are embedded in the concrete. Lifting and connection inserts are cast in the backup concrete. If the panel is to be insulated, the insulative core is installed between the stone and the concrete. Since the insulation acts as a bond breaker, the polyethylene sheet is omitted. The metal anchors must be long enough to pierce the insulation for proper embedment in the concrete. Bond breakers are used to compensate for the different volumetric changes of the stone and concrete caused by temperature, shrinkage, and creep. The marble or granite is usually from 1 to 1 1/2 in thick. (See Fig. 18.4.)

Figure 18.3 Pacific Gas and Electric headquarters, San Francisco, California. Granite-faced panels clad this landmark office structure.

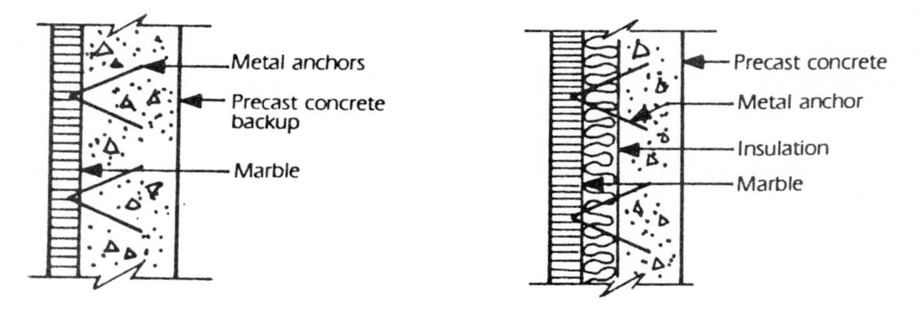


Figure 18.4 Details of panels faced with natural stone.

## Sandblasted Finish

Precast concrete units can be lightly sandblasted (see Fig. 18.5). The depth of sandblasting is usually sufficient only to expose sand particles and a portion of the coarse aggregate. To achieve greater depths, a light retarder is used along with the sandblasting. Depending upon the softness of the coarse aggregate, the sandblasting medium may also consist of walnut shells. Walnut shells will cut the cement and the sand but only slightly affect the coarse aggregate. The depth of sandblasting must be defined and worked out between the precaster and the architect prior to bidding. Sandblasting produces a lightly etched surface which exposes the aggregate and defuses light reflection.

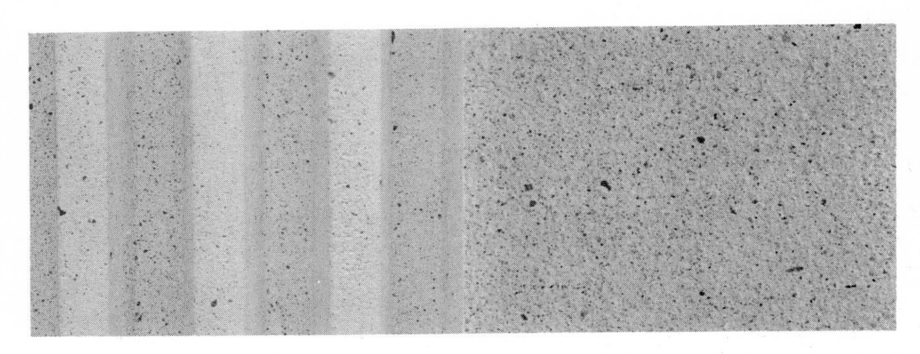

Figure 18.5 Cutter Laboratories, Berkeley, California. Light sandblasting presents a uniform and pleasant finish on the ribbed and smooth portions of the cladding panels on this project.

## Exposed Aggregate (Retarded) Finish

Exposed aggregate surfaces (see Fig. 18.6) are usually obtained by using retarders. The unit may be sandblasted afterward, but this is uncommon. If the surface to be exposed is in contact with the forms, the

Figure 18.6 Meridian Hill Park, Washington, D.C. This facade, constructed in 1920, demonstrates the durability and excellent weathering characteristics of exposed aggregate concrete. John Earley strictly controlled the water content and brushed the surfaces of the still green concrete to expose the aggregate.

retarder is brushed, rolled, or sprayed over a release agent previously applied to the mold surface. After removal from the form, the unhydrated surface mixture of cement paste and sand is washed with water and brushed off the unit. High-pressure water jets are also used. If the surface to be exposed is not in contact with the mold, the retarder is sprayed on after finishing with a steel trowel.

Retarders are available in light, medium, or heavy etch grades formulated to give various depths of exposure. Retarders prevent the set of the cement and deter the hardening process at the surface. If they are left on and not washed off, the cement will eventually set and cure; hence they must be washed off. The usual time is 48 to 72 hours for form-applied retarders and 24 hours for direct-applied retarders. However, some retarders may be left on a week or more before removal.

## Fractured Surface

A fractured surface can be obtained by placing cable or chains close together on a form bed. After the unit has been removed from the form, the cable or chain is pulled out, leaving an imprint along with fracturing between the ridges. Another method is to bush-hammer or chip the ridges to fracture the surface. When large ribs are cast in the face of the panel, these may be split in the yard by using a mason's chisel positioned at the base of the rib and struck with a heavy hammer or maul. This finish is called "split-rib." (See Fig. 18.7.) Care must be taken in forming this finish, in that the rib should be fractured by striking in only one direction so that light striking the panel from an angle results in a uniform shading effect. Another method of achieving a fractured surface appearance is to coat the smooth rib area with retarder prior to casting. Then by sandblasting the panel after stripping, a fractured-rib appearance is attained.

In modern practice all these surface finishes are usually achieved by using neoprene form liners.

## Sculptured Surface

Since concrete is plastic in the freshly mixed state, unlimited sculptural effects are achievable (see Fig. 18.8). Forms may be fabricated of wood or fiberglass. To create forms for complex shapes, wood, clay, plaster, or polystyrene patterns are made and sprayed with fiberglass resin and chopped fiberglass. The form may be reinforced or stiffened with wood, metal, or fiberglass ribs. Once completed, the form is sanded and gel-coated. One pattern may be used for more than one form. Wood and fiberglass forms may be used up to 50 times, depending on the shape. For greater numbers of casts, concrete or metal molds should be made.

**Figure 18.7** Alvarado sewage treatment plant, Union City, California. The split-rib effect is achieved economically by casting against a heavily textured form liner and then sandblasting the panel after stripping.

**Figure 18.8** Library stair tower, Civic Center, Inglewood, California. Sculptured forms combined with an exposed aggregate finish achieve the bold expression desired by the architect. *(Photograph by Eugene Phillips.)*

Prior to casting, the forms are coated with a release agent to ensure ease of removal. For a single nonrepeatable unit, polystyrene sheets can be carved into a negative pattern and used as a form liner. The polystyrene pattern is sprayed with a fiberglass resin to seal the surface, permitting ease of removal. In the stripping process, if the polystyrene sticks to the concrete, it is simply pulled off. One-time patterns are also made by casting impressions of sand patterns created by an artist.

### Acid-Etch Finish

When the desired architectural finish is a very light surface texture which just barely breaks the surface skin, an acid-etched finish may be selected. Acid etching consists of brushing a solution consisting of 1 part of muriatic acid to 2 or 3 parts of water on the surface of the concrete 2 or 3 days after casting. The surface is then cleaned off with water, resulting in a sand finish (see Fig. 18.9). A minor portion of the surface coarse aggregate may be revealed if a 100 percent acid solution is used. Acid etching results in a very uniform sanded appearance.

### Brick or Travertine Finish

Brick or travertine tiles from $^1/_2$ to $^3/_4$ in thick may be embedded in the face of a precast panel to provide the same effect that these materials present when installed on the jobsite (see Fig. 18.10). When used as brick facing elements with grouted joints, the tiles are laid down on a rubber form liner preindented with joints simulating raked joints. Next, a layer of grout is laid on the back of the tiles, using adequate vibration to assure that the grout is firmly deposited in the joints. Finally, after placement of reinforcing and positioning of connection hardware and handling inserts, the concrete backup is poured. When the tiles are laid up without joints, they are mechanically anchored to the precast concrete backup.

Brick tiles are manufactured with a positive dovetailed key to provide anchorage. Travertine tiles are drilled by the stonecutter with angled holes to receive special stainless-steel spring clips to anchor the tiles to the backup concrete. Subsequent cleaning is usually required after casting unjointed tiles. Brick-faced panels are usually cleaned by using high-pressure water jets.

### Form Liners

Form liners can be used to simulate wood boards, wood grain, small corrugations, rope, cable, or other small patterns. They may be neo-

Figure 18.9 Preparing an acid-etch finish. After the acid has been applied with brushes, a jet of water is used to clean the surface, revealing the sand.

prene, plastic, or metal. Neoprene form liners are preferred owing to ease of stripping, cutting, and reuse. Interesting effects can also be produced by placing various items such as beans or wood sticks on the molds. Beans will expand with the moisture and give an interesting texture to the concrete. Rock salt sprinkled on the bottom of a smooth mold also creates an interesting effect.

## Face Mixes

Often the cost of the coarse aggregates selected for exposed aggregate finishes is high when compared with the cost of locally available granitic, limestone, or river gravel coarse aggregates. When this situation arises in finish selection, the concrete may be cast in two different mixes, with the portion to be exposed with the expensive aggregate forming a thin outer layer (see Fig. 18.11). The depth of etch is directly related to the aggregate size. Where a light etch is desired the coarse aggregate size should not exceed 3/8 in. Where a medium etch is desired, the coarse aggregate size should be 5/8 or 3/4 in. For deep exposure a 1-in stone should be used. Except for very large coarse aggregate sizes, face mixes are usually 1 1/2 in thick. The architect should also be aware of variations caused by form configuration.

Figure 18.10 Mock-up of the Hibernia Bank, San Francisco, California. Brick tiles cast in a precast concrete manufacturer's plant. These panels were subsequently installed on this building at Front and California Streets.

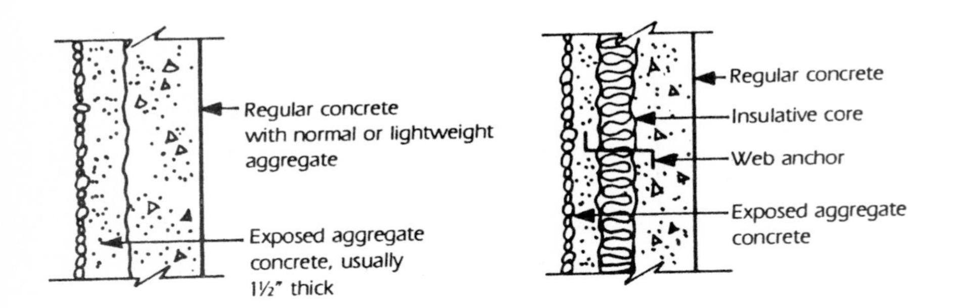


Figure 18.11 Concrete face mixes.

**Figure 19.1** Expansion of the Anheuser-Busch plant, Fairfield, California. A 14-in-square pile 110 ft long being lifted from the horizontal to the vertical by picking with both crane lines. By picking at three points as shown, approximately equal positive and negative bending moments, well below the value which would produce cracking, are induced in the pile.

Chapter

# 19

# Product Handling

The precast concrete manufacturer is responsible for handling precast and prestressed concrete units from casting to the time when they are erected and permanently built into the structure. When the precast manufacturer sells its product FOB jobsite, it is responsible to the point at which the buyer unloads the truck at the jobsite. In some instances for FOB jobsite contracts, the precaster may want to provide handling information to be used by the purchaser's erector even though the precaster's responsibility may not extend this far. In any event, proper handling of precast and prestressed units consists of the procedures and picking or support points to be used so as not to produce cracking. Product-lifting devices are also designed with sufficient factors of safety to guarantee shop and field personnel safety as well as product integrity. Design loads used to calculate handling loads and stresses include allowances for impact which reflect the magnitude of the dynamic loading to which the element is being subjected. This chapter will discuss handling devices and inserts, some of the variables in handling decisions made for stripping, storage, shipping, and erection, and, finally, a step-by-step procedure to use in analyzing handling.

## Lift Loops and Manufactured Inserts

Inserts are devices cast into the precast concrete unit which are used for handling the unit, connecting the unit to the building system, or both. When both handling and connection inserts are used, then dual usage is desirable to reduce cost. This is, however, often difficult to accomplish.

Strand lift loops are used only as lifting devices. Strand in short sections is readily available as waste from pretensioning operations.

As a lifting device, a short piece of looped strand is set in the concrete; the looped strand is usually shoved in to the required depth after the concrete has been screeded, thus not interfering with the screeding operation. When the precast section is thin and embedment is critical, the looped strand is usually anchored to the reinforcing cage with tie wire.

Where the precast concrete is to receive a poured-in-place concrete topping, the protruding strand can be either burned off or, depending on the topping thickness, left extended from the member. This is the usual case for beams, double T's, inverted beams, piling, and flat roof or floor panels. If the surface of the unit is to be exposed, then a lift loop is set in a pocket by using a polystyrene or wood blockout. After casting the concrete, the blockout is removed. After the precast unit has been erected, the lift loop is burned off and the pocket patched and finished to match the panel surface. This method is often used for wall panels; however, many precasters prefer to use manufactured inserts recessed for subsequent patching.

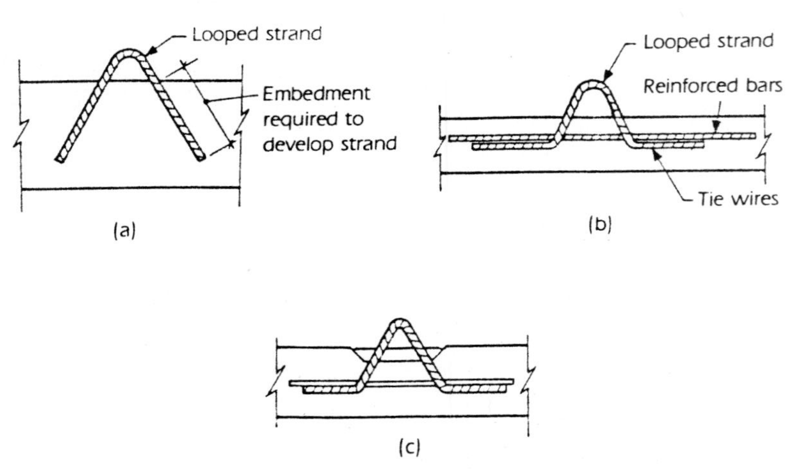


**Figure 19.2** Strand lift loops. (*a*) Looped strand cast in (no anchorage). (*b*) Looped strand tied in. (*c*) Looped strand placed in pocket in the concrete.

The size of strand lift loops is determined by calculating the load to the lifting point, including impact, and then applying a factor of safety of 4 to the ultimate capacity of the strand. For example, if waste 1/2-in-diameter $270^k$ strand were used as a lift loop, the ultimate capacity of the strand would be $0.153 \times 270 = 41.3^k$ kips. By applying a factor of safety of 4, the maximum safe load to the lift loop (including impact) is 10.3 kips.

The development length of the strand in the concrete may conservatively be taken as 50 diameters. In special cases, the shear cone method may be used to check the concrete strength of the lift loop assembly. This method is explained in the following paragraph on manufactured inserts.

Manufactured inserts are available for use with precast concrete.* The manufacturers' brochures list both the ultimate capacities and the safe working loads with a factor of safety of 4 applied. Two basic types of inserts are used for precast concrete: the coil-thread type and the standard-thread type (see Figs. 19.3 and 19.4).

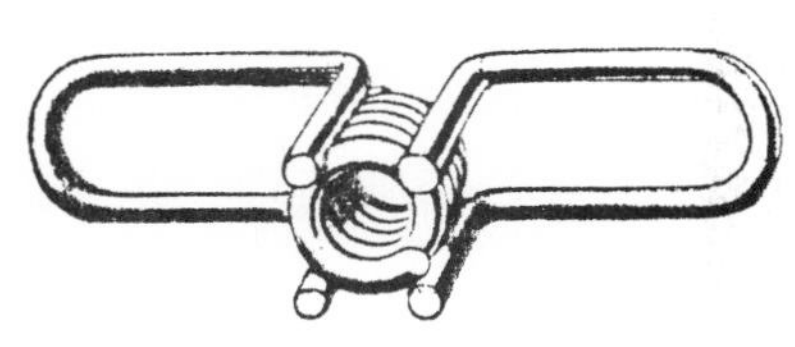

**Figure 19.3** Coil insert. It accepts coil-thread bolts.

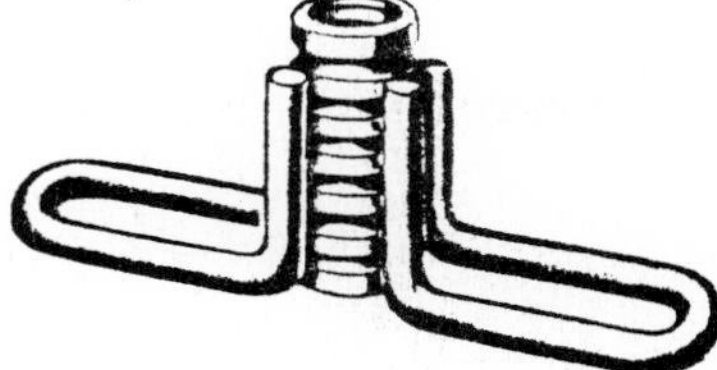

**Figure 19.4** Ferrule insert. It accepts standard-thread machine bolts.

A limited amount of technical information on inserts appears on page 6-10 of the *PCI Design Handbook*. More complete information is contained in product literature available from the various manufacturers of ferrule or coil inserts. Insert manufacturers have conducted tests on their products by using different concrete strengths. The results of these tests are contained in their literature.

Inserts are made in various diameters from 3/8 to 1 in or larger in some types. They are also made in various embedment configurations to maximize concrete pullout strength for thick as well as thin sections. When the insert is in full tension and there is sufficient concrete surrounding it, the shear pattern shown in Fig. 19.5 develops in the concrete.

However, if there is insufficient concrete, only a partial shear cone is developed. The shaded area in the cone in Fig. 19.6 shows the extent to which the area is reduced.

---

*Factors of safety may vary with actual application circumstances.

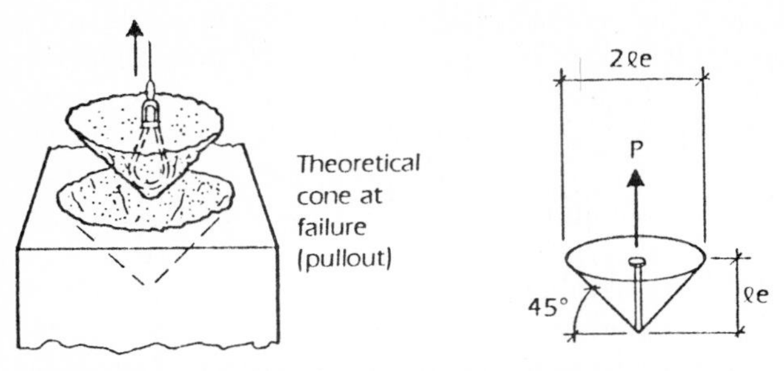


**Figure 19.5** Shear pattern developed in sufficient concrete.

$P_u = 1.4P$ ; $P_u = 2.8\, A_o \lambda \sqrt{f_c'}$ ; $\phi = 0.85$; $A_o = \pi \sqrt{2}\, \ell e^2$.

Charts in the *PCI Design Handbook* give design values for inserts (pages 6-53 to 6-58).

In the normal handling of precast concrete units, lifting slings are used

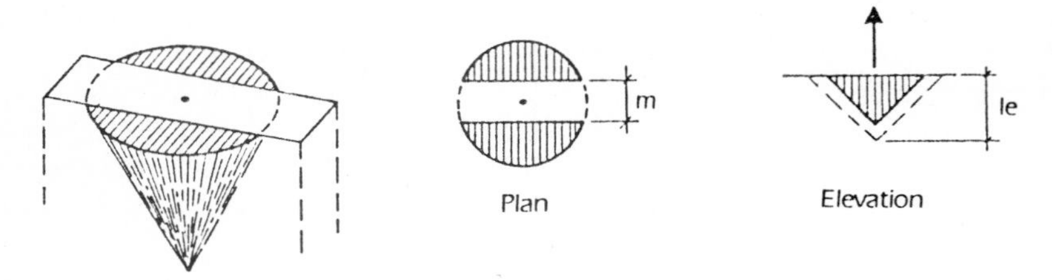


**Figure 19.6** Partial shear cone in insufficient concrete.

$$A_P = A_o - 2A_P \,;\, A_P' = \sqrt{2}\left[\ell e^2 \cos^{-1}\frac{(m)}{2e} - \frac{m}{2}\sqrt{\ell e^2 - \frac{m^2}{4}}\right]$$

which produce combined shear and tension on the insert. Also, when rotating a precast element into its final erected position a maximum critical condition of combined shear and tension will be produced. This combined loading condition should be checked for both the steel insert capacity and the concrete capacity in accordance with the interaction relationship given on page 6-9 of the *PCI Design Handbook.*

When the panel is lifted by an edge insert, the stress pattern shown in Fig. 19.7 occurs. The insert will tend to bend about *X;* the dotted line indicates the outline of probable concrete failure.

One final note of caution regarding lifting accessories: only strand or manufactured inserts are to be used; bent reinforcing bars are not to be used under any circumstances for lifting and handling precast concrete elements.

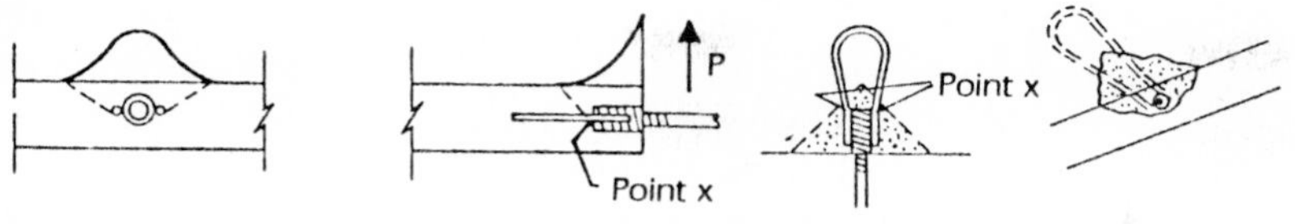


**Figure 19.7** Stress pattern in a panel lifted by an edge insert.

## Stripping

The relative ease (or difficulty) of the stripping operation varies with the shape of the member. During the stripping operation, care must be taken to keep moment, shear, and torsional stresses to a minimum. The design loads to consider during stripping are (1) dead load of the member, (2) suction (atmospheric pressure) between the form and the contact surface of the member, and (3) mechanical bonding between the form and some projection, indentation, or other surface variation in the member.

Wall panels usually require more consideration during the stripping operation than do other member shapes because (1) the stripping and handling loads perpendicular to the plane of the walls are much greater than the in-place wind and seismic loads, (2) wall panels are cast face down as shallow members with relatively little depth of section to develop resisting moments, (3) suctional loads tend to be relatively large, and (4) decorative-surface variations tend to increase the degree of mechanical bonding between the form and decorative surface cast against the form. In many instances, the reinforcement required for stripping and handling loads is more than adequate for the loads which must be resisted after the panel has been erected. The handling decision to be made for stripping is to pick the precast element at a sufficient number of points so as to keep tensile stresses within allowable limits and not to require reinforcing in excess of that required to satisfy service loads in the final installed condition.

The other structural shapes (beams, columns, etc.) are relatively easy to strip because (1) the stripping loads are usually less than the in-place loads for which the primary reinforcing is designed, (2) the sections are relatively deep and capable of developing large resisting moments, (3) suction is released when the prestress load is transferred to the concrete, (4) decorative and other surface variations tend to be minimal.

## Storage

The support system used in storage is very important. The effect of differential settlement or movement of two-point and three-point support systems in storage and shipping of prestressed precast concrete members is shown in Figs. 19.8 and 19.9.

A two-point simple support system is normally used to support all shapes of precast concrete members. If a support system of three or more points were used, one of the supports might settle and the resultant redistribution of stresses could severely damage the member. If a support in a two-point system settles, there is no redistribution of stresses, only a slight and negligible tilting of the supported member as shown in Fig. 19.8. However, permanent in-storage support systems are often built into solid poured foundations, as in the plants of many pile manufacturers, permitting multiple-point supports. An alternative method of support is to use continuous longitudinal supports, such as for thin solid prestressed slabs. In general, the method of support during storage will mirror the support system used for transportation. For example, if two points of support horizontally satisfy shipping handling stresses, then this is the logical method for storage. If long, thin panels must be shipped on edge, they will be stored this way also. Long prestressed concrete piles are stored and shipped on four points of support, with the shipping rig being composed of special pivot supports.

The fabricator's storage yard is valuable real estate. If the members are designed so that they can be stacked on top of one another, then some flexibility and economy of total production time might be realized in some situations. For example, stacking might allow the producer to

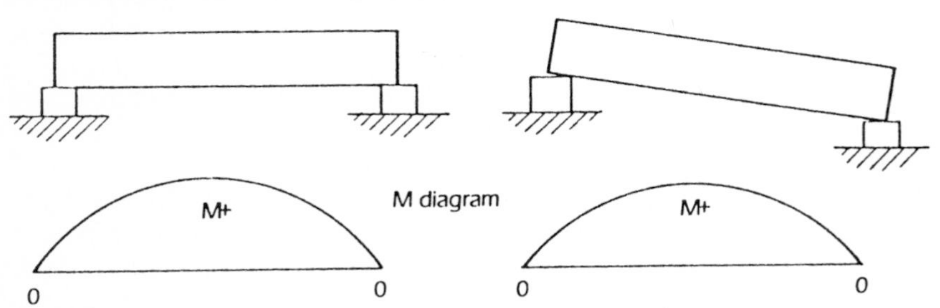


**Figure 19.8** Settlement in a two-point support system. There is no redistribution of moments due to settlement.

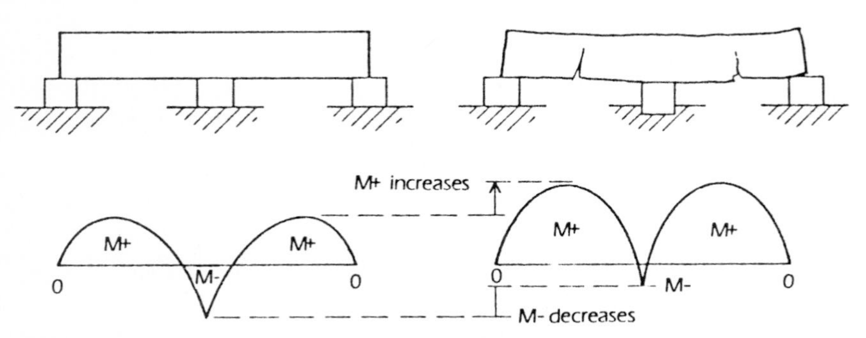


**Figure 19.9** Changes in moment due to settlement, which may cause cracking.

begin a continuous operational cycle at an early date, stack the members, and develop a readily available inventory from which to supply the erector quickly as necessary. When this is done, total construction time is governed more by the rate of erection than by the rate of fabrication. If the members cannot be stacked or if the fabricator does not have a sufficiently large storage yard, both the fabricator and the erector may have to run a slow and costly start-and-stop operation; i.e., fabricate some units, stop fabrication until the units are moved out of the way, start up again, etc.

Slabs, double T's, columns, and flat wall panels of the same shape (width and length) are usually stacked; single T's and deep girders are usually stored side by side (unstacked); and decorative-surface wall panels are usually stored in a vertical (tilted) position as separate units which are protected from contact with potentially damaging items. Prestressed flexural members should be blocked close to the ends to minimize the amount of camber growth in storage.

## Shipping

Members must be designed to resist the impact loads encountered while being transported to the jobsite. The manner of placement (stacked, side by side, or specially braced) and the shipping position (vertical, flat, or tilted) influence the magnitude and direction of these important design loads. Members are usually (1) loaded on the transporting vehicle (usually a tractor-drawn trailer) in the same position in which they are stored, (2) carried from the fabricating plant to the jobsite on the trailer, (3) lifted from the trailer and repositioned (reoriented) as necessary while still suspended from the crane, and (4) moved from the trailer to the final in-place position in the building, where the member is anchored, often temporarily, as quickly as possible. The lifting, reorienting, and temporary anchoring processes are repeated until the trailer is unloaded. Whenever possible, on-site storage and subsequent double handling operations are avoided. [If the transporting operation is defined as encompassing movement from the fabricating plant to the in-place position (with temporary connections), then any transporting system which requires on-site storage and relifting may be described as an inefficient transportation system. The exception to this statement is a pile-driving operation with several drivers spread out on the site.]

The shape of the member often determines how many members may be placed on one trailer. If the member shapes can be stacked, then each trailer may be able to carry a full trailer capacity load and the number of trips may be minimized. If the members cannot be stacked or if they require special bracing, then each trailer may not be able to carry a full load and more trips will have to be made. Normally, full trailer capacity can be achieved with all shapes except decorative-surface wall panels.

In summary, for shipping other than special situations such as the pivot rigs used for shipping piling, only two points of support are available on the truck. This makes shipping critical for handling in most cases.

## Erection

Handling in erection consists of rotating the precast element from the horizontal, or sideways, position into the vertical erected position for wall elements or merely of transferring from truck to final position for deck members. Prior thought should be given to the erection process in product design to assure that necessary erection loops or inserts are provided to facilitate the translation of precast wall elements into the vertical position. Care must also be taken to assure that erection slings do not create an angle greater than 45° so as not to induce excessive shear in erection inserts.

## Handling Procedures

The following procedure is useful as a checklist in designing precast and prestressed concrete elements for handling:

1. Calculate section properties of the cross section of the precast concrete element through the portion resisting bending.
2. Calculate member weight and the location of the center of gravity.
3. Check concrete stresses for handling during stripping, shipping, and erection. Concrete stresses during handling should be kept below the modulus of rupture ($f_r$), divided by a factor of safety (FS) of 1.5.

$$f_t = \frac{f_r}{\mathrm{FS}} = \frac{7.5\sqrt{f'_c}}{1.5} = 5\sqrt{f'_c}$$

where $f'_c$ is the concrete strength in the member at the time at which it is being analyzed.

For sand-lightweight concrete, multiply the above value by 0.85:

$$0.85 \times 5\sqrt{f'_c} = 4.25\sqrt{f'_c}$$

Use the following impact factors in calculating service load moments for the various conditions: *

*Other producers may use different values; these are not code requirements.

Stripping 1.5
Shipping 2.0
Erection 1.25

The following general equations (Figs. 19.10 through 19.20) are useful in determining design moments for various handling conditions:

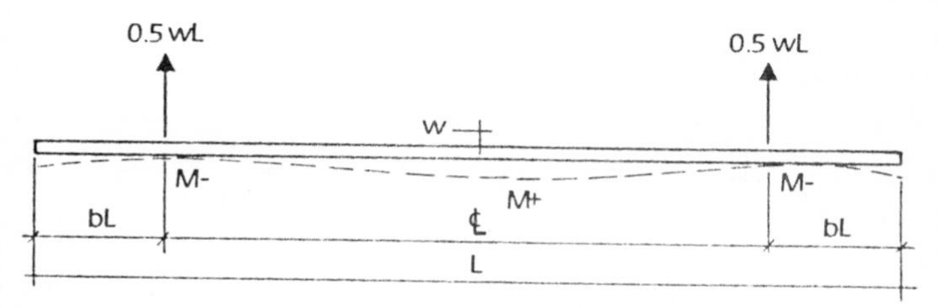


**Figure 19.10** General equation: two-point pick for stripping. $M^{+} = (2c - 1)\frac{wL^2}{8}$ $M^{-} = 0.5b^2wl^2$.

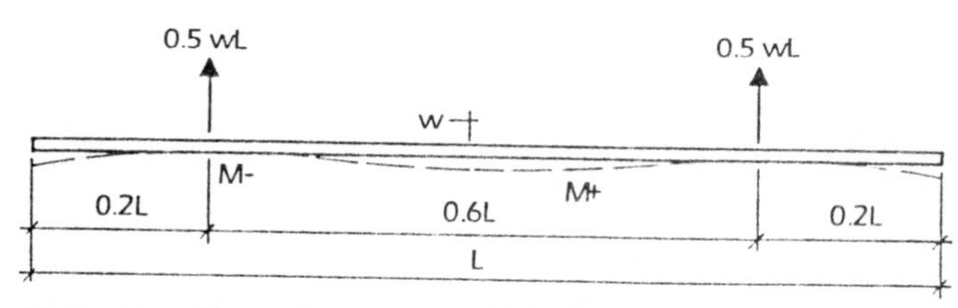


**Figure 19.11** Two-point pick for stripping and two-point support for shipping; equal maximum positive and negative bending moments. $M^{+} = M^{-} \cong 0.025\,wL^2$.

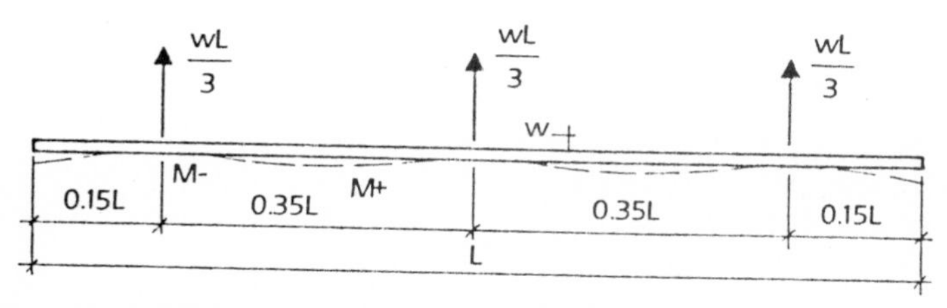


**Figure 19.12** Three-point pick for stripping; equal reactions. $M^{-} = 0.012wL^2$; $M^{+} = 0.006\,wL^2$.

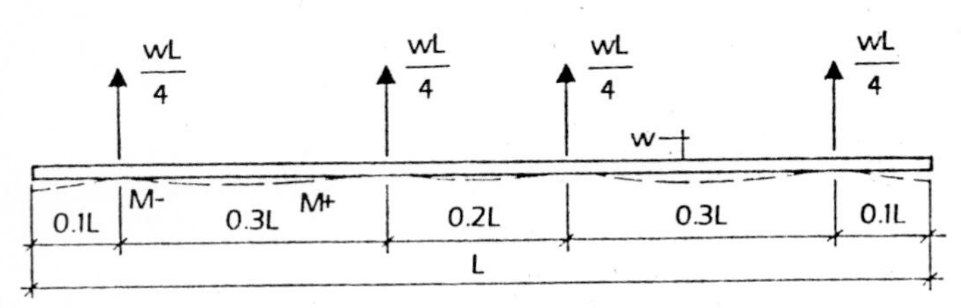


**Figure 19.13** Four-point pick for stripping; equal reactions; equal negative and positive bending moments. $M^+ = M^- = 0.0056\,wL^2$.

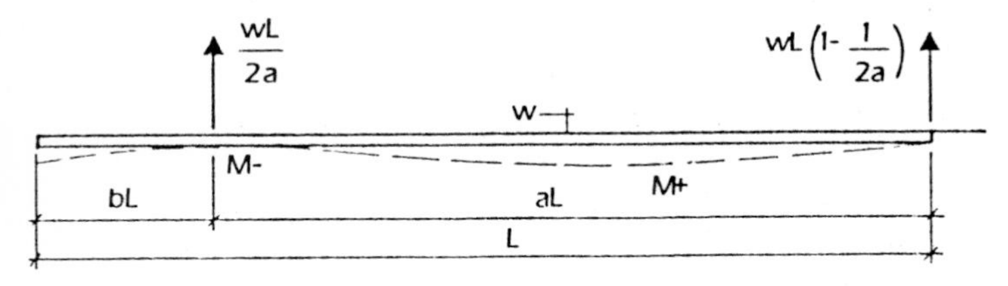


**Figure 19.14** General equation: two-point pick for erection. $M^- = 0.56\, b^2wL^2$ $M^+ = \left(1 - \frac{1}{2a}\right)^2 \frac{wL^2}{2}$.

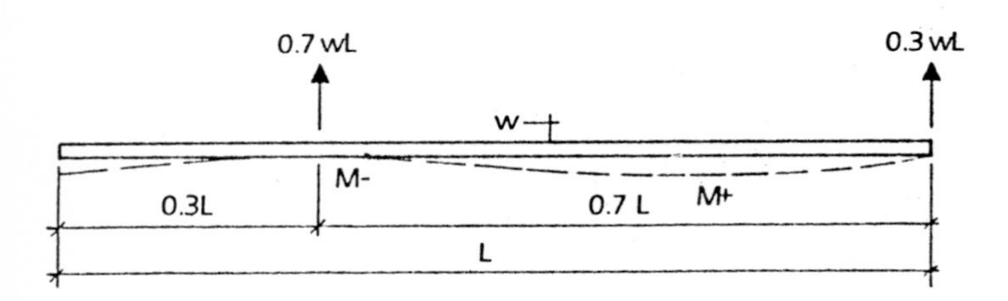


**Figure 19.15** Two-point pick for erection; equal negative and positive bending moments. $M^+ = M^- = 0.043\,wL^2$.

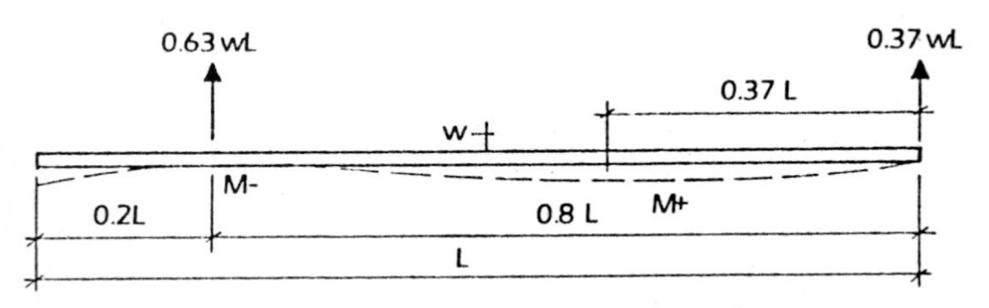


**Figure 19.16** Two-point pick for erection; lower pick point at bottom; two-point stripping insert location. $M^+ = 0.069\,wL^2$.

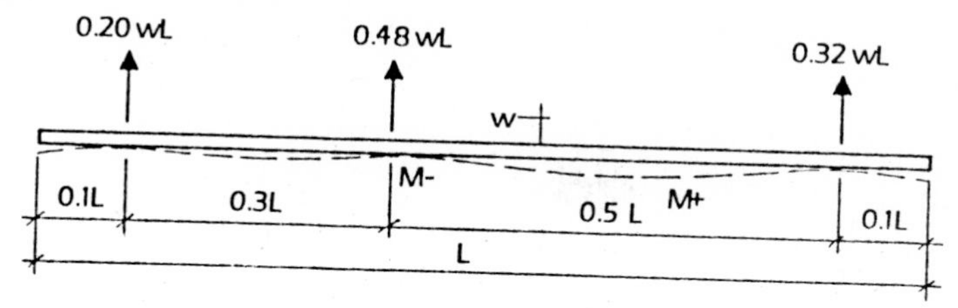


**Figure 19.17** Three-point pick for erection; four-point stripping insert locations. $M^+ = 0.019\,wL^2$; $M^- = 0.021wL^2$.

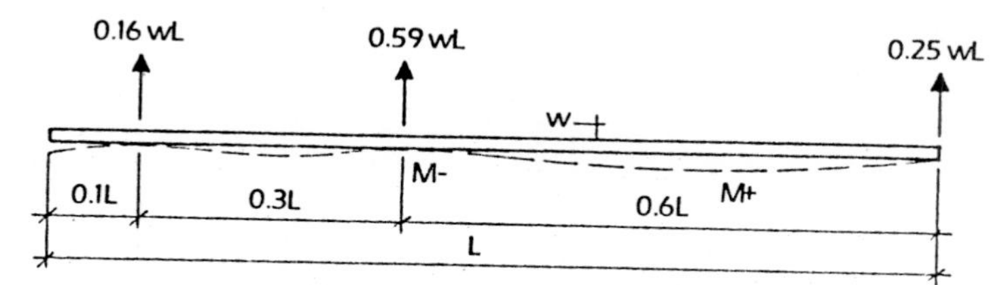


**Figure 19.18** Three-point pick for erection; at top and bottom four-point stripping insert locations. $M^+ = 0.030\,wL^2$; $M^- = 0.033wL^2$.

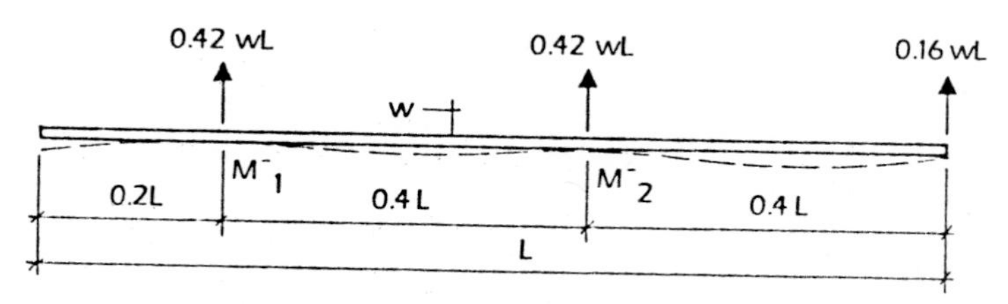


**Figure 19.19** Three-point pick for erection; equal lower-end reactions using two-point stripping insert locations.
$M^+ = 0.013wL^2$; $M_1^- = 0.02wL^2$; $M_2^- = 0.015wL^2$.

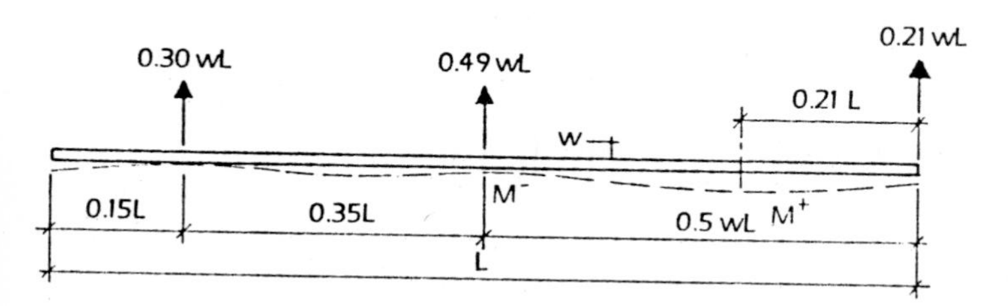


**Figure 19.20** Three-point pick for erection; top plus third-point stripping locations. $M^+ = 0.021wL^2$; $M^- = 0.023\,wL^2$.

**Figure 20.1** Dorothy Chandler Pavilion, Los Angeles Music Center. Erected in 1965, the architectural precast concrete cladding elements shown here demonstrate the beauty and permanence for which this wonderful material is known.

Chapter

# 20

# General Design Procedure for Architectural Precast Concrete Cladding Units

This chapter presents a logical procedure to use in sizing and detail-designing non-load-bearing architectural precast concrete cladding panels. It is written to provide a basis for decisions required in designing these elements, many of which must be made in the preliminary stages of the project.

I. *Preliminary size determination.* Try to keep panel weights below 11 tons per unit. This is compatible with the handling capacity of available erection equipment and will also facilitate vertical shipment on easel trailers, a common method for handling long, flat units that would be cracked if shipped flat or would require excessive reinforcing needed only for handling.

If possible, analyze the site access available for cranes and trucks. Tight sites or complicated spread-out buildings may require even further reducing of weights of units for feasible erection.

II. *Applicable design criteria*

A. *Uniform Building Code*

1. *Seismic.* $F_p = ZIC_pW_p$; Sec. 2312(*g*)2.

a. *Reinforcing design*

Wall panel element: $c_p = 0.75$ (applied normal to flat surface); $ZC_p$ in seismic Zone 4 = 0.3

*b*. *Connection design*: (UBC 2312 (*h*) 20)
Connector body: shall be designed for 1.33 times $F_p$
Connector fastener: shall be designed for 4 times $F_p$

Prior to 1979 building codes required an arbitrary value of $ZC_p$ in seismic Zone 4 = 2.0 to be applied in a lateral direction on the panel for connector design. Although this value is higher than the lateral force coefficients given above, codes did not make provisions to assure that the connector behaved in a ductile manner. Current code provisions for panel connections are intended to assure that distortion will be accommodated by the body of the connector and not by brittle failure of welds and inserts or studs embedded in the concrete. In this way, the connector body by yielding can form an energy-absorbing mechanism to accommodate extreme seismic conditions and large values of building drift (see Fig. 20.2). The 0.4-*g* loading to connector bodies is an elastic load, with

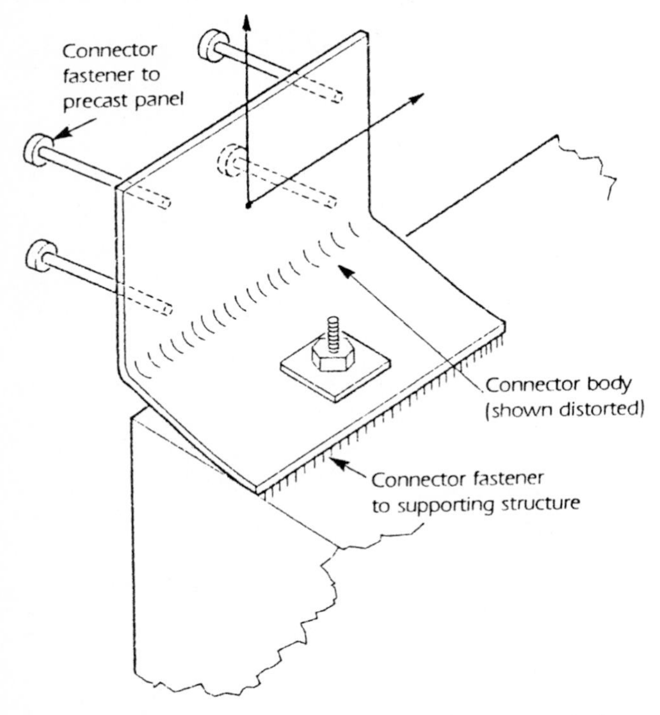


**Figure 20.2** Detail showing deformation of connector body during an extreme seismic disturbance. The connector body yields under extreme loading, thereby absorbing energy, accommodating building drift, and preventing buildup of high stress levels in connector fasteners.

the steel element being designed by the allowable-stress method. The 1.2-*g* load to fasteners is also considered an elastic load, with the fasteners being taken to ultimate strength or yield by using appropriate load factors and comparing this with the nominal strength of the fasteners, modified by appropriate capacity reduction factors.

Panel loadings are shown schematically in Fig. 20.3.

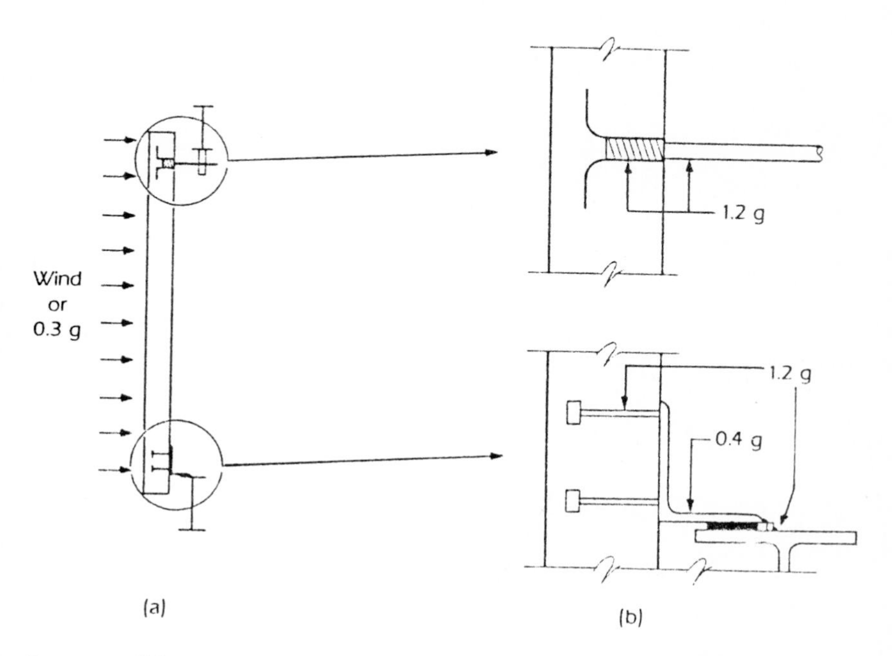


**Figure 20.3** Schematic presentation of panel loadings (Uniform Building Code). (*a*) Reinforcing design. (*b*) Connection design.

2. *Wind.* See Tables 23-F, 23-G, and 23-H and Fig. 1 of the Uniform Building Code.

B. *Concrete strength.* 5000 lb/in² minimum for durability.

C. *Cover.* 3/4 in minimum or as required to attain a fire rating greater than 2 hours. See Table 43-A and Sec. 2607 of the Uniform Building Code.

D. *Reinforcing*

Nos. 3 and 4 bars: ASTM A-615-40.

No. 5 and larger: ASTM A- 615-60.

Welded wire fabric: ASTM A-185.

All welded reinforcing bars should be A706 grade.

III. *Determine thickness of panel elements* .

A. As required for final design condition.

1. *Seismic*
2. *Wind*

B. As required for required minimum fire resistance ratings set forth in Table 43-B of the Uniform Building Code.

C. As determined by handling (shipping will usually be most critical).

Note that for panel thicknesses of less than 5 in one row of reinforcing should be used; for panel thicknesses 5 in or greater, two layers of reinforcing are desirable. Walls greater than 10 in shall have two layers.

IV. *Design panel reinforcing.*

A. Calculate section properties of the panel element, usually a cross section through the portion resisting bending.

B. Calculate panel weight and location of the center of gravity.

C. Analyze handling; determine reinforcing requirements for strip-striping, shipping, and erection. The formulas given are useful in determining design moments for various handling conditions.

Keep concrete stresses during handling below the modulus of rupture ($f_r$) divided by a factor of safety of 1.5;

i.e. $$\frac{7.5\sqrt{f_c}}{1.5} = 5\sqrt{f_c}$$

For sand-lightweight concrete, multiply the above value by 0.85.

Use the following impact factors in calculating service load moments for the various handling conditions: *

| | |
|---|---|
| Stripping | 1.5 |
| Shipping | 2.0 |
| Erection | 1.25 |

Summarizing the handling criteria for mild steel reinforced panels,

1. *Stripping*

$f'_{ci} = 2000 \text{ lb/in}^2$

$M_{\text{design}} = 1.5\,M_{\text{actual}}$ (1.5 is the factor of safety.)

$F_{t,\text{ allowable}} = 0.225$ ksi (normal weight )

* Other producers may use different values; these are not code requirements.

$$= 0.190\,\text{ksi}\quad(\text{lightweight})$$

$$A_s \text{ required} = \frac{M_{\text{design}}}{1.0d}$$

where $M$ is in ft · k, $d$ is in inches, and $A_s$ is in square inches.

2. *Shipping*

$$f'_{ci} = f'_c = 5000\,\text{lb/in}^2$$

$$M_{\text{design}} = 2.0M_{\text{actual}}$$

$$F_{t,\text{allowable}} = 0.350\,\text{ksi}\quad(\text{normal weight})$$

$$= 0.300\,\text{ksi}\quad(\text{lightweight})$$

$$A_s \text{ required} = \frac{M_{\text{design}}}{1.44d}\quad\text{for } f_y = 40\text{ ksi rebar}$$

$$A_s \text{ required} = \frac{M_{\text{design}}}{1.76d}\quad\text{for } f_y = 60\text{ ksi rebar or welded wire fabric*}$$

3. *Erection.*

$$M_{\text{design}} = 1.25M_{\text{actual}}$$

Other criteria the same as for shipping.

For prestressed concrete panels, keep maximum tensile stress below $3\sqrt{f_c}$.

D. Locate panel connection points to resist lateral and vertical forces. Check panel reinforcing for the most critical of the following conditions:

1. *Wind*
2. *Seismic.* 0.3 *g*
3. *Temperature steel requirements.* Sec. 2607 (n) of the Uniform Building Code. Adjust or increase the number of lateral panel connection points to optimize the amount of reinforcing used in the panel.

   In detailed connection design, only two bearing connections are used. As many lateral connections as required may be provided.

E. Check any bowing induced by thermal gradients, and provide ad-

*This is based on the working stress method and not the ultimate strength method of concrete design.

ditional lateral supports as required (usually critical for thin, flat panels only). Design charts are presented in Figs. 20.4 and 20.5.

V. *Design panel connections.*

A. Using fixed and lateral connection points previously selected, analyze the distribution of a unit load of $1^k$ applied in any direction that is at the center of gravity of the panel element; that is

$1^k$ down

$1^k$ longitudinal (left or right)

$1^k$ lateral (in or out)

Load analysis sketches are presented in Figs. 20.6, 20.7, and 20.8 to help in visualizing these load conditions on typical cladding panel configurations.

B. By the principle of superposition, determine the service load to the connector bodies and the service load to the connector fasteners, or

$P$ connector body = 1.0 $g$ down + 0.4 $g$ in any lateral direction
$P$ connector fastener = 1.0 $g$ down + 1.2 $g$ in any lateral direction

Multiplying by the actual panel weight $W$ will give the actual service load on the connection to be used in the design. The above service loads must be multiplied by approximate load factors in order to design the connections for the appropriate material capacities. From Sec. 2609 of the Uniform Building Code the required ultimate strength shall be

$$UP = 0.75(1.4D + 1.87E) = 1.05D + 1.4E$$

The Prestressed Concrete Institute also recommends that the following capacity reduction factors be used when the above ultimate-load criteria are followed:

$$\phi_c = 0.85 \text{ (concrete)}$$
$$\phi_s = 1.0 \text{ (steel)}$$

Chapter 6 of the *PCI Design Handbook* incorporates these values in the design charts for connectors as governed by concrete and steel strengths.

VI. *Other design criteria to be checked.*

---

* Note that the Uniform Building Code does not require designing for any component of vertical acceleration. However, high, narrow elements such as column covers on tall buildings will be subjected to vertical acceleration during an earthquake. For this reason, it is recommended to design the bearing connectors of these elements for some value of vertical acceleration, say, 0.3 to 0.5$g$, to be combined with gravity loading.

A. Assure that the movement of the building under wind or seismic loading can be accommodated in the connections and the joint widths and configurations selected [2 times story drift from the wind or 3/k times the seismic story drift or 1/2 in, whichever is greater (Sec. 2312, Uniform Building Code)].

B. Design any required temporary bracing as required by Occupational Safety and Health Administration (OSHA) regulations.

Long, flat cladding elements without intermediate supports will be subject to bowing due to temperature gradients such as those produced by hot sunlight on panels covering air-conditioned buildings. Conversely, cold exterior surfaces of flat cladding elements enclosing heated buildings would cause an inward movement. The magnitude of this movement which is acceptable is subject to designer judgment and the ability of glazing or window wall infill to accommodate the movement. Figure 20.4 estimates the amount of movement that would be encountered for a 60°F thermal gradient.

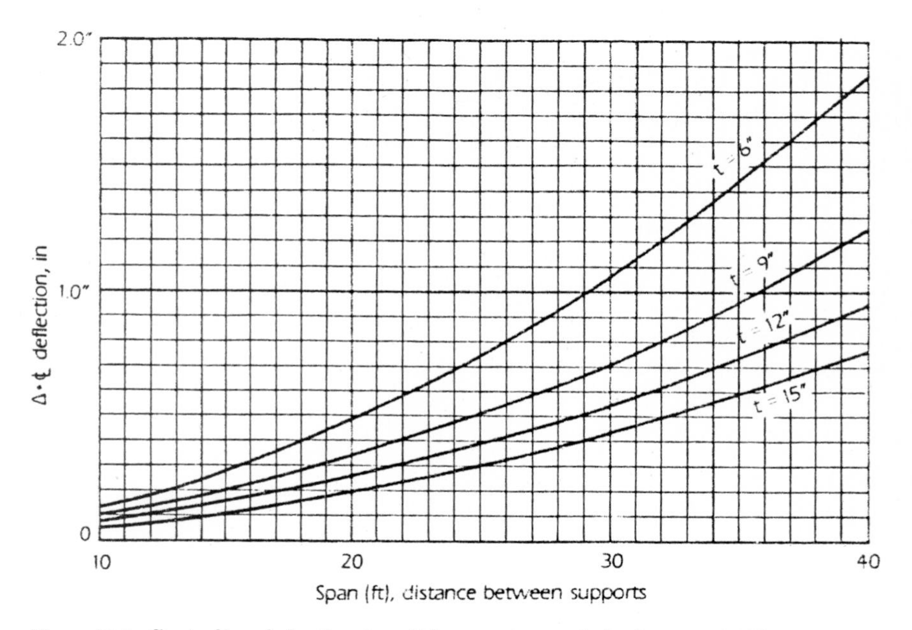


**Figure 20.4** Centerline deflection in solid precast spandrels due to a 60°F temperature differential between the outer and inner surfaces.

The diagram to the right of Fig. 20.5 shows the geometrical relationship which is the basis for the design graphs to the left of the figure and in Fig. 20.4. While being empirical and ignoring thermal lag due to the mass of the concrete, especially in thicker panels, it gives the designer an idea of the relative order of magnitude of movements and intermediate restraining forces. The graph of Fig. 20.5 furnishes restraining-force values which can be correlated to the specific design situation. Remember that if you restrain these precast elements at intermediate points, be sure that the connections can develop the forces imposed by thermal gradients.

Load analysis sketches for typical spandrels, two-story column cover units, and window panels are shown in Figs. 20.6, 20.7, and 20.8.

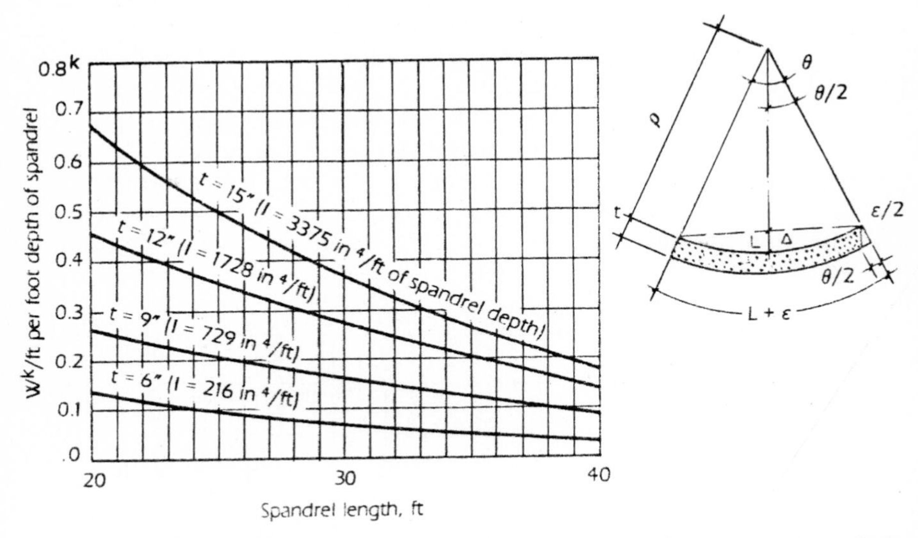


**Figure 20.5** Uniform restraining force required ($W$) in kips per foot per foot of spandrel depth for solid flat precast spandrels to resist force generated by a 60°F thermal gradient. $\epsilon = (6 \times 10^{-6})^{*}(L)(\Delta T)$; $\tan \theta/2 = \epsilon/2t$; $P = 360L/2\pi\theta$; $\Delta = \rho(1 - \cos\theta/2)$; $W = 384EI\,\Delta/5L^4$. (*Coefficient of expansion for hard-rock concrete.)

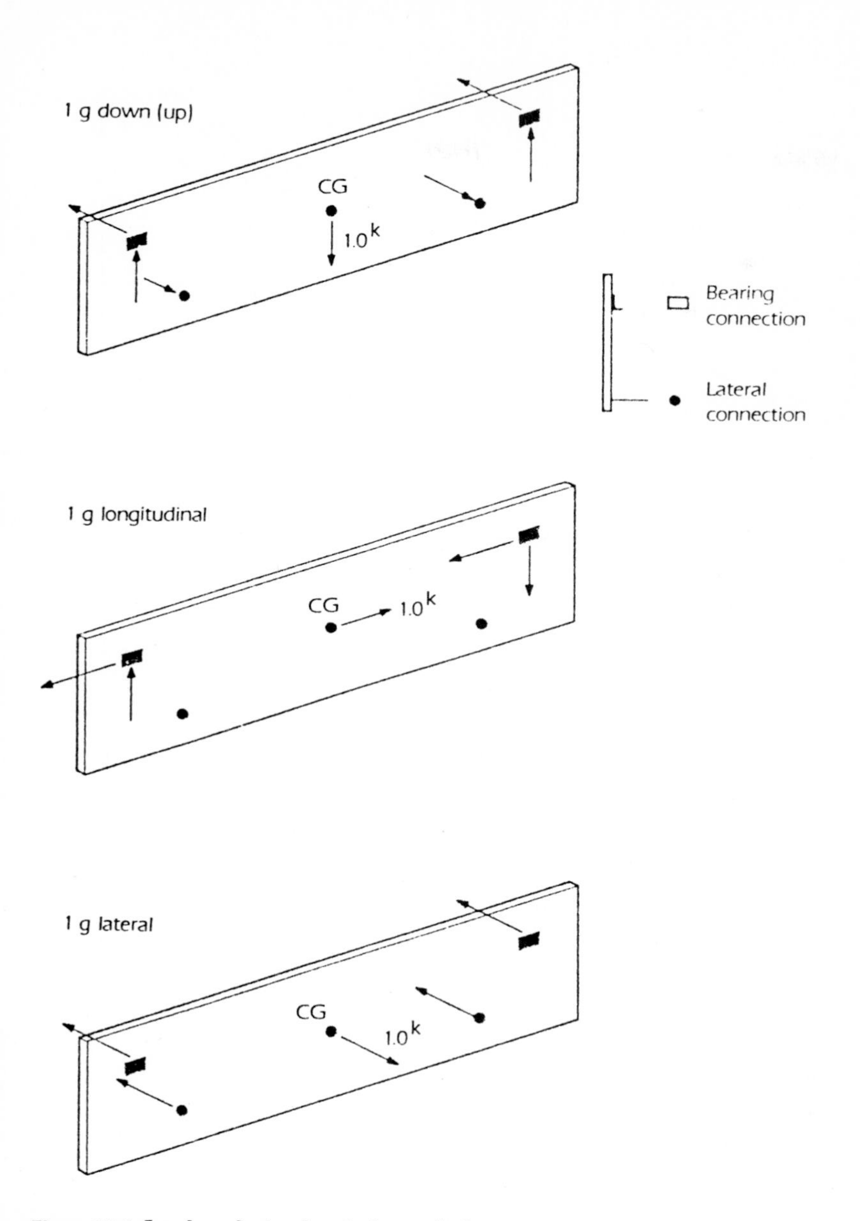


**Figure 20.6** Load analysis of typical spandrels.

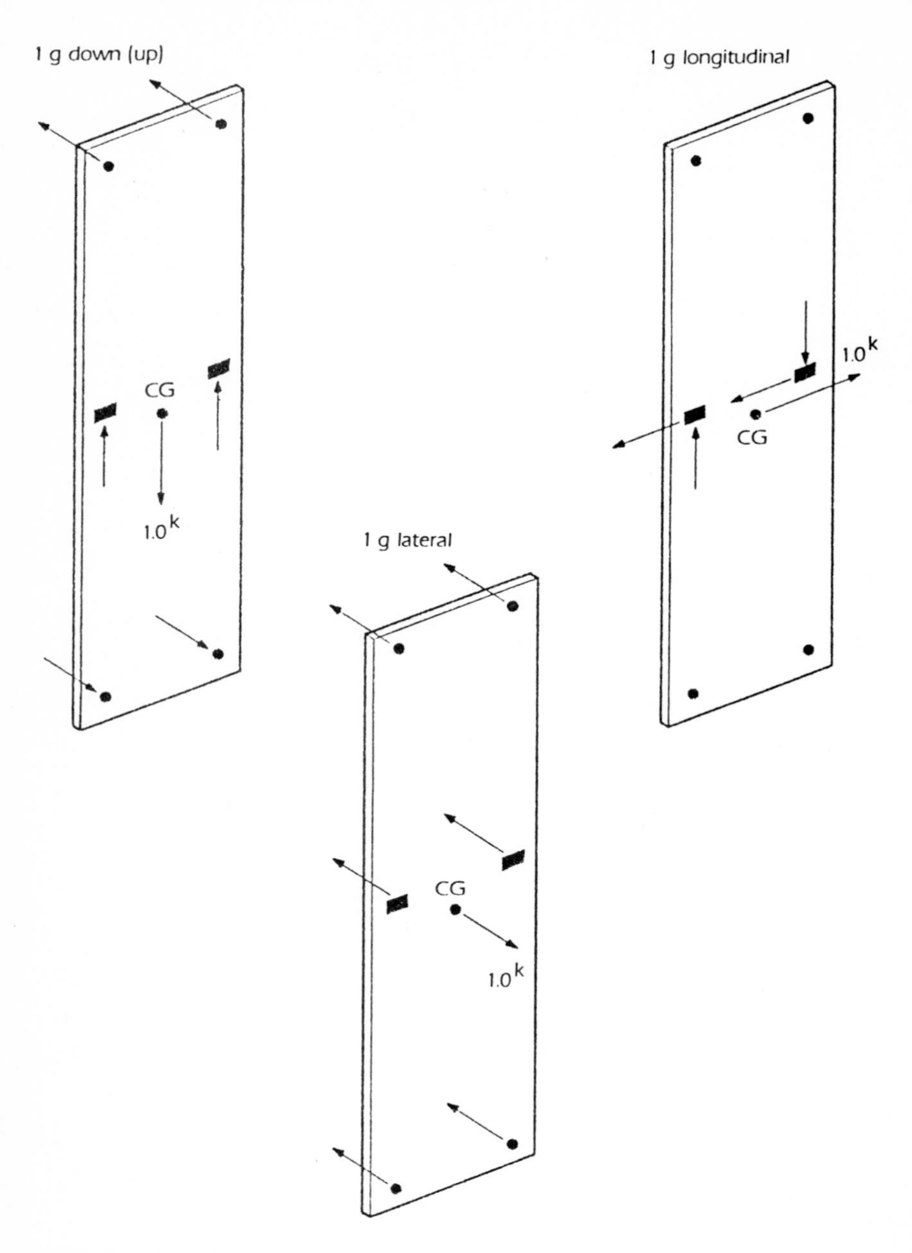


**Figure 20.7** Load analysis of typical two-story column cover units.

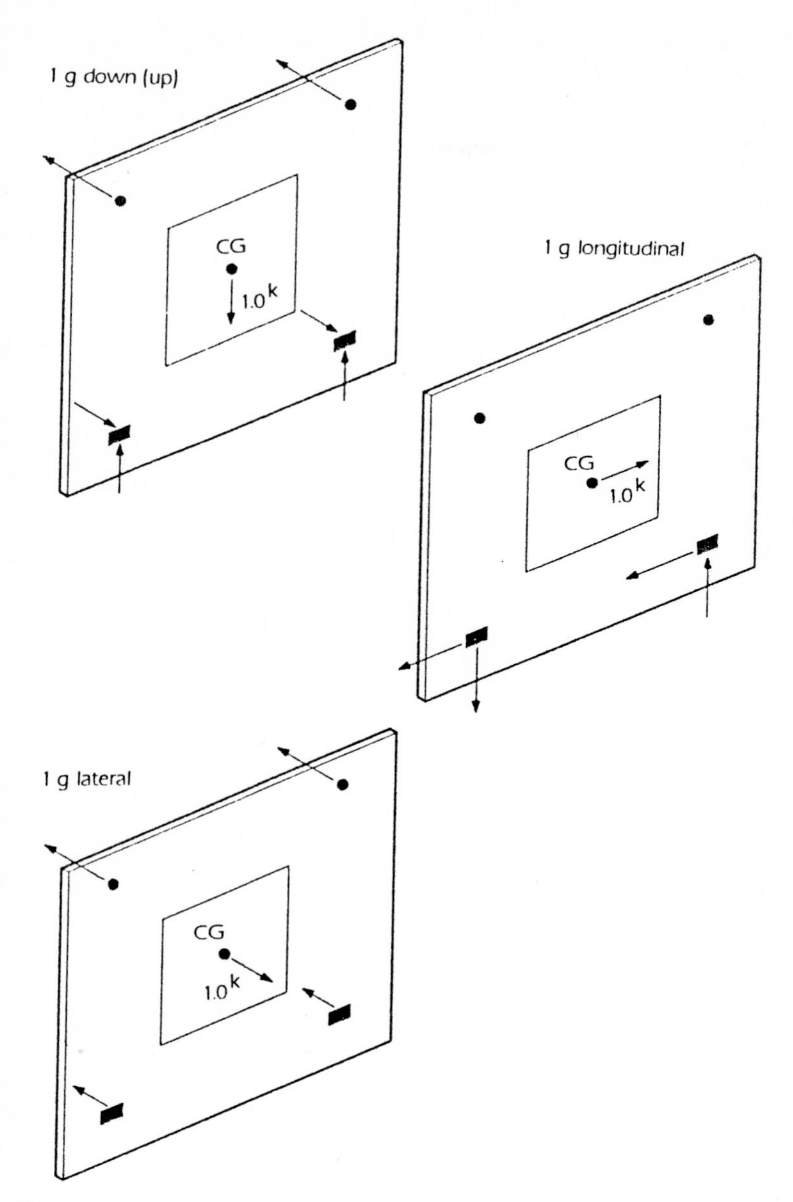


**Figure 20.8** Load analysis of typical window panels.

**Figure 21.1** Alvarado sewage treatment plant, Union City, California. A standard double T 8 ft wide by 24 in deep being erected on the roof of this project. There is standard production information for this shape both in the manufacturer's literature and in load tables in the *PCI Design Handbook*. In the absence of such information the design procedures presented in this chapter can save time and unnecessary trial designs.

Chapter

# 21

# General Design Procedure for Structural Prestressed Concrete Flexural Members

The intent of this chapter is to give the designer a step-by-step procedure to use in designing a prestressed concrete flexural member. It is presented in such a manner that the design may be developed from the beginning point where all that is known is the design span and the design loading. Normally, some of the preliminary rough design rules of thumb and preliminary design relationships may be skipped when manufacturer's load tables or the load-span tables given in Chap. 2 of the *PCI Design Handbook* are available. In all cases, however, when beginning the design of prestressed concrete flexural members, the known factors are:

1. Span
2. Superimposed dead and live loads and the loading arrangement
3. Strength of the concrete
4. Grade of the prestressing steel
5. Allowable concrete and strand stresses at release of prestress and in service
6. Approximate prestress losses

7. Camber or deflection requirements

The unknown factors are:

1. Size of the beam
2. Number of strands
3. Size of the strands
4. Strand profile and positioning in the member

The procedure gives a logical way to achieve the final design in as few steps as possible with a minimum of trial designs. In the absence of manufacturer's product literature, the preliminary depth calculation guidelines are used to select the size of the beam. Then, the strand information is developed with the balance of the preliminary design procedure. The rest of the process is self-explanatory and is given below:

## Design Procedure

### Notation

$f_b$ Bending stress or bending stress for bottom fiber
$f_t$ Tensile stress or bending stress for top fiber
$f'_{ci}$ Concrete strength at release of prestress
$f'_c$ 28-day concrete cylinder strength
$M_B$ Bending moment resisted by basic precast section
$M_C$ Bending moment resisted by composite section
*Basic section moduli (see Fig. 21.2)*
$S_t$ Section modulus at top fiber
$S_b$ Section modulus at bottom fiber
*Composite section moduli (see Fig. 21.3)*
$S_{tc}$ Section modulus at top fiber of precast section
$S_{bc}$ Section modulus at bottom fiber
$F_o$ Initial strand tension, kips
$F_i$ Prestressed force induced in member at release, kips
$F_f$ Final prestressed force induced in member after losses, kips
$f_{pu}$ Ultimate strength of prestressing steel
$f_{ps}$ Stress in prestress strand at nominal (ultimate) strength of member
$f_{se}$ Effective stress in prestressing steel after all losses have occurred
$e_c$ Strand eccentricity at centerline span

| | |
|---|---|
| $e_e$ | Strand eccentricity at the end of the member |
| $A_{ps}$ | Area of prestressing strand |
| $A$ | Area of prestressed concrete section |

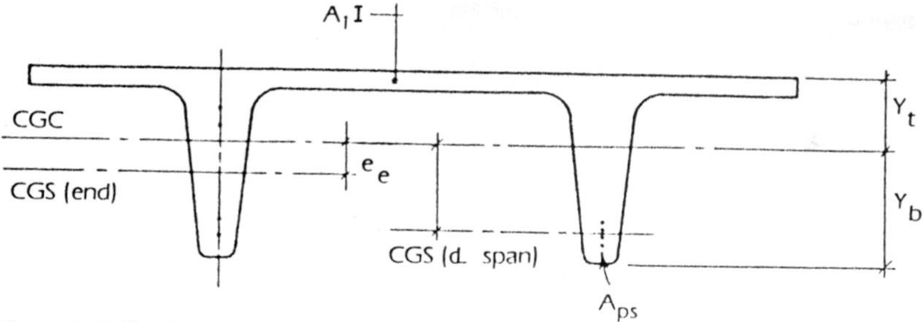


**Figure 21.2** Basic section. $S_t = I/y_t$. $S_b = I/y_b$.

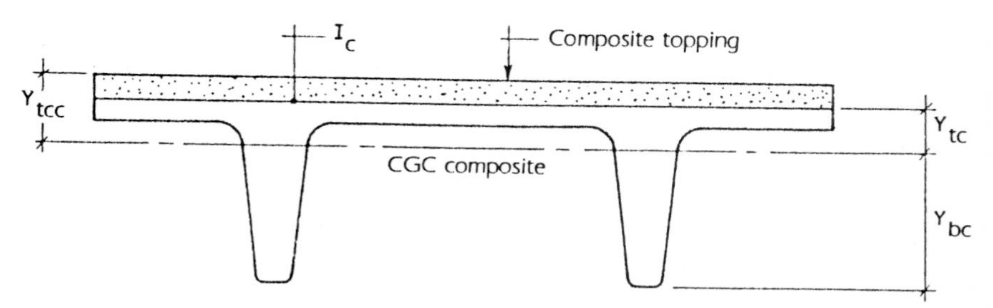


**Figure 21.3** Composite section $S_{tc} = I_c/y_{tc}$. $S_{bc} = I_c/y_{bc}$. $S_{tcc} = I_c/y_{tcc}$.

I. *Preliminary depth calculation.* Span-depth ratios should be as follows:

A. Single T
   1. Floor $L/D = 23$
   2. Roof $L/D = 29$

B. Double T
   1. Floor $L/D = 26$
   2. Roof $L/D = 34$

C. Single or double T (H20-44 highway loading)
   $L/D = 18$

D. Single or double T (plaza load: $S_{D+L} = 350$ lb/ft$^2$)
   $L/D = 12$

E. Hollow-core plank
   1. Floor $L/D = 40$
   2. Roof $L/D = 46$

F. Girders $L/D = 18$

II. Approximate $S_{bc}$ in$^3$ required = 6× $M_{SD+L}$(k-ft)

where $S_{D+L}$ is superimposed dead and live load.

III. Procedure

A. Calculate section properties for the trial section, if not already known.

B. Determine $f_b = \frac{M_B}{S_b} + \frac{M_C}{S_{bc}}$.

C. * If $f_b > f_t + f'_{ci}/2$, strands must be depressed. $\left(f_t = 6\sqrt{f'_c}\right)$

D. * If $f_b > f_t + f'_{ci}/2 < M_{BM}/6S_b + M_{TPG}/S_b + M_C/S_{bc}$, then a deeper section must be used.

For $f'_c = 5000\ \text{lb/in}^2\ \left(f'_{ci} = 3500\ \text{lb/in}^2\right)$, $f_t + f'_{ci}/2 = 2.2$ ksi.

For $f'_c = 6000\ \text{lb/in}^2\ \left(f'_{ci} = 4200\ \text{lb/in}^2\right)$, $f_t + f'_{ci}/2 = 2.6$ ksi.

E. Of course, manufacturer's load-span tables, where available, allow us to select the proper section without requiring the above preliminary trials.

F. Find the effective prestress force required. (At this point in the design, we know everything but the required effective prestress force. A very close approximation for prestress eccentricity *e* may be assumed based upon the strand profile and required concrete cover at the bottom of the member.)

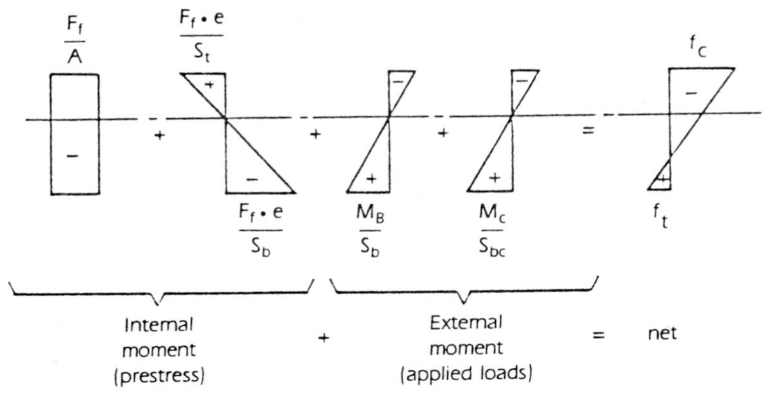


Figure 21.4 Prestress force requirements.

* See derivations at the end of the chapter.

Basic relationships (see Fig. 21.4)

$$f_{\text{top}} = -\frac{F_f}{A} + \frac{F_f e}{S_t} - \frac{M_B}{S_t} - \frac{M_C}{S_{tc}} = f_c \qquad (21.1)$$

$$f_{\text{bottom}} = -\frac{F_f}{A} - \frac{F_f e}{S_b} + \frac{M_B}{S_b} + \frac{M_C}{S_{bc}} = f_t \qquad (21.2)$$

Working with equation 21.2,

$$-F_f\left(\frac{1}{A} + \frac{e}{S_b}\right) = f_t - \frac{M_B}{S_b} - \frac{M_C}{S_{bc}}$$

$$F_f = \frac{\dfrac{M_B}{S_b} + \dfrac{M_C}{S_{bc}} - f_t}{\dfrac{1}{A} + \dfrac{e}{S_b}}$$

G. Select number of strands and check service load stresses.

H. When end stresses cannot be satisfied and it is not feasible to break bond on the strands at the ends or use top-end mild steel, recalculate the required prestress, substituting $M_{0.4L} = 0.96M_c$ (see Figs. 21.5 and 21.6),

$$e_{0.4L} = 0.90e_c \text{ for shallow drape}$$

$$e_{0.4L} = 0.80e_c \text{ for steep drape}$$

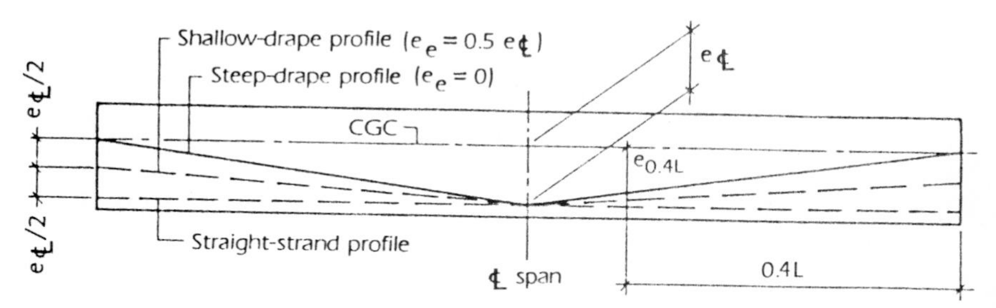


**Figure 21.5** Strand profiles. $e_{0.4L} = 0.9e_{℄}$ for shallow drape. $e_{0.4L} = 0.8e_{℄}$ for steep drape.

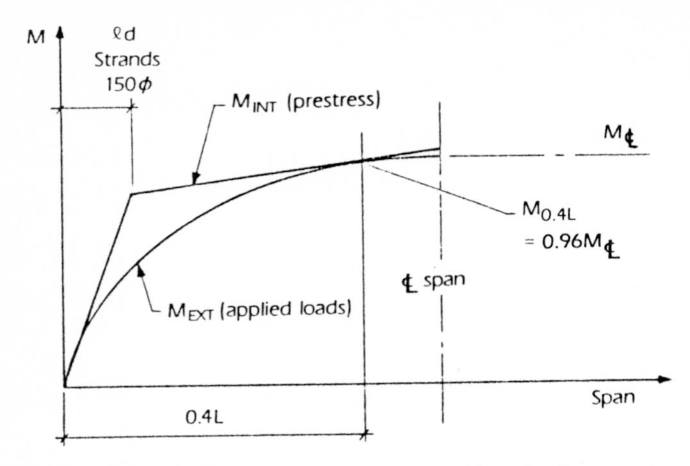


**Figure 21.6** Midpoint harped strands; uniform-load moment relationship.

IV. Detailed design (see Sec. 2618, Uniform Building Code). Check service load stresses at critical points:

A. *Concrete stresses*

1. At release

| | |
|---|---|
| Compression | $0.60f'_{ci}$ |
| Tension (except at ends) | $3\sqrt{f'_{ci}}$ |
| Tension at ends (simply supported members) | $6\sqrt{f'_{ci}}$ |

2. *Final design condition*

| | |
|---|---|
| Compression | $0.45f'_c$ |
| Tension (see Fig. 21.7) | |
| Hollow-core slabs | |
| Flat slabs | $6\sqrt{f'_c}$ |
| Beams and stemmed members | * $12\sqrt{f'_c}$ |

* When deflections are checked in accordance with bilinear strain relationships. (See *PCI Handbook*, page 4-44.)

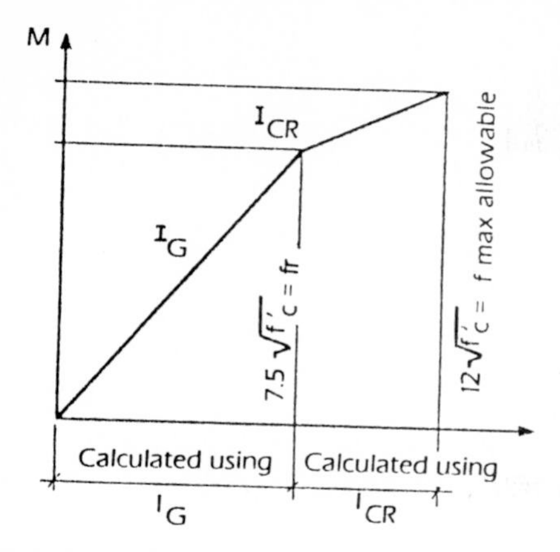


**Figure 21.7** Bilinear moment-deflection relationship.

3. *Stress losses.* Stress losses in pretensioned members result from four principal causes:

   a. *Elastic shortening.* This results from the deformation and resultant strain decrease when the prestress force $P$ is released into the cross-sectional area $A$.

   b. *Relaxation of steel.* There is a decrease in stress in the prestressing strand due to creep.

   c. *Creep of concrete.* A time-dependent phenomenon, this is the redistribution of the concrete molecules causing shortening under sustained load. Just as for relaxation of the steel, the magnitude of creep depends upon the stress level at the various positions in the cross section.

   d. *Shrinkage of concrete.* This is a shortening due to moisture loss from concrete.

For most typical design applications, stress losses at release can be conservatively taken as 10 percent of the initial prestress; final losses, at 78 percent for hard-rock concrete and 75 percent for sand-lightweight concrete. For members with high span-depth ratios and prestress levels or heavy sustained loads, a more detailed investigation should be carried out. One such method is given in the *PCI Design Handbook* on page 4-40. For low-relaxation strand the losses will be less.

B. *Strand stresses*

Initial strand stress: $0.70 f_{pu}$ for stress-relieved strand. For low-relaxation strand use $0.74 f_{pu}$.

1. *At release*

$$F_i = 0.9 \times 0.70 f_{pu} \times A_{ps}$$

2. *Final condition*

Hard-rock concrete (22 percent losses)

$$F_f = f_{se} \times A_{ps} \qquad F_f = 0.78 \times 0.70 f_{pu} \times A_{ps}$$

Lightweight concrete (25 percent losses)

$$F_f = 0.75 \times 0.70 f_{pu} \times A_{ps}$$

For 1.2-in-diameter-270-ksi strand:

$$(A_{ps} = 0.1531 \text{ in}^2\text{/strand})$$

$$F_o = 0.70 \times 0.1531 \times 270 = 28.9^k \text{ strand}$$

$$F_i = 0.9 \times 28.9 = \quad 26.0^k$$

or $F_f = 0.78 \times 28.9 = \quad 22.5^k$ (hard-rock concrete)

$F_f = 0.75 \times 28.9 = \quad 21.7^k$ (sand-lightweight concrete)

V. Final checks

A. Check stresses at critical points:

At centerline span for straight strands
At 0.4*L* for depressed strand

$$f_{top} = -\frac{F_f}{A} + \frac{F_f e}{S_t} - \frac{M_B}{S_t} - \frac{M_C}{S_{tc}}$$

$$f_{bottom} = -\frac{F_f}{A} - \frac{F_f e}{S_b} + \frac{M_B}{S_b} + \frac{M_C}{S_{bc}}$$

B. Check ultimate strength (see *PCI Design Handbook*, page 4-6).

C. Check shear-proportion web reinforcement (see *PCI Design Handbook*, page 4-21).

D. Check camber and deflection (see *PCI Design Handbook*, page 4-42).

## Taking Advantage of Inherent Capacity In Prestressed Concrete Designs Often Overlooked

This is possible without requiring high uneconomical concrete release strengths or final design strengths.

1. Use transformed section properties. This procedure can increase $I$ and $S$ by 10 percent or more for deep flexural members.

2. Break bond at the ends of some strands so that top-end tension $\left(> 6\sqrt{f'_c}\right)$ does not govern design or dictate a higher release strength.

CAUTION: Take strand development length into account when calculating flexural resistance and shear strength near the end of the member. Bottom mild steel, plus additional stirrups at the end, may be required.

3. Use mild steel or nonprestressed strand in bottom to satisfy ultimate-strength requirements when additional prestressing steel would result in excessively high release strengths.

4. Use strain compatibility in calculating $f_{ps}$ for use in ultimate-strength equation:

$$\phi A_{ps} f_{ps} (d - a/2) = M_u$$

5. Let bottom tension under all loads approach $12\sqrt{f'_c}$, consistent with allowable deflections. (Use bilinear moment-deflection relationship.)

6. Take allowable live-load reduction where permitted (see Uniform Building Code, Sec. 2306).

7. Use midpoint or third-point temporary shoring under long-span or heavily loaded composite flexural members to take advantage of the composite section to carry the deck dead loads as well as live loads; use top and bottom mild steel as required to increase $I$ and satisfy deflection or camber requirements.

8. Use unbonded tendons when release strengths are too high, member depth may not be increased, and deflection requirements prevent a partially prestressed design such as suggested in Par. 3. (Note that precautions to ensure satisfactory long-term performance in both normal and corrosive environments are given in a specification presented in the March--April 1985 issue of the *PCI Journal*.)

## Prestressed Concrete Flexural Design: Derivation of Preliminary Design Relationships for Service Load Stresses

### For straight strands

*At bottom-end release*

$$0.9\left(-\frac{F_o}{A} - \frac{F_o e}{S_b}\right) = -0.6 f'_{ci} \tag{21.1}$$

*At bottom centerline span: maximum service load condition*

$$0.78\left(-\frac{F_o}{A}-\frac{F_o e}{S_b}\right)+\frac{M_B}{S_b}+\frac{M_c}{S_{bc}}=f_t \tag{21.2}$$

$$-\frac{F_o}{A}-\frac{F_o e}{S_b}=-\frac{0.6}{0.9}f'_{ci}=\frac{f_t-\dfrac{M_B}{S_b}-\dfrac{M_c}{S_{bc}}}{0.78}$$

$$f_t+\frac{0.6\times 0.78}{0.9}f'_{ci}=\frac{M_B}{S_b}+\frac{M_c}{S_{bc}}$$

$$f_t+\frac{f'_{ci}}{1.923}=\frac{M_B}{S_b}+\frac{M_c}{S_{bc}}$$

$$f_t+\frac{f'_{ci}}{2}\cong\frac{M_B}{S_b}+\frac{M_c}{S_{bc}}$$

**For depressed strands: maximum harp (see Fig. 21.8)**

$$e=e_{\text{℄}} \qquad e_{\text{end}}=0 \qquad e_{0.4L}=0.8e \qquad M_{0.4L}=0.96M_{\text{℄}}$$

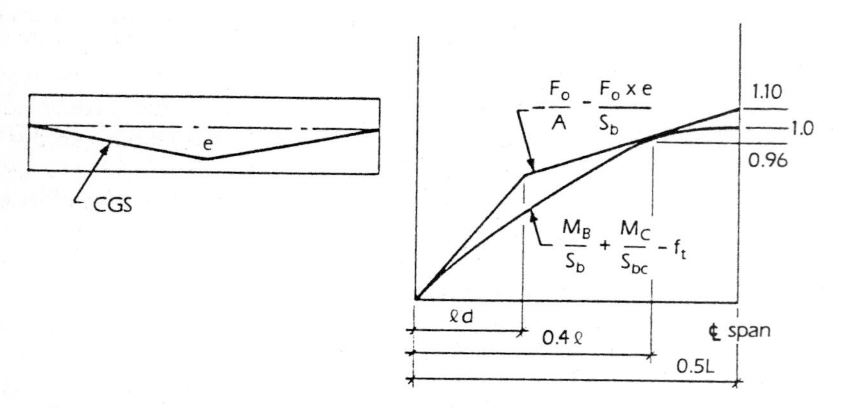


**Figure 21.8** Maximum harp for depressed strands.

*At centerline release (bottom)*

$$0.9\left(-\frac{F_o}{A}-\frac{F_o e}{S_b}\right)+\frac{M_{BM}}{S_b}+-0.6f'_{ci} \tag{21.3}$$

*At 0.4L (bottom): maximum service load condition*

$$\frac{0.78\left(-\frac{F_o}{A}-\frac{F_o e}{S_b}\right)}{1.1}+\frac{0.96 M_B}{S_b}+\frac{0.96 M_c}{S_{bc}}=f_t \tag{21.4}$$

$$-\frac{F_o}{A}-\frac{F_o e}{S_b}=-\frac{0.6}{0.9}f'_{ci}-1.111\frac{M_{BM}}{S_b}=1.410 f_t$$

$$-1.354\frac{M_B}{S_b}-1.354\frac{M_c}{S_b}$$

$$-0.667 f'_{ci}-1.111\frac{M_{BM}}{S_b}=1.140 f_t-1.354\frac{M_{BM}}{S_b}-1.354\frac{M_{TPG}}{S_b}$$

$$-1.354\frac{M_c}{S_{bc}}$$

$$\frac{0.243}{1.354-1.111}\frac{M_{BM}}{S_b}+1.354\frac{M_{TPG}}{S_b}+1.354\frac{M_c}{S_{bc}}$$

$$=1.410 f_t+0.667 f'_{ci}$$

Divide both sides by 1.382.

$$1.176\frac{M_{BM}}{S_b}+0.980\frac{M_{TPG}}{S_b}+0.980\frac{M_c}{S_{bc}}=1.020 f_t+0.483 f'_{ci}$$

$$\therefore \frac{M_{BM}}{6S_b}+\frac{M_{TPG}}{S_b}+\frac{M_c}{S_{bc}}\simeq f_t+\frac{f'_{ci}}{2}$$

# Part 3

# Component Design

**Figure 22.1** Transamerica Pyramid, San Francisco, California. Precast concrete cladding panels form the exterior of this San Francisco landmark. White cement and light-colored coarse and fine aggregates achieve harmony with the San Francisco tradition of being the alabaster city. The panels were lightly sandblasted in the plant and then erected on this 48-story 853-ft-high tower. In the background can be seen the tower crane used to erect precast panels on 601 Montgomery, under construction in the summer of 1979.

Chapter

# 22

# Solid Precast Concrete Wall Panels

Solid precast concrete wall panels are modular elements used to form the envelope of the building and to express the character of the exterior design. They fall into three general categories:

1. *Non-load-bearing cladding panels.* They are designed to support only their own weight and wind or seismic forces perpendicular to the panel. Cladding panels may be designed as spandrel panels, solid wall panels, or pierced window panels.
2. *Non-load-bearing shear walls.* They are designed to transfer lateral wind or seismic forces from horizontal diaphragms to the foundation or to another panel element.
3. *Load-bearing wall panels.* They are designed to support vertical loads from the building framing system. Load-bearing panels may also be designed to transmit lateral forces to the building foundations.

With precast concrete architects have the opportunity to bring out the full sculptural potential of the material and to freely express their design concepts for a building. The three-dimensional form of precast concrete is limited only by its function and the creativity of designers. Owing to its industrialized manufacturing process and the repetition resulting from correct modular design, forms which are not economically feasible with cast-in-place concrete become possible with precast concrete. Precast concrete is in reality the dream material for the architect.

## Advantages

Plant-cast precast concrete exterior panels have several advantages which justify their use:

1. *Fast construction.* Units are factory-cast on a daily cycle and stockpiled while other necessary site work and building framework are completed. Quick erection of precast panel units assures savings in interim financing costs, with early completion realized.
2. *Quality control.* Plant-cast units made in precision-made molds achieve close tolerances and uniform finish quality.
3. *Durability.* High-strength dense concrete guarantees "rock of ages" life expectancy.
4. *Finishes.* A variety of surface textures and finishes are attainable.
5. *Economy.* Modular design and multiuse of forms result in economy.
6. *Fire resistance.* Fire resistance is inherent in concrete construction.
7. *Low maintenance.* Proper selection of exposed aggregate finishes, coupled with proper panel design and drip details, eliminates long-term staining and streaking from accumulated surface dirt in the polluted air associated with city environments.
8. *Energy efficiency.* This type of construction affords the inherent thermal lag in concrete elements.

Exterior finishes are available in a wide range of textures from smooth to deeply exposed aggregate. Textures between these extremes are achieved by acid etching to lightly reveal the sand and aggregate at the surface, by sandblasting light or medium exposed aggregate, or by using retarders sprayed on the mold before casting. With this method, after stripping, the aggregate is revealed by washing the panel surface with high-pressure water hoses. The retarder material prevents cement hydration reactions from occurring at the surface, thereby allowing this powdery material to be washed away. Retarded finishes leave the coarse aggregate smooth, whereas sandblasting dulls the stone. The cement matrix may be white, gray, or colored by admixtures added to the mix. A smooth gray finish should be avoided on flat surfaces owing to the difficulty in preventing streaking and mottling of the surface, especially with steam curing. Textured surfaces may be obtained with form liners. (See Fig. 22.2.)

**Figure 22.2** Ambassador College Auditorium, Pasadena, California. Column covers, soffit panels, and parapet panels are of precast concrete with white cement and milky quartz aggregate with a medium sandblasted finish.

## Architectural Design Considerations

Before designing wall panels the architect should visit one or more precasters who produce wall panels or architectural precast concrete. He or she should visit their manufacturing plants as well as projects being erected, if possible. Designers can thus become familiar with the manufacturing process, including fabrication of the molds, problems in casting and finishing, methods of plant and jobsite handling, and methods used in connecting panels to the structure. This is very important for a full understanding of the material and its proper utilization.

In developing working drawings, the architect should work closely with the precaster. The engineering of the panels is done by the architect in conjunction with his or her structural engineer. The precaster will subsequently prepare shop drawings and necessary calculations to assure that the panels will be properly handled from manufacture through shipping to erection. The precaster will also design the handling and erection inserts. These shop drawings are submitted to the architect for shape approval prior to the manufacturing of molds. For large projects, a mock-up of a typical panel is first fabricated and approved by the architect as to finish and design. This mock-up serves as the standard throughout manufacture.

The precaster will either handle the erection of the units with its own crews or subcontract and supervise the erection. Hauling may also be subcontracted, but nonetheless the precaster is responsible for the panels until they are installed on the building and accepted by the architect through the general contractor.

In designing wall panels the following must be considered: (1) form design and fabrication, (2) concrete placement, (3) lifting the units from the forms (stripping), (4) plant handling during finishing and storing, (5) transporting the units to the jobsite, (6) erection, and (7) connection to the structure, including temporary bracing. In the manufacturer's analysis of the units, handling inserts and panel reinforcing are designed for (1) form removal, (2) in-plant handling, (3) transporting, (4) erection, (5) connection to the structure, and (6) in-place loading. Often the stresses on panels during form removal, handling, and erection exceed those caused by in-place loads. The inserts and reinforcing must be designed for each of the above situations. To reduce costs dual use of inserts is considered wherever possible.

### Form design

For simple shapes or in cases where there are many casts and the form cost is justified, steel forms are used. Where shapes are complex, wood, fiberglass, or concrete forms are used. With fiberglass forms, depending upon the complexity of the shape, up to 70 reuses are possible with minor rework. Most wall panels and architectural precast concrete units are cast in fiberglass or concrete forms. If the shape requires it, the form may be in parts that are assembled and disassembled for each cast. The primary consideration is the architect's design concept and how to accomplish this concept in the most efficient manner. The final design should consider maximum form reuse, keeping the number of different shapes to a minimum, ease of removal from the form, and the quality of the edges or changes in direction of the various planes desired.

The concrete unit is normally cast in a horizontal flat position with the exposed, textured, or sculptured face down and the flat inside face up. The vertical faces of the form must be sloped (must possess draft) for ease of removal from the mold. Where plug sections of forms or back formers are used, these must also be sloped both for ease of removal and to avoid pockets or voids when casting the concrete. An exception to this rule is made when the side forms are in removable sections. If the panel is pretensioned, consideration must be given to shortening occurring in the concrete during detensioning. Outside forms present few problems, but inside forms unless removed will cause the unit to crack during detensioning owing to binding on the forms. The magnitude of the draft is subject to the width of the section (surface friction versus strength of the section; see Fig. 22.3).

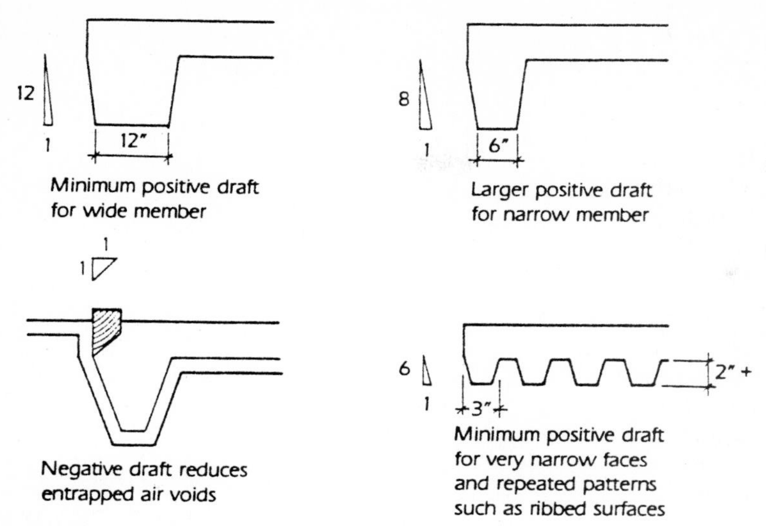


**Figure 22.3** Recommended minimum drafts.

Where edge grooves are required, removable side forms or expendable form materials are used. These are pulled out after the concrete unit has been removed from the basic form. (See Fig. 22.4.)

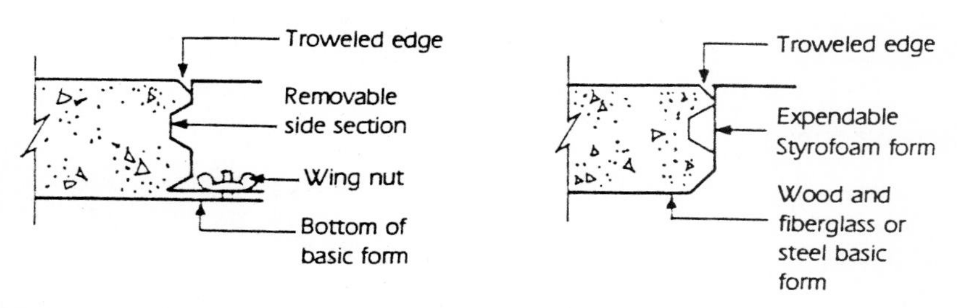


**Figure 22.4** Forming edge grooves.

In the final design of the shape, sharp edges and corners should be avoided and replaced with curved or chamfered edges because of possible edge damage in form removal and in handling. When the edge is sharp, only fine aggregate collects there and weakens the edge since larger aggregate is not present. Also, voids occur because of the plugging action of larger aggregate. As concrete shrinks during curing or otherwise changes volume, stresses build up at sharp corners, which can cause cracking as well as binding in the form. All inside corners should therefore have a radius or chamfer. The recommended minimum is $3/8$ in. (See Fig. 22.5.)

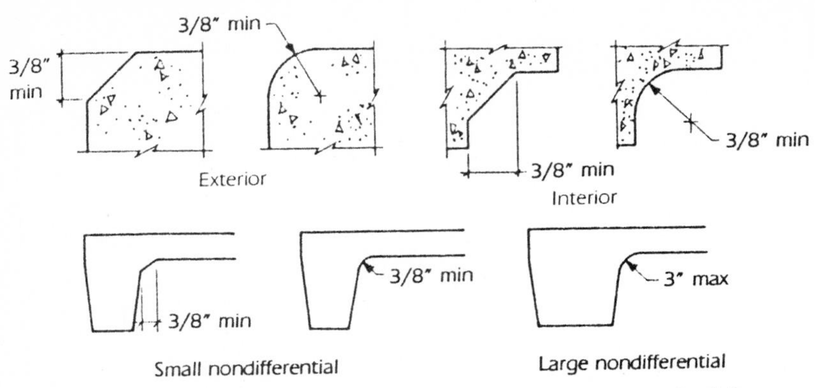


**Figure 22.5** Use of curved or chamfered edges. Note that as the mass of a rib becomes greater than that of the panel, the inside-corner transition should be increased.

The casting surface of the forms must be smooth and adequately reinforced to prevent form deflection. The forms are usually made up and coated with fiberglass to the shape of the desired section. Complex forms are made by a sprayed-on fiberglass-fiber reinforced coating applied to a pattern of the desired shape. The form is released from the pattern and hand-finished with a gel coat and sanded. It is also reinforced during and after the spray-up process. Another excellent method for making molds of complex shapes is to cast a concrete mold around a full-sized model of the desired panel.

## Concrete placement

The concrete strength used for nonprestressed wall panels is usually 5 ksi. This high strength results from the minimum cement content desirable from the standpoint of durability, usually 6 1/2 to 7 sacks per cubic yard. If the wall panel is pretensioned, sand-lightweight concrete of 5 ksi or normal-weight concrete of either 5 or 6 ksi is used. The maximum coarse aggregate size is 3/4 in. For exposed aggregate concrete, the coarse aggregate where the surface is to be exposed may range from 3/8 to 1/2 in, subject to the desired design effect. Usually the concrete is of the same mix throughout the panel. Where an expensive special exposed aggregate is to be used, two mixes are often used within the same panel (see Fig. 22.6). The special aggregate mix (face mix) is placed as a thin layer only in those areas to be exposed. While the face mix is still fresh, a more economical backup mix is poured, care being taken to blend the two to prevent separation due to laitance. When two different mixes are used, the designer should note the demarcation line on the drawings. The use of high-range water reducers (superplasticizers) is strongly recommended for all concrete mixes.

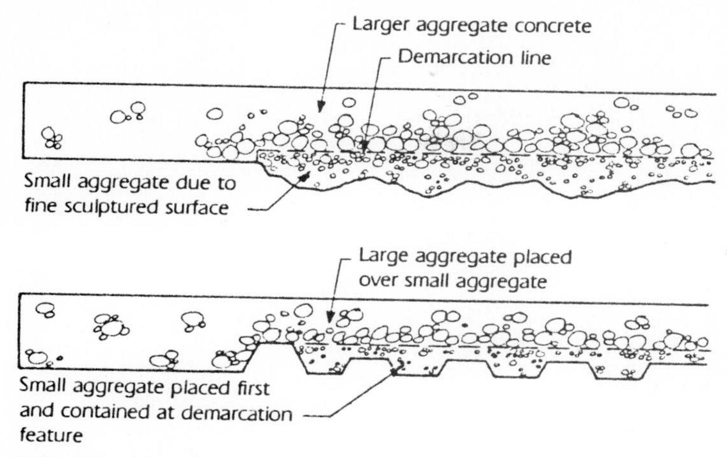


**Figure 22.6** Use of two mixes within a panel.

Where part of the surface is to receive a fine sculptured detail, two mixes can be used, with the mix in the sculptured area containing a smaller maximum coarse aggregate size (see Fig. 22.7). While concrete is being placed in the forms, either internal vibration with spud vibrators or external pneumatic mold vibrators are used to assure densely compacted concrete. (See also Fig 22.8.)

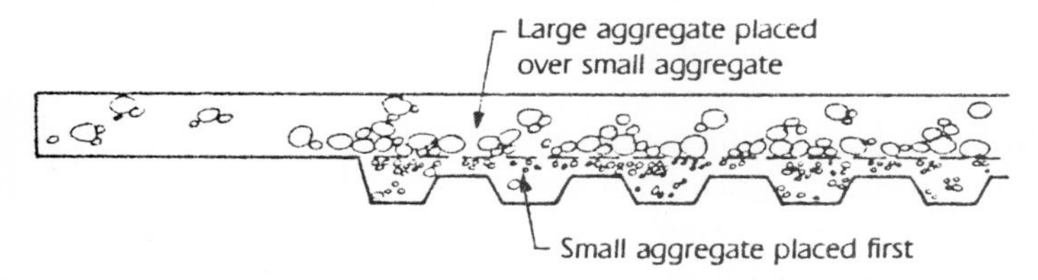


**Figure 22.7** Placement of two aggregate mixes in a panel.

### Removal from the form and handling

The primary consideration in the design of wall panels is the architect's design concept and how to accomplish this concept in the most efficient manner. This is the underlying principle behind the design of any wall panel. In finalizing the panel design, consideration must be given to the handling of the panel from stripping through erection at the jobsite. The stages consist of the position in which the panel is cast, how it is to be removed from the form, and how it is handled from finishing to storage, shipment, and erection. This sequence is studied, and the panel is structurally designed for each step and the inserts positioned accordingly. In each step, the panel is kept in equilibrium and balanced about

**Figure 22.8** Ambassador College Auditorium, Pasadena, California. Note the fine detail and excellent alignment of the precast column covers and soffit panels, achieved with quality precision molds and high vibration of low-slump concrete with small coarse aggregate. (*Photograph by Eugene Phillips.*)

its center of gravity. At no time during handling may the stresses in any portion exceed the stress which would produce cracking. Also, a factor of safety for impact and handling must be considered. The panel and each section are designed as an uncracked section, neglecting the contribution of the reinforcing steels. (See Chap. 19 for handling criteria.) Whenever possible, four pickup points are used in the horizontal flat position and two in the vertical position. Sometimes the size, weight, or configuration is such that more than four pickup points are necessary. In stripping the panel from the mold, the pickup points must be in balance with equal load to each sling; otherwise, the panel will bend or rotate and bind in the form with possible cracking and form distortion. In these situations specially designed jigs are required for stripping. (See Fig. 22.9.)

Some precasters will use spreader beams, traveling blocks, slings with fixed cable lengths, or vacuum lift systems for lifting the units out of the molds. Any openings or major projections must be designed so that beam sections within the panel can be used to transfer the panel weight without cracking to the pickup points. The beam section must be large enough to keep the stresses induced by handling, bending tension, and direct tension within allowable limits (see Fig. 22.10).

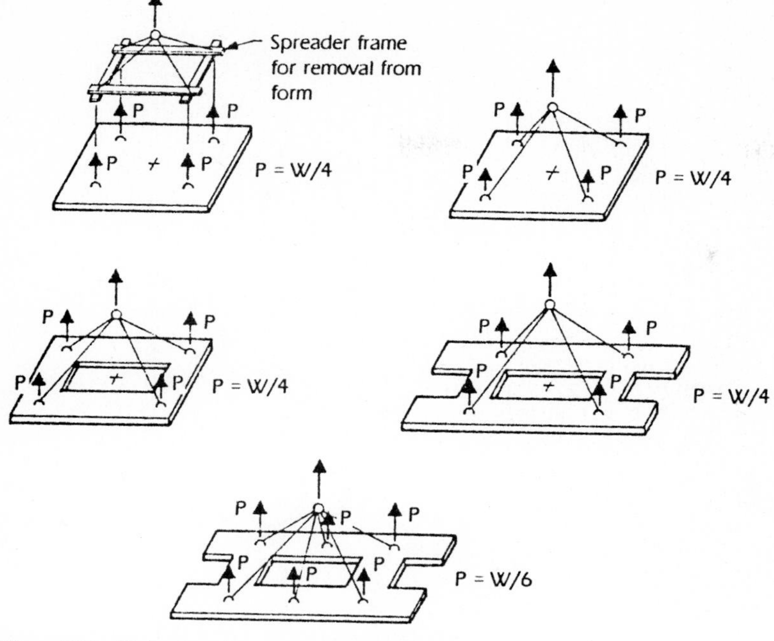


**Figure 22.9** Stripping the panel from the form.

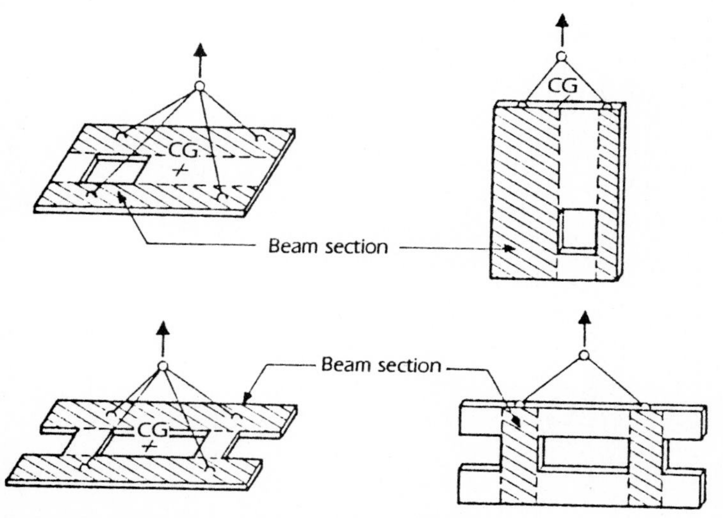


**Figure 22.10** Handling at beam sections of panels.

Some panel positions occurring during the handling process are shown in Fig 22.11.

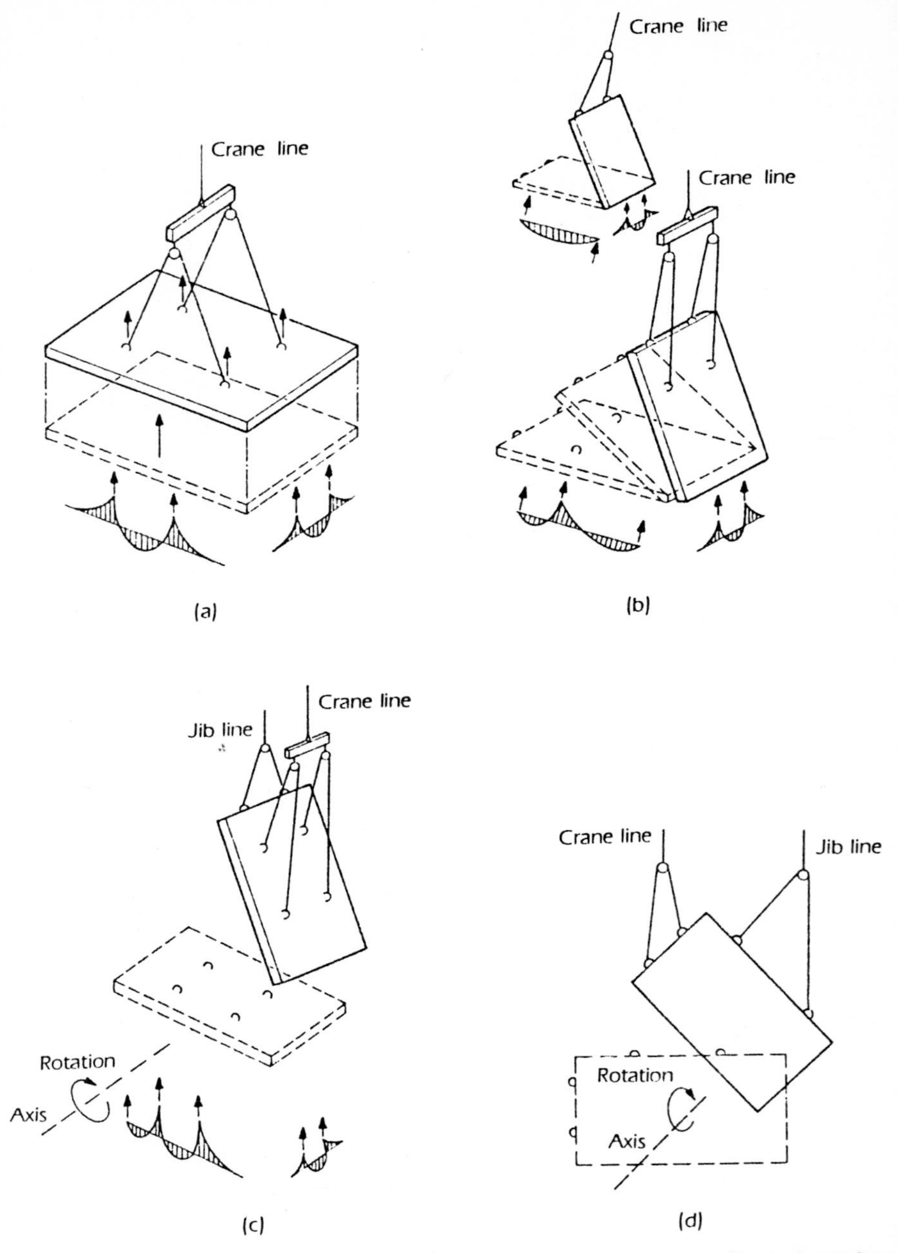


**Figure 22.11** Panel positions during handling. (*a*) Removal from form: sculptured panel. (*b*) Removal from form: plain panel tilted up. (*c, d*) Panel rotation.

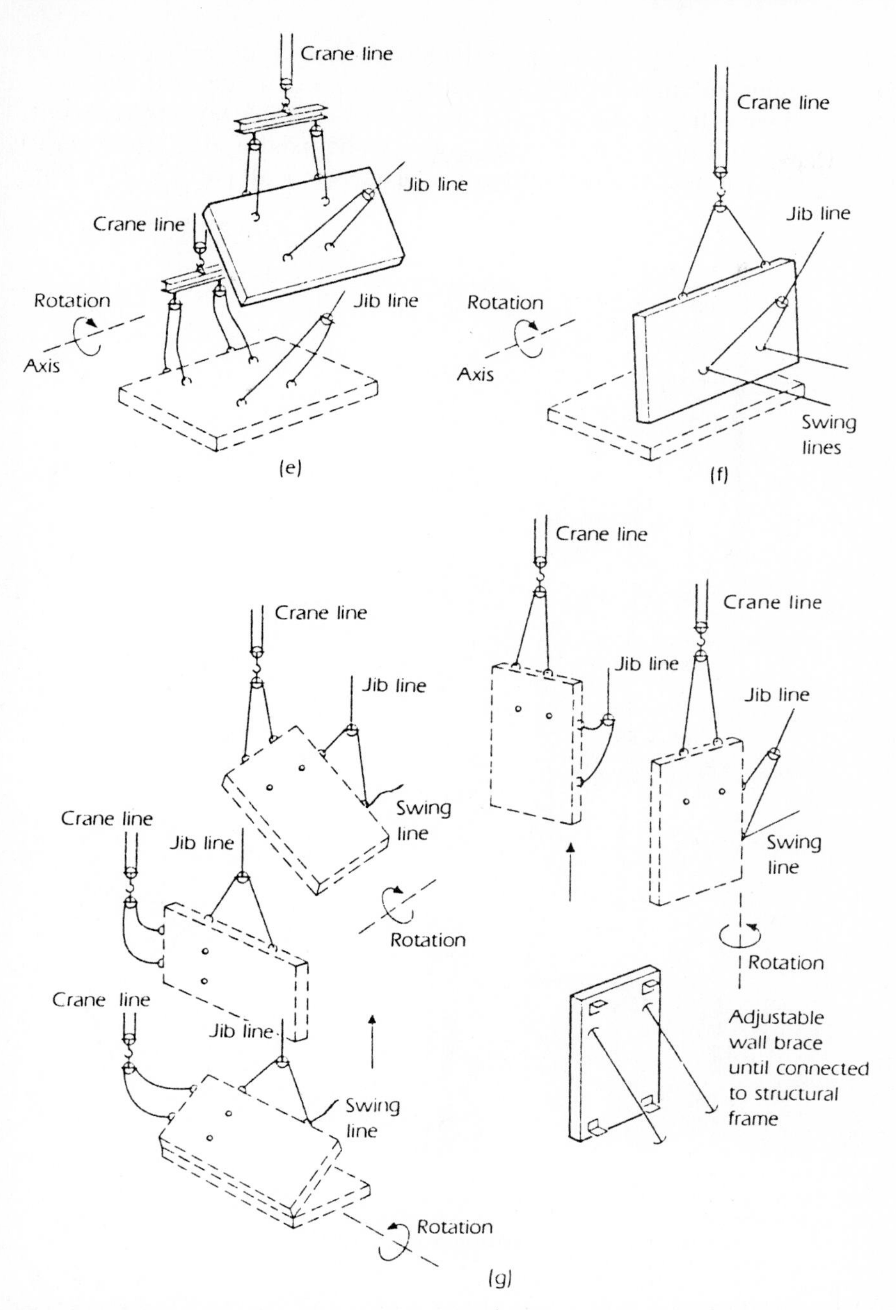


**Figure 22.11 (*continued*)** Panel positions during handling. (*e, f*) Panel rotation. (*g*) Panel erection.

In the remainder of this chapter the design process described in Chap. 20 is applied to various typical architectural and structural wall elements. Here will be demonstrated in practical example form the current philosophy and procedure recommended to be used in designing reinforcing and connections for solid precast concrete wall panels. (See Fig. 22.12.)

**Figure 22.12** Embarcadero Center, San Francisco, California. Four high-rise towers comprise this landmark complex, all clad with two-story precast concrete panels. Steel molds were used to produce these panels.

**Figure 22.13** 595 Market Street, San Francisco, California. Spandrels 34 ft by 5 ft 3 in deep band this 30-story building, completed in 1979. The old gives way to the new as the modern architecture of the financial district invades the Mission Street Redevelopment Area.

## Spandrel Panels

Spandrel panels are long, narrow horizontal elements that span from column to column at the elevation of the floor and roof planes. They enclose the floor framing system and fill the space between the finished floor to the bottom of the fenestration or to the parapet height, as in open parking structures. The spandrel may be a non-load-bearing cladding element or be designed to carry the tributary load from the floor system. The space between units is usually filled with glass or window units, as in office buildings, educational buildings, residential buildings, condominiums, apartments, and hotels. Spandrels are supported near or at their ends, on column corbels, or by edge beams or the floor system itself. One aspect of the structural design of spandrels often overlooked is the allowance for bowing due to thermal differentials between the outer and inner surfaces of the panel. The resulting movement should be able to be accommodated by the windows or other intermediate infill materials. Conversely, if intermediate lateral connectors are used, they should be able to develop the forces induced by restraint of the thermal gradient. The following design example demonstrates the development of intermediate connectors to restrain against the thermal forces induced. (See also Figs. 22.13 and 22.14.)

**Figure 22.14** 595 Market Street, San Francisco, California. A typical spandrel arrangement. Precast concrete was made with limestone aggregate and white cement, with a light sandblasted finish.

## Spandrel panel design

Design a non-load-bearing cladding spandrel for a high-rise office building in San Francisco. Use the Uniform Building Code. The panel cross section and elevation are shown in Fig. 22.15; $f'_c = 5000$ lb/in$^2$ hard-rock concrete.

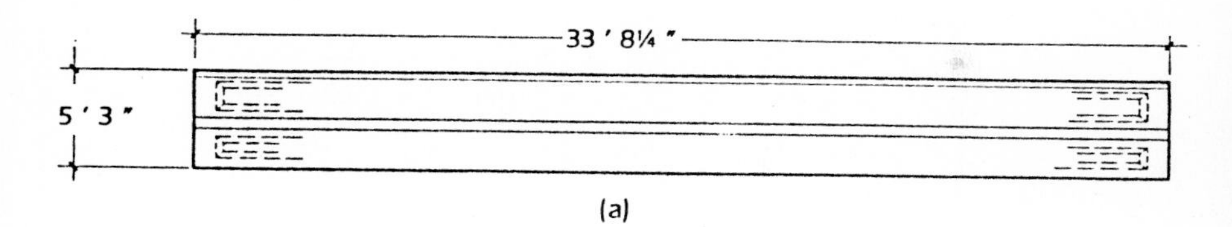


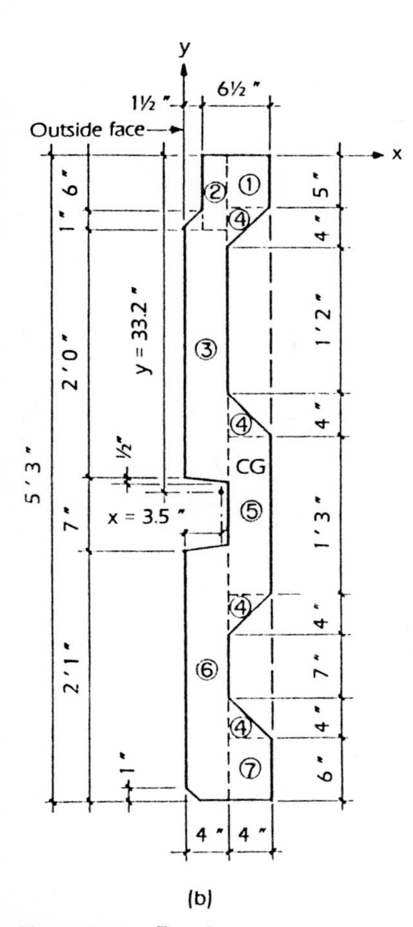


**Figure 22.15** Panel cross section (*a*) and elevation (*b*).

Locate the center of gravity. First, calculate $\bar{y}$.

| | Part | $A$ | $y_i$ | $Ay_i$ |
|---|---|---|---|---|
| 1 | 4 × 5 | 20 | 2.5 | 50 |
| 2 | 2.5 × 7 | 18 | 4.0 | 72 |
| 3 | 24 × 4 | 96 | 20.0 | 1,920 |
| 4 | 0.5 × 4 × 4 × 4 | 32 | 33.8 | 1,082 |
| 5 | 15 × 4 | 60 | 34.5 | 2,070 |
| 6 | 25 × 4 | 100 | 50.5 | 5,050 |
| 7 | 6 × 4 | 24 | 60.0 | 1,440 |
| Σ | | 350 | | 11,684 |

$$\bar{y} = \frac{11{,}684}{350} = 33.2 \text{ in.}$$

Then calculate $\bar{x}$.

| | Part | $A$ | $x_i$ | $Ax_i$ |
|---|---|---|---|---|
| 1 | | 20 | 6.0 | 120 |
| 2 | | 18 | 2.75 | 50 |
| 3 | | 96 | 2.0 | 192 |
| 4 | | 32 | 5.33 | 171 |
| 5 | | 60 | 6.0 | 360 |
| 6 | | 100 | 2.0 | 200 |
| 7 | | 24 | 6.0 | 144 |
| Σ | | 350 | | 1237 |

$$\bar{x} = \frac{1237}{350} = 3.5 \text{ in.}$$

**Section properties.** Determine section properties.

| Part | $\bar{x} - x_i$ | $(\bar{x} - x_i)^2$ | $A$ | $A(\bar{x} - x_i)^2$ | $I_o$ |
|---|---|---|---|---|---|
| 1 | 2.5 | 6.25 | 20 | 125 | 27 |
| 2 | 0.75 | 0.56 | 18 | 10 | 10 |
| 3 | 1.5 | 2.25 | 96 | 216 | 128 |
| 4 | 1.83 | 3.35 | 32 | 107 | 28 |
| 5 | 2.5 | 6.28 | 60 | 375 | 80 |
| 6 | 1.5 | 2.25 | 100 | 225 | 133 |
| 7 | 2.5 | 6.25 | 24 | 150 | 37 |

$$Iyy = \Sigma A(\bar{x} - x_i)^2 + I_o = 1651 \text{ in}^4.$$

$$S_{\text{outside}} = \frac{1651}{3.5} = 472 \text{ in}^3$$

$$S_{\text{inside}} = \frac{1651}{4.5} = 367 \text{ in}^3$$

Panel weight for normal-weight concrete is

$$\frac{352}{144} \times 0.150 = 0.370^{\text{k}}/\text{ft}$$

Use $0.40^{\text{k}}$/ft to account for end blocks.

**Handling and stripping.** Check handling and stripping. Try two-point pick at one-fifth points.

$$M^+ = M^- = 1.5 \times 0.025 \times 0.40 \times 33.7^2 = 17.0^{\text{k}\cdot\text{ft}}$$

$$f_t = \frac{17.0 \times 12}{367} = 0.556 \text{ ksi} > 5\sqrt{f'_{ci}} \text{ (too high)}$$

Therefore, use four-point pick.

0.1L | 0.3L | 0.2L | 0.3L | 0.1L
L

$$M^+ = M^- = 1.5 \times 0.0056 \times 0.40 \times 33.7^2 = 3.82^{\text{k}\cdot\text{ft}}$$

$$f_t = \frac{3.82 \times 12}{367} = 0.124 \text{ ksi} > 5\sqrt{f'_{ci}} \text{ (OK)}$$

$$A_s \text{ required} = \frac{3.82}{1 \times 5.8} = 0.66 \text{ in}^2$$

Six no. 3 bars are required in each face.

Store and ship in the vertical position since stresses are excessive with two-point support.

The bottom of the panel must be reinforced to satisfy shipping stresses. Block panel 3 ft 0 in in from each end.

$$M = 2.0 \times (2 \times 0.8 - 1) \times (0.125 \times 0.40 \times 33.7^2) = 68^{k \cdot ft}$$

$$A_s = \frac{68}{1.44 \times 60} = 0.79 \text{ in}^2$$

Use four no. 4 bars in bottom.

**Resistance of lateral forces.** Design panel element to resist lateral forces. (Assume that the panel connectors are 6 in from each end.)

If the wind loading is 30 lb/ft$^2$,

$$w = 0.030 \times 12.5 = 0.375^k/\text{ft}$$

$$M = 0.125 \times 0.375 \times 32.7^2 = 50.1^{k \cdot ft}$$

$$f_{t,\text{inside}} = \frac{50.1 \times 12}{367} = 1.64 \text{ ksi (too high)}$$

Try lateral connectors at midspan top and bottom.

$$M^- = 0.125 \times 0.375 \times 16.5^2 = 12.8^{k \cdot ft}$$

$$f_{t,\text{outside}} = \frac{12.8 \times 12}{472} = 0.325 \text{ ksi} < 5\sqrt{f'_c} \text{ (OK)}$$

$$A_s = \frac{12.8}{1.76 \times 5.8} = 1.23 \text{ in}^2$$

Six no. 4 bars are required in each face.

If the seismic design lateral force is 0.3 $g$,

$$w = 0.3 \times 0.40 = 0.120^k/\text{ft} < \text{wind (OK)}$$

For temperature steel requirements,

$$A_s = 0.002 \times 1/2 \times 352 = 0.35 \text{ in}^2 \text{ on each face} < \text{wind (OK)}$$

Check spandrel for bowing due to 60° thermal gradient. (Ignore beneficial effect of long-term differential shrinkage.)

$$I = \frac{1631}{5.25} = 314 \text{ in}^4/\text{ft}$$

$$\text{Equivalent thickness} \cong 7 \text{ in}$$

From charts for lateral support spacing of 16.5 ft, $\Delta = 0.3$ in, which is excessive. Therefore, use lateral supports at 8 ft 0 in ± on center, for which $\Delta \cong 0.08$ in. (OK)

Reanalyze wind-loading reinforcement requirements.

$$M = 0.11 \times 0.158 \times 8.0^2 = 1.11^{\text{k}\bullet\text{ft}}$$

$$A_s = \frac{1.11}{1.55 \times 5.8} = 0.13 \text{ in}^2$$

Therefore, handling governs.

Determine temperature steel requirements for transverse section.

$$A_s = 0.002 \times \frac{352}{5.25} = 0.13 \text{ in}^2/\text{ft}$$

Use no. 3 bars at 10 ft on center ($A_s = 0.11 \times 12/10 = 0.13 \text{ in}^2/\text{ft}$).

**Reinforcing summary.** Reinforcing is summarized in Fig. 22.16.

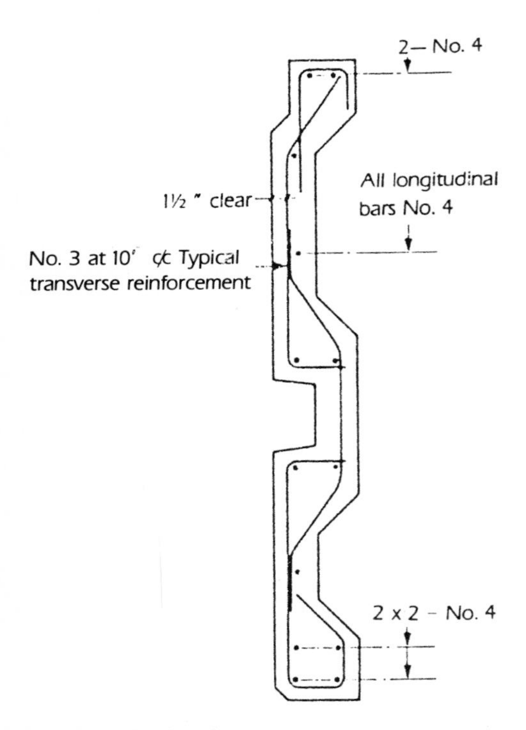


**Figure 22.16** Reinforcing summary.

**Connection design.** Distribution factors for loads to lateral connections are shown in Fig. 22.17.

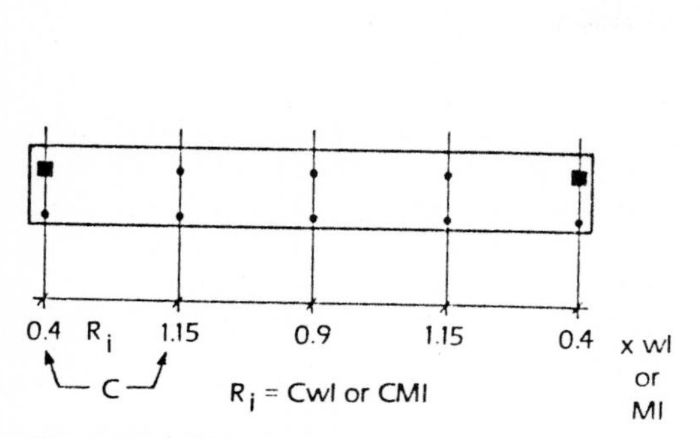


Figure 22.17 Distribution factors for loads to lateral connections.

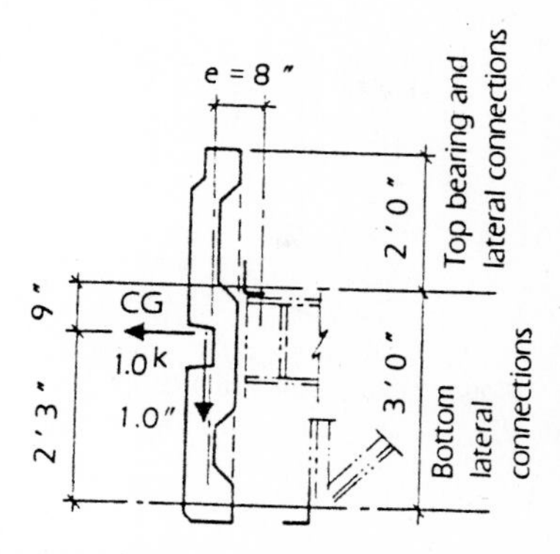


Figure 22.18 Connection arrangement.

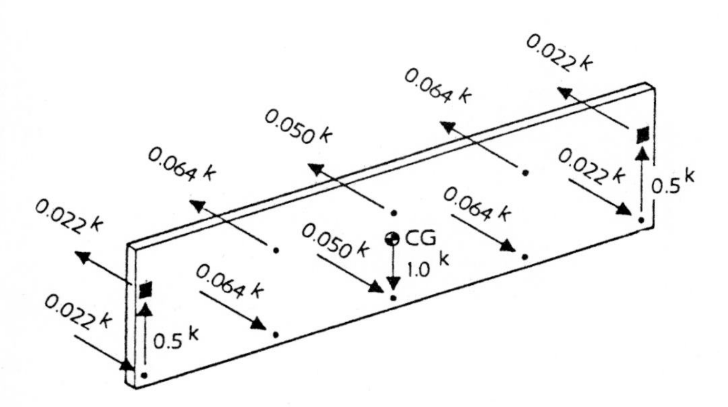


Figure 22.19 Analyze $1.0^k$ down (gravity).

$$M = \frac{1.0 \times 8}{33.7} = 0.237^{k \cdot in}/\text{ft}$$

$$\ell = 8.4 \text{ ft}$$

| $C$ | $CM$ | $H = \frac{CM\ell}{3 \times 12}$ |
|---|---|---|
| 0.4 | $0.80^{k \cdot in}$ | $0.022^k$ |
| 0.9 | $1.79^{k \cdot in}$ | $0.050^k$ |
| 1.15 | $2.29^{k \cdot in}$ | $0.064^k$ |

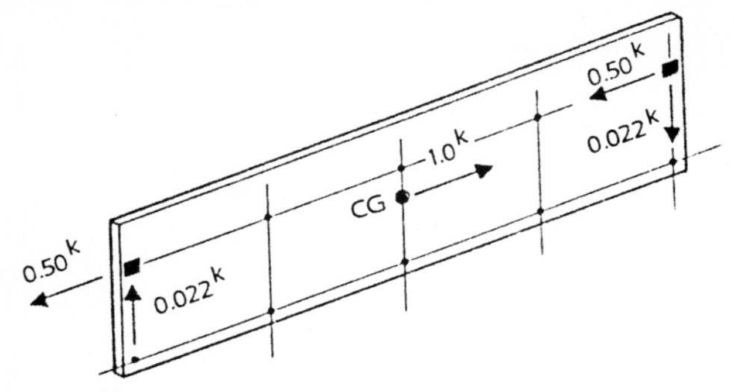


**Figure 22.20** Analyze $1.0^k$ longitudinal. $\Sigma M_{CG} = 0$. $2R_v \times 16.8 = 1 \times 0.75$. $R_v = 0.022^k$.

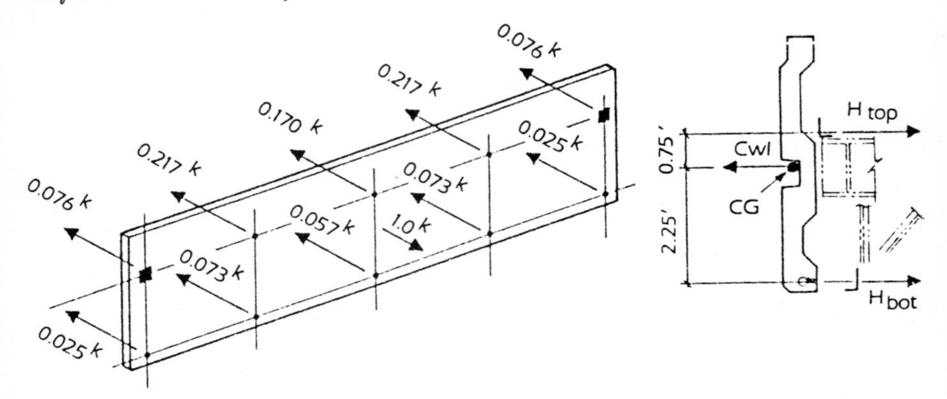


**Figure 22.21** Analyze $1.0^k$ lateral. $M_{top} = 3/4\, cw\ell$ ; $M_{bottom} = 1/4\, cw\ell$ . (See Fig. 22.17 for distribution factors.)

$$w = \frac{1.0}{33.7} = 0.003^k/\text{ft} \qquad \ell = 8.4\ \text{ft}$$

| $C$ | $Cw\ell$ | $H_{top}$ | $H_{bottom}$ |
|---|---|---|---|
| 0.4 | $0.101^k$ | $0.076^k$ | $0.025^k$ |
| 0.9 | $0.227^k$ | $0.170^k$ | $0.057^k$ |
| 1.15 | $0.290^k$ | $0.217^k$ | $0.073^k$ |

Panel weight $W = 0.4^k/\text{ft} \times 33.7\,\text{ft} = 13.5^k$. (Glazing weight is neglected for simplicity.)

**Connector fastener; bottom lateral connection.** Design this fastener. (See Fig. 22.22.)

$$W\ \left(1.0^{k}\ \text{down} + 1.2^{k}\ \text{in}\right) = P_{\text{critical}}$$

or

$$\left(13.5^{k}\right)(0.064 + 1.2 \times 0.073) = 2.04^{k}\ \text{compression}$$

$$\left(13.5^{k}\right)(0.064 + 1.2 \times 0.073) = 0.32^{k}\ \text{tension}$$

A 5/8 -in ϕ ferrule loop insert is OK. $P_{\text{allow}} = 3.0^{k}$.

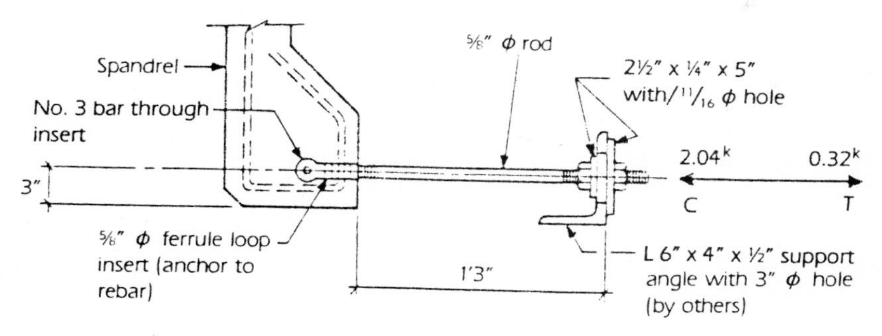


**Figure 22.22** Bottom lateral connection.

Calculate size of threaded rod stock required as governed by compression (body of connector).

$$\left(13.5^{k}\right)(0.064 + 0.4 \times 0.073) = 1.26^{k}\ \text{compression}$$

Use 5/8-in rod.

| ϕ nom II | ϕ net II | A net IIII | $I$ in$^4$ * | $r = \sqrt{\frac{I}{A}}$ | $\frac{L}{r} = 200$, $Fa = 8.27$ $L_u$ | $P_{\text{allow}}$ | $\frac{L}{r} = 100$, $Fa = 16.47$ $L_u$ | $P_{\text{allow}}$ | $\frac{L}{r} = 50$, $Fa = 22.60$ $L_u$ | $P_{\text{allow}}$ | Tension $F_t = 26.70$ $P_t$ † |
|---|---|---|---|---|---|---|---|---|---|---|---|
| 1/2 | 0.42 | 0.1385 | 0.0015 | 0.10 in | 20 in | 1.2ᵏ | 10 in | 2.3 | 5 in | 3.2 | 3.7 |
| 3/4 | 0.63 | 0.3079 | 0.0077 | 0.16 in | 32 in | 2.5 | 16 in | 5.1 | 8 in | 6.9 | 8.2 |
| 1 | 0.83 | 0.5410 | 0.0233 | 0.21 in | 42 in | 4.5 | 21 in | 8.9 | 10 in | 12.2 | 14.4 |
| 1¼ | 1.08 | 0.9161 | 0.0668 | 0.27 in | 54 in | 7.6 | 27 in | 15.0 | 14 in | 20.7 | 24.4 |

NOTE: $Fy$ = 38 ksi (ASTM A-307); $k$ = 1.0. Allowable stresses are increased by 33 percent.
*$I = 0.049087d^4$.
† Allowable insert value will govern.

**Connector fastener; top lateral connection.** Design this fastener. (See Fig. 22.23.)

$$W\left(1.0^{k}\,\text{down} + 1.2^{k}\text{ in and out}\right) = P_{\text{critical}}$$

$$\left(13.5^{k}\right)(0.64 + 1.2 \times 0.217) = 4.37^{k}\text{ tension}$$

$$\left(13.5^{k}\right)(0.64 - 1.2 \times 0.217) = 2.65^{k}\text{ compression}$$

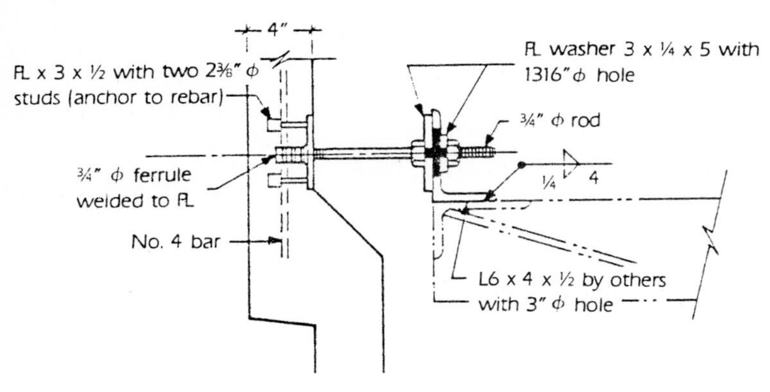


**Figure 22.23** Top lateral connection.

Design connection to panel.

$$UP = 0.75(1.4D + 1.7L + 1.87E)$$

$$= 0.75(1.4 + 0.064 + 1.87 \times 1.2 \times 0.217)\left(13.5^{k}\right) = 5.82^{k}$$

or

$$UP = 0.9D + 1.43E$$

$$= (0.9 \times 0.064 + 1.43 \times 1.2 \times 0.217)\left(13.5^{k}\right) = 5.80^{k}$$

Try plate $5 \times 3 \times 3/8$ with two $3/8$-in $\phi \times 2\,1/2$-in-long studs.
As governed by concrete strength (see *PCI Design Handbook*, page 6-54):

$$\ell_e = 2.5\text{ in} \quad y = 0 \quad x = 4\text{ in} \quad \therefore \phi P_c = 13^{k}$$

As governed by stud steel capacity (see *PCI Design Handbook*, page 6-52):

$$\phi P_s = 2 \quad 6.0 = 12.0^{k}\text{ governs.}$$

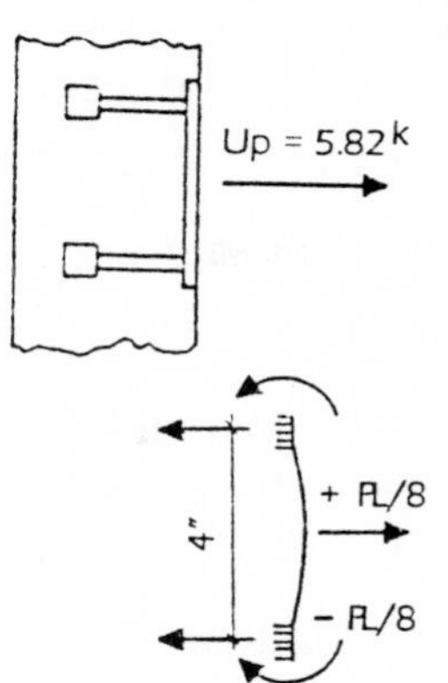


℄ thickness required:

Use $f_b = 0.9F_y = (0.9)(36$ $= 32.4$ ksi

$$i \;= \frac{M}{S} = \frac{PL/8}{1/6\,b\,t^2}$$

$$32.4 = \frac{(1/8)\,(4.37)\,(4)}{(1/6)\,(3)\,t^2} \qquad t^2 = 0.135 \quad t = 0.37\text{ in}$$

Use 3/8-in ℄.
For the rod connector t ℄, use 3/4 -in ferrule welded to ℄ ($P_T = 8.2^k$).
Check punching shear $C = 2.65^k$.

$$UP = 0.75\,(1.4 \times 0.064 \; - 1.87 \times 1.2 \times 0.217)\,(13.5^k)$$

$$v_{bo} = \frac{C}{\phi b_o d} = \frac{4.02}{(0.85)\,(\;\times 9 + 2 \times 7)\,(4)}$$

$$= 0.037\text{ ksi}$$

$$v_c = 4\sqrt{f_c'} = 0.283\text{ k i (OK)}$$

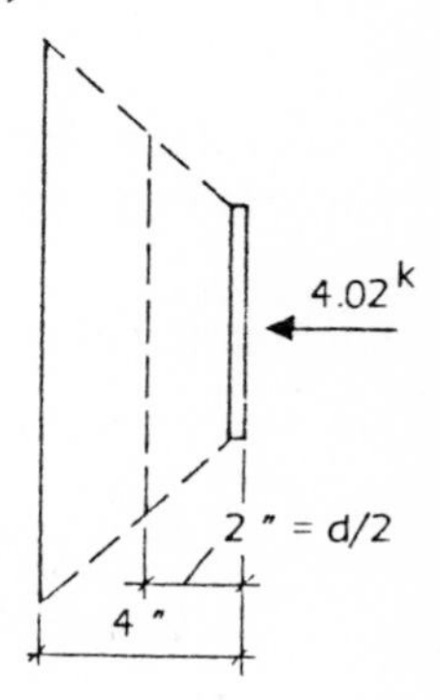


Check weld; ∠ 6 × 4 1/2 to curb ∠ cast in floor slab. Use E70 XX elect odes (see *AISC Manual for Steel Construction*, 8th :dition, page 4-76).

$$P = 4.37^k \quad \ell = 4\text{ in} \quad C_1 = 1.0 \quad K = 0$$

$$a = 4.5\text{ in} \quad \therefore a = \frac{4.5}{4} = 1.125$$

$$C = 0.375$$

$$b = \frac{P}{CC_1\ell} = \frac{4.\;7}{(0.375)\;1)(4)} = 2.91$$

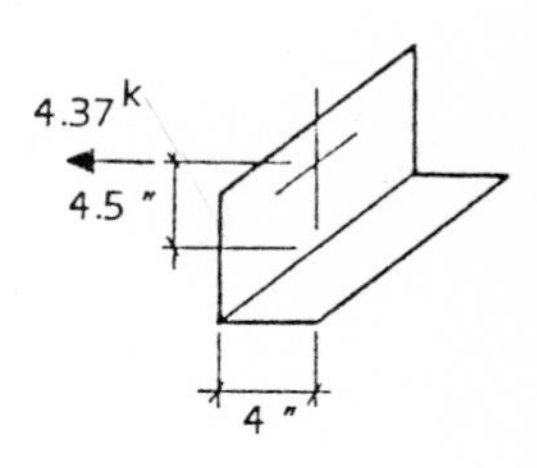


Therefore, use 1/4-ir welds. (A 33 percent increase in weld stress has been allowed but not aken.)
Design connector b ly; ∠ 6 × 4 × $t$.

$$P_{\text{critical}} = W\left(1.0^{\text{k}} \text{ down} + 0.4^{\text{k}} \text{ in and out}\right)$$
$$= \left(13.5^{\text{k}}\right)(0.064 + 0.4 \times 0.217) = 2.03^{\text{k}}$$

Try 6-in-long angle (assume $K = 0.8$ in).

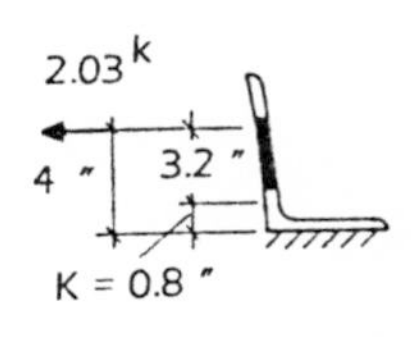


$$F_b = \frac{M}{S} = \frac{(2.03)(3.2)}{1/6 \times 6 \times t^2} = 1.33 \times 0.6 \times 36$$

$t^2 = 0.23$

$t = 0.48$ in

Use $\angle$ $6 \times 4 \times 1/2 \times 0$ ft 6 in long.

Check angle stress for $P = 1.2\,g$.

$$f_b = \frac{4.37 \times 3.2}{1/6 \times 6 \times 0.5^2} = 55.9 \text{ ksi} > F_y \text{ (OK)}$$

Therefore, the body of the connector exhibits ductile behavior at maximum design force on the connector fastener.

**Top bearing connection.** Design top bearing connection.

Force to connector body:

$$P_{\text{critical}} = W\left(1.0^{\text{k}} \text{ down} + 0.4^{\text{k}} \text{ longitudinal}\right)$$
$$= \left(13.5^{\text{k}}\right)(0.5 + 0.4 \times 0.022) = 6.87^{\text{k}}$$
$$= 0.022 \times 13.5 = 0.30^{\text{k}}$$

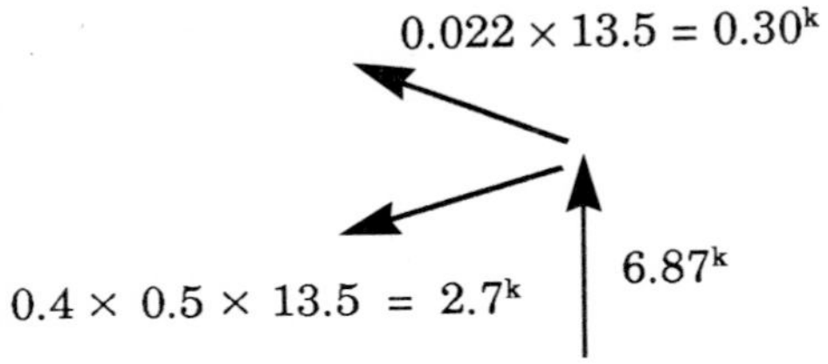

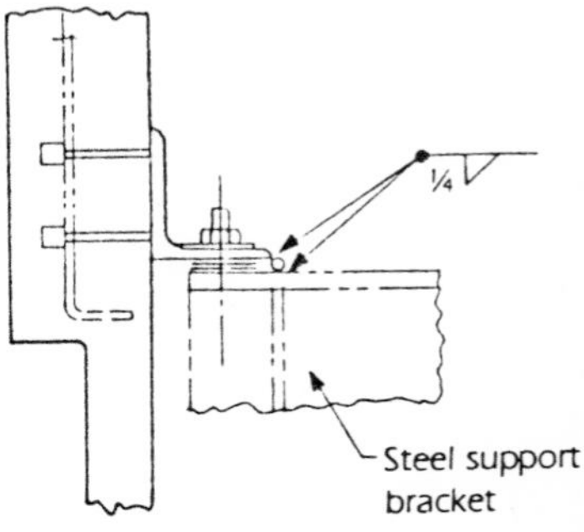

$0.022 \times 13.5 = 0.30^{\text{k}}$

$0.4 \times 0.5 \times 13.5 = 2.7^{\text{k}}$

$6.87^{\text{k}}$

¼

Steel support bracket

$$\text{or } P_{\text{critical}} = W\left(1.0^{\text{k}} \text{ down} + 0.4^{\text{k}} \text{ out and in}\right)$$
$$= \left(13.5^{\text{k}}\right)(0.022 + 0.4 \times 0.076) = 0.71^{\text{k}}$$

$0.5 \times 13.5^k = 6.75^k$ $0.71^k$

Determine bearing-angle thickness. Try 9-in-long L leg. (Assume $k = 1.1$ for 6-in leg.)

$$F_b = \frac{M}{S} = \frac{(6.87)(2.9)}{1/6 \times 9 \times t^2} = 28.7 \qquad t_{\text{required}} = 0.68 \text{ in}$$

Use $\angle$ 8 × 6 × 3/4 × 9 in long.
Design connection to panel.

$$UP = 0.75(1.4D + 1.87E)$$
$$= 0.75(1.4 \times 0.5 + 1.87 \times 1.2 \times 0.022)(13.5) = 7.6^k$$
$$UP = 0.9D + 1.43E$$
$$= (0.9 \times 0.5 + 1.43 \times 1.2 \times 0.022)(13.5) = 6.6^k$$

$0.022 \times 1.4 \times 13.5 = 0.42^k$ $(UH)$

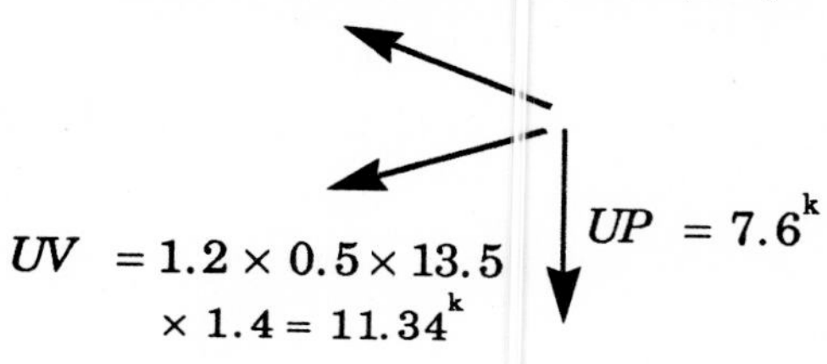

$UP = 7.6^k$

$$UV = 1.2 \times 0.5 \times 13.5 \times 1.4 = 11.34^k$$

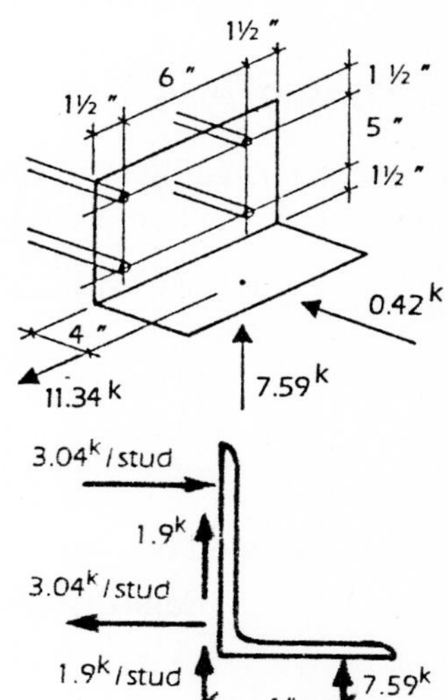


Try four studs.
1. Effect of $UP = 7.6^k$.

$$2T_1 \times 5 = 7.6 \times 4$$
$$T_1 = 3.04^k/\text{stud}$$
$$R_{yv} = \frac{7.6}{4} = 1.9^k/\text{stud}$$

2. Effect of eccentric shear; $UV = 11.34^k$.
Due to shear:

$$R_{xv} = \frac{11.34}{4} = 2.84^k/\text{stud}$$

3″ Ry R
2.5″ Rx
e = 4″
11.34k

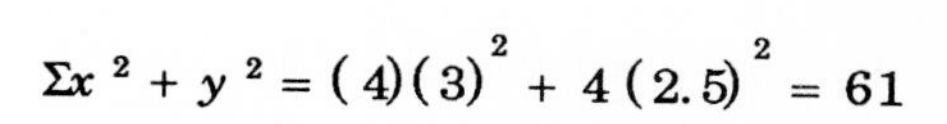

$$\Sigma x^2 + y^2 = (4)(3)^2 + 4(2.5)^2 = 61$$

Due to torsion :

$$R_{xT} = \frac{Pey.}{\sum x^2 + y^2} = \frac{(11.34)(4)(2.5)}{61} = 1.86^{k}/\text{stud}$$

$$R_{yT} = \frac{Pex}{\sum x^2 + y^2} = \frac{(11.34)(4)(3)}{61} = 2.23^{k}/\text{stud}$$

Effect of eccentric shear causing bending ($e$ = 4 in):

$$2T_2 \times 6 = (11.34)(4) \quad T_2 = 3.78^{k}/\text{stud}$$

$$R = \sqrt{(R_{xv} + R_{xT})^2 + (R_{yv} + R_{vT})^2}$$

$$= \sqrt{(2.84 + 1.86)^2 + (1.90 + 2.23)^2} = 6.26^{k}/\text{stud}$$

The critical design condition is $\begin{cases} V_u = 6.26^{k}/\text{stud} \\ P_u = 3.04 + 3.78 = 6.82^{k}/\text{stud} \end{cases}$

Try four 3/4 -in $\phi \times$ 6-in-long studs.

As governed by stud capacity (see *PCI Design Handbook*, 3d edition, pages 6-9, 6-52, and 6-53):

$$\phi P_s = 23.9^{k}/\text{stud} \quad \phi V_s = 19.9^{k}/\text{stud} \quad (\phi = 1.0)$$

$$\frac{1}{\phi}\left[\left(\frac{P_u}{P_s}\right)^2 + \left(\frac{V_u}{V_s}\right)^2\right] = \frac{1}{0.85}\left[\left(\frac{6.82}{23.9}\right)^2 + \left(\frac{6.26}{19.9}\right)^2\right]$$

$$= 0.08 + 0.10 = 0.18 < 1.0 \text{ (OK)}$$

As governed by concrete capacity:

Edge distance $d_e$ = 3 in* $f'_c$ = 5000 lb/in$^2$

$$\phi P_c = 16.5^{k}/\text{stud} \quad \phi V_e = 9.4^{k}/\text{stud}$$

where $\phi = 0.85$ ($P_O = 1.94$; $V_0 = 11.1$).

$$\frac{1}{\phi}\left[\left(\frac{P_u}{P_c}\right)^2 + \left(\frac{V_u}{V_c}\right)^2\right] = \frac{1}{0.85}\left[\left(\frac{6.82}{19.4}\right)^2 + \left(\frac{6.26}{11.1}\right)^2\right] = 0.51 < 1.0 \text{ (OK)}$$

* Pullout only.

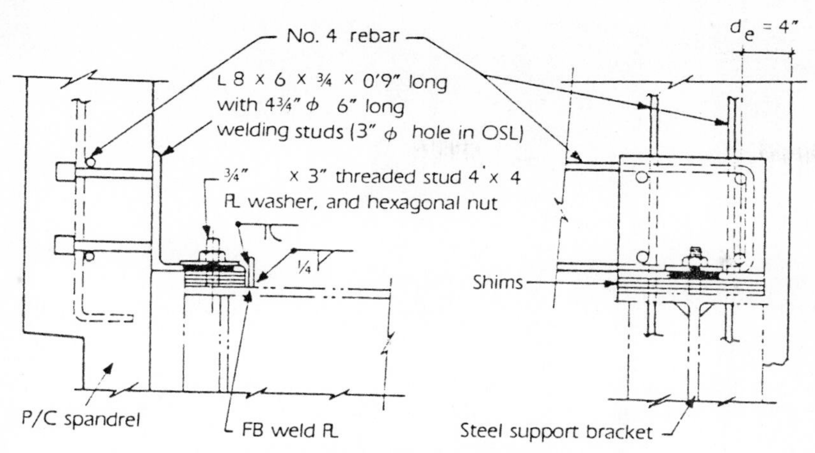


**Figure 22.24** Detail of bearing connection.

Therefore, use four $^3/_4$-in $\phi \times$ 6-in-long studs. See Fig. 22.24 for bearing connection detail.

## Column Cover Units

While spandrel units give the exterior of the building a banded appearance, column cover units emphasize the verticality of the structure (see Figs. 22.25 and 22.26). These units are made in multistory lengths and sometimes are prestressed concentrically to keep handling stresses within allowable limits and to achieve economy in the reinforcing of the member. A key element in the location of the bearing connections of column cover units is to position these connections as closely as possible to the center of gravity of the units. In this manner, the counterbalancing moment of forces required to counteract longitudinal seismic effects on the column cover is minimized. Another aspect of structural design of column cover connector bearing assemblies is that there is often a considerable distance between the attachment point on the column cover and the support location on the building frame, requiring a torsion-resistant steel assembly to form the connection bracket. There are a variety of ways to "hide" the structural column with the column cover unit, and attachments may be made to the flanges of the column section or to the adjacent floor framing (see Fig. 22.27). One arrangement is demonstrated in the following design example.

**Figure 22.25** Ambassador College Auditorium, Pasadena, California. The column cover units shown here were erected progressively from the ground level, with bottom "blind" connections and top welded connections being made to the structural steel columns as erection progressed. See the typical connections presented at the end of this chapter. *(Photograph by Eugene Phillips.)*

**Figure 22.26** 333 Market Street, San Francisco, California. Two-story V-shaped column cover units clad this 33-story building, emphasizing the vertical mode and forming a pleasing contrast with continuous window wall and glazing infill.

**Figure 22.27** Transamerica Pyramid, San Francisco, California. Architectural precast concrete column covers clad the complex structural steel framing at the building's base, providing beauty as well as protection from corrosion and fire.

### Column cover design

Design a two-story column cover unit for a high-rise office building in San Francisco. Use the Uniform Building Code. The column cover cross section and elevation are shown in Fig. 22.28; $f'_c$ = 5000 lb/in² hard-rock concrete.

Weight of cover:

$$A_c = (2)(3.08 - 0.37)(0.37) = 2.01 \text{ ft}^2$$

$$A_{\text{diaphragm}} = \frac{(2.33)^2}{2} = 2.72 \text{ ft}^2$$

$$V = (2.01)(27) + (2.72)(2)(0.42) = 57 \text{ ft}^3$$

$$W = (57)(0.150) = 8.6^k$$

$$w = \frac{8.6}{27} = 0.32^k/\text{ft}$$

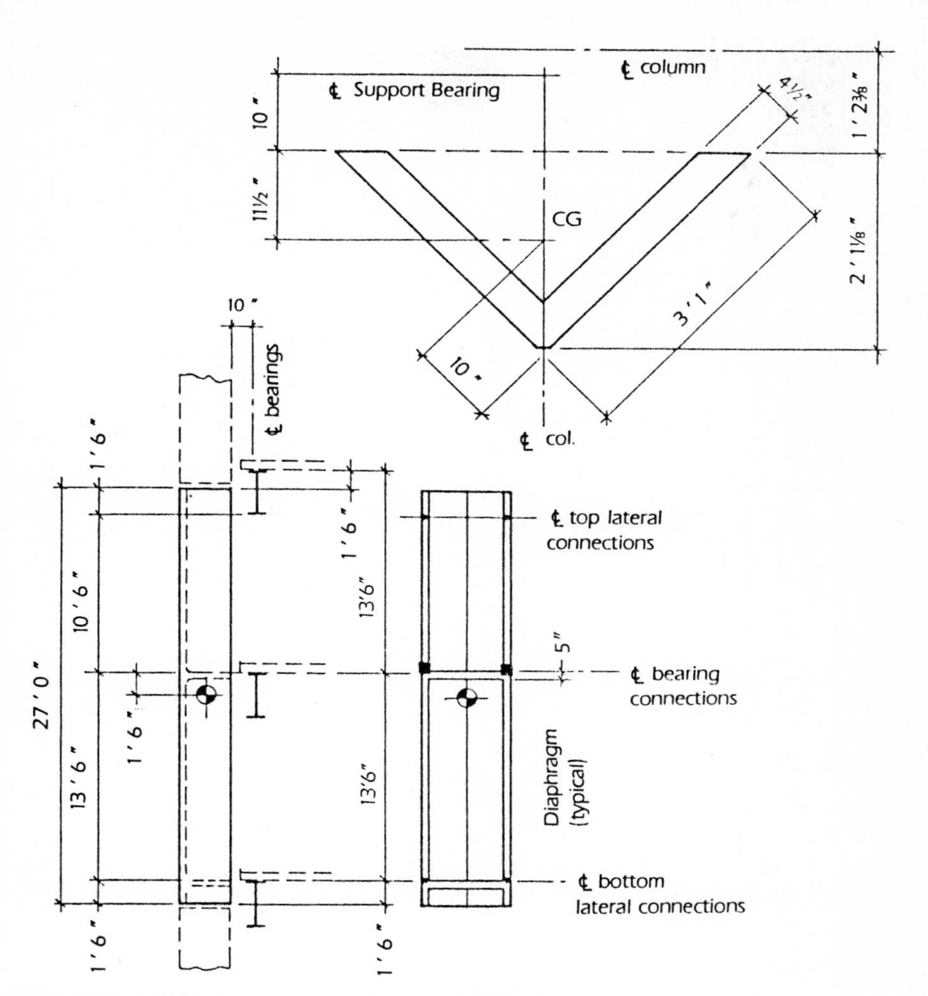


Figure 22.28 Column cover cross section and elevation.

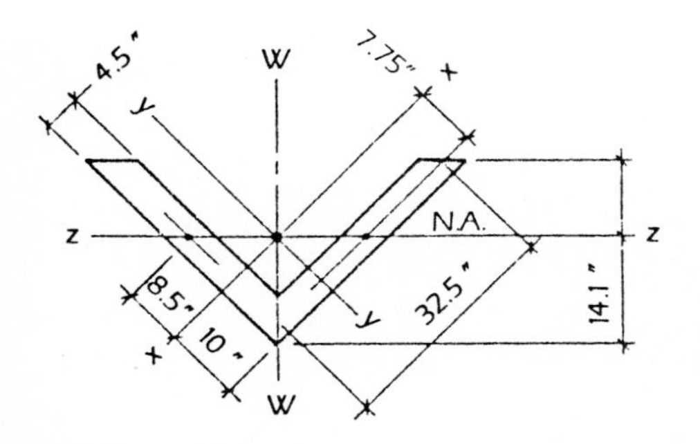


Figure 22.29 Section of column cover.

**Section properties.** Determine section properties (see Fig. 22.29). See *AISC Manual for Steel Construction*, 8th edition, page 6-25. (See also Figs. 22.30 and 22.31.)

$$I_x = I_y$$

$$= \left(\tfrac{1}{3}\right)\left[(4.5)(34.75 - 10)^3 + (34.75)(10)^3 - (30.25)(10 - 4.5)^3\right]$$

$$= \left(\tfrac{1}{3}\right)(68{,}224 + 34{,}750 - 5033) = 32{,}647 \text{ in}^4$$

$$K = \pm\frac{(30.25)(34.75)(30.25)(34.75)(4.5)}{(4)(34.75 + 30.25)} = \pm 19{,}125 \text{ in}^4$$

$$I_z = I_x \sin^2\theta + I_y \cos\theta + K \sin^2\theta$$

$$= (2)(32{,}647)(\sin 45°)^2 + (-19{,}125)(\sin 90°)$$

$$= 32{,}647 - 19{,}125 = 13{,}520 \text{ in}^2$$

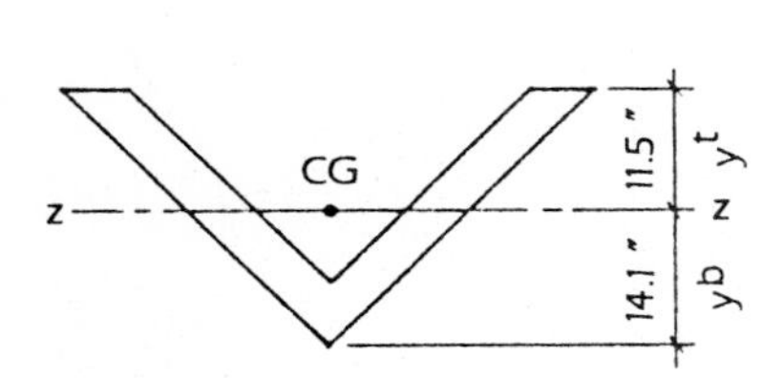


Figure 22.30 Stripping about $z$-$z$ axis.

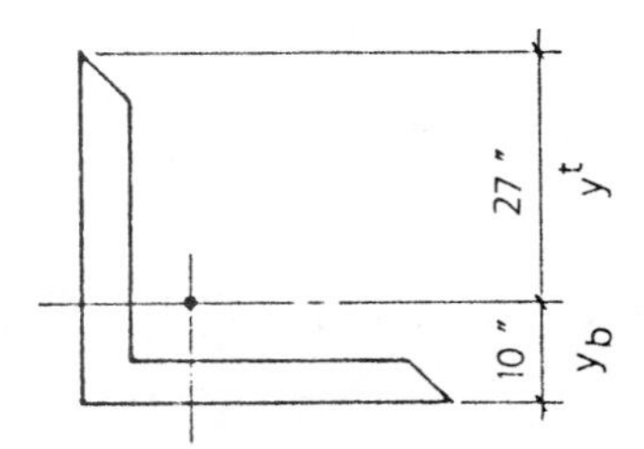


Figure 22.31 Storing and shipping about $x$-$x$ axis.

$$I_z = 13{,}520 \text{ in}^4$$

$$S_t = \frac{13{,}520}{11.5} = 1176 \text{ in}^3$$

$$S_b = \frac{13{,}520}{14.1} = 959 \text{ in}^3$$

$$I_x = 32{,}647 \text{ in}^4$$

$$S_t = \frac{32{,}647}{27} = 1209 \text{ in}^3$$

$$S_b = \frac{32{,}647}{10} = 3265 \text{ in}^3$$

**Stripping.** Use two-point pick at one-fifth points.

$$M = 1.5 \times 0.025 \times 0.32 \times 27^2 = 8.75^{\text{k}\cdot\text{ft}}$$

$$f_t = \frac{8.75 \times 12}{959} = 0.109 \text{ ksi} < 5\sqrt{f'_{ci}} \text{ (OK)}$$

$$A_s \text{ required} = \frac{8.75}{1.0 \times 20} = 0.44 \text{ in}^2$$

Therefore, use two no. 4 bars top and bottom.

**Shipping.** Design as follows.

$$M = 2.0 \times 0.025 \times 0.32 \times 27^2 = 11.66^{k \cdot ft}$$

$$f_t = \frac{11.66 \times 12}{1209} = 0.116 \text{ ksi} < 5\sqrt{f'_c} \text{ (OK)}$$

$$A_s \text{ required} = \frac{11.66}{1.44 \times 33} = 0.25 \text{ in}^2$$

Therefore, use one no. 4 bar + welded wire fabric (WWF) as required.

**Erection.** Try lifting at one end only.

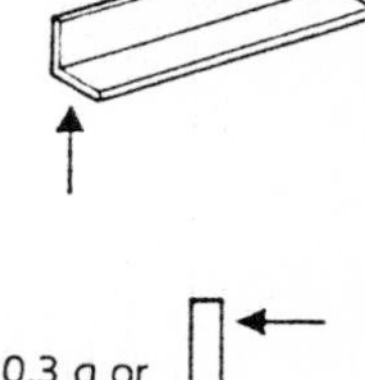

$$M = 1.25 \times 0.125 \times 0.32 \times 27^2 = 36.5^{k \cdot ft}$$

$$f_t = \frac{36.5 \times 12}{3265} = 0.134 \text{ ksi} < 5\sqrt{f'_c} \text{ (OK)}$$

$$A_s \text{ required} = \frac{36.5}{1.4 \times 33} = 0.77 \text{ in}^2$$

Therefore, use three no. 4 bars + WWF.

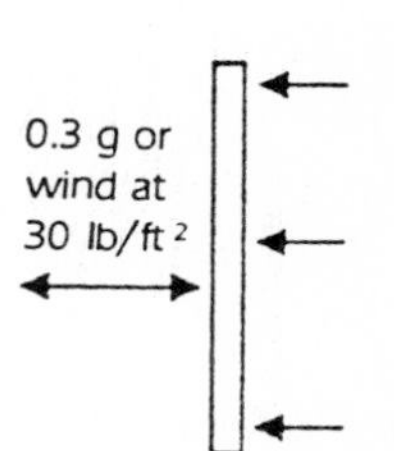


**Service load condition.** Design as follows.

$$\text{Seismic: } 0.3g = 0.3 \times 0.32 = 0.096^k/\text{ft}$$

$$\text{Wind: } 0.030 \times 4.3 = 0.129^k/\text{ft}$$

$$M = 0.125 \times 0.129 \times 13.5^2 = 2.94^{k \cdot ft}$$

Therefore, this is less critical than handling.
Temperature: $A_s = (0.0018)(4.5)(12) = 0.10 \text{ in}^2/\text{ft}$ each way.
Use $6 \times 6$ $W5.5 \times W5.5$ WWF ($A_s = 0.11 \text{ in}^2/\text{ft}$).

**Diaphragm reinforcing.** Design as follows.

$$A_s = (0.0025)(5)(12) = 0.15 \text{ in}^4$$

Use two no. 4 bars each way.

**Reinforcing summary.** Reinforcing is summarized in Fig. 22.32.

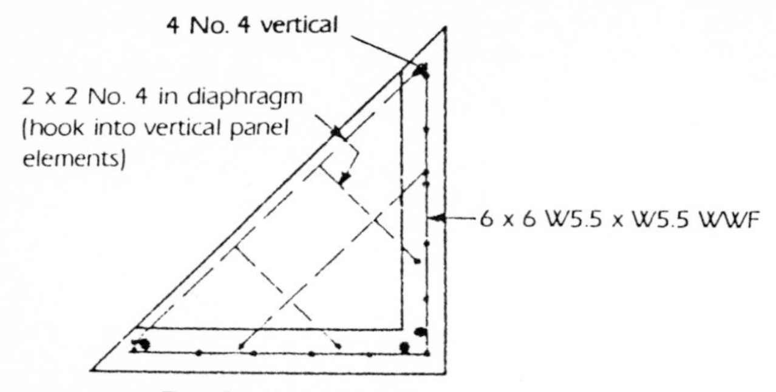


**Figure 22.32** Reinforcing summary.

**Connection design.** Distribution factors for loads to connectors are shown in Figs. 22.33 and 22.34.

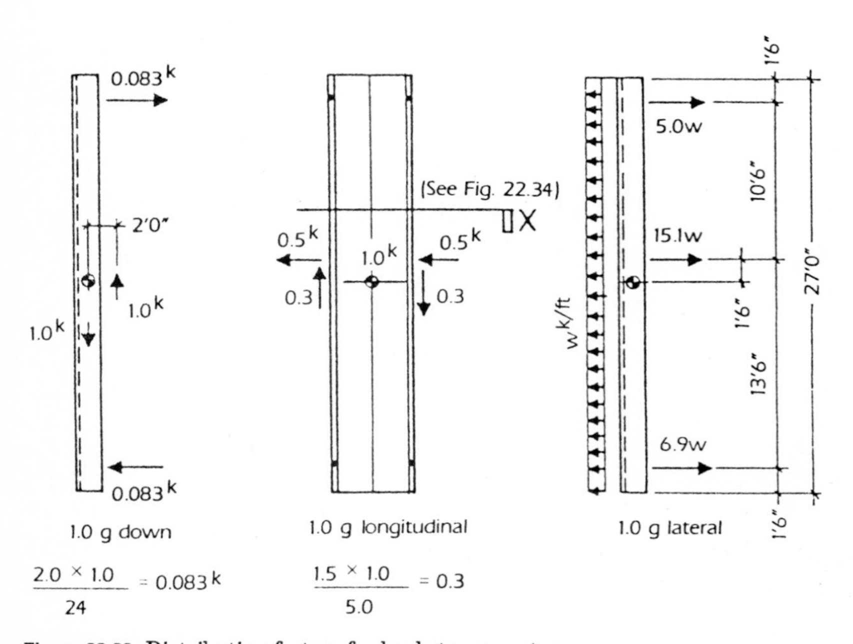


**Figure 22.33** Distribution factors for loads to connectors.

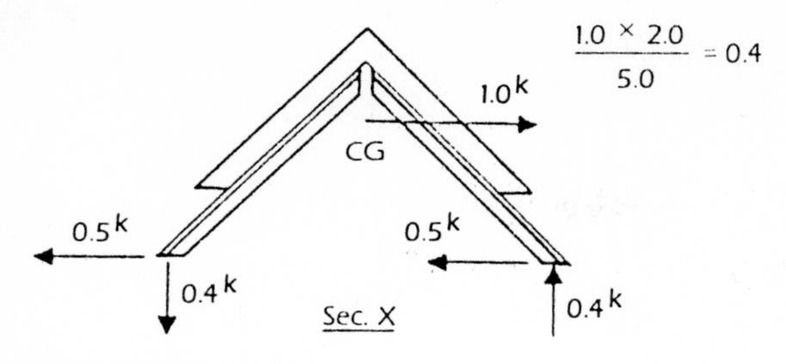


**Figure 22.34** Section $X$.

**Connector fastener; bottom and top lateral connection.** Design this fastener (see Fig. 22.35).

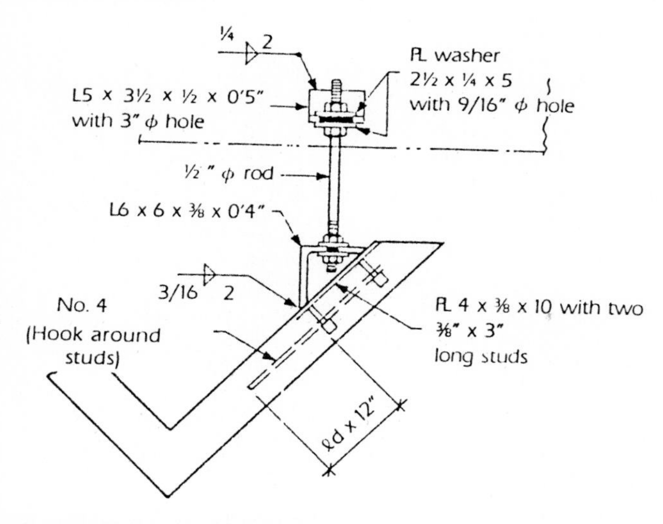


**Figure 22.35** Lateral connection.

$$W\ \left(1.0^{k}\ \text{down} + 1.2^{k}\ \text{in and out}\right) = P_{\text{critical}}$$

$$8.6^{k} \times 0.088 + 1.2 \times 6.9 \times 0.32^{k}/\text{ft} = 0.71 + 2.65 = 3.36^{k}$$

or $$\frac{3.36}{2} = 1.68^{k}/\text{connector}$$

Rod size:

$$A_{\text{required}} = \frac{1.68}{20} = 0.084\ \text{in}^2$$

Use $\frac{1}{2}$-in $\phi$ rod ($A_s$ = 0.139 in²).
Angle on panel:

$$M = \frac{P\!\!\!\!L}{4} = \frac{1.68 \times 6}{4} = 2.52^{\text{k}\bullet\text{in}}$$

$$S = \frac{M}{1.33F_b} = \frac{2.52}{22 \times 1.33} = 0.086 \text{ in}^3$$

By using 6-in L leg 4 in long,

$$1/6 \times 4 \times t^2 = 0.086$$

$$t = 0.36 \text{in}$$

Use L 6 × 6 × 3/8 × 0 ft 4 in.

**Insert ℄ in panel.** Design as follows.

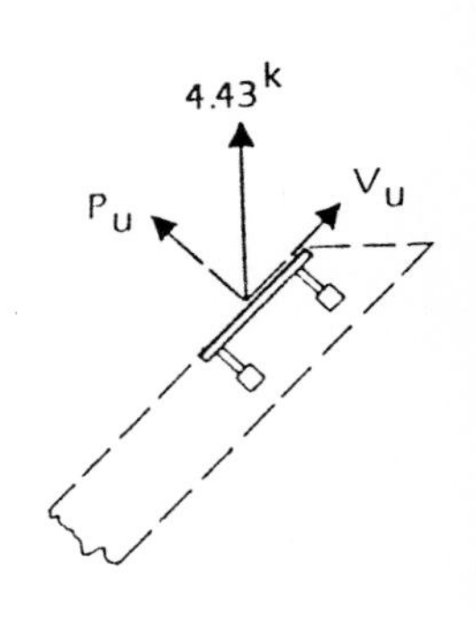


$$UP = 0.75(1.4D + 1.7L + 1.87E)$$

$$= (0.75)(1.4 \times 0.71 + 1.87 \times 2.65) = 4.46^{\text{k}}$$

or $\frac{4.46}{2} = 2.23^{\text{k}}$/ connector

$$UP = 0.9D + 1.43E = (0.9)(0.71)$$

$$+ (1.43)(2.65) = 4.43^{\text{k}}$$

or $\frac{4.43}{2} = 2.22^{\text{k}}$/ connector

$$V_u = P_u = 0.707 \times 2.23^{\text{k}} = 1.58^{\text{k}}$$

Try two $\frac{3}{8}$-in $\phi$ × 3-in-long studs ($d_e$ = 2 in). See *PCI Design Handbook*, pages 6-9, 6-53, and 6-53.

As governed by stud capacity:

$$\phi V_s = 5.0^{\text{k}}/\text{ stud}$$

$$\phi P_s = 6.0^{\text{k}}/\text{ stud}$$

$$\left(\frac{P_u}{P_s}\right)^2 + \left(\frac{V_u}{V_s}\right)^2 = \left(\frac{0.5 \times 1.58}{5.0}\right)^2 + \left(\frac{0.5 \times 1.58}{6.0}\right)^2 = 0.04 < 1.0 \text{ (OK)}$$

As governed by concrete capacity:

$$\phi V_c = 1.5^{\text{k}}/\text{ stud } \left(V_c = 1.76^{\text{k}}\right)$$

$$\phi P_c = 4.9^{\text{k}}/\text{ stud } (P_c = 5.76)$$

$$\frac{1}{\phi}\left[\left(\frac{P_u}{P_c}\right)^2 + \left(\frac{V_u}{V_c}\right)^2 = \left(\frac{0.5\times 1.58}{5.76}\right)^2 + \left(\frac{0.5\times 1.58}{1.76}\right)^2\right]$$

$$= 0.26 < 1.0\,(\text{OK})$$

Insert ℙ thickness: use 3/8-in ℙ × 4 in × 10 in.

$$\text{Weld: } \frac{1.68}{3\times 0.93} = 0.62 \text{ in}$$

Use two 3/16-in fillet welds each side and each end.

**Angle bracket to beam.** Design as follows.

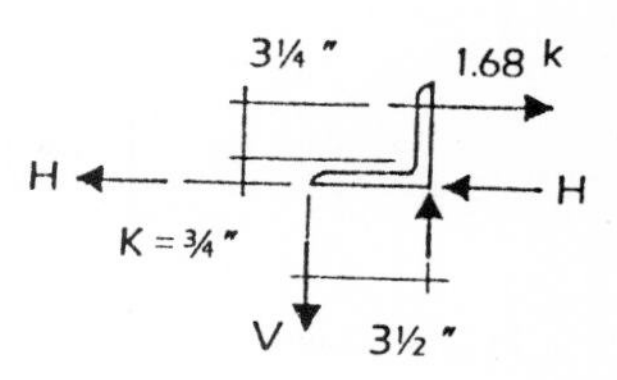


$$M = 1.68\times 3.25 = 5.46^{\text{k}\cdot\text{in}}$$

$$S = \frac{5.46}{22\times 1.33} = 0.19 \text{ in}^3$$

$$1/6\times 5\times t^2 = 0.19 \quad \therefore t = 0.48$$

Use L $5\times 3\,1/2\times 1/2$.

$$\text{Weld: } H = 0.84^{\text{k}} \quad V = 1.92^{\text{k}} \quad R = \sqrt{H^2 + V^2} = 2.1^{\text{k}}$$

By using 1/4-in welds, $L = 2.1/(4\times 0.93\times 1.33) = 0.4$ in.
Use 2-in-long welds centered on heel and toe of angle.

**Load-bearing connection.** Design connection (gravity plus longitudinal seismic govern). See Fig. 22.36.

**Bracket section.** Design moments.

$$M_v = 5.35\times 21 = 112^{\text{k}\cdot\text{in}}$$

$$M_H = 2.20\times 21 = 46^{\text{k}\cdot\text{in}}$$

$$S_{\text{required}} = \frac{112+46}{1.33\times 24} = 4.95 \text{ in}^3$$

Check condition with dead load only:

$$M_v = 4.30\times 21 = 90^{\text{k}\cdot\text{in}}$$

$$S_{\text{required}} = \frac{90}{24} = 3.75 \text{ in}^3$$

Use tube support $4\times 4\times 0.375$ ($S = 5.10$ in$^3$).

**Connector fasteners.** Design as follows. (Bracket reactions are shown in Figs. 22.37 and 22.38.)

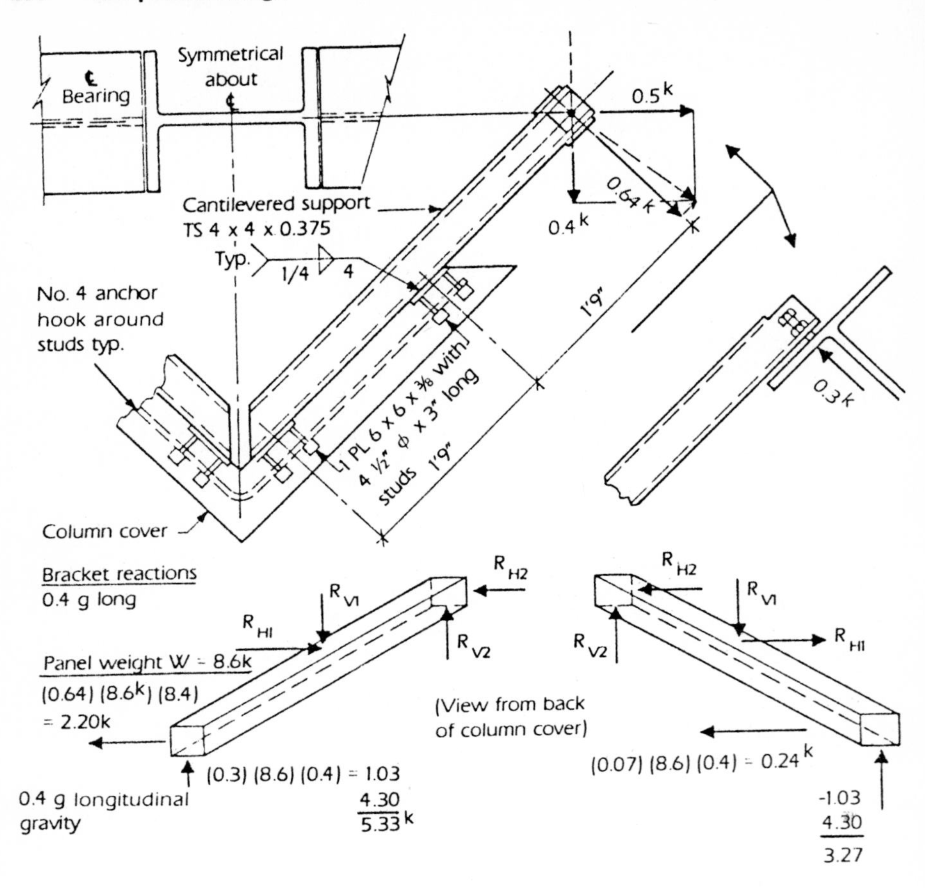


**Figure 22.36** Load-bearing connection; 0.4 $g$ longitudinal. $R = \sqrt{0.4^2 + 0.5^2} = 0.64^k$. $\tan\theta = 0.4/0.5; \therefore \theta = 39°$. $0.64\cos(45 - 39) \cong 0.64^k$. $0.64\sin 6° = 0.07^k$.

Left cantilever:

$$R_{V1} = \frac{42}{21} \times 5.33 = 10.66^k$$

$$R_{V2} = 10.66 - 5.33 = 5.33^k$$

$$R_{H1} = \frac{42}{21} \times 2.20 = 4.40^k$$

$$R_{H2} = 4.40 - 2.20 = 2.20^k$$

Right cantilever:

$$R_{V1} = \frac{42}{21} \times 3.27 = 6.54^k$$

$$R_{V2} = 6.54 - 3.27 = 3.27^k$$

$$R_{H1} = \frac{42}{21} \times 0.24 = 0.48^k$$

$$R_{H2} = 0.48 - 0.24 = 0.24^k$$

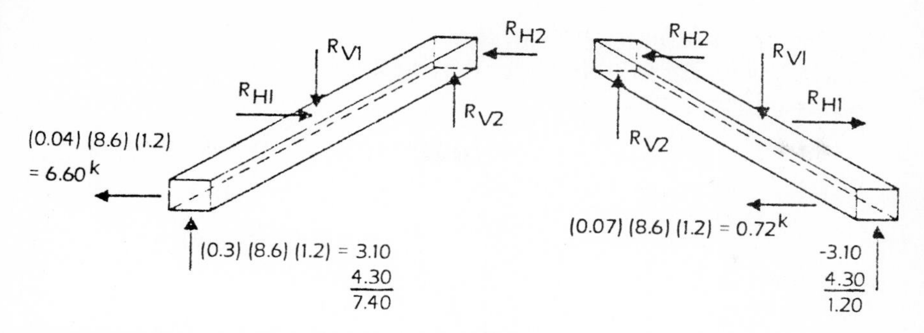


**Figure 22.37** Bracket reaction; 1.2 $g$ longitudinal.

Left cantilever:

$$R_{V1} = \frac{42}{21} \times 7.40 = 14.80^{k}$$

$$R_{V2} = 14.80 - 7.40 = 7.40^{k}$$

$$R_{H1} = \frac{42}{21} \times 6.6 = 13.20^{k}$$

$$R_{H2} = 13.20 - 6.60 = 6.60^{k}$$

Right cantilever:
Not critical

$$UV_{\text{critical}} = 0.75(14D + 1.7L + 1.87E)$$

$$= (0.75)\left(1.4 \times \frac{42}{21} \times 4.3 + 1.87 \times \frac{42}{21} \times 3.10\right) = 17.7^{k}$$

$$UH_{\text{critical}} = 0.75 \times 1.87 \times 13.20 = 18.5^{k}$$

Panel weight $W = 0.32^{k}$/ft.

$$P = (0.5)(15.1)\left(0.32^{k}/\text{ft}\right)(1.2) = 2.90^{k}$$

$$R_{H} = 0.707 \times 2.90 = 2.05^{k}$$

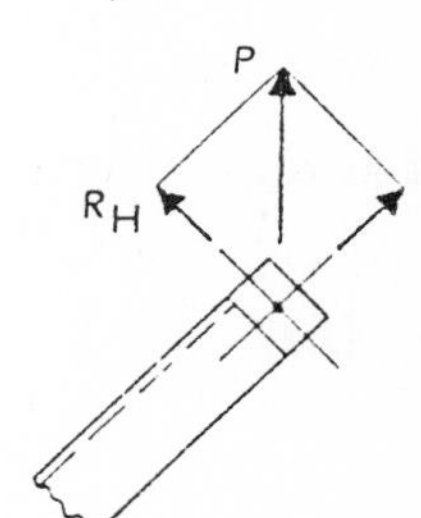


**Figure 22.38** Bracket reaction;1.2$g$ lateral.

Therefore, 1.2 $g$ lateral is not as critical as 1.2 $g$ longitudinal.

**Bracket insert ℙ's.** Design as follows.

$$UV = 17.7^{k} \quad UP = 18.5^{k}$$

Try 6 × 6 × 3/8 with four 1/2-in $\phi$ × 3-in-long studs (see Fig. 22.39). See *PCI Design Handbook*, pages 6-9, 6-52, and 6-53.

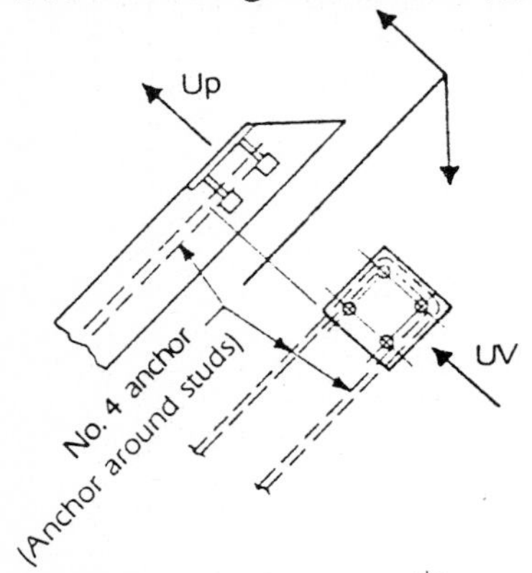


**Figure 22.39** Bracket insert ℙ's.

As governed by stud capacity:

$$\phi V_s = 8.8^k/\text{stud}$$

$$\phi P_s = 10.6^k/\text{stud}$$

$$\left(\frac{P_u}{\phi P_s}\right)^2 + \left(\frac{V_u}{\phi V_s}\right)^2$$

$$= \left(\frac{18.5}{4 \times 10.6}\right)^2 + \left(\frac{17.7}{4 \times 8.8}\right)^2$$

$$= 0.19 + 0.25 = 0.44 < 1 \quad \text{(OK)}$$

As governed by concrete capacity:

$\phi V_c$:

$$\left.\begin{array}{lll} 2 \text{ at } d_e = 2 \text{ in} & 2 \times 1.5^k = 3.0^k \\ 2 \text{ at } d_e = 5 \text{ in} & 2 \times 9.4^k = 18.8^k \end{array}\right\} 21.8^k = \phi V_c \quad V_c = \frac{21.8}{0.85} = 25.6^k$$

$\phi P_c$:

$$\left.\begin{array}{lll} 2 \text{ at } d_e = 2 \text{ in} & 2 \times 6.1^k = 12.2^k \\ 2 \text{ at } d_e = 5 \text{ in} & 2 \times 9.1^k = 18.2^k \end{array}\right\} 30.4^k = \phi P_c \quad P_c = \frac{30.4}{0.85} = 35.8^k$$

$$\frac{1}{\phi}\left[\left(\frac{P_u}{P_c}\right)^2 + \left(\frac{V_u}{V_c}\right)^2\right] = \frac{1}{0.85}\left[\left(\frac{18.5}{35.8}\right)^2 + \left(\frac{17.7}{25.6}\right)^2\right] = 0.88 < 1.0 \quad \text{(OK)}$$

Use ℙ 6 × 6 × 3/8 with four 1/2-in $\phi$studs 3 in long.

**Weld tube support to insert.** See Fig. 22.40.

$$\frac{14.8 \times 2}{4} = 7.4^k$$

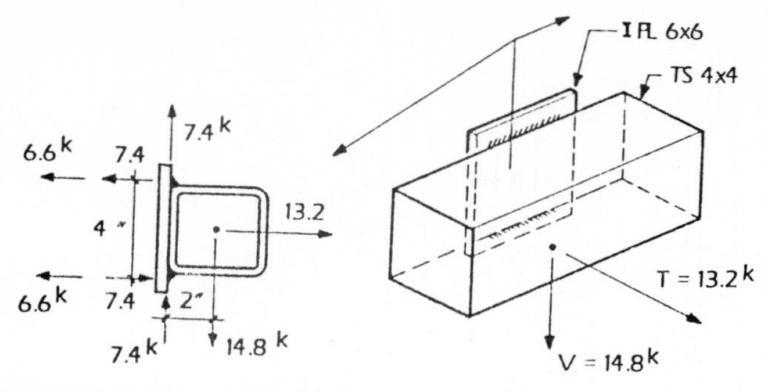


**Figure 22.40** Tube support welded to insert ℙ.

Resultant to top weld: $\sqrt{7.4^2 + 14^2} = 15.83^k$.
By using 1/4-in-welds,

$$L = \frac{15.83}{1.33 \times 4 \times 0.93} = 3.2\,\text{in}$$

Use 4-in, 1/4-in weld.

**Leveling bolt at bracket support.** Design as follows.

$$P = 4.3^k$$

Use 3/4-in $\phi$ bolt 4 in long. $\left(P_{\text{allow}} = 8.8^k\right)$.
See *AISC Manual for Steel Construction*, page 4-3.

**Weld at bracket support.** Handle as follows:

$$T = 6.6^k \quad L = \frac{6.6}{1.33 \times 0.93 \times 4} = 1.33$$

Use 3-in, 1/4-in weld each side with backing bar.

**Rebar anchor into slab.** See Fig. 22.41.

$$A_s = \frac{6.6^k}{20 \times 1.33} = 0.24\ \text{in}^2$$

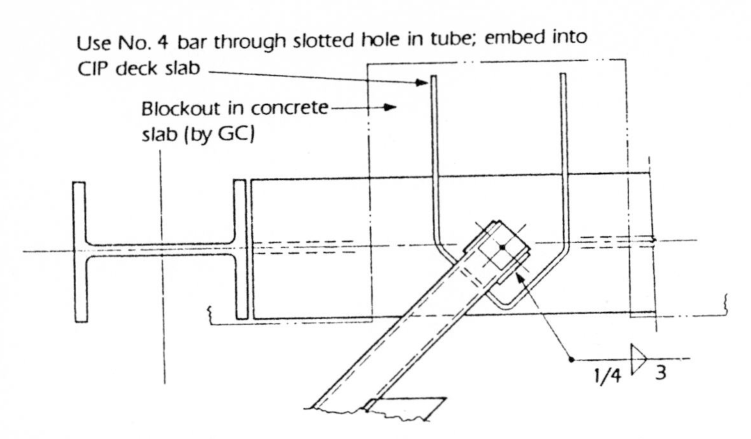


Figure 22.41 Rebar anchor embedded into slab.

## Window Panel Units (Pierced Wall Panels)

These units are used as either load-bearing or non-load-bearing cladding units for buildings, with the load-bearing applications normally limited to a four-story, or 50-ft, height. However, large panel building systems for housing have employed these elements as shear panels in seismic zones for many years in Europe. In California, window panel units have been used principally as cladding on multistory structures, with lateral and vertical load resistance being supplied by the building frame (see Figs. 22.42 and 22.43). The openings in the precast units are normally cast with a polyvinylchloride reglet, which serves to receive a zipper-type neoprene gasket and subsequent glazing in the field. Problems with trades on site have traditionally prevented glazing from being installed in the precast concrete fabricator's yard. Often, the shape of the panel is designed so that the projecting elements are used as sunshades, thus reducing solar heat gain and glare. Panel sizes are limited by the usual shipping and erection constraints. Multistory cladding elements with widths of 8 to 12 ft are subject to seismic drift limitations on the design of the top and bottom connectors. For this reason, in seismic zones the tendency is toward single-story, single-opening punched units. Multiwindow units are subject to shipping limitations and require a project large enough to justify special tilt truck frames to ship units up to 14 ft in width. Pierced panels are seldom prestressed owing to the difficulty in achieving concentric stressing and elimination of consequent bowing due to eccentric prestressing. The following example demonstrates the proper structural design of the reinforcing and connections of a typical single-opening window panel.

**Figure 22.42** 601 Montgomery, San Francisco, California. Pierced window panels are the traditional solution to high-rise exteriors in precast concrete. Uniformity of finish and a high degree of quality control are exemplified in these single-unit plant-cast cladding panels.

**Figure 22.43** 601 Montgomery, San Francisco, California. Only precision molds and factory-casting techniques can provide the degree of quality demonstrated by these precast concrete cladding units.

### Window panel design

Design an architectural precast concrete window panel for a high-rise office building in San Francisco. Use the Uniform Building Code. The panel cross section and elevation are shown in Fig. 22.44; 5000 lb/in$^2$ hard-rock concrete will be used for these pierced non-load-bearing cladding elements.

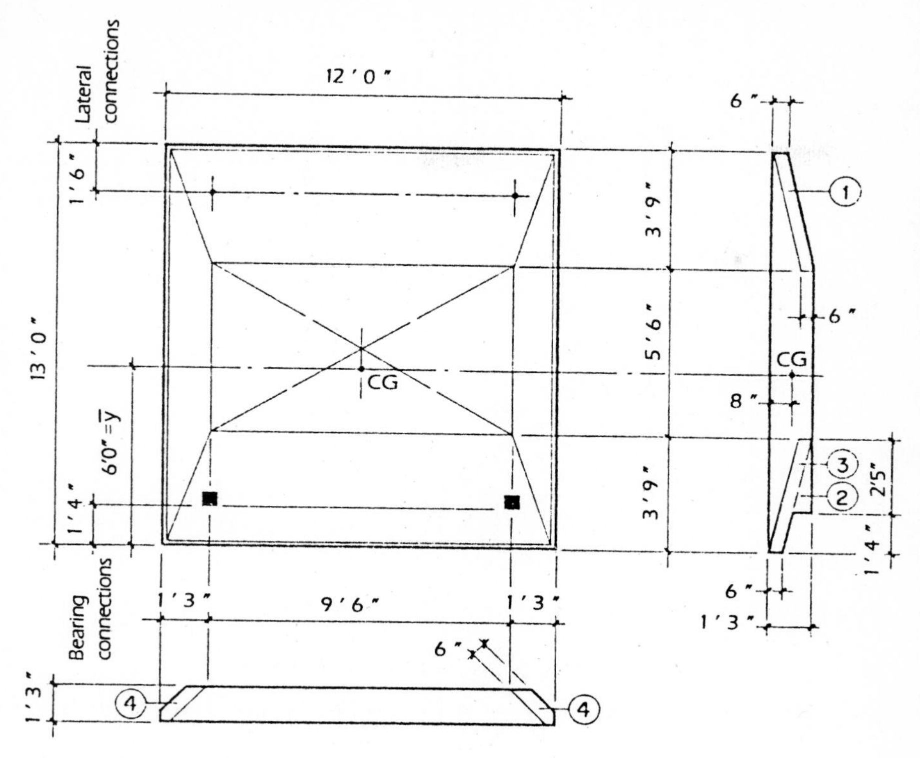


**Figure 22.44** Panel cross section and elevation.

**Center of gravity.** Locate the center of gravity.

| Part | | $A$ | × | $L$ | = volume | × | $y_{bot}$ | $= V_y$ |
|---|---|---|---|---|---|---|---|---|
| 1 | 6 × 45 | 270 | | 144 | 38,880 | | 11.13 | 432,734 |
| 2 | 1/2 × 6 × 29 | 87 | | 144 | 12,528 | | 2.14 | 26,810 |
| 3 | 6 × 45 | 270 | | 144 | 38,880 | | 1.88 | 73,094 |
| 4 | 2 × 6 × 29 | 228 | | 111 | 25,308 | | 6.50 | 164,502 |
| | Σ | | | | 115,596 | | | 697,140 |

$$\text{Weight} = \frac{115{,}596}{1728} \times 0.150 = 10.0^{k} \qquad y_{bot} = \frac{697{,}140}{115{,}596} = 6.08\,\text{ft}$$

**Handling.** Check handling.

Strip and final support points (see Fig. 22.45).

For section properties of vertical mullions, see *AISC Manual for Steel Construction*, 7th edition, page 6-25.

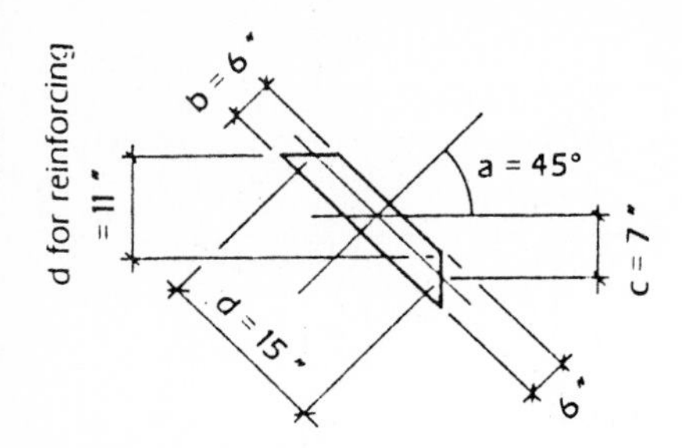


**Figure 22.45** Mullion cross section.

$$S = \frac{bd\left(b^2 \sin^2 a + d^2 \cos^2 a\right)}{6(b \sin a + d \cos a)}$$

$$S = \frac{(6)(15)\left[(6^2)(0.707^2) + (15^2)(0.707)\right]}{6[(6)(0.707) + (15)(0.707)]} = 131 \text{ in}^3$$

$I = 1.5 \quad A = 6 \times 15 = 90 \text{ in}^2$

$$\text{Span length} = 9 \text{ ft } 3 \text{ in} \quad w = \frac{90}{144} \times 0.150 = 0.094^{k}/\text{ft}$$

$$M = 1.5 \times 0.125 \times 0.094 \times 9.25^2 = 1.51^{k\cdot ft} \quad (18^{k\cdot in})$$

$$f_t = \frac{18}{131} = 0.137^{k}/\text{in}^2 < 5\sqrt{f'_{ci}} = 5\sqrt{2000} = 0.225 \quad (\text{OK})$$

Therefore, handling is not critical. To simpify off-loading at tight access points at the jobsite in midcity streets, it may be desirable to ship these panels vertically on easel trailers. This decision would also dictate the mode in which the panels would be stored.

**Reinforcing for stripping.** Quantity of reinforcing required to satisfy stripping:

$$A_s = \frac{1.51}{1.0 \times 11} = 0.14 \text{ in}^2$$

Therefore, use one no. 4 bar top and bottom.

**Service load condition.** Wind on vertical mullion governs.

$$w = 0.030 \times 6.0 = 0.180^{k}/\text{ft}$$

$$M = 0.125 \times 0.180 \times 9.25^2 = 1.93^{k}/\text{ft}$$

$$f_t = \frac{1.93 \times 12}{131} = 0.176^{k}/\text{in}^2 = 5\sqrt{f'_c} \quad (\text{OK})$$

$$A_s = \frac{1.93}{1.44 \times 11} = 0.12 \text{ in}^2$$

Therefore, use one no. 4 bar top and bottom.

**Temperature reinforcing.** Analyze as follows.

$$A_s = 0.0025 \times 90 \times 0.33 \text{ in}^2$$

Therefore, use one no. 4 bar top and bottom in mullion.
Sill: $A_s = 0.0025 \times 400 = 100 \text{ in}^2$.
Therefore, use six no. 4 bars ($A_s = 1.20$).
Head: $A_s = 0.0025 \times 270 = 0.68 \text{ in}^2$.
Therefore, use four no. 4 bars ($A_s = 0.80$).
Transverse: $A_s = 0.0025 \times 6 \times 12 = 0.18 \text{ in}^2$.
Therefore, use no. 4 bar at 12 in on center ($A_s = 0.20$).

**Reinforcing summary.** Reinforcing is summarized in Fig. 22.46.

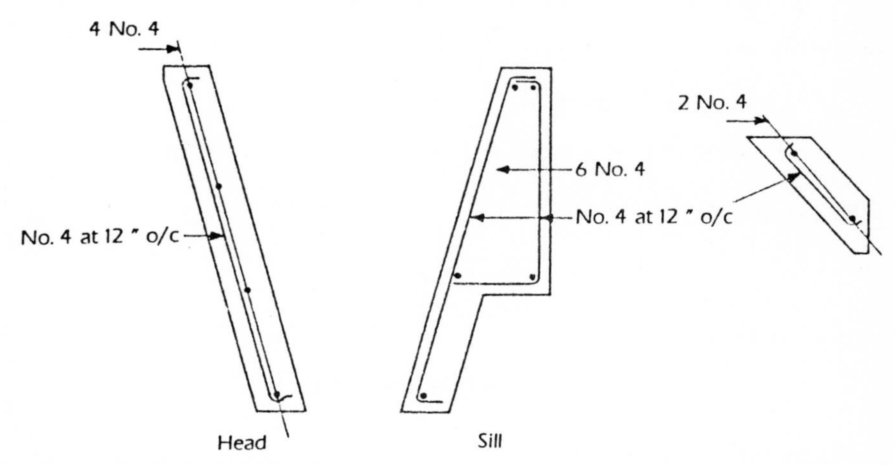


**Figure 22.46** Reinforcing summary.

**Connection design.** Connection design is illustrated in Figs. 22.47 and 22.48.

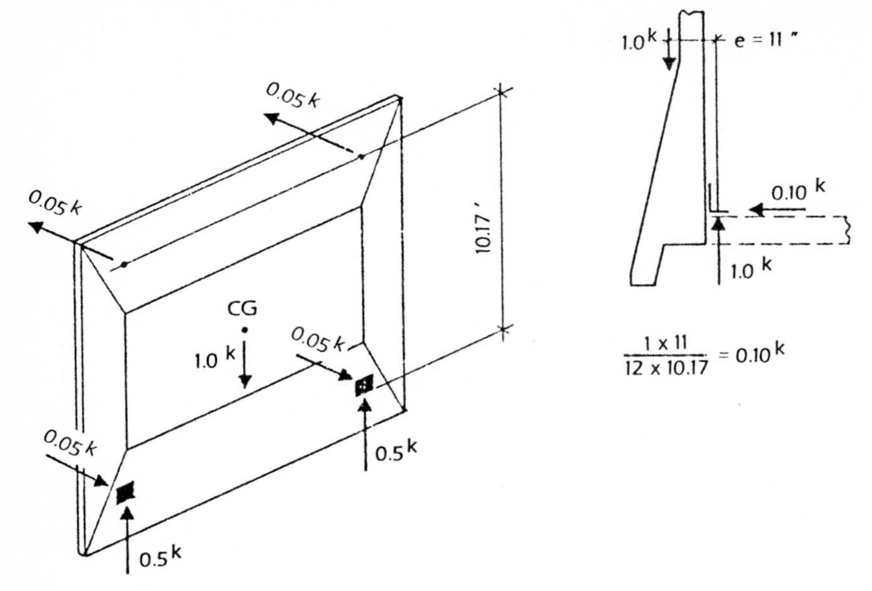


**Figure 22.47** Connection design: 1.0 $g$ down (gravity).

**Connector fastener; top lateral connection.** See Fig. 22.49.

$$W \left(1.0^{k} \text{ down} + 1.2^{k} \text{ in and out}\right) = P_{\text{critical}}$$

$$10.0(0.05 + 1.2 + 0.23) = 3.26^{k} \quad L_u = 16 \text{ in}$$

Rod size:

$$A_{\text{required}} = \frac{3.26}{20} = 0.163 \text{ in}^2$$

Use 3/4-in $\phi$ rod.

$$P_{\text{allow}} = 5.1^{k} \text{ compression;} \quad 8.2^{k} \text{ tension}$$

Insert ℙ in panel head:

$$UP = 0.75(1.4D + 1.7L + 1.87E)$$

$$= 0.75(1.4 \times 0.5 + 1.87 \times 1.2 \times 2.3) = 4.4^{k}$$

or

$$UP = 0.9D + 1.43E$$

$$= 0.9 \times 0.5 + 1.43 \times 1.2 \times 2.3 = 4.4^{k}$$

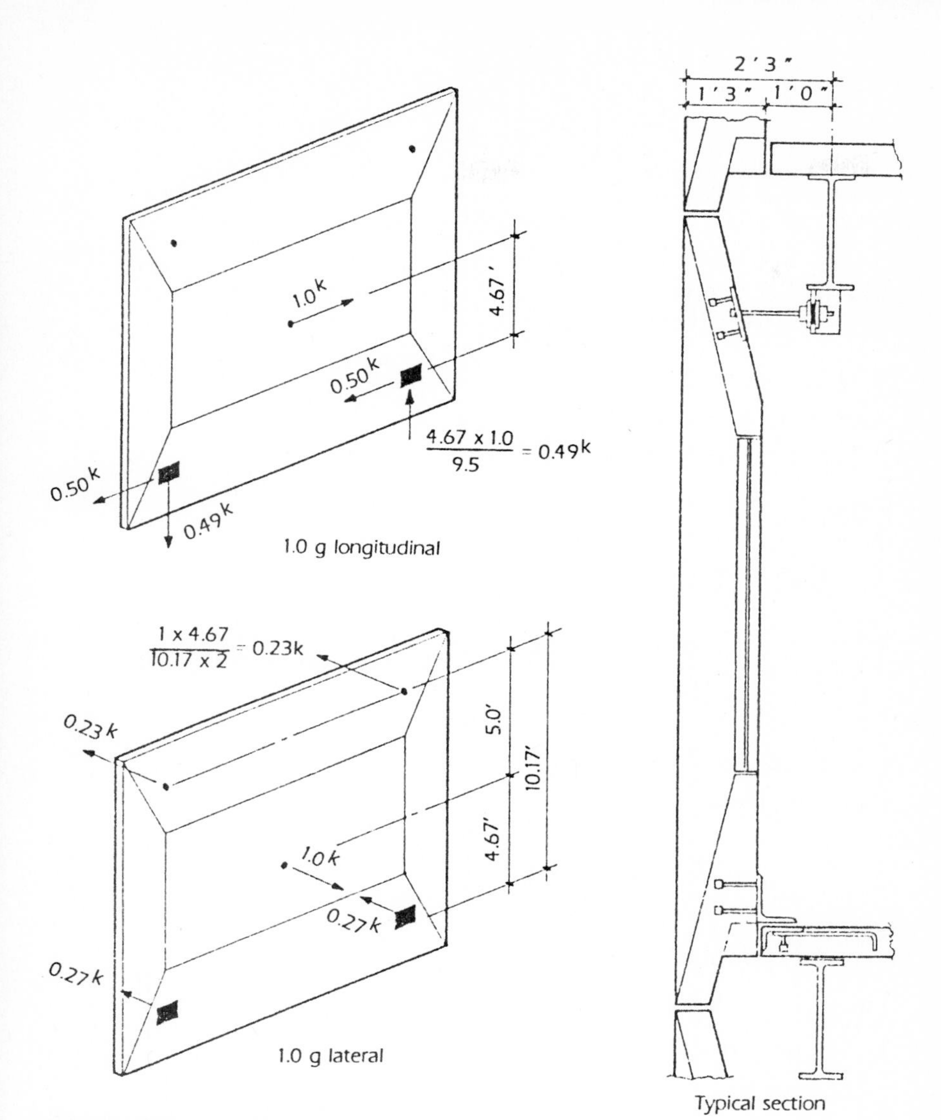


**Figure 22.48** Connection designs.

Try ℙ 6 ×3 ×3/8 with two 3/8-in ϕ ×4-in-long studs. (See *PCI Design Handbook*, 3d edition, pages 6-9, 6-52, and 6-53.)

$$\phi P_c = 2 \times 6.1 = 12.2^k$$

$$P_s = 2 \times 6 = 12^k$$

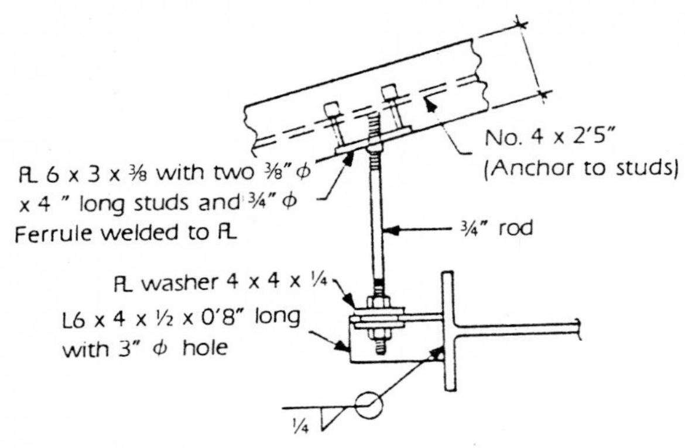


**Figure 22.49** Top lateral connection.

PL thickness required:

$$F_b = \frac{M}{S} = 0.9F_y = \frac{PL/8}{1/6bt^2}$$

$$(0.9)(36) = \frac{(3.26)(5)(0.125)}{(1/6)(3)(t^2)} \qquad t^2 = 0.126 \quad t = 0.35 \text{ in}$$

Use 3/8-in PL.
Punching shear:

$$v_{bo} = \frac{C}{b_o d} = \frac{4.4}{(0.85)(42)(6)} = 0.021 \text{ ksi} < 4\sqrt{f'_c} \quad \text{(OK)}$$

**Connector body at WF spandrel.** See Fig. 22.50.

$$W \ (1.0\,g \ \text{down} + 0.4\,g \ \text{in and out})$$

$$= 10.0(0.05 + 0.4 \times 0.23) = 1.42^{\text{k}}$$

Try L $6 \times 4$ with 3-in ϕ hole.
L thickness required:

$$F_b = \frac{M}{S}$$

$$1.33 \times 22 = \frac{1.42 \times 4.5}{1/6 \times 5 \times t^2}$$

$$t^2 = 0.26 \quad t = 0.51 \text{ in}$$

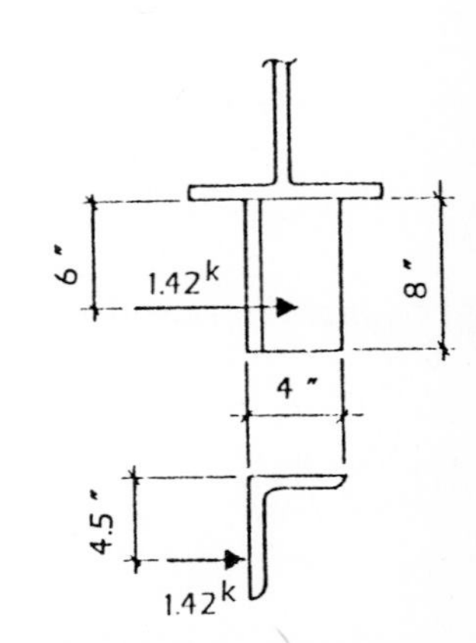


**Figure 22.50** Connector body.

Use L 6 × 4 × 1/2 × 0 ft 8 in long with 3-in ϕ hole.
Weld required to bottom flange of wide-flange (WF) spandrel:

$$W \ (1.0\, g \text{ down} + 1.2\, g \text{ lateral}) = 3.26^{k}$$

$$M = (3.26)(6) = 19.56^{k\cdot in}$$

Section modulus of weld group:

| Part | L | X | Lx |
|---|---|---|---|
| 4 in | 8 | 2 | 16 |
| 6 in | 12 | 0 | 0 |
| Σ | 20 | | 16 |

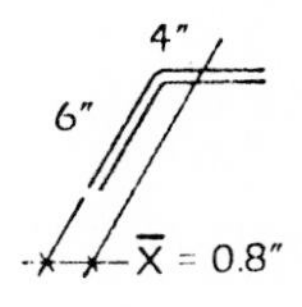


$$x = \frac{16}{20} = 0.8 \text{ in}$$

$$S_{min} = \frac{30}{3.2} = 9.4 \text{ in}^2$$

| Part | $L_o$ | $Ld^2$ |
|---|---|---|
| 4 in | 10.7 | 11.5 |
| 6 in | 0 | 7.7 |

$$\Sigma I_x = 30 \text{ in}^3$$

$$f_{max} = \frac{V}{A} + \frac{M}{S} = \frac{3.26}{20} + \frac{19.56}{9.4} = 2.24^{k}/\text{in}$$

Use ¼-in weld$(0.93 \times 4 = 37.2^{k}/\text{in})$.

**Bearing connection.** Design bearing connection (see Fig. 2.51).

$$\text{Force to connector body} = W \ (1.0\, g \text{ down} + 0.4\, g \text{ longitudinal})$$
$$= 10.0(0.50 + 0.4 \times 0.49) = 6.96^{k}$$

Use L 8× 6× 3/4.
Determine angle length required.

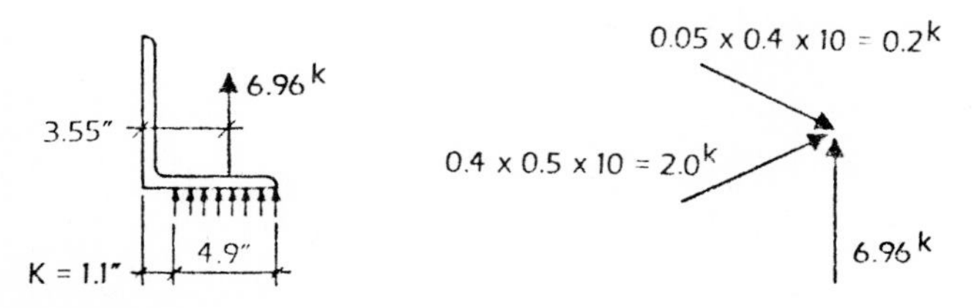


**Figure 22.51** Bearing connection.

$$M = (6.96)\left(\frac{4.9}{2}\right) = 17.05^{\text{k}\cdot\text{in}}$$

$$f_b = \frac{M}{S} = \frac{17.05}{1/6 \times L \times (0.75)^2} = 28.7 \quad \therefore L = 6.34 \text{ in}$$

Use L $8 \times 6 \times 3/4 \times 0$ ft 8 in long.

**Connection to panel.** Design connection (see Fig. 22.52).

$$UP = 0.75(1.4D + 1.7L + 1.87E)$$

$$= 0.75(1.4 \times 0.5 + 1.87 \times 1.2 \times 0.49)(10) = 13.5^{\text{k}}$$

or

$$UP = 0.9D + 1.43E = (0.9 \times 0.5 + 1.43 \times 1.2 \times 0.49)(10)$$

$$= 12.9^{\text{k}}$$

$$UH = 1.4 \times 0.05 \times 10 = 0.7^{\text{k}}$$

$$UV = 1.4 \times 0.50 \times 1.2 \times 1.0 = 8.4^{\text{k}}$$

Try four studs.

1. Effect of $UP = 13.5^{\text{k}}$.

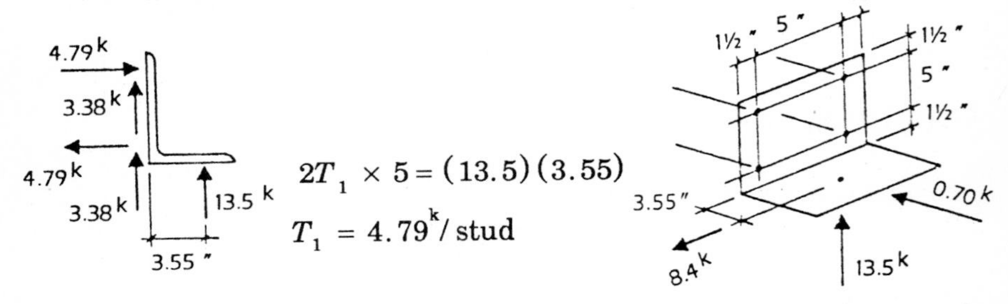


$$2T_1 \times 5 = (13.5)(3.55)$$

$$T_1 = 4.79^{\text{k}}/\text{stud}$$

**Figure 22.52** Connection to panel.

$$R_{yv} = \frac{13.5}{4} = 3.38^{\text{k}}/\text{stud}$$

2. Effect of eccentric shear; $UV = 8.4^{\text{k}}$. See Fig. 22.53.
   Due to bending ($e = 3.55$ in):

$$2T_2 \times 5 = (8.4)(3.55)$$

$$T_2 = 2.98^{\text{k}}/\text{stud}$$

Due to shear:

$$R_{xy} = \frac{8.4}{4} = 2.1^k/\text{stud}$$

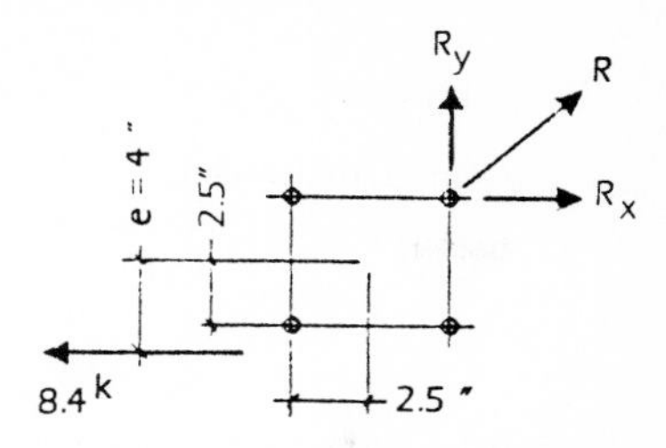


**Figure 22.53** Effect of eccentric shear. $\Sigma x^2 + y^2 = (4)(2.5)^2(2) = 50$.

Due to torsion:

$$R_{xt} = \frac{Pey}{\Sigma x^2 + y^2} = \frac{(8.4)(4)(2.5)}{50} = 1.68^k/\text{stud}$$

$$R_{yt} = \frac{Pex}{\Sigma x^2 + y^2} = \frac{(8.4)(4)(2.5)}{50} = 1.68^k/\text{stud}$$

$$R = \sqrt{(R_{xv} + R_{xT})^2 + (R_{yv} + R_{yT})^2}$$

$$= \sqrt{(2.1+1.68)^2 + (3.38 + 1.68)^2}$$

$$= 6.31^k/\text{stud}$$

The critical design condition is

$$V_u = 6.31^k/\text{stud}$$

$$P_u = 4.79 + 2.98 = 7.77^k/\text{stud}$$

Try four 5/8-in $\phi$ studs 6 in long.

As governed by stud capacity (see *PCI Design Handbook*, 3d edition, pages 6-9, 6-52, and 6-53):

$$\phi P_s = 16.6^k/\text{stud} \qquad V_s = 13.8^k/\text{stud}$$

$$\left(\frac{P_u}{\phi P_s}\right)^2 + \left(\frac{V_u}{\phi V_s}\right)^2 = \left(\frac{7.77}{16.6}\right)^2 + \left(\frac{6.31}{13.8}\right)^2$$

$$= 0.22 + 0.21 = 0.43 < 1.0 \quad \text{(OK)}$$

As governed by concrete capacity:

$$d_e = 8 \text{ in (shear)} \quad d_e = 2.5 \text{ in (pullout)} \quad f'_c = 5000 \text{ lb/in}^2$$

$$\phi P_c = 13.0^{k}/\text{stud} \quad P_c = \frac{13.0}{0.85} = 15.3^{k}/\text{stud}$$

$$\phi V_c = 14.7^{k}/\text{stud} \quad V_c = \frac{14.7}{0.85} = 17.3^{k}/\text{stud}$$

$$\frac{1}{\phi}\left[\left(\frac{P_u}{P_c}\right)^2 + \left(\frac{V_u}{V_c}\right)^2\right] = \frac{1}{0.85}\left[\left(\frac{7.77}{15.3}\right)^2 + \left(\frac{6.31}{17.3}\right)^2\right] = 0.46 < 1.0 \text{ (OK)}$$

Therefore, use four 5/8-in $\phi$ studs 6 in long.

**Design weld of bearing angle to insert ℄ in floor slab.** See Fig 22.54. 1.2 $g$ longitudinal governs.

$$V = 1.2 \times 0.5 \times 10 = 6.0^{k}$$

$$M = 6 \times 6 = 36^{k \cdot in}$$

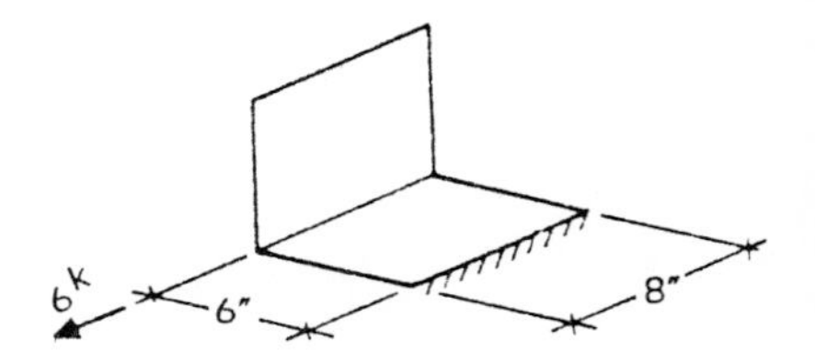


**Figure 22.54** Weld of bearing angle to insert ℄. Note that loading eccentricity may require the use of a channel or angle section in lieu of the flat bar shape shown.

Section modulus of weld:

$$S = 1/6 \times 8^2 = 10.67 \text{ in}^2$$

$$f = \frac{V}{L} + \frac{M}{S} = \frac{6}{8} + \frac{36}{10.67} = 4.12^{k}/\text{in}$$

$$\text{Weld } D = \frac{4.12}{0.93} = 4.4$$

Use 5/16-in weld.

**Bearing connection detail.** Bearing connection is shown in detail in Fig. 22.55.

## Panel Erection

In Figs. 22.56, 22.57, and 22.58 panels are being erected at 601 Montgomery, San Francisco.

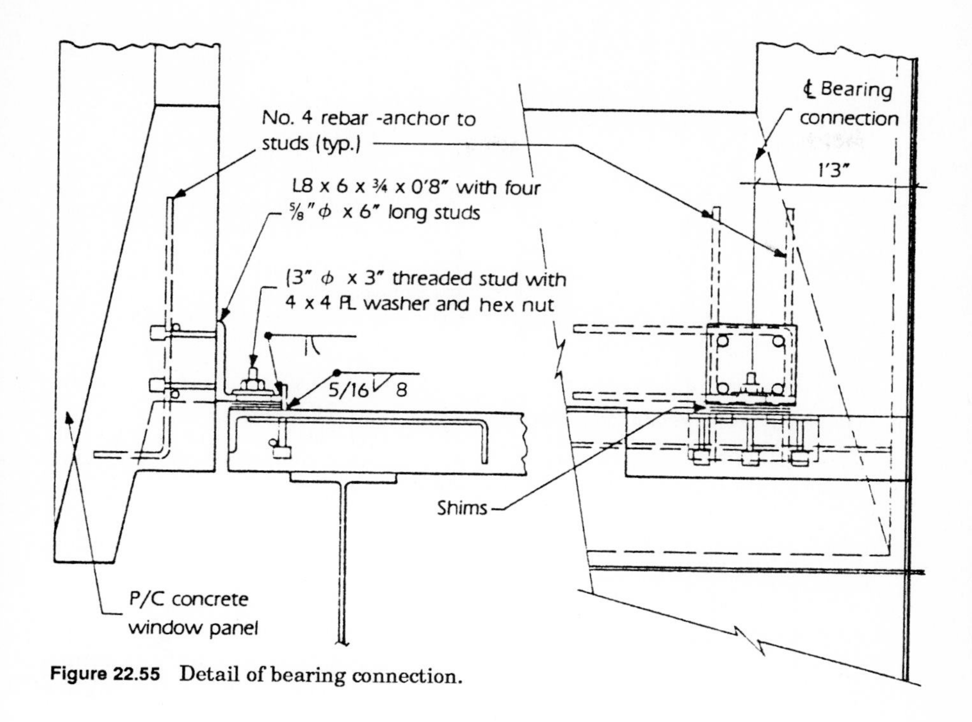

**Figure 22.55** Detail of bearing connection.

**Figure 22.56** Panel being lowered into position by roof-mounted tower crane.

Figure 22.57 An iron worker adjusts the top in-and-out connection as the panel is aligned.

Figure 22.58 Bottom bearing connection with an oversized hole to allow for the field-placing tolerance required for positioning the threaded stud.

## Solid Wall Panels: Narrow, High Panels.

Solid wall panels may be designed as cladding elements or as load-bearing, shear-resisting elements in the building (see Fig. 22.60). They also lend themselves to design as prestressed concrete elements and may be cast in the form of standard structural shapes such as single or double T's. Sandwich panel design is very advantageous, especially with the high cost of energy required to heat and cool building interiors. This design aspect of solid panels is covered in Chap. 23. Panels may be interconnected horizontally and vertically when they are used to transmit vertical and lateral loads to another panel element or to the foundation. When they are used as cladding elements, the joints between elements are left open, with subsequent caulking installed in the field. They may be used to enclose an entire wall or be spaced with areas of fenestration in between. Normally, these panels are placed vertically one or two stories high, with 8 ft being used as a common panel width and building module. Units 8 ft wide may be shipped without special permits.

Two aspects of design of non-load-bearing high wall panels merit additional explanation. First, the thickness of these solid, relatively long units is such that it is usually more economical to store and ship the units edgewise so as to negate the requirement for having additional reinforcement or prestressing required only for controlling stresses associated with flat storage and shipment. The other unusual aspect of the design of high panels is that the top lateral connections (the bottom connections are usually always the bearing connections) must be designed to restrain the panel in the longitudinal direction as well as in the lateral (in-and-out) direction. However, this top connection must also be designed to accommodate building drift under extreme seismic conditions (see Fig. 22.59). This is achieved by designing the connector body to behave elastically at $0.4g$ but to yield at forces approaching $1.2g$. The following example demonstrates these items.

**Figure 22.59** Cutter Laboratories, Berkeley, California. Parapet panels, spandrels, column cover panels, and 24-ft-high unbraced wall panels completely clad this steel-framed building. Permanence, speed of construction, and attractive appearance, as well as low initial cost, influence owners to select precast concrete cladding units.

**Figure 22.60** Alvarado sewage treatment plant, Union City, California. These wall panel units transmit seismic lateral forces to the foundation, as well as providing support for the double-T roof.

### Narrow, high wall panel design

Design a high cladding panel for the exterior of a steel-framed laboratory building in Berkeley, California. Use the Uniform Building Code. No intermediate support points are available, and the top connections must be designed to resist the effect of seismic forces parallel to as well as perpendicular to the plane of the panel; $f'_c$ = 5000 lb/in² hard-rock concrete. Panel cross section and elevation are shown in Fig. 22.61.

**Handling.** For handling calculations use

$$A = 5 \times 12 = 60 \text{ in}/\text{ft}^2$$

$$S = 1/6 \times 12 \times 5^2 = 50 \text{ in}^3/\text{ft}^2$$

$$\text{Weight} = \frac{5.5}{12} \times 0.50 = 0.070^k/\text{ft}^2$$

Panel weight:

$$(0.07)(8)(24) = 13.5^k \ (W)$$

or

$$\frac{15.5}{24} = 0.56^k/\text{ft} \ (W)$$

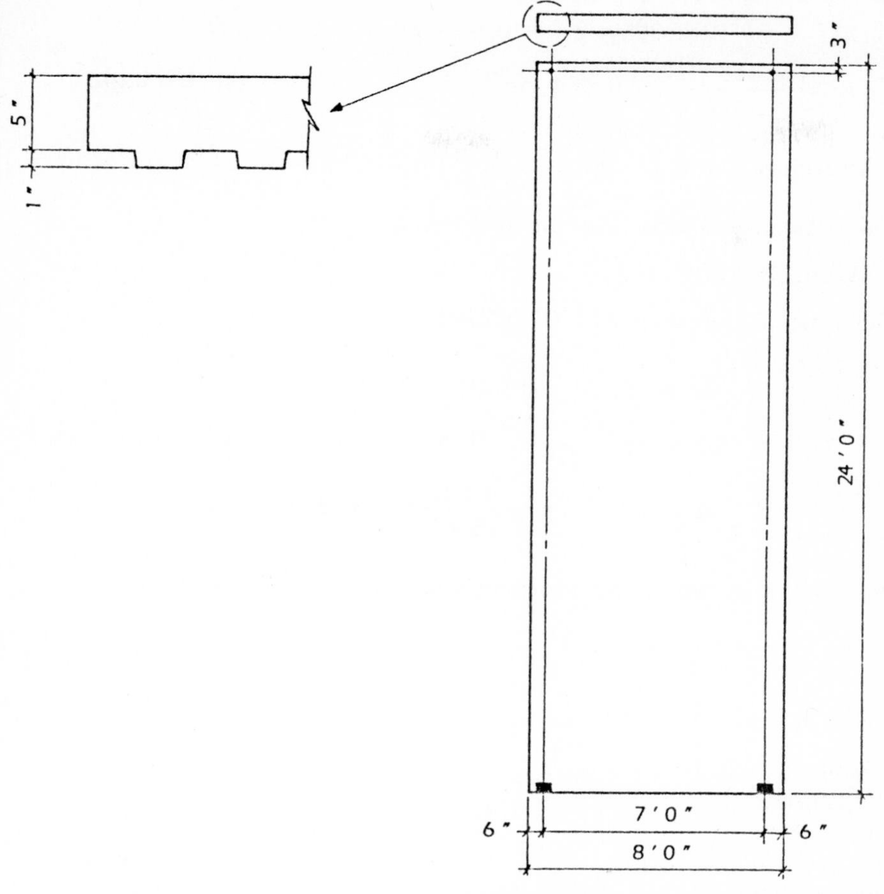


**Figure 22.61** Panel cross section and elevation.

**Stripping.** Try four-point pick.

$$M^{+} = M^{-} = 0.0056 \times 1.5 \times 0.07 \times 24^{2} = 0.339^{\text{k}\cdot\text{ft}}/\text{ft}$$

$$f_t = \frac{M}{S} = \frac{0.339 \times 12}{50} = 0.081^{\text{k}}/\text{in}^2 \quad (\text{OK})$$

$$A_s \text{ required} = \frac{0.339}{(1)(2.5)} = 0.136 \text{ in}^2/\text{ft}$$

or $\quad 8 \times 0.136 = 1.088 \text{ in}^2$ per 8-ft panel

Required: $4 \times 4$ $W4.0 \times W4.0$ + two no. 3 bars longitudinal ($A_s = 1.18$ in$^2$).

**Shipping.** Try two points flat.

L/10 3L/10 2L/10 3L/10 L/10

$$M^+ = M^- = 0.025 \times 2.0 \times 0.07 \times 24^2 = 2.016^{\text{k}\cdot\text{ft}}/\text{ft}$$

$$f_t = \frac{2.016 \times 12}{50} = 0.484^{\text{k}}/\text{in}^2 > 5\sqrt{f'_c} = 0.350 \text{ ksi (too high)}$$

Therefore store and ship panel vertically.

**Final design condition.** Handle as follows:

Wind: 15 lb/ft²

Seismic: $0.3\,g = (0.3)(70) = 21 \text{ lb/ft}^2$ $\therefore$ governs.

$$M = 0.125 \times 0.021 \times 24^2 = 1.512^{\text{k}\cdot\text{ft}}/\text{ft}$$

$$f_t = \frac{1.512 \times 12}{50} = 0.363 \simeq 5\sqrt{f'_c} \text{ (OK)}$$

$$A_s \text{ required} = \frac{1.512}{1.76 \times 2.5} = 0.344 \text{ in}^2/\text{ft}$$

**Reinforcing summary.** Reinforcing is summarized in Fig. 22.62.

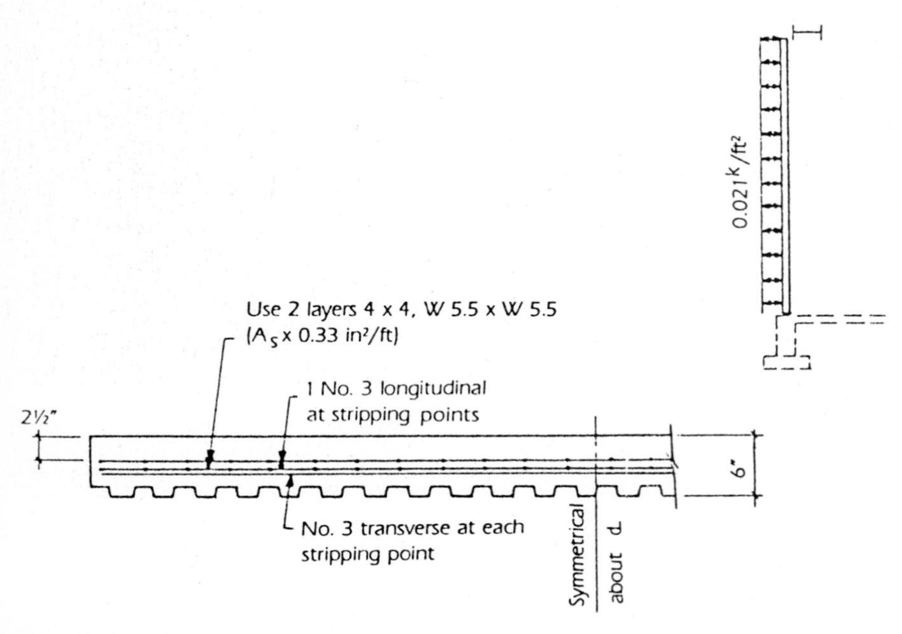


**Figure 22.62** Reinforcing summary.

**Schematic handling summary.** Handling is summarized schematically in Figs. 22.63, 22.64, 22.65, and 22.66.

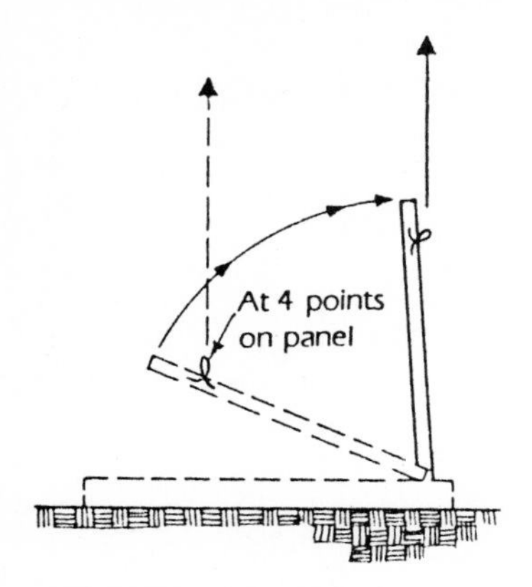


**Figure 22.63** Stripping: tilt up at four stripping points.

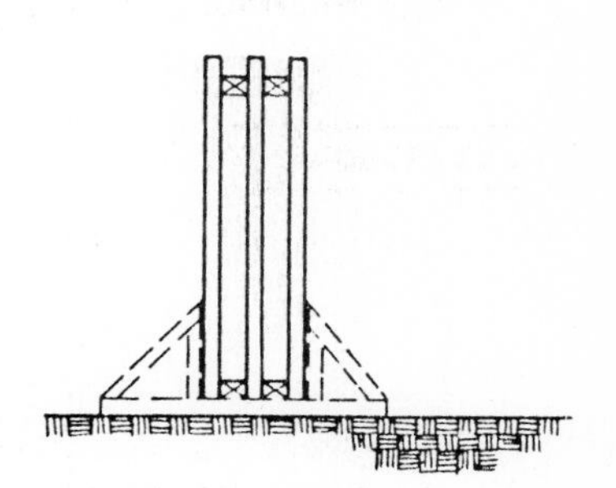

**Figure 22.64** Storage: store vertically; handle at side inserts.

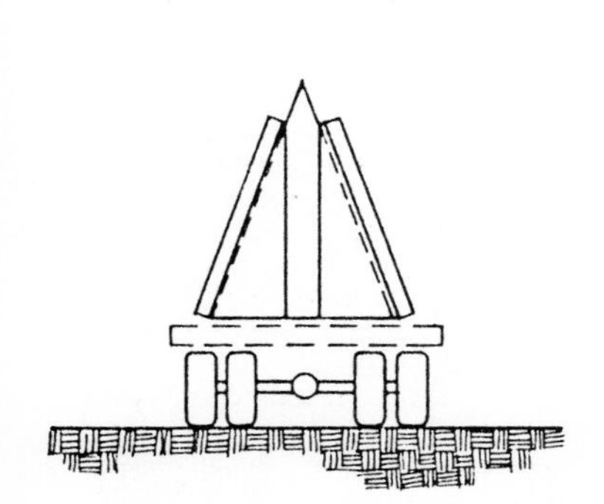

**Figure 22.65** Shipping: easel trailer.

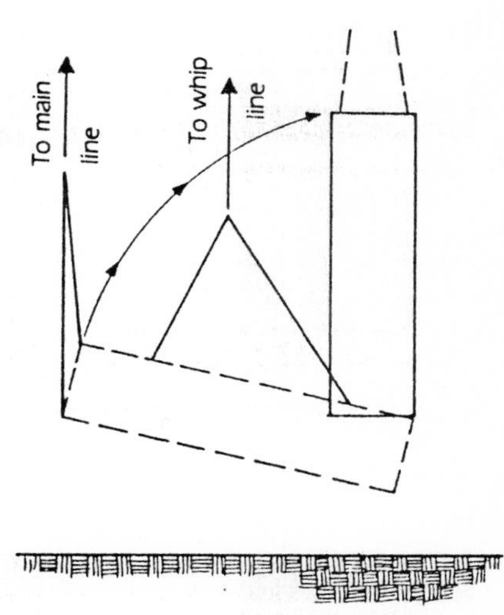


**Figure 22.66** Erection: tilt up as shown.

**Connection design.** See Fig. 22.67. Design top and bottom connections to behave elastically under 0.4 $g$ longitudinal loading but to yield and absorb energy at 1.2 $g$ loading.

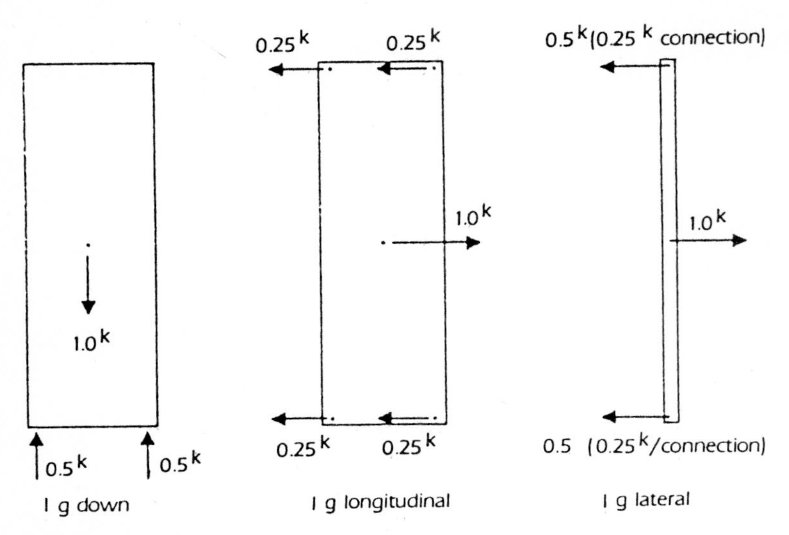


**Figure 22.67** Connection design.

**Bottom Bearing Connection.** Design as shown in Fig. 22.68.

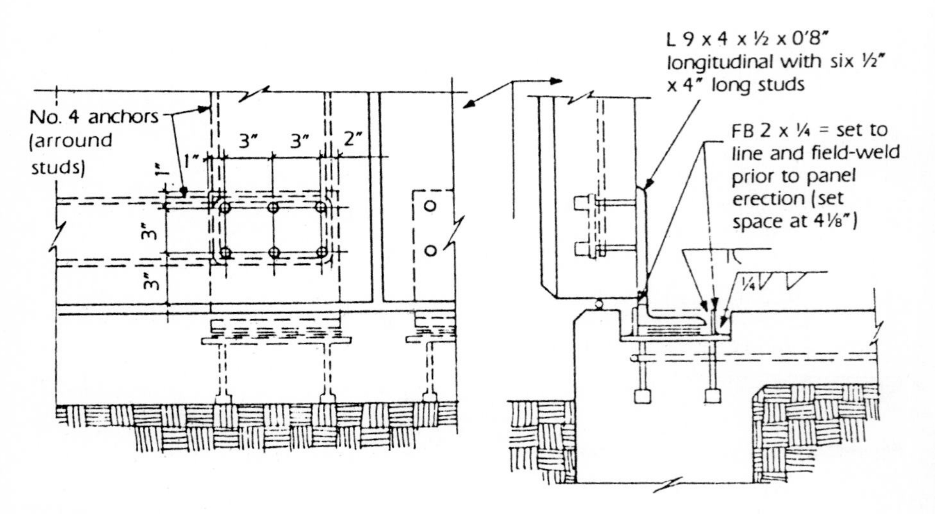


**Figure 22.68** Bottom bearing connection.

$$UP = 1.4 \times 0.5 \times 13.5^{k} = 9.45^{k} \ (1.0\,g \ \text{vertical})$$

$$UH = 1.4 \times 0.25 \times 13.5^{k} \times 1.2 = 5.67^{k} \ (1.2\,g \ \text{longitudinal})$$

1. Effect of $UP = 9.45^{k}$. See Fig. 22.69.

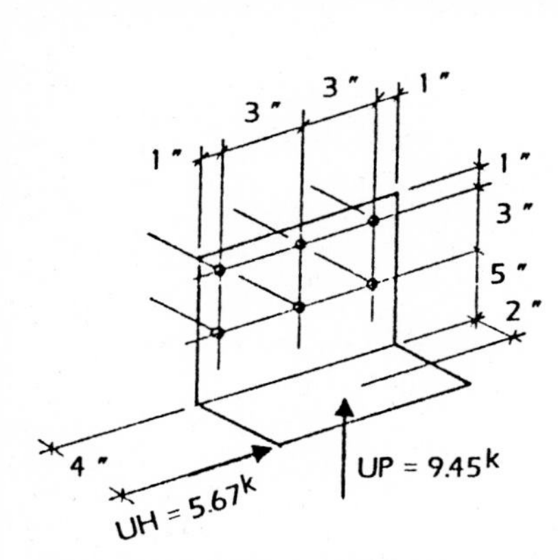


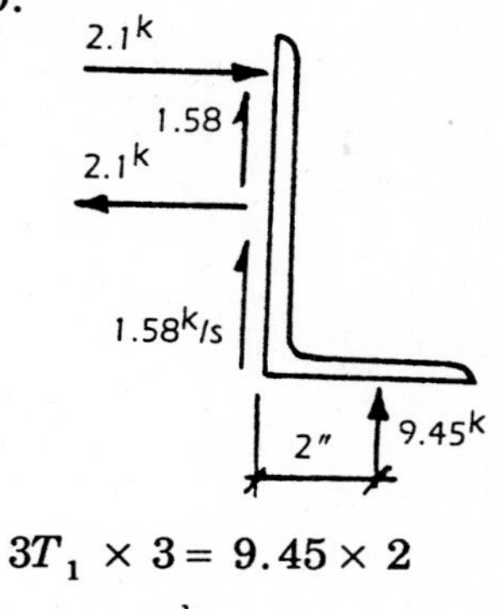


Figure 22.69 Effect of $UP = 9.45^{k}$.

$$3T_1 \times 3 = 9.45 \times 2$$

$$T_1 = 2.1^{k}/\text{stud}$$

$$R_{yv} = \frac{9.45}{6} = 1.58^{k}/\text{stud}$$

2. Effect of eccentric shear; $UH = 5.67^{k}$.
Due to bending ($e = 4$ in; see Fig. 22.70):

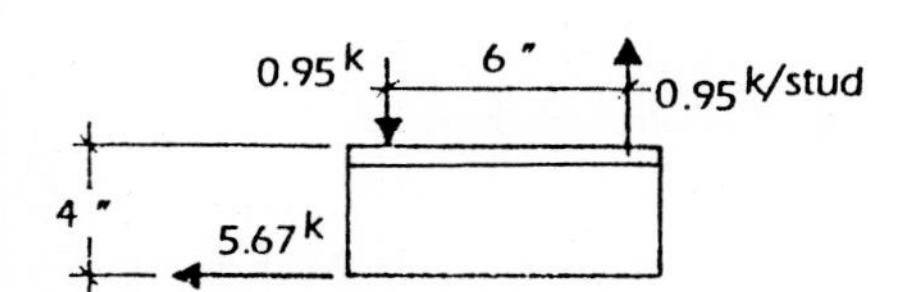


Figure 22.70 Eccentric shear due to bending.

$$4T_2 \times 6 = 5.67 \times 4$$

$$T^{2} = 0.95^{k}/\text{stud}$$

Due to shear:

$$R_{xv} = \frac{5.67}{6} = 0.95^{k}/\text{stud}$$

Due to torsion (see Fig. 22.71):

$$R_{xT} = \frac{Pey}{\Sigma x^2 + y^2} = \frac{(5.67)(6.5)(1.5)}{49.5} = 1.12^{k}/\text{stud}$$

$$R_{yT} = \frac{Pex}{\Sigma x^2 + y^2} = \frac{(5.67)(6.5)(3)}{49.5} = 2.24^{k}/\text{stud}$$

$$R = \sqrt{(R_{xv} + R_{xT})^2 + (R_{yv} + R_{yT})^2}$$

$$R = \sqrt{(0.95 + 1.2)^2 + (1.58 + 2.24)^2} = 4.3^k/\text{stud}$$

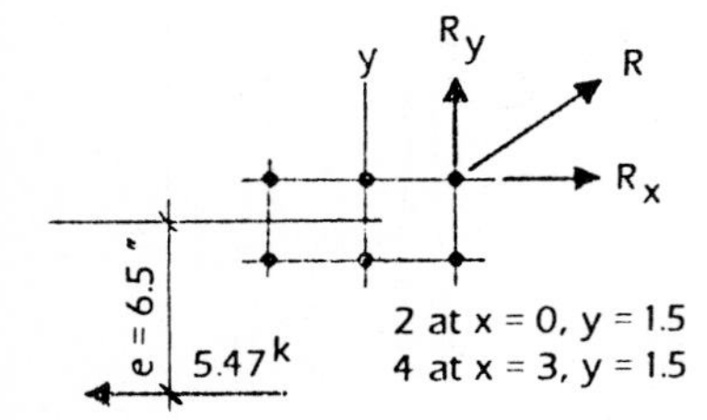


**Figure 22.71** Eccentric shear due to torsion. $\Sigma x^2 + y^2 = 2(2.25) + 4(9 + 2.25) = 49.5$.

Try six ½-in $\phi$ studs×4 in long ($L_e = 4$ in; $d_e = 3$ in).

$$V_u = 4.3^k/\text{stud}$$

$$P_u = T_1 + T_2 = 2.1 + 0.95 = 3.05^k/\text{stud}$$

See *PCI Design Handbook*, 3d edition, pages 6-9, 6-52, and 6-53.
As governed by stud capacity:

$$\phi P_s = 10.6^k/\text{stud} \quad \phi V_s = 8.8^k/\text{stud}$$

$$\left(\frac{P_u}{\phi P_s}\right)^2 + \left(\frac{V}{\phi V_s}\right)^2 = \left(\frac{3.05}{10.6}\right)^2 + \left(\frac{4.36}{8.8}\right)^2 = 0.08 + 0.24$$
$$= 0.32 < 1.0 \text{ (OK)}$$

As governed by concrete capacity $(\phi = 0.85;\ L_e = 4 \text{ in};\ f_c' = 5000)$:

$$\phi P_c = 11.3^k/\text{stud} (d_e = 3 \text{ in}) \quad \phi V_c = 7.7^k/\text{stud} (d_e = 4.5 \text{ in})$$

$$P_c = 13.3^k/\text{stud},\ V_c = 9.1^k/\text{stud for } \phi = 0.85$$

$$\frac{1}{\phi}\left[\left(\frac{P_u}{P_c}\right)^2 + \left(\frac{V_u}{V_c}\right)^2\right] = \frac{1}{0.85}\left[\left(\frac{3.05}{13.3}\right)^2 + \left(\frac{4.36}{9.1}\right)^2\right] = 0.34 < 1.0 \text{ (OK)}$$

Therefore, use six ½-in × 4-in-long studs as shown.

Check 1.2 $g$ out on connection:

$$UH = 1.4 \times 1.2 \times 0.25 \times 13.5 = 5.67^{k}$$

$$5.67 \times 8 = 3T \qquad \therefore T = 15.12^{k} \text{ or } 5.04^{k}/\text{stud}$$

$$\phi P_s = 10.6^{k} \quad \text{(OK)}$$

Check angle leg at 0.4 $g$ out:

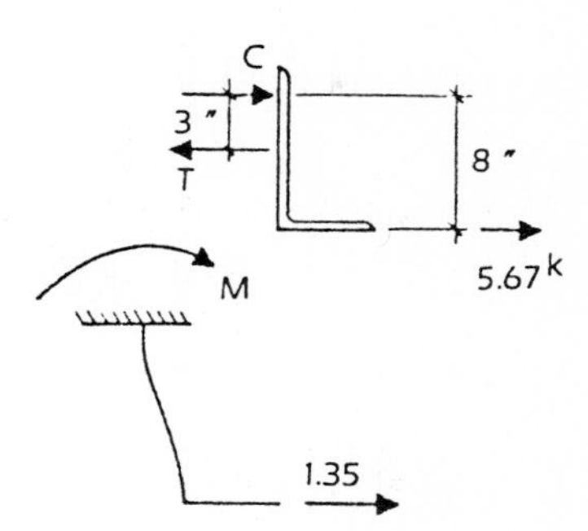


$$H = 0.4 \times 0.25 \times 13.5 = 1.35^{k}$$

$$M = (1.35)(5) = 6.75^{k\cdot in}$$

$$f = \frac{M}{S} = \frac{6.75}{1/6 \times 8 \times 0.5^2}$$

$$= 20.3 \text{ ksi} \quad \text{(OK)}$$

Check force in outstanding leg of L 9 × 4 × 1/2 when panel rotates under building drift at 1.2 $g$. Ascertain that connection is developed.

$$M_p = 3F$$

$$\text{At yield: } Z = \frac{bd^2}{4} = \frac{(8)(0.5)^2}{4} = 0.5 \text{ in}^2$$

$$ZF_y = M_p$$

$$(36)(0.5) = 3F \qquad \therefore F = 6^{k} \quad \text{(OK)}$$

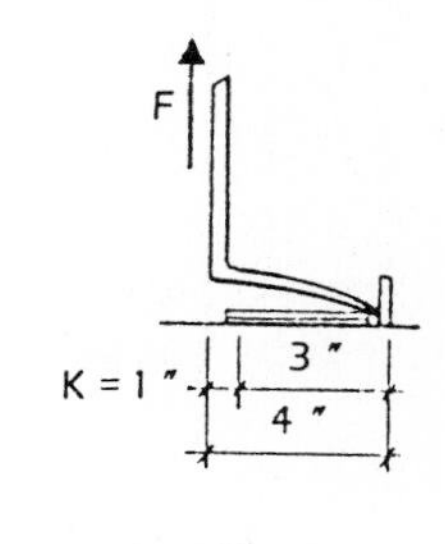


Design weld ∠9 × 4 × 1/2 to insert ℙ in cast-in-place floor:

$$M = (5.67)(4) = 22.68^{k}$$

$$(UH \text{ at } 1.2\,g = 5.67^{k})$$

$$S = 1/6 \times 8^2 = 10.67$$

$$f = \frac{22.68}{10.67} = 2.13^{k}/\text{in}$$

$$D = \frac{2.13}{0.93} = 2.28$$

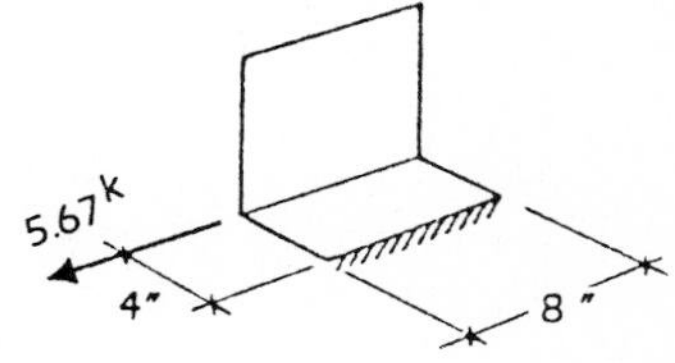

Use ¼-in weld.

**Top connection.** See Fig. 22.72.

$$1.2\ g\ \text{ lateral} = (1.2)(0.25)(13.5) = 4.05^{k}$$

$$UP = 1.4 \times 4.05 = 4.67^{k}$$

Use two ½-in $\phi$ × 4-in-long studs plus L 4 × 4 × 3/8.

$P_u = V_u = 0.707 \times 5.67 \times 0.5 = 2.00^{k}$/stud (concrete capacity governs)

$$d_e = 3 \text{ in} \quad L_e \cong 3 \text{ in} \quad \phi = 0.85$$

$$\phi P_c = 7.2^{k} \quad P_c = 8.5^{k}$$

$$\phi V_c = 3.4^{k} \quad V_c = 4.0^{k}$$

$$P_u = V_u = 0.707 \times 5.67 \times 0.5 = 2.00^{k}/\text{stud}$$

$$\frac{1}{0.85}\left[\left(\frac{2.0}{8.5}\right)^2 + \left(\frac{2.0}{3.4}\right)^2\right] = 0.36 < 1.0 \text{ (OK)}$$

Threaded rod size:

$$A_{\text{required}} = \frac{4.05}{1.33 \times 20} = 0.15 \text{ in}^2$$

Use ⅝-in $\phi$ rod.

Tube size:

$$M = (4.05)(9) = 36.45^{k \cdot ft}$$

$$S = \frac{36.45}{1.33 \times 24} = 1.14 \text{ in}^3$$

Use tube support $4 \times 4 \times 0.25$ ($s = 4.00$ in$^3$).

Panels being erected are shown in Fig. 22.73.

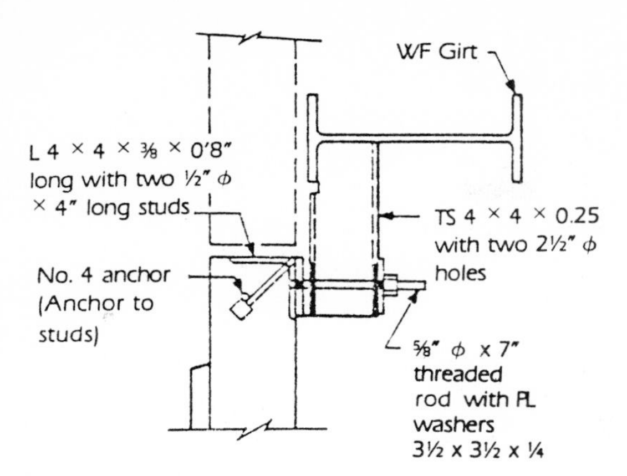


**Figure 22.72** Detail of top connection.

**Figure 22.73** 601 Montgomery, San Francisco, California. Panel erection with a roof-mounted tower crane allows uninterupted erection of precast concrete cladding panels despite limited access at the site and congested midcity streets.

**Figure 22.73** ***(continued)*** 601 Montgomery, San Francisco, California. Panel being hoisted into position.

## Prestressed Solid Wall Panels

Solid precast concrete wall panels are often prestressed whereas pierced panels are not. This is due to the interference of the openings with the prestressing strands, the tendency of the opening surrounds to bind to the form after release of prestress but before stripping, and the possibility of undesired bowing induced by eccentric prestressing. Prestressed panels are usually flat, ribbed, or made from standard structural shapes such as the double T. The panels may be single-thickness or of sandwich construction. Sandwich panels usually have the structural wythe prestressed, with the wythe covering the insulation being isolated, or "floating." Flat sandwich panels with both outer and inner wythes prestressed are designed by taking into account the temperature-gradient difference between the two wythes and the effect of differential drying shrinkage and the resultant induced stresses and bowing. These members are often designed as composite elements, with horizontal shear transfer being designed in the ties between the outer and inner wythes. The usual selections of finishes are available, with the conditional requirement that transverse ribs should be avoided to prevent binding in the forms resulting from prestress elastic-shortening strains. Prestressed panels are made in a long line similar to the process used for flexural deck members, with associated efficiencies in plant labor.

An advantage of prestressing is that it is a more economical way to reinforce the wall panel as compared with mild steel reinforcing.

**Figure 22.74** Levitz furniture warehouse, Concord, California. Double-T wall panels 24 and 40 ft high support vertical loads and resist lateral loads from earthquake and wind.

Prestressing also helps to keep handling stresses within allowable limits. Moreover, for units where the average effective prestress exceeds 225 lb/in$^2$, the minimum amounts of reinforcing specified for mild steel reinforced walls are waived, and ties around the prestress strands may be omitted unless they are required for shear. Prestressing also increases the ultimate flexural capacity of the wall, and in most applications bending governs the design and not the axial load capacity of the element.

In designing these walls, the fire-resistive requirements often dictate the flange thickness of stemmed or double-T units. Thicknesses required for various fire ratings are shown below:

| Fire rating, hours | 1 | 2 | 3 | 4 |
|---|---|---|---|---|
| Sand-lightweight concrete, in | 3 | 3 3/4 | 4 3/4 | 5 1/2 |
| Hard-rock concrete, in | 3 1/2 | 5 | 6 1/4 | 7 |

Of course, the proper cover distances for reinforcing and strand, as listed in Table 43A of the Uniform Building Code, would be observed also, in addition to the overall required wall or flange thickness. The moment magnification method given in the code is sufficient analysis for slenderness for $k\ell/r \leq 100$. For a slenderness ratio greater than 100, the $P$-$\Delta$ effect shall be checked in accordance with Sec. 2610 of the Uniform Building Code.

Load-bearing prestressed concrete wall panels are normally designed to exhibit no tension under full-service-load conditions. For tall, slender panels, the seismic lateral loading condition of $0.3g$, combined with dead plus live load, will normally govern. The wall element will also have to fall within the envelope created for combined ultimate axial and bending indicated by the interaction diagrams for the particular section and reinforcing under consideration. Often overlooked in the design of prestressed panels are the minimum eccentricities specified in Sec. 2610 of the Uniform Building Code.

The design procedure for analysis of a prestressed concrete wall panel is as follows:

1. Select flange or wall thickness based upon the required fire rating.
2. For preliminary design, select a section for which $k\ell/r \leq 100$. Preliminary sections can be rectangular or standard structural shapes.
3. Summarize the loads to the panel, both vertical and lateral.
4. Determine the most critical ultimate loading condition:

$$1.4D + 1.7L$$

$$0.75(1.4D + 1.7L + 1.7W)$$

$$0.75(1.4D + 1.7L + 1.87E)$$

5. Refer to interaction diagrams, such as those in the 3d edition of the *PCI Design Handbook* on pages 2-51 through 2-53 to select a trial section.
6. Select prestressing to provide a minimum of 225 lb/in²; check for zero tension under full-service loading. The cover over the strand should conform to that required for the fire rating or weather exposure conditions as outlined in Sec. 2607 of the Uniform Building Code.
7. Evaluate slenderness effects.
8. Check handling.
9. Construct interaction diagrams; ascertain that the trial section satisfies ultimate strength design criteria.

The following design example demonstrates the use of this procedure in designing a load-bearing double-T wall panel in an industrial building.

### Double-T wall panel design

Design a load-bearing wall panel 8 ft 0 in wide for an industrial building in Fire Zone 1 (1-hour rating required for the wall). Its physical geometry is shown in Fig. 22.75.

**Preliminary design.** From Table 43-B of the Uniform Building Code, 3½ in of Grade A hard-rock concrete is required for a 1-hour rating. Therefore, use 3½-in-thick double-T flange.

**Loads and Moments per 8-ft-Wide Double-T Panel**

| | | |
|---|---|---|
| Long-span joists | $\frac{20/\text{ft}}{4} \times \frac{60}{2} \times 8$ | = 1200 lb |
| Metal deck | $3\ \text{lb/ft}^2 \times 30 \times 8$ | = 720 lb |
| Built-up roof and insulation | $6\ \text{lb/ft}^2 \times 30 \times 8$ | = 1440 lb |
| Sprinklers | $5\ \text{lb/ft}^2 \times 30 \times 8$ | = 1200 lb |
| Mechanical and miscellaneous | $5\ \text{lb/ft}^2 \times 30 \times 8$ | = 1200 lb |
| Live load | $16\ \text{lb/ft}^2 \times 30 \times 8$ | = 3840 lb |
| Σ | $40\ \text{lb/ft}^2$ | 9600 lb |

$$\text{Estimated T weight: } 60\ \text{lb/ft}^2 \times 8 \times 29 = 13{,}920\ \text{lb}$$

$$UP_{\text{top}} = (1.4)(5.760) + (1.7)(3.840) = 14.6^{\text{k}}$$

$$UP_{\text{bottom}} = 14.6 + (1.4)(13{,}920) = 34.1^{\text{k}}$$

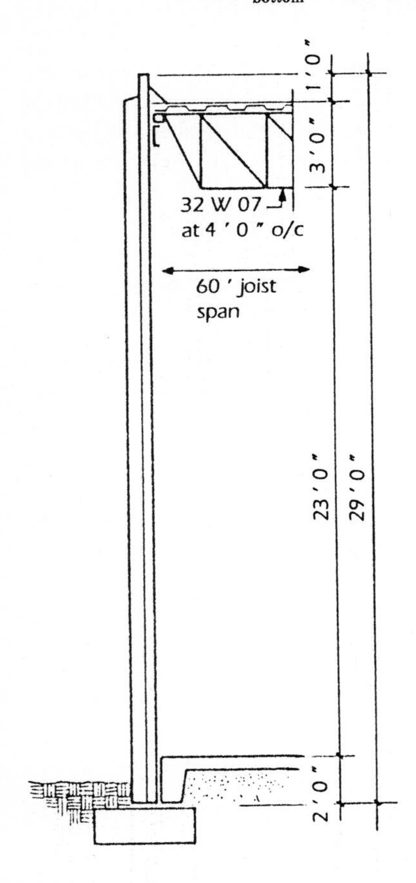


**Figure 22.75** Geometry of a double-T wall panel.

Wind:

$$1/8 \times 0.015 \times 8 \times 26^2 = 10.1^{\text{k}\cdot\text{ft}}$$

Seismic (0.3 $g$ ; Uniform Building Code, Table 23-J):

$$1/8 \times 0.3 \times 0.060 \times 8 \times 26^2 = 12.2^{\text{k}\cdot\text{ft}}$$

Loading conditions may be summarized as follows:

1. $D + L$.
   At top:

$$UP = 14.6^{\text{k}}$$

$$UM = (14.6)\left(\frac{4}{12}\right) = 4.9^{\text{k}\cdot\text{ft}}$$

(Assume $e = 4$ in.)

At base (assume 50 percent fixed):

$$UP = 34.1$$

$$UM \cong 4.9 \times 0.5 = 2.5^{\text{k}\cdot\text{ft}}$$

2. $D + L + W$ (maximum at midheight).

$$UP = 0.75\left(\frac{14.6 + 34 \text{ in}}{2}\right) = 18.3^{\text{k}}$$

$$UM = 0.75 \times 1.7 \times 10.1 = 3/4 \times 0.75 \times 4.9 = 10.3^{\text{k}\cdot\text{ft}}$$

3. $D + L + E$ (maximum at midheight).

$$UP = 18.3^{\text{k}}$$

$$UM = 0.75 \times 1.7 \times 1.1 \times 12.2 + 3/4 \times 0.75 \times 4.9 = 19.9^{\text{k}\cdot\text{ft}}$$

Condition 3 governs. From Fig. 2.6.3 of the 3d edition of the *PCI Design Handbook*, a 12-in-deep section with a 2-in flange is ample.

Try a 10-in-deep panel (see Fig. 22.76).

$$f'_c = 5000 \text{ lb/in}^2$$
$$E_c = 4300$$

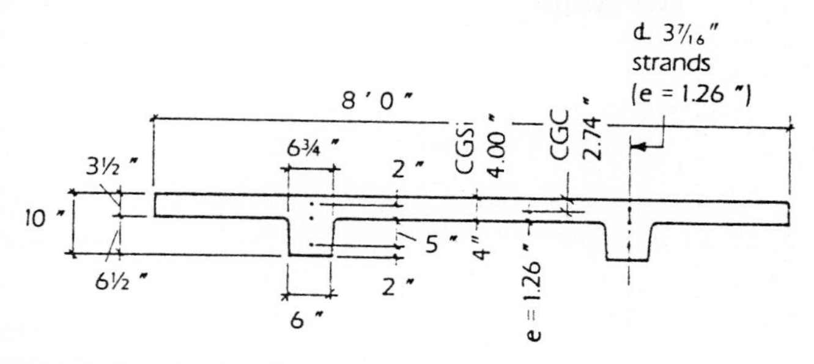


**Figure 22.76** Panel 10 ft deep.

Section properties:

| Part | $A$ | $Y_t$ | $A_{yt}$ | $I_o$ | $Ad^2$ |
|---|---|---|---|---|---|
| 3.5 × 96 | 336 | 1.75 | 588 | 348 | 329 |
| 2 × 6³ × 6.5 | 83 | 6.75 | 560 | 292 | 1335 |
| | 419 | | 1148 | $\Sigma I_0 + Ad^2$ = 2300 in⁴ | |

$$y_t = \frac{1148}{419} = 2.74 \text{ in} \qquad S_t = 839 \text{ in}^3 \qquad (\text{Weight} = 55 \text{ lb/ft}^2)$$
$$y_b = 7.26 \text{ in} \qquad S_b = 317 \text{ in}^3$$

Check prestressing (minimum required = 225 lb/in², per Uniform Building Code, Sec. 2618). (See *PCI Design Handbook*, 3d edition, pages 11-16 and 4-39.)

$$F_{f,\,av} = (0.225)(419) = 94.3^k \qquad \text{For } 7/16\text{-in } \phi \text{ } 270^k \text{ strands}$$

$$N = \frac{94.3}{16.9} = 5.58 \qquad F_f = (0.78)(21.7) = 16.9^k/\text{strand}$$

Use six 7/16 -in $\phi$ 270$^k$ strands ($e$ = 1.26 in).

Service load condition: $D + L + E$ (see *PCI Design Handbook*, page 4-52).

$$f_{\text{outside}} = \frac{(16.9)(6)}{419} - \frac{(101.4)(1.26)}{317}$$

$$+ \frac{\left[\left(\frac{3}{4}\right)(4.9) + 12.2\right](12)}{317} - \frac{9.6 + 0.5 \times 13.9}{419}$$

$$= -0.242 - 0.403 + 0.601 - 0.039 = -0.083 \text{ ksi (OK)}$$

Check average prestress induced:

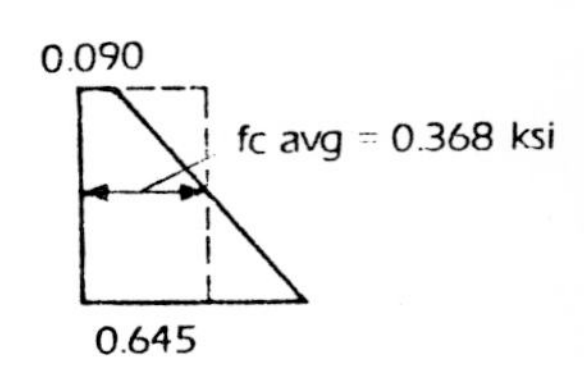


$$f_{\text{inside}} = -\frac{101.4}{419} + \frac{(101.4)(1.26)}{839} = -0.090$$

$$f_{\text{outside}} = -\frac{101.4}{419} + \frac{(101.4)(1.26)}{317} = -0.645$$

**Handling.** Check handling.

Stripping: use impact factor $I = 1.5$.
Try two-point pick at one-fifth points.

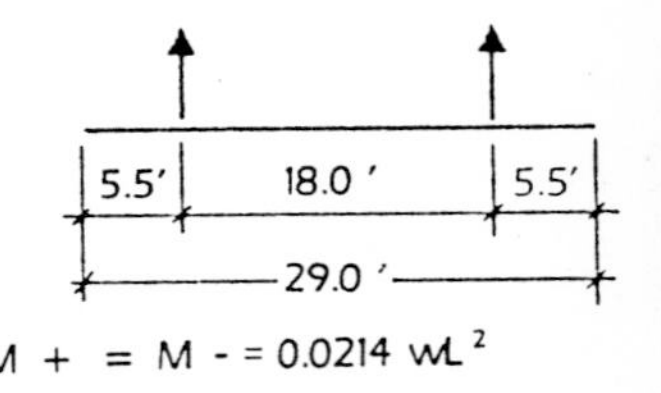


$$M^{+} = M^{-} = (0.0214)(0.44)(1.5)(29^2) = 11.9^{\text{k}\cdot\text{ft}}$$

$$f_{\text{top}} = -0.090 + \frac{(11.9)(12)}{859} = +0.080 \text{ ksi}$$

$$\text{Maximum allowable} = \frac{f_r}{\text{FS}} = \frac{(7.5)(\sqrt{3500})}{1.65} = +0.270 \text{ ksi (OK)}$$

Therefore strip, store, and ship by supporting on dunnage at one-fifth points.

**Evaluate slenderness effects of panel or column.** See Uniform Building Code, sec. 2610; *PCI Design Handbook*, 3d edition, page 3-22.

Radius of gyration of panel:

$$r = \sqrt{\frac{I}{A}} = \sqrt{\frac{2300}{419}} = 2.34 \text{ in}$$

For semifixed base, $K = 1.0$ (see *AISC Manual for Steel Construction*, page 5-124).

$$\frac{K\ell}{r} = \frac{(25)(12)}{2.34} = 128 > 100$$

Therefore, the $P$-$\Delta$ effect must be analyzed according to the Uniform Building Code, sec. 2610. For the analysis include effect of the panel being erected 2 in out of plumb. (See Fig. 22.77.)

$$\Delta = \frac{5wL^4}{384EI} = \frac{(5)(0.202)(25^4)(1728)}{(384)(4300)(2300)} = 0.18\text{in}$$

$$UP = (18.5)(0.18/12) = 0.3^{\text{k}\cdot\text{ft}}$$

Additional moment from 2-in out-of-plumbness:

$$UM^{2''e} = (14.6)(2/12) = 2.4^{\text{k}\cdot\text{ft}}$$

Final magnified ultimate design moment:

$$\Sigma UM = 19.9 + 0.5 + 2.4 = 22.6^{\text{k}\cdot\text{ft}}$$

$$UP = 18.3^{\text{k}}$$

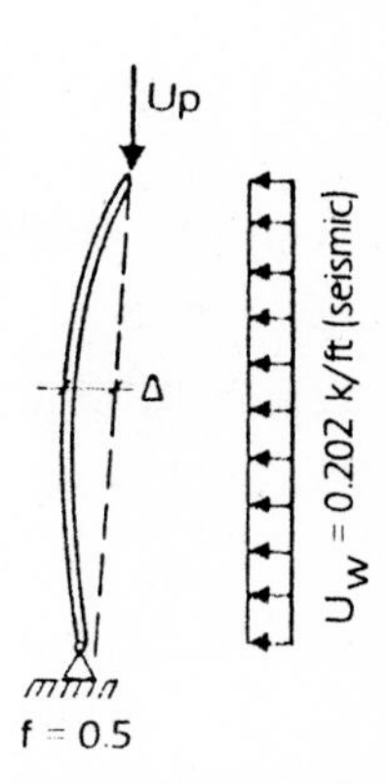


**Figure 22.77** Effect of panel being erected 2 in out of plumb.

Construct interaction diagram.

1. Determine $M_o$ ($\phi P_n = 0$). See *PCI Design Handbook*, pages 4-6, 4-48, and 4-63.

$$f_{se} = 147 \text{ ksi} \quad cwp = (1.06)\left(\frac{4 \times 0.115}{96 \times 5.5}\right)\left(\frac{270}{5}\right) = 0.05 < 0.08$$

$$\therefore f_{ps}/f_{pu} = 0.98 \quad \therefore f_{ps} = 264.6 \text{ ksi}$$

$$a/2 = \frac{A_{ps}\, f_{ps}}{1.7 f'_c\, b} = \frac{(0.46)(264.6)}{(1.7)(5)(96)} = 0.15 \text{ in}$$

$$\phi M_o = 0.85\,\phi A_{ps}\, f_{ps}\,(d - a/2)$$

$$= (0.85)(0.9)(0.46)(264.6)(5.5 - 0.15) = 498^{\text{k}\cdot\text{in}}\ (41.5^{\text{k}\cdot\text{ft}})$$

2. Determine point where $\phi = 0.7$.

$$0.7P_n = 0.1 f'_c\, Ag = (0.1)(5)(419) = 210^{\text{k}} \quad (P_n = 300^{\text{k}})$$

3. Determine another point near end of curve. See *PCI Design Handbook*, page 4-49.

Set $a = 0.5\text{in.} \;\therefore c = \dfrac{a}{\beta_1} = \dfrac{0.5}{0.8} = 0.625 \text{ in};\; y' = \dfrac{a}{2} = 0.24 \text{ in}$

$A_{comp} = (0.625)(96) = 60 \text{ in}^2 \quad d = 5.5 \text{ in} \quad 0.85f'_c = 4.25 \text{ ksi}$

$$\varepsilon_s = \left[\frac{f_{se}}{E_s} - \frac{0.003}{c}(c - d')\right] = \left[\frac{147}{27{,}500} - \frac{0.003}{0.625}(0.625 - 1.5)\right]$$

$$= 0.00955 \text{ in/in}$$

$$\varepsilon_{ps} = \left[\frac{f_{se}}{E_s} - \frac{0.003}{c}(d - c)\right] = \left[\frac{147}{27{,}500} - \frac{0.003}{0.625}(5.5 - 0.625)\right]$$

$$= 0.02875 \text{ in/in}$$

$f_s = \varepsilon_s E_s = (0.00955)(27{,}500) = 262.6 \text{ ksi}$

$f_{ps} = 0.98 f_{pu} = 264.6 \text{ ksi}$

(See *PCI Design Handbook*, page 11-18.)

$$0.85\phi P_n = 0.85\phi[(60)(4.25) - (2)(0.115)(262.6) - (4)(0.115)(264.6)] = 62.0\phi^{\text{k}}$$

By interpolation between $\phi = 0.9$ at $P_n = 0$ and $\phi = 0.7$ at $P_n = 300^{\text{k}}$, $\phi$ at $P_n = 62^{\text{k}} = 0.74$. Therefore, $\phi P_n = (0{,}74)(62) = 45.9^{\text{k}}$

$$0.85\phi M_n = (0.85)(0.74)[(255)(2.74 - 0.25) - (60.4)(274 - 1.50) + (121.7)(5.50 - 2.74)] = 564^{\text{k}\cdot\text{in}}\left(47.0^{\text{k}\cdot\text{ft}}\right)$$

**Draw interaction diagram.** See Fig. 22.78. (See also *PCI Design Handbook*, page 2-51.) Section is adequate since all loading points fall within the boundary envelope.

**Design flange reinforcing.** See Uniform Building Code, sec. 2614; *PCI Design Handbook*, pages 11-22 and 11-23.

$$A_s \text{ horizontal (transverse)} = 0.0020Ag = (0.0020)(3.5)(12) = 0.084 \text{ in}^2 \;(W2.9 \text{ at } 4 \text{ in})$$

$$A_s \text{ vertical (longitudinal)} = 0.0012Ag = (0.0012)(3.5)(12) = 0.05 \text{ in}^2 \;(W2 \text{ at } 4 \text{ in})$$

**Reinforcing summary**: 8-ft double-T wall panel. See Fig. 22.79.

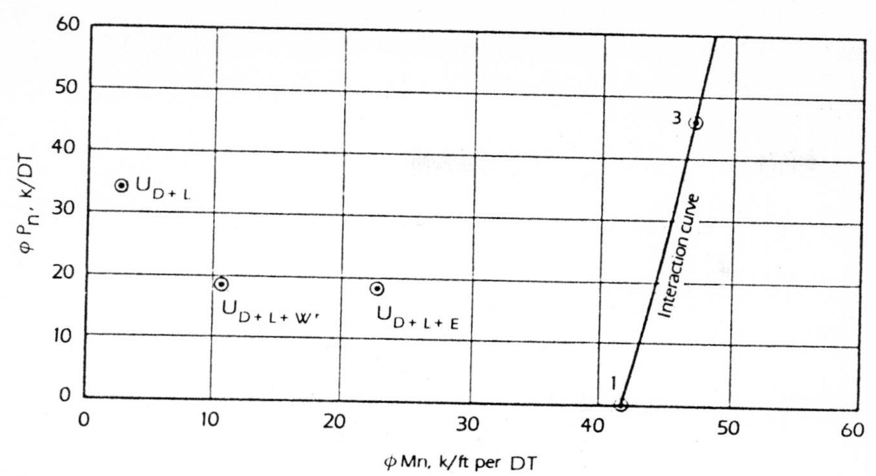


Figure 22.78 Interaction diagram.

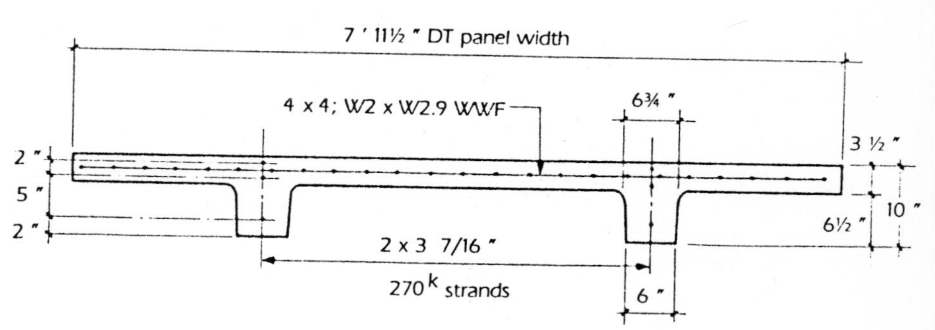


Figure 22.79 Reinforcing summary: 8-ft double-T wall panel.

## Connection Design for Lateral Load-Resisting Solid Wall Panels

In 1969 static testing was carried out in a program cosponsored by the Prestressed Concrete Manufacturers Association of California and performed by San Jose State University. These tests verified the capacity of standard shear connections used to develop the horizontal shear transfer between flanges of double-T members used as floors or walls. The following details demonstrate the use of these connectors for the example double-T wall panel problem presented in the preceding pages. The other connection details show the details recommended at the roof and at the foundation.

### Design of double-T wall panel flange weld plates

Design plates to develop a factored seismic design base shear of 1.54 kips per lineal foot of wall.

**Design shear strength of concrete in double-T flange.** See Fig. 22.80. (See *PCI Design Handbook*, page 3-62.)

$$v_{ru} = 2\phi\sqrt{f'_c}\,t$$
$$= (2)(0.85)(0.71)(3.5) = 420 \text{ lb/in}$$

or
$$5.04^{k}/\text{ft} > 1.54^{k}/\text{ft} \quad (\text{OK})$$

By using A706 no. 5 bars ($f_y = 60$),

$$C_u = T_u = \phi f_y A_s$$
$$= (0.9)(60)(0.31) = 16.7^{k}/\text{bar}$$
$$v_{ru} = (C_u + T_u)\cos 45°$$
$$= (16.7 + 16.7)(0.707) = 23.6^{k}$$
$$s = 23.6/1.54^{k}/\text{ft} = 15.3 \text{ ft spacing required}$$

Use three connections per panel joint (see Fig. 22.81).

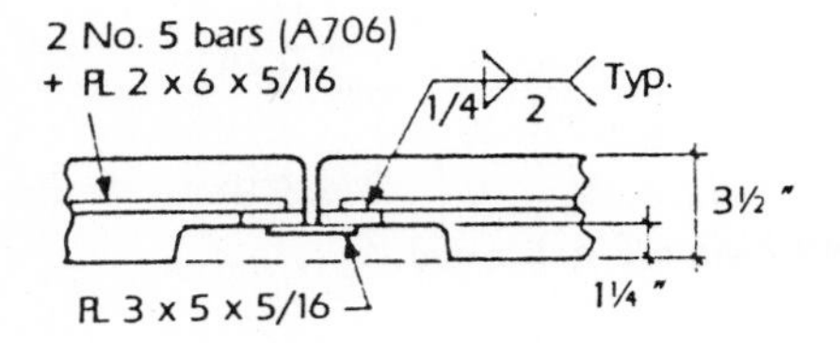


**Figure 22.80** Double-T flange connector.

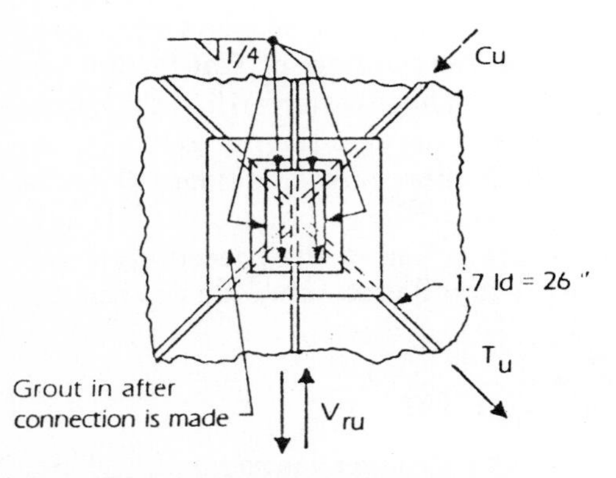


**Figure 22.81** Rear elevation of double -T flange connector.

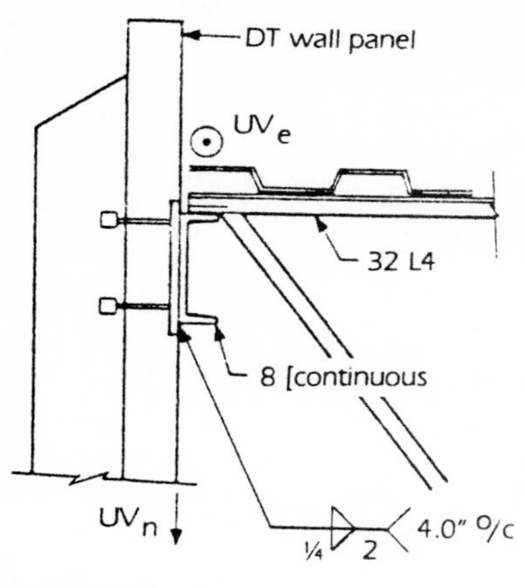


**Figure 22.82** Top connection.

**Load to continuous channel chord.** See Fig. 22.82.

$$UV_e = 1.54^k/\text{ft} \times 4\,\text{ft} = 6.2^k \text{ connection}$$

$$UV_n = 14.6/2 = 7.3^k/\text{connection}$$

Use PL $3 \times 10 \times 5/16$ + two $1/2$-in $\phi \times$ 5-in-long studs $\left(\Sigma\phi V_c = 18.8^k\right)$ at 4 ft 0 in on center at stems. (See *PCI Design Handbook*, page 6-53.)

Required length of ¼-in weld: $\dfrac{9.6}{0.93 \times 4 \times 1.33} = 1.94$ in

Use 2-in-long welds top and bottom.

**Double-T wall panel base connection**

$$UV_e = 1.54 \times 8 \times 12.3^k/\text{T}$$

Required length of ¼-in weld (see *PCI Design Handbook*, page 6-53):

$$\frac{12.3}{0.93 \times 4 \times 1.33} = 2.49$$

Use 3-in-long weld.
Connect at one stem per double T. (See Fig. 22.83.)

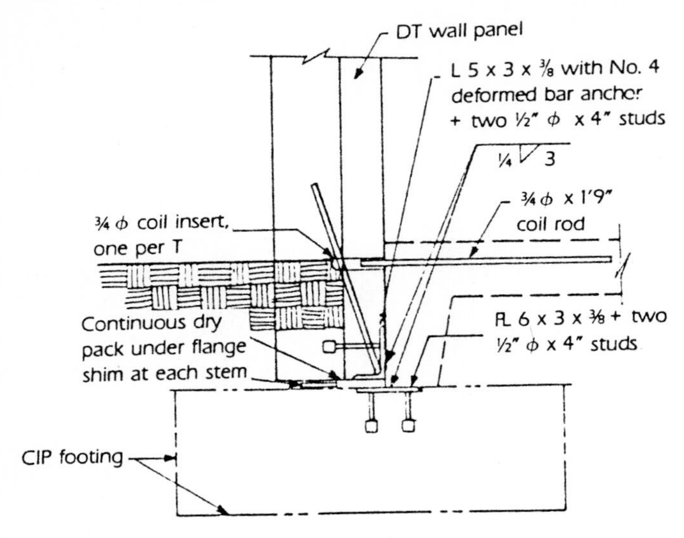


**Figure 22.83** Base connection.

### Connection details.

Typical connection details are depicted in Figs. 22.84 through 22.98.

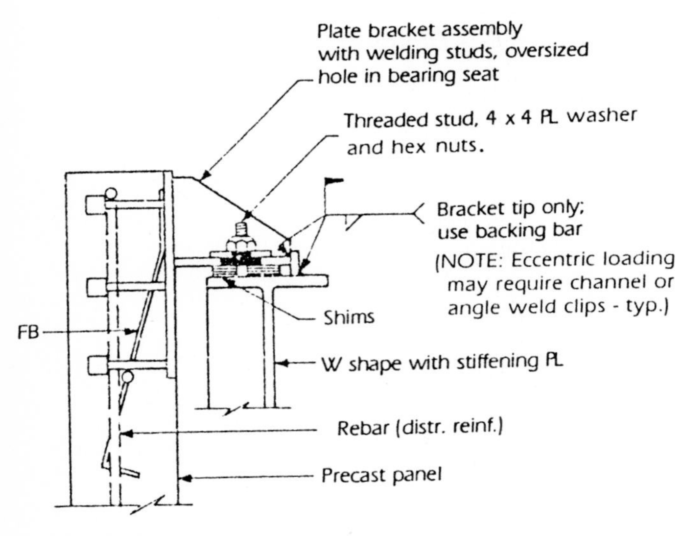


**Figure 22.84** Top bearing (hanger) connection.

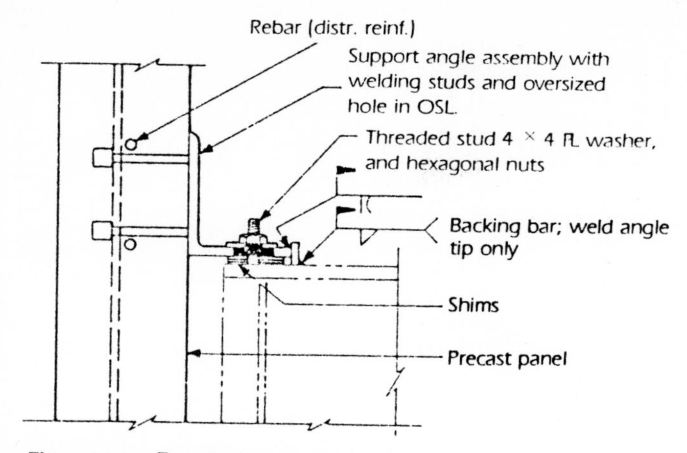


**Figure 22.85** Detail of bearing connection.

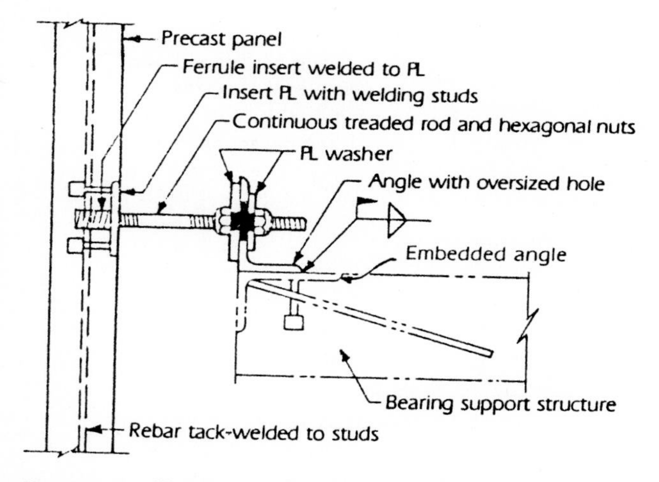


**Figure 22.86** Thin-panel lateral connection.

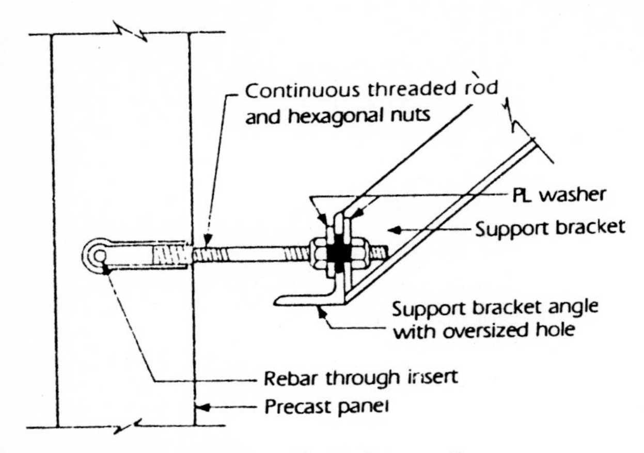


Figure 22.87 Thick-panel lateral connection.

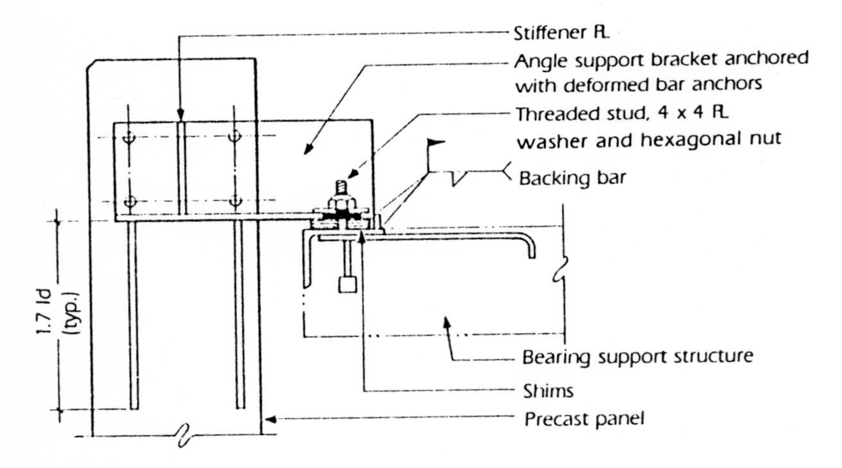


Figure 22.88 Bearing connection: embedded structural shape.

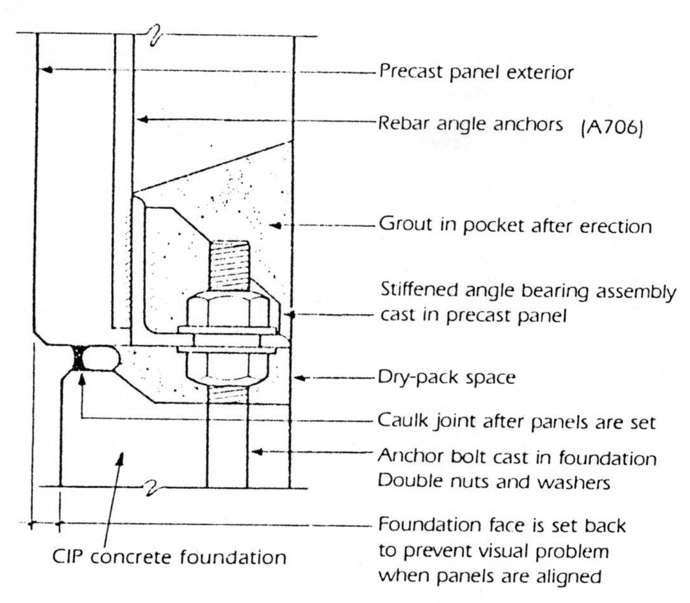


**Figure 22.89** Hidden panel bearing connection.

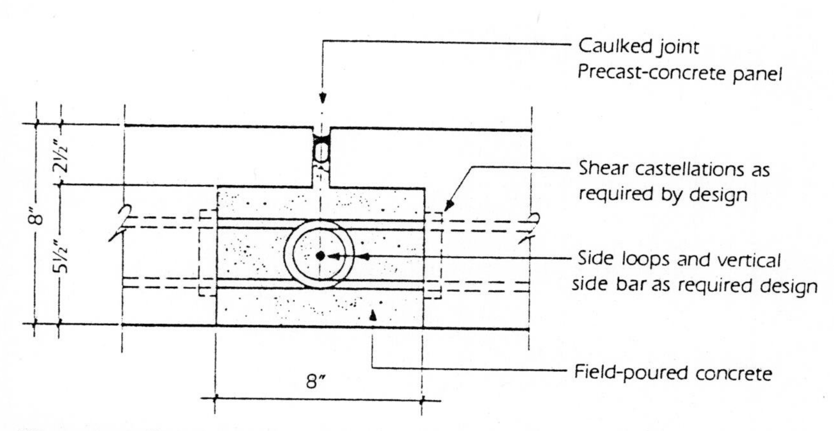


**Figure 22.90** Fixed connection: shear wall vertical joint in lateral load-resisting precast wall construction.

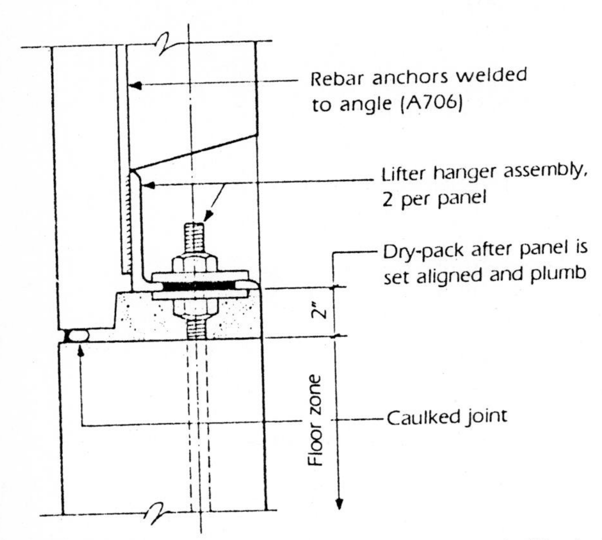


**Figure 22.91** Shear wall: horizontal joint detail. Vertical section at lifter hanger; lateral load-resisting precast wall construction.

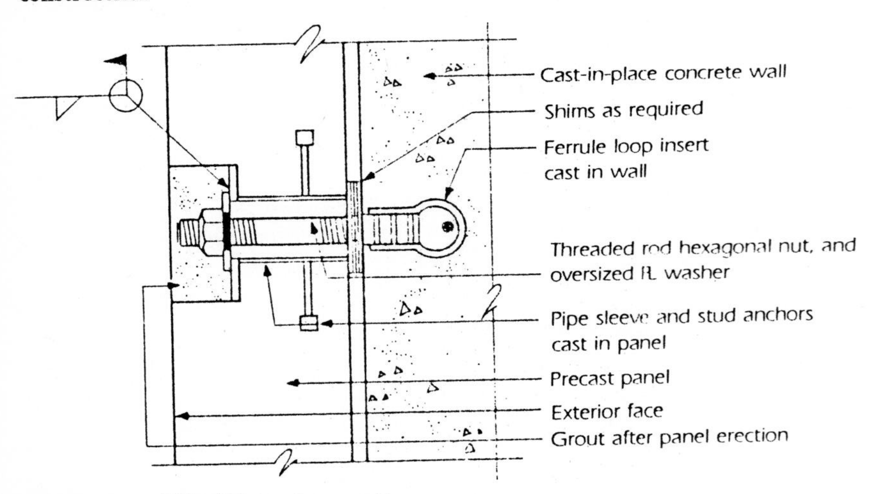


**Figure 22.92** "Blind" lateral connection.

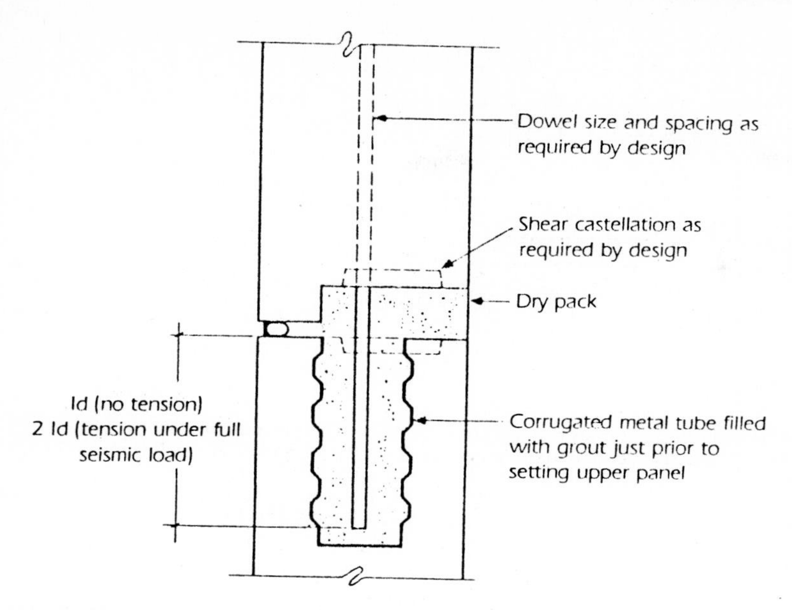


**Figure 22.93** Vertical section at dowel; lateral load-resisting precast wall construction.

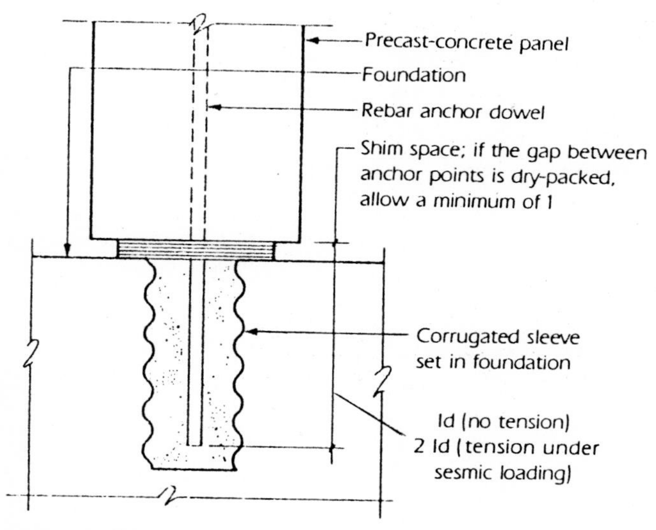


**Figure 22.94** Bottom "blind" bearing connection.

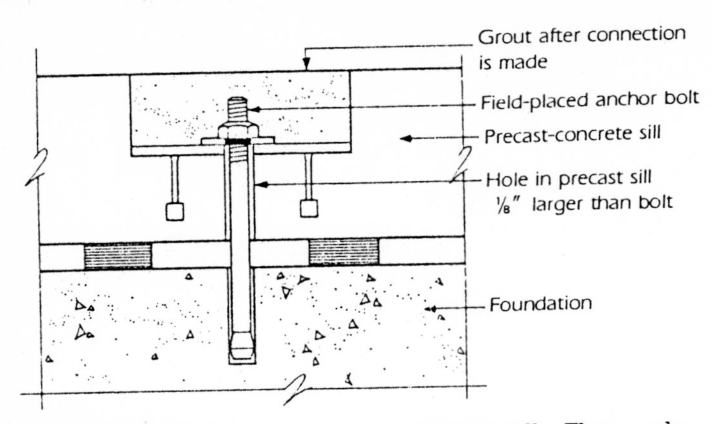


**Figure 22.95** Blind connection for precast concrete sills. The procedure is as follows: (1) Set shim stacks to proper elevation. (2) Set and align precast sill. (3) Field-drill bolt hole in foundation concrete (the hole size is the same as the bolt diameter). (4) Install the proper type of expansion bolt that is strong in tension such as the Kwik-Bolt, Phillips, Parabolt, or Wej-It. (5) Patch hole in precast sill finish to match balance of panel.

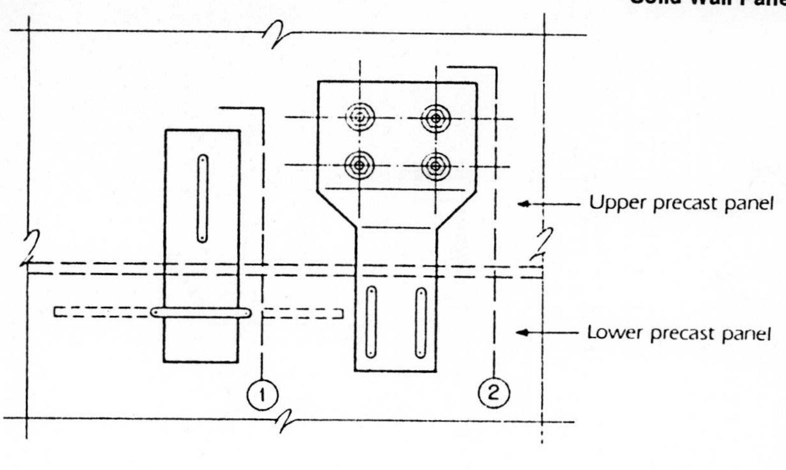


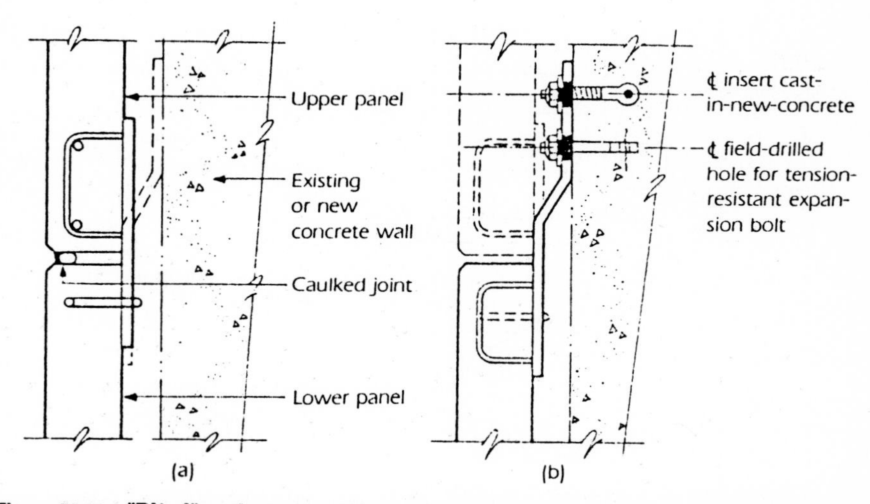


**Figure 22.96** "Blind" multistory cladding panel connection. (*above*) Upper and lower panels. (*below*) (*a*) Section (1) of bottom lateral connection. (*b*) Section (2) of top bearing connection.

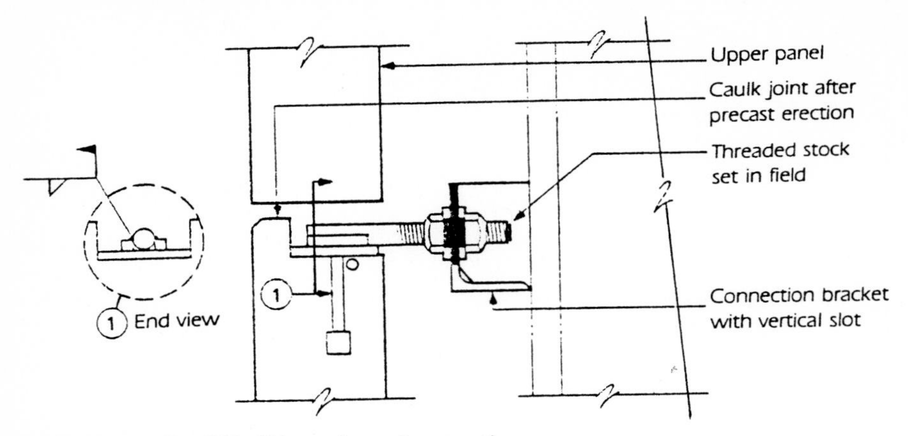


**Figure 22.97** Top "blind" lateral panel connection.

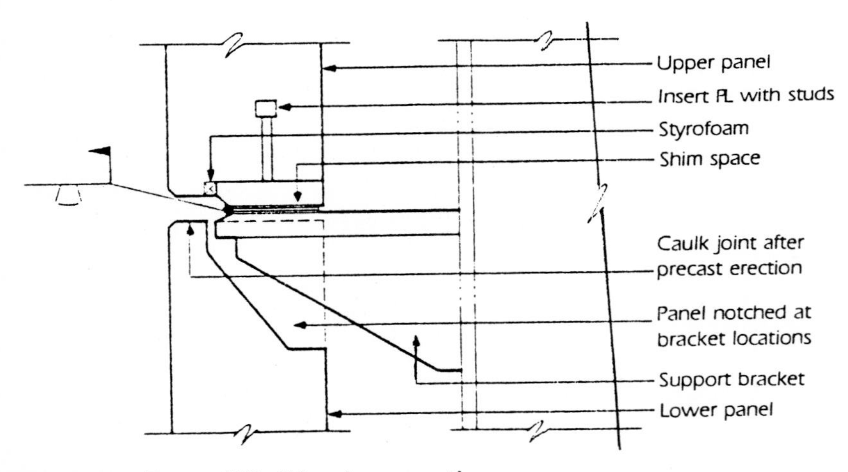


**Figure 22.98** Bottom "blind" bearing connection.

**Figure 23.1** Contra Costa County Detention Facility, Martinez, California. Non-load-bearing sandwich panels 10 in thick clad the top two floors of this project, shown while under construction; 4 in of polyurethane insulation was sandwiched between a 2-in outer skin and a 4-in inner skin. The two concrete wythes are interconnected by concrete ribs to cause them to act as one unit in resisting transverse forces.

Chapter

# 23

# Precast Concrete Sandwich Panels

In view of the need to conserve energy and the energy regulations being implemented at state and national levels, precast concrete sandwich panels offer an ideal building solution. With sandwich panels a durable surface is provided on both sides of the panels. The surface may be smooth on both sides, or have an exposed aggregate, or sandblasted or patterned exterior surface. The concrete surface offers resistance to weather, fire, and vandalism and is easy to maintain. With sandwich panels project construction time can be reduced, resulting in savings to the owner. The $U$ (thermal transmission) value of a 6-in solid normal-weight concrete wall is 2.38. For a sandwich panel consisting of 2 in of concrete for one wythe, $1\frac{1}{2}$ in of polyurethane, and 5 in of concrete for the other wythe, the $U$ value is 0.10. Sandwich panels can be effectively used for commercial, educational, governmental, industrial, and residential buildings. In designing with sandwich panels, architects still have the range of form, pattern, surface texture, and color that they have with solid noninsulated precast concrete panels. (See Figs. 23.1 and 23.11.)

Sandwich panels may be designed as load-bearing or non-load-bearing elements. They may be designed to function as either composite or noncomposite panels where design temperatures differ on either side of a wall. A noncomposite panel is one in which one wythe is nonstructural and is usually from $1\frac{1}{2}$ to $2\frac{1}{2}$ in thick. The outer wythe is supported from the structural wythe with flexible hangers and, as a floating wythe, is free to react to temperature and other volumetric changes. Its main purpose is to protect the insulative core. A composite panel is one in which the two wythes act as a unit to resist transverse forces and the wythes are connected with concrete ribs or steel shear connectors.

Various types of sandwich panels are shown in Figs. 23.2 through 23.7.

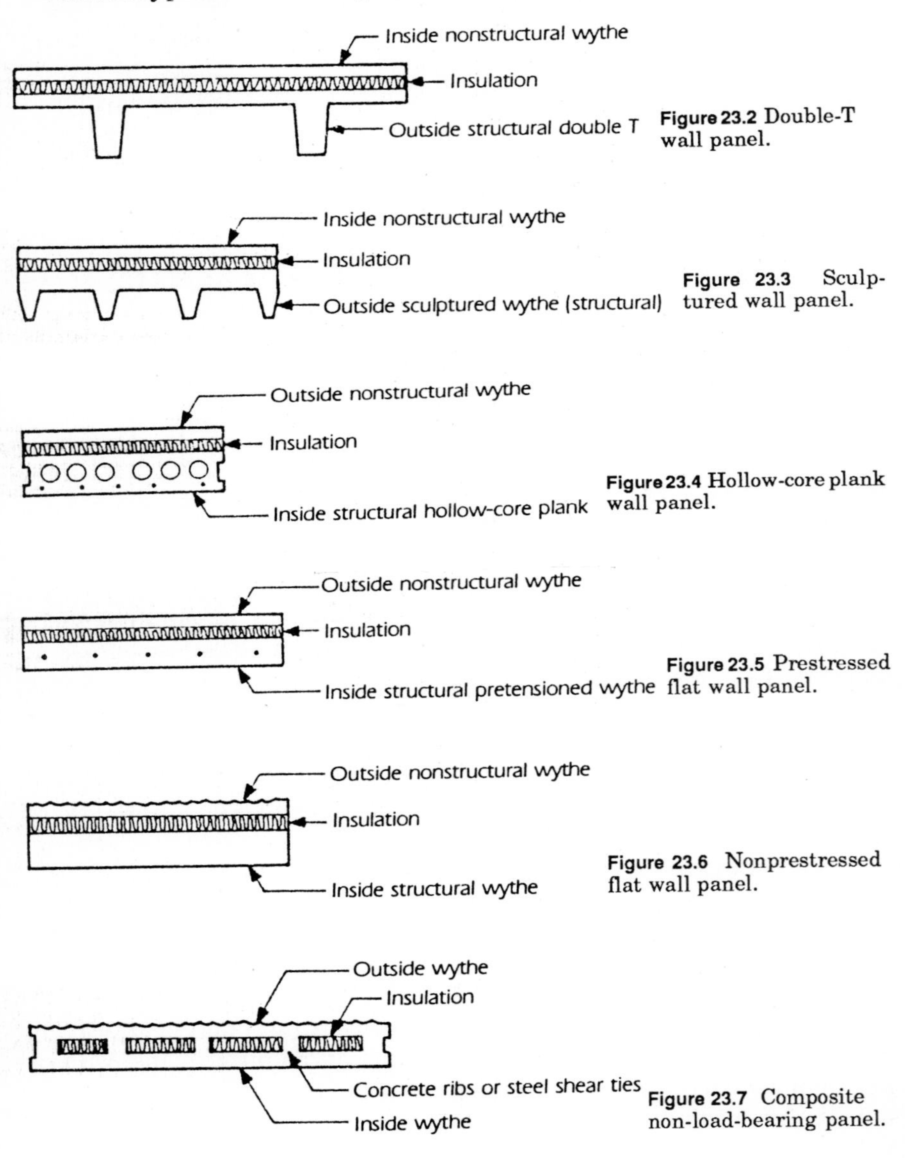


**Figure 23.2** Double-T wall panel.

**Figure 23.3** Sculptured wall panel.

**Figure 23.4** Hollow-core plank wall panel.

**Figure 23.5** Prestressed flat wall panel.

**Figure 23.6** Nonprestressed flat wall panel.

**Figure 23.7** Composite non-load-bearing panel.

## Insulation

The type of insulation used is subject to the $U$ value desired, wall thickness, cost, and type of panel. Polystyrene has an $R$ value (thermal resistance) of 4.00 per inch of thickness, polyurethane has an $R$ value of 6.25 per inch of thickness, and fiberglass board has an $R$ value of 4.00 per inch of thickness.

$$U = \frac{1}{R}$$

$$= \frac{1}{R_{\text{outside air}} + R_{\text{exterior concrete}} + R_{\text{insulation}} + R_{\text{interior concrete}} + R_{\text{inside air}}}$$

To reduce the required wall thickness, polyurethane would be the better insulation. Cost is affected by the current economic market, the thickness, and the $R$ value desired. For composite panels, insulation with a high bond surface is desired but with a minimum water absorption potential from the fresh concrete. For noncomposite panels a bond breaker sheet or a treated surface adjacent to the floating wythe is necessary to prevent bonding of the concrete to the insulation.

## Noncomposite Sandwich Panels

With a sandwich panel the outside wythe is subject to volumetric changes which differ from those of the inside wythe. These volumetric changes result from differential temperature changes between the two wythes, differential temperature changes due to time, shrinkage, lateral load stresses, and creep. For noncomposite panels, one wythe is made thin, from 1 1/2 to 2 1/2 in thick with an optimum thickness of 2 in. This wythe is attached to the other wythe with a combination of flexible and nonflexible ties. The flexible ties allow the floating wythe to accommodate movements from its different volumetric changes while the nonflexible tie anchors or fixes one portion of the floating wythe relative to the other wythe. The other wythe is the structural wythe and resists the forces placed on the panel due to stripping, plant handling, transportation, erection, and in situ loadings. This structural wythe is the thicker wythe, its thickness being determined by the stresses placed upon it and by building code requirements. Usually the thinner wythe is placed on the outside and the structural wythe on the inside. An exception to this rule occurs when the structural wythe is made with a pretensioned double-T section. In colder climates, attention should be given to avoidance of any pockets that could trap water and result in freezing water cracking the wythes.

The flexible tie may be one or several of the types shown in Figs. 23.8 and 23.9. The nonflexible ties or hangers may be one of the types shown in Fig. 23.10. The ties or hangers are galvanized or stainless steel and are either no. 2 reinforcing steel or 6-gauge wire. A 6- gauge wire tie will provide a tensile strength of about 1400 lb. The ties are usually spaced at 4 ft on centers each way. The insulative core runs out to the panel edges, thus allowing the floating wythe to move independently of the structural wythe. The insulation is either covered with a bond-breaking sheet or spray-coated with a liquid bond-breaking compound.

### Lifting stresses

Double-T and other structural ribbed panels are usually cast with the floating wythe on top and in a horizontal position, while flat and sculp-

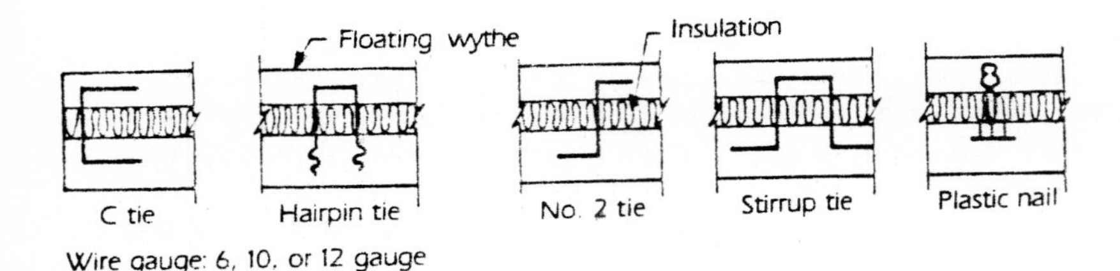


**Figure 23.8** Types of flexible ties.

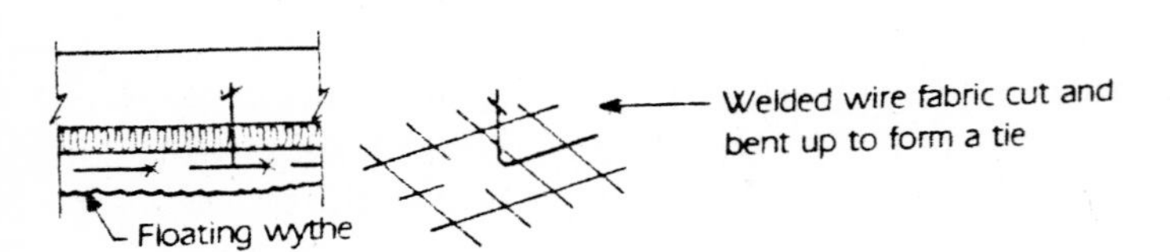


**Figure 23.9** Tie formed of welded wire fabric.

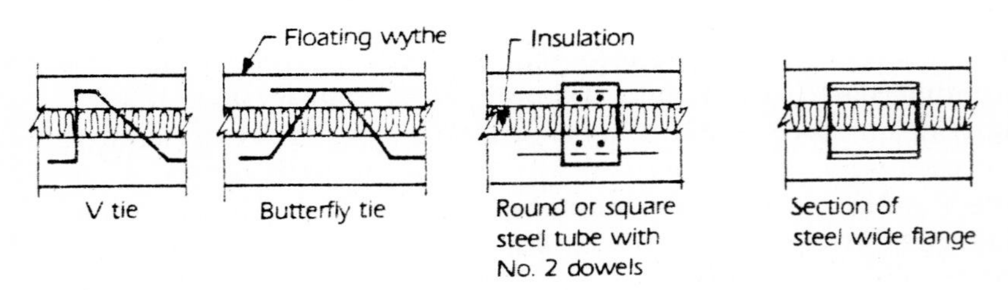


**Figure 23.10** Nonflexible ties or hangers.

tured architectural panels are usually cast with the floating wythe down. Since the floating wythe is nonstructural, all the lifting stresses are resisted by the structural wythe. The structural wythe is designed on the concept of an uncracked section with extreme fiber in tension not exceeding $f_{t,\,\text{design}} = \left(7.5\sqrt{f'_c}\right)/1.5$, where 1.5 is a factor of safety and $f'_c$ is the concrete compressive strength at the time of lifting. It should also be large enough so that frictional resistance of the floating panel does not induce strain which would cause cracking of the floating panel where it may be in tension. Here the precast manufacturer's experience is an important factor. The value of $f_{t,\,\text{actual}} = \dfrac{MC}{I}$, where $I$ and $C$ are for the structural wythe and the moment $M$ is based upon the total panel weight, including an allowance for impact.

### Lifting, handling, and erection inserts and connections

The location and special detailing of these inserts depend upon which wythe of the panel is cast up, which wythe is placed to the outside, whether the panel is bottom - or side-supported and load-bearing or non-load-bearing, the panel story height, and the method of plant handling, transportation, and erection. If the inserts pierce the floating wythe, they should be isolated so that they do not bear against the floating wythe. Styrofoam is usually placed around the insert at the floating wythe. However, one or several inserts may be bonded to both the floating wythe and the structural wythe where hanger ties are used or at corbels. This is the point where the two wythes are fixed together. If prestressing strands are used as lifting bales, the strands can be burned off after the panel has been erected and the remaining holes grouted. A preformed pocket is usually provided at these points if they are in the exposed face of the panel or are exposed to the weather.

### Floating-wythe reinforcement

The Uniform Building Code has no provisions for reinforcement in the floating wythes of sandwich panels. For walls in general the minimum requirements for welded wire fabric are 0.0020 $bt$ horizontal and 0.0012 $bt$ vertical (Sec. 2614, Uniform Building Code). For a 2-in wythe the horizontal minimum requirement is $0.0020 \times 12 \times 2 = 0.048$ in$^2$/ft ($6 \times 6 - W4.0 \times W4.0$ provides 0.08 in$^2$/ft).

Several types of noncomposite panels and details are shown in Figs. 23.12 through 23.19.

**Figure 23.11** Contra Costa County Detention Facility, Martinez, California. The precast concrete sandwich panels provide energy savings, security, and economy as well as an attractive appearance, allowing the building to blend harmoniously with the surrounding neighborhood.

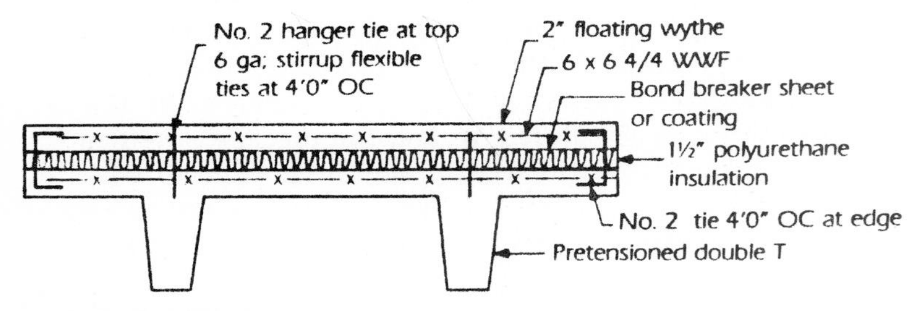


**Figure 23.12** Double-T sandwich panel.

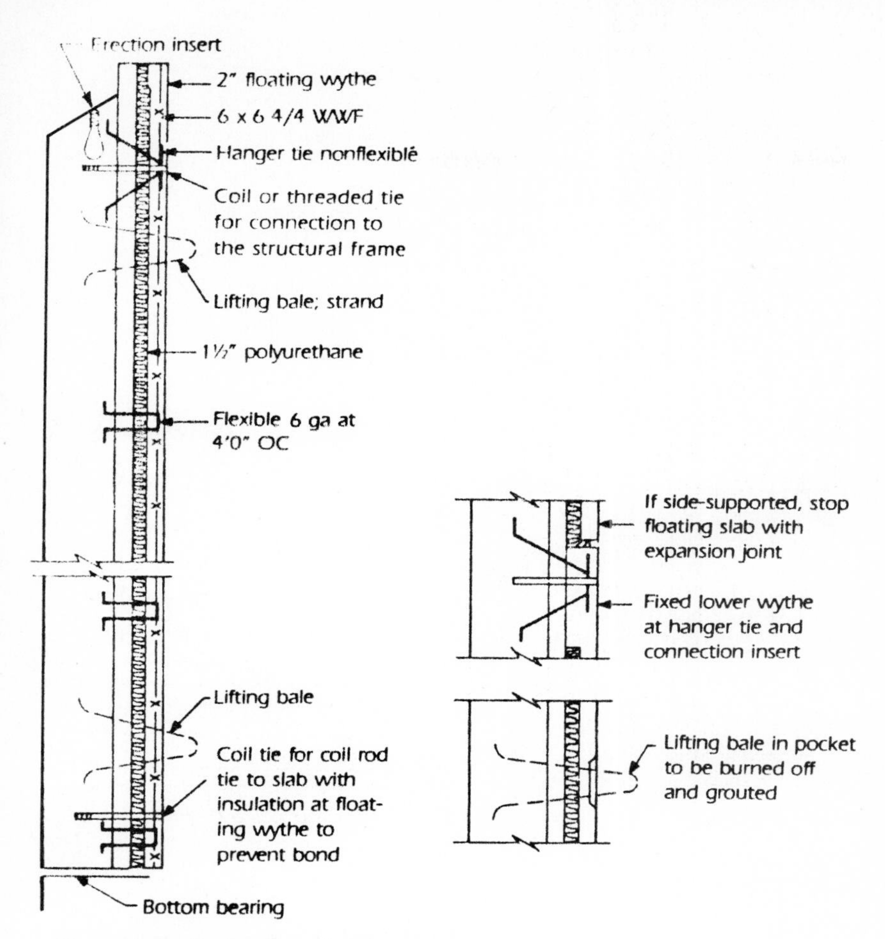


**Figure 23.13** Double-T sandwich panel.

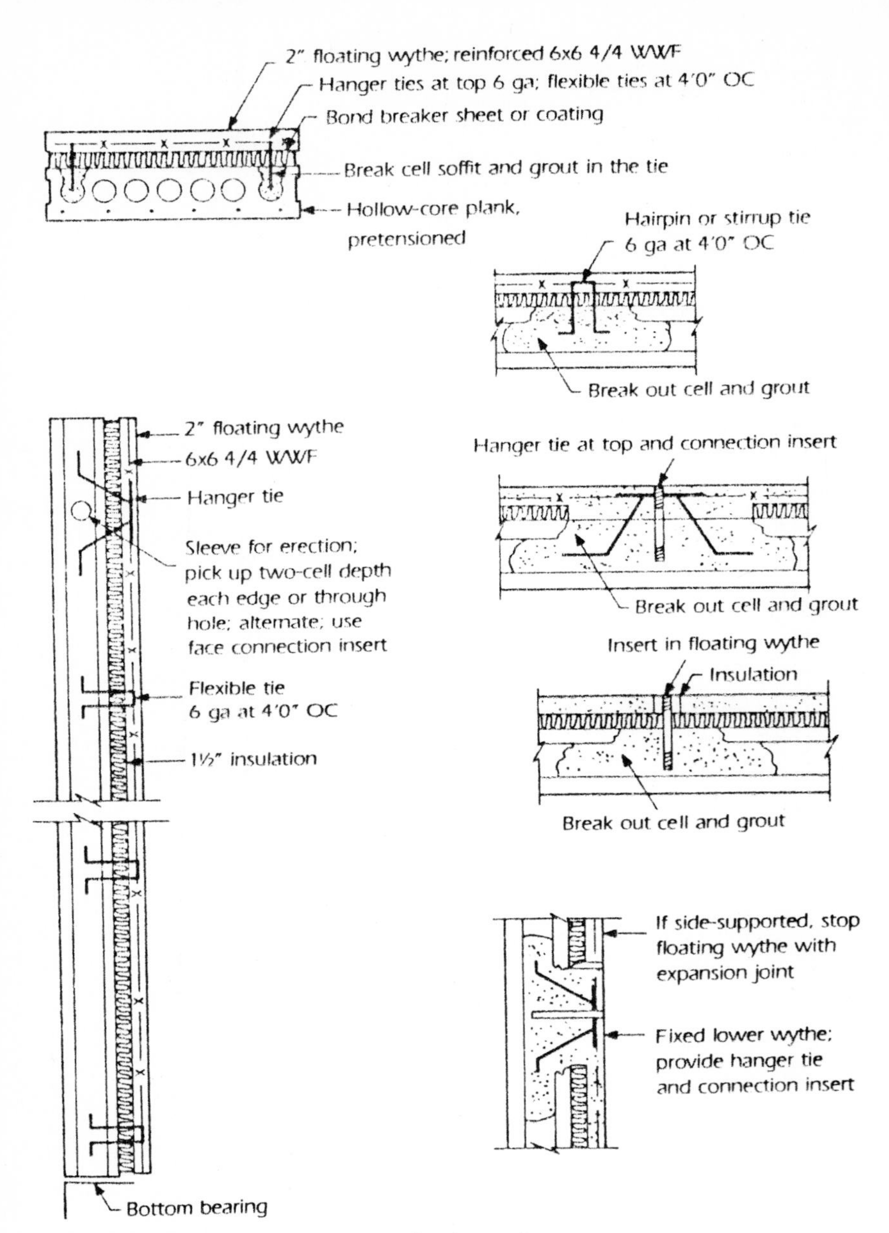


**Figure 23.14** Hollow-core plank sandwich panel.

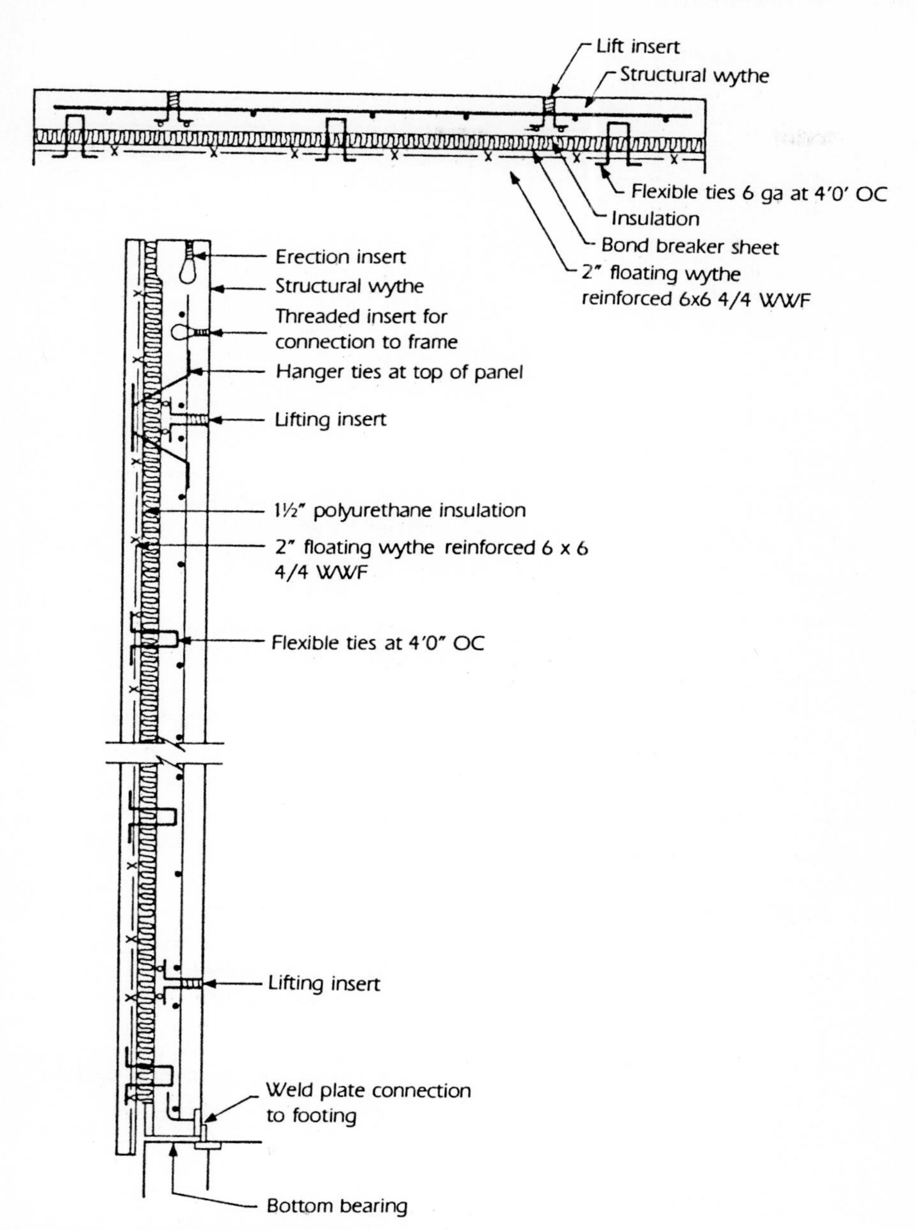


**Figure 23.15** Flat sandwich panel.

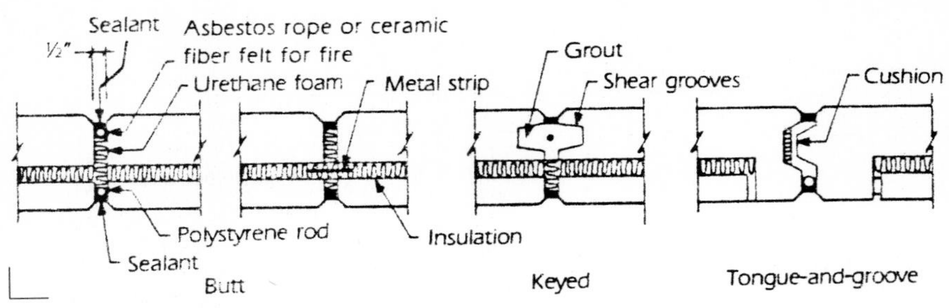


**Figure 23.16** Vertical joints.

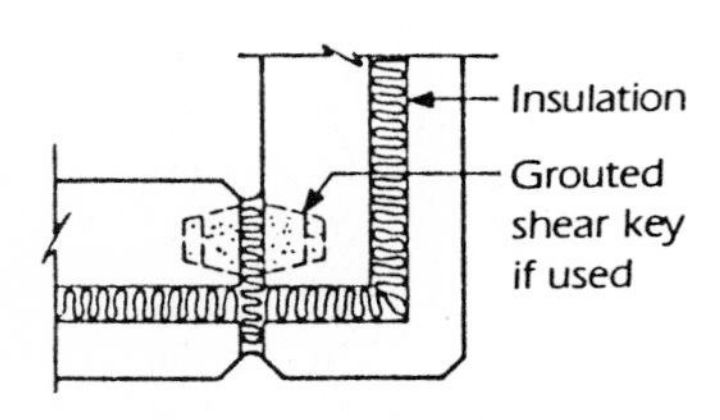


**Figure 23.17** Corner joint.

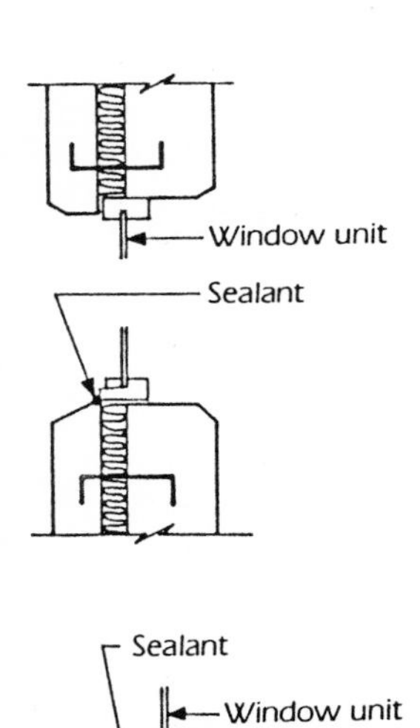


**Figure 23.18** Window openings.

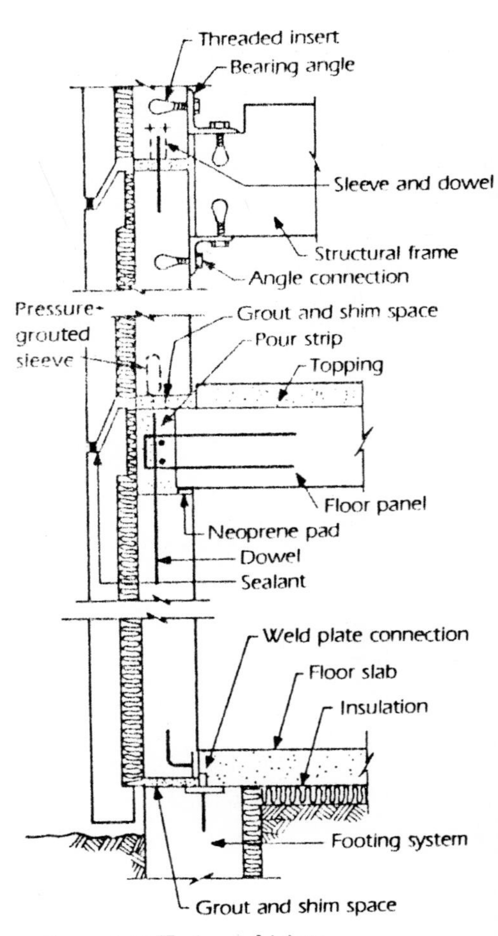


**Figure 23.19** Horizontal joints.

## Composite Panels

A composite panel is one in which the two wythes are designed to act as a unit in resisting transverse or vertical forces. The edges are usually solid concrete, and the wythes are connected with solid concrete ribs, steel ribs, or a combination of concrete and steel. The ribs are designed to resist the imposed horizontal shear forces. If the temperature and humidity difference between the outside and the inside surfaces of the composite panel is zero or small, then the volumetric changes due to temperature, shrinkage, and creep will be the same for both wythes and they will work together with little differential movement. However, as the volumetric changes become different between the two wythes, the effect will be warping or bowing of the panel. The advantage of composite panels is that they have higher strength than equivalent concrete noncomposite panels. Composite panels are usually non-load-bearing and, as such, resist transverse forces only. The shear ties or connecting ribs may be made of concrete, trussed steel joint reinforcement as used in masonry construction, trussed steel studs of many different patterns, or expanded metal lath embedded in a concrete rib. If the panel is pretensioned, then both wythes are pretensioned and the ribs are concrete. The panel edges are usually solid concrete; however, panels have been designed without concrete solid edges but with metal stud or trussed reinforcement as shear ties. (See Figs. 23.20 and 23.21.)

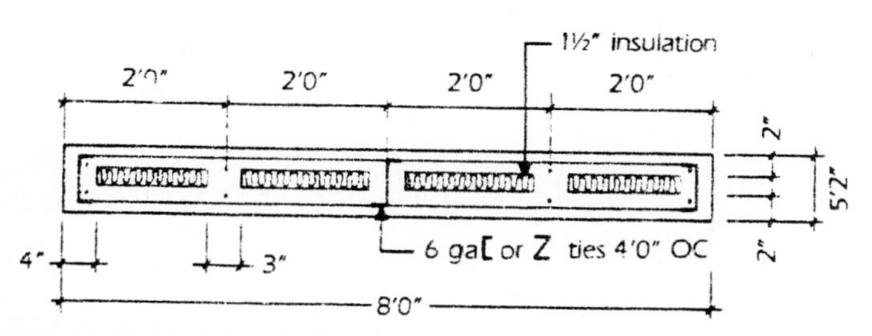


**Figure 23.20** Panel without metal shear reinforcement.

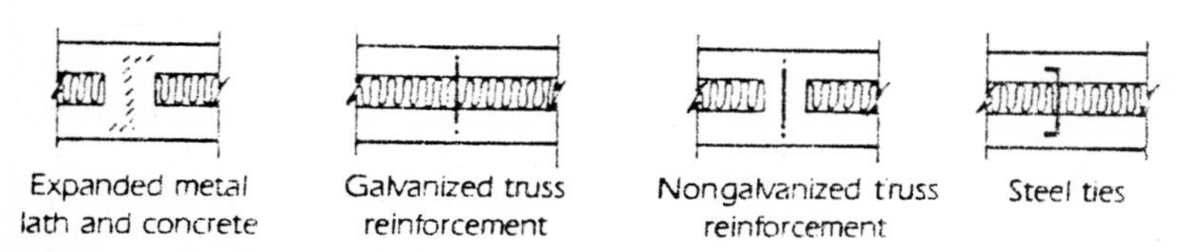


**Figure 23.21** Panels with metal shear reinforcement.

The spacing of the metal ties is often governed by the standard supplied widths of the insulation sheets.

### Thermal-bridge effect

The use of concrete ribs and edges and metal shear ties reduces the thermal efficiency of the panels. They produce thermal bridges (heat flow paths through the panel). The approximate width of the thermal bridge is shown in Fig. 23.22.

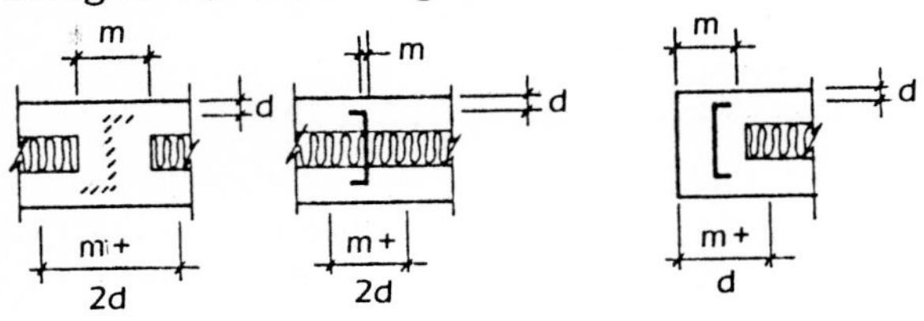


**Figure 23.22** Thermal bridge. $m$ = width of concrete rib or the metal shear tie; $d$ = distance from panel surface to metal tie.

$$U_{av} = \frac{U_{\text{insulated area}} + U_{\text{thermal-bridge area}}}{\text{total panel area}} + \text{mass coefficient}$$

The mass coefficient is used to calculate the total heat loss due to the heat storage factor of the wall mass but is omitted when comparing wall $U$ values.

### Insulation

Insulation may be polystyrene, polyurethane, or fiberglass board. Again, type and thickness are determined by the desired $R$ value, the cost, and the panel thickness. For composite panels the insulation should not be coated or have a bond breaker sheet since as much bond as possible should be developed between the concrete and the insulation in order to improve the composite action.

### Design stresses

In calculating the value of $f_t = \frac{MC}{I}$ for composite panels, the value of $I$ must be corrected because of the stiffness behavior of the shear ties or ribs. The theoretical value of $I$ is

$$I_t = \frac{b}{12}\left(d_o^3 - d_i^3\right)$$

See also Fig. 23.23.

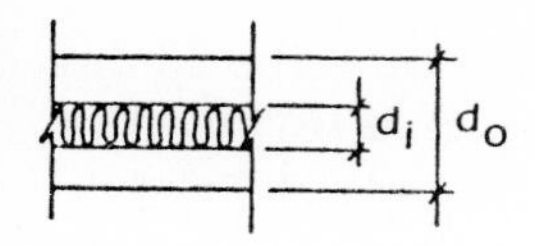


**Figure 23.23** Calculating design stress. $I_{\text{corrected}} = CI_t$; $C$ = correction factor.

Tests conducted by the Portland Cement Association have indicated the following percentages of rigidity to that of $I_{\text{theoretical}}$ for composite panels (correction coefficients):

| | |
|---|---|
| 22 percent | Wythes connected by rigid board insulation or by metal connectors with no shear value |
| 39 percent | Panels with solid concrete periphery edges only |
| 50 percent | Panels with metal truss members but no concrete edges |
| 70 percent | Panels with metal truss or steel stud shear connectors and concrete edges |

The above values are in the direction of loading of the metal shear connectors. If the panel has solid pickup points and solid edges and the width of the panel is small (8 ft or less), then composite action can be assumed in the transverse direction for the transverse bandwidth without the use of metal shear ties in the transverse direction. Thus, the two wythes will work together to distribute the lifting stresses both ways. If the panel has no solid concrete edges, then transverse shear connectors should be provided in the transverse bandwidth at lifting points. The transverse bandwidth is limited to the lesser of 10 times the total panel thickness, the distance from the pickup point to the closest edge, or half the spacing between the pickup points. The value of $C$ is usually taken from 40 to 60 percent, subject to the panel design. The design value of $f_t = \left(7.5\sqrt{f'_c}\right)/\text{FS}$, where FS (factor of safety) is usually 1.5. Handling design is done in the manner specified in Chap. 19, taking into account the reduced value of $I$ and the normal impact factors. (See Fig. 23.24.) If sand-lightweight concrete is used, multiply $f_t$ by 0.85.

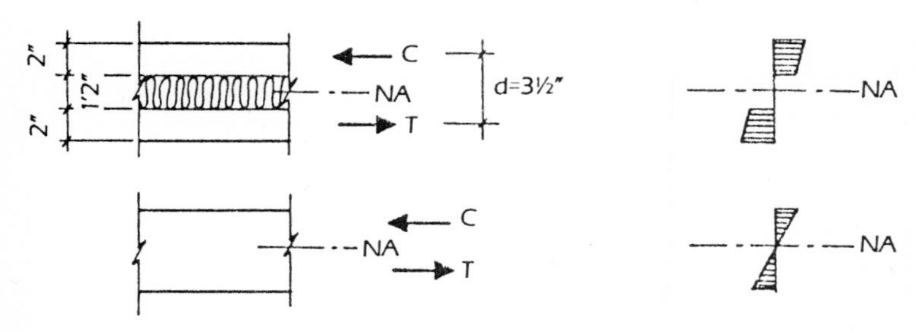


**Figure 23.24** Composite panel equivalent stiffness.

The concrete strength at pickup is usually 2000 lb/in$^2$, and the 28-day strength is usually 5000 lb/in$^2$ or greater. The value of $M$ (moment) is determined in the same way as for nonsandwich panels and is dependent upon the location of the lifting and erection inserts.

### Wythe reinforcement

The reinforcing steel in the wythe has little influence on the lifting and handling operation; hence the panels are designed as uncracked sections with the limiting factor being the tensile stress in the concrete. However, the code-required amount of reinforcing steel is provided to satisfy handling operations, service loads, and volumetric changes. The code minimum is set primarily to control volumetric changes, but lifting stresses are usually the most critical stresses to which the panels are ever subjected.

The code reinforcement requirement for welded wire fabric is 0.0020-$bt$ in the horizontal direction and 0.0012$bt$ in the vertical direction. For 40-ksi-yield reinforcing bars the values are 0.0025 in the horizontal direction and 0.0015 in the vertical direction. With composite panels this can be divided proportionately between the two wythes. For pretensioned panels the strands are balanced and placed in the center of the wythes. In analyzing axial loads, the full section is used to determine the radius of gyration. However, composite panels are not recommended for use as load-bearing panels owing to volumetric strain differences between the wythes. Where they have been used as load-bearing elements, the axial load is usually carried by one wythe.

### Pickup points

Where pickup points or inserts are provided, the panel is usually solid, This solid area is made large enough both to enclose the insert and to resist the imposed shear forces (see Fig. 23.25).

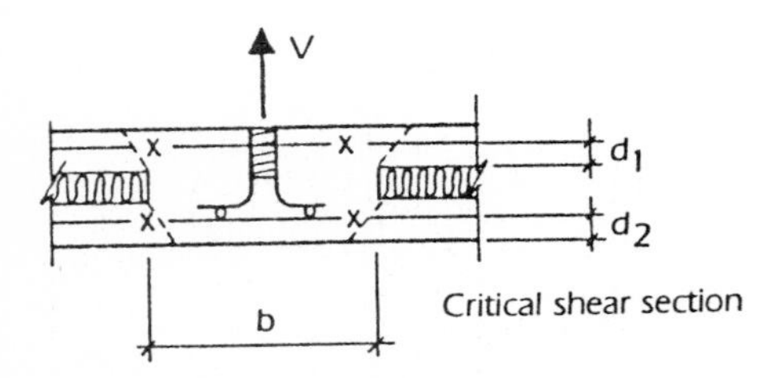


**Figure 23.25** Design for pickup points. $V_c = V/b_o d$, where $V$ = reaction at insert, $b_o$ = 4$b$ (if square), and $d$ = sum of the distances from reinforcing steel to the extreme compression fiber of the wythe.

### Waterproofing and sealant details

These details are shown in Fig. 23.26.

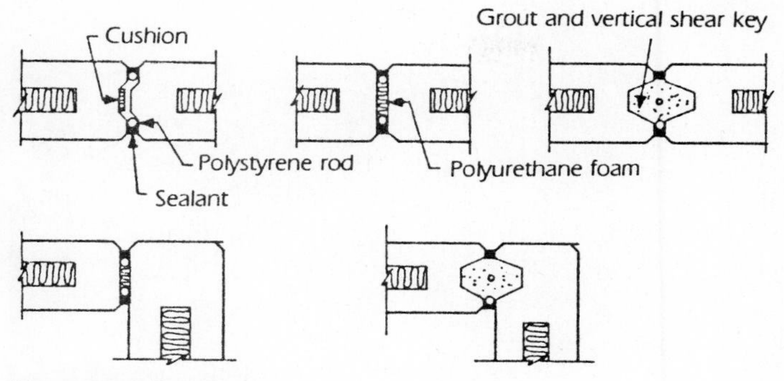


**Figure 23.26** Panel joints.

**Figure 24.1** Bonaventure Hotel, Los Angeles, California. 684,000 $ft^2$ of prestressed concrete flat slabs were used for the floors in this 35-story multitower complex. The structural steel erector installed both the steel framing and the precast prestressed flat slabs in one continuous, uninterrupted operation.

Chapter

# 24

# Prestressed Concrete Solid Flat Slabs

Solid prestressed concrete flat slabs are used for floor or roof members, load-bearing or non-load-bearing wall panels, and spandrel panels or in any situation in which flat slabs in one direction may be used. They may be designed as either solid or insulated units. As deck members, they may be designed as simple spans or continuous spans with or without shoring and use a field-poured cast-in-place concrete topping. They may also be designed as untopped deck elements, with due attention being given to the connections of the elements to ensure satisfactory performance of the diaphragm under seismic loading conditions. Since these elements are relatively thin sections, careful attention should also be given to the camber induced by prestressing as it relates to in-storage and in situ conditions of slab deadweight and possible differential camber. These elements are used as floors in motel and hotel construction (see Fig. 24.2), intermediate spanning slabs in spread T or spread channel floor systems and offices and garages, stay-in-place deck panels in bridge construction, and walls for industrial buildings.

Figure 24.2 Bonaventure Hotel, Los Angeles, California. Shown here are floors in various stages of construction. The longest span was 24 ft, with temporary midspan shoring. Prestressed concrete slabs 3 1/2 in thick with a 2 1/2 in-thick cast-in-place concrete topping formed the entire floor assembly, with the underside of the slab being sprayed with acoustical textured paint to form the finished ceiling.

Solid prestressed concrete slabs are usually fabricated in depths ranging from 3 to 6 in thick. For depths of 6 in or greater it is usually more efficient to use hollow-core slabs or shallow double-T units. The width of the units is restricted by shipping and by the width of the precaster's casting beds. Optimum slab widths range from 8 to 12 ft. Casting beds in the factory may vary from 200 ft to as much as 800 ft in length. Many units are cast in a long line simultaneously, with bulkhead separators between units, or in a continuous pouring operation, with individual units being saw-cut to length after the concrete has reached its required release strength. Untopped units are designed only as simple spans. Topped units are designed as simple spans or as continuous units with temporary midspan shoring. The minimum thickness of units is sometimes governed by such nonstructural considerations as fire-resistive requirements or sound transmission criteria. Table 24.1 indicates maximum spans for various slab thickness and design conditions, and Table 24.2 shows fire rating requirements.

**Figure 24.3** Bonaventure Hotel, Los Angeles, California. Plant casting allowed close control over tolerances. Here we see slabs stockpiled in storage awaiting shipment. Openings for mechanical risers and vertical ventilator ducts were cast in the slabs during fabrication. The 16-gauge metal edge former and Unistrut assembly eliminated perimeter edge forming and provided an instant attachment point for the window wall.

**TABLE 24.1 Preliminary Solid-Slab Design Criteria ($S_{LL}$ = 100 lb/ft²)**

| Slab thickness, in | Maximum simple span without topping, ft | Maximum simple span with 2 1/2 in of hard-rock concrete topping, ft | Maximum continuous span with 2 1/2 in of hard-rock concrete topping and temporary midspan shoring, ft* |
|---|---|---|---|
| Lightweight concrete slab; 1 1/8-in cover over strands | | | |
| 3 1/2 | . . . † | 14 1/2 | 24 |
| 4 | 14 | 16 | 26 |
| 4 1/2 | 15 1/2 | 17 1/2 | 28 1/2 |
| 5 | 17 | 19 | 31 |
| 5 1/2 | 18 1/2 | 20 1/2 | 33 1/2 |
| 6 | 20 | 22 | 36 |

**TABLE 24.1 (*Continued*) Preliminary Solid-Slab Design Criteria ($S_{LL}$ = 100 lb/ft²)**

| Slab thickness, in | Maximum simple span without topping, ft | Maximum simple span with 2 1/2 in of hard-rock concrete topping, ft | Maximum continuous span with 2 1/2 in of hard-rock concrete topping and temporary midspan shoring, ft* |
|---|---|---|---|
| Hard-rock concrete slab; 1 1/8-in cover over strands | | | |
| 4 | . . . † | 15 | 25 |
| 4 1/2 | 14 1/2 | 16 1/2 | 27 1/2 |
| 5 | 16 | 18 | 30 |
| 5 1/2 | 17 1/2 | 19 1/2 | 32 1/2 |
| 6 | 19 | 21 | 35 |

*Longer spans may require third-point shoring.
†Usually not permitted because of fire rating requirements.

**TABLE 24.2 Fire Rating Requirements for Solid Prestressed Concrete Flat Slabs (Uniform Building Code)**

| Item | Fire rating, h | | | |
|---|---|---|---|---|
| | 1 | 2 | 3 | 4 |
| Minimum composite floor slab thickness of hard-rock concrete, in | 3 1/2 | 4 1/2 | 5 1/2 | 6 1/2 |
| Minimum composite floor slab thickness of lightweight concrete, in | 3 | 4 | 4 1/2 | 4 |
| Unrestrained rating | | | | |
| Required cover over strand for hard-rock concrete, in | 1 | 1 1/2 | 2 | 2 1/2 |
| Required cover over strand for lightweight concrete, in | 3/4 | 1 1/8 | 1 1/2 | 1 7/8 |
| Restrained rating | | | | |
| Required cover over strand for hard-rock concrete, in | 3/4 | 3/4 | 1 | 1 1/4 |
| Required cover over strand for lightweight concrete, in | 3/4 | 3/4 | 3/4 | 1 |

## Long-Term Cambers and Deflections

Initial cambers and deflections are magnified with time owing to several factors, among which are loss of prestress force and strength gain of concrete, as well as plastic deformation of the reinforced-concrete complex with time, or creep. The third edition of the *PCI Design Handbook*, page 4-46, presents a chart with suggested multipliers to use in calculating long-term cambers and deflections, taking into account the above factors. Calculate initial camber and member weight deflection by using the initial concrete modulus of elasticity and initial prestress force; always apply the multipliers listed for these values. Calculation of movements occurring due to loads applied in the final erection condition are naturally made by using full values of the modulus of elasticity. These factors will give the total approximate expected cambers and deflections for the condition being analyzed and not just the additional effect, as is given in some codes and publications. See Fig. 24.4.

To satisfy energy requirements, roof panels are usually covered with rigid insulation prior to applying the roofing material. Sandwich panels consisting of two flat slabs with an insulation core between them may be used for both wall and roof panels. In this case one of the slabs is usually the structural slab, and the other serves as a protective-finish surface covering. This method also overcomes problems due to bowing caused by differential thermal expansion between the outer and inner surfaces of the panel.

## Analysis of Topped Prestressed Concrete Flat Slabs

### Design example: unshored flat floor slab

Design a precast pretensioned floor slab for a midrise office building project. Slabs are to be 12 ft wide and span 16 ft. They are to be fabricated with sand-lightweight concrete and must have a 2-hour unrestrained fire rating. The slab concrete strength at release is 3500 lb/in$^2$; at 38 days, 5000 lb/in$^2$. Topping will be normal-weight concrete ($f_c' = 3000$ lb/in$^2$); $1/2$-in $\phi - 270^k$ prestressing strands will be used.

From preliminary design charts a 4-in-thick section is chosen with $2\frac{1}{2}$-in cast-in-place topping (see Fig. 24.4).

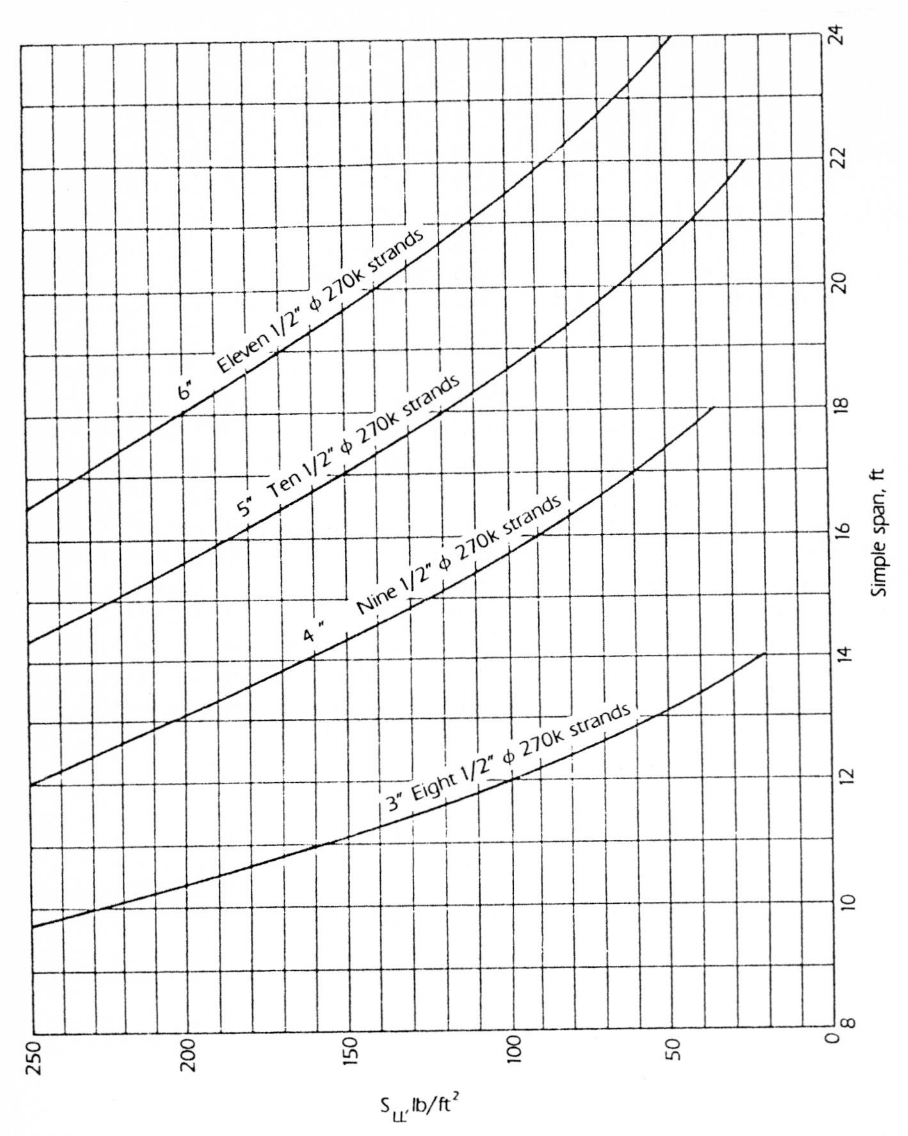


**Figure 24.4** Load table: solid prestressed concrete flat slab plus 2 1/2-in cast-in-place topping (prestressing shown is for a 12-ft-wide slab). $f'_c$ = 5000 lb/in² for a lightweight concrete slab and 3000 lb/in² for hard-rock concrete topping. 1 1/8-in cover over strand (2-hour unrestrained rating).

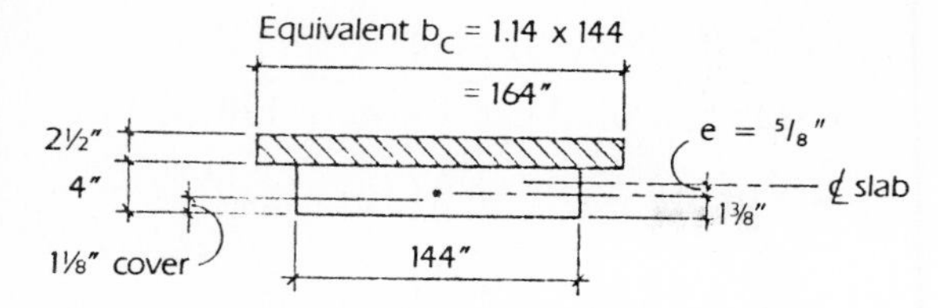


**Figure 24.5** Cross section through slab and topping.

$$\frac{E_{\text{topping}}}{E_{\text{precast}}} = \frac{3.3}{2.9} = 1.14$$

Weight of slab = 38 lb/ft²

Weight of topping = 32 lb/ft²

Live load = 50 lb/ft²

**Section properties.** Determine section properties.

Basic section:

$$A = (144)(4) = 576 \text{ in}^2$$
$$I = (1/12)(144)(4^3) = 768 \text{ in}^4$$
$$S_b = S_t' = (\tfrac{1}{2})(768) = 384 \text{ in}^3$$

Composite section:

| Part | $A$ | $y_{tcc}$ | $ay_{tc}$ | $I_c$ | $Ad^2$ |
|---|---|---|---|---|---|
| Basic | 576 | 4.50 | 2592 | 768 | 1049 |
| Topping | 410 | 1.25 | 513 | 213 | 1480 |
| Total | 986 | | 3105 | | |

$\Sigma I_c = 3510.$

$$y_{tcc} = \frac{3105}{986} = 3.15 \text{ in}$$
$$y_{tc} = 0.65 \qquad S_{tc} = 5400 \text{ in}^3$$
$$y_{bc} = 3.35 \qquad S_{bc} = 1048 \text{ in}^3$$

**Loads and moments.** For a 12-ft-wide slab.

Slab weight $M_S = (1/8)(12)(0.038)(16^2)(12) = 175^{k \cdot in}$

Topping $M_C = (1/8)(12)(0.032)(16^2)(12) = 147^{k \cdot in}$

Live load $M_L = (1/8)(12)(0.050)(16^2)(12) = 230^{k \cdot in}$

**Allowable stresses.** Stresses are as follows:

Concrete:

At transfer $(f'_{ci} = 3500 \text{ lb/in}^2)$

Compression: $0.6 \times 3500 = 2100 \text{ lb/in}^2$

Tension: $3 \times \sqrt{3500} = 177 \text{ lb/in}^2$

Final design $(f'_c = 5000 \text{ lb/in}^2)$

Compression: $0.45 \times 5000 = 2250 \text{ lb/in}^2$

Tension: $6 \times \sqrt{5000} = 425 \text{ lb/in}^2$

Strand:

At transfer (assume 10 percent losses):

$f_i = (0.9)(0.7)(270) = 170$ ksi

Final design (assume 20 percent losses):

$f_i = (0.8)(0.7)(270) = 151$ ksi

**Prestress force.** Calculate prestress force required.

$$F_f = \frac{\dfrac{M_B}{S_B} + \dfrac{M_{BC}}{S_{BC}} - f_t}{\dfrac{1}{A} + \dfrac{e}{S_B}} = \frac{\dfrac{175 + 147}{384} + \dfrac{230}{1048} - 0.425}{\dfrac{1}{576} + \dfrac{0.625}{384}} = 188^k$$

$$N = \frac{188}{151 \times 0.1531} = 8.13$$

Use eight ½-in ϕ-270k strands

Use initial tension of $\dfrac{188}{(8)(0.153)(270)(0.8)} = 0.71 f_{pu}$

$$f_i = (0.9)(0.71)(270) = 173 \text{ lb/in}^2 \quad f_f = (0.8)(0.71)(270) = 153 \text{ ksi}$$

$$F_i = (8)(173)(0.1531) = 212^{\text{k}} \quad F_f = (8)(153)(0.1531) = 187^{\text{k}}$$

At release:

$$-\frac{F_i}{A} \pm \frac{F_i e}{S_B} \pm \frac{M_S}{S_B}$$

At end:

$$f_t = -\frac{212}{576} + \frac{(212)(0.625)}{384} = -0.23 \text{ (compression)}$$

$$f_b = -0.368 - 0.345 = -0.713 \text{ (compression)}$$

At centerline:

$$f_t = -0.368 + 0.345 - \frac{175}{384} = -0.479 \text{ (compression)}$$

$$f_b = -0.368 - 0.345 + 0.456 = -0.257 \text{ (compression)}$$

**Final design condition under all loads.** At centerline of span:

$$f_t = -\frac{187}{576} + \frac{(187)(0.625)}{384} - \frac{322}{384} - \frac{230}{5400} = -0.902 < 0.45 f_c' \quad \text{(OK)}$$

$$f_b = -0.324 - 0.304 + 0.839 + \frac{230}{1048} = +0.430 \simeq 6\sqrt{f_c'} \quad \text{(OK)}$$

**Ultimate strength.** Check ultimate strength.

$$\rho_p = \frac{A_{ps}}{bd} = \frac{(3)(0.1531)}{(144)(5.13)} = 0.001658$$

$$f_{ps} = f_{pu}\left(1 - 0.5\rho_p \frac{f_{pu}}{f_c'}\right) = 270\left(1 - 0.5 \times 0.001658 \times \frac{270}{5}\right)$$

$$= 257.9 \text{ ksi}$$

$$a/2 = \frac{A_{ps} f_{ps}}{1.7 f_c' b} = \frac{(8)(0.1531)(257.9)}{(1.7)(3)(144)} = 0.48 \text{ in}$$

$$M_u = \phi A_{ps} f_{ps} (d - a/2) = (0.9)(8)(0.1531)(257.9)(5.13 - 0.43)$$

$$= 1336^{\text{k}\cdot\text{in}}$$

$$UM = 1.4M_D + 1.7M_L = (1.4)(175 + 147) = (1.7)(230)$$
$$= 842^{k \cdot in} < M_u \quad (OK)$$

**Transverse steel in precast slab.** For temperature and handling.

$$A_s = (0.002)(4)(12) = 0.096 \text{ in}^2/\text{ft required}$$

Use no. 3 bars at 12 in on center transverse.

**Shear.** Check shear.

$$v_u = \frac{UV}{bd} = \frac{(1.4)(0.07)(8)(12) + (1.7)(0.05)(8)(12)}{(144)(5.13)} = 0.024 \text{ ksi}$$

$$v_c = 0.85\phi\, 2\sqrt{f'_c} = \frac{(0.85)(0.85)(2)(\sqrt{5000})}{1000} = 0.102 \text{ ksi} \quad (OK)$$

**Deflection.** Check deflection.
Instantaneous camber:

$$\Delta_{P/S} \uparrow = \frac{F_i e L^2}{8EI} = \frac{(212)(0.63)(16^2)(144)}{(8)(2400)(768)} = 0.33 \text{ in} \uparrow$$

Deflection of slab self-weight (assume slab is blocked in storage 1 ft from each edge):

$$\Delta_S \downarrow = \frac{5wL^4}{384EI} = \frac{(5)(0.038)(12)(14^4)(1728)}{(384)(2400)(768)} = 0.21 \text{ in} \downarrow$$

By using a creep factor of 1.8, when the slab reaches the jobsite, it will exhibit a net upward deflection of

$$1.8 \times (0.33 - 0.21) = 0.21 \text{ in} \uparrow$$

When wet topping is poured, the slab will deflect an additional

$$\Delta_{topping} \downarrow = \frac{(5)(0.032)(12)(16^4)(1728)}{(384)(2900)(768)} = 0.25 \text{ in} \downarrow$$

Long-term effect of topping weight (*additional* creep deflection):

$$1.5 \times \frac{(5)(0.032)(12)(16^4)(1728)}{(384)(3100)(3510)} = 0.05 \text{ in} \downarrow$$

Instantaneous live-load deflection:

$$\Delta_{LL} = \frac{(5)(0.05)(12)(16^4)(1728)}{(384)(3100)(3510)} = 0.08 \text{ in} \downarrow$$

Maximum live-load deflection allowable from American Concrete Institute 318-83, Table 9.5 (b):

$$\frac{L}{480} = \frac{16 \times 12}{480} - 0.40 \text{ in} > \Delta_{LL} \quad (\text{OK})$$

Net condition under full loading:

$$+\uparrow \Sigma\Delta = +0.21 - 0.25 - 0.05 - 0.08 = -0.17 \text{ in} \downarrow$$

**Handling.** Check handling conditions (see Fig. 24.6).

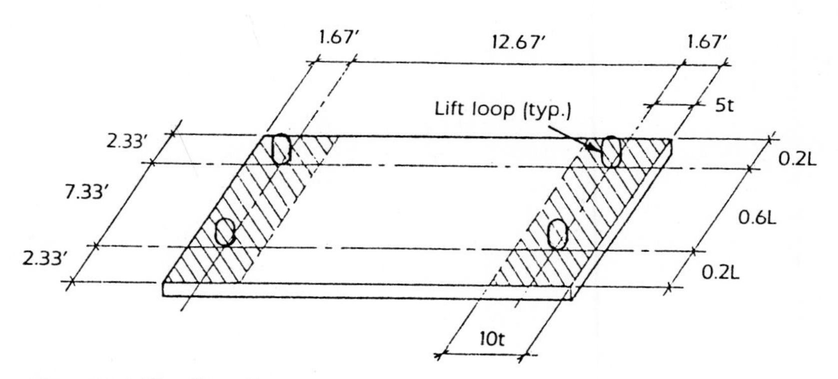


**Figure 24.6** Handling diagram.

Stripping (use impact factor $I$ = 1.5):

$$M^- = (0.056)\left(\frac{1.67^2}{2}\right)(12) = 0.93^{\text{k}\cdot\text{in}}/\text{ft}$$

$$M^+ = (0.056)\left(\frac{12.67^2 \times 12}{8}\right) - 0.93 = 12.55^{\text{k}\cdot\text{in}}/\text{ft}$$

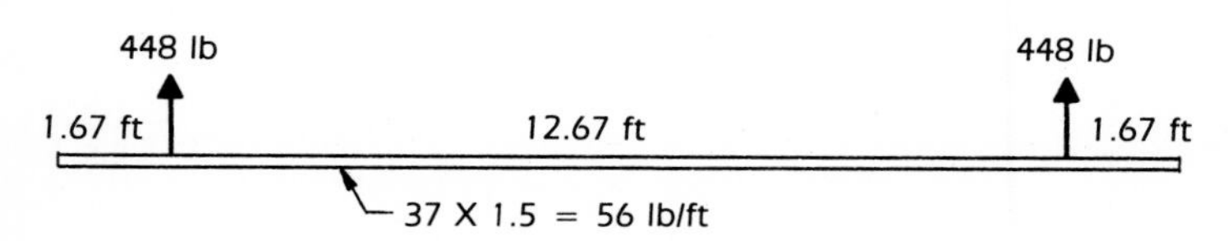

Check tensile stresses.
At cantilever:

$$f_{top} = -\frac{F_i}{A} + \frac{F_i e}{S_T} = \frac{M_{strip}}{S_T} = -\frac{17.7}{48} + \frac{(17.7)(0.63)}{32} + \frac{0.93}{32}$$

$$= -0.569 + 0.348 + 0.030 = +0.009 \text{ ksi (OK)}$$

At midspan:

$$f_{bottom} = -\frac{17.7}{48} - \frac{(17.7)(0.63)}{32} + \frac{12.55}{32} = -0.325 \text{ ksi (OK)}$$

Allowable tensile stress at stripping:

$$F_T = \frac{f_{ri} \times 0.85}{\text{FS}} = \frac{7.5\sqrt{3500} \times 0.85}{2.0} = +0.189 \text{ ksi}$$

Transverse condition (assume transverse beam 40 in wide):

$$M^{+} = M^{-} = \frac{(0.444)(2.33^2)}{2} = 14.46^{\text{k·in}}$$

$$f_t = \frac{M}{S} = \frac{14.46}{32 \times 3.33} = +0.136 \text{ ksi} < 0.189 \text{ (OK)}$$

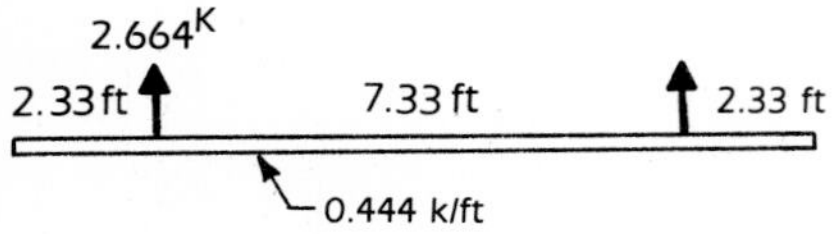

2.664^K
2.33 ft 7.33 ft 2.33 ft
0.444 k/ft

**Shipping.** Block at lifting points with transverse dunnage; use $I = 2.0$. By inspection from longitudinal stripping condition, this is not critical.

**Figure 24.7** Palo Alto Financial Center. This totally precast concrete building features the use of exterior and deck elements acting compositely with cast-in-place concrete. Here we see 2 3/4-in-thick solid prestressed lightweight concrete slabs being erected on temporary midspan shoring. The final design span is 18 ft, with shoring being removed after 2 3/4-in of lightweight concrete topping has been cured.

**Figure 24.8** Wesley Towers housing for the elderly, Campbell, California. Solid prestressed concrete slab 4 in thick being erected on precast concrete "voided" bearing walls. Temporary midspan shoring is used until the cast-in-place topping has been poured and cured, thus providing a 24-ft-clear-span floor.

**Design example: shored solid-slab system**

Design a solid prestressed concrete slab to span 28 ft between masonry bearing walls for a motel project. Tempoary midspan shoring will be provided. A 2-hour fire rating is required. (See also Figs. 24.7 and 24.8.) From Table 43-A of the Uniform Building Code, in order to achieve a 2-hour rating $1\frac{1}{2}$-in of cover is required for hard-rock concrete, and $1\frac{1}{8}$-in of cover is required for lightweight concrete. By referring to preliminary load-span design tables, a 4 in-thick hard-rock concrete slab with $2\frac{1}{2}$ in of cast-in-place concrete topping, spanning continuously between supports, is selected. (Note that only a slight increase in maximum span is afforded by using lightweight aggregate concrete.)

**Section properties per foot of width.** See Fig 24.9.

Basic section:

$$f'_c = 5000 \text{ lb/in}^2 \qquad f'_{ci} = 3500 \text{ lb/in}^2$$

$$A = 48 \text{ in}^2 \qquad I_B = 64 \text{ in}^4$$

$$Y_b = Y_t = 2 \text{ in} \qquad S_b = S_t = 32 \text{ in}^3$$

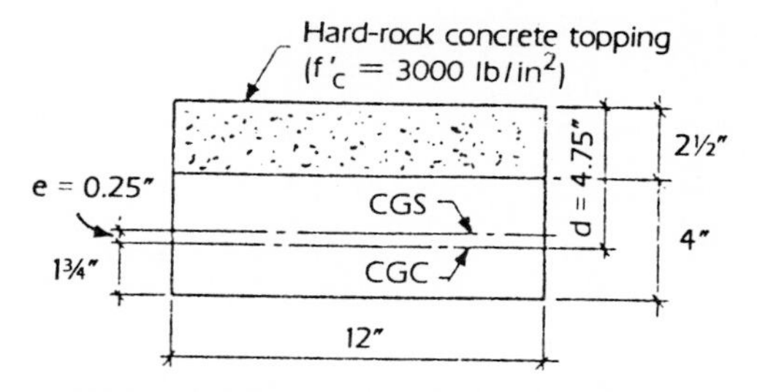


**Figure 24.9** Design for shored solid-slab system.

Composite section:

$$\frac{E_c}{E_B} = \frac{3300}{4500} = 0.77$$

Equivalent composite topping width = $0.77 \times 12 = 9.24$ in

| Part | $A$ | $Y_{tcc}$ | $Ay_{tcc}$ | $I_c$ | $Ad^2$ |
|---|---|---|---|---|---|
| Precast | 48 | 4.5 | 216 | 64 | 48 |
| Composite, 9.24 × 2.5 in | 23 | 1.25 | 29 | 12 | 116 |
| Σ | 71 | | 245 | | |

$\Sigma I = I_c = 240 \text{ in}^4$.

$$Y_{tcc} = \frac{245}{71} = 3.5 \text{ in}$$

$$Y_{bc} = 3.0 \text{ in} \qquad S_{bc} = \frac{240}{3} = 80 \text{ in}^3$$

Service loads:

| | |
|---|---|
| 4-in prestressed solid slab | 50 lb/in² |
| 2 1/2 -in cast-in-place concrete topping | 32 lb/in² |
| Partitions and miscellaneous | 15 lb/in² |
| Live load | 40 lb/in² |
| Temporary construction load ($S_{D+L}$)* | 100 lb/in² |

1. Elastic analysis: service load design
   (*a*) Maximum positive-moment condition
      (1) Weight of prestressed slab plus wet topping (see Fig. 24.10).

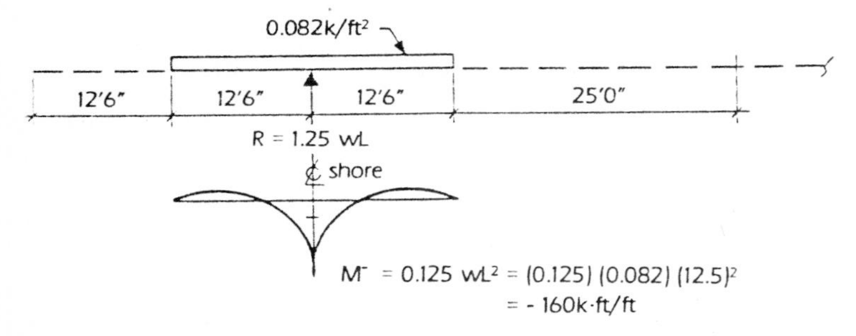


**Figure 24.10** Design for service load.

(2) Effect of shore removal. Place two bearing walls 12 ft 6 in apart at ends of building and at either side of expansion joints in order to minimize positive moment in first 25 ft interior span. (See Fig. 24.11.)

*OSHA regulations.

+

Moment sign convention

$$K_{AB} = \frac{80}{12.5} = 6.4$$

$$K_{AB} = \frac{3}{4} \times 6.4 = 4.8$$

$$K_{BC} = \frac{80}{25} = 3.2$$

$$\text{FEM} = \frac{PL}{8} = \frac{(1.28)(25)}{8} = 4.0^{\text{k}\cdot\text{ft}}/\text{ft}$$

$1.25wL = (1.25)(0.082)(12.5) = 1.28^{\text{k}}$

$I = 80$ constant

*A* *B* *C* *D*

12 ft 6 in 25 ft 0 in 25 ft 0 in

| | | | | | | | |
|---|---|---|---|---|---|---|---|
| K | 4.8 | 6.4 | 3.2 | 3.2 | 3.2 | 3.2 | 3.2 |
| DF ($K/\Sigma K$) | 1.0 | 0.67 | 0.33 | 0.5 | 0.5 | 0.5 | 0.5 |
| FEM* | 0 | 0 | +4.00 | -4.00 | 0 | 0 | 0 |
| Balance | | -2.68 | -1.32 | +2.00 | +2.00 | | |
| Carryover | | | +1.00 | -0.66 | | +1.00 | |
| Balance | | -0.67 | -0.33 | +0.33 | +0.33 | -0.50 | -0.50 |
| Carryover | | | +0.16 | -0.16 | -0.25 | +0.16 | |
| Balance | 0 | -0.11 | -0.05 | +0.21 | +0.20 | -0.08 | -0.08 |
| Adjusted moment | | -3.46 | +3.46 | -2.28 | +2.28 | +0.58 | -0.58 |

*FEM = fixed-end moment.

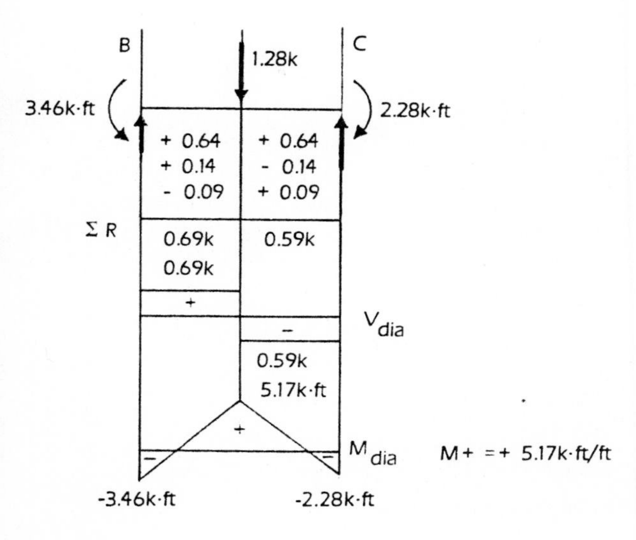


**Figure 24.11** Effect of shore removal.

(3) Superimposed dead and live load on first 25-ft interior span.

$$\text{FEM} = \frac{wL^2}{12} = 0.055 \times \frac{25^2}{12} = 2.86^{\text{k}\cdot\text{ft}}/\text{ft}$$

0.055 k/ft²

| | A | B (12 ft 6 in) | C (25-ft 0-in) | | D (25 ft 0 in) | | |
|---|---|---|---|---|---|---|---|
| K | 4.8 | 6.4 | 3.2 | 3.2 | 3.2 | 3.2 | 3.2 |
| DF ($K/\Sigma K$) | 1.0 | 0.67 | 0.33 | 0.5 | 0.5 | 0.5 | 0.5 |
| FEM* | 0 | 0 | +2.86 | -2.86 | 0 | 0 | 0 |
| Balance | | -1.92 | -0.94 | +1.43 | +1.43 | | |
| Carryover | | | +0.72 | -0.47 | | +0.72 | |
| Balance | | -0.48 | -0.24 | +0.24 | +0.23 | -0.36 | -0.36 |
| Carryover | | | +0.12 | -0.12 | -0.18 | +0.12 | |
| Balance | | -0.08 | -0.04 | +0.15 | +0.15 | -0.06 | -0.06 |
| Adjusted moment | 0 | -2.48 | +2.48 | -1.63 | +1.63 | +0.42 | -0.42 |

*FEM = fixed-end moment.

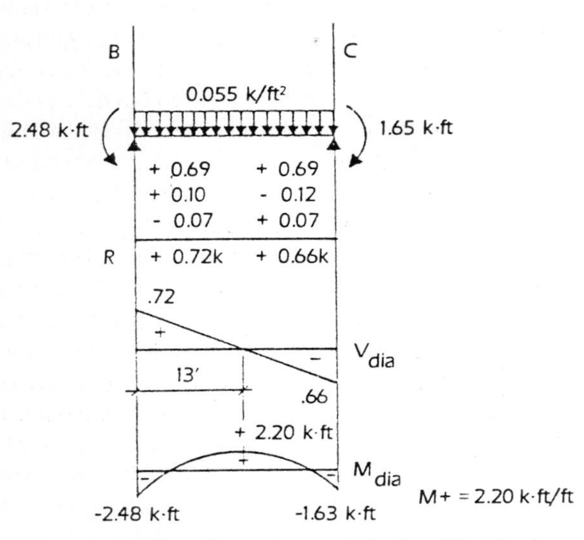


**Figure 24.12** Effect of superimposed dead and live load.

Prestress force required to develop positive moment:

$$F_f = \frac{\dfrac{M_B}{S_b} + \dfrac{M_C}{S_{bc}} - f_t}{\dfrac{1}{A} + \dfrac{e}{S_b}} = \frac{\dfrac{-1.60 \times 12}{32} + \dfrac{(5.17 + 2.20)(12)}{80} - 0.425}{\dfrac{1}{48} + \dfrac{0.25}{32}}$$

$$= \frac{-0.600 + 1.106 - 0.425}{0.02083 + 0.00781} = \frac{0.081}{0.029} = 2.8^{\text{k}}/\text{ft (small)}$$

Therefore ultimate strength analysis governs strand requirement.

(*b* ) Maximum negative-moment condition in slab*

12'6" 12'6"

℄ shore

$$M^{-} = (0.125)(0.100)(12.5^2) = 1.95^{\text{k·ft}}/\text{ft}$$

Prestress force required to develop negation moment (see *PCI Design Handbook*, 3d edition, page 11-15):

Use allowable tension equal to modulus of rupture.

$$f_r = 7.5\sqrt{f'_c} = 0.530 \text{ ksi}$$

$$f_t = -\frac{F_f}{A} + \frac{F_f \times e}{S_t} + \frac{M}{S_t}$$

$$f_t = \left(\frac{1}{A} - \frac{e}{S_t}\right) = \frac{M}{S_t} - f_t$$

$$F_f = \frac{\dfrac{M}{S_t} - f_t}{\dfrac{1}{A} - \dfrac{e}{S_t}} = \frac{\dfrac{(1.95)(12)}{32} - 0.530}{\dfrac{1}{48} - \dfrac{0.25}{32}} = \frac{0.731 - 0.530}{0.02083 - 0.00781}$$

$$= \frac{0.201}{0.013} = 15.5^{\text{k}}/\text{ft}$$

*OSHA regulations.

For a 12-ft-wide slab the required number of ½-in ϕ270$^{k}$ strands is (*PCI Design Handbook*, page 11-16)

$$A_{ps} = 0.1531 \quad F_o = 28.9^{k}/\text{strand} \quad F_f = 0.78 \times 28.9 = 22.5^{k}/\text{strand}$$

$$N = \frac{15.5 \times 12}{22.5} = 8.3 = 9 \text{ strands required}$$ (*PCI Design Handbook*, page 4-39)

2. Ultimate strength design

$$UM^{+} \cong (1.4)(5.17) + (1.4)(15/55)(2.2) + (1.7)(40/55)(2.2)$$
$$= 7.24 + 0.84 + 2.72$$
$$= 10.8^{k\cdot ft}/\text{ft of slab}$$

Determine prestressing requirements (*PCI Design Handbook*, page 4-61).

$$K_u = \frac{12{,}000\,UM}{bd^2} = \frac{(12{,}000)(10.8)}{(12)(4.75^2)} = 479$$

For $f'_c = 3000$ and $K_u = 480$:

$\overline{w}_p = 0.228 < 0.300$ [OK; Uniform Building Code, Sec. 2618(i)]

$$A_{ps} = \overline{w}_p bd\, f'_c / f_{pu}$$
$$= (0.228)(12)(4.75)(3/270) = 0.144 \text{ in}^2/\text{ft}$$

$$N = \frac{0.144 \times 12}{0.1531} = 11.3$$ (*PCI Design Handbook*, page 11-16)

Use twelve ½-in ϕ - 270$^{k}$ strands.

Check deflections (*PCI Design Handbook*, page 4-46):

$$\Delta \cong 0.008\frac{PL^3}{EI} + 0.004\frac{wL^4}{EI}$$

where $p = 1.28^{k}$(shore)

$w = 0.015^{k}/\text{ft}\ (s_{DL})$

$$E = \frac{(4300)(4) + (3300)(2.5)}{6.5} = 3900 \text{ ksi}$$

$$\Delta = \left\{ 0.008\left[\frac{(1.28)\left(25^{3}\right)}{(3900)(240)}\right] + 0.004\left[\frac{(0.015)\left(25^{4}\right)}{(3900)(240)}\right]\right\}(1728)$$

$$= (0.000171 + 0.000025)(1728) = 0.034 \text{ in}$$

$$\Delta_{LT} \cong 2.5 \times 0.34 = 0.85 \text{ in}$$

$$\frac{L}{240} = \frac{25 \times 12}{240} = 1.25 \text{ in (OK; Uniform Building Code, Table 23 D)}$$

Check horizontal shear between slab and topping (*PCI Design Handbook*, page 4-28; see shear diagrams for critical loading conditions):

Positive moment region:

$$UV^{+} = (1.4)(0.65) + (1.4)\left(\frac{15}{55}\right)\left(\frac{0.69}{2}\right) + (1.7)\left(\frac{40}{55}\right)\left(\frac{0.69}{2}\right) = 1.47^{k}$$

$$\phi V_u h = \phi \times 80bvd = \frac{(0.85)(80)(12)(4.75)}{1000} = 3.88^{k} > UV \quad \text{(OK)}$$

Negative moment region:

$$UV^{-} = (1.4)(0.69) + (1.4)\left(\frac{15}{55}\right)(0.60) + (1.7)\left(\frac{40}{55}\right)(0.60) = 1.93^{k} > UV \quad \text{(OK)}$$

Therefore roughen top of slab to satisfy horizontal shear.
Check vertical shear:

$$v_u \max = \frac{UV^{-}}{bd} = \frac{1.93}{(12)(4.75)} = 0.034 \text{ in}$$

$$1/2\phi v_c = 1/2 \times 0.85 \times 2\sqrt{5000} = 0.060 \text{ ksi} > v_u \quad \text{(OK)}$$

No web reinforcing is required.
Check handling of 25-ft-long slabs; shipping governs.
Try two-point support transverse.

I = 2.0

W = 2.0 × 0.05 = 0.1k/ft²

0.2L 0.6L 0.2L

$$M^- = M^+ = 0.025 \times 0.1 \times 25^2 = 1.56^{\text{k·ft}}/\text{ft}$$

$$f^+ = \frac{M}{S} = \frac{1.56 \times 12}{32} = 0.585^{\text{k}}/\text{in}^2 \text{ (excessive)}$$

Use continuous longitudinal dunnage on truck:

$$M = 0.025 \times 0.1 \times 12^2 = 0.36^{\text{k·ft}}/\text{ft}$$

$$f_{\text{transverse}} = \frac{0.36 \times 12}{32} = 0.135 \text{ ksi}$$

$$f_{T,\text{ allowable}} = \frac{f_r}{1.5} = \frac{0.530}{1.5} \; 0.350 \text{ ksi (OK)}$$

Determine negative moment requirements in topping over supports.

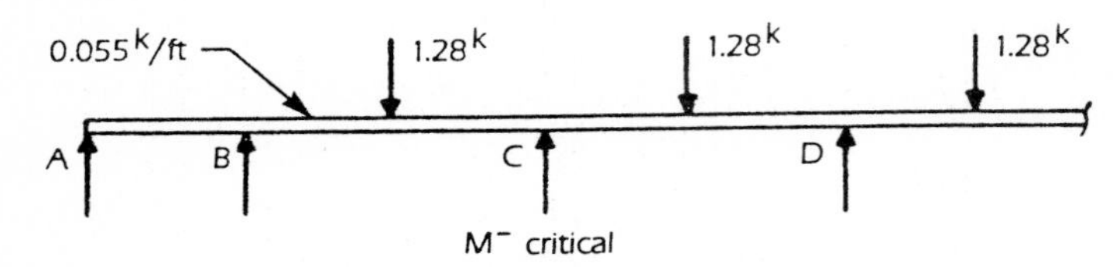


By using moment distribution $M^-_{\text{max}}$ is found to occur at support $C$.

$$M_C^- = \frac{\text{shore load } 50 + L}{0.135PL + 0.088wL^2} = \frac{\text{shore load } 50 + L}{4.33 + 3.04} = 7.37^{\text{k·ft}}/\text{ft}$$

$$UM^- = (1.4)(4.33) + (1.4)(15/50)(3.04) + (1.7)(40/50)(3.04)$$
$$= 11.5^{\text{k·ft}}/\text{ft}$$

By using ASTM 615-60 reinforcement, from *ACI Design Handbook,*

$$A_s = \frac{UM}{a_u d} = \frac{11.5}{4.18 \times 5.5} = 0.50 \text{ in}^2/\text{ft, less topping mesh area}$$

Use no. 6 bars at 12 in on center; use 5000 lb/in² grout between slabs at walls.

Transverse temperature steel:

$$0.002 \times 12 \times 4 = 0.096$$

Use no. 3 bars at 12 in on center transverse.

Reinforcing of the solid-slab system is summarized in Fig. 24.13.

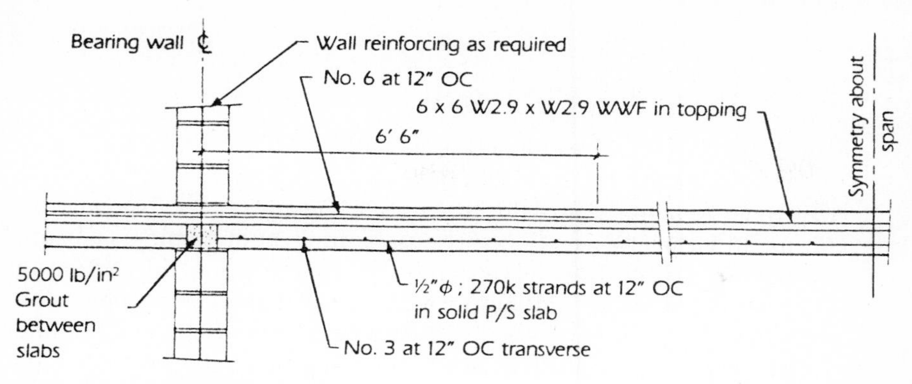


**Figure 24.13** Reinforcing summary.

## Analysis of Untopped Prestressed Concrete Flat Slabs

Untopped prestressed concrete slabs are a very economical solution for floors in hotels and housing or in conjunction with spread systems in parking structures. Other types of structures such as offices use cast-in-place topping to surround cast-in electrical conduit and communication services in addition to performing the structural function, thereby normally precluding the use of untopped slabs. When untopped slabs are used, the boundaries should be tied into supporting members or chord beams by poured-in-place concrete and bars or the strand left exposed beyond the ends of the members. It is recommended that shear keys between adjacent slabs be wide enough to allow the grouted-joint top surface to form a gradual change in elevation between adjacent slabs exhibiting a small degree of differential camber. This keyway should also be castellated to develop horizontal shear forces developed by the composite untopped slab diaphragm. A recommended joint detail and typical forces acting on the untopped slab diaphragm are shown below.

In analyzing these diaphragms, the following assumptions are made (see Fig. 24.15):

1. The slab diaphragm is assumed to be a rigid plane that is fully supported laterally.
2. The diaphragm transfers lateral shears to stiffening elements (shear walls, frame elements, etc.) which provide overall structural stability.
3. A properly designed slab diaphragm resists both shear and flexural stresses.

The three types of loads shown above that are resisted by the untopped diaphragm and its joints are:

$v$ Horizontal shear parallel to the grouted slab joint. $v_s$ is calculated by the relationship

$$v_s = \frac{V_{ay}}{I_b} \quad \text{(unit shear stress, ksi)}$$

with the slab diaphragm acting as a beam.

$V_e$ Horizontal shear parallel to the end of the slab at the bearing. $V_e$ is usually the end reaction of the slab diaphragm acting as a beam where $V_e$ is the drag force being taken into a lateral stiffening element. The force $V_e$ is usually transferred from the untopped slab diaphragm by bars or strand protruding beyond the ends of the slabs, providing a positive mechanical load transfer at the interface.

$V_p$ Vertical shear perpendicular to the grouted slab joint. $V_p$ is a result of load transfer from a point loading from one side of the slab joint.

The shear castellation may be conservatively estimated to have the following capacity:

Cross-sectional area of castellation:

$$(2.5)(1.5)(0.5) = 1.875 \text{ in}^2$$

Using 3000 lb/in² grout, and an ultimate bearing capacity of $0.7 f'_c = (0.7)(3000)$, or 2100 lb/in², the ultimate resisting load per key is

$$(1.875)(2100) = 3939 \text{ lb/key}$$

or

$$1.938^k/\text{key} \times \frac{12}{10} = 4.73^k/\text{ft ultimate horizontal}$$

shear resistance of castellated shear key

Preliminary load tables are shown for untopped slabs fabricated in both hard-rock and lightweight concrete. In most applications the as-built condition will permit these slabs to be considered as being restrained. Finally, in designing these slabs the prestress force should be such that the instantaneous camber at release is 1.1 times the instantaneous dead-load deflection produced by the slab self-weight. The following example serves to demonstrate that proper design of untopped slabs is a result of a camber analysis, with the load-carrying capacity being determined afterward and compared with the capacity required. This procedure will assure successful performance in the finished structure, minimizing the potential of problems resulting from differential camber and creep sag occurring with time.

In lieu of providing protruding end steel from the slabs and forming the castellated shear key, which may make smaller projects excessively costly, the diaphragm may be formed by using intermediate and boundary reinforcing calculated by the use of the shear friction concept as covered by the Uniform Building Code, Sec. 2611. The use of this type of reinforcing to maintain the integrity of diaphragms composed of untopped slab elements subjected to seismic (cyclic) loading has been substantiated by testing. This shear friction reinforcing is in addition to any reinforcing required to satisfy flexural loading conditions.

**Figure 24.14** Los Angeles County Public Social Services parking garage, Long Beach, California. Untopped prestressed concrete slabs 5 in thick span between double-T channels that span 63 ft between supports. Shear castellations in the plank sides and a positive tie into the field-poured concrete over the channel ensure that the completed floor acts as a rigid diaphragm to transmit lateral forces.

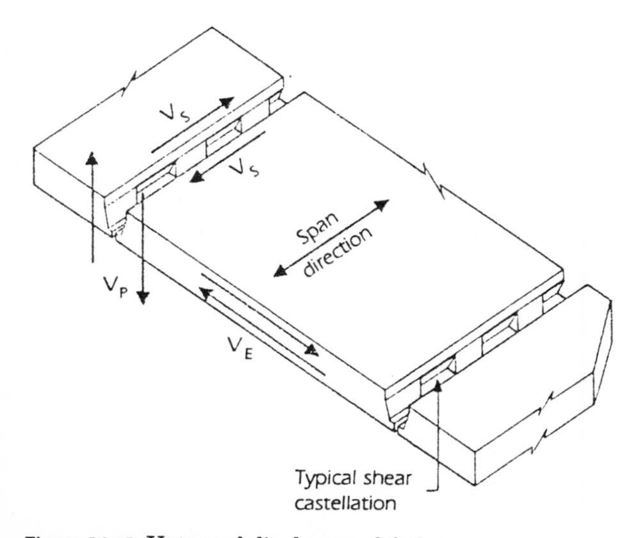


**Figure 24.15** Untopped diaphragm slab forces.

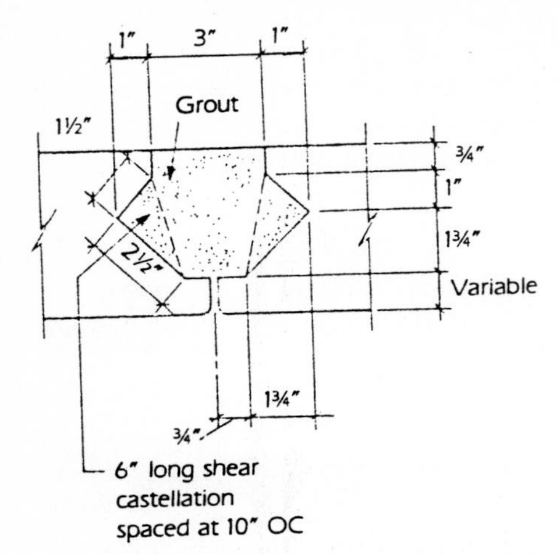


**Figure 24.16** Detail of castellated lateral slab joint.

**Figure 24.17** Los Angeles County Public Social Services parking garage, Long Beach, California. Untopped slabs 5 in thick being erected. The slabs span 16 ft between supports. The prestressing strands were allowed to extend beyond the ends of the slabs to form a positive tie with the cast-in-place concrete over the prestressed channels.

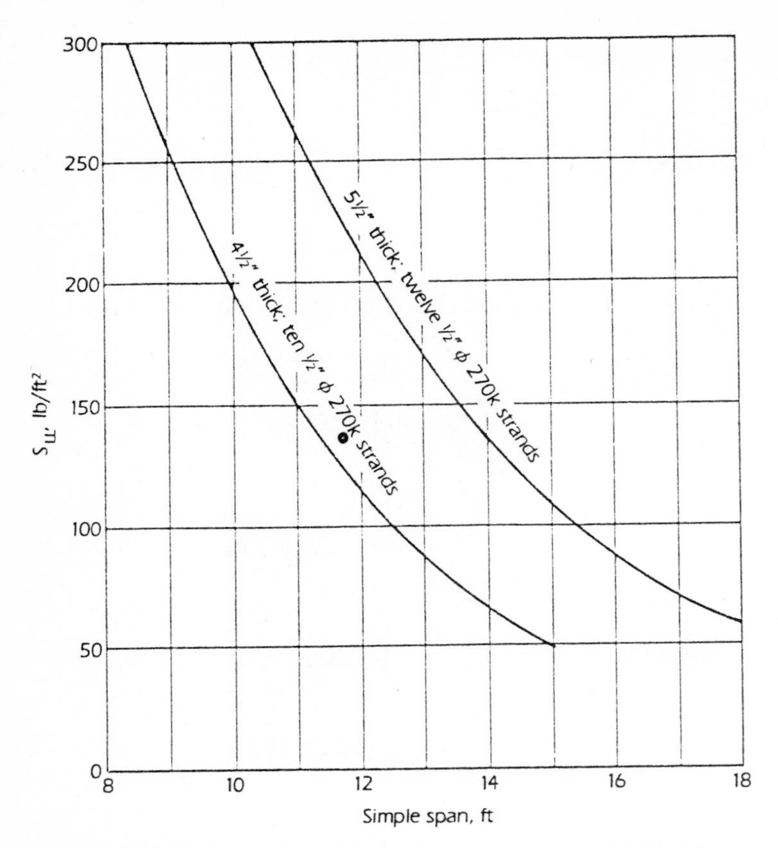


**Figure 24.18** Untopped solid slabs 12 ft wide. $f'_c$ = 5000 lb/in² hard-rock concrete. 1 1/2-in cover over strand (2-hour unrestrained rating).

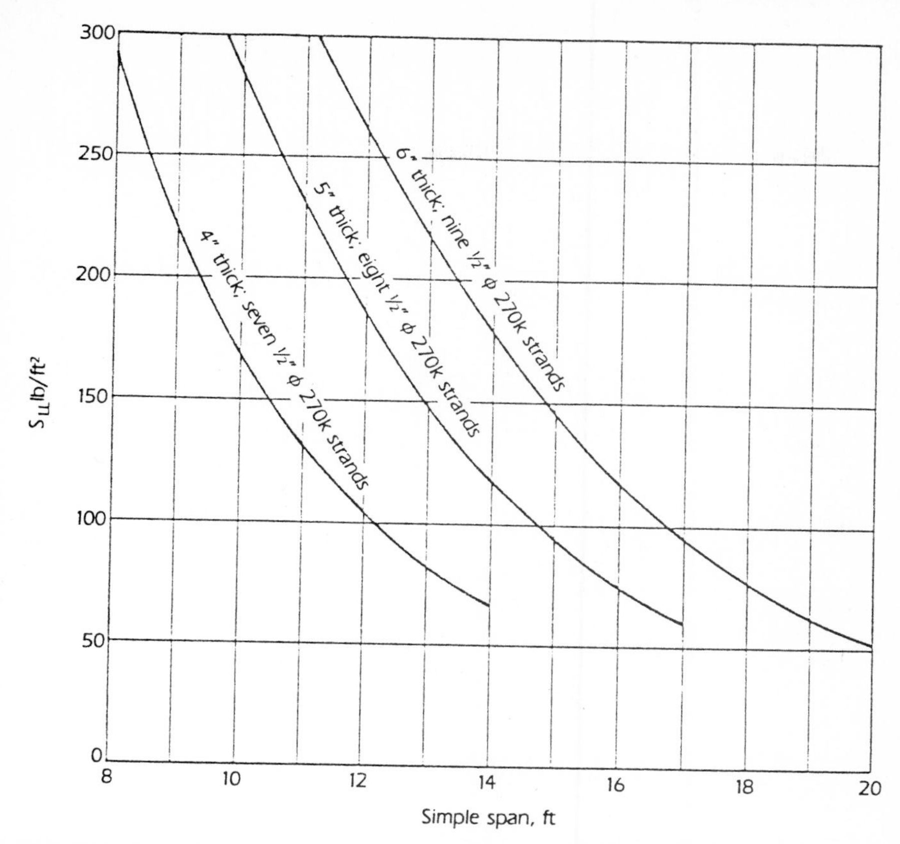


**Figure 24.19** Untopped solid slabs 12 ft wide. $f'_c$ = 5000 lb/in² lightweight concrete. 1 1/8-in cover over strand (2-hour unrestrained rating).

## Design example: untopped prestressed flat slab

Design an untopped slab for a parking structure spanning 17 ft between spread channels. The area over the channels will subsequently be filled with cast-in-place topping flush with the tops of the untopped slabs. Therefore the slabs may be considered to be restrained. A 2-hour rating is required. Use 1/2-in $\phi$-270$^k$ strands. (See Fig. 24.18.)

From preliminary design charts a 5-in-thick section is selected. Cover 3/4 in thick is required. See Fig. 24.20.

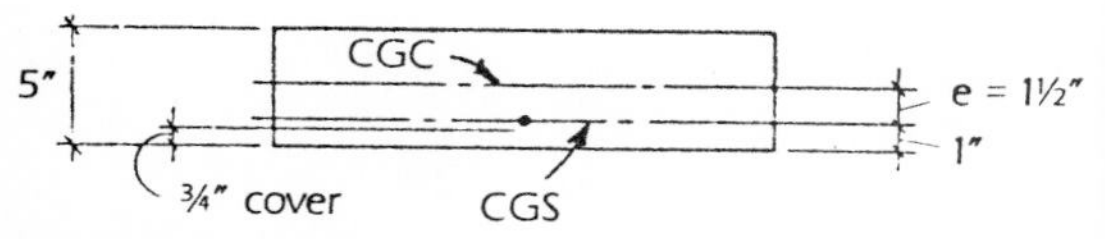


**Figure 24.20** Cross section through untopped prestressed flat slab.

$S = 50\ \text{in}^3$ $\qquad A = 60\ \text{in}^2/\text{ft}$

$I = 125\ \text{in}^4$ $\qquad \text{Weight} = 63\ \text{lb/ft}^2$

$f'_{ci} = 3500\ \text{lb/in}^2$ $\qquad E_{ci} = 3600$

$f'_c = 5000\ \text{lb/in}^2$ $\qquad E_c = 4300$

$S_{LL} = 50\ \text{lb/ft}^2$

Camber control:

$$\Delta_i\ \text{slab weight} = \frac{5wL^4}{384E_{ci}I} = \frac{(5)(0.063)(17^4)(1728)}{(384)(3600)(125)} = -0.26\ \text{in}\downarrow$$

Therefore, required $\Delta\ P/S \uparrow = 1.1 \times 0.26 = +0.29\ \text{in}\uparrow$

Calculate $F_i$ required:

$$\Delta P/S = \frac{F_i eL^2}{8E_{ci}I}$$

$$0.29 = \frac{F_i\,(1.5)(17^2)(144)}{(8)(3600)(125)}$$

$$F_i = 16.7^k/\text{ft}$$

For a 12-ft-wide slab,

$$N = \frac{16.7 \times 12}{0.9 \times 28.9} = 7.7$$

Use eight ½-in $\phi$-$270^k$ strands.

$$F_i = (0.9)(8)(28.9) = 208^k$$

$$F_f = (0.78)(8)(28.9) = 180^k$$

Check final stresses (use 22 percent losses):

$$M_{slab} = 1/8 \times 0.063 \times 12 \times 12 \times 17^2 = 328^{k\cdot in}$$

$$M_{SLL} = 1/8 \times 0.050 \times 12 \times 12 \times 17^2 = 260^{k\cdot in}$$

$$UM = (1.4)(328) + (1.7)(260) = 901^{k\cdot in}$$

$$f_{\text{bottom}}^{\mathbb{C}} = -\frac{180}{720} - \frac{(180)(1.5)}{600} + \frac{328 + 260}{600}$$

$$= -0.250 - 0.450 + 0.980 = +0.280 \text{ ksi} < 6\sqrt{f'_c} \quad \text{(OK)}$$

$$\left(6\sqrt{f'_c} = +0.425 \text{ ksi}\right)$$

At release:

$$f_{\text{top}}^{\text{end}} = -\frac{208}{720} + \frac{(208)(1.5)}{600}$$

$$= -0.289 + 0.520 = +0.231 < 6\sqrt{f'_{ci}}$$

$$6\sqrt{f'_{ci}} = +0.354 \text{ksi}$$

Check ultimate strength:

$$\rho_p = \frac{A_{ps}}{bd} = \frac{(8)(0.1531)}{(144)(4)} = 0.002126$$

$$f_{ps} = f_{pu}\left(1 - 0.5\,\rho_p\,\frac{f_{pu}}{f'_c}\right) = (270)\left(1 - 0.5 \times 0.002126 \times \frac{270}{5}\right)$$

$$= 254.5 \text{ ksi}$$

$$\frac{a}{2} = \frac{A_{ps} f_{ps}}{1.7 f'_c b} = \frac{(8)(0.1531)(254.5)}{(1.7)(5)(144)} = 0.25 \text{ in}$$

$$M_u = \phi A_{ps} f_{ps}\left(d - \frac{a}{2}\right) = (0.9)(8)(0.1531)(254.5)(4 - 0.25)$$

$$= 1052^{\text{k}\bullet\text{in}} > UM \quad \text{(OK)}$$

Check camber and deflection in final long-term condition with creep:

$$\Delta P/S_f = \frac{(208)(1.5)(17^2)(144)}{(8)(3600)(1500)} \times 2.45\,(\text{CF}) = +0.74 \text{ in} \uparrow$$

$$\Delta_{\text{slab }F} = 0.26 \text{in} \times 2.70 = \quad -0.70 \text{ in} \downarrow$$

$$-0.04 \text{ in} \uparrow \quad \text{(OK)}$$

$$\Delta_{SLL} = \frac{50}{63} \times 0.26 \text{ in} = 0.21 \text{ in}$$

$$\frac{L}{240} = \frac{12 \times 17}{240} = 0.85 \text{ in} > 0.21 \text{ (OK)}$$

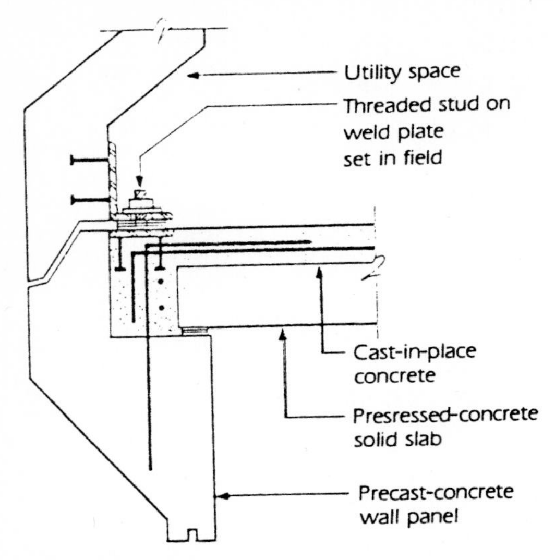


**Figure 24.21** Connection detail.

## Connection Details

Connection details are shown in Figs. 24.21 through 24.38.

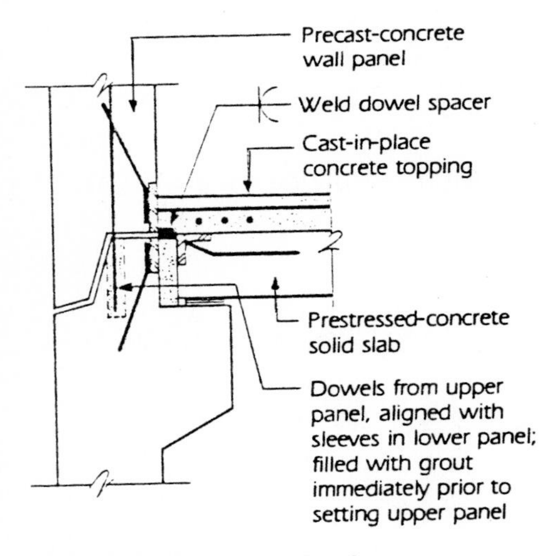


**Figure 24.22** Connection detail.

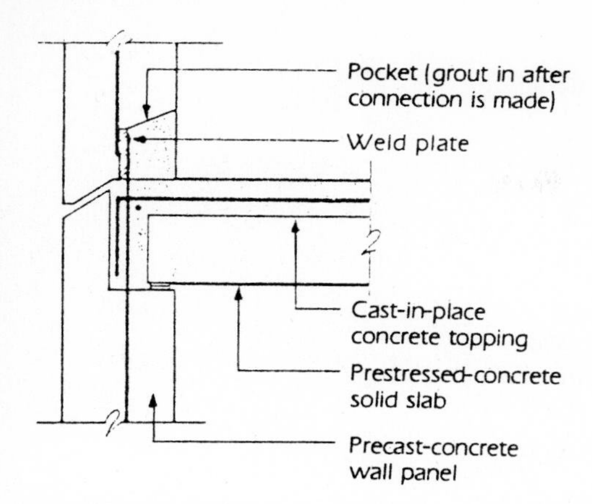


**Figure 24.23** Detail of cast-in-place concrete diaphragm at exterior wall; shear wall horizontal joint detail.

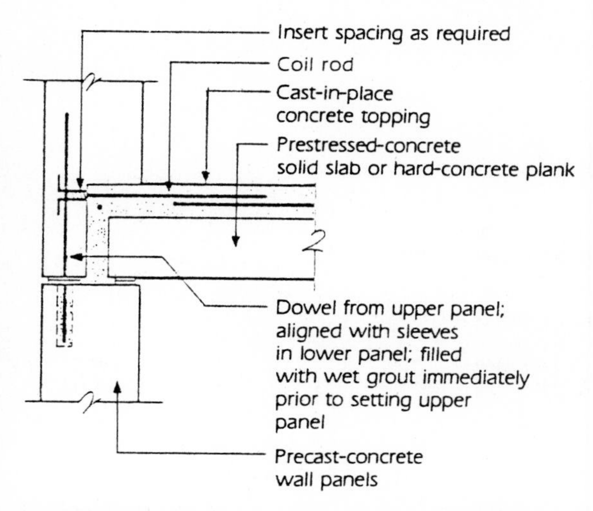


**Figure 24.24** Detail of cast-in-place concrete diaphragm at exterior wall; shear wall horizontal joint detail.

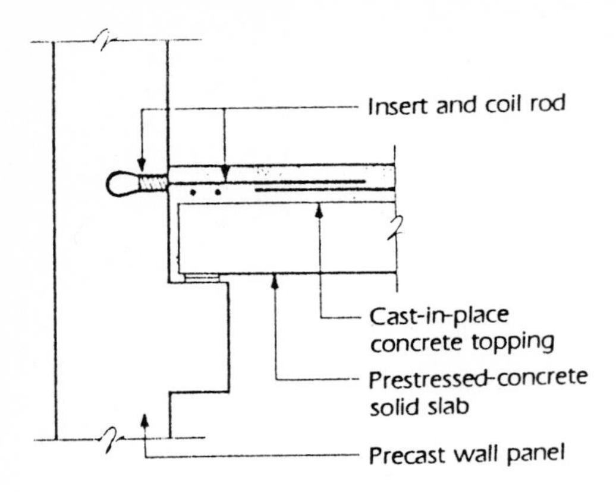


**Figure 24.25** Cast-in-place concrete diaphragm at exterior wall.

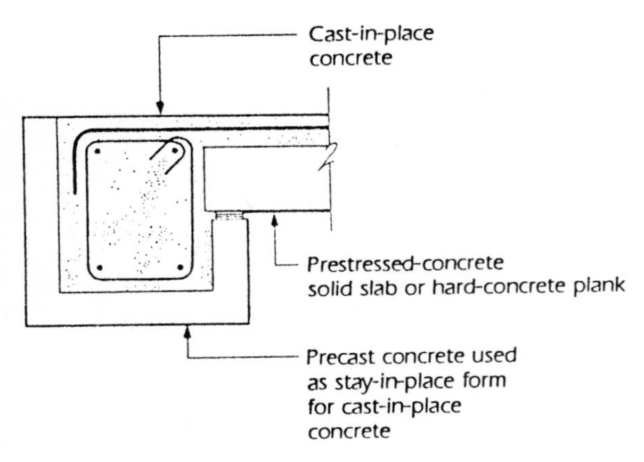


**Figure 24.26** DMRF spandrel detail.

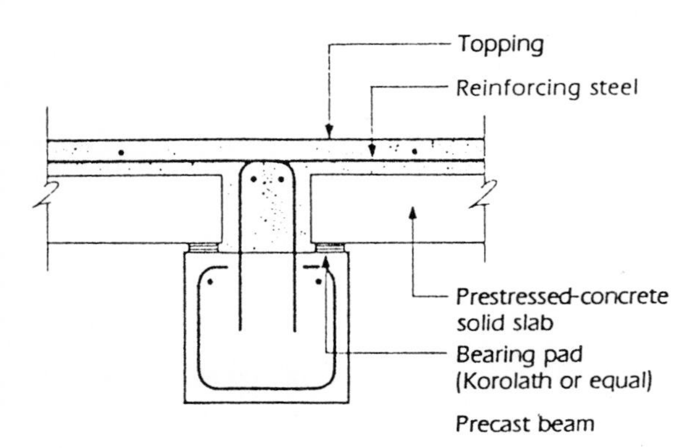


**Figure 24.27** Connection detail.

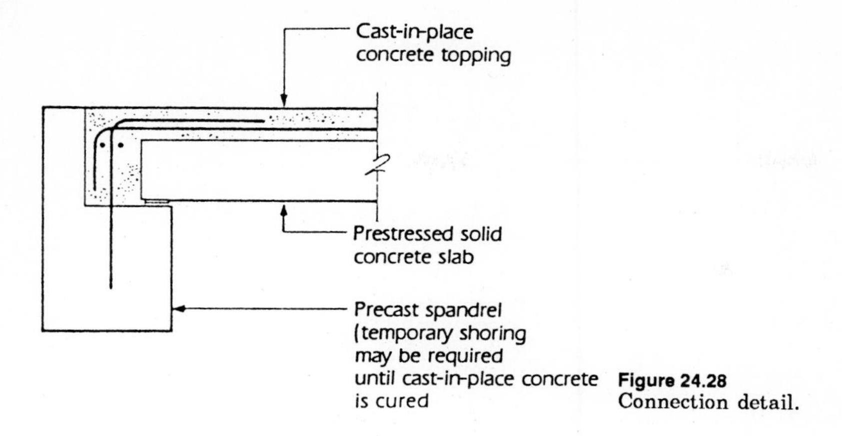


**Figure 24.28** Connection detail.

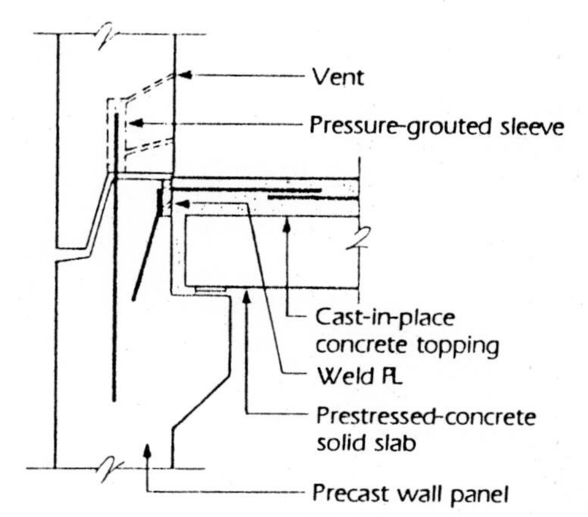


**Figure 24.29** Shear wall horizontal joint detail.

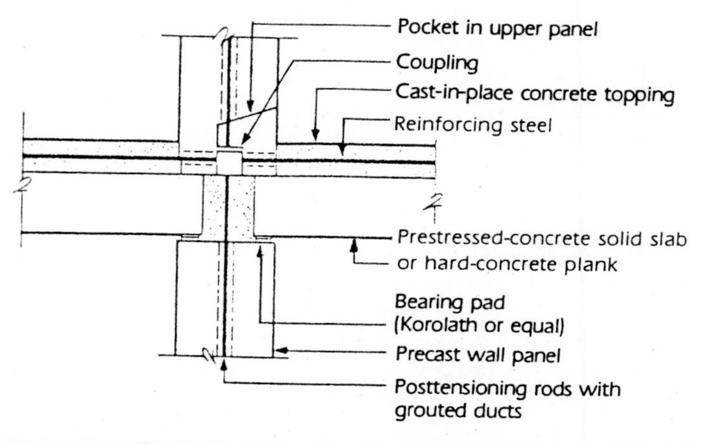


**Figure 24.30** Vertically posttensioned shear wall: horizontal joint detail.

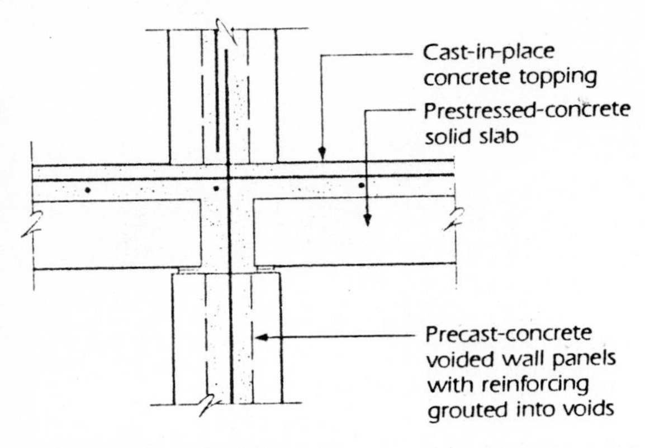


**Figure 24.31** Voided shear wall panel: horizontal joint detail.

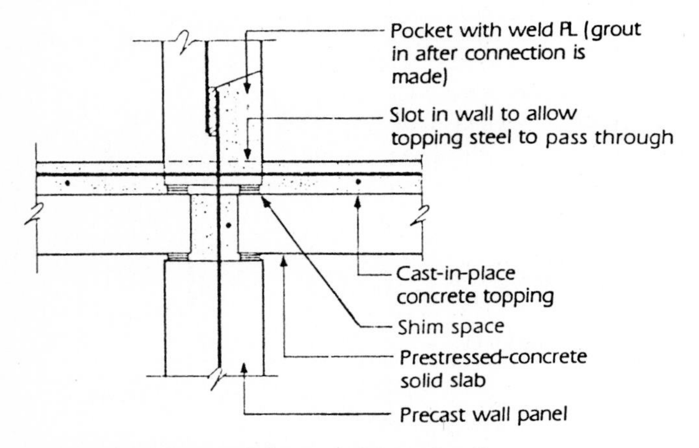


**Figure 24.32** Shear wall horizontal joint detail.

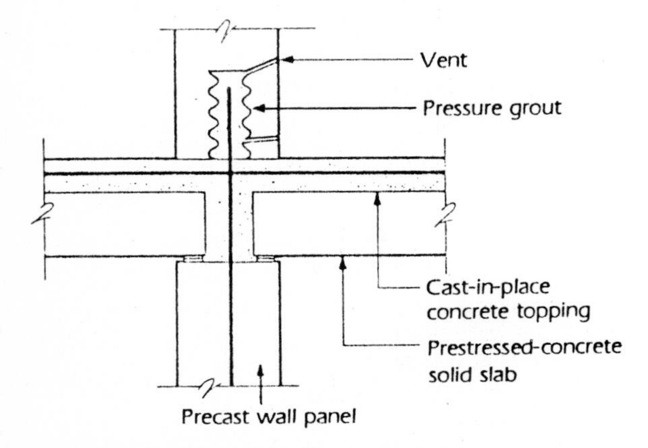


**Figure 24.33** Shear wall horizontal joint detail.

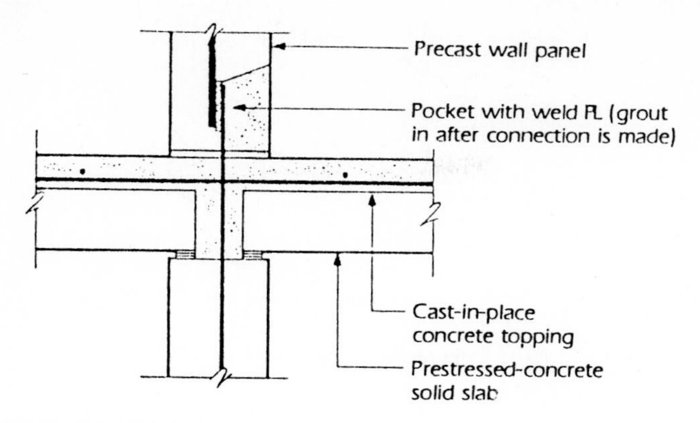


**Figure 24.34** Shear wall horizontal joint detail.

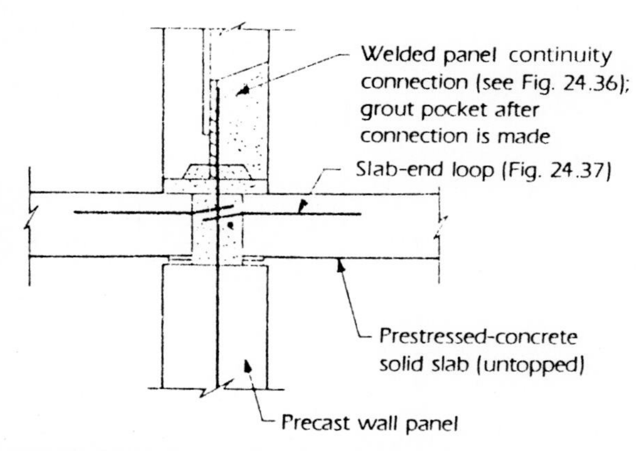


**Figure 24.35** Untopped diaphragm wall panel: horizontal joint detail.

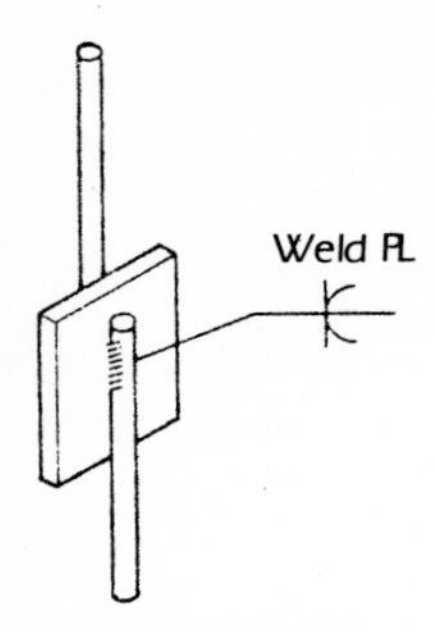


**Figure 24.36** Joint detail. This is preferred because it provides construction tolerance.

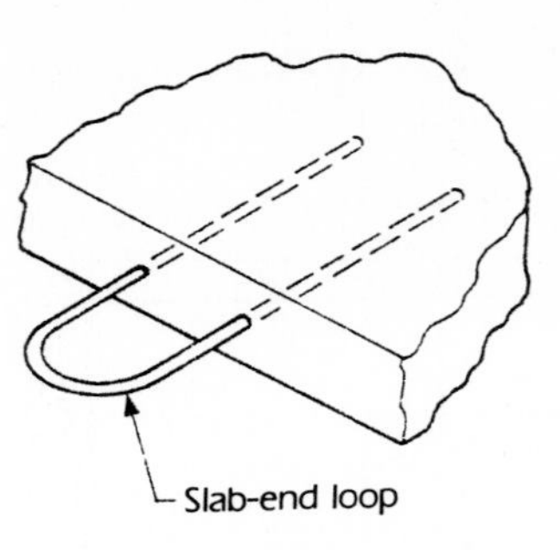


**Figure 24.37** Butt joint detail.

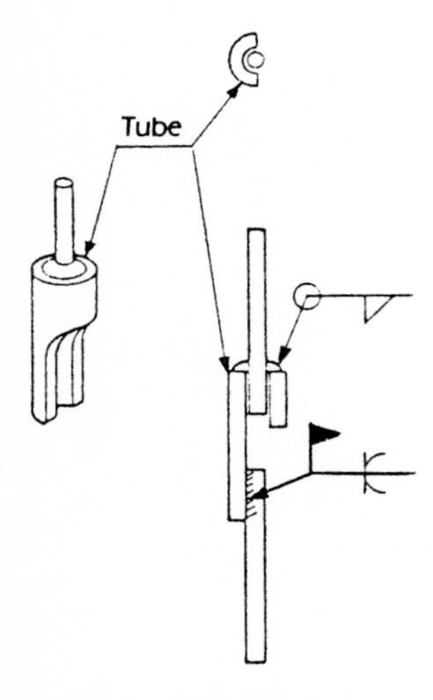


**Figure 24.38** Joint detail. This detail has very little allowance for construction tolerance.

**Figure 24.39** Bonaventure Hotel, Los Angeles, California. Tower cranes were used to erect structural steel and prestressed concrete flat slabs and perform hoisting for other trades working on the building. Two levels of steel framing were erected, then two levels of prestressed slabs, etc.

Figure 25.1 Heritage House, Concord, California. 118,000 ft$^2$ of 8-in Spancrete hollow-core planks form the floors of this eight-story residence for senior citizens. One entire floor was erected each week on masonry bearing walls. This fast construction allowed the building to be completed 2 months ahead of schedule. The hollow-core planks also provide inherent fire protection and prevent the passage of both impact noise and sound transmission through the floor.

Chapter

# 25

# Hollow-Core Planks

Hollow-core planks are voided precast pretensioned concrete deck units used mainly to form floors and roofs in buildings. They can also be used as noninsulated or insulated wall panels in either load-bearing or non-load-bearing structural functions. These units are made by commercially franchised processes using specialized forming machinery. The six principal processes produced in the United States are:

*Dynaspan.* This is made in 8-ft widths by a slipforming process with low-slump concrete. Each slab has 14 cores.

*Flexicore.* This wet-cast product is cast in 2-ft widths in 60-ft-long pans. Voids are formed with deflatable rubber tubes.

*Span-Deck.* This wet-cast product is cast in two sequential operations, with the second operation slipforming 8-ft-wide planks with rectangular voids.

*Spancrete.* This is made in 40-in-wide units by tamping an extremely dry mix with three sequential sets of tampers compacting the mixes around slipforms.

*Spiroll.* This extruded product is made in 4-ft-wide units with round voids formed by augers which are a part of the casting machine.

*Dy-Core.* This extruded 4-ft-wide product is made by compressing zero-slump concrete into a solid mass by a set of screw conveyors in the extruder. High-frequency vibration combined with compression around a set of dies in the forming chamber of the machine produces the plank with oblate or octagonal-shaped voids.

Hollow-core slabs are made in 6-, 8-, 10-, and 12-in depths and are used in spans from 18 to 42 ft. They are used as floors or roofs in hotels, motels, offices, shopping malls, department stores, schools, hospitals, and multifamily housing, as decks in parking structures, and as infill systems spanning between spread channels. They may be erected side by side or be spread apart a distance of 2 or 3 ft with the space in between spanned with metal deck. These spread systems are covered with cast-in-place concrete topping to form an integral diaphragm and achieve the required fire rating. In certain applications, such as roofs or floors in multifamily housing, the units may be installed without a cast-in-place concrete topping provided that the completed diaphragm maintains its integrity in withstanding design seismic stresses.

Fire ratings of up to 4 hours are achieved by providing adequate cover over the strands, using cast-in-place concrete topping and lightweight aggregate concrete. Untopped hollow-core units 8 in thick meet HUD sound transmission criteria for floors in multifamily, multistory construction.

## Manufacturing

The two hollow-core-plank processes typical of this type of precast element are Span-Deck and Spancrete. In the Span-Deck process, two separate casting machines are used. The first machine (bottom casting machine) lays down a $1^3/_4$-in-thick soffit of fluid workable concrete around the tensioned prestress strands (usually $^1/_2$-in-diameter strands). Then the second machine (top casting machine) deposits the balance of the concrete section around temporary rectangular voids formed with pea gravel deposited from inside a slipform. This second machine has two large hoppers, one for the pea gravel and one for the concrete. Then the freshly cast slabs are steam-cured, sawed to length, and transported to a void material recovery area where the pea gravel is dumped out of the voids by tilting the planks on end. In the wet-cast Span-Deck process, side rails with a shear key former keep the wet concrete inside the bed. These side rails pivot down prior to sawing and stripping. Large preformed openings may be made in this product to accommodate mechanical risers or heating, ventilating, and air-conditioning (HVAC) ducts. Insert plates and top reinforcing may be placed in the wet concrete. The side rails may be altered to form a castellated shear key in the plank (see Chap. 24, Figs. 24.14, 24.15, and 24.16). Span-Deck is cast on a daily cycle in beds that are 400 to 600 ft long.

The Spancrete process uses a very dry mix which may vary in aggregate quality in each layer that is deposited. These dry mixes are placed in three layers by the successive compaction of tampers which are part of the machine. The top two layers are tamped around and above slipforms, leaving the voids in the product. One pass is made each day on 750-ft-long beds. The product is stack-cast (one layer is cast directly on the preceding layer). The beds are cast up to five stacks high, and the product is ambient-cured for 1 week. Water is also run through the voids in the initial curing phase. The shear key is slipformed by the side of the machine. Large openings for shafts, HVAC ducts, or mechanical risers may be saw-cut in the plant. Owing to the nature of the dry-mix process, it is not possible to cast plates or rebar in the slabs. Top strand may be placed subject to design limitations. Spancrete is tensioned with ¼-in-diameter through ½-in-diameter strands.

In both processes, the top finish may be roughened sufficiently to satisfy Sec. 2617 of the Uniform Building Code. If the planks are to be used untopped, as in roof construction, they should have a smooth top finish. Other advantages of hollow-core planks are the use of the voids as electrical raceways and, in specially designed instances, as HVAC ducts. Hangers for suspended ceilings or pipes may be placed by the respective trades in the field in the plank joints prior to grouting. Small openings (less than 10 by 10 in) should be core-drilled by the respective trades in the field.

## Attachment of Partitions

When hollow-core planks are used in housing or motel applications where the underside of the plank forms the finished ceiling, partition attachment details should allow for movement of the plank due to temperature changes and creep and shrinkage effects on camber and plank self-weight deflection. The use of proper attachment details will ensure that unsightly gaps do not form at the tops of the partitions or, conversely, will prevent cracking of partitions due to long-term downward movement of the floor system. Some suggested details are shown in the section "Connection Details" at the end of the chapter.

## Hollow-Core-Plank Production

The production of hollow-core planks is illustrated in Figs. 25.2 and 25.3.

**Figure 25.2** Spancrete production beds showing the casting machine and the long-line stack-casting method used.

**Figure 25.3** Span-Deck line and casting machines. The bottom casting machine shown in the foreground places soffit concrete. The top casting machine in the background follows, slipforming the balance of the section, which contains temporary aggregate void-forming material.

## Design

Figures 25.4 and 25.5 are preliminary design charts for topped and untopped hollow-core plank. Preliminary design information for hollow-core plank can also be found in the third edition of the *PCI Design Handbook*, pages 2-29 to 2-34. Figure 25.6 is a sample of a detailed load table for a specific manufacturer's hollow-core-plank system.

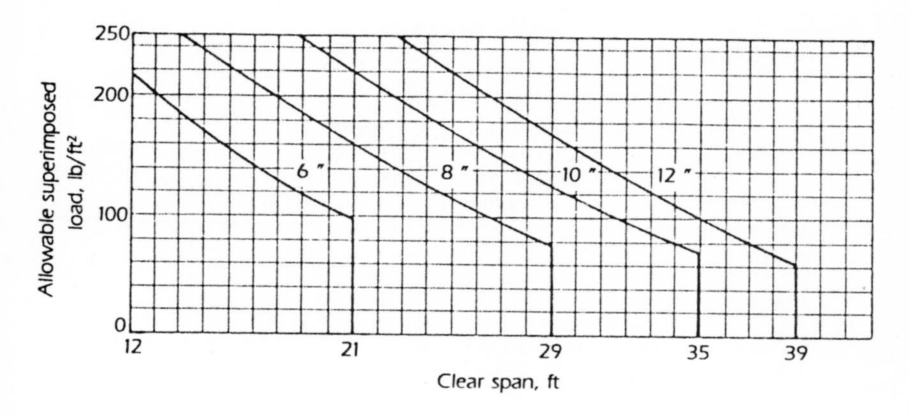


**Figure 25.4** Untopped hollow-core plank.

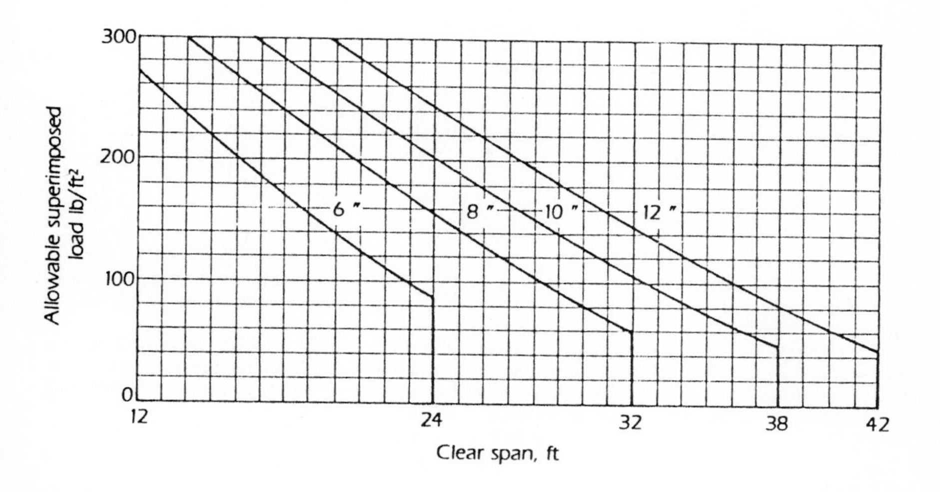


**Figure 25.5** Hollow-core plank with 2 1/2 in of composite topping.

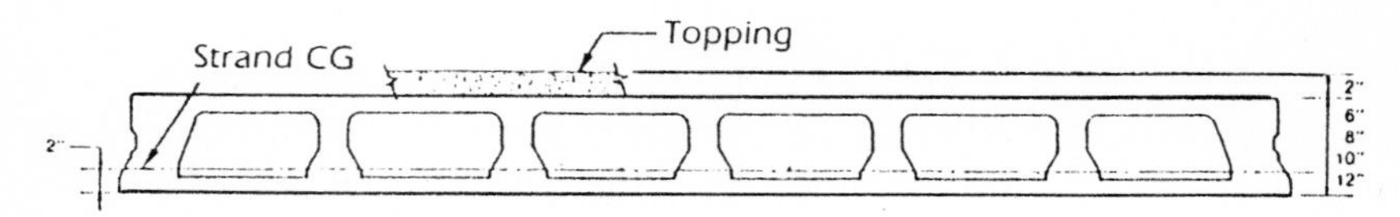


**Figure 25.6** Span-deck load table.

| Section Properties | | | | |
|---|---|---|---|---|
| 8' 0" × | Bare | | Composite | |
| 8" | $A$ = 370.78 in$^2$<br>$I_p$ = 3090 in$^4$<br>$y_b$ = 3.91 in<br>$f'_c$ = 5000 lb/in$^2$ | $S_b$ = 791 in$^3$<br>$S_t$ = 754 in$^3$<br>$e$ = 1.91 in<br>$Wt$ = 55 lb/in$^2$ | $I_c$ = 5895 in$^4$<br>$Y_{cb}$ = 5.36 in<br>$f'_c$ = 3000 lb/in$^2$<br>(topping only) | $S_{ct}$ = 1271 in$^3$<br>$S_{cb}$ = 1099.0 in$^3$<br>$Wt_c$ = 80 lb/in$^2$ |

Explanation of chart

1. ½ in-ϕ-270$^k$ strands are used in all cases. Initial prestress force is 28.9$^k$ per strand.
2. $M_u$ is the ultimate resisting moment, in k· ft per plank.
3. Other notation is explained below:

$A$ area of precast section
$W_{tp}$ weight of precast section
$W_{tc}$ weight of precast section and topping
$I_p$ moment of inertia of precast section
$I_c$ moment of inertia of composite section
$S_b$ bottom-section modulus of precast section
$S_t$ top-section modulus of precast section
$S_{cb}$ bottom-section modulus of composite section
$S_{ct}$ top-section modulus of composite section
$e$ eccentricity of prestressing strands
$y_b$ distance from centroid to bottom of precast section
$Y_{cb}$ distance from centroid of composite section to bottom of composite section

NOTES

1. Design criteria: ACI 318-83.
2. The safe superimposed load is a uniformly distributed load consisting of the live load plus any dead load that is in addition to the weight of the plank or the weight of the plank and topping.
3. Concentrated loads should be transposed into an equivalent uniform load (in pounds per square foot) that produces a moment equal to that of the concentrated load. This equivalent uniform load should be added to any other uniform superimposed dead load and live load before entering tables and choosing strand pattern.
4. Values to the left and below the solid stepped line are controlled by shear.
5. Shaded areas to the right side of chart indicate spans for which deflections must be checked, i.e., flat roofs, etc.
6. Maximum theoretical safe load values are 125 lb/ft$^2$ for untopped plank and 250 lb/ft$^2$ for composite plank with a field-applied reinforced-concrete structural topping. Untopped plank is normally used for live loads asociated with residential applications. Topping is recommended for industrial, commercial, and institutional floor plank applications.

| | No. of strands | $M_u$ ft•k | Simple span with 2-in concrete topping (composite unit) | | | | | | | | | | | | | | | | | | | | |
|---|---|---|---|---|---|---|---|---|---|---|---|---|---|---|---|---|---|---|---|---|---|---|---|
| | | | 14 | 15 | 16 | 17 | 18 | 19 | 20 | 21 | 22 | 23 | 24 | 25 | 26 | 27 | 28 | 29 | 30 | 31 | 32 | 33 | 34 |
| 8" | 5 | 112 | | 228 | 193 | 163 | 138 | 117 | 100 | 84 | 71 | 59 | 49 | 40 | | | | | | | | | |
| | 6 | 132 | | | 238 | 203 | 174 | 150 | 129 | 111 | 95 | 81 | 69 | 59 | 49 | 41 | | | | | | | |
| | 7 | 151 | | | 244 | 225 | 208 | 181 | 157 | 136 | 118 | 102 | 89 | 77 | 66 | 56 | 46 | 35 | | | | | |
| | 8 | 170 | | | 246 | 227 | 210 | 195 | 182 | 160 | 140 | 123 | 107 | 94 | 82 | 70 | 58 | 47 | 36 | | | | |
| | 9 | 187 | | | 248 | 229 | 212 | 197 | 184 | 172 | 161 | 142 | 125 | 110 | 97 | 83 | 70 | 58 | 47 | 37 | | | |
| | 10 | 204 | | | 250 | 231 | 214 | 199 | 185 | 173 | 162 | 152 | 142 | 126 | 112 | 96 | 82 | 69 | 57 | 47 | 37 | | |
| | 11 | 220 | | | | 232 | 215 | 200 | 187 | 174 | 163 | 153 | 144 | 136 | 125 | 108 | 93 | 79 | 67 | 56 | 46 | 37 | |
| | 12 | 235 | | | | 234 | 217 | 202 | 188 | 176 | 165 | 155 | 145 | 137 | 129 | 120 | 104 | 90 | 77 | 65 | 55 | 45 | 36 |

* Reprinted by permission from Blakeslee Prestress, Inc., Branford, Conn.

**Hollow-core plank: design example 1**

Design a hollow-core plank with 3 1/2 in of hard-rock composite topping to span 28 ft 8 in in a department store floor. A fire rating of 2 hours is required. Use hard-rock concrete for the hollow-core plank. $f'_c$ (precast) = 5000 lb/ in$^2$; $f'_c$ (topping) = 3000 lb/ in$^2$. (See Fig. 25.7.)

From a local manufacturer's load-span tables, an 8-ft-wide by 8-in-deep section is selected. Note that the load tables indicate that 8-in-deep Span-Deck hollow-core plank with 2 in of hard-rock concrete composite topping carries a superimposed load of 90 lb/ft$^2$ for a 29-ft span. With the 3 1/2-in-thick topping, as specified in the design description, this system should carry the superimposed loading of 75 lb/ft$^2$ of live load and 25 lb/ ft$^2$ of partitions and mechanical systems.

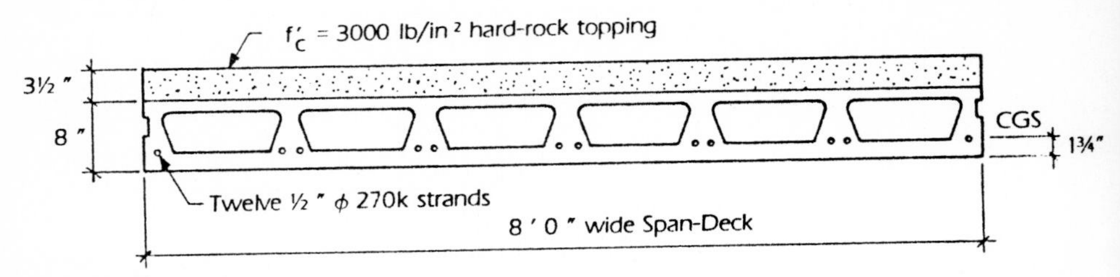


**Figure 25.7** Hollow-core-plank design.

Note also that because the manufacturing process used for forming the voids results in ribs slightly thicker than the theoretical dimensions shown, the design assigns a slightly higher value to the area of the cross section, based upon actual plant measurements.

**Section properties.** Determine section properties.

Basic section (concrete weight 150 lb/ft³):

Actual area: $A = 390 \text{ in}^2$ Weight $= 51 \text{ lb/ft}^2$

$I_B = 3217 \text{ in}^4$ $s_t = 775 \text{ in}^3$

$y_t = 4.26 \text{ in}$ $s_b = 860 \text{ in}^3$

$y_b = 3.74 \text{ in}$

$e = 3.74 - 1.75 = 1.99 \text{ in}$

Composite section (with 3½-in cast-in-place hard-rock topping):

| Part | $A$ | $y_{tcc}$ | $Ay_{tcc}$ | $I_o$ | $Ad^2$ |
|---|---|---|---|---|---|
| 8-in Span-Deck | 390 | 7.76 | 3026 | 3217 | 2246 |
| 3½-in CIP: 0.77 × 3.5 × 96 = | 259 | 1.75 | 453 | 264 | 3575 |
| Composite section | 649 | | 3479 | | |

$$\Sigma I_0 + Ad^2 = I_c = 9102 \text{ in}^4.$$

$$y_{tcc} = \frac{3479}{649} = 5.36 \text{ in}$$

$y_{tc} = 1.86 \text{ in}$ $s_{tc} = 4894 \text{ in}^3$

$y_{bc} = 6.14 \text{ in}$ $s_{bc} = 1482 \text{ in}^3$

**Loads and moments.** For an 8-ft-wide slab:

To basic section :

| | k/ft | $M_c$ k·in | ×LF | $=UM_{₵}$ |
|---|---|---|---|---|
| 8-in Span-Deck | $1.5 \times 0.408 \times 28.67^2 =$ | 503 | 1.4 | 705 |
| 3 1/2-in CIP topping | $1.5 \times 0.350 \times 28.67^2 =$ | 432 | 1.4 | 605 |
| | | $M_B = 935^{k \cdot in}$ | | |

To composite section :

| | k/ft | $M_c$ k·in | ×LF | $=UM_{₵}$ |
|---|---|---|---|---|
| $S_{DL}$ (25 lb/ft²) | $1.5 \times 0.200 \times 28.67^2 =$ | 246 | 1.4 | 345 |
| $S_{LL}$ (75 lb/ft²) | $1.5 \times 0.600 \times 28.67^2 =$ | 740 | 1.7 | 1258 |
| | | $M_C = 986^{k \cdot in}$ | | $UM = 2913^{k \cdot in}$ |

**Stresses.** Stresses are as follows (see *PCI Design Handbook*, 3d edition, page 4-16).

Prestress force required :

$$F_f = \frac{\dfrac{M_B}{S_b} + \dfrac{M_c}{S_{bc}} - f_t}{\dfrac{1}{A} + \dfrac{e}{S_b}} = \frac{\dfrac{935}{860} + \dfrac{986}{1482} - 6\sqrt{5000}}{\dfrac{1}{390} + \dfrac{1.99}{860}}$$

$$= \frac{1.087 + 0.665 - 0.425}{0.0002564 + 0.002314} = \frac{1.327}{0.004878} = 272^k$$

Using 1/2-in $\phi$-270$^k$ strands,

$$A_{ps} = 0.1531 \text{ in}^2$$

$$F_O = 28.9^k \text{strand}(0.7 f_{pu})$$

$$F_i = 0.9 \times 28.9 = 26.0^k/\text{strand}$$

$$F_f = 0.78 \times 28.9 = 22.5^k/\text{strand}$$

$$N = \frac{272}{22.5} = 12.09$$

Use twelve ½-in $\phi$-270$^k$ strands.

Check end stresses at release:

$$f_{top} = \frac{(12)(26)}{390} + \frac{(12)(26)(1.99)}{755} = -0.800 + 0.822$$

$$= +0.0022 \quad (OK)$$

$$f_{bottom} = -\frac{(12)(36)}{390} - \frac{(12)(26)(1.99)}{860} = 0.800 - 0.722$$

$$= -1.522 \quad (OK)$$

$$(< 0.6 f'_{ci} \;, \therefore OK)$$

Check final stresses at centerline span:

$$f_{top} = -\frac{(12)(22.5)}{390} + \frac{(12)(22.5)(1.99)}{755} = \frac{935}{755} - \frac{986}{4894}$$

$$= -0.692 + 0.712 - 1.238 - 0.201 = 1.419 < 0.45 f'_c \quad (OK)$$

$$f_{bottom} = -\frac{(12)(22.5)}{390} - \frac{(12)(22.5)(1.99)}{860} + \frac{935}{860} + \frac{986}{1482}$$

$$= -0.692 - 0.625 + 1.087 + 0.665 = +0.435 \cong 6 f'_c \quad (OK)$$

**Ultimate strength.** Check ultimate strength. (See *PCI Design Handbook*, 3d edition, pages 4-6 and 4-63.)

$$f_{se} = 147 \text{ ksi}$$

$$C\bar{\omega}p = C \frac{A_{ps}}{bd} \times \frac{f_{pu}}{f'_c} = 1.0\left(\frac{12 \times 0.1531}{96 \times 9.75}\right)\left(\frac{270}{3}\right) = 0.18$$

From the chart read $f_{ps}/f_{pu} = 0.96$; therefore, $f_{ps} = 0.96 \times 270 = 259.2$ ksi.

$$a/2 = \frac{A_{ps} f_{ps}}{1.7 f'_c b} = \frac{(1.837)(259.2)}{(1.7)(3)(96)} = 0.97 \text{ in}$$

$$M_u = \phi A_{ps} f_{ps} (d - a/2)$$

$$= (0.9)(1.837)(259.2)(9.75 - 0.97)$$
$$= 3763^{\text{k·in}} > UM = 2913^{\text{k·in}} \text{ (OK)}$$

**Camber and deflection.** Check camber and deflection.

$$e = 1.99 \text{ in} \quad E_{ci} = 3600 \text{ ksi} \quad E_c = 4300 \text{ ksi}$$

Camber (see *PCI Design Handbook*, page 4-73):

$$\Delta_{P/S} \uparrow = \frac{F_i e L^2}{8E_{ci} I_b} = \frac{(312)(1.99)(28.67^2)(144)}{(8)(3600)(3217)} = 0.79 \text{ in}$$

Slab self-weight deflection:

$$\Delta_{\text{slab}} \downarrow = \frac{5wL^4}{384E_{ci} I_b} = \frac{(5)(0.408)(28.67^4)(1728)}{(386)(3600)(3217)} = 0.54 \text{ in}$$

Topping weight deflection:

$$\Delta_{\text{topping}} \downarrow = \frac{5wL^4}{384E_c I_b} = \frac{(5)(0.350)(28.67^4)(1728)}{(386)(4300)(3217)} = 0.38 \text{ in}$$

Superimposed loads:

$$\Delta_{SD+SL} \downarrow = \frac{5wL^4}{384E_c I_c} = \frac{(5)(0.800)(28.67^4)(1728)}{(384)(4300)(9102)} = 0.31 \text{ in}$$

**Creep.** Determine creep effects (see *PCI Design Handbook*, page 4-46).

At erection :

| | $\Delta_i \times CF =$ | $\Delta_{\text{long-term}}$ |
|---|---|---|
| Camber | 0.79 × 1.80 = | +1.42 |
| Slab weight | 0.54 × 1.85 = | - 1.00 |
| Σ basic section | | +0.42 in ↑ |
| Topping weight | | - 0.38 ↓ |
| Σ at topping pour | | +0.04 in ↑ |

Long-term as-built condition:

| | $\Delta_i \times CF =$ | $\Delta_{\text{long-term}}$ |
|---|---|---|
| Camber | 0.79 × 2.20 = | +1.74 |
| Slab weight | 0.54 × 2.40 = | - 1.30 |
| Topping weight | 0.38 × 2.30 = | - 0.87 |
| Σ long-term composite | | - 0.43 in |

**Allowable deflection.** Check $\Delta_{SD+SL}$ allowable (see *PCI Design Handbook*, page 4-42; Uniform Building Code, Table 23-D).

$$\frac{L}{360} = \frac{28.67 \times 12}{360} = 0.96 \text{ in} > 0.31 \text{ in} \quad \text{(OK)}$$

**Shear.** Check horizontal and vertical shear.

Horizontal shear between plank and topping (see *PCI Design Handbook*, page 4-28):

$$UV = [(1.4)(0.408 + 0.350 + 0.200) + (1.7)(0.600)]\,\frac{28.67}{2} = 33.8^k$$

$$\phi V_{nh}{}^2 = \phi 80\, b_v d = (0.85)(80)(96)(9.75)/1000 = 63.6^k$$

Therefore, roughen top of plank to satisfy horizontal shear requirement. (Broom finish or equivalent roughness satisfies code requirements.)

Vertical shear (see Fig. 25.8; *PCI Design Handbook*, page 4-66):

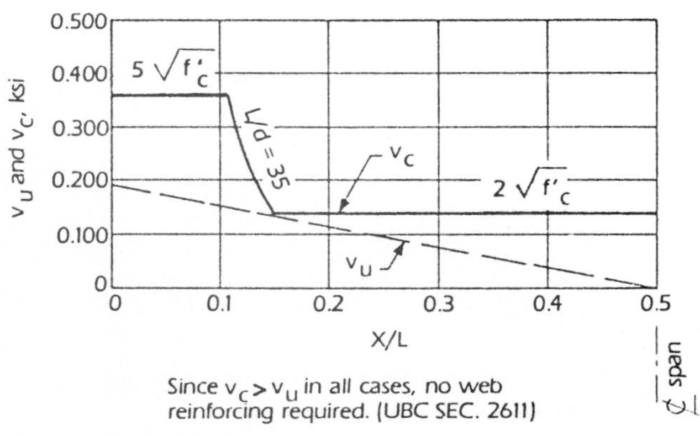


**Figure 25.8** Vertical shear.

**Figure 25.9** Sears-Eastridge Mall, San Jose, California. 8-in-thick Spancrete spanning 28 ft. A 2½-in-thick cast-in-place concrete topping is subsequently placed to form the completed diaphragm and take lateral loads to shear walls. Precast concrete beams and columns complete the building frame.

**Figure 25.10** Sears-Eastridge Mall, San Jose, California. 8-in-thick Spancrete planks being erected two at a time to maximize crane efficiency.

$$b_w d \cong (21.1)(9.75) = 206\,\text{in}^2$$

$$v_{u,\,\text{support}} = \frac{UV}{\phi b_w d} = \frac{33.8}{(0.85)(206)} = 0.19\ \text{ksi}$$

$$\frac{L}{D} = \frac{28.67 \times 12}{9.75} = 35$$

**Hollow-core plank: design example 2**

Design a hollow-core plank with 2 1/2 in of hard-rock concrete composite topping to span 37 ft on center bearings over parking and to support motel construction. $f'_c = 4000\ \text{lb/in}^2$; $f'_{ci} = 3500\ \text{lb/in}^2$. A 3-hour rating is required.

Use lightweight concrete with 1 1/2 in of cover to achieve the desired fire rating, from charts in the *PCI Design Handbook* (3d edition, page 2-32). A 10-in section should work.

**Section properties.** Determine section properties.

Basic section (see *PCI Design Handbook*, page 2-36):

$A = 272\ \text{in}^2$

$y_t = 5.09$ in

$y_b = 4.91$ in

$e = 4.91 - 1.69 = 3.22$ in

$I_b = 2790$ in

$s_t = 585\ \text{in}^3$

$s_b = 604\ \text{in}^3$

Weight $= 65\ \text{lb/ft}^2$

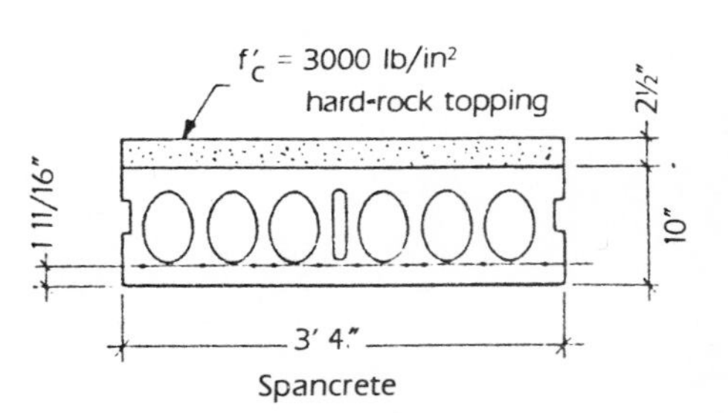


Composite section:

$$I_c = 6611\,\text{in}^4 \quad y_{tc} = 3.02\,\text{in} \quad s_{tc} = 2186\,\text{in}^3$$

$$y_{bc} = 6.98\,\text{in} \quad s_{bc} = 947\,\text{in}^3$$

**Loads and moments.** Determine loads and moments.

To basic section:

| | k/ft | $M_c$ k in | × LF | = $UM_{℄}$ |
|---|---|---|---|---|
| 10-in lightweight Spancrete | $1.5 \times 0.217 \times 37^2 =$ | 446 | 1.4 | 624 |
| Cast-in-place topping | $1.5 \times 0.104 \times 37^2 =$ | 214 | 1.4 | 299 |
| | | 660 k·in | | |

To composite section:

| | k/ft | $M_c$ k·in | × LF | = $UM_{\mathcal{C}}$ |
|---|---|---|---|---|
| $S_{DL}$ $(15\ \text{lb/ft}^2)$ | $1.5 \times 0.050 \times 37^2 =$ | 103 | 1.4 | 144 |
| $S_{LL}$ $(40\ \text{lb/ft}^2)$ | $1.5 \times 0.133 \times 37^2 =$ | 273 | 1.7 | 464 |
| | | 376 k·in | $\Sigma UM_{\mathcal{C}}$ | = 1531 k·in |

**Stresses.** Stresses are as follows.

Prestress force required (see *PCI Design Handbook*, page 4-16):

$$F_f = \frac{\dfrac{M_s}{S_b} + \dfrac{M_c}{S_{bc}} f_t}{\dfrac{1}{A} + \dfrac{e}{s_b}} = \frac{\dfrac{660}{604} + \dfrac{376}{947} - 0.425}{\dfrac{1}{272} + \dfrac{3.2}{604}} = \frac{1.065}{0.00901} = 118.2^k$$

Using $^3/_8$-in $\phi$-270$^k$ strands (see *PCI Design Handbook*, pages 11-16 and 4-39),

$$A_{ps} = 0.085$$

$$F_o = 16.1^k/\text{strand}$$

$$F_i = 0.09 \times 16.1 = 14.5^k/\text{strand}$$

$$F_f = 0.75 \times 16.1 = 12.1^k/\text{strand}$$

$$N = \frac{118.2}{12.1} = 9.77$$

Therefore, use ten $^3/_8$-in $\phi$-270 strands.

Check end stresses at release ( − = compression ; + = tension ; see *PCI Design Handbook*, page 11-15):

$$f_{\text{top}} = -\frac{145}{272} + \frac{(145)(3.22)}{585} = 0.533 + 0.798 = +0.265$$

$$< 0.6\sqrt{f'_{ci}}\ \text{(OK)}$$

$$f_{\text{bottom}} = -\frac{145}{272} - \frac{(145)(3.22)}{604} = -0.533 - 0.773$$

$$= -1.306\ \text{ksi} < 0.6 f'_{ci}\ \text{(OK)}$$

Check service load stresses at centerline span:

$$f_{top} = -\frac{121}{272} + \frac{(121)(3.22)}{585} - \frac{660}{585} - \frac{376}{2186}$$

$$= -0.455 + 0.666 - 1.138 - 0.172 = -1.089 < 0.45f'_c \quad \text{(OK)}$$

$$f_{bottom} = -\frac{121}{272} - \frac{(121)(3.22)}{604} + \frac{660}{604} + \frac{376}{947}$$

$$= -0.445 - 0.645 + 1.093 + 0.397 = +0.400 < 6f'_c \quad \text{(OK)}$$

**Ultimate strength.** Check ultimate strength (see *PCI Design Handbook*, pages 4-6 and 4-63).

$$f_{se} = 142 \text{ ksi}$$

$$C\bar{\omega}p = 1.0\left(\frac{0.85}{40 \times 10.81}\right)\left(\frac{270}{3}\right) = 0.177$$

From the chart read $f_{ps}/f_{pu} = 0.96$; therefore, $f_{ps} = 259.2$ ksi.

$$a/2 = \frac{A_{ps}f_{ps}}{1.7f'_c b} = \frac{(0.85)(259.2)}{(1.7)(3)(40)} = 1.08 \text{ in}$$

$$M_u = \phi A_{ps} f_{ps}(d - a/2) = (0.9)(0.85)(259.2)(10.81 - 1.08)$$

$$= 1929^{k\cdot in} > UM \quad \text{(OK)}$$

**Horizontal shear.** Check horizontal shear between plank and topping (see *PCI Design Handbook*, page 4-28).

$$UV \ [(1.4)(0.217 + 0.104 + 0.050) + (1.7)(0.133)]\ \frac{37}{2} = 13.8^k$$

$$\phi V_{nh} = \phi\, 80b_v d = (0.85)(80)(40)(10.81)/1000 = 29.4^k > UV$$

Therefore, broom top finish or equivalent is sufficient to assure shear transfer.

**Camber and deflection.** Check camber and deflection. See *PCI Design Handbook*, pages 4-42 and 4-46.

$$\Delta_{i,\text{ prestress}} = 1.61 \text{ in} \uparrow \qquad \Delta_{i,\text{ Spancrete}} = -1.28 \text{ in} \downarrow$$

$$\Delta_{\text{topping}} = -0.57 \text{ in} \downarrow \qquad \Delta_{SDL} = -0.03 \text{ in} \downarrow$$

$$\Delta_{SLL} = -0.09 \text{ in} < L/480 = 0.93 \text{ in} \quad \text{(OK)}$$

At erection:

$$\Delta = +(1.80)(1.61) - (1.85)(1.28) = +0.53\,\text{in} \uparrow$$

All sustained loads:

$$\Delta = +(2.20)(1.61) - (2.40)(1.28) - (2.3)(0.57) - (3.0)(0.03) = -0.93\,\text{in} \downarrow$$

**Vertical shear.** Check vertical shear (see Fig. 25.11; *PCI Design Handbook*, page 4-66).

$$b_w d = (13)(10.81) = 141$$

$$v_{u\text{, support}} = \frac{UV}{\phi b_w d} = \frac{13.8}{(0.85)(141)} = 0.115\,\text{ksi}$$

$$L/d = \frac{(37)(12)}{10.81} = 40$$

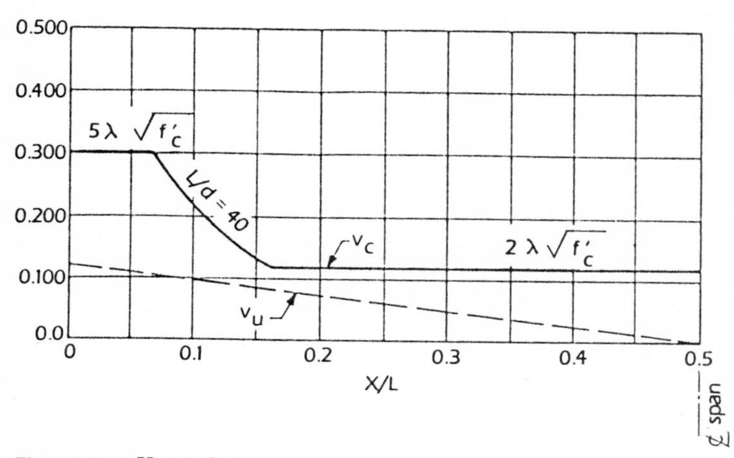


**Figure 25.11** Vertical shear.

Since $v_c > v_u$ in all cases, no web reinforcing is required (Uniform Building Code, Sec. 2611).

**Reinforcing summary: 10-in lightweight Spancrete.** See Fig. 25.12.

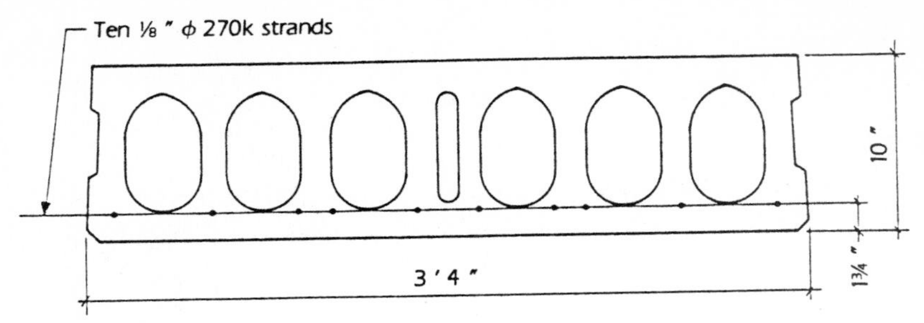


**Figure 25.12** Reinforcing summary: 10-in lightweight Spancrete.

## Spread Plank Systems

Sometimes it is economical to spread hollow-core planks apart and span the resulting gap with metal deck. Figure 25.13 shows a typical installation during construction, followed by a design example using 10-in Spancrete hollow-core planks.

### Design example: spread hollow-core-plank system

Design a spread plank system for a department store floor. A fire rating of 2 hours is required. Design span = 26.5 ft. Planks are spread 2 ft 10 in apart. (See Fig. 25.14.)

**Section properties.** Determine section properties.

Basic section:

$A = 272$ in $\quad I_b = 2970\ \text{in}^4 \quad$ Weight $= 85\ \text{lb/ft}^2$

$y_t = 5.09$ in $\quad s_t = 585\ \text{in}^3 \quad e = 3.16$ in

$y_b = 4.91$ in $\quad s_b = 604\ \text{in}^3 \quad E_{3000}/E_{4000} = 3.3/3.8 = 0.87$

Composite section:

| Part | $A$ | $y_{tcc}$ | $Ay_{tcc}$ | $I_o$ | $Ad^2$ |
|---|---|---|---|---|---|
| 10-in Spancrete | 272 | 8.59 | 2336 | 2970 | 2786 |
| 4.5-in CIP: 0.87 × A 4.5 = | 141 | 2.25 | 317 | 238 | 1399 |
| 3.5-in CIP: 0.87 × A 3.5 = | 116 | 1.75 | 203 | 118 | 1545 |
| Composite section | 529 | | 2856 | | |

**Figure 25.13** Sears building, Modesto, California. Spancrete planks 10 in thick spanning 28 ft 8 in and spread 34 3/4 in apart. Cast-in-place concrete topping 2 1/2 in thick is placed over the planks, and the metal deck spanning the gaps is covered with 4 1/2 in of cast-in-place concrete. The spread system also makes vertical penetrations for mechanical and electrical systems much simpler.

$$\Sigma I_0 + Ad^2 = I_c = 9038 \text{ in}^4.$$

$$y_{tcc} = 5.4 \text{ in} \quad y_{tc} = 1.9 \text{ in} \quad s_{tc} = 4757 \text{ in}^3$$

$$y_{bc} = 8.1 \text{ in} \quad s_{bc} = 1116 \text{ in}^3$$

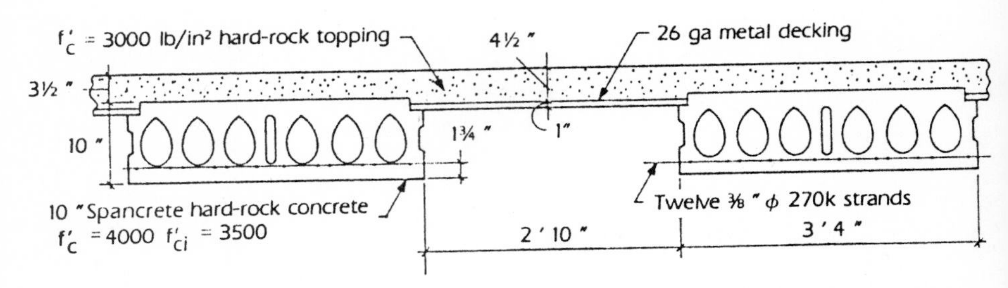


**Figure 25.14** Design data for 10-in-spread Spancrete system. $s_{LL} = 75 \text{ lb/ft}^2$; $s_{bl} = 25 \text{ lb/ft}^2$.

**Loads and moments.** Determine loads and moments.

To basic section:

| | | | | |
|---|---|---|---|---|
| 10-in Spancrete | $1.5 \times 0.283 \times 26.5^2 =$ | 298 | 1.4 | 417 |
| CIP topping | $1.5 \times 0.326 \times 26.5^2 =$ | 343 | 1.4 | 481 |
| | | $641^{\text{k in}}$ | | |

To composite section:

| | | | | |
|---|---|---|---|---|
| $S_{DL}$ | $1.5 \times 0.154 \times 26.5^2 =$ | 162 | 1.4 | 227 |
| $S_{LL}$ | $1.5 \times 0.463 \times 26.5^2 =$ | 488 | 1.7 | 830 |
| | | $650^{\text{k-in}}$ | | $1955^{\text{k in}}$ |

**Stresses.** Stresses are as follows.

Prestress force required (see *PCI Design Handbook*, 3d edition, page 4-16):

$$F_f = \frac{\dfrac{M_B}{S_b} + \dfrac{M_c}{S_{bc}} - f_t}{\dfrac{1}{A} + \dfrac{e}{s_b}} = \frac{\dfrac{641}{604} + \dfrac{650}{1116} - 0.425}{\dfrac{1}{272} + \dfrac{3.16}{604}} = \frac{1.219}{0.00891} = 136.8^{\text{k}}$$

Using $^3/_8$-in $\phi$-270$^k$ strands (see *PCI Design Handbook*, pages 4-39 and 11-16),

$$A_{ps} = 0.085 \text{ in}^k$$

$$F_o = 16.1^k/\text{strand}$$

$$F_i = 14.5^k/\text{strand}$$

$$F_f = 0.78 \times 16.1 = 12.6^k/\text{strand}$$

$$N = \frac{136.8}{12.6} = 10.9$$

Use twelve $^3/_8$-in $\phi$ - 270$^k$ strands.

Check end stresses at release (– = compression; + tension; see *PCI Design Handbook*, page 11-15):

$$f_{\text{top}} = -\frac{174}{272} + \frac{(174)(3.16)}{585} = -0.640 + 0.940$$

$$= +0.300 < 6\sqrt{f'_{ci}} \quad \text{(OK)}$$

$$f_{\text{bottom}} = -\frac{174}{272} - \frac{(174)(3.16)}{604} = -0.640 - 0.910$$

$$= -1.550 < 0.6 f'_{ci} \quad \text{(OK)}$$

Check final stresses at centerline span:

$$f_{\text{top}} = -\frac{151.2}{272} + \frac{(151.2)(3.16)}{585} - \frac{641}{585} - \frac{650}{4757}$$

$$= -0.556 + 0.817 - 1.096 - 0.137 = -0.972 < 0.45 f'_c \quad \text{(OK)}$$

$$f_{\text{bottom}} = -\frac{151.2}{272} - \frac{(151.2)(3.16)}{604} + \frac{641}{604} + \frac{650}{1116}$$

$$-0.556 - 0.791 + 1.061 + 0.582 = +0.296 < 6\sqrt{f'_c} \quad \text{(OK)}$$

**Ultimate strength.** Check ultimate strength. (See *PCI Design Handbook*, pages 4-6 and 4-63.)

$$f_{se} = 147.4 \text{ ksi}$$

$$C\overline{w}p = 1.0\left(\frac{1.02}{74 \times 11.75}\right)\left(\frac{270}{3}\right) = 0.106$$

From the chart read $f_{ps}/f_{pu} = 0.975$; therefore, $f_{ps} = 263.3$ ksi.

$$a/2 = \frac{A_{ps}f_{ps}}{1.7f'_c b} = \frac{(1.02)(263.3)}{(1.7)(3)(74)} = 0.71 \text{ in}$$

$$M_u = \phi A_{ps}f_{ps}(d - a/2) = (0.9)(1.02)(263.3)(11.75 - 0.71)$$

$$= 2668^{\text{k}\cdot\text{in}} > UM \quad \text{(OK)}$$

**Deflections.** Check deflections. (See *PCI Design Handbook*, pages 4-42 and 4-46.)

At erection:

$$\Delta = 0.63 \text{ in} \uparrow$$

All sustained loads:

$$\Delta_{LT} = 0.18 \text{ in} \downarrow$$

$$\Delta_{LL} = 0.15 \text{ in} \downarrow \text{ (OK)}$$

**Shear.** Check horizontal and vertical shear.

Horizontal shear between plank and topping (see *PCI Design Handbook*, page 4-28):

$$UV = [(1.4)(0.283 + 0.326 + 0.154) + (1.7)(0.463)]\frac{26.5}{2} = 24.6^{\text{k}}$$

$$\phi V_{nh} = \phi 80\, b_v d = (0.85)(80)(38)(11.75)/1000 = 30.4^{\text{k}} \text{ (OK)}$$

Therefore, broom finish or equivalent on top of plank is required.

Vertical shear (see Fig. 25.15; *PCI Design Handbook*, page 4-66):

$$b_w d = (13)(11.75) = 153 \text{ in}^2$$

$$v_{u\text{, support}} = \frac{UV}{\phi b_w d} = \frac{24.6}{(0.85)(153)} = 0.189 \text{ ksi}$$

$$\frac{L}{D} = \frac{26.5 \times 12}{11.75} = 27$$

Since $v_c > v_u$ in all cases, no web reinforcing is required (Uniform Building Code, Sec. 2611).

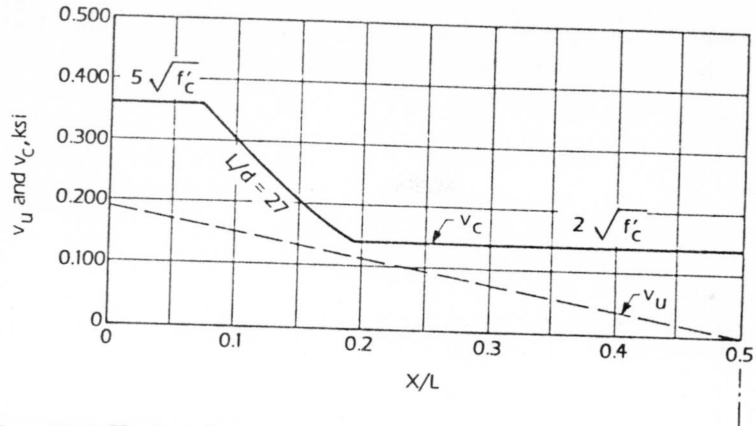


**Figure 25.15** Vertical shear.

## Additional Design Considerations

In designing untopped hollow-core plank, as discussed in Chap. 24, section "Analysis of Untopped Prestressed Concrete Flat Slabs," the prestress force should be such that the instantaneous camber at release is 1.1 times the instantaneous dead-load deflection produced by the slab self-weight. Proper design of untopped hollow-core systems is a result of camber analysis with the load-carrying capacity being determined afterward and then compared with that required. This will assure that differential camber (the difference in in situ elevations of adjacent units) will be minimized. In situations where heavy superimposed loads are intended to be carried on untopped hollow-core slabs, several field expedients serve to level adjacent planks exhibiting excessive camber differential until grouting can be performed. These are:

1. Use of temporary wooden wedges driven into shear keys
2. 4- by 4-in fish plates top and bottom connected by bolts through the shear keys
3. Placement of weights or water-filled drums on high planks

Untopped diaphragm action may be achieved by the use of edge weld plates cast into wet-cast hollow-core-plank products. Dry-mix hollow-core-plank products may use intermediate and boundary reinforcing calculated by the use of the shear friction concept to form the diaphragm. The use of this concept, as permitted by the Uniform Building Code, Sec. 2611, has been substantiated by testing.

The manufacturing process also affects the design of cantilever reinforcing. Wet-cast products can have top bars cast into the section. Dry-cast products use top strand and auxiliary mild steel cast in shear keys for situations where heavy point loadings occur on the cantilever ends.

**Figure 25.16** After curing, hollow-core planks are saw-cut to specified lengths and placed in storage. Odd-width planks are cut lengthwise. In the Span-Deck process shown here, the side rails rotate down and out of the way, allowing full access for the saw. (*Photograph by Ted Gutt.*)

## Connection Details

A common situation in the use of hollow-core planks is bearing the planks on masonry walls which also serve as the lateral-force-resisting system in the building. Close attention must be paid to assure that the transfer of lateral forces at the floor zone is not impaired by poor detailing at this critical interface. The details shown in Fig. 25.17 should be used. For further connection details, see Figs. 25.19 through 25.33.

**Figure 25.17** Typical interior bearing detail for hollow-core plank.

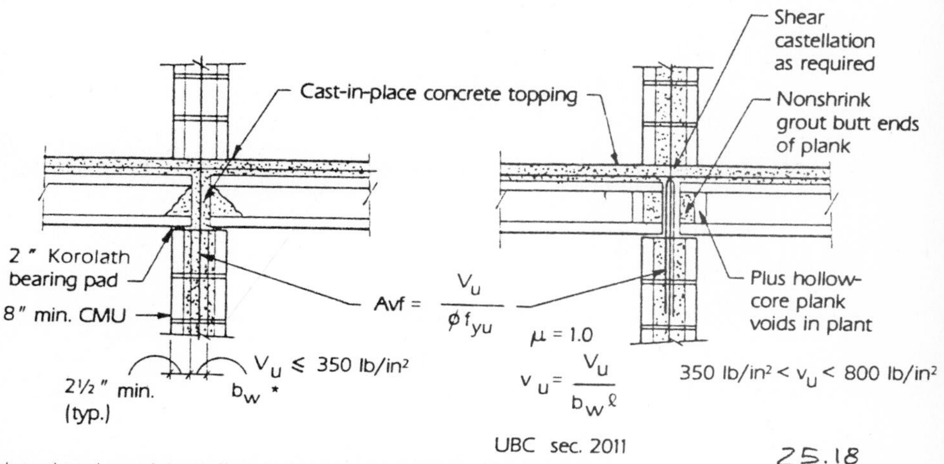


**Figure 25.18** Hollow-core planks in storage awaiting shipment. Plant fabrication is performed at the same time site work is done, thereby telescoping total project construction time. Note the preformed opening in the plank in the foreground.

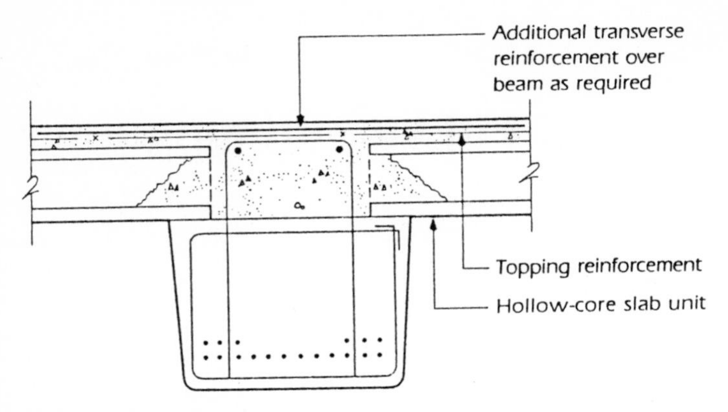


**Figure 25.19** Prestressed concrete girder.

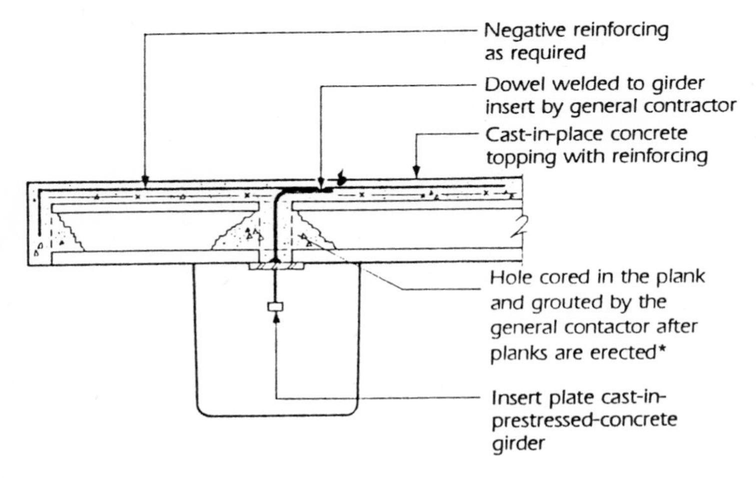


*Note that for Span-Deck process openings may be performed in the plant.

**Figure 25.20** Typical hollow-core-plank cantilever detail.

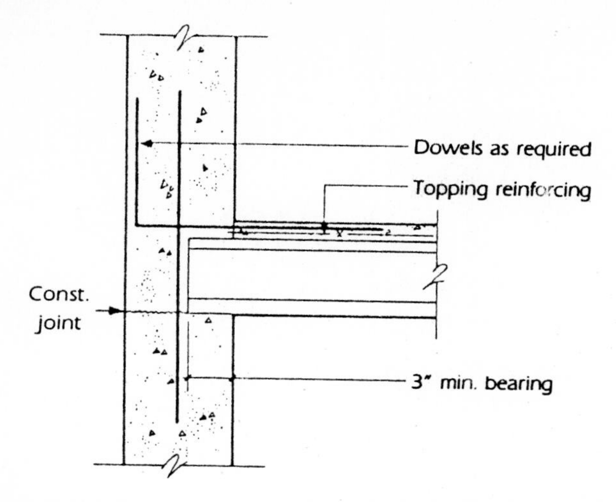


**Figure 25.21** Cast-in-place concrete exterior wall bearing detail.

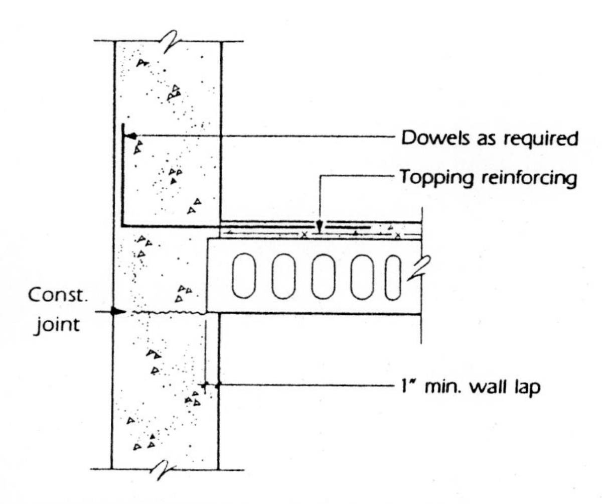


**Figure 25.21** ***(continued)*** Cast-in-place concrete exterior wall–nonbearing direction.

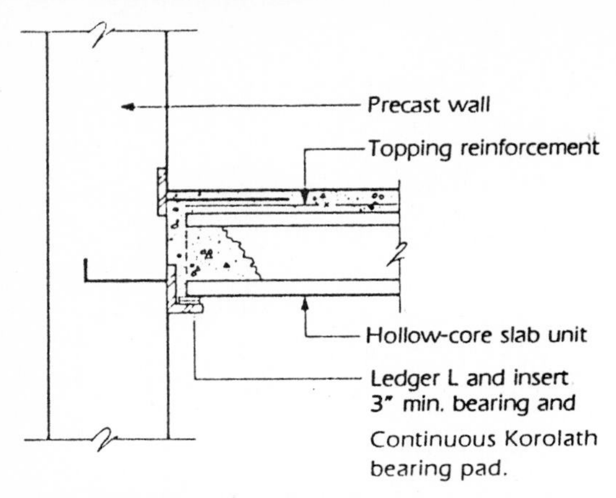


Figure 25.22 Precast wall unit with ledger support angle.

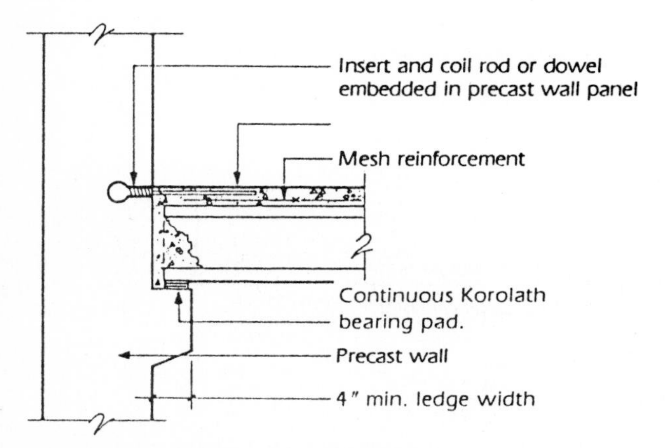


Figure 25.23 Precast concrete wall with corbel.

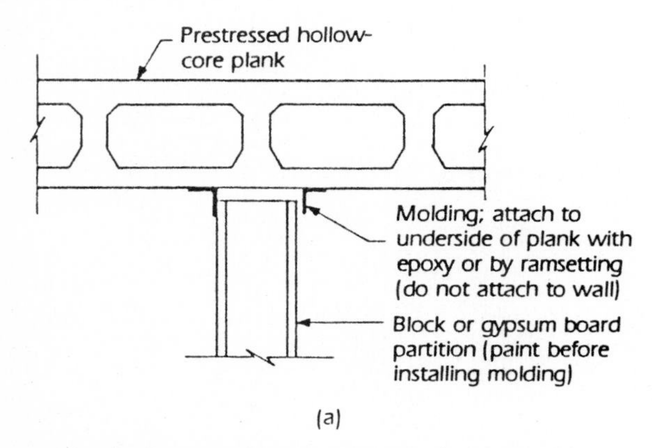


Figure 25.24 Partition attachment details. (*a* ) Option 1.

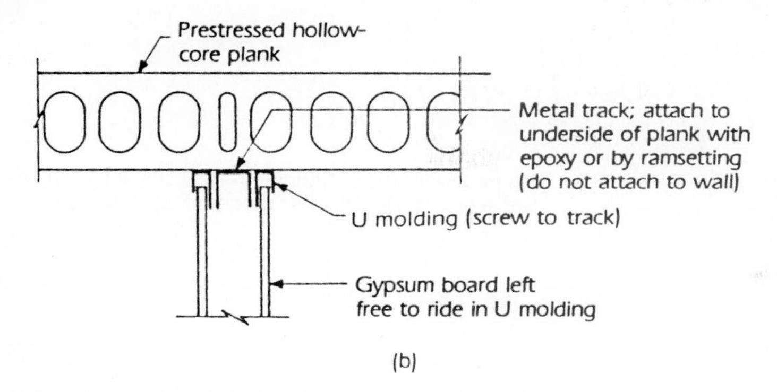


**Figure 25.24** ( *Continued* ) Partition attachment details. (*b* ) Option 2.

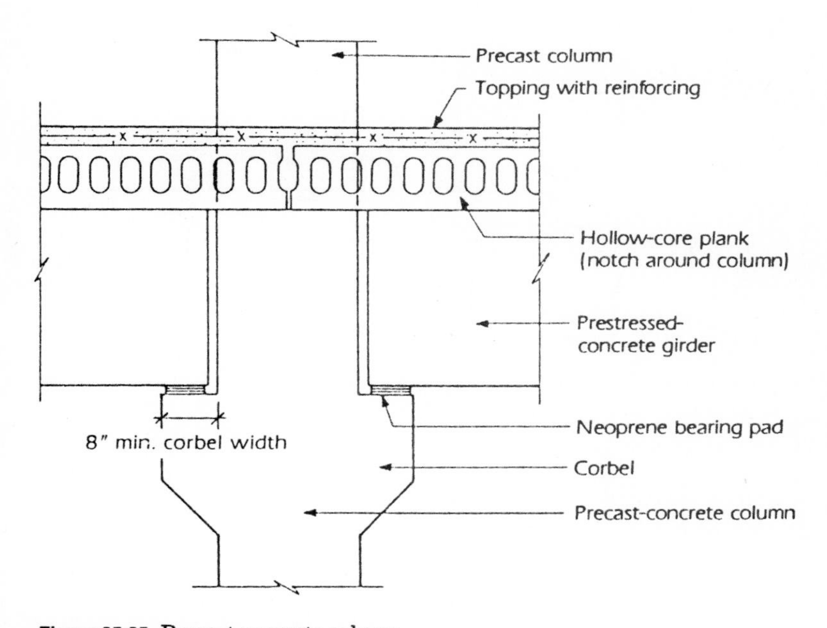


**Figure 25.25** Precast concrete column.

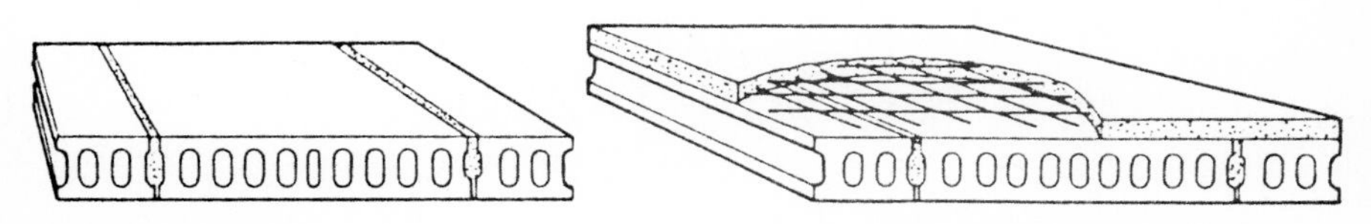

**Figure 25.26** Grouted joints.

**Figure 25.27** Topping with reinforcing.

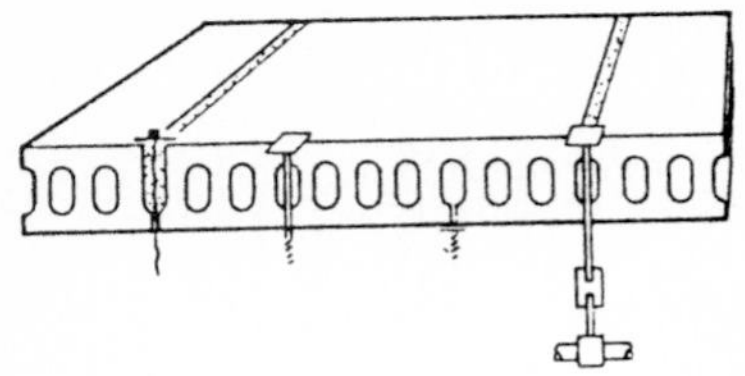

**Figure 25.28** Place hanger inserts for suspended ceilings and electrical or mechanical systems.

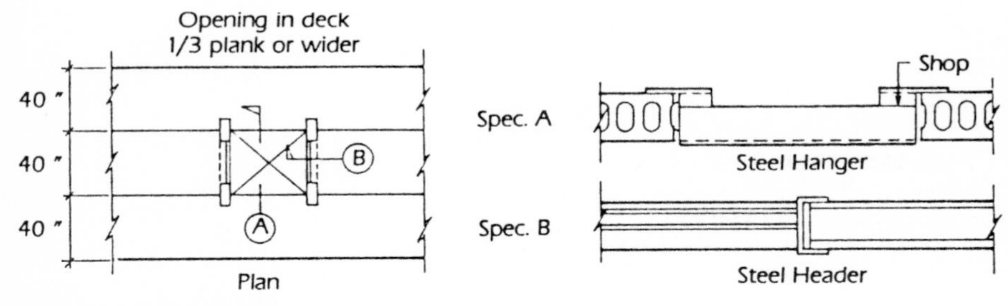


**Figure 25.29** Typical header detail.

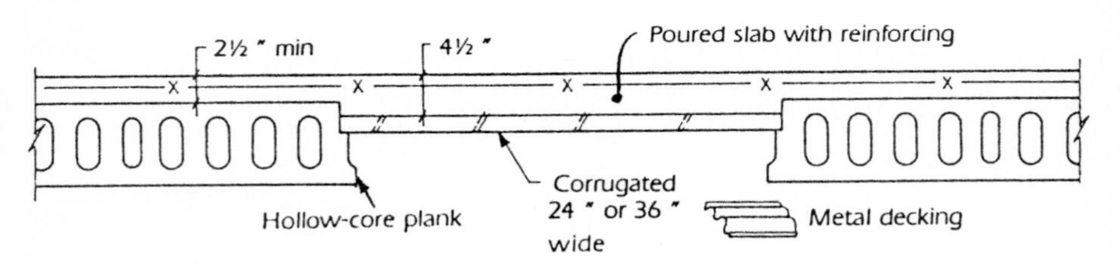


**Figure 25.30** Typical spread system.

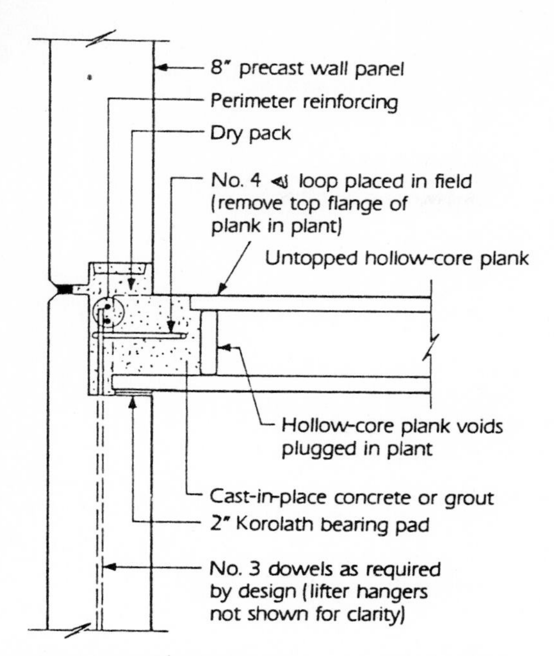


**Figure 25.31** Untopped hollow-core-plank diaphragm: exterior bearing connection.

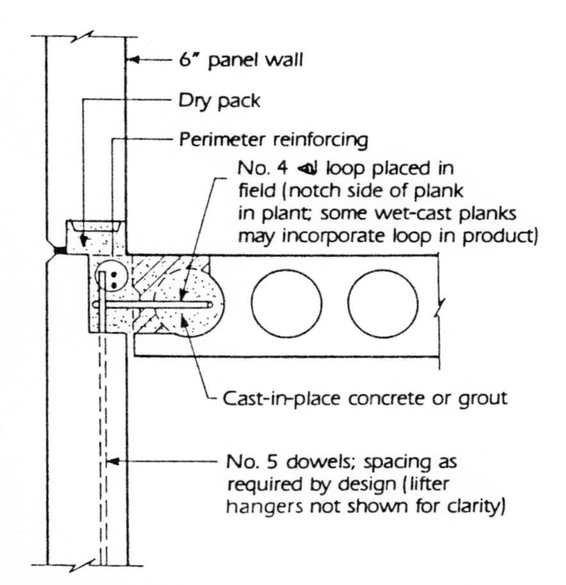


**Figure 25.32** Untopped hollow-core-plank diaphragm: exterior nonbearing lateral-load-resisting connection.

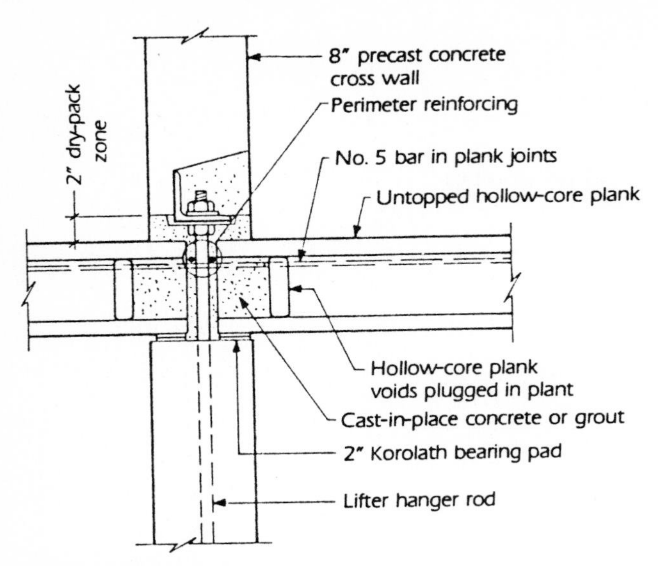


**Figure 25.33** Untopped hollow-core-plank diaphragm: interior bearing (platform) connection.

Production of Span-Deck is shown in Fig. 25.34.

**Figure 25.34** Span-Deck production operation. (*above*) The bottom (soffit) casting machine receiving concrete from a specialized yard delivery truck called a sidewinder. (*below*) The bottom casting machine as seen from the rear with the $1^3/_4$-in-thick soffit placed and ready for the subsequent top casting operation. (*Photographs by Ted Gutt.*)

**Figure 26.1** Alvarado sewage treatment plant, Union City, California. Double T's up to 70 ft long were provided for this complex of many buildings. Inverted-T girders and vertical- and lateral-load-resisting wall panels were also plant-precast and then installed, quickly enclosing these buildings.

Chapter

# 26

# Prestressed Concrete Double T's

The double T has become the mainstay of the prestressed concrete industry. It is used as a deck member in floors and roofs of offices, shopping malls, department stores, and schools and in parking structures. Deeper sections are used to form clear-span roofs over gymnasiums and swimming pools. In roof construction, rigid insulation can be incorporated into the flanges in the plant. The double T can also be used as a wall panel in commercial and industrial buildings. It can be used to resist vertical as well as lateral loads in these applications. Wall panel units can be solid units as well as insulated sandwich panel units. Windows and door openings may be placed in the flange areas between the stems.

In this chapter, we shall deal with double T's used as flexural deck members. As floor units, double T's usually receive a field-poured cast-in-place concrete topping, which provides a smooth, level floor surface and serves as the horizontal diaphragm. Roof units may be designed without cast-in-place concrete topping, in which case the units are connected by flange weld plates, thus forming the diaphragm.

## Availability

Double T's are made in various widths, with 8- and 12-ft-wide units being the most frequently used (see Figs. 26.2 and 26.3). They are made in depths from a minimum of 10 in to a maximum of 41 in. The standard 2-in flange thickness may be increased by altering the height of the side rail. Double-T spans of up to 90 ft are possible in the deeper sections.

**Figure 26.2** Heavy-stemmed double T's 8 ft wide and 32 in deep in storage in the plant. Prestressed concrete double T's are fabricated and stockpiled while site work is done, thereby telescoping total project time. Note the proper dunnage, clean access, and well-ordered rows of product.

**Figure 26.3** Union Bank – Oceangate parking structure, Long Beach, California. Double T's 12 ft wide and 36 in deep, spread apart 4 ft, provide the clear-span deck for this three-level parking structure.

**Figure 26.3** (*Continued*) Union Bank – Oceangate parking structure, Long Beach, California. The area between the spread T's was formed and poured at the same time that the cast-in-place concrete topping was poured. Long-span, column-free parking bays are afforded by the 64-ft-span double T's.

## Manufacturing

Double T's are cast specifically for individual projects and are not stockpiled prior to order. They are cast in long lines ranging from 300 to 600 ft. Many units are fabricated in the same bed, with bulkhead separators being placed between the individual units. Units are cast in steel molds which have variable stem widths, since the form sidewalls slope approximately $^{1}/_{2}$ in per foot of depth. Various depths of T's are fabricated by casting in temporary concrete soffit liners or by using preformed special metal soffit liners. Odd-width units can be made by blocking out portions of the overhanging flange. Blockouts can be placed on the bed to create openings in the unit, eliminate one flange, stop the stem short of a flange, stop the flange short of a stem, or create single T's (see Fig. 26.4). Openings can also be placed in the stem of the member up to one-third the total member depth with no decrease in flexural strength. These openings should be located to clear the prestressing strands, optimally in the upper portion of the web. For topped designs, the top surface of the T should be roughened sufficiently to develop horizontal shear in accordance with the requirements of the Uniform Building Code. A broom finish on the top surface satisfies these roughness requirements. Untopped roof T's should have their top surfaces steel-trowel-finished.

Concrete strengths for hard-rock concrete are 5000 and 6000 lb/in$^2$; 5000 lb/in$^2$ sand-lightweight concrete also is used when the additional cost of the lightweight aggregate is offset by savings in some other area, such as shipping or erection. Prestressing strand is usually $^1/_2$ in in diameter, 270-ksi grade. Economical designs usually specify harped strand. This is accomplished either by depressing down with special devices from the top of the bed after tensioning the strand or by using hold-down devices fastened to anchors built in under the bed. In the latter case, the hold-down devices are positioned prior to tensioning the strand. Welded wire fabric is used for flange reinforcing. Either welded wire fabric or mild steel stirrups, fabricated in advance into cages, are used as stem reinforcing.

**Figure 26.4** Alvarado sewage treatment plant, Union City, California. Single T's 4 ft wide were made in the double-T mold by installing a temporary continuous flat bar former down the centerline of the bed. This was done to accommodate the 4-ft module used to lay out the bays in the building.

## Erection Considerations

For plant handling and erection, lifting bales consisting of loops made up from prestress strand are embedded in each end of each stem. After the double T's have been erected, these may be burned off for uptopped units or left cast in the concrete topping. Access for trucks into each bay of the structure is essential when erecting these large units, which can weigh 15 tons or more. Often prudent architects and their engineers consult with precast concrete operations personnel early in a project to determine the best way to design the long-span members and not overlook practical details that could save money in the installation of the prestressed concrete components. See Figs. 26.5 and 26.6.

**Figure 26.5** J. C. Penney store, Santa Rosa, California. Prestressed concrete shallow double T's, inverted-T girders, and precast concrete columns form the basic building frame for the floors and roof of this department store, with lateral forces being transmitted to the masonry infill walls. Quick erection saves construction time, and the precast decks form an instant working platform for other trades.

**Figure 26.6** Kaiser Permanente plant expansion, Cupertino, California. Double T's form the walls, floors, and roof of this complicated industrial building. Interior precast wall panels, columns, and beams complete the range of precast concrete products used in this totally precast concrete structure. Note the bracing used to stabilize the structure during erection. See photograph below: the large openings were preplanned and formed in the plant in these 80-ft-span double T's.

## Design Considerations

The span of the double T's is dependent upon architectural considerations, building module, transportation, and economics. The architectural considerations are use of the enclosed space, building shape, and design effect (aesthetics). Certain types of buildings such as parking structures have set column spacings and clear-span requirements which give optimum parking arrangements and maximum number of spaces. For hotels and apartment buildings, the size and arrangement of the room layouts will dictate the location of structural supports, which in turn will dictate the span of the double T. Office and industrial occupancies require large clear spans. For industrial or engineering research occupancies, the design live loads also have an effect on spans.

The module to be selected in relation to the T span is chosen to give the maximum number of units of the same length as well as to be compatible with the wall panel size or module. The effect of cantilevers also is taken into consideration. Usually the building is designed around an 8-ft module in both directions, being a multiple of the T width or bay spacing. It is desirable but not essential that all the T units be of the same length.

To reduce costs it is desirable to keep the spans within the truck and load lengths permitted without special permits. For double T's this results in a T length of approximately 60 ft. Longer lengths may be transported at a small premium in shipping cost. Since economics is always a factor, a cost study is often made to determine which system as well as span-length module would be the most desirable: hollow-core plank with shorter spans versus long-span double T's; solid double-T system versus spread channel system with pretensioned flat panels.

Untopped diaphragms may be formed in two ways. The first uses flange weld plates spaced as required by design to develop diaphragm integrity. Care must be taken at boundaries and interfaces with lateral-load-resisting systems to adequately transfer seismic diaphragm forces and provide structural redundancy at these points. The second method uses cast-in-place concrete edge beams and intermediate and boundary reinforcing calculated by the use of the shear friction theory to form the diaphragm. The use of this concept as permitted by the Uniform Building Code, Sec. 2611, has been substantiated by testing. Untopped diaphragms are formed with double-T units on roofs and, in certain specialized instances, in parking structures. The flanges are thickened to 4 or 4 1/2 in and are interconnected with special welded connections (see section "Connection Design"). Other connection conditions also are shown for these untopped structures.

Preliminary design charts for double-T units commonly made by precasters are shown in Figs. 26.7 and 26.8. The charts in the third edition of the *PCI Design Handbook*, pages 2-6 through 2-24, also are useful. Designers should always contact prestressed concrete manufacturers in the general vicinity of the project to ascertain the specific details and sections available. A sample of a specific manufacturer's cross section and product literature is shown in Fig. 26.9, and an example double-T design calculation is presented.

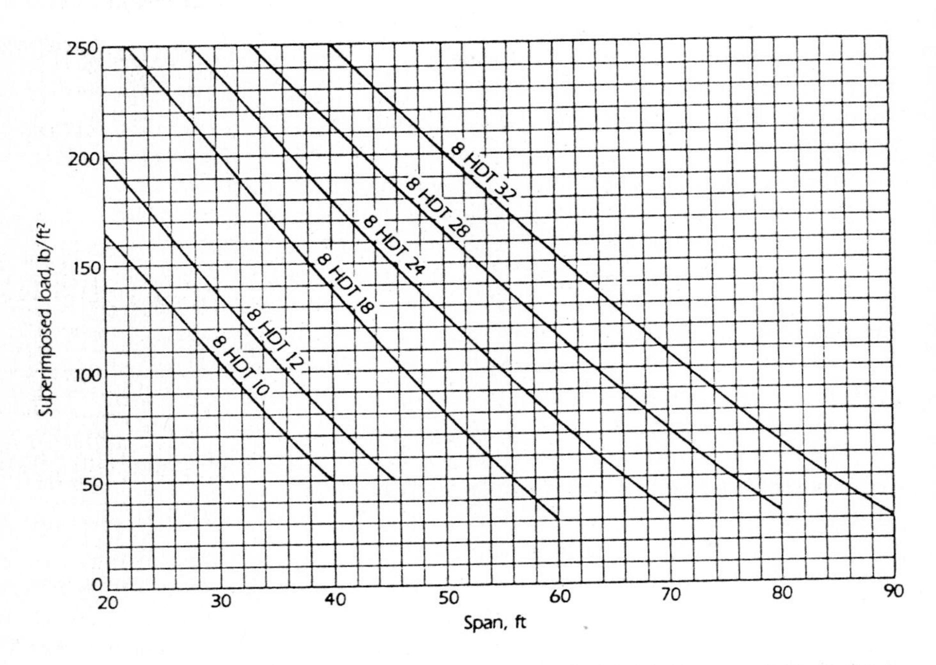


**Figure 26.7** Preliminary load table for heavy-stemmed double T's (bottom width, ≥ 6 in). All double T's are fabricated with hard-rock concrete ($f'_c$ = 5000 lb/ in$^2$); 2 1/2 in of hard-rock concrete topping is cast in place. The T designation 8HDT means "8-ft-wide heavy-stemmed double T 24 in deep."

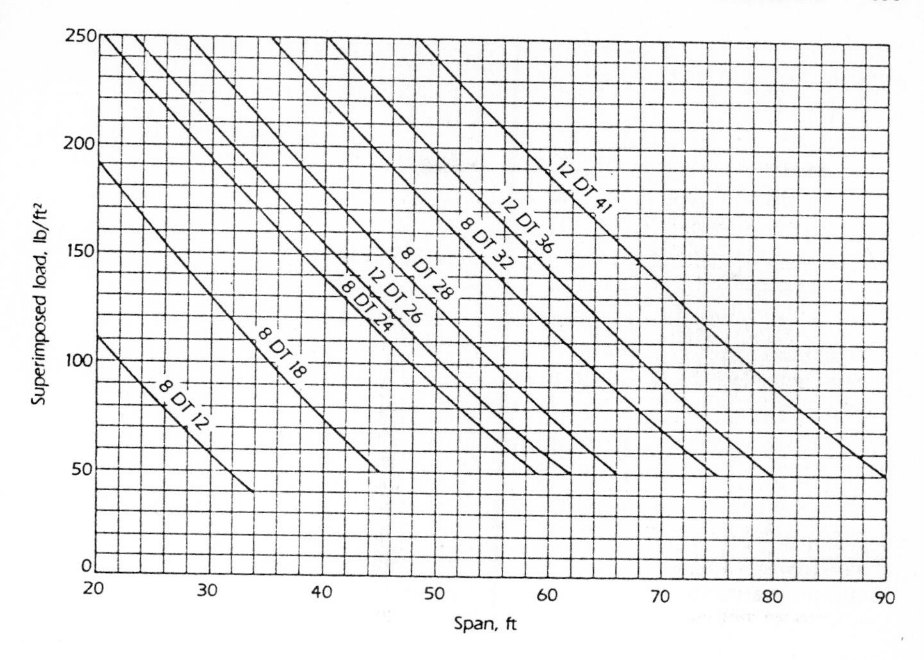


**Figure 26.8** Preliminary load table for normal-stem-width double T's (bottom stem width, < 4 3/4 in for 8-ft-wide T's, 7 1/4 in for 12-ft-wide T's). All double T's are fabricated with hard-rock concrete ($f'_c$ = 5000 lb/in$^2$) ; 2 1/2 in of hard-rock concrete topping is cast in place. The T designation 12DT36 means "12-ft-wide double T 36 in deep."

### Double-T design

Design an 8-ft-wide double T for a parking structure (see Fig. 26.10). The span is 58 ft long; live load is 50 lb/ft$^2$. Use hard-rock concrete: $f'_c = 5000$ lb/in$^2$; $f'_{ci} = 3500$ lb/in$^2$.

From the *PCI Design Handbook* (3d edition; page 2-16), a 24 in-deep-section is selected. From the manufacturer's product literature shown in Fig. 26.9, specific section properties for a 24-in double T are noted.

Basic section:

| | |
|---|---|
| $A = 444$ in$^2$ | Weight = $0.480^k$/ft |
| $I_b = 24{,}237$ in$^4$ | $S_t = 3245$ in$^3$ |
| $y_t = 7.47$ in | $S_b = 1466$ in$^3$ |
| $y_b = 16.53$ in | |

Section properties

$A = 444.5$ in$^2$

Weight

44.4 lb/ft$^2$ (lightweight at 115 lb/ft$^3$)

57.9 lb/ft$^2$ (hard rock at 150 lb/ft$^3$)

Without topping: $I = 24{,}237$ in$^4$

$C_t = 7.47$ in  $K_t = 3.29$ in  $S_t = 3245$ in$^3$

$C_b = 16.53$ in  $K_b = 7.30$  $S_b = 1466$ in$^3$

Composite with 2-in topping slab: $I = 33{,}914$ in$^4$

$C_t = 6.91$ in  $S_t = 4905$ in$^3$

$C_c = 4.91$ in  $S_c = 6902$ in$^3$

$C_b = 19.09$ in  $S_b = 1776$ in$^3$

Composite with 2½-in topping slab: $I = 36{,}205$ in$^4$

$C_t = 6.91$ in  $S_t = 5238$ in$^3$

$C_c = 4.41$ in  $S_c = 8208$ in$^3$

$C_b = 19.59$ in  $S_b = 1848$ in$^3$

Fire rating

2 hours with 2-in lightweight topping

2 hours with 2½-in hard-rock topping

Recommended uses

Parking structures with 2½-in hard-rock topping:

Spans to 62 ft with 50 lb/in$^2$ unreduced live load

Spans to 64 ft with 50 lb/ft$^2$ live load reduced to 40 lb/ft$^2$

Roof:

Spans to 64 ft with or without topping

Floor:

Spans to 58 ft with 2-in topping

Spans to 53 ft with 2½-in topping

Load assumptions

Roof:

18 lb/ft$^2$ superimposed dead load

20 lb/ft$^2$ live load to 50-ft span

16 lb/ft$^2$ live load over 50-ft span

Floor:

30 lb/ft$^2$ superimposed dead load

50 lb/ft$^2$ live load reduced for area

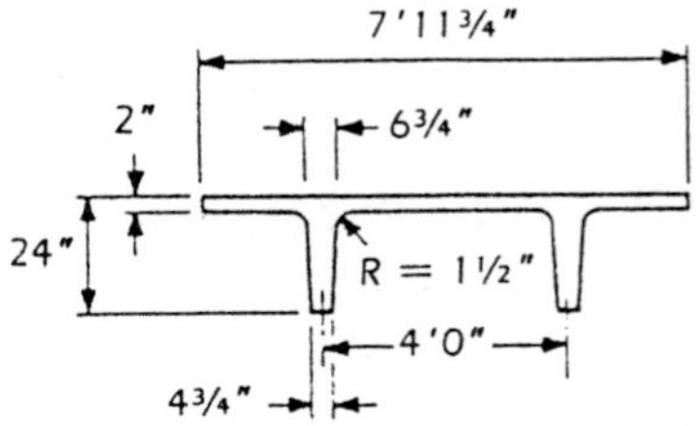

7′11¾″
2″
6¾″
24″
R = 1½″
4′0″
4¾″

**Figure 26.9** Data for 8-ft lightweight Kabo-Karr double T × 24 in. (*Courtesy of Kabo-Karr, L. A. Compton Group, Visalia, California.*)

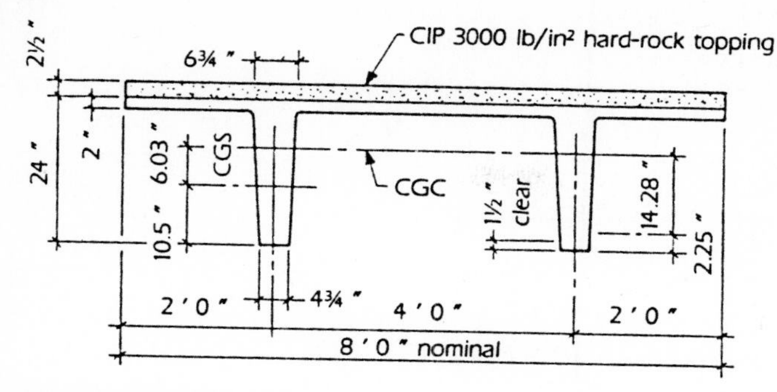


**Figure 26.10** Double T design.

Composite section:

$$I_c = 36{,}205\,\text{in}^4 \quad S_{tc} = 8208\,\text{in}^3$$
$$y_{tc} = 4.41\,\text{in} \quad S_{bc} = 1848\,\text{in}^3$$
$$y_{bc} = 19.59\,\text{in}$$

Strand data for 1/2-in $\phi$ -270$^k$ strand (see *PCI Design Handbook*, page 11-16) :

$$f_{pu} = 270\,\text{ksi} \quad A_{ps} = 0.1531^2/\text{strand}$$

| | |
|---|---|
| Initial jacking force | $F_o = 0.7A_{ps}f_{ps} = 28.9^k/\text{strand}$ |
| At release | $F_i = 0.9 \times F_o = 26.0^k/\text{strand}$ |
| Final design (22 percent losses) | $F_f = 0.78 \times F_o = 22.5^k/\text{strand}$ |

**Loads and moments.** Determine loads and moments (see *PCI Design Handbook*, page 4-39).

To basic section:

| | k/ft | $M_c^{\text{k·in}}$ | × LF | = $UM_c$ |
|---|---|---|---|---|
| Double T | $1/8 \times 0.480 \times 58^2 \times 12 =$ | 2422 | 1.4 | 3391 |
| CIP | $1/8 \times 0.250 \times 58^2 \times 12 =$ | 1262 | 1.4 | 1766 |
| | | 3684$^{\text{k·in}}$ | | |

To composite section :

| | k/ft | $M_c^{\text{k·in}}$ | × | LF | = | $UM_{\mathbb{C}}$ |
|---|---|---|---|---|---|---|
| Live load | $1/8 \times 0.400 \times 58^2 \times 12 =$ | $2018^{\text{k·in}}$ | | 1.7 | | 3431 |
| | | | | $\Sigma UM$ | = | $8588^{\text{k·in}}$ |

**Strand harping:** Determine whether or not strands must be depressed.

$$f_b = \frac{M_B}{S_b} + \frac{M_c}{S_{bc}} = \frac{3684}{1466} + \frac{2018}{1848} = 3.6 \text{ ksi}$$

$$f_t = + \frac{f_c'}{2} = 6\sqrt{5000} + \frac{3500}{2} = 2.2 \text{ ksi} < f_b$$

Therefore, use depressed strands (shallow drape).

$$M_{0.4L} = 0.96M_{\mathbb{C}}$$
$$e_{0.4L} \cong 0.93e_{\mathbb{C}}$$

**Stresses.** Determine stresses.
Required prestress force:

$$F_f = \frac{\dfrac{M_B}{S_b} + \dfrac{M_c}{S_{bc}} - f_t}{\dfrac{1}{A} + \dfrac{e}{S_b}} = \frac{\dfrac{(0.96)(3684)}{1466} + \dfrac{(0.96)(2018)}{1848} - 6\sqrt{5000}}{\dfrac{1}{444} + \dfrac{(0.93)(14.28)}{1466}}$$

$$= \frac{2.412 + 1.048 - 0.425}{0.00225 + 0.00906} = \frac{3.035}{0.01131} = 268^{\text{k}}$$

$$N = \frac{268}{22.5} = 11.9$$

Therefore, use ½-in $\phi$-$270^{\text{k}}$ strands.
Check stresses at end at release (see *PCI Design Handbook*, pages 4-16 and 11-15):

$$f_{\text{top}} = - \frac{(12)(26)}{444} + \frac{(312)(6.03)}{3245}$$

$$= -0.703 + 0.580 = -0.123 \text{ ksi composite}$$

$$f_{\text{bottom}} = -0.703 - \frac{(312)(6.03)}{1466}$$

$$= -0.703 - 1.283 = -1.986 < 0.6f'_{ci} \quad (\text{OK})$$

Check bottom tension under all loads at $0.4L$:

$$e_{0.4L} = (0.8)(14.28 - 6.03) + 6.03 = 12.63 \text{ in}$$

$$f_{\text{bottom}} = -\frac{(12)(22.5)}{444} - \frac{(270)(12.63)}{1466} + \frac{(0.96)(3684)}{1466} + \frac{(0.96)(2018)}{1848}$$
$$= -0.608 - 2.326 + 2.412 + 1.048$$
$$= +0.526 \text{ ksi}$$

$$7.5\sqrt{f'_c} = +0.530 \text{ ksi} \quad (\text{OK})$$

**Ultimate strength.** Check ultimate strength (see *PCI Design Handbook*, page 4-6).

1. Approximate code method (Uniform Building Code, Sec. 2618)

$$\rho_p = \frac{A_{ps}}{bd} = \frac{(12)(0.1531)}{(96)(22.6)} = 0.000847$$

$$f_{ps} = f_{pu}\left(1 - 0.5\rho_p \frac{f_{pu}}{f'_c}\right) = 270\left(1 - 1.05 \times 0.000847 \times \frac{270}{3}\right)$$

$$= 259.7 \text{ ksi}$$

$$a/2 = \frac{A_{ps}f_{ps}}{1.7f'_c b} = \frac{(1.837)(259.7)}{(1.7)(3)(96)} = 0.97 \text{ in}$$

$$M_u^{0.4L} = \phi A_{ps} f_{ps}(d - a/2) = (0.9)(1.837)(259.7)$$

$$(22.6 - 0.97) = 9287^{\text{k}\cdot\text{in}}$$

$$UM^{0.4L} = 0.96 \times 8588 = 8244^{\text{k}\cdot\text{in}} < M_u \quad (\text{OK})$$

2. Using strain compatibility (see *PCI Design Handbook*, pages 4-6 and 4-63) and Fig. 3.9.5,

$$f_{se} = 147 \text{ ksi}$$

$$C\overline{w}p = C\frac{A_{ps}}{bd} \times \frac{f_{pu}}{f'_c} = 1.0\left(\frac{1.837}{96 \times 22.6}\right)\left(\frac{270}{3}\right) = 0.076$$

From the chart,

$$\frac{f_{ps}}{f_{pu}} = 0.98; \therefore f_{ps} = (0.98)(270) = 264.6 \text{ ksi}$$

$$a/2 = \frac{(1.837)(264.6)}{(1.7)(3)(96)} = 0.99 \text{ in}$$

$$M_u^{0.4L} = (0.9)(1.837)(264.6)(22.6 - 0.99) = 9454^{\text{k}\cdot\text{in}}$$

**Shear.** Check horizontal and vertical shear.

Horizontal shear between precast T and topping (see *PCI Design Handbook*, page 4-28):

$$UV = [(1.4)(0.480 + 0.250) + (1.7)(0.400)]\,\frac{58}{2} = 49.4^{\text{k}}$$

$$\phi V_{nh} = \phi\, 80\, b_v d = (0.85)(80)(96)(22.6)/1000 = 147.5^{\text{k}}$$

Therefore, roughen top of T; no ties are required. [This is analogous to Uniform Building Code, Sec. 2617, where $v_{dh} = (UV/\phi b_v d) \leq 80$ lb/in². Therefore, roughen top only.]

Check vertical shear; proportion web reinforcement. Note certain provisions of the Uniform Building Code :

1. Web reinforcing may be omitted where $v_u < 0.5v_c$.
2. Maximum spacing shall be the smallest of:

   *a.* $0.75h = (0.75)(24) = 18$ in

or *b.* 24 in

or *c.* $\dfrac{A_v f_y}{50b_w} = \dfrac{(0.22)(40{,}000)}{(50)(11.5)} = 15.3$ in, which governs

Construct shear diagram. (See Fig. 26.11; *PCI Design Handbook*, page 4-67.)

$$d = 0.8h = 19.2 \text{ in}$$

$$v_{u,\text{ support}} = \frac{UV}{\phi bd} = \frac{49.4}{(0.85)(11.5)(19.2)} = 0.263 \text{ ksi}$$

$$\frac{L}{D} = \frac{58 \times 12}{19.2} = 36$$

$$5\sqrt{f_c'} = 0.350 \text{ ksi}$$

$$2\sqrt{f_c'} = 0.140 \text{ ksi}$$

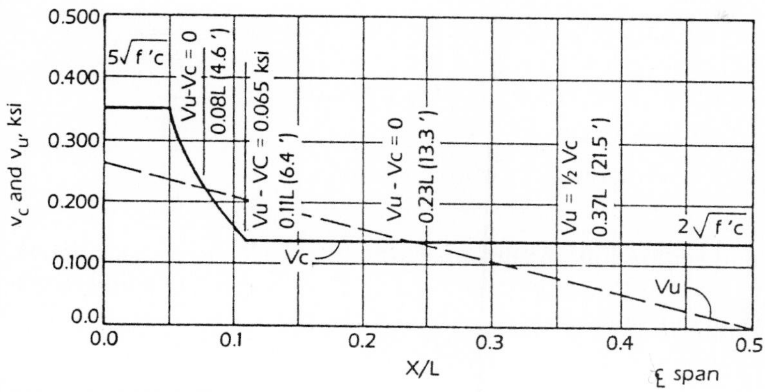


**Figure 26.11** Shear diagram.

**Shear reinforcement:** See Fig. 26.12.

Determine minimum web reinforcement requirements (see *PCI Design Handbook*, page 4-71).

Using no. 3 stirrups (one row of bars for each T stem),

$$b_w d = (11.5)(19.2) = 220 \text{ in}^2$$

$$A_v = 0.22 \text{ in}^2$$

$$A_{ps} = 1.837 \text{ in}^2$$

$A_{v \text{ provided}}$ no. 3 bars at 15 in on center = $0.22 \times 12/15 = 0.176$ in² (OK)

(Use no. 3 stirrups, one row per T stem.)

Determine web reinforcing required at maximum value of $(v_u - v_c)\,b_w = (0.065)(11.5) = 0.748$ (see *PCI Design Handbook*, page 4-72):

For no. 3 bars at 12 ($A_v = 0.22$) $\quad (v_u - v_c)b_w = 0.733^k/\text{in}$ (OK)

For $4 \times 4 - W\,5.5 \times W\,5.5$ $\quad (v_u - v_c)b_w = 0.810^k/\text{in}$ (OK)

Select either option (see *PCI Design Handbook*, page 11-22).

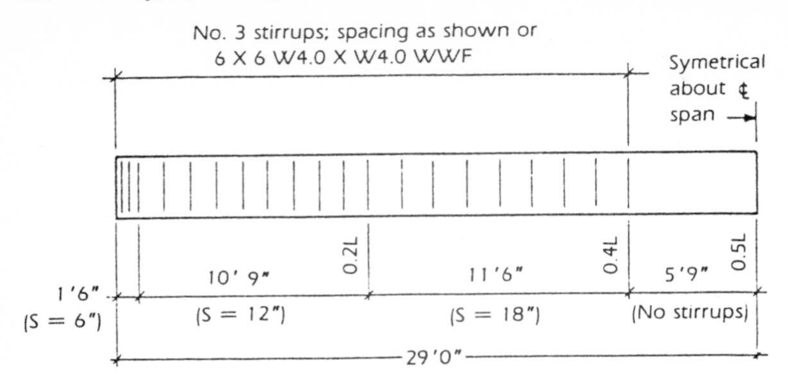


**Figure 26.12** Web reinforcement. $S$ = stirrup spacing.

Design end reinforcing in bottom of stem (see *PCI Design Handbook,* page 4-46). Take full ultimate moment at 25 $\phi$ length of strand development. (See Fig. 26.13.)

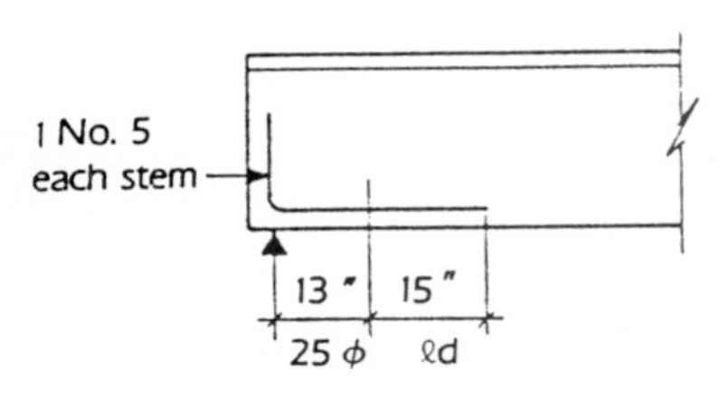


**Figure 26.13** End reinforcing.

$$UM^{1.1'} \cong (49.4)(1.1) = 54.3^{\text{k} \cdot \text{ft}}$$

$$A_s \text{ required} = \frac{UM}{a_u d} \cong \frac{54.3}{4 \times 24}$$

$$= 0.57 \text{ in}^2$$

Use two no. 5 ($A_s = 0.62$ in$^2$).

Development length of no. 5 bar $Ld = 15$ in (see *PCI Design Handbook,* page 11-20). Note that this reinforcing may be deleted when a properly designed end-bearing connection assembly is used (see Fig. 26.15).

Design flange reinforcing.
Temperature steel:

$$A_s = 0.0018 \times 12 \times 2 = 0.043 \text{ in}^2/\text{ft each way}$$

Temporary construction load:

$$S_{DL+LL} = 100\,\text{lb/ft}^2 \text{ (OSHA requirements; flange cantilever governs)}$$

$$M = \frac{0.100 \times 20.5^2}{2 \times 144} = 0.15^{\text{k}\cdot\text{ft}}/\text{ft}$$

$$A_s = \frac{M}{ad} \cong \frac{0.15}{1.76 \times 1} = 0.085\ \text{in}^2/\text{ft transverse}$$

Use 4 × 4 – W 1.4 × W 2.9 welded wire fabric (longitudinal; $A_s$ = 0.043 in²/ft) (transverse; $A_s$ = 0.087 in²/ft) (see *PCI Design Handbook*, page 11-22).

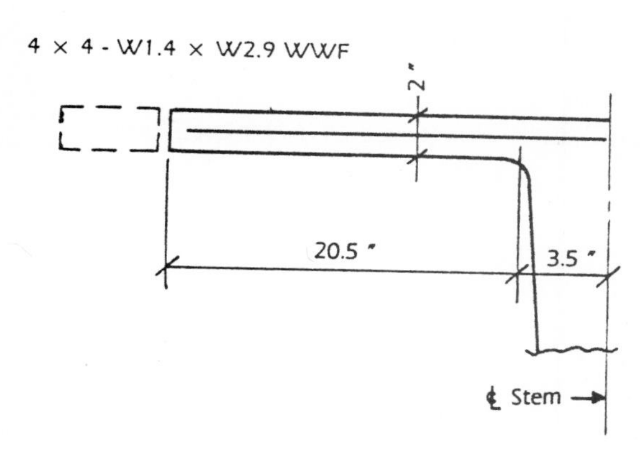


**Camber and deflection.** Check camber and deflection (see *PCI Design Handbook*, pages 4-73 and 11-15).

1. Determine instantaneous values of camber (+) and deflection (-).
   Camber :

$$e_e = 6.03\ \text{in} \quad e' = 14.28 - 6.03 = 8.25\ \text{in} \quad E_{ci} = 3600\ \text{ksi}$$

$$\Delta_{PS} \uparrow = \frac{F_i e_e L^2}{8E_{ci} I_b} + \frac{F_i e' L^2}{12 E_{ci} I_b} = \frac{(312)(6.03)(58^2)(144)}{(8)(3600)(24{,}237)}$$

$$+ \frac{(312)(8.25)(58^2)(144)}{(12)(3600)(24{,}237)}$$

$$= 1.31 + 1.19 = +2.5$$

Double-T weight deflection:

$$\Delta_{DT} \downarrow = \frac{5wl^4}{384E_{ci}I_b} = \frac{(5)(0.480)(58^4)(1728)}{(384)(3600)(24{,}237)} = -1.40 \text{ in}$$

Topping weight deflection:

$$\Delta_{\text{topping}} \downarrow = \frac{5wl^4}{384EI_b} = \frac{(5)(0.250)(58^4)(1728)}{(384)(4300)(24{,}237)} = -0.61 \text{ in}$$

Live-load deflection:

$$\Delta_{LL} \downarrow = \frac{5wl^4}{384EI_c} = \frac{(5)(0.400)(58^4)(1728)}{(384)(4300)(36{,}205)} = -0.65 \text{ in}$$

2. Determine long-term effects of creep (see *PCI Design Handbook*, page 4-46).

At erection:

| | $\Delta_i$ | × | CF | $\Delta_{\text{long-term}}$ |
|---|---|---|---|---|
| Camber | +2.50 | × | 1.80 = | +4.50 |
| Deflection (double-T-weight) | -1.40 | × | 1.85 = | -2.60 |
| Σ | | | | +1.90 in ↑ |

Final condition :

| | $\Delta_i$ | × | CF | $\Delta_{\text{long-term}}$ |
|---|---|---|---|---|
| Camber | +2.50 | × | 2.20 = | +6.00 |
| Deflection (double-T weight) | -1.40 | × | 2.40 = | -3.36 |
| Deflection (topping) | -0.61 | × | 2.30 = | -1.40 |
| Σ | | | | +1.24 in ↑ |
| Deflection (live-load) | | | | -0.65 |
| | | | | +0.59 in ↑ |

**Allowable $\Delta_{LL}$.** Check allowable $\Delta_{LL}$. [See ACI 318-83, Table 9.5(b); Uniform Building Code, Table 23-D.]

$$\frac{L}{360} = \frac{58 \times 12}{360} = 1.93 \text{ in} \quad (\text{OK})$$

**Reinforcing summary: 8DT24.** See Fig. 26.14.

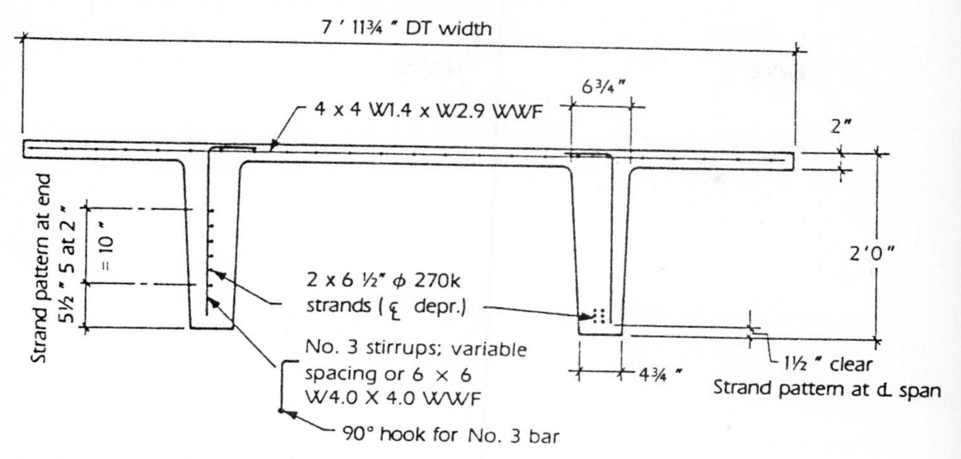


**Figure 26.14** Reinforcing summary: 8DT24.

## Connection Design

It can be stated axiomatically that the bottoms of double T's and other stemmed prestressed concrete members are *never* welded at their supports; they are left to "float" free on neoprene to allow long-term creep and shrinkage movements to be accommodated without causing distress in the prestressed concrete elements and their connections.

### End-bearing assembly design

Design end-bearing assembly for stem of double T (see Fig. 26.15).

$$R_D = 10.6^{\text{k}} \quad R_L = 5.8^{\text{k}} \rightarrow UV = 24.7^{\text{k}}$$

Evaluate $N_u$ (bearing on neoprene pad); see *PCI Design Handbook*, page 6-18):

$$N_u = F_S = f_f A_b = (0.1)(4)(5) = 2.0^{\text{k}}$$

$$A_n = \frac{N_u}{\phi f_y} = \frac{2.0}{(0.85)(60)} = 0.04 \text{ in}^2$$

**Figure 26.15** End-bearing assembly for stem of double T.

Determine $A_{cr}$ and $A_{vf}$:

$$A_{cr} = \frac{bw}{\sin 20°} = \frac{(4)(5)}{0.342} = 58 \text{ in}^2$$

$$\mu_e = \frac{1000\lambda A_{cr}\mu}{V_\mu} = \frac{(1000)(58)(1.4)}{24{,}700} = 3.3 < 3.4$$

$$A_{vf} = \frac{V_\mu}{\phi f_y \mu_e} = \frac{24.7}{(0.85)(60)(2.3)} = 0.15 \text{ in}^2$$

$A_{vf} + A_n = 0.15 + 0.04 = 0.19 \text{ in}^2 \Rightarrow$ use one no. 4 anchor (A706)

$$A_{sh} = \frac{(A_{vf} + A_n)f_y}{\mu_e f_{ys}} = \frac{(0.19)(60)}{(3.4)(60)} = 0.06 \text{ in}^2$$

$\Rightarrow$ one no. 3 bar required

**Neoprene bearing pad design**

Design a neoprene bearing pad to accommodate long-term movement and bearing pressure on stem of double T. Double T's are erected 45 days after casting.

$R_{D+L} = 10.6 + 5.8 = 16.4^{k}/\text{stem}$

Bearing pressure $f_b = \dfrac{V}{wb}\ \dfrac{16.4}{(4)(5)} = 0.820$ ksi

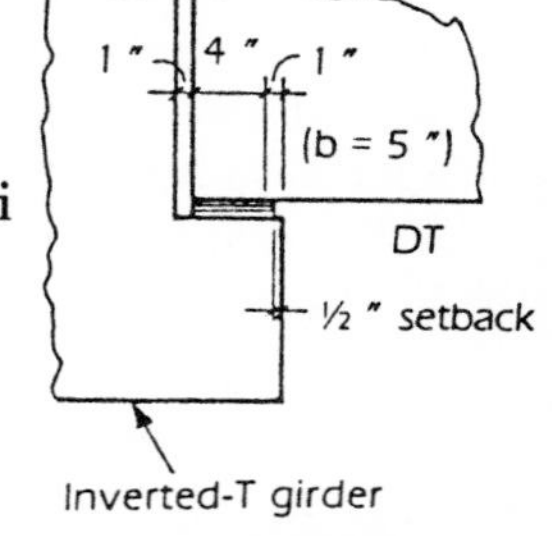


**Figure 26.16** Neoprene bearing pad.

Calculate long-term movement due to creep and shrinkage (see *PCI Design Handbook*, page 3-8):

$f'_{ci} = 3500$ lb/in²
Average P/S = 1000 lb/in²
V/S = 444/288 = 1.5

| | Long-term | – | 45 days | = | Net |
|---|---|---|---|---|---|
| Creep strain | 315 | | 155 | | 159 |
| Shrinkage strain | 560 | | 251 | | 309 |
| Total | | | | | $468 \times 10^{-6}$ in/in |

$$\Delta = 468 \times 10^{-6} \times 29 \times 12 = 0.16 \text{ in maximum strain in pad}$$

$$\text{Recommended } t = 1.4\,\Delta = (1.4)(0.16) = 0.22 \text{ in}$$

Try 1/4-in pad (see *PCI Design Handbook*, page 6-16):

$$\text{Shape factor} \frac{wb}{2(w+b)t} = \frac{(4)(5)}{(2)(4+5)(1/4)} = 4.4$$

$$f_b \text{ allowable} = 4DS \text{ or } 800 \text{ lb/in}^2 \; (4 \times 60 \times 4.4 = 1056 \text{ lb/in}^2)$$

Use pad 4½ × 5 × ¼ in; 60 Durometer chloroprene (AASHTO specification).

### Connection details

Connection details are presented in Figs. 26.17 through 26.29.

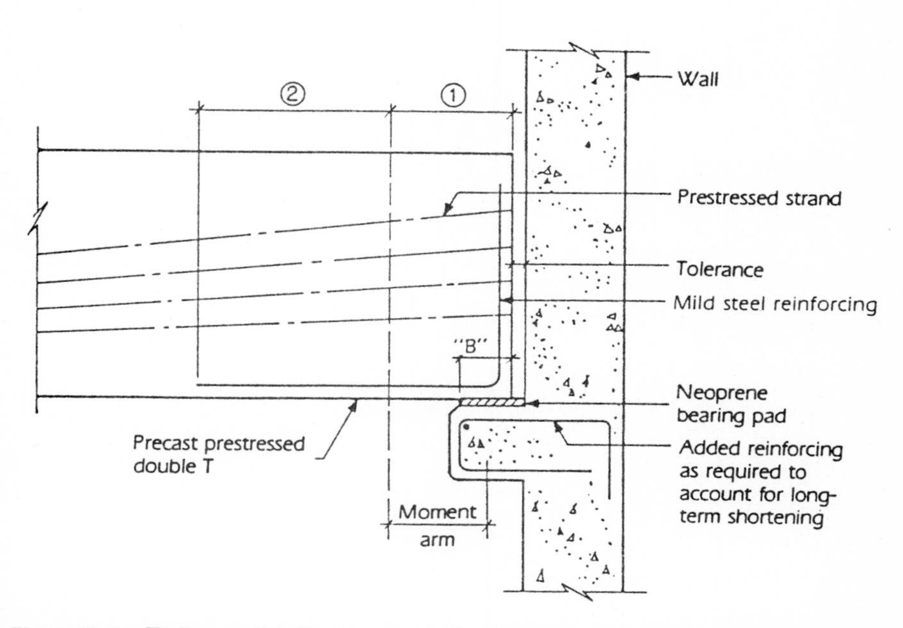


**Figure 26.17** End zone detail. Zone 1: to be designed as reinforced concrete since prestressing force is not fully transferred to concrete. Zone 2: anchorage distance for mild steel reinforcing in fully prestressed concrete. Bearing dimension *B* (bearing length) is determined by allowable bearing load, prestressed member width, and tolerances. Mild steel reinforcing may be deleted when end reinforcing is designed in accordance with Chap. 6 of the third edition of the *PCI Design Handbook*.

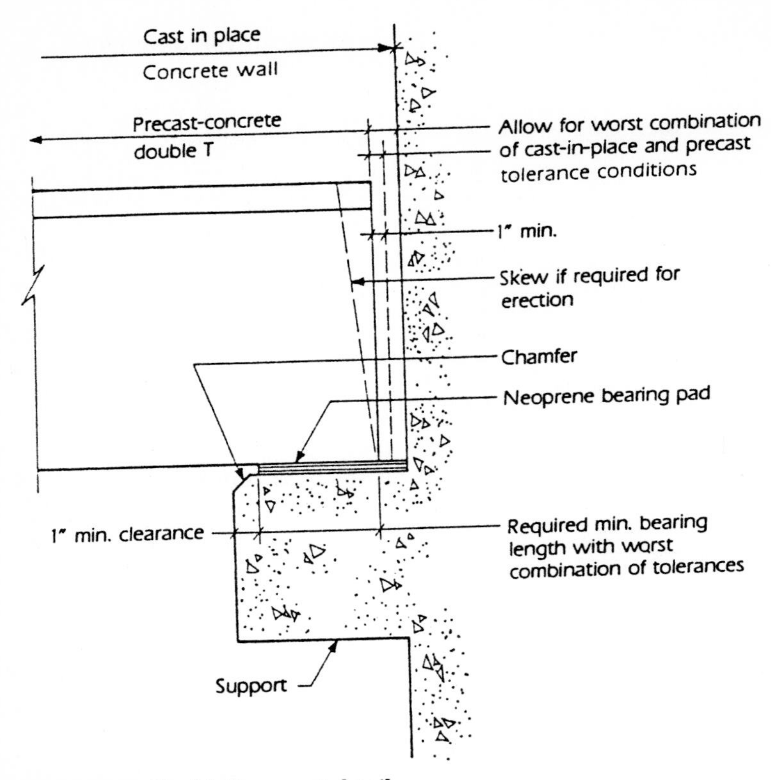


**Figure 26.18** Double-T support detail.

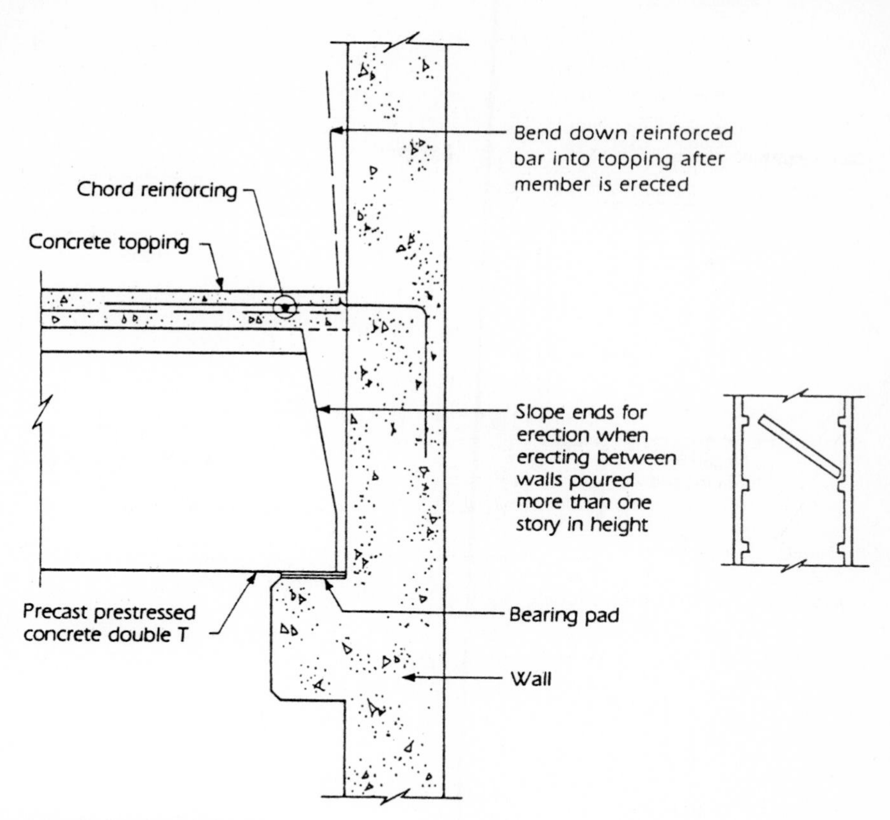


**Figure 26.19** Wall bearing.

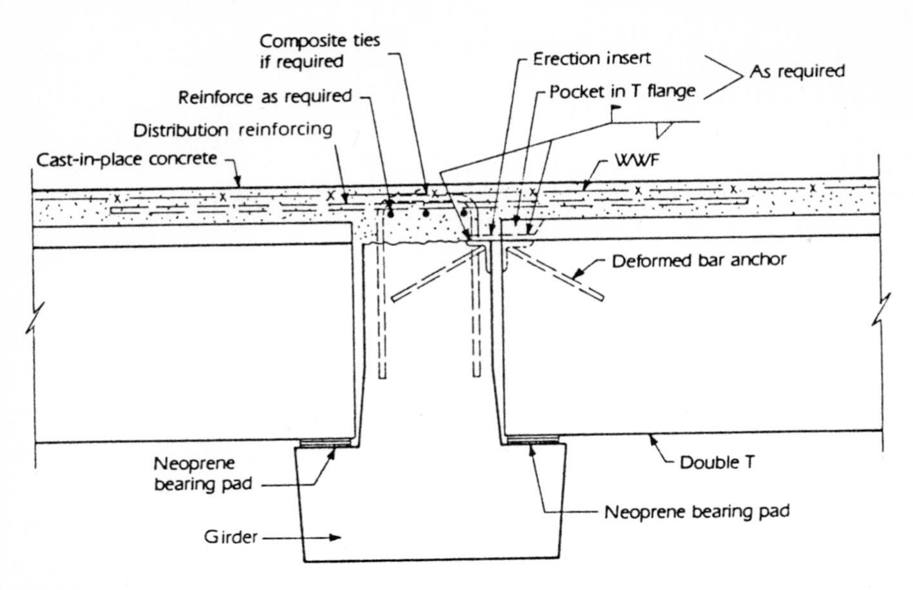


**Figure 26.20** Double T to girder.

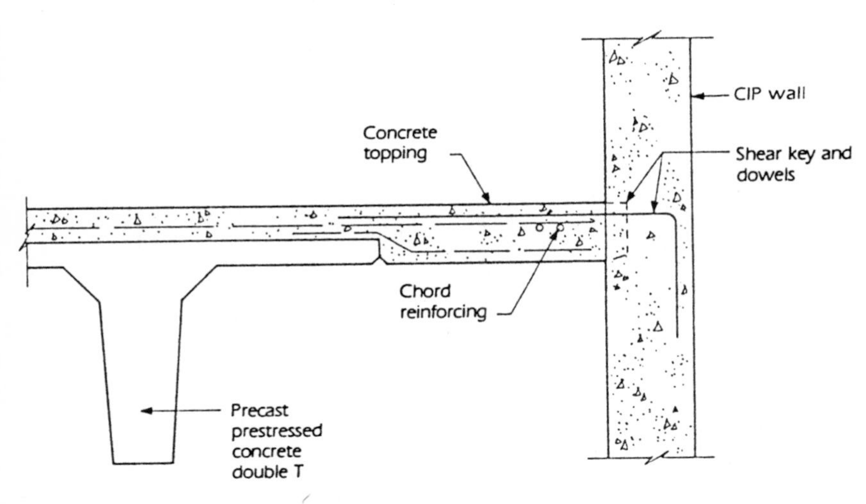


**Figure 26.21** Member parallel to wall.

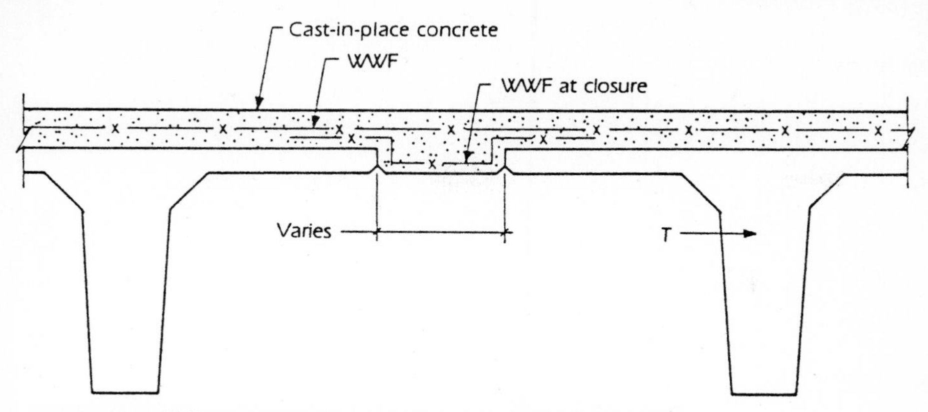


**Figure 26.22** T flange connection closure for spread double-T system.

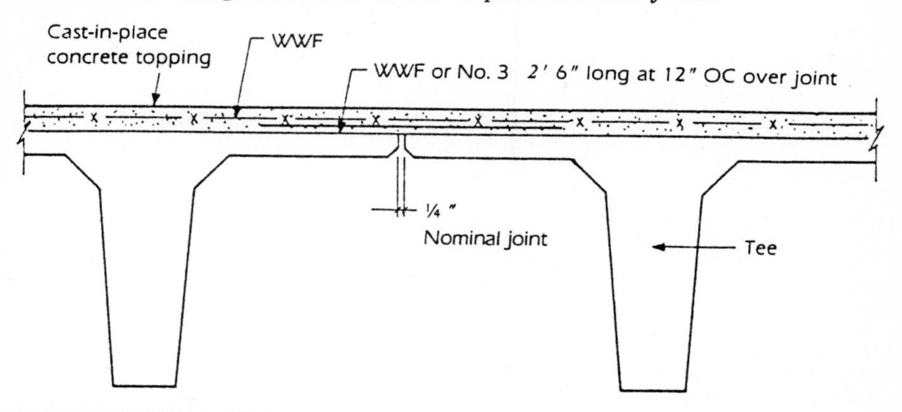


**Figure 26.23** Typical T flange connection.

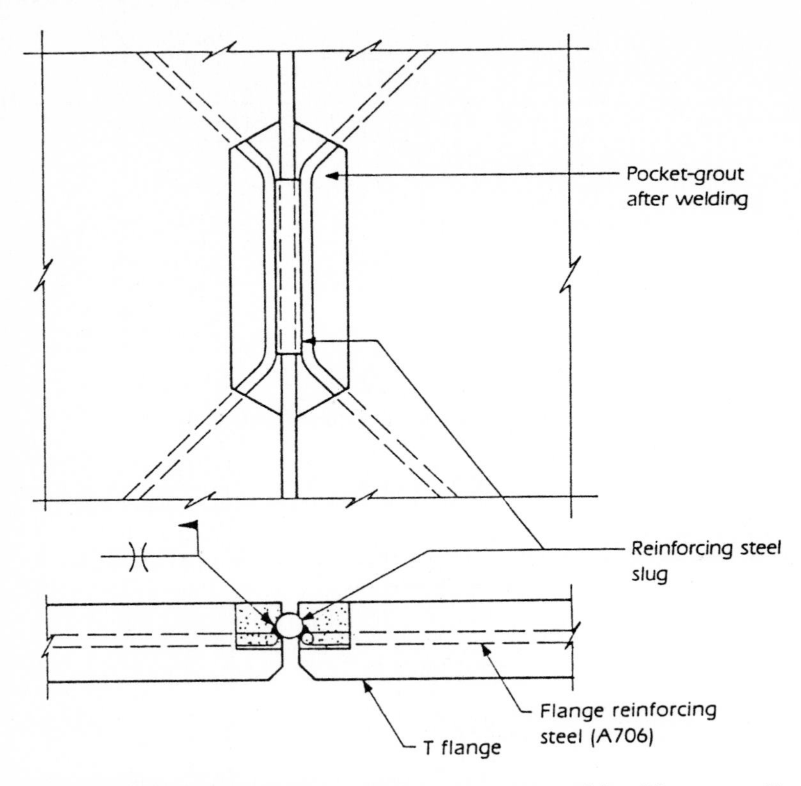


**Figure 26.24** T flange seismic connection; no topping slab. Floor or wall diaphragm connection.

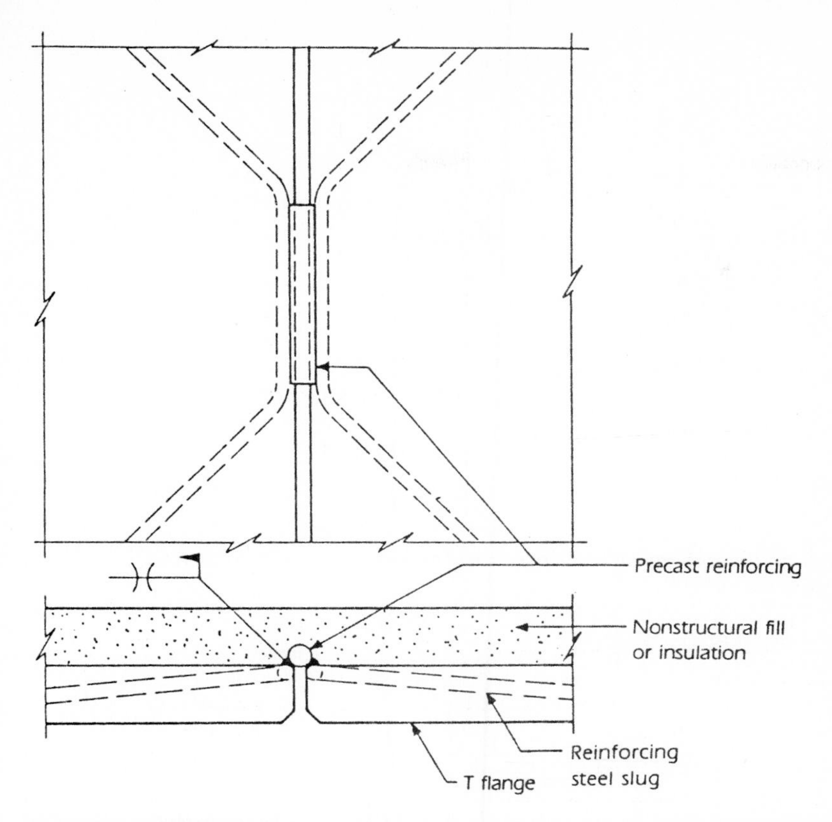


**Figure 26.25** T flange seismic connection; nonstructural fill. Roof diaphragm detail.

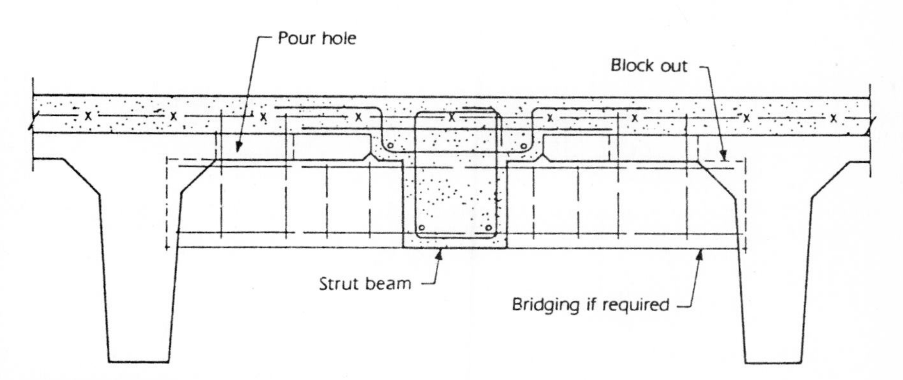


**Figure 26.26** T flange connection; seismic strut.

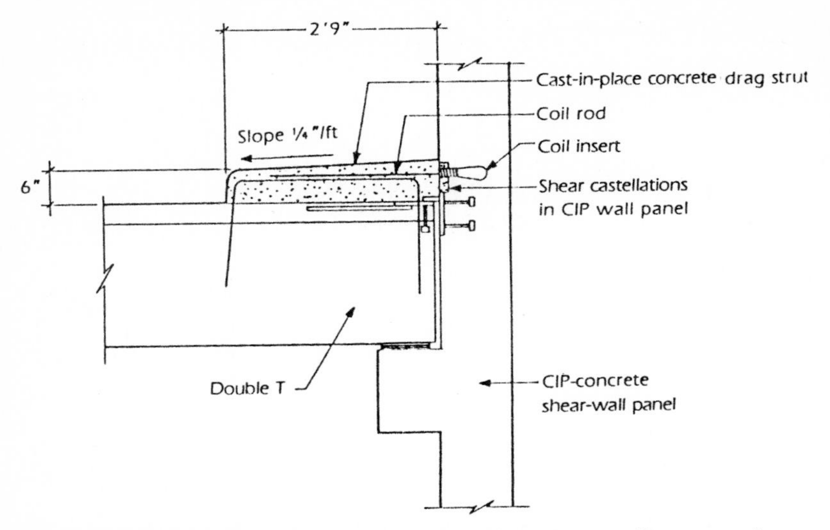


**Figure 26.27** Diaphragm connection at cast-in-place shear wall; untopped parking garage details.

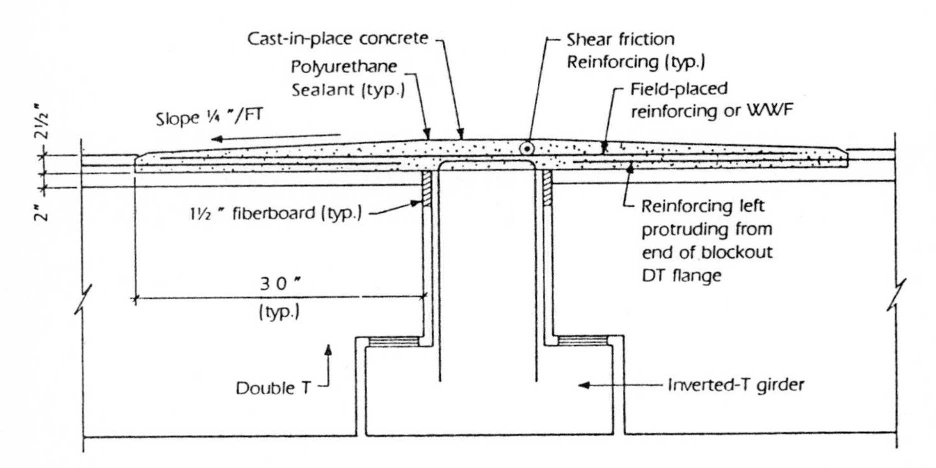


**Figure 26.28** Diaphragm connection at double-T and inverted-T girders; untopped parking garage details.

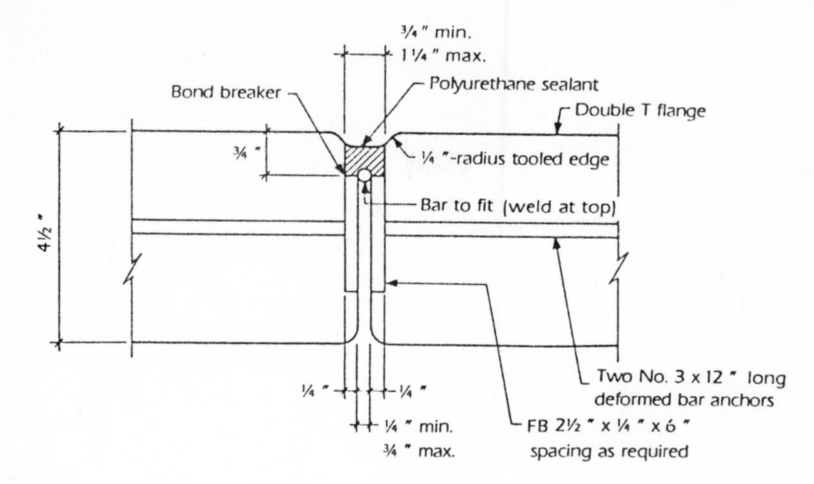


**Figure 26.29** Double-T flange connection detail; untopped parking garage details.

A double T being erected is shown in Fig. 26.30.

**Figure 26.30** Alvarado sewage treatment plant, Union City, California. A double T 8 ft wide by 24 in deep being erected. The specifications called for special paint on the underside of the T. The painting contractor found it economically feasible to do this work in the fabricator's plant while the T's were in storage, thus eliminating on-site painting and saving time.

**Figure 27.1** The Broadway, La Jolla, California. Prestressed concrete inverted-T girders being erected on two-story precast concrete columns. The ironworker on the ladder is positioning a neoprene bearing pad on the column corbel in preparation for the next girder. The bottoms of these girders are left "floating" on elastomeric bearing pads so that long-term creep and shrinkage movements may be accommodated.

Chapter

# 27

# Prestressed Concrete Girders

Plant-cast prestressed concrete girders are normally used to support other prestressed concrete deck members, such as solid slabs, hollow-core planks, or double T's, and to deliver vertical loads to precast concrete columns or wall systems. They are classified in three general categories: rectangular, ledger, and inverted-T. They are made in depths up to 48 in, although deeper sections can be precast subject to handling and erection limitations. The larger sizes may be voided to reduce weight. Some manufacturers also have forms for I shapes, which have more efficient section properties than the rectangular shape. Voids, where used, are formed by hollow tubes, cardboard box forms, plywood liners, or polystyrene blocks or by using removable steel mandrels.

Standard sizes and cross sections, subject to individual manufacturer mold inventories, are as follows :

1. Rectangular girder: from 12 in wide by 18 in deep to 36 in wide by 48 in deep
2. Ledger girder: from 12 in wide by 18 in deep to 30 in wide by 48 in deep
3. Inverted-T girder: from 12 in wide by 18 in deep to 24 in wide by 48 in deep

Widths and depths within the minimum-maximum ranges usually vary in 2-in increments.

The widths given for ledger girders and inverted-T girders are the stem width at the top; standard ledge dimensions are 6 in wide by 12 in deep. Sometimes girders are made with pockets sized and located to correspond with double-T framing. These are often used in bakeries and food-processing plants, where it is mandatory that no ledges be present to catch and collect dust. They are also used as parking garage transfer girders where the area between the T stem and the pocket is subsequently dry-packed to gain the additional web width and consequently increased load-carrying capacity. Precast prestressed concrete girders are used as framing components in parking structures, shopping malls, department stores, office buildings, commercial buildings, subterranean parking decks, reservoir covers, schools, and hospitals—anywhere where precast concrete construction is feasible.

## Manufacturing

Girders are cast either in permanent-type beds, where the forms are fixed in place, or in "universal beds," where transportable molds are set up for the specific project being manufactured. The stressing lines are usually 200 to 400 ft long. Some manufacturers have pile lines with beds 800 ft long or more in which girders are cast. Where universal beds are used with transportable forms, there is usually no strand-depressing capability. Plants with fixed girder forms often have strand hold-down anchors cast below the pallet of the girder form as part of the bed. Fixed side forms exhibit a side slope (draft) which facilitates stripping. The draft is usually $^3/_8$ or $^1/_2$ in/ft. Girder forms with movable sides obviously need not have draft. These forms are also the most efficient from the standpoint of the initial costs involved in bed setup for individual project cross sections. Concrete strengths range to 6000 lb/in$^2$ for hard-rock concrete and 5000 lb/in$^2$ for lightweight concrete; $^1/_2$-in-diameter 270-ksi strand is usually used for prestressing. (See Figs. 27.2 through 27.5.)

**Figure 27.2** Girder casting bed. Shown here is a 430-ft-long girder line. One side is movable in 2-in increments to permit casting girders in widths up to 24 in. The side forms shown can produce girders to 3 ft in depth. Side rail extensions increase the maximum girder depth to 4 ft. Note the stressing abutment in the foreground with drillings for $\frac{1}{2}$-in-diameter strand spaced 2 in apart each way.

**Figure 27.3** Rectangular girders: 14- by 31-in prestressed concrete girders in storage for the Bedford parking structure in Beverly Hills, California. The girders are 65 ft long.

Figure 27.4 Ledger girders in storage.

Figure 27.5 Inverted-T girders awaiting shipment. These 64-ft-span girders are destined for service in a parking structure.

## Erection Considerations

In planning projects with large prestressed concrete girders, anticipated jobsite conditions and erection requirements should be investigated early. In general, any project containing precast concrete components weighing in excess of 10 tons should be analyzed on an individual basis, often by consulting with prestressed concrete manufacturing operations and erection personnel. Another practical consideration that often causes problems in the field is the introduction of torsional stresses in ledger and inverted-T girders as deck members are erected on them. Often temporary bracing and shoring are required to resist these forces, which sometimes can cause extremely large movements in the partially erected building frame. Sometimes it is prudent to provide for temporary bolted or welded connections between T and beam or between beam and column or wall to resist the problems caused by erection torsional

**Figure 27.6** The Broadway, La Jolla, California. Completed precast concrete frame consisting of columns, ledger girders, inverted-T girders, and double T's. A double-nut connection to the column base plate was used to plumb the columns. The base plates were then grouted, giving sufficient rigidity to the two-story columns to negate any requirement for additional bracing. Top connections were made at the ledger-column juncture to prevent torsional rotation when the T's were erected.

**Figure 27.7** The Broadway, La Jolla, California. An inverted-T girder with roughened top and extended ties to develop horizontal shear with the cast-in-place concrete topping in accordance with the Uniform Building Code, Secs. 2617 and 2611. Note also the double T's framing into the girder. Also demonstrated are the two degrees of roughness required to satisfy UBC requirements. The broom finish on the T will satisfy horizontal shear stresses to 80 lb/in$^2$. Greater stresses than this will require a roughness providing a modulus of amplitude of $^1/_4$ in, as demonstrated by the girder top.

forces. These temporary connections can usually be removed once all the members have been erected, permanent connections made, and cast-in-place concrete topping and closures poured. (See Fig. 27.6.)

## Design Considerations

Owing to the manufacturing conditions mentioned above, it is usually wise to design girders in straight strands while allowing precast manufacturers with strand-depressing capabilities to redesign the girders to take advantage of the inherent economy in their particular operation. When designing with straight strands, end stresses often become critical. How this situation is handled in design is demonstrated in the following design example. Preliminary load tables for noncomposite girders are contained in the third edition of the *PCI Design Handbook*, pages 2-44 through 2-46. However, it is often just as convenient to use the approximate formulas given in the beginning of the design example. Another option often overlooked in design, especially in large projects, is the material savings gained by utilizing temporary midpoint or third-point shoring, especially for longer spans. This allows the bulk of the load to be carried by the composite section, as opposed to unshored designs, where the basic section uses the major portion of its capacity in carrying supported dead loads from precast concrete deck members and cast-in-place concrete topping.

Introducing continuity at intermediate supports results in increased positive-moment capacity. The resulting tradeoff in negative reinforcing, however, often negates any economic gains in the prestressed concrete girder design.

One last item that is often overlooked in design is the use of transformed girder section properties in calculating deflections and verifying service load performance. Deep, heavily prestressed girders with top reinforcing both in the basic section and in the composite pour exhibit transformed section properties considerably higher than those indicated by the bare concrete cross sections.

See also Fig. 27.7 for a discussion on horizontal shear transfer in composite construction.

### Girder design

Design an inverted-T girder to support the double T's in the parking structure referred to in Chap. 26, subsection "Double-T Design." (See Fig. 27.8.)

Select a girder span compatible with the parking layout and the double-T width.

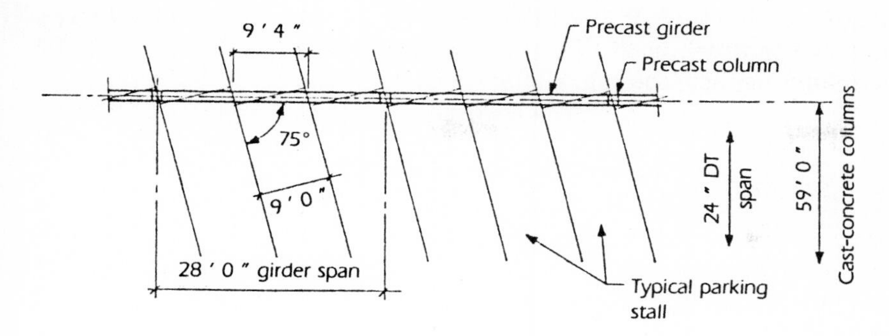


**Figure 27.8** Inverted-T girder design.

**Preliminary design.** Try 28-ft on-center column spacing (girder span approximately 27 ft).

Approximate loads to beam:

| | | | |
|---|---|---|---|
| Reduced live load | 0.6 × 0.050 × 59 | = | 1.77^k/ft |
| Double-T weight | 0.060 × 58 | = | 3.48 |
| Topping weight | 0.031 × 59 | = | 1.83 |
| Estimated beam weight | | = | 0.60 |
| $\Sigma w$ | | | 7.68^k/ft |

When preliminary load-span tables are not available, the following formulas are useful for selecting trial sizes:

$$\text{Noncomposite girder: } h = \sqrt{\frac{2.5M_T}{b}}$$

$$\text{Composite girder: } h = \sqrt{\frac{1.7M_T}{b}}$$

where $h$ = precast beam depth, in

$M_T$ = total moment, in · kips

$b$ = beam stem width, in

For $b$ = 12 in,

$$h = \sqrt{\frac{(1.7)(1.5)(7.68)(27^2)}{12}} = 34.5 \text{ in}$$

**Girder concrete:** $f'_c = 5000 \text{ lb/in}^2$; $f'_{ci} = 3500 \text{ lb/in}^2$ (hard-rock concrete).

From the chart on page 2-46 of the *PCI Design Handbook* (3d edition),

24IT36 carries 7.73$^k$/ft (by interpolation). From the manufacturer's product literature, a 261T34 is available with the following specific section properties (see Fig. 27.9).

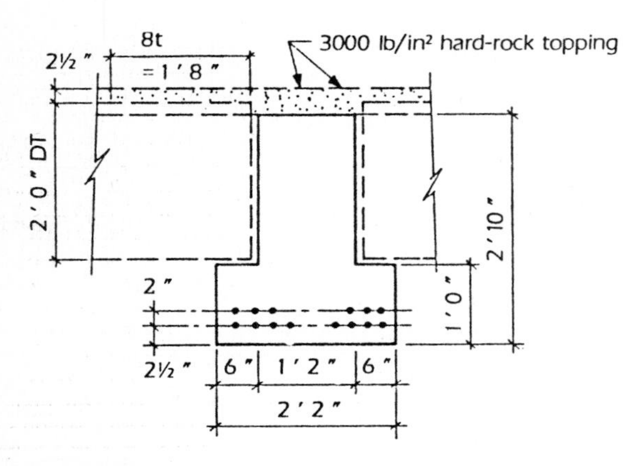


**Figure 27.9** Prestressed concrete girder section.

Basic section:

$$A = 620 \text{ in}^2 \qquad \text{Weight} = 0.65^k/\text{ft}$$

$$I_b = 60{,}960 \text{ in}^4$$

$$y_t = 19.6 \text{ in} \qquad s_t = 3117 \text{ in}^3$$

$$y_b = 14.4 \text{ in} \qquad s_b = 4220 \text{ in}^3$$

Composite section :

$$I_c = 174{,}850 \text{ in}^4$$

$$y_{tc} = 11.1 \text{ in} \qquad s_{tc} = 15{,}646 \text{ in}^3$$

$$y_{bc} = 22.9 \text{ in} \qquad s_{bc} = 7660 \text{ in}^3$$

Strand data for 1/2 -in $\phi$- 270$^k$ strand (see *PCI Design Handbook*, page 11-16):

$$F_o = 28.9^k/\text{strand} \qquad F_i = 26.0^k/\text{strand}$$

$$F_f = 22.5^k/\text{strand} \qquad A_{ps} = 0.1531 \text{ in}^2$$

**Loads and moments.** Determine loads and moments.

To basic section :

| | k/ft | | $M_{\mathrm{\mathcal{C}}}^{\mathrm{k \cdot in}}$ | × LF | = *UM* |
|---|---|---|---|---|---|
| Double T | $1.5 \times 3.48 \times 27.0^2$ | = | 3805 | 1.4 | 5328 |
| Topping | $1.5 \times 1.83 \times 27.0^2$ | = | 2001 | 1.4 | 2802 |
| Beam weight | $1.5 \times 0.65 \times 27.0^2$ | | 711 | 1.4 | 995 |
| | | | 6517 | | |

To composite section :

| | | | | |
|---|---|---|---|---|
| Reduced live load | $1.5 \times 1.77 \times 27.0^2$ | $= 1935^{\mathrm{k \cdot in}}$ | 1.7 | 3290 |
| | | | $\Sigma\,\mathrm{UM}_c$ = | $12{,}415^{\mathrm{k \cdot in}}$ |

**Stresses.** Determine stresses.

Required prestress force; straight strands ($e \cong 14.4 = 3.5 = 10.9$ in):

$$F_f = \frac{\dfrac{M_B}{S_b} + \dfrac{M_c}{S_{bc}} - f_t}{\dfrac{1}{A} + \dfrac{e}{S_b}} = \frac{\dfrac{6517}{4220} + \dfrac{1935}{7660} - 0.530}{\dfrac{1}{620} + \dfrac{10.9}{4220}}$$

$$= \frac{1.544 + 0.253 - 0.530}{0.00161 + 0.00258} = \frac{1.267}{0.00419} = 302^{\mathrm{k}}$$

$$N = \frac{302}{22.5} = 13.4$$

Therefore, use fourteen 1/2-in $\phi$-270$^{\mathrm{k}}$ strands.

Check stresses: $F_i = (14)(26.0) = 364^{\mathrm{k}}$ (see *PCI Design Handbook*, pages 4-16 and 11-15).

Top end at release:

$$f_{\mathrm{top}} = -\frac{364}{620} + \frac{(364)(10.9)}{3117} = -0.587 + 1.272 = +0.685\,\mathrm{ksi} > 6\sqrt{f'_{ci}}$$

$$f_{\text{bottom}} = -\frac{364}{620} + \frac{(364)(10.9)}{4220} = -0.587 - 0.940$$

$$= -1.527 \text{ ksi} < 0.6\sqrt{f'_{ci}} \quad \text{(OK)}$$

Top-end steel is required to resist full tension force (see **Fig. 27.10**).

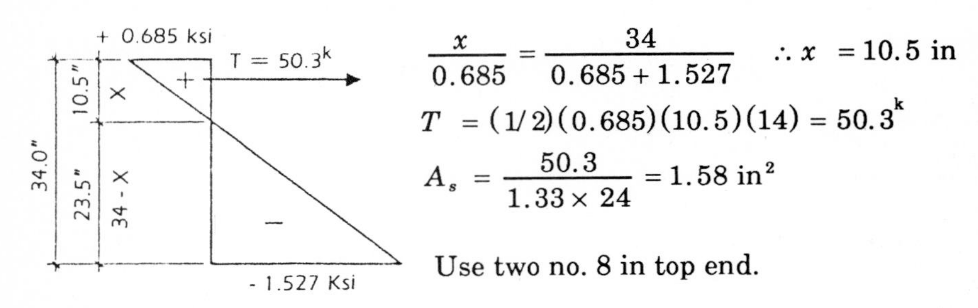


$$\frac{x}{0.685} = \frac{34}{0.685 + 1.527} \qquad \therefore x = 10.5 \text{ in}$$

$$T = (1/2)(0.685)(10.5)(14) = 50.3^{k}$$

$$A_s = \frac{50.3}{1.33 \times 24} = 1.58 \text{ in}^2$$

Use two no. 8 in top end.

**Figure 27.10** Stress diagram at end of girder.

Determine where top-end steel is no longer required, that is, where the beam weight moment reduces top tension to $6\sqrt{f_{ci}'}$.

Compression required from beam weight

$$= 0.685 - 6\sqrt{3500} = 0.685 - 0.355 = 0.330^{k}/\text{in}^2$$

$$\frac{M_{BM}}{S_t} = 0.330 \text{ ksi minimum required}$$

0.65^k/ft, x, 9^k, 27′, 9^k

At centerline span:

$$\frac{748}{3117} = 0.240 \text{ ksi} < 0.330 \text{ ksi}$$

Therefore, run top steel continuously since top tension is always greater than $6\sqrt{f'_{ci}}$.

Check bottom tension under all loads: $F_f = (14)(22.5) = 315$.

$$f^{\text{¢}}_{\text{bottom}} = -\frac{315}{620} - \frac{(315)(10.9)}{4220} + \frac{6517}{4220} + \frac{1935}{7660}$$

$$-0.508 - 0.814 + 1.544 + 0.253 = +0.475^{k}/\text{in}^2 < 7.5\sqrt{f'_c} \quad \text{(OK)}$$

**Ultimate strength.** Check ultimate strength (see *PCI Design Handbook,* pages 4-6 and 4-65).

$$f_{se} = 147\text{ ksi} \quad A_{ps} = (14)(0.1531) = 2.143\text{ in}^2$$

$$C\overline{\omega}p = C\,\frac{A_{ps}}{bd} \times \frac{f_{pu}}{f_c'} = (1.0)\left(\frac{2.143}{54 \times 35}\right)\left(\frac{270}{3}\right) = 0.10$$

From the chart read $f_{ps}/f_{pu} = 0.97$; therefore $f_{ps} = (0.97)(270) = 261.9$ ksi.

$$a/2 = \frac{A_{ps}f_{ps}}{1.7f_c'\,b} = \frac{(2.143)(261.9)}{(1.7)(3)(54)} = 2.04\text{ in}$$

Assume that topping and T flange act compositely so that compression block depth $a$ is within flange. Therefore,

$$M_u = \phi A_{ps}f_{ps}(d - a/2) = (0.90)(2.143)(261.9)(35.0 - 2.04)$$

$$= 16{,}649^{\text{k}\cdot\text{in}} > UM = 12{,}415^{\text{k}\cdot\text{in}} \quad (\text{OK})$$

**Shear.** Determine tie requirements to transfer horizontal shear force between precast and cast-in-place concrete (see *PCI Design Handbook,* page 4-28).

$$T = A_{ps}f_{ps} = (2.143)(261.9) = 561^{\text{k}}$$

$$l_{vh} = \frac{27 \times 12}{2} = 162\text{ in} \quad f_y = 40\text{ ksi}$$

$$C = (0.85)f_{cc}'\,A_{\text{top}} = (0.85)(3)(2.5 \times 40 + 4.5 \times 14) = 416^{\text{k}} = F_h$$

$$\text{Maximum allowable without ties} = 40\phi b_h l_{vh}$$

$$= (40)(0.85)(14)(162/1000)$$

$$= 77^{\text{k}} < F_h$$

Therefore, ties are required.

$$\mu_e = \frac{1000\lambda^2 b_v l_{vh}}{F_h} = \frac{(1000)(1.0^2)(14)(162)}{416{,}000} = 5.45$$

$$A_{cs} = \frac{F_h}{\mu_e f_y} = \frac{416}{(0.85)(5.45)(40)} = 2.25\text{ in}^2$$

Minimum requirements:

$$1.\ A_{cs,\min} = \frac{120 b_v l_{vh}}{f_y} = \frac{(120)(14)(162)}{40{,}000} = 6.8 \text{ in}^2$$

$$\text{or } 1.33 \times 2.23 = 2.99 \text{ in}^2 \text{ (controls)}$$

$$2.\ A_{cs,\min} = \frac{50\, b_v vh}{f_y} = \frac{(50)(14)(162)}{40{,}000} = 2.84 \text{ in}^2$$

Use no. 4 ties.

$$s = \frac{(13.5)(12)(0.40)}{2.99} = 21.7 \text{ in}$$

Use 18 in maximum. (Maximum spacing is $4 \times 4.5 = 18$ in. )

Calculate horizontal shear steel requirements using the Uniform Building Code:

$$v_{dh} = \frac{v_u}{\phi b_v d} = \frac{153}{(0.85)(14)(35)} = 0.367 \text{ ksi} > 0.350 \text{ (Sec. 2617)}$$

$$A_{vf} = \frac{V_u}{\phi f_y \mu} = \frac{153}{(0.85)(40)(1.0)} = 4.50 \text{ in}^2 \text{ [Sec. 2611(p)]}$$

$$s = \frac{163 \times 0.40}{4.50} = 14.4 \text{ in (use 15 in)}$$

**Web reinforcing.** Determine web reinforcing requirements (see Fig. 27.11; also *PCI Design Handbook*, page 4-66).

$$b_w d = (14)(35) < 490 \text{ in}^2$$

$$v_u \text{ support} = \frac{UV}{\phi bd} = \frac{153}{(0.85)(490)} = 0.367 \text{ ksi}$$

$$\frac{L}{D} = \frac{27 \times 12}{35} = 9.5 \text{ in} \qquad A_{ps} = 2.143 \text{ in}^2$$

By inspection, minimum shear reinforcement suffices. From the chart on page 4-71 of the *PCI Design Handbook*, $A_{v,\min} = 0.10 \text{ in}^2/\text{ft}$.

By using no. 4 stirrups,

$$s_{\max} = \frac{A_y f_y}{50 b_w} = \frac{(0.40)(40{,}000)}{(50)(14)} = 22.9 \text{ in}$$

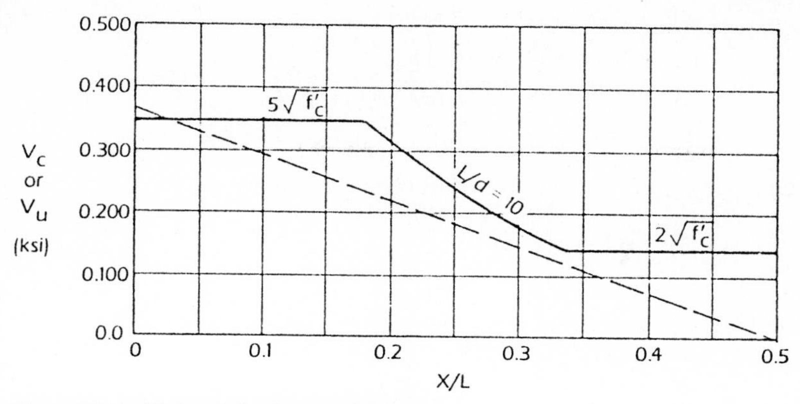


**Figure 27.11** Web reinforcing requirements.

Use no. 4 bars at 15 in on center; extend into topping.

**Camber and deflection.** Check camber and deflection (see *PCI Design Handbook*, pages 4-73 and 11-15).

$$\text{Camber: } \Delta_{P/S,i} = \frac{F_i e l^2}{8EI_b} = \frac{(364)(10.9)(27^2)(144)}{(8)(3600)(60{,}960)} = +0.24 \text{ in}$$

$$\text{Beam weight: } \Delta_{BM,i} = \frac{5wl^4}{384EI} = \frac{(5)(0.65)(27^4)(1728)}{(384)(3600)(60{,}960)} = -0.04 \text{ in}$$

$$\text{Double-T weight: } \Delta_{DT,i} = \frac{3.84}{0.65} \times 0.035 = -0.19 \text{ in}$$

$$\text{Topping: } \Delta_{\text{topping},i} = \frac{1.83}{0.65} \times 0.035 = -0.10 \text{ in}$$

Final condition:

| | $\Delta_i$ | × | CF | = | $\Delta_{\text{long-term}}$ |
|---|---|---|---|---|---|
| Camber | +0.24 | × | 2.2 | = | +0.53 |
| Double T and beam | -0.23 | × | 2.4 | = | -0.55 |
| Topping | -0.10 | × | 2.3 | = | -0.23 |
| Σ Long-term movement | | | | | -0.25 in ↓ |

$$\Delta_{LL} = \frac{(5)(1.77)(27^4)(1728)}{(384)(4300)(174{,}850)} = 0.03 \text{ in} \qquad (\text{OK})$$

**Reinforcing summary; 26IT34.** Reinforcing is summarized in Fig. 27.12.

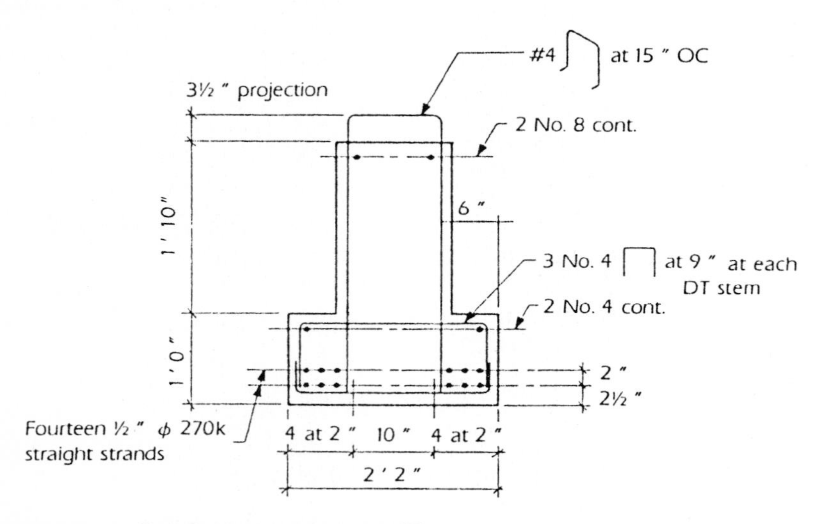


**Figure 27.12** Reinforcing summary: 26IT34.

See installation of inverted-T girders and supporting double T's in Fig. 27.13.

## Connection Details

Special attention should be paid to the design and detailing of corbels and ledges owing to the high stress concentrations at these critical interfaces. Floating neoprene bearings under girders assure relief of creep and shrinkage movement stresses.

**Figure 27.13** Alvarado sewage treatment plant, Union City, California. Inverted-T girders supported on cast-in-place concrete bents and supporting double T's up to 70 ft long. Note the clean lines and close dimensional control afforded by the plant-cast precast concrete members.

### Neoprene bearing pad design

Design a neoprene bearing pad for an inverted-T beam where it bears on the column corbel shown. Girders are erected 45 days after casting. The column size is 14 × 24 in.

$$R_{D+L} = 83 + 25 = 108^{k}$$

$$\text{Bearing pressure } f = \frac{V}{wb}$$

$$= \frac{108}{(8)(14)} = 0.964\,\text{ksi}$$

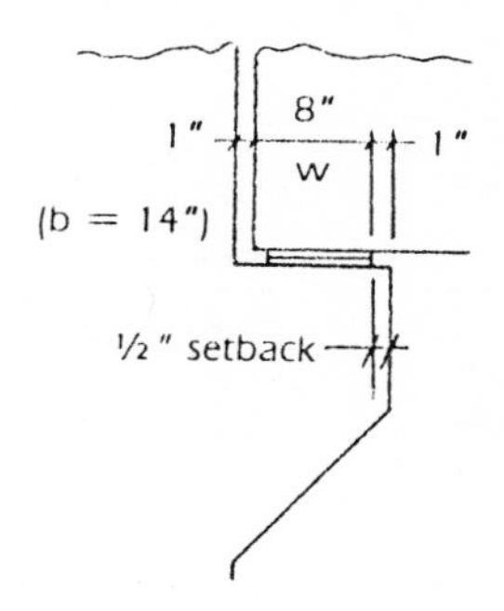

Calculate long-term movement due to creep and shrinkage (see *PCI Design Handbook*, page 3-8).

$$\Delta = 574 \times 10^{-6} \times 14 \times 12 = 0.10 \text{ in}$$

$$\text{Recommended } t = 1.4\,\Delta = (1.4)(0.10) = 0.14 \text{ in}$$

Try 3/8-in pad (see *PCI Design Handbook*, page 6-16):

$$\text{Shape factor} = \frac{wb}{2(w + b)t} = \frac{(8)(14)}{(2)(8 + 14)(2/8)} = 6.8$$

$$f_{b\text{, allowable}} = 1000 + 100S < 1500(1000 + 680 = 1680 \text{ lb/in}^2)$$

Use pad 8 1/2 × 14 × 3/8 in; 80 Durometer (random fiber reinforced pad).

### Ledge reinforcing design

Design ledge reinforcing for the inverted-T girder shown where it supports the double T's (see Fig. 27.14; also *PCI Design Handbook*, page 6-32).

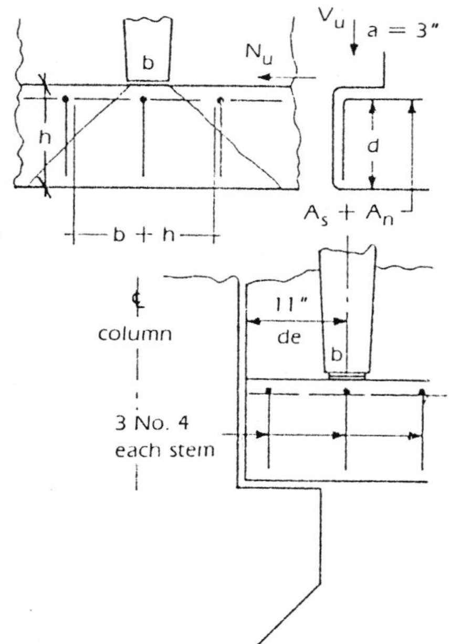


**Figure 27.14** Ledger reinforcing for inverted-T girder.

$$N_u = 0.2 \qquad V_u = 4.9^k \quad V_u = 24.7^k$$

$$b + h = 17 \text{ in} \qquad S = 48 \text{ in} > b + h$$

$$\phi V_n = 3\phi h \sqrt{f_c'}\,(2lp + b + h)$$

$$= (3)(0.85)(12)(0.071)(2 \times 6 + 17) = 62.7^k$$

or

$$\phi V_n = \phi h \sqrt{f_c'}\,(2lp + b + h + 2d_e)$$

$$= (0.85)(12)(0.071)(2 \times 6 + 17 + 2 \times 11) = 36.8^k$$

$$> V_u \quad (\text{OK})$$

Design ledge reinforcing.

$$\text{Minimum reinforcing: } \frac{200(b + h)d}{f_y}$$

$$= \frac{200(17)(10)}{40,000} = 0.85 \text{ in}^2$$

From Table 6.20.20 on page 6-63 of the *PCI Design Handbook*, for 6-in ledge $b = 6$ in and $h = 12$ in:

$$\text{For } V_u = 24.7^k, \; A_{s1} = \text{two no. 4}$$

$$\text{For } N_u = 4.9^k, \; A_{s2} = \text{one no. 4}$$

Use $A_s + A_n$ = three no. 4 bars; or three no. 4 bars at 9 in ($A$ = 0.60 in$^2$).

### Connection details

Connection details are presented in Figs. 27.15 through 27.20.

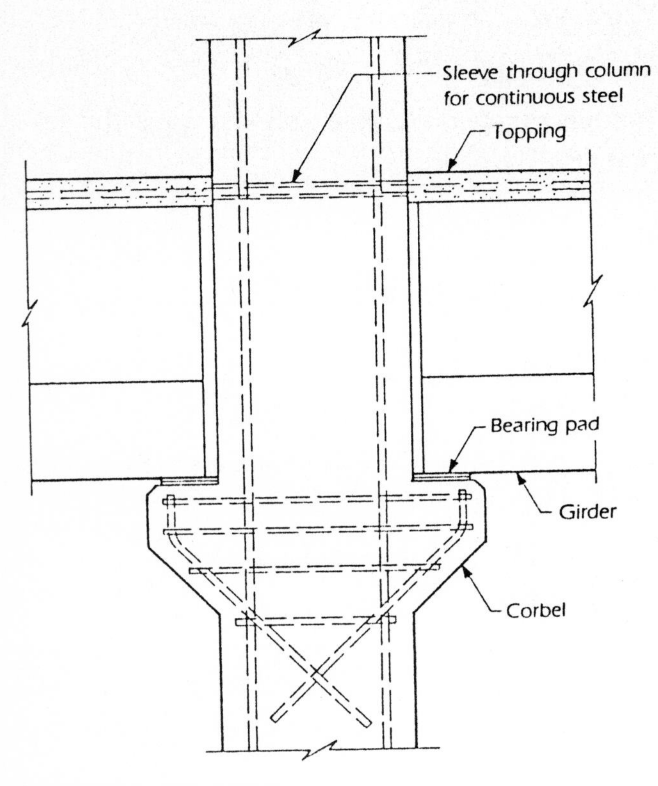


**Figure 27.15** Girder-column.

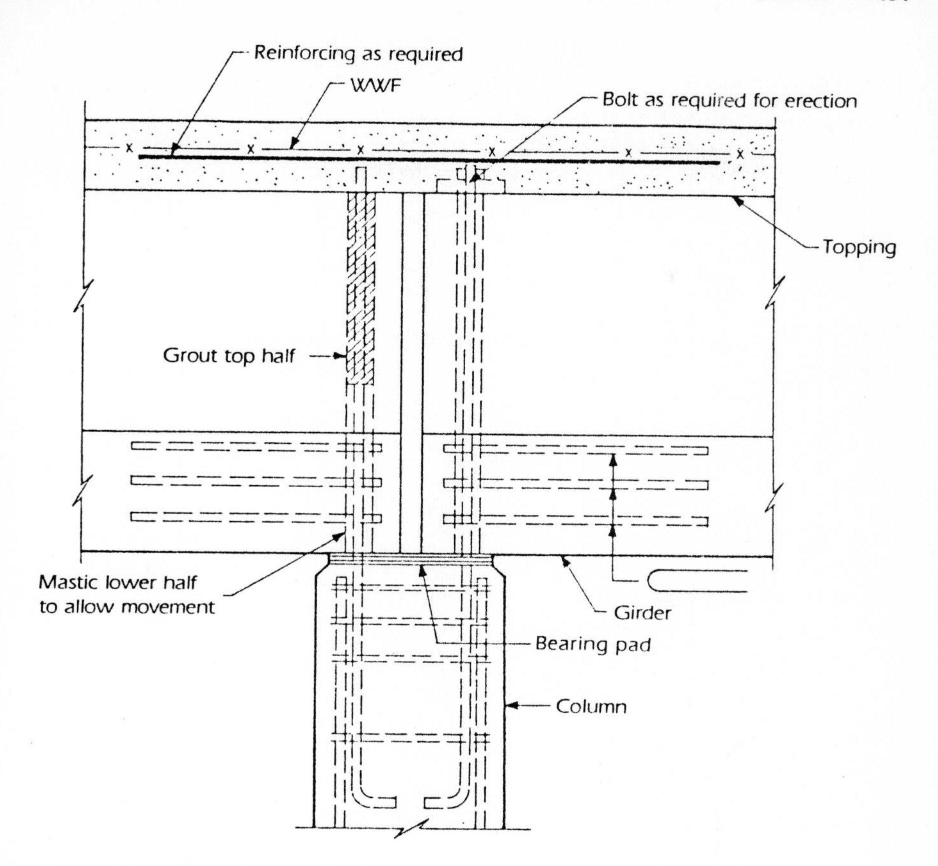


**Figure 27.16** Girder to column top.

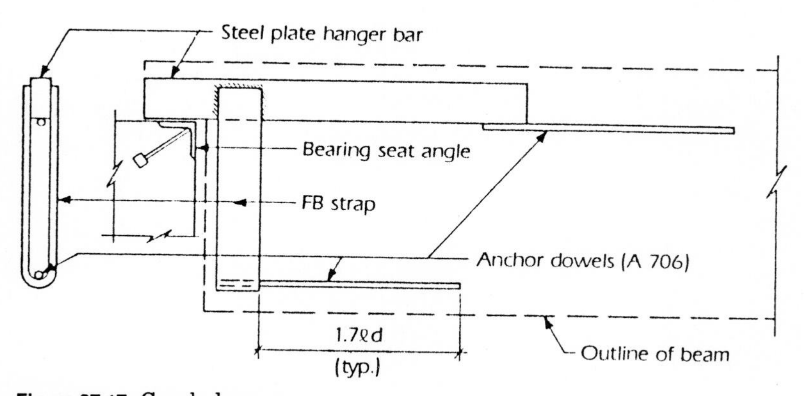


**Figure 27.17** Cazaly hanger.

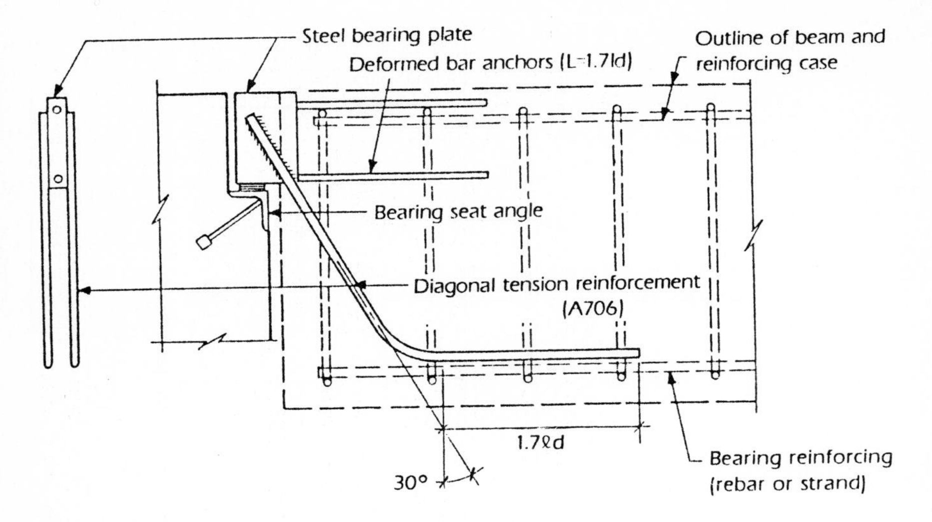


Figure 27.18 Knife-blade connection reinforcing.

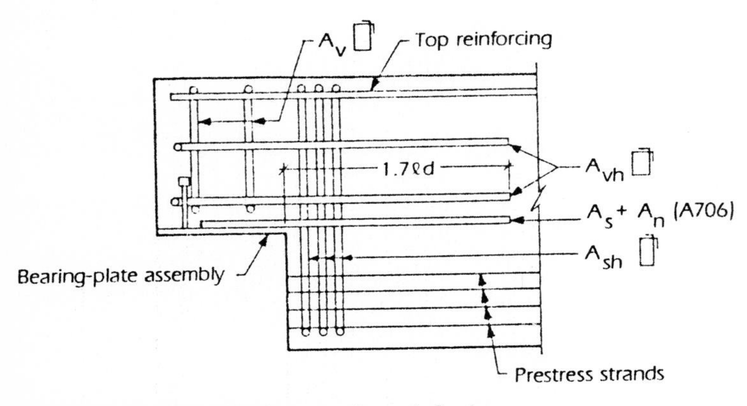


Figure 27.19 Dapped-end connection reinforcing.

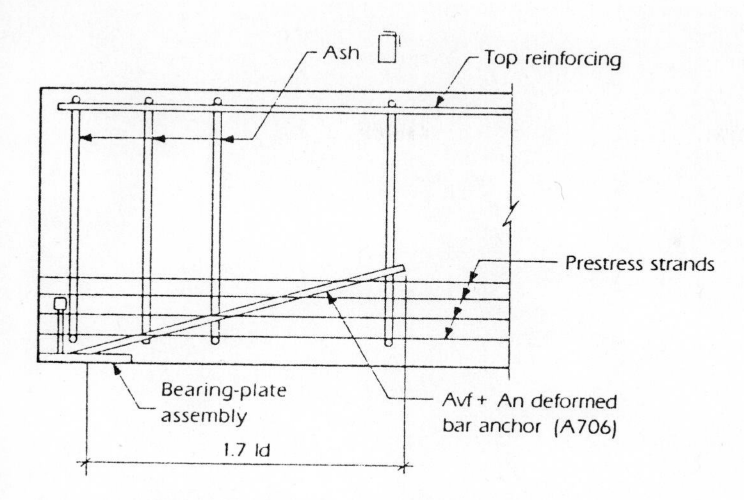


**Figure 27.20** Plain-end reinforcing and connection detail.

**Figure 28.1** Classroom building, University of California at Davis. Prestressed concrete columns 87 ft long form the vertical-load-carrying frame for this nine-story building. Other precast components are exposed aggregate wall panel units, precast channel floor units, and single-T roof members seen cantilevering out above.

Chapter

# 28

# Precast Concrete Columns

Precast concrete columns may be made in many different cross sections, but square or rectangular cross sections are standard because precast concrete manufacturers have these types of forms available in their inventories. Standard column sizes range from a minimum of 10 by 10 in square to a maximum of 24 by 24 in square. Rectangular columns are available consistent with the size of the pallet or soffit form standard in a particular manufacturer's mold inventory. For larger quantities of the same type of column, say, over 1000 ft of product on a given project, the selection of available mold cross section is not as critical as for smaller quantities, for which standard column form sizes should definitely be ascertained in the planning stages of the structure. Precast concrete columns can be designed in lengths consistent with handling requirements at the plant and the jobsite. Shipping will usually limit the maximum length of the column, since only two points of support normally are feasible. However, special rigs are available with double-pivot supports that will permit handling long pieces. These rigs are used to haul piling in excess of 100 ft in length. Splices may be used to cut down the length of individual pieces to be handled yet satisfy final design situations in multistory structures. Another way to keep handling stresses within allowable limits is by prestressing. This will allow transportation of longer pieces yet reduce tension under lateral loads induced by trucking as well as by service loads in the completed structure.

**Figure 28.2** Union Bank--Oceangate Building, Long Beach, California. Precast concrete columns form the exterior architecture as well as performing a load-bearing function in this structure. In addition, the beam-column connection was made by using details conforming to code provisions for cast-in-place concrete moment-resisting ductile frames. The columns are designed and detailed to conform to these code requirements.

## Production

Standard column cross-sectional dimensions are determined by the location of the standard corbel in the side forms. Standard column heights (measured in the corbel width dimension) are 12, 16, 20, and 24 in. Standard column widths (plan dimension of the mold soffit corbel or pallet) are from 10 to 24 in, varying in 2-in increments. To make a corbel on the third side, a box is built across the top of the mold, and the corbel is cast at the same time the balance of the column is cast. For corbels on the fourth side of the column (bottom side as cast), a second cast is normally performed around a steel bracket welded to an insert plate cast in the bottom of the column. In performing a second-cast operation in a precast plant, the primary-cast shape, in this case a precast column, is stripped, moved to a finishing area, and turned over 180° so that the bottom side as cast is then on top. Then a form is placed around the corbel location, and the corbel on the fourth side is cast. For very large projects, it is more economical to cut a hole in the soffit liner and introduce a bottom form for the corbel on the fourth side of the column so that the entire member is cast in one operation, thereby eliminating double handling in the plant. (See Fig. 28.3.)

**Figure 28.3** Precast column production. Parking garage columns are shown in storage in a fabricator's yard. Note the corbels on three sides and keyed base detail for socketed connection in footing in lieu of the customary steel base plate design.

## Design

The state of the art of the use of precast and prestressed concrete columns is such that these elements normally are used only as vertical-load-resisting elements in a building unless they are built into part of the lateral-force-resisting system of the building, such as a boundary element at the end of a cast-in-place concrete shear wall. As such, due consideration should be given to the column's ability to accommodate seismic drift in the connections at the base and other points where it is tied into the structure. Moreover, the column should not be inadvertently built into the structure in such a manner that it contributes stiffness not accounted for in the lateral analysis of the building, resulting in undesirable cracking or failure. The proper use of bolted base plate connections (as opposed to socketed-type footings) and free-bottom neoprene bearings at beam-corbel interfaces will usually allow sufficient drift movements to be accommodated without distress, especially in relatively stiff shear wall buildings.

The Uniform Building Code gives some basic requirements for the reinforcing of columns. Among them are:

| | |
|---|---|
| 2610(j) | Limits of vertical reinforcement (minimum and maximum) |
| 2618($\ell$) | Waiver of minimum vertical reinforcement for prestressed columns with effective prestress exceeding 225 lb/in$^2$ |
| 2618($\ell$), 2607(k) | Lateral reinforcing: mild steel ties |
| 2618($\ell$), 2607(k), 2610(j) | Lateral reinforcing: mild steel spirals |

**Figure 28.4** Union Bank Building, Long Beach, California. Precast concrete columns form a part of a ductile moment-resisting frame in this 16-story building. The main reinforcing was cast into the columns and extended out into U-shaped precast spandrels around the perimeter at each floor. Cast-in-place concrete ductile moment-resistant frame provisions of the Uniform Building Code were used in detailing the rebar and cast-in-place concrete closure pours. The floor system was formed with spread single T's and hollow-core plank infill.

Both prestressed and mild steel reinforced columns may be designed with mild steel ties as lateral reinforcement. In designing and producing large quantities of columns, it is more economical to prestress these members than to fabricate them in all-mild-steel units. The principal saving is due to material efficiency resulting from the use of prestressing strand over reinforcing bars and to saving in plant labor. For a given column cross section and desired ultimate capacity at the balance point (axial load capacity at simultaneous ultimate strain of concrete and yielding of tension steel) it is considerably more advantageous to use strand in lieu of costly reinforcing bars. At present this cost advantage is more than 400 percent. In terms of plant labor, long-line prestressing effects optimum efficiencies in preliminary and casting labor. Column ties are "hung" in bundles off tensioned strand in the beds and wired in position. This compares favorably with the secondary operation required for fabricating cages and subsequent double handling of these reinforcing units.

In reinforced-concrete column design, for a given nominal superimposed load the greatest economy is achieved with the cross section resulting from the use of the highest-strength concrete attainable under plant conditions and with the minimum allowable vertical reinforcing ($A_s = 0.01 A_g$). Selecting 6000 lb/in² hard-rock concrete combined with the fact that most applications are for pin-ended columns results in designs where slenderness does not govern the design of the column. Preliminary designs using this strength of concrete and 2 percent of vertical steel with minimum column eccentricity usually result in optimum selections. Slenderness effects are covered in Sec. 2610($\ell$) of the Uniform Building Code. Sometimes, as in exterior parking garage columns, an appreciable bending moment is induced owing to the summation of eccentric loadings delivered to the column corbels at each parking level. Table 3.4.1 on page 3-21 of the third edition of the *PCI Design Handbook* is helpful quickly in determining the sum of the induced moment from this unbalanced loading. The resulting maximum-moment condition may also be determined by the use of moment distribution.

An unfortunate aspect of current code provisions makes the use of spiral reinforcing for pin-ended prestressed concrete columns uneconomical, in that relatively large spiral bar diameters and closely spaced pitches result from the use of code equation 10-5. For this reason, mild steel ties are recommended in all cases for plant-cast precast and prestressed concrete columns.

The following design example demonstrates the procedure for selecting an economical column shape and shows how to construct the interaction diagram for the design chosen.

### Column design

Design a precast concrete column supporting four levels of interior girders as shown for the parking structure in the subsection "Girder Design" in Chap. 27.

Calculate loads to column:

| | | $P^k$ | × | LF | = | $UP^k$ |
|---|---|---|---|---|---|---|
| Double T's | $0.060^k/\text{ft}^2 \times 58 \times 28 \times 4 =$ | 390 | × | 1.4 | = | 546 |
| CIP topping | $0.031^k/\text{ft}^2 \times 59 \times 28 \times 4 =$ | 205 | × | 1.4 | = | 287 |
| Girder weight | $0.65^k/\text{ft} \times 27 \times 4 =$ | 70 | × | 1.4 | = | 98 |
| Reduced live load | $0.40 \times 0.050 \times 59 \times 28 \times 4 =$ | 132 | × | 1.7 | = | 225 |
| | | | | | $\Sigma UP$ | $1156^k$ |

**Preliminary design.** When preliminary column tables or interaction charts, such as those in the *CRSI Handbook*, are not available, the following formulas are useful in selecting trial sizes (see *PCI Design Handbook*, 3d edition, page 2-49):

| $f'_c = 5000$ lb/in$^2$ | $f'_c = 6000$ lb/in$^2$ | |
|---|---|---|
| $A_{st} = 0.02A_g$ | $A_{st} = 0.02\,A_g$ | |
| $e = 0.1t$ | $f_y = 60$ ksi | $e = 0.1t$ |
| $A_g = \dfrac{UP}{2.9}$ | $A_g = \dfrac{UP}{3.4}$ | |

The most economical cross section will result from using the highest-strength concrete attainable for a given steel percentage.

For our problem use:

$$f'_c = 6000 \text{ lb/in}^2 \qquad f_y = 60 \text{ ksi}$$

$$\therefore A_g = \frac{1156}{3.4} = 340 \text{ in}^2 \qquad E_s = 29{,}000$$

Try a 14- by 24-in column with four no. 11 bars (two each way). See Fig. 28.5.

**Interaction diagram.** Construct an interaction diagram.

Determine the following parameters:

$$\beta_1 = 0.85 - 0.10 = 0.75$$

$$d = 24 - 2.5 = 21.5 \text{ in} \qquad d' = 2.5 \text{ in}$$

$$y_t = 12 \text{ in} \qquad 0.85 f'_c = 5.1 \text{ ksi}$$

$$A_g = 24 \times 14 \times 336 \text{ in}^2$$

$$A_s = A'_s = 3.14 \text{in}^2$$

*Step 1.* Determine $P_o$ from Fig. 4.7.1(c) on page 4-48 of the *PCI Design Handbook.*

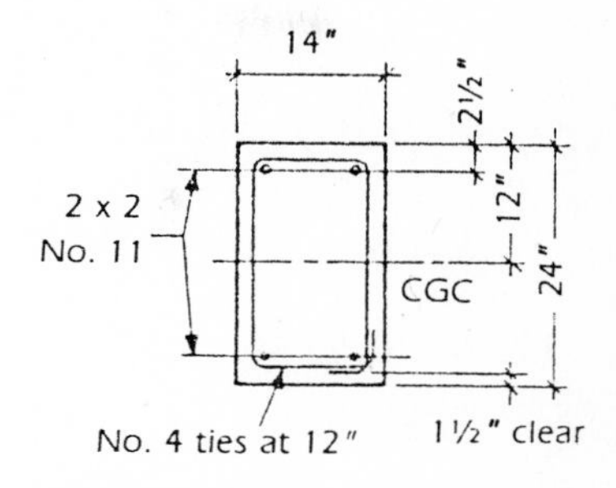


**Figure 28.5** 14- by 24-in column: cross-section detail.

$$\phi P_o = \phi\left[0.85f'_c \left(A_g - A'_s - A_s\right) + \left(A'_s + A_s\right) f_y\right]$$

$$= 0.70\,[5.1\,(336 - 6.28) + (6.28)\,(60)] = 1441^{k}$$

*Step 2.* Determine $P_{nb}$ and $M_{nb}$ from Fig. 4.7.1 (d) of the handbook.

$$c = \frac{0.003d}{0.003 + f_y / E_s} = \frac{(0.003)\,(21.5)}{0.003 + \dfrac{60}{29{,}000}} = 12.72 \text{ in}$$

$$f'_s = E_s \left[\frac{0.003}{c}\,(c - d')\right] \leq 60 \text{ ksi}$$

$$= 29{,}000\left[\frac{0.003}{12.72}\,(12.72 - 2.5)\right] = 69.9 \text{ ksi} \quad \therefore f'_s = f_y = 60 \text{ ksi}$$

$$A_{\text{comp}} = ab = \rho_1 cb = (0.75)\,(12.72)\,(14) = 133.6 \text{ in}^2$$

$$y' = a/2 = \frac{(0.75)\,(12.72)}{2} = 4.77 \text{in}$$

$$\phi P_{nb} = \phi\left[\left(A_{\text{comp}} - A'_s\right)\,0.85f'_c + A'_s f'_s + A_s f_y\right]$$

$$= 0.70[(133.6 - 3.14)\,(5.1) + (3.14)\,(60) - (3.14)(60)] = 460^{k}$$

$$= 460^{k}$$

$$\phi M_{nb} = \phi P_{nb} e = (0.70)\,[(665)\,(12 - 4.77)$$

$$+ (188.4)\,(21.5 - 12) + (188.4)\,(12 - 2.5)]$$

$$= 0.70(4808 + 1790 + 1790) = 5872^{\text{k}\cdot\text{in}}\left(490^{\text{k}\cdot\text{ft}}\right)$$

*Step 3.* Determine $M_o$. (Neglect compression reinforcement.)

$$a \quad \frac{A_s f_y}{0.85 f'_c b} = \frac{(3.14)(60)}{(0.85)(6.0)(14)} \quad 2.64 \text{ in}$$

$$M_o = A_s f_y (d - a/2) - (3.14)(60)\left(21.5 - \frac{2.64}{2}\right) = 3802^{\text{k}\cdot\text{in}}$$

$$\phi = 0.9 \text{ at } \phi P_n = 0 \qquad M_o = (0.9)\left(\frac{3802}{12}\right) = 285^{\text{k}\cdot\text{ft}}$$

For $\phi = 0.7$, $M_o = 222^{\text{k}\cdot\text{ft}}$

$$\text{at } \phi = 0.7 \qquad P_n = 0.1 f'_c A_g = (0.1)(6)(336) = 202^{\text{k}}$$

Maximum design load $0.8\phi P_o = (0.8)(1441) = 1153^{\text{k}}$ gives curve point at $e = 0.1t$. Therefore, $\phi M_n = (1153)(2.4/12) = 231^{\text{k}\cdot\text{ft}}$.

*Step 4.* Calculate an intermediate curve point in the compression range ($a > a_{\text{bal}} < 0.75d$). [See Fig. 4.7.1(a) on page 4-48 of the *PCI Design Handbook*.]

Set $a = 16$ in. $\quad \therefore c = 16/0.75 = 21.3$ in

$A_{\text{comp}} = ab = (16)(14) = 224 \text{ in}^2$

$$f_s = E_s \left[\frac{0.003}{c}(c - d')\right] = 29{,}000\left[\frac{0.003}{21.3}(21.3 - 2.5)\right] = 76.7^{\text{k}} \text{ ksi}$$

(Use 60 ksi)

$$f_s = E_s \left[\frac{0.003}{c}(d - c)\right] = 29{,}000\left[\frac{0.003}{21.3}(21.5 - 21.3)\right] = 1 \text{ ksi}$$

$$\left(y' = a/2 = 60^{\text{k}}\right)$$

$$\phi P_n = 0.7[(224 - 3.14)(5.1) + (3.14)(60) - 3]$$
$$= 0.7(1126 + 188 - 3) = 918^{\text{k}}$$

$$\phi M_n = 0.7[(1126)(12 - 8) + (3.14)(1)(21.5 - 12)$$
$$+ (3.14)(60)(12 - 2.5)] = 4427^{\text{k}\cdot\text{in}} \left(369^{\text{k}\cdot\text{ft}}\right)$$

**Drawing the interaction diagram.** See Fig. 28.6.

For an unsupported height of 9 ft 0 in, $KL/r = (9\times 12)/4.2 = 26$, which is less than than 34. Therefore, slenderness effects need not be considered [Uniform Building Code, Sec. 2610 (1)].

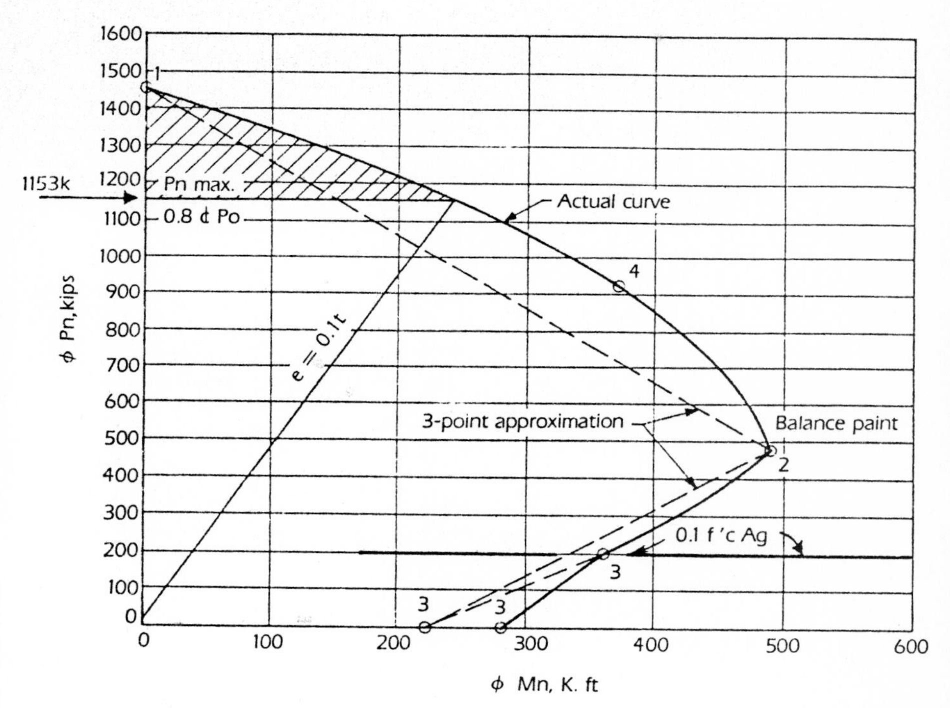


**Figure 28.6** Interaction diagram.

## Connection Design

Below are shown example calculations for a typical column corbel and for the design of a steel column base plate. In designing precast column base plates, two conditions of loading should be investigated. A temporary erection loading exists where the ungrouted base plate is subjected to the column weight plus the erection loads. The final condition shows the dead plus live loading on the grouted base plate. Page 6-21 of the third edition of the *PCI Design Handbook* gives additional information on base plate design.

### Reinforcing for a column corbel

Design reinforcing for a column corbel supporting an inverted-T girder (see Fig. 28.7). Loads are shown below.

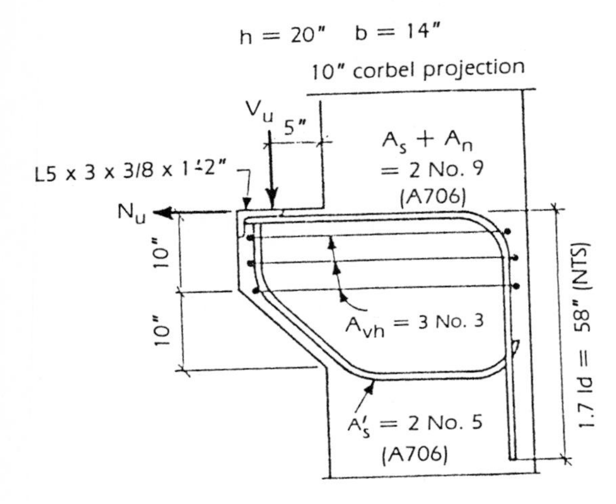


**Figure 28.7** Reinforcing for a column corbel supporting an inverted-T girder.

$$V_u = 1.4 \times 83 + 1.7 \times 25 = 159^{\mathrm{k}}$$

$$N_u = 31^{\mathrm{k}} \text{ (given)}$$

Determine $A_s + A_n$ (see Table 6.20.20 on page 6-65 of the *PCI Design Handbook*).

For $b = 14$ in and $h = 20$ in, $V_n = 190^{\mathrm{k}}$; using two no. 9 bars, ($A_s$ = 2.0 in²).

Determine $A_{vh}$.

$$\mu e = \frac{1000bh\,\mu}{V_u} = \frac{(1000)(14)(20)(1.4)}{(159{,}000)} = 2.47$$

$$A_{vh} = \frac{V_u}{3\phi f_{yv}\,\mu_e} = \frac{159}{(3)(0.85)(40)(2.47)} = 0.63 \text{ in}^2$$

$$A_{vh,\min} = \frac{40bh}{f_{yv}} = \frac{(40)(20)(14)}{40{,}000} = 0.28 \text{ in}^2$$

Use three no. 3 bars ($A_v = 0.66 \text{ in}^2$).

**Column base plate**

Design a column base plate for the 14- by 24-in precast column shown to adequately transmit bearing to a cast-in-place concrete pier with $f'_c$ = 3000 lb/in$^2$. (A36 steel is to be used.) See Fig. 28.8.

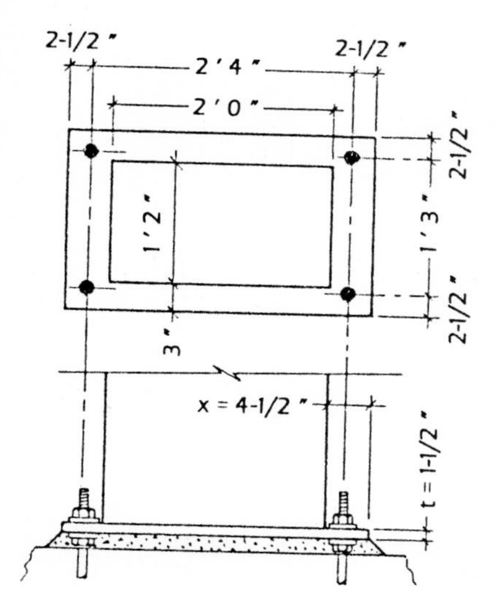


**Figure 28.8** Design of a column base plate.

**Allowable bearing pressure.** See Uniform Building Code, Sec. 2610(p); *PCI Design Handbook*, page 6-21.

$$F_{bu} = 0.85\phi f'_c = (0.85)(0.70)(3) = 1.785 \text{ ksi}$$

$$A_{\text{required}} = \frac{UP}{F_p} = \frac{1156}{1.785} = 648 \text{ in}^2$$

Use ℙ 20 × 33 = 660 in$^2$.

**Base plate thickness.** See *PCI Handbook*, page 6-62.

$$t = x_o\sqrt{\frac{2f_{by}}{\phi f_y}}$$

$$= 4.5\sqrt{\frac{(2)(1.785)}{(0.9)(36)}} = 1.49 \text{ in}$$

Use ℙ 20 × 1½ × 2 ft 9 in.

Bracing and subsequent construction of a classroom building are shown in Fig. 28.9.

Figure 28.9 Classroom building, University of California at Davis. The top photograph is the bracing scheme used to stabilize the nine-story prestressed concrete columns during construction. The 26 columns subsequently received precast concrete infill panels, single T's at the roof, prestressed channel deck elements, and precast sunshades to complete the structure, shown during panel erection in the lower photograph.

**Connection details**

Connection details are shown in Figs. 28.10 through 28.21.

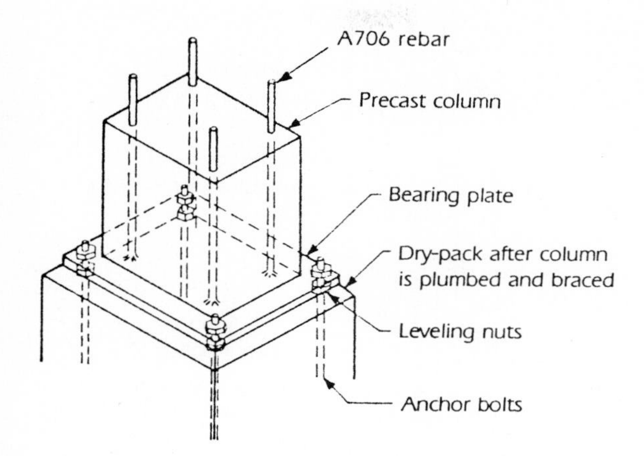


**Figure 28.10** Bolted column base (exposed).

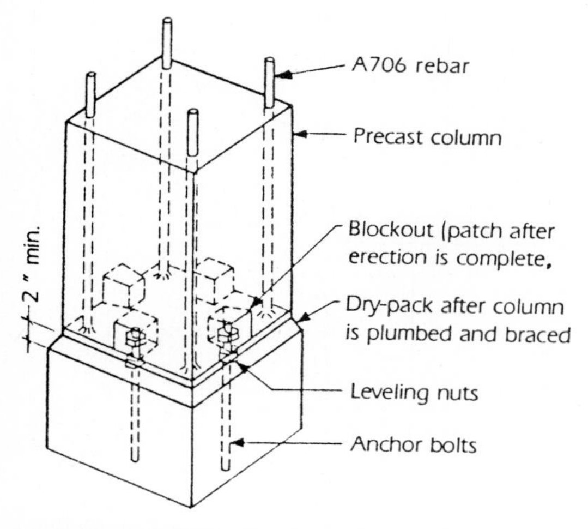


**Figure 28.11** Bolted column base (hidden).

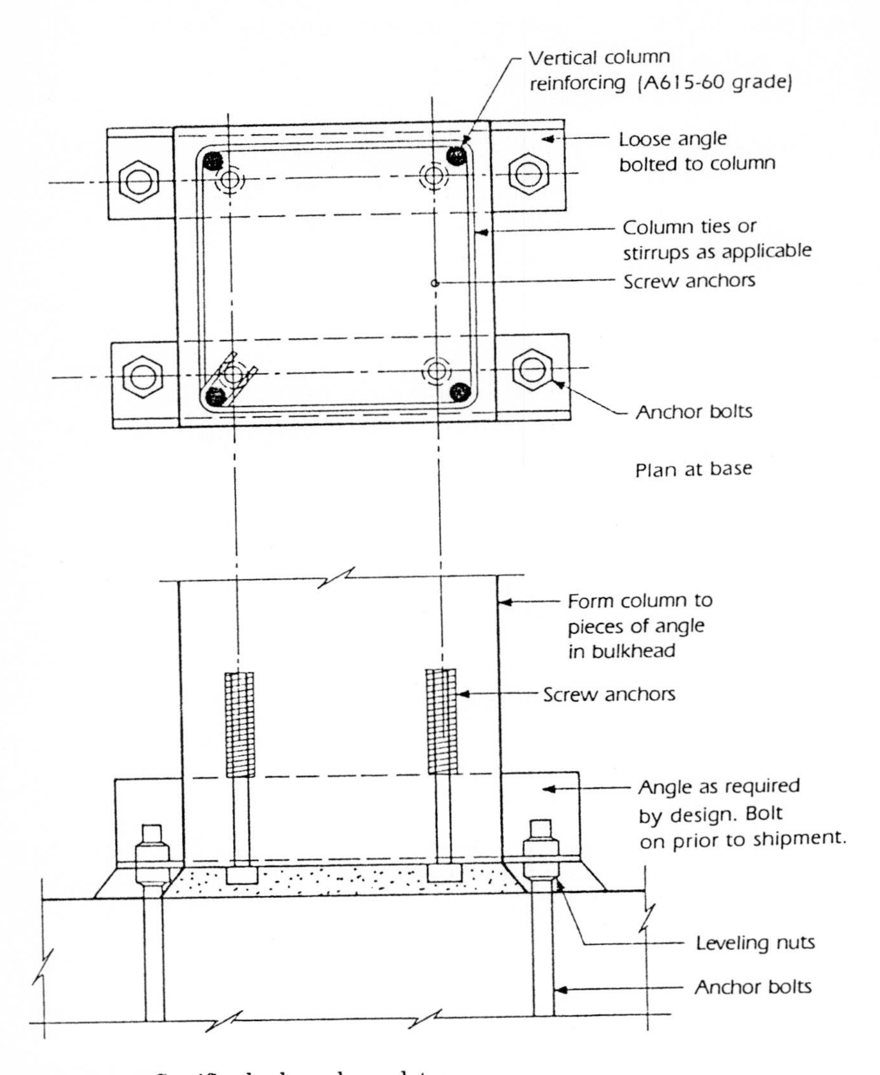


**Figure 28.12** Semifixed column base plate.

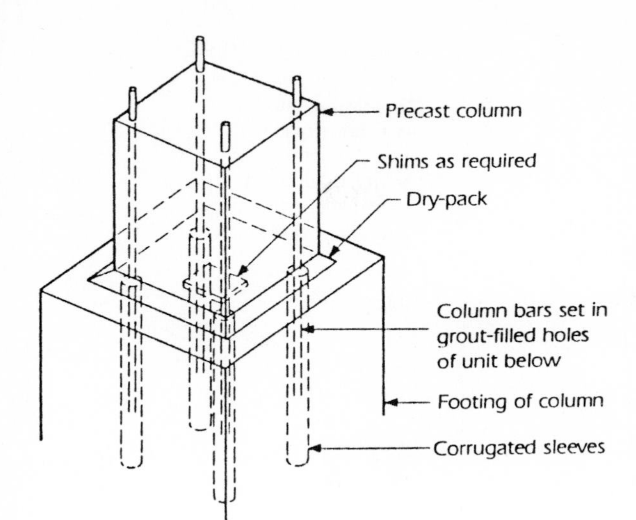


**Figure 28.13** Doweled column base.

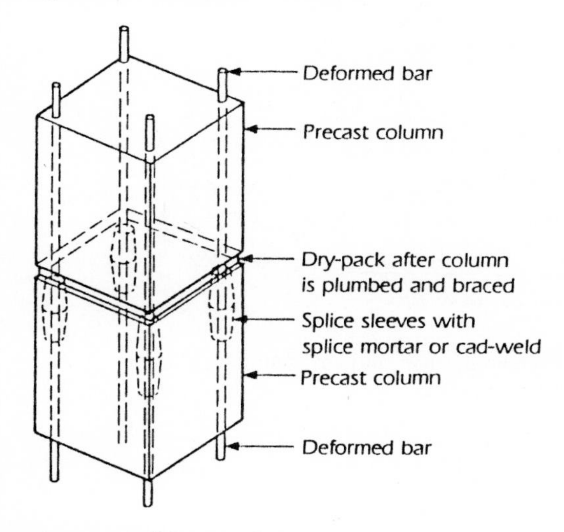


**Figure 28.14** Column splice.

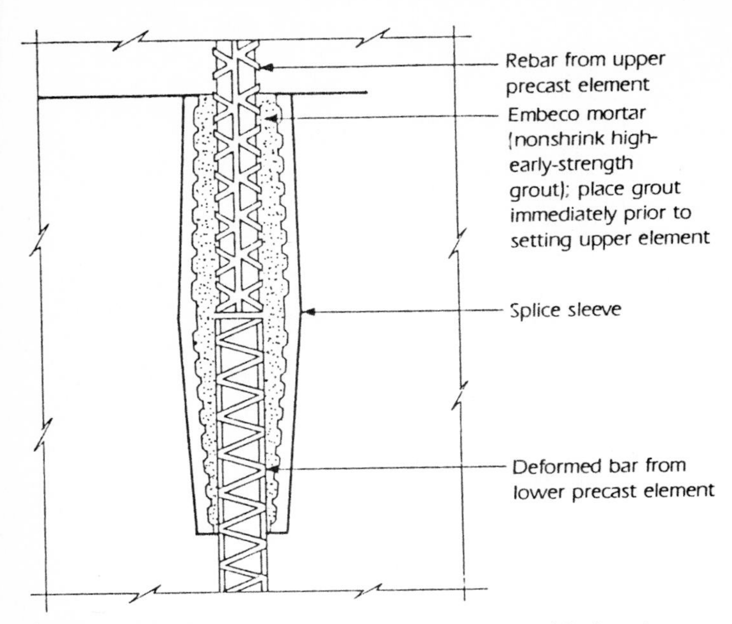


**Figure 28.15** NMB splice sleeve system, manufactured for bar sizes no. 5 through no. 18.

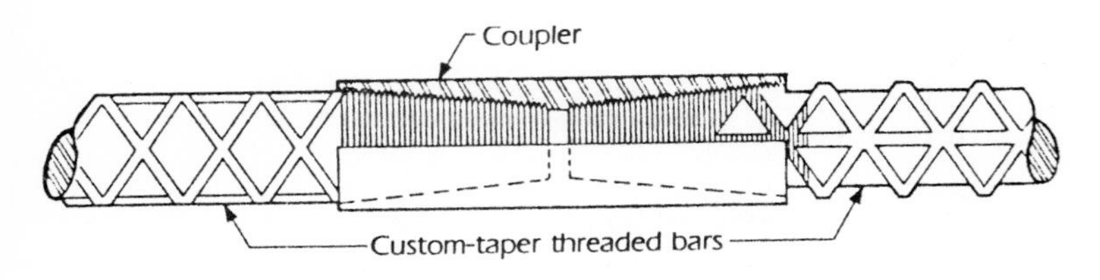


**Figure 28.16** Fox-Howlett no-slip reinforcing bar coupler, manufactured for no. 11, no. 14, and no. 18 bars.

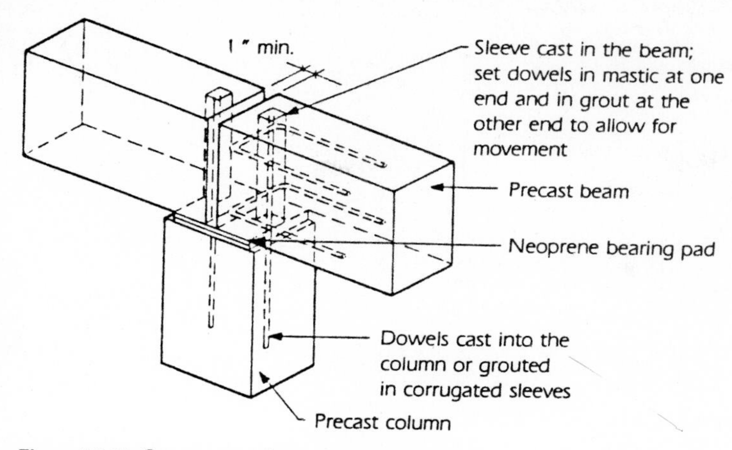


**Figure 28.17** One-story column-beam connection.

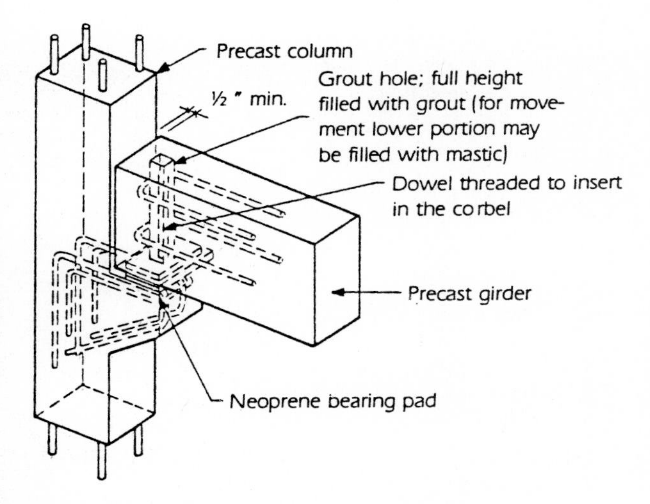


**Figure 28.18** Hinged column-girder connection.

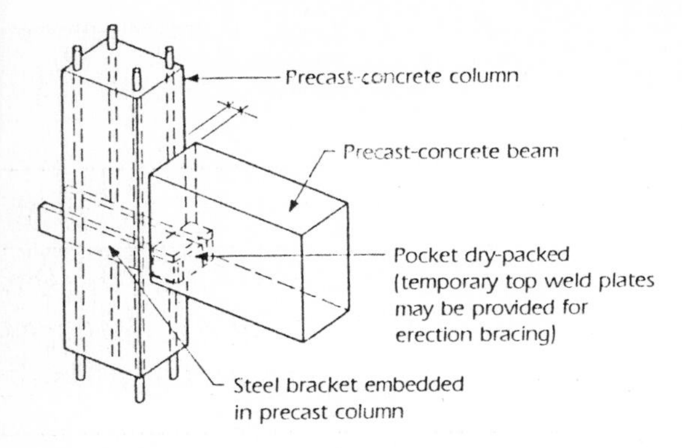


**Figure 28.19** Connection from beam to column.

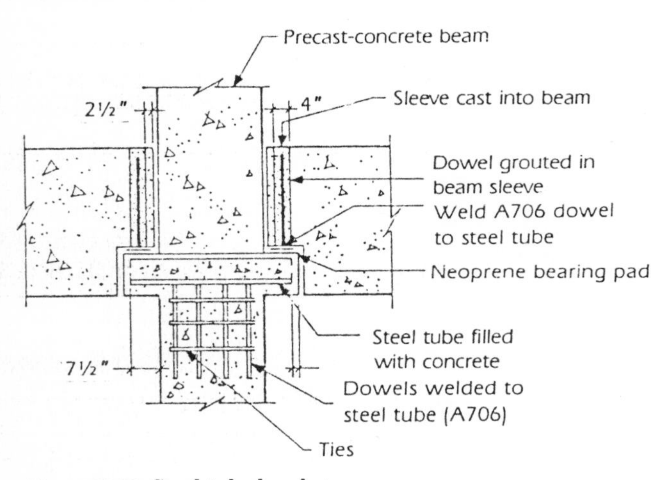


**Figure 28.20** Steel tube bracket.

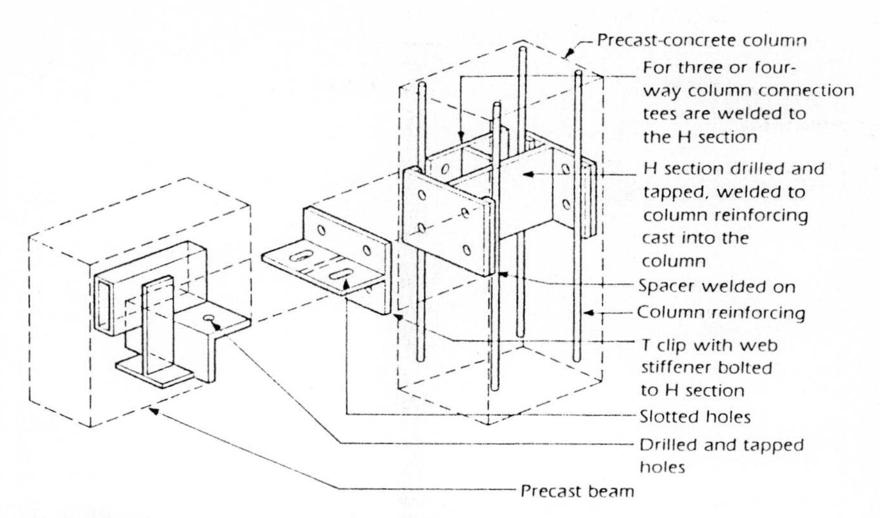


**Figure 28.21** Steel insert column-beam connection. With this connection the column can be cast in a continuous beam or piling form.

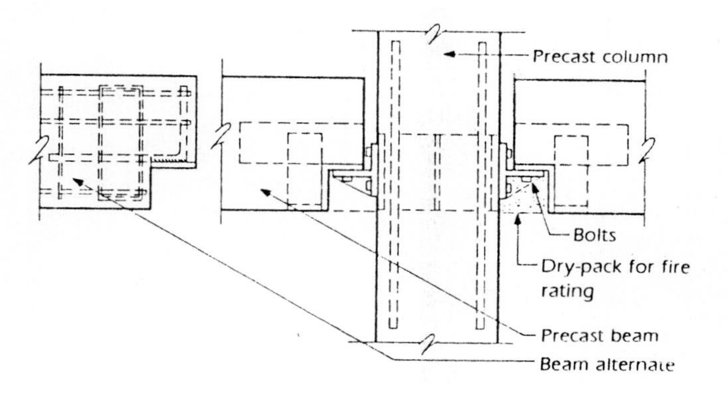

Figure 29.1 Golden Gateway III, San Francisco, California. Prestressed concrete piling 14 in square being driven in downtown San Francisco. Note the cast-in plate in the piling units in the foreground, which form a part of the Dyn-A-Splice mechanical pile splice. This eliminated the problem with transporting long units on midcity streets. First, the lower 80-ft section was driven; then the upper 40-ft section was spliced, and driving continued immediately.

Chapter

# 29

# Prestressed Concrete Bearing Piles

When vertical loads delivered to the base of a building or a bridge structure cause allowable soil pressures to be exceeded under reasonably sized spread footings or mat foundations, a bearing-pile foundation system is used. Sometimes, the proximity of bedrock to the ground surface will result in economy over spread footings even when soil bearing capacities are satisfied. Where end-bearing conditions are not attainable, piles are driven until sufficient skin friction is developed to safely support the applied loads. Prestressed concrete piles, having a large constant surface area, develop frictional resistance in a shorter distance than other types of piling. Friction piles are most commonly used in silty sand, clay-silt, and clay-sand types of soils. The use of piling dates back to before the Roman era, and evidence of such use is worldwide. Prestressed concrete piling is also immune from fungus attack, marine borers, termites, corrosion, and fire. When considering the proven splicing methods available, prestressed concrete piling can be driven to any length. For these reasons, most piling used for foundations is pretensioned concrete piling. The advantage of prestressing is that compressive stresses are introduced which counteract tension stresses resulting from handling, stress waves produced by driving, or tension stresses due to buckling or eccentricity during driving. A complete overview of the use of concrete piling in general is given in the ACI Committee 543 report entitled "Recommendations for Design, Manufacture, and Installation of Concrete Piles." A wealth of practical information on the use of prestressed concrete bearing piles is given in Prof. Ben C. Gerwick, Jr. 's book *Construction of Prestressed Concrete Structures* (Wiley, New York, 1971).

The principal competing types of piling systems are structural steel, timber, and concrete-filled tapered metal casing piles. In addition to cost and construction time savings, the following advantages are also offered by prestressed concrete piling:

1. *High load-carrying capacity.* Because of superior strength, high axial loads, well within the allowable limits of regulating codes can be carried by prestressed piles to the extent allowed by soil conditions. Higher loads per pile mean fewer piles, smaller footings, and, consequently, in most instances lower costs per ton carried.

2. *Durability.* The dense, high-quality concrete, maintained permanently under stress, is relatively free from shrinkage and other cracks and is impervious to water. Experience plus accelerated tests has proved pretensioned concrete piles to be exceptionally durable under even the most severe conditions of exposure.

3. *Ease of handling.* Fewer picking points and the great strength of pretensioned piles greatly facilitate transportation and handling and contribute to lower driving costs.

4. *Ability to take hard driving.* The ability to withstand driving allows the use of heavier hammers for driving through denser soils, thus permitting the development of adequate bearing in the soil.

5. *Greater column strength.* The long-column behavior of prestressed concrete piling is excellent and thus permits high axial loads even when the piles are also subject to eccentricities or lateral forces and moments. When bending resistance is critical, pile capacity can be increased by increasing effective prestress up to the recommended maximum.

6. *Resistance to uplift.* Pretensioned piles can be utilized very effectively in tension where uplift must be resisted, and transfer of tension to the pile can be easily accomplished.

7. *Quality control.* Close supervision by experienced personnel of materials and workmanship in a centrally controlled plant ensures a

high-quality product with 6000 lb/in$^2$ concrete. Guaranteed strengths to 8000 lb/in$^2$ may be provided at slight additional cost.

8. *Economy.* Prestressed concrete piles, when properly used, have shown substantial economic advantages for pile foundations.

9. *Availability.* Piling sizes can obviously vary with design selection and local availability of standard sizes. Some of the standard sizes available in most areas are :

| | |
|---|---|
| Solid square | 12, 14, 16, and 20 in |
| Solid octagonal | 18 and 24 in |

Some manufacturers make a larger number of different types and sizes, including hollow sections and large-diameter hollow-cylinder piles. Product information on a more extensive range of prestressed concrete piling is shown on the J. H. Pomeroy & Co. piling property sheet (Fig. 29.2).

## Fabrication

The advantage of pretensioning is that the piling can be quickly and efficiently produced in continuous steel forms with steam curing on a daily cycle. The piling is cast in lines often 800 ft or more in length. The square sizes are often cast in "gang" molds, with several lines being cast side by side to achieve more efficient utilization of bed labor. Octagonal or round shapes are made by using molds with the upper portions hinged to swing out of the way during strand placement and positioning of spiral reinforcement. Hollow sections are formed by using hollow tubes which are cast in or by employing a moving mandrel which forms the inner void. As with other precast concrete products, vertical sides of molds will have a slight draft on the order of ½ in/ft to facilitate stripping. Normally 6000 lb/in$^2$ concrete is used for piling for more cost-efficient design even though required release strengths rarely exceed 3000 lb/in$^2$. Piling to be used in extremely corrosive environments, as in tidal areas with heavy concentrations of salts, is made with cement-rich concrete mixes containing 7½ or 8 sacks of cement per cubic yard of concrete. In addition, as for all structural precast or prestressed concrete products, the water-cement ratio should be kept below 0.40. Corrosion-resistant piling also exhibits increased cover, on the order of 2½ to 3 in, over the reinforcing assembly. In some instances a 3- to 7-day moist or water cure is required to assure durability. (See Figs. 29.3, 29.4, and 29.5 for some views of plant production.)

# J. H. POMEROY & CO., INC.

P. O. Box 300, Petaluma, California 94953 Telephone (707) 763-1918

## PROPERTIES OF PRETENSIONED PRESTRESSED CONCRETE PILES

| PILE SIZE DIAMETER | SHAPE | SOLID OR HOLLOW | Ac | WEIGHT PLF | NUMBER OF STRANDS | EFFECTIVE PRESTRESS (TO NEAREST 5 PSI) | I | I/C | r | PERIMETER | ALLOWABLE MOMENT 300 PSI TENSION | ALLOWABLE MOMENT 600 PSI TENSION | ALLOWABLE LOADS BASED ON f'c 6000 PSI | ALLOWABLE LOADS BASED ON f'c 7000 PSI |
|---|---|---|---|---|---|---|---|---|---|---|---|---|---|---|
| INCHES (1) | | (2) | SQ. IN. (3) | LBS. (4) | PER PILE (5) | PSI (6) | IN.$^4$ | IN.$^3$ | IN. | INCHES | KIP INCHES (7) | KIP INCHES (7) | TONS (8) | TONS (8) |
| 10'' | SQUARE | SOLID | 98 | 105 | 4—7/16" | 760* | 790 | 158 | 2.84 | 38 | 167 | 215 | 87 | 103 |
| 12'' | SQUARE | SOLID | 142 | 152 | 5—½" | 830 | 1,664 | 277 | 3.42 | 46 | 313 | 396 | 125 | 148 |
| 14'' | SQUARE | SOLID | 194 | 209 | 6—½" | 730 | 3,112 | 445 | 4.00 | 54 | 458 | 592 | 172 | 204 |
| 15'' | OCTAGONAL | SOLID | 186 | 196 | 6—½" | 760 | 2,765 | 368 | 3.86 | 50 | 390 | 500 | 165 | 195 |
| 16'' | SQUARE | SOLID | 254 | 273 | 9—½" | 835 | 5,344 | 668 | 4.59 | 62 | 758 | 958 | 222 | 264 |
| 18'' | OCTAGONAL | SOLID | 268 | 288 | 9—½" | 790 | 5,705 | 634 | 4.61 | 60 | 691 | 881 | 236 | 281 |
| 18'' | SQUARE | SOLID | 322 | 346 | 11—½" | 805 | 8,597 | 955 | 5.17 | 70 | 1,055 | 1,342 | 283 | 336 |
| 20'' | SQUARE | SOLID | 398 | 428 | 13—½" | 770 | 13,146 | 1,315 | 5.75 | 78 | 1,407 | 1,801 | 353 | 418 |
| 20'' | SQUARE | 11'' H. C. | 303 | 326 | 10—½" | 775 | 12,427 | 1,243 | 6.40 | 78 | 1,336 | 1,709 | 268 | 318 |
| 24'' | SQUARE | 14'' H. C. | 418 | 450 | 13—½" | 730 | 25,490 | 2,124 | 7.81 | 94 | 2,188 | 2,825 | 372 | 440 |
| 36'' | ROUND | 26'' H. C. | 487 | 524 | 17—½" | 820 | 60,016 | 3,334 | 11.10 | 113 | 3,734 | 4,735 | 428 | 508 |
| 48'' | ROUND | 38'' H. C. | 675 | 726 | 24—½" | 835 | 158,222 | 6,593 | 15.31 | 151 | 7,483 | 9,460 | 592 | 703 |
| 54'' | ROUND | 44'' H. C. | 770 | 829 | 28—½" | 855 | 233,409 | 8,645 | 17.41 | 170 | 9,985 | 12,578 | 673 | 800 |

(1) Nominal pile-size.
(2) Holes for hollow core piles are circular.
(3) Reduction in area for chamfers on square piles has been taken into account.
(4) Tables are based on concrete of 155 lb./cu. ft. density.
(5) Based on 1/2'' and 7/16'' diameter ASTM A416 Grade 270 strands with ultimate strengths of 41,300 lbs. and 31,000 lbs., respectively. If different diameter strand is used, the number of strands per pile should be increased or decreased, in accordance with strand manufacturer's tables, to provide approximately the same minimum effective prestress shown in the table.
(6) Effective prestress assumes a uniform distribution of strands resulting in a uniform prestress. Piles marked with * have effective prestress based on 60% of ultimate strand strength. All other piles have effective prestress based on initial prestress in strand of 70% of ultimate minus losses of 35,000 psi.
(7) Allowable bending moments listed are for a permissable concrete tensile stress of 300 psi with an effective prestress as given in the table, no external axial load, f'c=6000 psi and assuming a modulus of rupture of 600 psi. Allowable moments for earthquake or similar transient loads are based on a tension of 600 psi. Piles with both axial load and bending should be analyzed considering the effect of the sustained external load. When bending resistance is critical, the allowable moment may be increased by using more strands to raise the effective prestress to a maximum of 0.2 f'c psi.
(8) Allowable design loads are based on the accepted formula of N=Ac (0.33 f'c - 0.27 fpe), and are computed for f'c =6000 psi and 7000 psi. For concrete strength in excess of 7000 psi, consult our Plant Engineering Department for information on practicability and economics.

**Figure 29.2** J. H. Pomeroy & Co. piling property sheet.

**Figure 29.3** A view of 16-in-square piling gang forms. In the foreground is the multiple-strand stressing jack assembly used in the plant. The vertical wide-flange beams are abutments designed to withstand the stressing force of almost 15 tons per strand.

**Figure 29.4** Cylinder pile production. Shown is the production line for 54-in-diameter hollow-cylinder piling for the Dumbarton Bridge project crossing lower San Francisco Bay. The spiral reinforcing is being spaced according to design; the top portion of the form is being placed in the background. A total of 246 piles, each 90 ft long, were produced for this project and shipped to the bridge site by barge.

**Figure 29.5** Square pile production. These 800-ft-long lines in the manufacturer's plant contain 16-in-square piling for the Embarcadero Center 4 project in San Francisco, California. Note the lift loops cast in for stripping and handling. The gaps in the foreground are bulkheads between units with space to permit the strands to be burned to release the prestress force into the member.

## Handling

Owing to the extreme length of prestressed concrete piling (lengths in excess of 120 ft are not uncommon) particular care and planning are required to assure that allowable tensile stresses are not exceeded during stripping, transportation, and "tripping up" into the vertical position in the field. The usual impact factors are applied to service load moments, and actual calculated tensile stresses are compared with allowable stresses, usually the modulus of rupture of the concrete divided by 1.5. See Chap. 19 for various picking points and specifics on product handling. See also Fig. 29.6.

**Figure 29.6** Anheuser-Busch plant expansion, Fairfield, California. Special four-point pivot support rigs are used to ship these 110-ft-long sections of 14-in-square piling. Note the special spreader beams used to off-load and handle sections at the site.

## Installation

When driving into hard strata or rock, a steel tip consisting of a wide flange or H section is cast into the end of the prestressed concrete pile. The standard tip is the flat concrete surface with chamfered edges. High capacity requires that the final seating be done with a large hammer. Hammers delivering an energy of 30,000 to 60,000 ft · lb per blow are generally employed for developing design load capacities of 150 to 250 tons. These piles must be seated to or driven into a satisfactory bearing stratum as determined by the soils engineer. To ensure penetration and to prevent the excessive absorption of driving energy from upper strata, special methods must sometimes be employed, depending on the characteristics of the overlying soil. Jetting, including pilot jetting, internal jetting, and external jetting during driving, is often effective and practical. Predrilling with a wet drill, with or without the aid of a bentonite slurry to hold the hole, is also practical. (See Fig. 29.7.)

## Design Considerations for Pile Placement and Driving

The minimum center-to-center spacing of piles not driven in rock (friction piles) should not be less than twice the average diameter of a round or octagonal pile, or less than 1.75 times the diagonal dimension of a square pile, or less than 2 ft 6 in. For sand or clay or clay-sand soil, spacing should be increased by 10 percent for each interior piling in a group. This increase need not exceed 40 percent. Piling driven to rock (bearing piles) should have a minimum center-to-center spacing of not

**Figure 29.7** Stoneridge overcrossing, Pleasanton, California. The 70-ft-long sections of 12-in-square prestressed concrete piling shown here are being driven at a 20° batter.

less than twice the average diameter of round piles, or less than 1.7 times the diagonal dimension of rectangular piles, or less than 2 ft 0 in. A column or pier supported by piles, unless connected to permanent construction which provides adequate lateral support, should rest on not less than three piles. When the supporting capacity of a single row of piles is adequate for a wall, effective measures should be taken to provide for eccentricity and lateral force, or the piles should be driven alternately in lines spaced at least 1 ft apart and located symmetrically under the center of gravity of the carried loads. The normal tolerance for placing prestressed concrete piling is within ± 6 in from the theoretical location shown on the drawings. It has been found from experience that heavier rams and shorter strokes reduce driving stresses. The use of soft wood cushion blocks 6 to 12 in thick will reduce driving stresses by one-half or more. These compress during driving and thus do not adversely affect the transmission of energy during hard driving. New blocks should be used for each pile. Tests conducted in Sweden show that new-type hammers with special impact-absorbing caps have even better, more uniform force-delivering characteristics. These hammers reduce the impact force in the pile and impart more energy than conventional wood caps. Thus, fewer blows are needed to install the pile. In soft or irregular driving it is desirable to reduce the velocity of the blow to minimize the magnitude of the rebounding tension wave.

**Figure 29.8** Anheuser-Busch plant expansion, Fairfield, California. A three-point pick is used to lift a 100-ft-long pile into the vertical position and place it into the leads of the pile-driving rig for driving.

## Splicing

Often it is advantageous to make piling in shorter lengths and splice the piles together in the field. Splices are used when:

- The ordered length of a prestressed concrete pile is insufficient to obtain the specified bearing value.
- The estimated length of a prestressed concrete pile cannot be economically or feasibly transported or safely handled in the driver.
- Engineering opinions vary on estimated pile lengths derived from analysis of widely spaced borings.
- There are nonuniform underground conditions.
- Experience in local areas indicates erratic lengths which could not be forecast with accuracy.

When splicing is used, individual pile lengths are kept below 60 to 80 ft. A splice must be equal to or greater than the piles it connects in load-carrying capacity, moment resistance, shear, and uplift (if applicable). The splice must be durable, economical, and quickly connected so as to allow driving to be continued as soon as possible. Two types of splices have proved to be very effective. Both have had their capacities verified by actual full-scale testing.

1. *Cement-doweled splice.* This splice is made with the upper pile section cast with protruding rebar dowels and the lower section having sleeves cast in the top of the pile. After the lower section has been driven down, the upper section is installed but separated from the lower section by a 1/2 -in shim stack at the center of the pile. Then a steel splice boot with pouring pockets is clamped around the sections to be spliced. Plasticized cement is poured into the pockets to seal the joint. In 10 to 15 minutes, the plasticized cement has developed a compressive strength of 5000 lb/in$^2$; the boot is removed, and driving can continue. (Epoxy can also be used. The epoxy is poured into the sleeves prior to lowering the upper section, which displaces epoxy into the joint between sections, where it is retained by a form clamped around the perimeter of the piles.) A practical note: the dowel lengths should vary to facilitate stabbing during mating of the pile sections. This splice is sometimes specified where corrosive conditions prohibit use of the bolted splice.

2. *Dyn-a-Splice (bolted splice).* This splice is made up of upper and lower steel plates attached to threaded rebar cast in the pile sections. The threaded rebar is fastened to the plate section with special lock bolts, which fix the plates in exact mated positions. The splice is completed in the field by hammering in wedges which interlock the bolts at each corner of the pile. This splice is preferred over the cement-doweled splice since less time is lost in making the connection. See Fig. 29.9.

**Figure 29.9** Dyn-A-Splice mechanical pile splice. Note how the upper and lower sections are quickly mated in the field. The center-aiding pile automatically centers the upper unit over the driven lower section. (*Photograph courtesy of A-Joint Corp., North Brunswick, New Jersey.*)

## Connection of Pile to Cap or Footing

The design varies in individual cases. The connection may be designed as a hinge to eliminate moment, or a moment connection may be required. Frequently a satisfactory connection is obtained by extending the prestressing strands into the footing or capping beam. The length of strand required may be exposed during cutoff of the pile to grade or by allowing for the extra length during manufacture. An effective moment connection can be made by embedment of the pile head into the cap by about 2 ft. At this depth the full moment resistance of the pile can be developed.

Where the thickness of the pile cap will not permit embedment of strands, the connection is made by mild steel dowels. Where the length of piles can be predetermined within 1 or 2 ft, the dowels can be either (1) fully cast into the head of the pile and exposed during cutoff to grade, (2) left projecting at the head and a specially notched head or follower used during driving, or (3) inserted after driving and grouted into holes formed in the head during manufacture or drilled in after driving. Where the pile length is variable, it is often more economical and practical to drill the holes after cutoff. However, during manufacture ties must be cast into the pile along a sufficient length to be effective on the dowels in their eventual position. The preferred method for connecting piles to pile caps is to use pile anchor dowels anchored to sleeves (or holes drilled into the pile) with neat cement paste. The diameter of the holes shall be $\frac{1}{2}$ to $\frac{1}{4}$ in larger than the outside diameter of the pile anchor dowels. The cement paste is placed into the holes before the dowels are inserted so that no voids remain. The dowels are left undisturbed until the paste has hardened.

## Reinforcing and Jet Pipe Details

Typical reinforcing and jet pipe details are shown in Figs. 29.10 and 29.11.

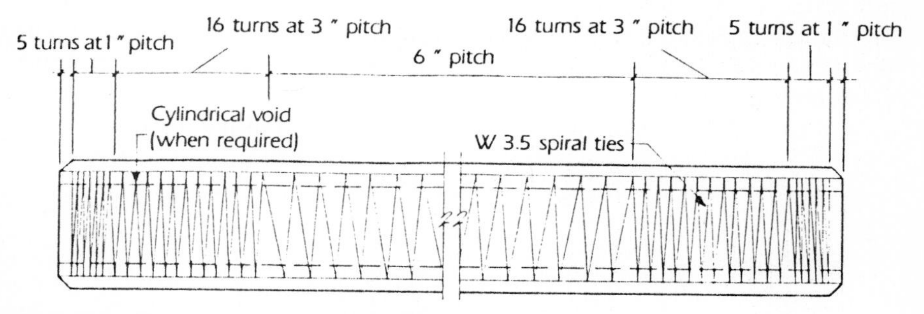


**Figure 29.10** Typical elevation: non-lateral-load-resisting pile.

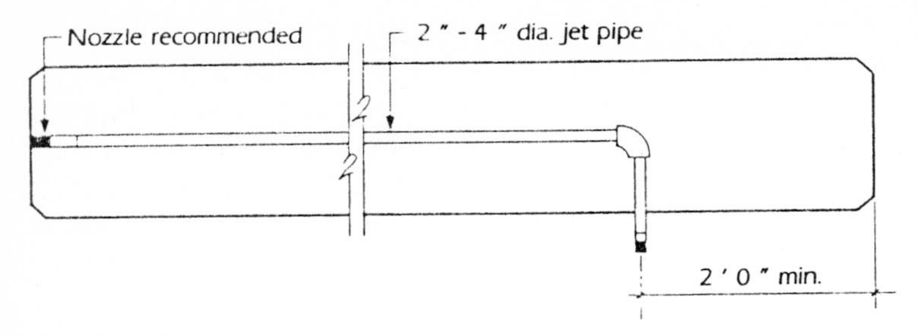


Figure 29.11 Jet pipe details (spiral not shown).

## Design Considerations

The soils engineer plays an important part in the design of prestressed concrete bearing piles. He or she will recommend expected lengths of pile that will be required to develop required frictional resistance, minimum pile stiffness needed to develop fixity at a specified depth below the surface, and any imposed curvatures resulting from layered soil mass movements occurring during seismic disturbances.

Building code requirements for prestressed concrete piling are given in the Uniform Building Code, Sec. 2909, where a formula is presented for the allowable compressive stress in the concrete due to externally applied axial loads: $f_c = 0.33f'_c - 0.27f_{ps}$. This formula, being somewhat arbitrary, should be checked against the specific stress conditions imposed by axial and bending loads and compared with allowable stresses given in the Uniform Building Code, Sec. 2618. Also, load combinations with seismic or wind conditions are permitted a 33 percent increase in these allowable stresses. The code also indicates minimum effective prestress of 700 lb/in$^2$ required for piling greater than 50 ft in length. The J. H. Pomeroy & Co. piling property sheet (Fig. 29.2) gives allowable concentric loads for the empirical formula from Sec. 2909 of the Uniform Building Code. These values are helpful in preliminary design. Page 2-55 of the third edition of the *PCI Design Handbook* gives similar information. In final design, the interaction diagram for the specific section being considered is drawn based upon design criteria outlined in the *PCI Design Handbook*, pages 4-47 through 4-49. The actual ultimate load combinations are plotted on the resulting diagram to see whether or not they fall within the required envelope. Following the design example presented below, another J. H. Pomeroy chart (Fig. 29.16) shows interaction diagrams drawn for several standard pile sizes.

Another area where reinforcement requirements for piling differ is the required lateral reinforcing: the Uniform Building Code, Sec. 2909, indicates the requirements for steel wire spiral lateral reinforcing varying from 5-gauge for 24-in-diameter piles and smaller to 3-gauge 1/4-in-diameter) for piles larger than 24 in in diameter. Spacing requirements are also given for this reinforcing. Recent studies and tests on piling to develop data on reinforcing required to enable prestressed concrete piling to withstand imposed curvatures from extreme (catastrophic) seismic conditions indicate that the diameter of spiral reinforcing should be larger and the spacing (pitch) of the spiral smaller, similar to the lateral reinforcement provisions for spiral reinforcing in concrete columns in order to develop ductility. In other words, spiral reinforcing for these extreme design situations should be about 0.7 percent where normal pile lateral reinforcing is about 0.1 percent. Additional mild steel may also be required to develop additional ultimate moment capacity required at the pile–pile cap interface. (See below, "Seismic Design of Prestressed Concrete Piling.")

Previously, the requirements recommended for piling to be designed in corrosive or extremely saline environments were covered. These included additional cover, cement-rich mixes coupled with a low water-cement ratio, and additional curing requirements. Allowable tensions in these areas should be held to zero, or the modulus of rupture divided by a factor of safety of 2.5 at the most. This item is often overlooked in some codes.

### Prestressed concrete bearing-pile design

Design prestressed concrete piling to support a parking garage shear wall and column foundation. Figure 29.12 shows the structural configuration and loads delivered to the pile cap. The soils engineer's report recommends using piling with a minimum stiffness ($EI$) of $7 \times 10^9$ lb in$^2$. On the basis of this recommendation the pile will develop a point of fixity 12 ft below the bottom of the pile cap. It is estimated that 70 ft of piling will be required to develop the required horizontal lateral resistance. The pile–pile cap connection should be designed to provide fixity.

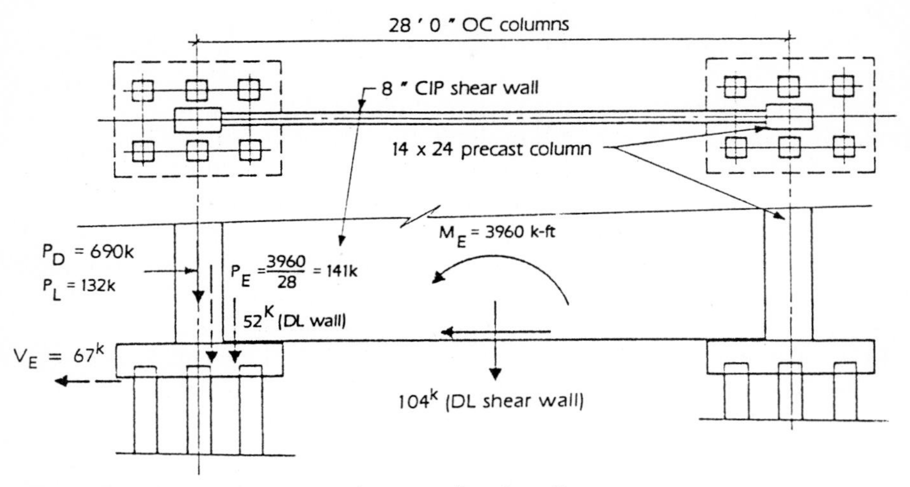


Figure 29.12 Design of prestressed concrete bearing pile.

**Service load conditions.** See Fig. 29.13.

$$P_{D+L+E} = 1015^{\text{k}} \quad V_E = 67^{\text{k}}$$

$$M_E^- = (0.3)(12)(67) = 241^{\text{k}\cdot\text{ft}}$$

$$M_E^+ = (0.1)(12)(67) = 80^{\text{k}\cdot\text{ft}}$$

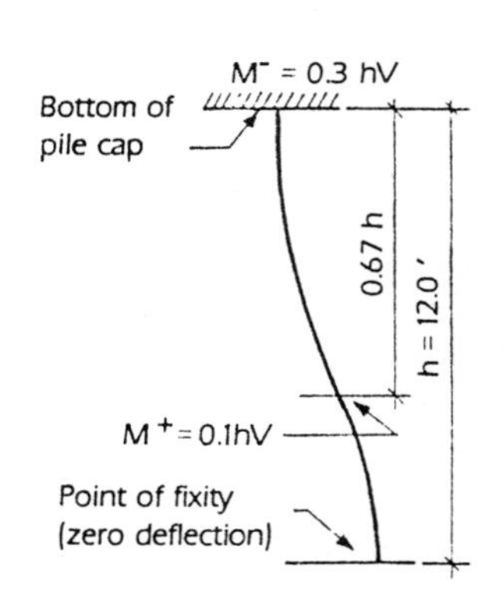


Figure 29.13 Elastic curve at top of pile.

or $$P_{D+L} = 874^{\text{k}}$$

whichever combination is more critical.

The first load combination qualifies for a one-third increase in allowable stresses [Uniform Building Code, Sec. 2302(d)].

**Ultimate load conditions.** Determine ultimate load conditions.

$$U_{D+L+E}\begin{cases} UP = (0.75)(1263 + 1.87 \times 141) = 1145^{k} \\ UV = 0.75 \times 1.87 \times 67 = 94^{k} \\ UM^{-} = 0.75 \times 1.87 \times 241 = 338^{k\cdot ft} \end{cases}$$

$$U_{D+L} = (1.4)(742) + (1.7)(132) = 1263^{k}$$

**Preliminary design.** Use the vertical load condition to choose a trial pile group. (See *PCI Design Handbook*, 3d edition, page 2-55.)

$$\frac{D+L}{4} = \frac{874}{4} = 219^{k}/\text{pile}\ (110\ \text{tons})$$

Try four 12-in-square piles $\left(f'_c = 6000\,\text{lb/in}^2\right)$.

**Seismic condition.** Investigate service load condition for seismic loading. − = compression; + = tension. (Work with J. H. Pomeroy pile design charts, Fig. 29.2.)

$$f^{+} = -\frac{P}{A} + \frac{M_{\text{ext}}}{S} - \frac{P_{\text{ext}}}{A}$$

$$f^{+} = -0.785 + \frac{241 \times 12}{277 \times 4} - \frac{1015}{142 \times 4} - 0.785 + 2.610 - 1.787$$

$$= +0.038\ (\text{OK})$$

$$f^{-} = -0.785 - 2.610 - 1.787 = 5.182\ \text{ksi}$$

Too high! $\left(1.33 \times 0.45 f'_c = 3.59\ \text{ksi.}\right)$ See *PCI Design Handbook*, page 11-15.)

Try six 12-in-square piles with six 7/16-in ϕ-$270^{k}$ strands (*PCI Design Handbook*, pages 4-39 and 11-16).

$$F_f = (6)(21.7)(0.78) = 101.6^{k}$$

$$f^{-} = \frac{101.6}{142} - \frac{241 \times 12}{277 \times 6} - \frac{1015}{142 \times 6} = -0.715 - 1.740 - 1.191 = -3.646$$

$$f'_c,\ \text{required} = \frac{3.646}{1.33 \times 0.45} = 6.078$$

Specify 6100 lb/in².

# J. H. POMEROY & CO., INC.

P. O. Box 300, Petaluma, California 94953
Telephone (707) 763-1918

## PRETENSIONED PRESTRESSED CONCRETE PILES
## PILE INTERACTION CURVES

SOLID SQUARE SHAPE PILES
10", 12", 14", 16", and 18"

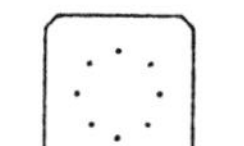

The interaction curves shown hereon are presented as an aid to the Engineer when designing structures using the ultimate strength method.

The curves are derived from a computer solution in accordance with ACI's Building Code. A rectangular stress block was assumed in the concrete with concrete strain limited to .003 in/in. Steel stresses are based on average values of stress strain curves for 1/2 inch and 7/16 inch diameter ASTM A416 grade 270 strands. Concrete Strength is 6,000 psi in 28 days.

A strength reduction factor $\phi$ of 0.7 has already been included in the diagrams for axial loads in the range from 100% to 10% of ultimate axial load. Below 10%, the $\phi$ factor increases linearly to a maximum of 0.9 for the case of pure flexure and combined tension and bending.

For section properties, effective prestress, allowable working design loads and other technical information for these piles see Pomeroy's "Properties of Pretensioned Prestressed Concrete Piles" sheet.

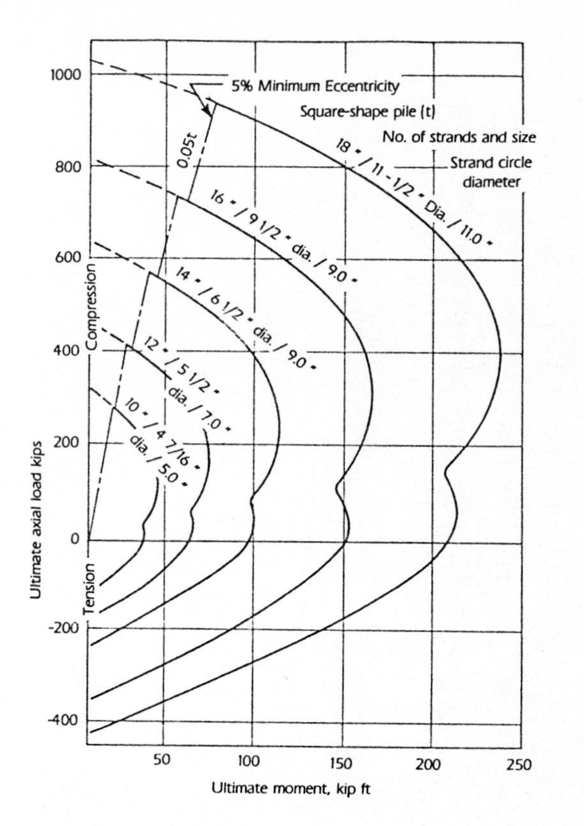


Figure 29.14 J. H. Pomeroy & Co. pile interaction curves.

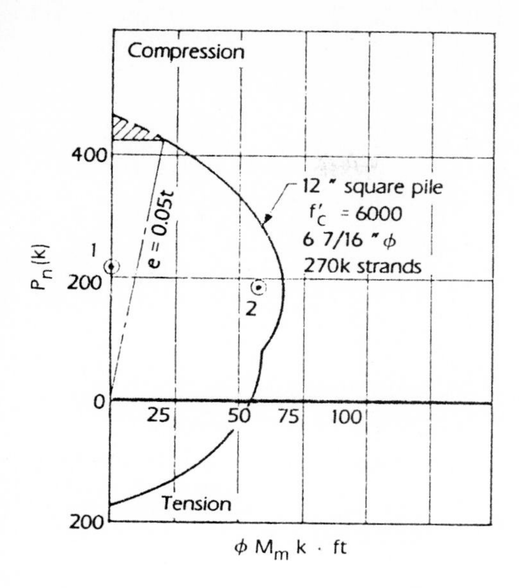


**Figure 29.15** Interaction diagram for ultimate loading conditions. (*From J. H. Pomeroy & Co. chart, Fig. 29.14.*)

**Ultimate strength.** Check ultimate strength for each pile (see *PCI Design Handbook*, page 2-48).

$$1.\ UP_{D+L} = \frac{1263}{6} = 211^{k}$$

$$2.\ \text{Seismic}\begin{cases} UP_E = \dfrac{1145}{6} = 191^{k} \\ UM_E = \dfrac{338}{6} = 56^{k\cdot ft} \end{cases}$$

Both conditions fall within the envelope of the interaction diagram.

**Handling.** Check handling stresses for 70-ft-long pile.

Stripping:

$$W = 1.5\,(\text{impact factor}) \times 0.150 = 0.225^{k}/\text{ft}$$

$$M = (0.0214)(0.225)(70^2)(12) = 283^{k\cdot in}$$

$$\text{At release } F_i = (0.9)(6)(21.7) = 117.2^{k}$$

$$f_t = \frac{117.2}{14.2} + \frac{283}{277} = -0.825 + 1.020 = +0.195\ \text{ksi}$$

$$F_{t,\ \text{allowable}} = \frac{F_r}{\text{FS}} = \frac{7.5\sqrt{3500}}{1.5} = +0.296\ \text{ksi}\ \ (\text{OK})$$

Storage: Block at four points.
Shipping:

$$w = 2.0 \text{ (impact factor) } \times 0.150 = 0.300^{k}/\text{ft}$$

4-point store and ship

0.1L 0.3L 0.2L 0.3L 0.1L

Try supporting at two points. $M^{+} = M^{-} = 0.0056\ wL^2$

$$M = (0.0214)(0.3)(70^2)(12) = 377^{k \cdot in}$$

$$f_t = -0.715 + \frac{377}{277} = +0.645 \text{ ksi}$$

Too high! Ship on special rigs supporting the piles at four points.
Erection:

$$w = 1.25 \times 0.150 = 0.188^{k}/\text{ft}$$

Try a three-point pick.

$$M = (0.02)(0.188)(70^2)(12) = 221^{k \cdot in}$$

$$f_t = -0.715 + \frac{221}{277} = +0.082 \text{ ksi (OK)}$$

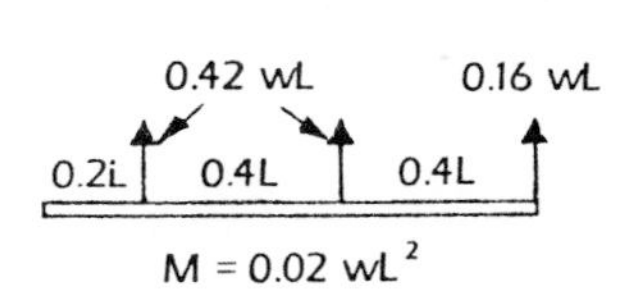


Erect as shown.

Mild steel is required at the end of the pile to develop the bending moment at the pile cap.

$$A_s = \frac{UM}{a_u d} = \frac{56}{4 \times 6} = 2.33 \text{ in}^2$$

Use four no. 7 bars $(A_s = 2.40 \text{ in}^2)$.
Use neat cement grout in sleeves cast in pile.

**Reinforcing summary: 12-in-square prestressed concrete pile.** See Fig. 29.16.

$$f'_{ci} = 3500 \text{ lb/in}^2$$

$$f'_c = 6100 \text{ lb/in}^2$$

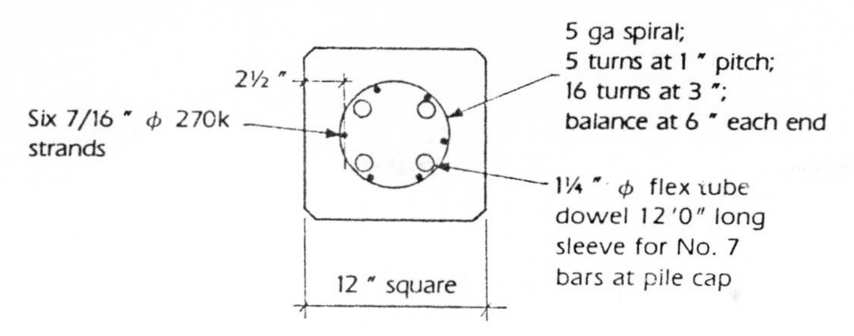


**Figure 29.16** Reinforcing summary.

## Seismic Design of Prestressed Concrete Piling*

Based upon current knowledge, the following conclusions may be drawn regarding the design for pile bending during earthquakes.

1. The pile-bending problem is one of geometry and ductility rather than moment capacity.
2. Pile foundations should be keyed into the soil.
3. Pile–pile cap embedments should develop strength in tension in addition to providing resistance to punching shear through the cap.
4. Spiral confinement reinforcing should be provided at the potential plastic-hinge region at the pile–pile cap interface for cantilevered and unrestrained pile foundations. The amount of confinement reinforcing should be that required for ductile moment-resisting frame columns.

### Recommended seismic design provisions for prestressed piling

The dual-response criteria for designing structures for earthquakes are presented in the recommended lateral force requirements of the Structural Engineers Association of California. Prestressed concrete piling should be designed for serviceability in a moderate earthquake, in addition to being capable of developing ductility at critical points of potential plastic hinging in order to dissipate energy generated by a maximum probable earthquake and provide safety against pile failure.

To ensure serviceability and ductility in prestressed concrete piles in seismic areas, the following design provisions are recommended.

---

*Reprinted by permission from the Prestressed Concrete Institute; D. A. Sheppard, "Seismic Design of Prestressed Concrete Piling," *Journal of the Prestressed Concrete Institute*, March–April 1983.

1. **Design for serviceability.** Provide reinforcement for shear and flexure as indicated by factored loads from ACI 318-83, Sec. 9.2.2:

$$U = 0.75(1.4 + 1.7L + 1.87E) \tag{29.1}$$

$$U = 0.9D + 1.43E \tag{29.2}$$

As an alternative, provide reinforcement as indicated by dynamic analysis using appropriate linear elastic response spectra.

2. **Design for ductility.*** Use confinement reinforcing to develop ductility in the following critical locations for various pile foundation arrangements:

*a.* Restrained piling where

$$M_u < 0.20\phi M_n$$

and

$$P_e < 0.3f_c'A_g$$

and those areas of embedded piling below the ductile range where not subjected to curvatures from layered soil movements during an earthquake:

$$\rho_s \geq 0.003$$

*b.* Embedded portions of piling subjected to curvatures from layered soil movements during an earthquake :

$$\rho_s \geq 0.014$$

*c.* Restrained piling where

$$M_u > 0.2\,\phi M_n$$

$\rho_s \geq 0.021$ for entire ductile range (see Fig. 29.18).

*d.* Unrestrained piling and cantilevered freestanding piling (see Figs. 29.19 and 29.20).

(1) $\rho_s \geq 0.021$ throughout entire ductile range.

(2) At the potential plastic-hinge region immediately below the pile cap, for a distance of two pile diameters, provide spiral reinforcing in accordance with the following equations:

*Not applicable to hollow-core piling sections.

$$\rho_s = 0.45\frac{f_c'}{f_{ysp}}\left(\frac{A_g}{A_c} - 1\right)\left(0.5 + 1.25\frac{P_e}{\phi f_c' A_g}\right) \tag{29.3}$$

but not less than

$$\rho_s = 0.12\frac{f_c'}{f_{ysp}}\left(0.5 + 1.25\frac{P_c}{\phi f_c' A_g}\right) \quad (29.4) \tag{29.4}$$

Confinement reinforcing has been referred to in two ways in the technical literature (see Fig. 29.17). Gerwick refers to spiral percentage as a function of nominal reinforcing area to cross-sectional area. ACI 318-83 defines the spiral reinforcing index ($\rho_s$) as the ratio of the volume of the spiral reinforcing to the volume of the confined concrete core. The latter definition is the more precise one and will be used for these recommendations.

When it is not possible to incorporate the above amounts of confine-

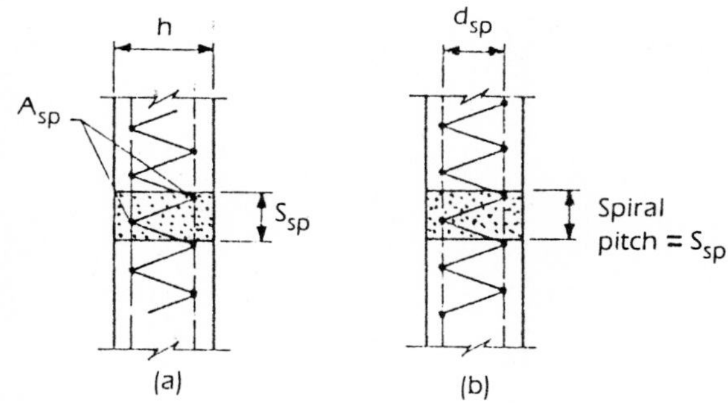


**Figure 29.17** Comparison of spiral confinement technology. (*a*) Percent spiral $= \frac{2A_{sp}}{hS_{sp}}$. [*Spiral confinement percentage (Gerwick).*] (*b*) $\rho_s$(spiral reinforcement index). $\rho_s$ = (vol. spiral)/(vol. core). $\rho_s = \frac{\pi d_{sp} A_{sp}}{\pi d_{sp}^2 \times S_{sp}} = \frac{4A_{sp}}{d_{sp} S_{sp}}$. [*Spiral confinement ratio (ACI).*]

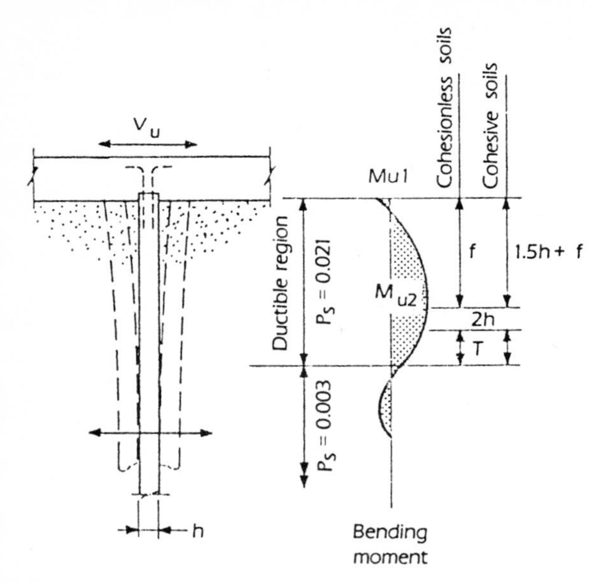


**Figure 29.18** Restrained long, slender piles in cohesive or cohesionless soils.

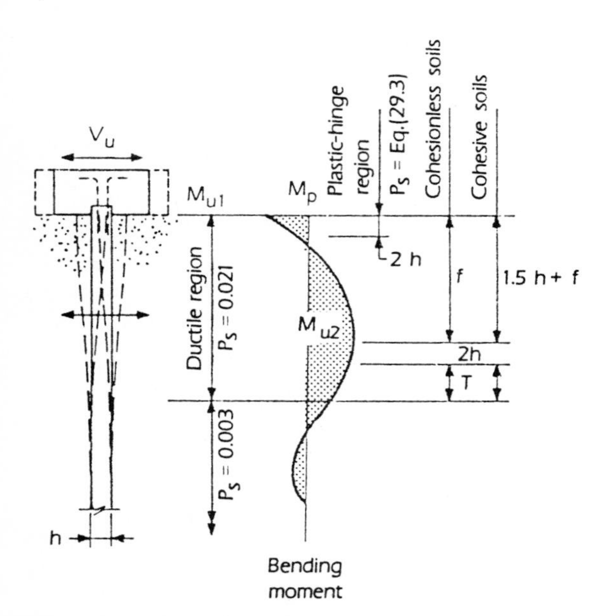


**Figure 29.19** Unrestrained long, slender piles in cohesive or cohesionless soils in which the pile cap is free to translate.

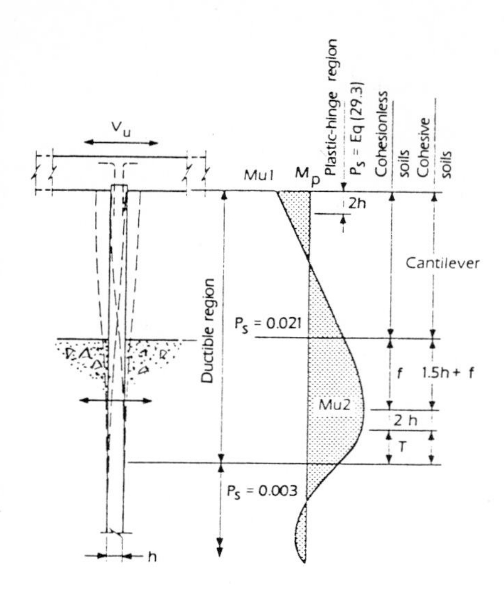


**Figure 29.20** Cantilevered freestanding piles, unrestrained at top and free to translate.

ment in smaller precast pile sections, a cast-in-place concrete pile buildup may be used to satisfy the requirements (see Fig. 29.21). Where minimum cover requirements are 1 1/4 in (32 mm) clear (ACI 318, Sec. 7.7.2), a secondary overlapping spiral section may be used in the plastic-hinge region to satisfy the above requirements (see Fig. 29.22).

The formulas in Subsec. *d*(2) are the relationships given in the New Zealand Code of Practice for Concrete Structures, which offset a degree of conservatism demonstrated by ACI 318-83 (Secs. 10.9.3 and A.6.5.2) for lower levels of axial load typical of actual prestressed concrete pile designs (see also ACI 318-83 Commentary, Sec. A.6.5.2).

The commentary to the New Zealand code also contains a good discussion of moment curvature exhibited by confined concrete at various axial load levels and how the current relationships for calculating confinement steel do not reflect the actual ductility requirements of the concrete section but are formulas which calculate the confinement required to sustain a given axial load after the concrete cover has spalled off.

In their book *Reinforced Concrete Structures* (Wiley, New York, 1975), Robert Park and Thomas Paulay also indicate how the ACI confinement relationships may be overly conservative when used to calculate the required confinement for smaller cross sections. However, the existing relationships are used as conservative interim provisions until more realistic procedures for calculating confinement reinforcing based upon actual moment-curvature analyses are developed.

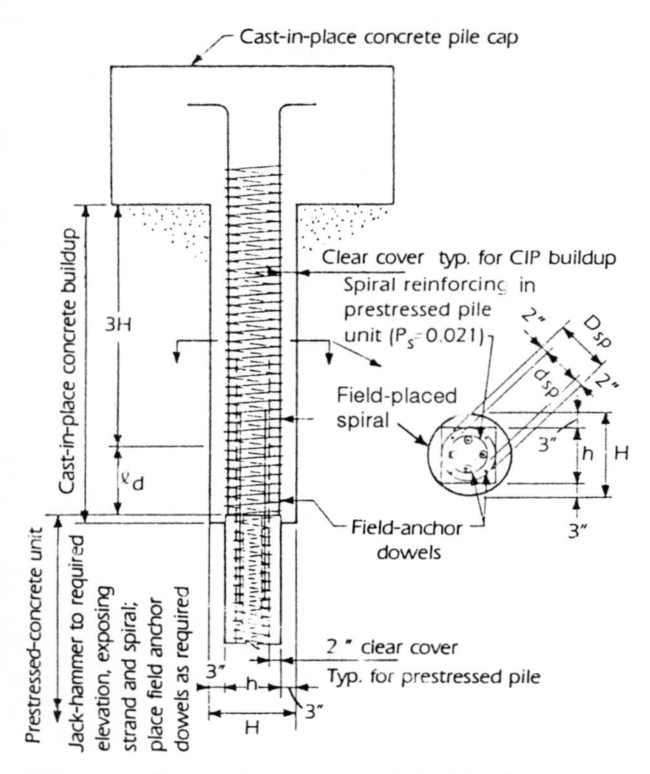


**Figure 29.21** Cast-in-place concrete pile buildup detail.

Charts of spiral confinement sizes and spacings for various standard sizes of prestressed concrete piling and values of $\rho_s$ are presented in Tables 29.1, 29.2, 29.7, and 29.8.

Also included in the detailing recommendations are tables showing spiral confinement provided by using ASTM A-648 wire (hard-drawn steel wire for prestressed concrete pipe; Tables 29.3 through 29.6). By using these high-strength wires, the total ductility requirements of Eq. (29.3) can be incorporated entirely in the precast pile section without the necessity of using a cast-in-place concrete pile buildup or an outer spiral cage in the pile at the plastic-hinge region.

ASTM A-648 wire is available in sizes up to W7.5 (7.9-mm $\phi$) and exhibits bending properties which permit it to be rolled into spiral coils. Testing conducted in Japan shows the value of very high strength wire confinement [yield strengths of up to 200 ksi (1378 MPa)] to improve the curvature ductility of prestressed piling.

For wire size conversion see Table 29.9.

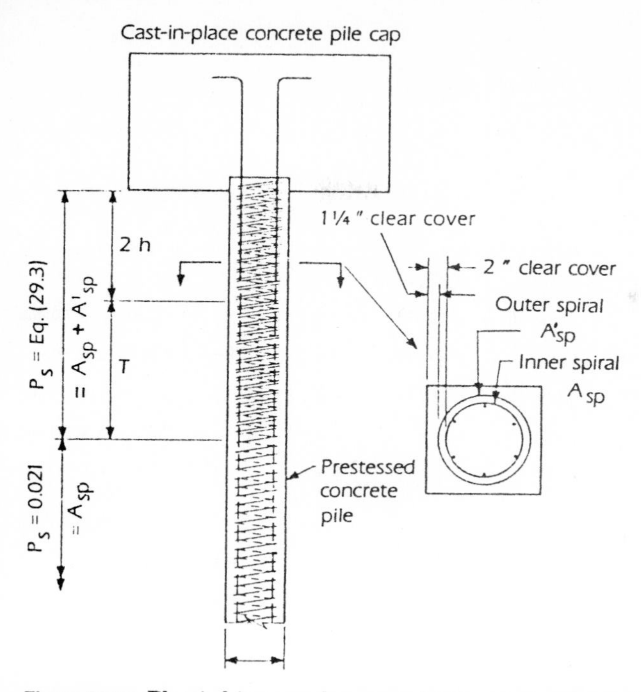


**Figure 29.22** Plastic-hinge confinement detail: overlapping spiral.

**TABLE 29.1 Plastic-Hinge Confinement: Overlapping Spiral for Various Prestressed Piling Sections (U. S. Units)**

| Pile size, $h$, in | $d_{sp}$, in | $D_{sp}$, in | Inner spiral, $\rho_s = 0.021$, $A_{sp}$, in$^2$ | $A_g$, in$^2$ | $A_c$ (typ.), in$^2$ | $\rho_s$, Eq. (29.3) | $A_{sp} + A'_{sp}$, in$^2$ | Outer spiral, $A'_{sp}$ |
|---|---|---|---|---|---|---|---|---|
| 12 square | 8 | 9 | 0.085 | 142 | 57 | 0.048 | 0.22 | W14 @ 2 in |
| 14 square | 10 | 11 | 0.11 | 194 | 87 | 0.039 | 0.22 | W11 @ 2 in |
| 16 square | 12 | 13 | 0.13 | 254 | 123 | 0.034 | 0.22 | W9 @ 2 in |
| 18 square | 14 | 15 | 0.15 | 322 | 165 | 0.030 | 0.23 | W8 @ 2 in |
| 20 square | 16 | 17 | 0.17 | 398 | 214 | 0.028 | 0.24 | W7 @ 2 in |
| 24 square | 20 | 21 | 0.22 | 574 | 330 | 0.024 | 0.25 | W4 @ 2 in |
| 15 octag. | 11 | 12 | 0.12 | 186 | 104 | 0.025 | 0.15 | W3.5 @ 2 in |
| 18 octag. | 14 | 15 | 0.15 | 268 | 165 | 0.020 | 0.15 | |
| 24 octag. | 20 | 21 | 0.22 | 473 | 330 | 0.014 | 0.15 | |

$f'_e = 6$ ksi; $f_{ysp} = 70$ ksi (ASTM $A$ - 82); $P_e = 0.2 f'_c A_g$.

NOTE: The current ACI code limitation on the value of $f_{ysp}$ to 60 ksi (413 MPa) is ignored in these tables.

**TABLE 29.2. Plastic-Hinge Confinement Overlapping Spiral for Various Prestressed Concrete Piling Sections (S.I. Units)**

| Pile size, $h$, mm | $d_{sp}$, mm | $D_{sp}$, mm | Inner spiral, $\rho_s = 0.021$, $A_{sp}$, $mm^2$ | $A_g$, (typ.), $mm^2$ | $A_c$, (typ.), $mm^2$ | $\rho_s$, Eq. (29.3) | $A_{sp} + A'_{sp}$, $mm^2$ | Outer spiral, $A'_{sp}$ |
|---|---|---|---|---|---|---|---|---|
| 305 square | 203 | 229 | 54.8 | $0.92 \times 10^5$ | $0.37 \times 10^5$ | 0.048 | 141.9 | 10.7 mm $\phi$ @ 50 mm |
| 356 square | 254 | 279 | 71.0 | $1.25 \times 10^5$ | $0.56 \times 10^5$ | 0.039 | 141.9 | 9.5 mm $\phi$ @ 50 mm |
| 406 square | 305 | 330 | 83.9 | $1.64 \times 10^5$ | $0.79 \times 10^5$ | 0.034 | 141.9 | 8.6 mm $\phi$ @ 50 mm |
| 457 square | 356 | 381 | 96.8 | $2.08 \times 10^5$ | $1.06 \times 10^5$ | 0.030 | 148.4 | 8.1 mm $\phi$ @ 50 mm |
| 508 square | 406 | 432 | 109.7 | $2.57 \times 10^5$ | $1.38 \times 10^5$ | 0.028 | 154.8 | 7.6 mm $\phi$ @ 50 mm |
| 610 square | 508 | 533 | 141.9 | $3.70 \times 10^5$ | $2.13 \times 10^5$ | 0.024 | 161.3 | 5.7 mm $\phi$ @ 50 mm |
| 380 octag. | 280 | 305 | 77.4 | $1.20 \times 10^5$ | $0.67 \times 10^5$ | 0.025 | 96.8 | 5.4 mm $\phi$ @ 50 mm |
| 457 octag. | 356 | 381 | 96.8 | $1.73 \times 10^5$ | $1.06 \times 10^5$ | 0.020 | 96.8 | |
| 610 octag. | 508 | 533 | 141.9 | $3.05 \times 10^5$ | $2.13 \times 10^5$ | 0.014 | 96.8 | |

$f'_c = 41.3$ MPa; $f_{ysp} = 482$ MPA; $P_c = 0.2 f'_c A_g$.

**TABLE 29.3 ASTM A-648 Hard-Drawn Steel Wire for Prestressed Concrete Pipe (Class III; U.S. Units)**

| Size, diameter | Ultimate tensile strength, ksi | Approximate yield strength, $f_{ysp}$, ksi |
|---|---|---|
| 8-ga/0.162-in W2 | 262 | 198 |
| 6-ga/0.192-in W2-9 | 252 | 191 |
| 1/4-in 0.250-in W5 | 240 | 182 |
| 5/16-in/0.312-in W7.5 | 221 | 167 |

**TABLE 29.4 ASTM A-648 Hard-Drawn Steel Wire for Prestressed Concrete Pipe (Class III; S.I. Units)**

| Size, diameter, mm | Ultimate-tensile strength, MPa | Approximate yield strength, $f_{ysp}$, MPa |
|---|---|---|
| 4.1 | 1810 | 1364 |
| 4.9 | 1740 | 1316 |
| 6.4 | 1650 | 1253 |
| 7.9 | 1520 | 1151 |

**TABLE 29.5 Ductile Confinement Chart for ASTM A-648 Steel Wire (U.S. Units)**

| Pile size, $h$, mm | $d_{sp}$, in | $A_g$, in$^2$ | $A_c$, in$^2$ | W 7.5 wire: $\rho_s$, Eq. (29.3) | W 7.5 wire: $S_{sp}$, in | W 5 wire: $\rho_s$, Eq. (29.3) | W 5 wire: $S_{sp}$ (typ.), in |
|---|---|---|---|---|---|---|---|
| 12 square | 8 | 142 | 57 | 0.020 | 1 7/8 | 0.018 | 1 3/8 |
| 14 square | 10 | 194 | 87 | 0.017 | 1 3/4 | 0.015 | 1 5/16 |
| 16 square | 12 | 254 | 123 | 0.014 | 1 3/4 | 0.013 | 1 1/4 |
| 18 square | 14 | 322 | 165 | 0.013 | 1 5/8 | 0.012 | 1 3/16 |
| 20 square | 16 | 398 | 214 | 0.012 | 1 1/2 | 0.011 | 1 1/8 |
| 24 square | 20 | 574 | 330 | 0.010 | 1 1/2 | 0.009 | 1 1/8 |
| 15 octag | 11 | 186 | 104 | 0.011 | 2 1/2 | 0.010 | 1 3/4 |
| 18 octag | 14 | 268 | 165 | 0.008 | 2 1/2 | 0.008 | 1 3/4 |
| 24 octag | 20 | 473 | 330 | 0.006 | 2 1/2 | 0.005 | 2 |

$f'_c = 6$ ksi. $\quad {}^*\rho_s = 0.45 \dfrac{f'_c}{f_{ysp}} \left( \dfrac{A_g}{A_c} - 1 \right) \left( 0.5 + 1.25 \dfrac{P_c}{\phi f'_c A_g} \right).$

2-in cover. $\quad p_e = 0.2 f'_c A_g$.

**TABLE 29.6 Ductile Confinement Chart for ASTM A-648 Steel Wire (S.I. Units)**

| Pile size, $h$, mm | $d_{sp}$, mm | $A_g$, mm$^2$ | $A_c$, mm$^2$ | 7.9-mm wire: $\rho_s$, Eq. (29.3) | 7.9-mm wire: $S_{sp}$, mm | 6.4-mm wire: $\rho_s$, Eq. (29.3) | 6.4-mm wire: $S_{sp}$ (typ.), mm |
|---|---|---|---|---|---|---|---|
| 305 square | 203 | $0.92 \times 10^5$ | $0.37 \times 10^5$ | 0.020 | 48 | 0.018 | 35 |
| 356 square | 254 | $1.25 \times 10^5$ | $0.56 \times 10^5$ | 0.017 | 44 | 0.015 | 33 |
| 406 square | 305 | $1.64 \times 10^5$ | $0.79 \times 10^5$ | 0.014 | 44 | 0.013 | 32 |
| 457 square | 356 | $2.08 \times 10^5$ | $1.06 \times 10^5$ | 0.013 | 41 | 0.012 | 30 |
| 508 square | 406 | $2.57 \times 10^5$ | $1.38 \times 10^5$ | 0.012 | 38 | 0.011 | 29 |
| 610 square | 508 | $3.70 \times 10^5$ | $2.13 \times 10^5$ | 0.010 | 38 | 0.009 | 29 |
| 380 octag | 280 | $1.20 \times 10^5$ | $0.67 \times 10^5$ | 0.011 | 64 | 0.010 | 44 |
| 457 octag | 356 | $1.73 \times 10^5$ | $1.06 \times 10^5$ | 0.008 | 64 | 0.008 | 44 |
| 610 octag | 508 | $3.05 \times 10^5$ | $2.13 \times 10^5$ | 0.006 | 64 | 0.005 | 51 |

$f'_c = 41.3$ MPa. $\quad {}^*\rho_s = 0.45 \dfrac{f'_c}{f_{ysp}} \left( \dfrac{A_g}{A_c} - 1 \right) \left( 0.5 + 1.25 \dfrac{P_e}{\phi f'_c A_g} \right).$

50-mm cover. $\quad P_e = 0.2 f'_c A_g$.

NOTE: The current ACI code limitation on the value of $f_{ysp}$ to 60 ksi (413 MPa) is ignored in these tables.

**TABLE 29.7 Spiral Confinement for Various Prestressed Concrete Piling Sections Using ASTM A-82 Wire (U.S. Units)**

| Spiral arrangement for various values of $\rho_s$ | | | | | Cast-in-place concrete pile buildup | | | | | |
|---|---|---|---|---|---|---|---|---|---|---|
| Pile size, $h$, in | $d_{sp}$, in | $\rho_s = 0.003$ | $\rho_s = 0.014$ | $\rho_s = 0.021$ | $H = h + 6$ in | $A_g$, in | $A_c$, in | $D_{sp}$, in | $\rho_s$* | $A_{sp}$ |
| 12 square | 8 | W3.5 @ 6 in | W6 @ 2 in | W8.5 @ 2 in | 18 ϕ | 254 | 113 | 12 | 0.020 | W12 @ 2 in |
| 14 square | 10 | W3.5 @ 4 ½ in | W7 @ 2 in | W11 @ 2 in | 20 ϕ | 314 | 154 | 14 | 0.017 | W12 @ 2 in |
| 16 square | 12 | W3.5 @ 3 ½ in | W8.5 @ 2 in | W13 @ 2 in | 22 ϕ | 380 | 201 | 16 | 0.014 | W12 @ 2 in |
| 18 square | 14 | W4 @ 3 ½ in | W10 @ 2 in | W15 @ 2 in | 24 ϕ | 452 | 254 | 18 | 0.013 | W12 @ 2 in |
| 20 square | 16 | W5 @ 4 in | W11 @ 2 in | W17 @ 2 in | 26 ϕ | 531 | 314 | 20 | 0.011 | W11 @ 2 in |
| 24 square | 20 | W5 @ 3 in | W14 @ 2 in | W22 @ 2 in | 30 ϕ | 707 | 452 | 24 | 0.009 | W11 @ 2 in |
| 15 octag. | 11 | W3.5 @ 4 in | W8 @ 2 in | W12 @ 2 in | 21 ϕ | 346 | 177 | 15 | 0.015 | W12 @ 2 in |
| 18 octag. | 14 | W4 @ 3 ½ in | W10 @ 2 in | W15 @ 2 in | 24 ϕ | 452 | 254 | 18 | 0.013 | W12 @ 2 in |
| 24 octag. | 20 | W5 @ 3 in | W 14 @ 2 in | W 22 @ 2 in | 30 ϕ | 707 | 452 | 24 | 0.009 | W11 @ 2 in |

$f'_c = 6$ ksi. $f_{ysp} = 70$ ksi (ASTM A-82).

2-in cover.

NOTE: The current ACI code limitation on the value of $f_{ysp}$ to 60 ksi (413.4 MPa) is ignored in these tables.

$$*\rho_s = 0.45 \frac{f'_c}{f_{ysp}} \left( \frac{A_g}{A_c} - 1 \right) \left( 0.5 + 1.25 \frac{P_e}{\phi f'_c A_g} \right).$$

$f'_c = 3$ ksi.

$f_{ysp} = 70$ ksi (ASTM A-82).

$P_e = 0.2 f'_c A_g$.

3-in cover.

**TABLE 29.8 Spiral Confinement for Various Prestressed Concrete Piling Sections Using ASTM A-82 Wire (U.S. Units)**

| Spiral arrangement for various values of $\rho_s$ | | | | | Cast-in-place concrete pile buildup | | | | | |
|---|---|---|---|---|---|---|---|---|---|---|
| Pile size, $h$, mm | $d_{sp}$, mm | $\rho_s = 0.003$ | $\rho_s = 0.014$ | $\rho_s = 0.021$ | $H = h + 6$, mm | $A_g$ (typ.), mm | $A_c$, mm | $D_{sp}$, mm | $\rho_s$* | $A_{sp}$ |
| 305 square | 203 | 5.4 mm @ 150 | 7.0 mm ϕ @ 50 | 8.3 mm ϕ @ 50 | 457 ϕ | $1.64 \times 10^5$ | $0.73 \times 10^5$ | 305 | 0.020 | 9.9 mm @ 50 mm |
| 356 square | 254 | 5.4 mm @ 114 | 7.6 mm ϕ @ 50 | 9.5 mm ϕ @ 50 | 508 ϕ | $2.03 \times 10^5$ | $0.99 \times 10^5$ | 356 | 0.017 | 9.9 mm @ 50 mm |
| 406 square | 305 | 5.4 mm @ 90 | 8.3 mm ϕ @ 50 | 10.3 mm ϕ @ 50 | 560 ϕ | $2.45 \times 10^5$ | $1.30 \times 10^5$ | 406 | 0.014 | 9.9 mm @ 50 mm |
| 457 square | 356 | 5.7 mm @ 90 | 9.0 mm ϕ @ 50 | 11.1 mm ϕ @ 50 | 610 ϕ | $2.92 \times 10^5$ | $1.64 \times 10^5$ | 457 | 0.013 | 9.9 mm @ 50 mm |
| 508 square | 406 | 6.4 mm @ 100 | 9.5 mm ϕ @ 50 | 11.8 mm ϕ @ 50 | 660 ϕ | $3.43 \times 10^5$ | $2.03 \times 10^5$ | 508 | 0.011 | 9.5 mm @ 50 mm |
| 610 square | 508 | 6.4 mm @ 75 | 10.7 mm ϕ @ 60 | 13.4 mm ϕ @ 50 | 762 ϕ | $4.56 \times 10^5$ | $2.92 \times 10^5$ | 610 | 0.009 | 9.5 mm @ 50 mm |
| 380 octag. | 280 | 5.4 mm @ 100 | 8.1 mm ϕ @ 50 | 9.9 mm ϕ @ 50 | 533 ϕ | $2.23 \times 10^5$ | $1.14 \times 10^5$ | 380 | 0.015 | 9.9 mm @ 50 mm |
| 457 octag. | 356 | 5.7 mm @ 90 | 9.0 mm ϕ @ 50 | 11.1 mm ϕ @ 50 | 610 ϕ | $2.93 \times 10^5$ | $1.64 \times 10^5$ | 457 | 0.013 | 9.9 mm @ 50 mm |
| 610 octag. | 508 | 6.4 mm @ 75 | 10.7 mm ϕ @ 50 | 13.4 mm ϕ @ 50 | 762 ϕ | $4.56 \times 10^5$ | $2.92 \times 10^5$ | 610 | 0.009 | 9.5 mm @ 50 mm |

$f'_c = 41.3$ MPa. $f_{ysp} = 482$ MPa (ASTM A-82).

50-mm clear cover.

$$^*\rho_s = 0.45 \frac{f'_c}{f_{ysp}} \left( \frac{A_g}{A_c} - 1 \right) \left( 0.5 + 1.25 \frac{P_e}{\phi f'_c A_g} \right)$$

$\rho_c = 0.2 f'_c A_g$. $f'_c = 20.7$ MPa.

$f_{wsp} = 482$ MPa. 76-mm clear cover.

**TABLE 29.9. Wire Size Conversion Chart**

| Old designation | Diameter, in | Area, $in^2$ | W, wire size | Diameter, in | SI units | |
|---|---|---|---|---|---|---|
| | | | | | Diameter, mm | Area, $mm^2$ |
| 5 | 0.207 | 0.034 | | | | |
| | | 0.035 | W3.5 | 0.211 | 5.4 | 22.6 |
| 4 | 0.225 | 0.040 | W4 | 0.225 | 5.7 | 25.8 |
| | | 0.045 | W4.5 | 0.240 | 6.1 | 29.0 |
| 3 | 0.244 | 0.047 | | | | |
| | | 0.050 | W5 | 0.252 | 6.4 | 32.3 |
| 2 1/2 | 0.253 | 0.050 | | | | |
| | | 0.055 | W5.5 | 0.264 | 6.7 | 35.5 |
| 2 | 0.265 | 0.055 | | | | |
| 1 1/2 | 0.273 | 0.059 | | | | |
| | | 0.060 | W6 | 0.276 | 7.0 | 38.7 |
| 1 | 0.283 | 0.063 | | | | |
| | | 0.065 | W6.5 | 0.287 | 7.3 | 41.9 |
| 0 1/2 | 0.295 | 0.068 | | | | |
| | | 0.070 | W7 | 0.298 | 7.6 | 45.2 |
| 0 | 0.307 | 0.074 | | | | |
| | | 0.075 | W7.5 | 0.309 | 7.8 | 48.4 |
| | | 0.080 | W8 | 0.319 | 8.1 | 51.6 |
| | | 0.085 | W8.5 | 0.328 | 8.3 | 54.8 |
| 2/0 | 0.331 | 0.086 | | | | |
| | | 0.090 | W9 | 0.338 | 8.6 | 58.1 |
| | | 0.095 | W9.5 | 0.347 | 8.8 | 61.3 |
| | | 0.100 | W10 | 0.356 | 9.0 | 64.5 |
| 3/0 | 0.363 | 0.103 | | | | |
| | | 0.110 | W11 | 0.374 | 9.5 | 71.0 |
| | | 0.120 | W12 | 0.390 | 9.9 | 77.4 |
| 4/0 | 0.394 | 0.122 | | | | |
| | | 0.130 | W13 | 0.406 | 10.3 | 83.9 |
| | | 0.140 | W14 | 0.422 | 10.7 | 90.3 |
| 5/0 | 0.431 | 0.146 | | | | |
| | | 0.150 | W15 | 0.437 | 11.1 | 96.8 |
| | | 0.160 | W16 | 0.451 | 11.5 | 103.2 |
| 6/0 | 0.462 | 0.168 | | | | |
| | | 0.170 | W17 | 0.465 | 11.8 | 109.7 |
| | | 0.180 | W18 | 0.478 | 12.1 | 116.1 |
| 7/0 | 0.490 | 0.189 | | | | |
| | | 0.190 | W19 | 0.491 | 12.5 | 122.6 |
| | | 0.200 | W20 | 0.504 | 12.8 | 129.0 |
| | | 0.220 | W22 | 0.529 | 13.4 | 141.9 |
| | | 0.240 | W24 | 0.553 | 14.0 | 154.8 |
| | | 0.260 | W26 | 0.575 | 14.6 | 167.8 |
| | | 0.280 | W28 | 0.597 | 15.2 | 180.7 |
| | | 0.310 | W31 | 0.628 | 16.0 | 200.0 |

Further details are presented in Figs. 29.23 through 29.28.

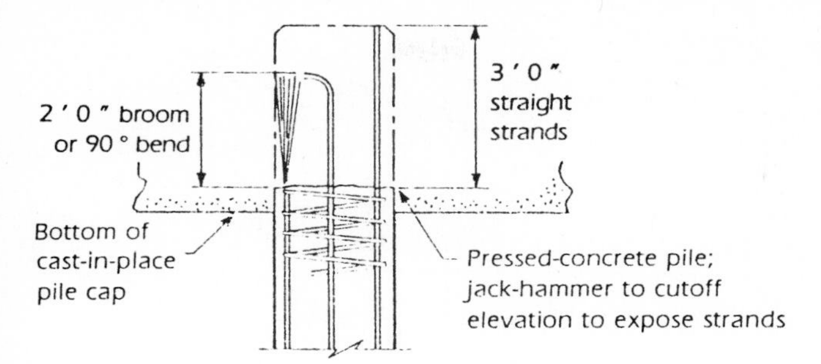


**Figure 29.23** Prestressed concrete pile showing fixed pile–pile cap connection using exposed strands with a diameter of 1/2 in (12.7 mm) or less.

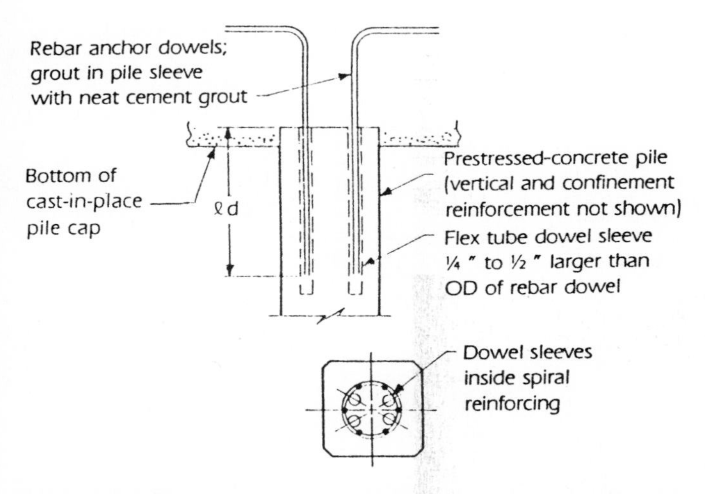


**Figure 29.24** Fixed pile–pile cap connection using mild steel dowels.

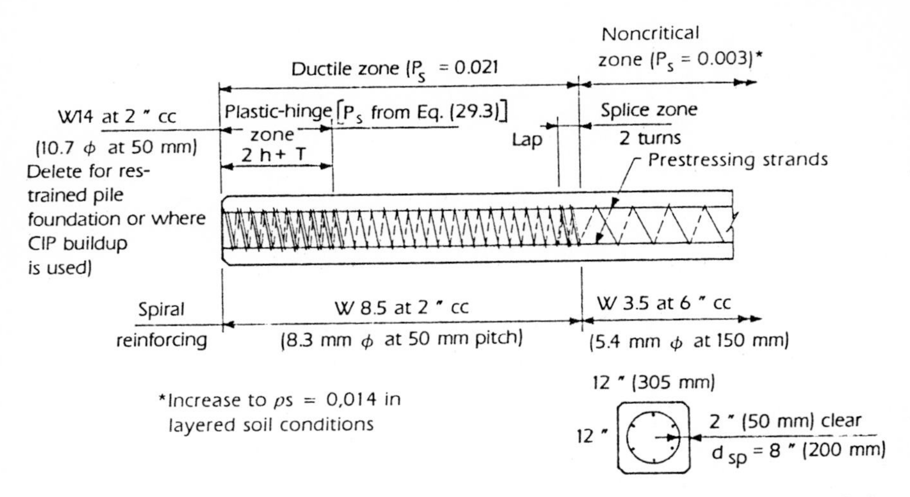


**Figure 29.25** Seismic-resistant pile 12 in (305 mm) square (Uniform Building Code, seismic Zone 4).

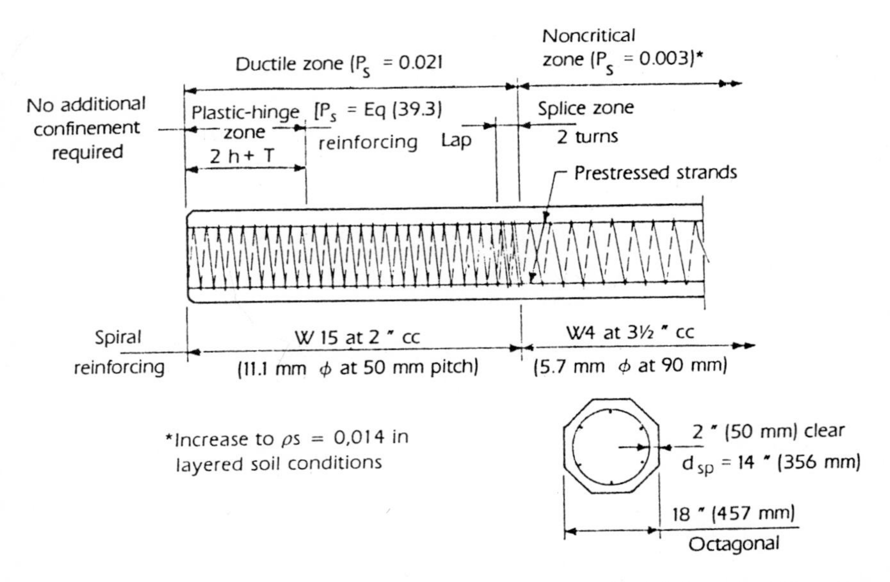


**Figure 29.26** Octagonal 18-in (457-mm) seismic-resistant pile (Uniform Building Code, seismic Zone 4).

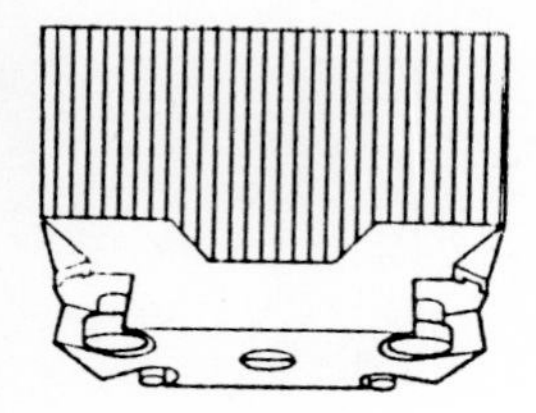

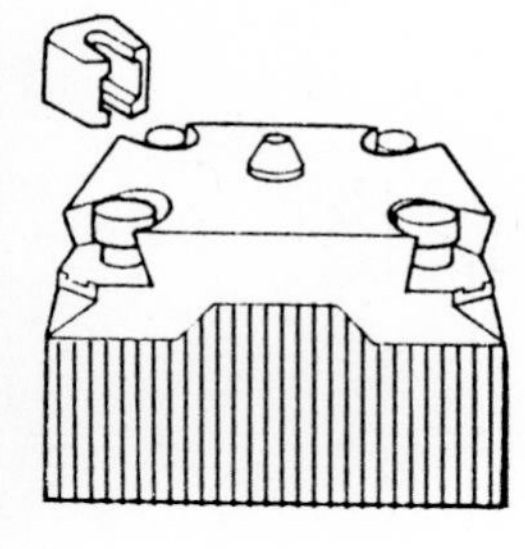

**Figure 29.27** Dyn-A-Splice mechanical splice.

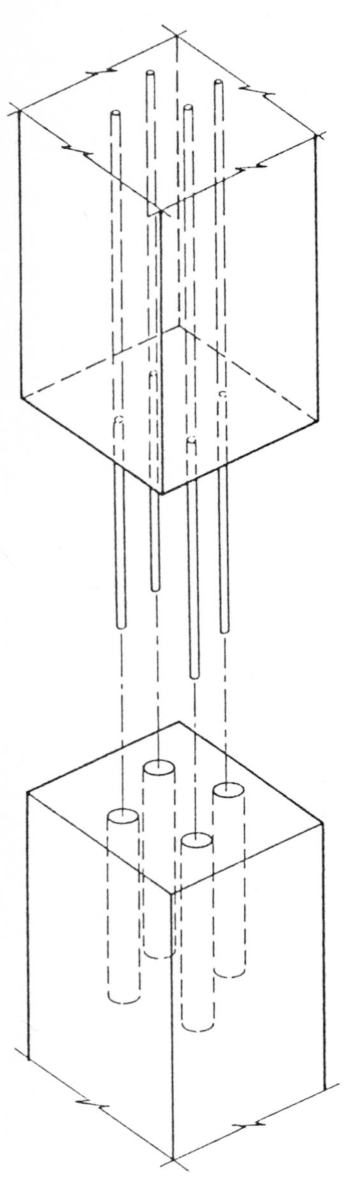

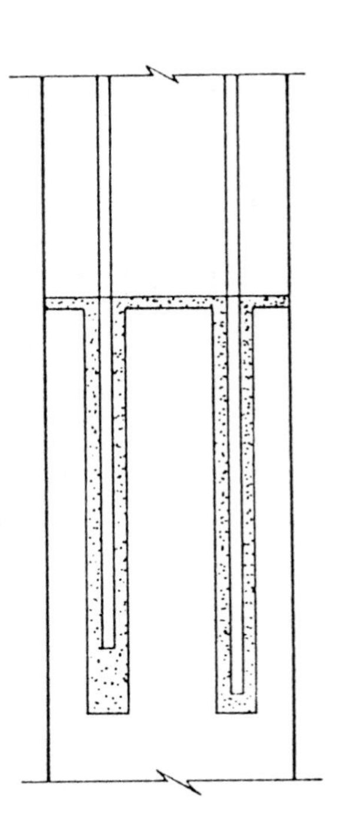

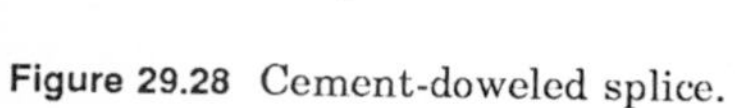

**Figure 29.28** Cement-doweled splice.

**Figure 30.1** Oyster Point Marina breakwater, South San Francisco, California. Prestressed concrete sheet pile units 60 ft long are being slid into place in soft bay mud. The 14-in-thick units were shipped by barge from the fabricator's yard. Installation equipment is also barge-mounted.

Chapter

# 30

# Prestressed Concrete Sheet Piling

Prestressed concrete sheet piling units are used to retain earth or other materials, as in bulkhead walls for marine applications, or in special instances, as in subterranean building foundation walls. They are also installed as breakwaters to protect harbors and marinas. In applications where they support vertical loads of any appreciable magnitude, they must be tied back or braced at sufficient intervals so that the bending stresses do not limit the axial load-carrying capacity of the pile. Interaction diagrams for prestressed concrete sheet piling are constructed in a similar manner as for prestressed concrete bearing piles or columns. Various cross sections have been used for sheet pile applications, but we shall discuss here only the rectangular shape as a standard product. The sections are normally made with a tongue-and-groove interlock to facilitate installation. Groove-and-groove configurations, where the area between the grooves is grouted above the waterline after installation, also are fabricated. The area below the waterline usually transitions to the tongue-and-groove detail on one side, once again for installation reasons. For more detailed practical information on sheet piles, see the excellent text by Prof. Ben C. Gerwick, Jr., *Construction of Prestressed Concrete Structures* (Wiley, New York, 1971).

**Figure 30.2** Huntington Harbor bulkhead wall, Huntington Beach, California. Prestressed concrete sheet pile units 40 ft long are stockpiled in the plant awaiting delivery.

## Availability

Solid rectangular units are fabricated in depths to 12 in. Depths greater than 12 in are normally voided to reduce weight in handling. Widths are selected subject to a weight limitation per pile of 12 tons unless a thorough investigation of on-site handling and installation limitations is performed. This limit, although admittedly somewhat arbitrary, assures that shipping will be optimized, in that two pieces may be handled per load. Prestressed concrete sheet piles are fabricated by all manufacturers of prestressed concrete products. Some manufacturers have universal prestressing lines with flat soffits specifically devoted to the production of rectangular sheet pile units. Others may temporarily convert their double-T or single-T beds into flat soffits by filling the stems with concrete. (See Fig. 30.3.)

## Installation Considerations

Sheet piling installation in the water usually employs a temporary wood or steel wale alignment system. Either wood or steel wales are supported on one end by the section of wall already driven and on the open end by a temporary steel or concrete pile driven on line and subsequently removed and leapfrogged ahead as wall installation proceeds. The wales are connected to the wall by large C clamps or with bolts through holes provided in the concrete units for this use and also to hold up steel forming brackets used to construct the cast-in-place concrete coping.

**Figure 30.3** Oyster Point Marina breakwater, South San Francisco, California. Prestressing bed with the previous day's pour awaiting quality control inspection and placement into storage. Note the slight corner bevel to facilitate installation. Note also the movable side rails with a keyway shape.

Driving or jetting proceeds with the tongue edge proceeding first in the direction of the driving. The bottoms of the piles are sloped so that as the pile is driven, it is forced into the portion of the wall already installed, assuring plumbness. Without this precaution, the units will tend to "walk," or experience a tendency for the bottom of the unit to splay out. The amount of slope depends upon the soil and method of installation. For units being jetted in sand, a full 45° bevel is usually provided. For units driven in silt or clay, only a slight corner bevel is required to keep the bottom edge in against the previously driven section. If hard driving is encountered or if the installation is in soft mud, too great a bevel could cause excessive shear key friction and impede driving. Given the manner in which the groove and tongue are related to the direction of driving and the design of the beveled bottom, the installation sequence of the wall must be known beforehand for the units to be properly designed. (See Figs. 30.4 and 30.5.)

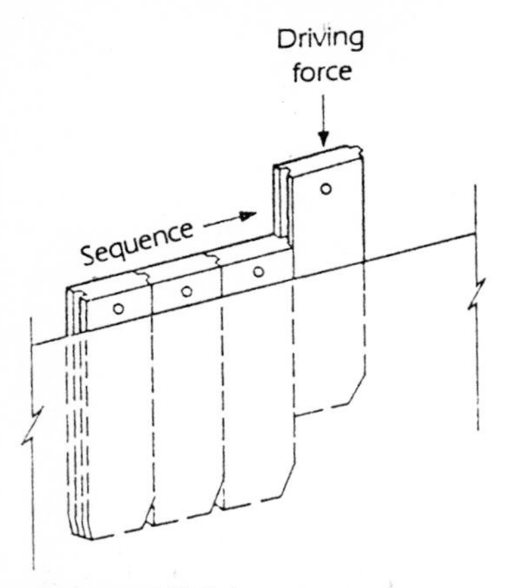


Figure 30.4 Driving sequence.

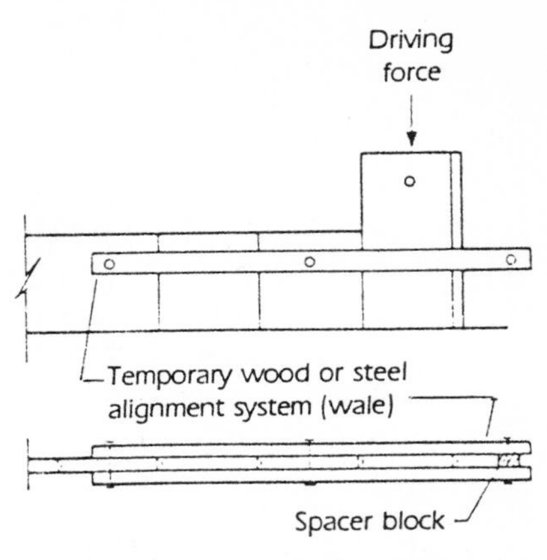


Figure 30.5 Detail of driving sequence.

Jetting consists of using water pumped at high pressure through holes cast in the pile units to install the piles in granular soils. It involves special techniques that are essential for efficient production in the field. The ducts cast in the sheet piles should be of sufficient size and number to accommodate an adequate volume of water in the pile. A minimum diameter of 2 1/2 in is recommended, with 4 in preferred; 6 to 8 percent of the gross pile cross-sectional area should be provided as jet tube opening. The bottom of the tube should be constricted to increase water velocity at discharge from the bottom of the pile, thereby increasing the movement of sand from the driving zone. Sufficient two- or three-stage pumps are required to develop the head and discharge essential for good efficient installation. Avoid the use of fragile polyvinyl chloride couplers at the top of the pile tube for providing a connection to the pump hoses. Sometimes a gang of pipes is connected to a manifold and placed over the pile to aid in jetting (external jetting). Sometimes in silty sands the problem is not in getting the pile down but rather in keeping the pile from sinking too far before sufficient backflow occurs around the driven unit to develop side friction to maintain position. A cast-in-place concrete coping is usually poured to serve as an anchorage beam for tieback systems and to dress up any vertical irregularities of as-driven pile elevations. As mentioned above, a 1 1/2 - in diameter hole placed 15 in down from the top of the pile serves to hold coping form brackets in position. The brackets will have a long vertical slot in the supporting

**Figure 30.6** Huntington Harbor bulkhead wall, Huntington Beach, California. A 12-ton sheet pile unit being positioned for jetting. An 85-ton crane on shore was used to hold the sheet piles during installation. A small barge-mounted crane was used to hold the auxiliary jetting equipment. The main jet pump was connected to hoses feeding water directly to pipes cast in the pile.

angle to adjust to final installed pile elevations. Wall installation should be within an allowable alignment tolerance of ±2 in. (See Fig. 30.6.)

## Sheet Pile Walls

An aspect of construction of sheet pile walls is the difficulty in maintaining exact horizontal control. For this reason, if the wall has changes in alignment or closes upon itself, it is advisable to make up the corner units later on the basis of field measurements. The disadvantage of this procedure is that additional plant and field setup costs are involved and that time is lost in waiting for these special units. An alternative is forming and pouring in place the corner or closure piece required for the wall at these points.

## Typical Sheet Piling Conditions

Some typical conditions peculiar to sheet piling are shown in Figs. 30.7, 30.8, and 30.9.

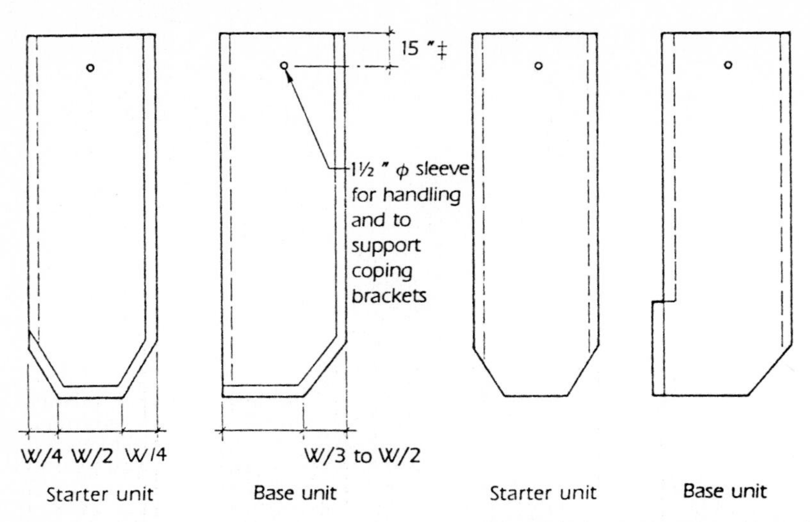


**Figure 30.7** Tongue and groove.

**Figure 30.8** Groove-and-groove combination.

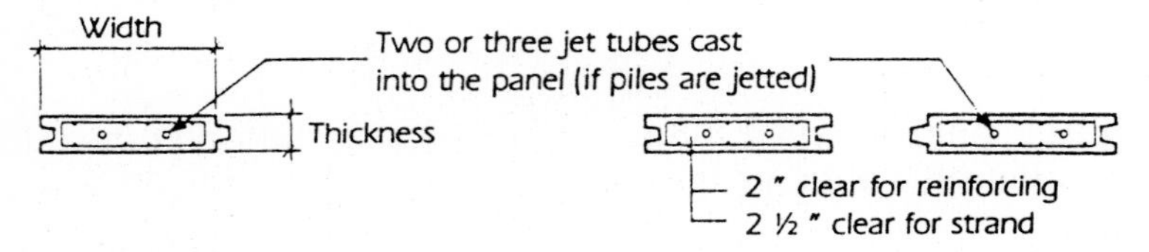


**Figure 30.9** Sheet pile cross sections.

## Coping

A cast-in-place concrete coping is usually formed after the wall installation is complete. This coping dresses up the top of the wall and hides minor variations in alignment. It also serves as a carrying beam to distribute tieback loads from wall active pressure. See Fig. 30.10.

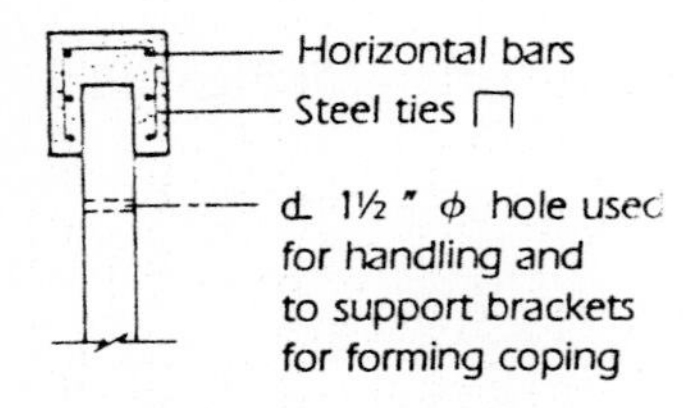


**Figure 30.10** Concrete cap.

### Typical installations

Typical installations are shown in Figs. 30.11 through 30.16.

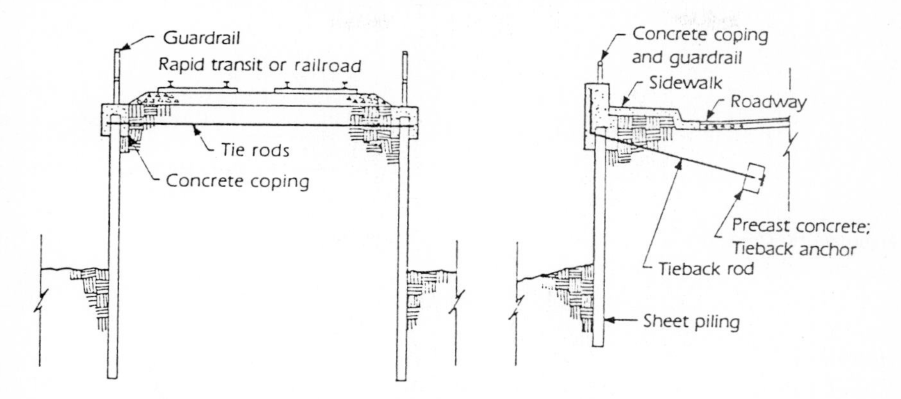


**Figure 30.11** Raised rapid transit or highway bed.

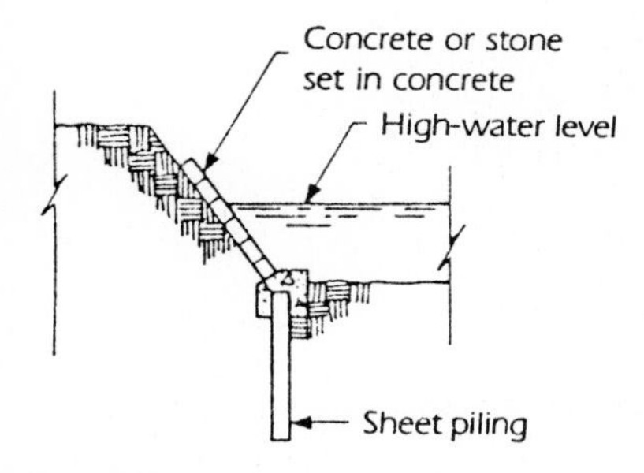


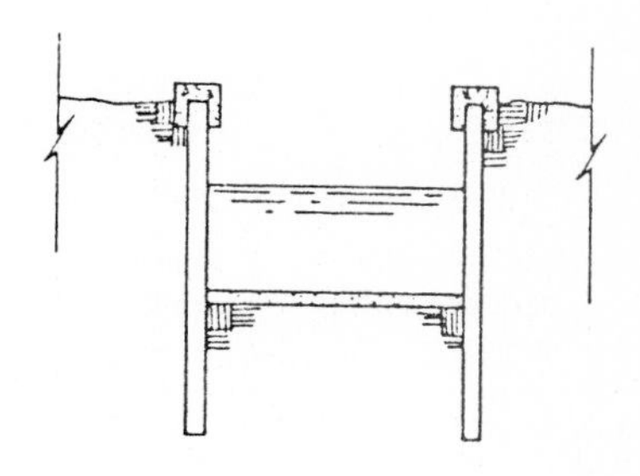

**Figure 30.12** Canal.

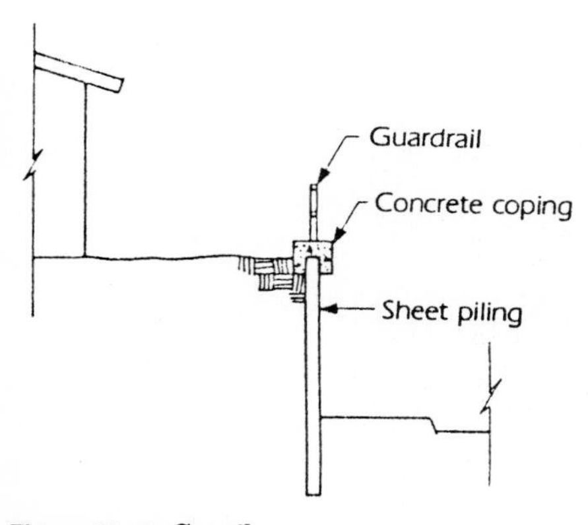


**Figure 30.13** Cantilever.

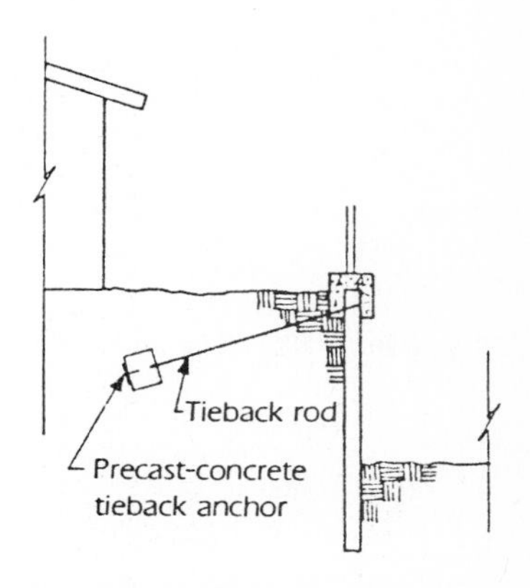


**Figure 30.14** Tieback system.

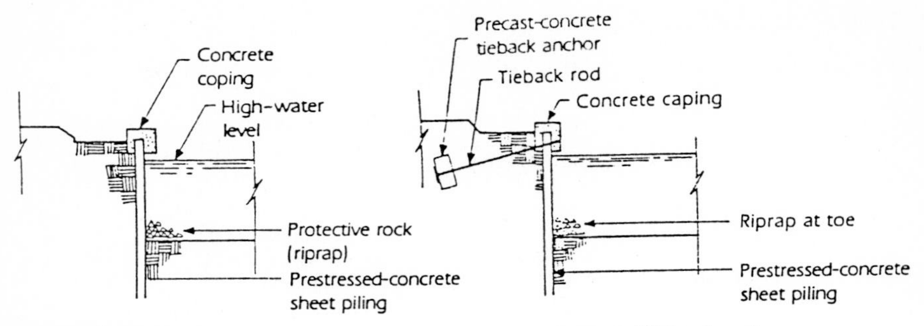


**Figure 30.15** Cantilever.

**Figure 30.16** Tieback rods.

### Handling

Sheet piling units are handled by lifting at cast-in strand lift loops or manufactured inserts (see Fig. 30.17). Typical handling conditions are shown in Figs. 30.18, 30.19, and 30.20.

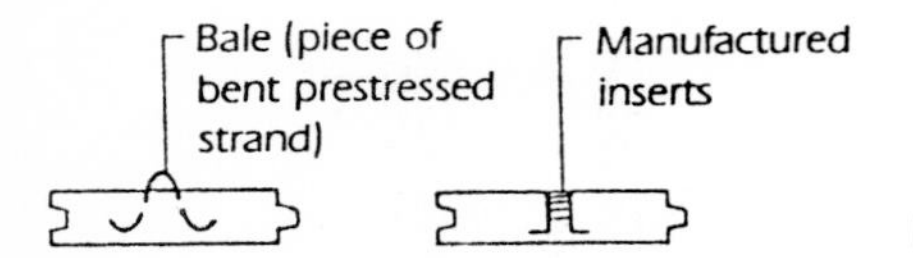


**Figure 30.17** Handling embedments.

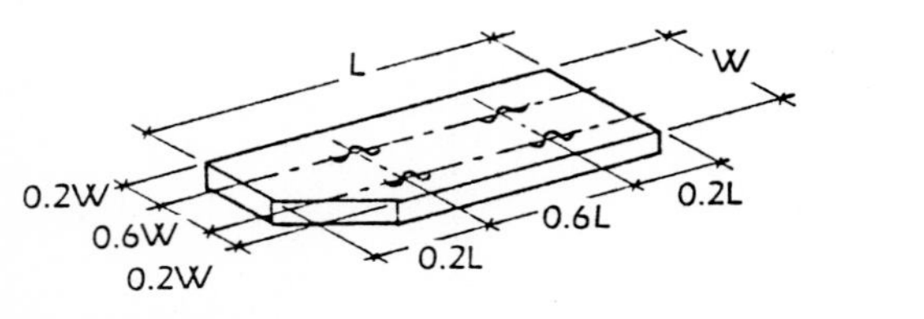


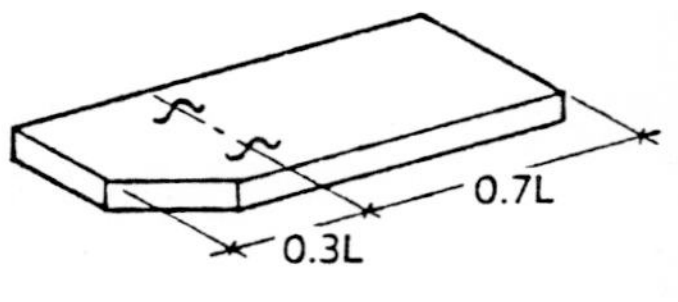


**Figure 30.18** Stripping and plant handling (equal positive and negative mements).

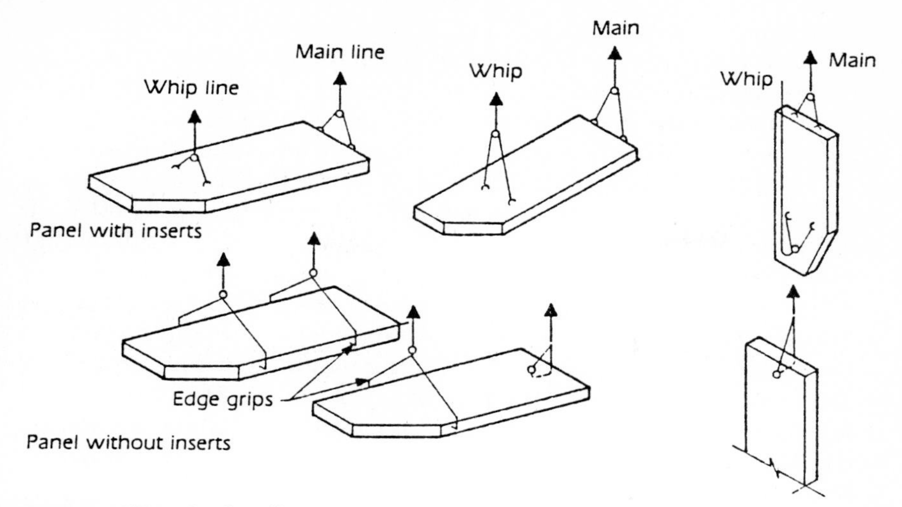


**Figure 30.19** Erection handling.

**Figure 30.20** Huntington Harbor bulkhead wall, Huntington Beach, California. A 12-ton sheet pile unit being tripped up into the vertical position from a flatbed trailer. Note the hoses attached to jet pipes cast in pile, which will subsequently be hooked up to jet pumps on an offshore barge.

## Design Considerations

A thorough knowledge of the soil conditions at the site is essential for sheet piles to be correctly designed. Not only should the soils report indicate the active and passive soil pressures to be used in the design, but it should also indicate the method of installation dictated by the soil present. If the soil is fine sand with a minimum of clay or silt, jetting is usually permitted. In addition, the soils report will give the soil pressure diagram to be used in the design. If the piles are to be driven, then a minimum concentric prestress of 700 lb/in$^2$ is indicated, and spiral or closed-loop ties should be used for lateral reinforcing, with the spacing being very close at the ends, as for prestressed concrete bearing piles. If, however, jetting is indicated, then the minimum prestress will be that required to satisfy bending and handling, and a certain amount of eccentricity is permitted, usually limited by that value producing an instantaneous camber of 1 in. The strands will be placed in the bottom of the member as cast, thereby facilitating handling and also resulting in the smooth soffit side of the member being adjacent to the backfill material behind the wall after installation, which minimizes soil friction and consequent buildup of active soil pressure. The *PCI Design Handbook* on page 2-56 has preliminary design charts to assist the designer in selecting a pile thickness from a known value of service load moment. The balance of the sheet pile design proceeds in a similar manner as for bearing piles in satisfying elastic design criteria for horizontal loading due to handling and vertical loading during installation and in satisfying both elastic and ultimate design requirements for both bending and axial forces resulting from the in situ condition. The following design example gives a practical demonstration of the principles discussed here.

### Prestressed concrete sheet pile design

Design a cantilever prestressed concrete sheet pile bulkhead wall along a tidal flood control channel. The soils engineer has recommended active and passive soil pressures and has furnished a soil pressure diagram to be used in the design. The native soil is fine sand; uniformly graded granular backfill will used behind the wall. The highest water table behind the wall is at an elevation of 0 ft 0 in. The wall is also to be designed to resist the lateral load effect of a 200 lb/ft surcharge acting at 10 ft from the outside face of the wall, which the soils report indicates is equivalent to a lateral load of 410 lb/ft acting 5 ft down from the top of the wall. See Fig. 30.21.

**Additional design criteria.** The cover over strand is 2½ in; the cover over rebar, 2 in. The allowable concrete tension is $3\sqrt{f'_c}$; $f'_c = 6000$ lb/in$^2$; $f'_{ci} = 3500$ lb/in$^2$. The minimum cement content for durability is seven sacks per cubic yard. The piles will be installed by jetting.

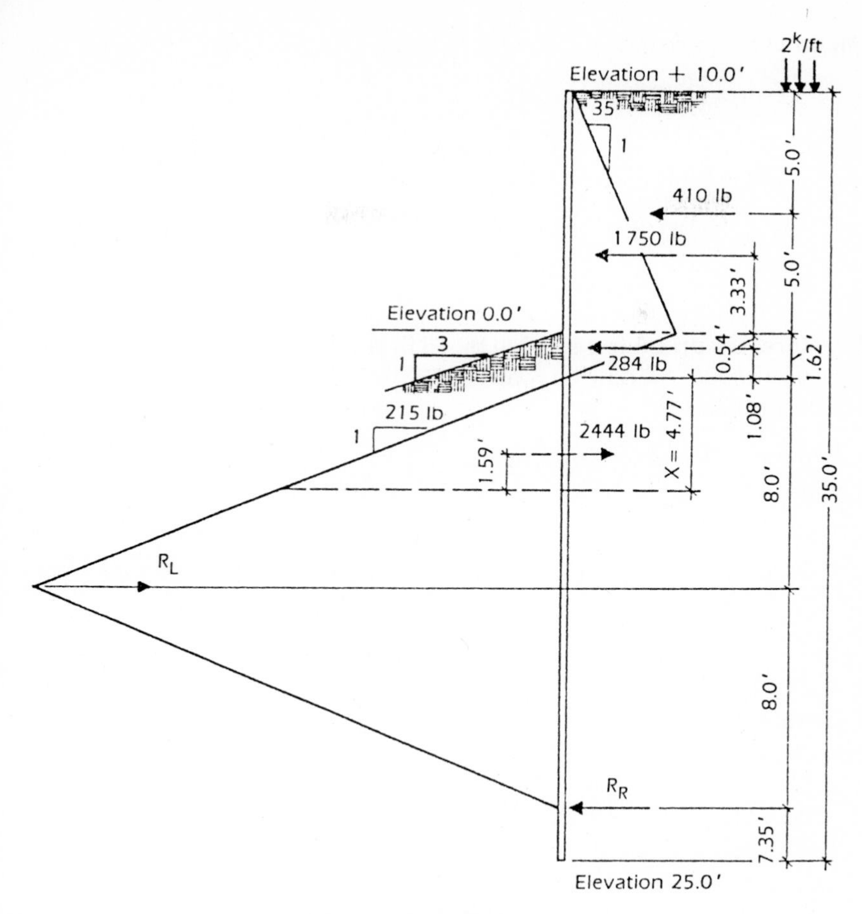


**Figure 30.21** Loading diagram per foot of wall.

**Factor of safety.** Check factor of safety for stability.

$$\Sigma MR_R = 0$$

$$8R_L = (284)(17.08) + (1750)(20.95) + (410)(22.62)$$

$$R_L = 6348 \text{ lb}$$

$$\text{Available } R_L = 215 \times \frac{16^2}{4} = 13{,}760 \text{ lb}$$

$$\text{FS} = \frac{13{,}760}{6348} = 2.17 \text{ (OK)}$$

**Maximum bending moment.** Determine maximum bending moment in pile.

Point of zero shear:

$$410 + 1780 + 284 = 215\frac{x^2}{2} \qquad \therefore x = 4.77\text{ ft}$$

$$M_{max} = (410)(11.39) + (1750)(9.72) + (284)(6.39)$$

$$- (2444)(1.59) = 19{,}609^{\text{lb}\cdot\text{ft}}/\text{ft}$$

Therefore, try unit 5 ft 0 in wide by 10 in thick (11 tons each); see *PCI Design Handbook*, 3d edition, page 2-57.

Section properties:

$A = 600\text{ in}^2 \quad \text{Weight} = 0.63^{k}/\text{ft}$

$I = 5000\text{ in}^4$

$S = 1000\text{ in}^3$

$d = 7.25\text{ in} \quad d' = 2.75\text{ in}$

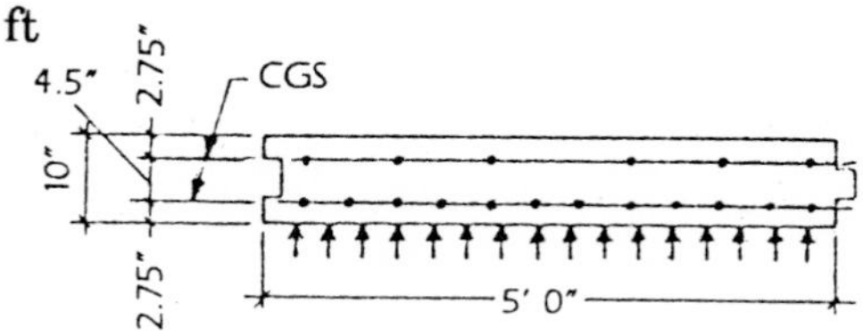

**Prestressing.** Determine the number of strands required to satisfy service loads (assume $e = 0.7$ in).

$$F_f = \frac{\dfrac{M}{S} - f_t}{\dfrac{1}{A} + \dfrac{e}{S}} = \frac{\dfrac{19.61 \times 5 \times 12}{1000} - 3\sqrt{6000}}{\dfrac{1}{600} + \dfrac{0.7}{1000}} = \frac{1.176 - 0.232}{0.00237} = 398^{k}$$

By using $\frac{1}{2}$-in $\phi$-270$^k$ strands, $F_f = 0.78 \times 28.9 = 22.5^{k}/\text{strand}$ (see *PCI Design Handbook*, page 11-16).

$$N = \frac{398}{22.5} = 17.68$$

Therefore, use eighteen $\frac{1}{2}$-in $\phi$-270$^k$ strands.

Try 12-in inside face and 6-in outside face.

$$e = \frac{(12)(7.25) + (6)(2.75)}{18} - 5.0 = 0.75\text{ in} > e \ \text{(assumed) (OK)}$$

**Compression.** Check the final design condition.

$$f_c = \frac{(22.5)(18)}{600} + \frac{(22.5)(18)(0.75)}{1000} - \frac{(19.61)(5)(12)}{1000}$$

$$= -0.675 + 0.303 - 1.176 = 1.548 < 0.45\sqrt{f'_c} \text{ (OK)}$$

**Ultimate strength.** Check ultimate strength (see *PCI Design Handbook*, pages 4-6, 4-63, and 11-15).

$$C\bar{\omega}p = 1.13\frac{(12)(0.1531)(270)}{(60)(7.25)(6)} = 0.215 \quad f_{se} = 147 \text{ ksi}$$

From the chart $f_{ps}/f_{pu} = 0.95$.
Therefore, $f_{ps} = (0.95)(270) = 256.5$ ksi

$$a/2 = \frac{A_{ps}f_{ps}}{1.7f'_c b} = \frac{(1.837)(256.5)}{(1.7)(6)(60)} = 0.77 \text{ in}$$

$$M_u = A_{ps}f_{ps}(d - a/2) = (0.9)(1.837)(256.5)(7.25 - 0.77) = 2748^{k \cdot in}$$

$$UM_H = 1.7 \times 5 \times 19.61 \times 12 = 2000^{k \cdot in} < M_u \text{ (OK)}$$

**Shear.** Check shear.

$$UV = 1.7 \times 5 \times (0.410 + 1.750 + 0.248) = 20.8^k$$

$$v_u = \frac{20.8}{(0.85)(60)(0.8)(10)} = 0.051\text{ksi} < \tfrac{1}{2} v_c$$

∴ No web reinforcing is required.

**Camber at the time of driving.** See *PCI Design Handbook*, pages 4-46 and 4-73.

$$\Delta_{P/S}\uparrow = \frac{F_i el^2}{8E_i I} \times \text{CF} = \frac{(468)(0.75)(35^2)(144)}{(8)(3600)(5000)} \times 1.80 = 0.77 \text{ in}$$

This is satisfactory for a jetted pile.

**Deflection.** Check deflection at top in installed condition. Assume fixity at 10 ft below the channel bottom [Uniform Building Code, Sec. 2908 (d)].

$$\Delta = [(2.05)(15^2)(60 - 15) + (8.75)(13.33^2)(60 - 13.33) + (1.42)(9.46^2)(60 - 9.46) - (17.4)(2.8^2)(60 - 2.8)] \left[\frac{1728}{(6)(4700)(5000)}\right]$$

$$\Delta = 1.13 \text{ in}$$

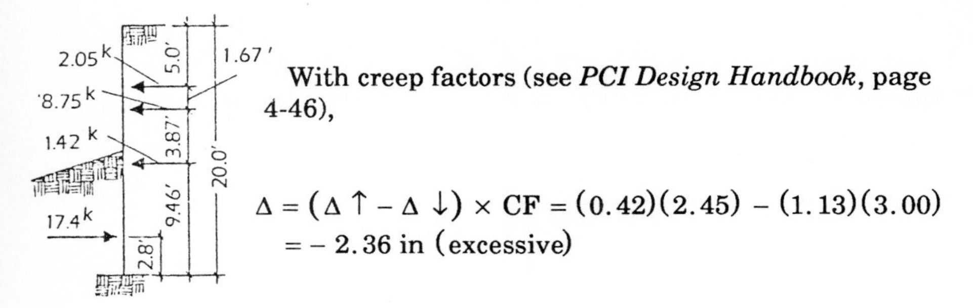


With creep factors (see *PCI Design Handbook*, page 4-46),

$$\Delta = (\Delta\uparrow - \Delta\downarrow) \times CF = (0.42)(2.45) - (1.13)(3.00)$$
$$= -2.36 \text{ in (excessive)}$$

Therefore, use a thicker and narrower pile; try a section measuring 12 in by 4 ft 0 in.

Section properties:

$A = 576 \text{ in}^2$

$\text{Weight} = 10.5 \text{ tons each} = 0.6^{k}/\text{ft}$

$I = 6912 \text{ in}^4$

$S = 1152 \text{ in}^3$

$d = 9.25 \text{ in}$

$d' = 2.75 \text{ in}$

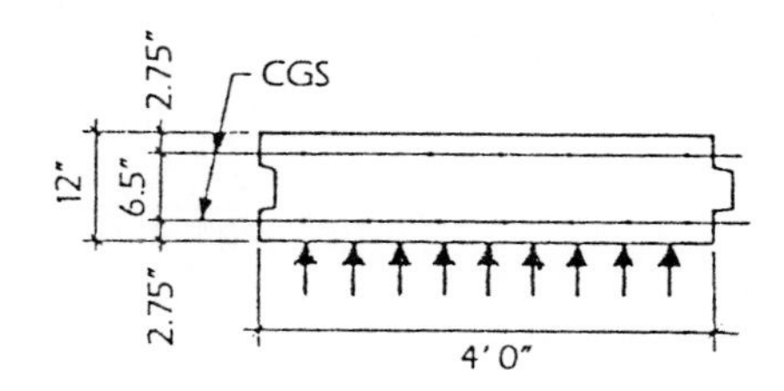


**Recalculated loads and moments.** These are as follows:

$$M = 4 \times 19.61 = 78.4^{k\cdot ft} \left(941^{k\cdot in}\right)$$

$$UM = 1.7 \times 941 = 1600^{k\cdot in}$$

$$UV = 4/5 \times 20.8 = 16.64^{k}$$

**Service load strand requirement.** Assume $e = 0.6$ in. (See *PCI Design Handbook*, page 11-15.)

$$F_f = \frac{\frac{941}{1152} - 0.232}{\frac{1}{576} + \frac{0.6}{1152}} = 259^k \quad N = \frac{259}{28.9 \times .78} = 11.5$$

Use seven 1/2 -in $\phi$ strands in bottom and five in top ($e$ = 0.54 in).

$$f_t = -\frac{271}{576} - \frac{(271)(0.54)}{1152} + \frac{941}{1152} = -0.470 - 0.127 + 0.817$$

$$= +0.220 < 3\sqrt{f_c'} \quad \text{(OK)}$$

**Ultimate strength.** Check ultimate strength. (See *PCI Design Handbook*, pages 4-6 and 4-63).

$$Cwp = 1.13\frac{(7)(0.1531)(270)}{(48)(9.25)(6)} = 0.123$$

$$\therefore f_{ps} = (0.96)(270) = 259.2 \text{ ksi}$$

$$a/2 = \frac{(1.072)(259.2)}{(1.7)(6)(48)} = 0.57 \text{ in}$$

$$M_u = (0.9)(1.072)(259.2)(9.25 - 0.57) = 2170^{\text{k·in}} > UM \quad \text{(OK)}$$

**Shear.** Check shear.

$$v_u = \frac{16.64}{(0.85)(48)(0.8)(12)} = 0.042 \text{ ksi} < 1/2 v_c$$

Therefore, no web reinforcing is required.

**Camber.** Check camber. (See *PCI Design Handbook*, pages 4-46 and 4-73.)

$$\Delta_{P/S} \uparrow = \frac{(312)(0.54)(35^2)(144)}{(8)(3600)(6912)} = 0.15 \text{ in} \times 1.8 = 0.27 \text{ in} \quad \text{(OK)}$$

**Deflection at top in installed condition.** See above, subsection "Deflection."

$$\Delta \downarrow = \left(\frac{4}{5}\right)\left(\frac{5000}{6912}\right) 1.13 = 0.65 \text{ in} \quad \text{(OK)}$$

**Handling.** Check handling.

0.2L 0.6L 0.2L

$M^{+} = M^{-} = 0.0214\ wL^2$

Stripping: Try two-point pick. Impact factor = 1.5.

$$M_s = 1.5 \times 0.0214 \times 0.6 \times 35^2 \times 12 = 283^{\text{k}\cdot\text{in}}$$

$$f_t^{+} = -\frac{312}{576} + \frac{(312)(0.54)}{1152} + \frac{283}{1152} = -0.542$$

$$+ 0.127 + 0.245 = -0.151\,\text{ksi (OK)}$$

Shipping: ($I = 2.0$) Try two-point support.

$$f_t^{+} = \frac{271}{576} + \frac{(271)(0.54)}{1152} + \frac{283 \times 4/3}{1152}$$

$$= -0.470 + 0.127 + 0.327$$

$$= -0.016\ \text{ksi (OK)}$$

Erection: Try lifting at ends.

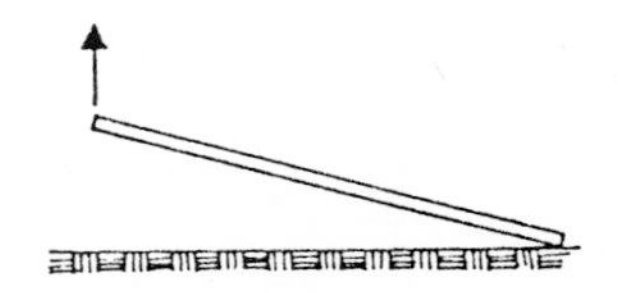

$$I = 1.25$$

$$M = (1.25)(0.125)(0.6)(35^2)(12) = 1378^{\text{k}\cdot\text{in}}$$

$$f_t^{+} = -\frac{271}{576} - \frac{(271)(0.54)}{1152} + \frac{1378}{1152}$$

$$= -0.470 - 0.127 + 1.196 = 0.599\,\text{ksi (too high)}$$

$$F_{t\text{, allowable}} = \frac{7.5\sqrt{f'_c}}{\text{FS}} = \frac{0.580}{1.5} = +0.352\,\text{ksi}$$

Check erection at bottom stripping point and erection loop.

$$M = (1.25)(0.0685)(0.6)(35^2)(12) = 755^{\text{k}\cdot\text{in}}$$

$$f_t^{+} = -0.470 - 0.127 + \frac{755}{1152} = +0.058\,\text{ksi (OK)}$$

0.8L 0.2L

$M^{+} = 0.0685\ wL^2$

**Handling summary.** Strip, store, and ship by handling at loops cast in $0.2L$ points. Erect at lower stripping loops and top erection loops.

**Design note.** The method of installation determines the design philosophy used in reinforcing the pile. If this pile were installed by driving to bearing, then the prestressing would be concentric and would have spiral confinement reinforcing or closed-loop stirrups arranged as in standard bearing piles.

**Reinforcing summary:** 12-in by 4-ft prestressed concrete sheet pile. See Fig. 30.22.

$$f'_c = 6000\ \text{lb/in}^2$$

$$f'_{ci} = 3500\ \text{lb/in}^2$$

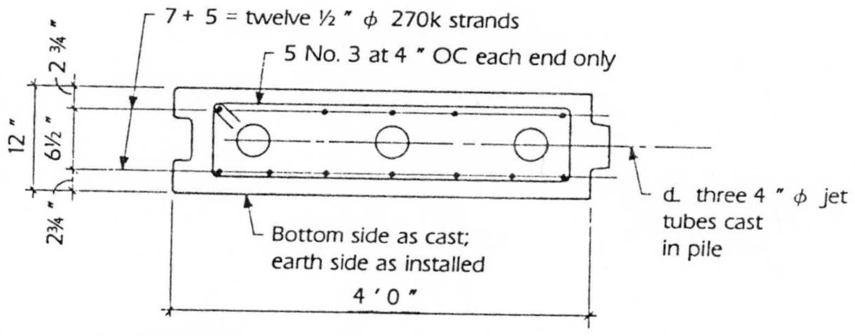


**Figure 30.22** Reinforcing summary.

**Figure 31.1** Candlestick Park, South San Francisco, California. Precast L-shaped stadium slabs were furnished in 1961 for the original stadium construction and again in 1973 for the conversion to allow the park to be used for football as well as baseball. The 28-ft-long prestressed concrete units were cast with the vertical riser in the down position and then rotated 180° prior to shipping.

Chapter

# 31

# Miscellaneous Structural Shapes

In addition to the standard sections discussed in previous component design chapters, two other elements have been prominent in the development of the plant-cast precast concrete industry. These elements are the single T and the stadium slab. While each has had extensive use, their lack of general marketing applications has prevented them from becoming truly standard sections.

## Single T

The single T, originally referred to as the Lin T (after T. Y. Lin), was originally the standard long-span deck element in both the eastern and the western United States. The single T was originally used in parking structures (see Fig 31.2), with the units set one next to the other, with each having its own supporting column, or spread apart, supporting cast-in-place concrete or hollow-core slab units spanning in between. Often the cast-in-place slab infill design solution was posttensioned transversely to the direction of the T span. The other principal use of the single T was in achieving long spans over gymnasiums and pools in school construction.

However, in recent years the relaxing of fire requirements in this type of construction in building codes, coupled with the lack of awareness of specifiers and buyers of the other features and benefits of precast prestressed concrete construction, has led to a reduction in this area of product application. Nonetheless, many manufacturers have single-T forms, and this shape can be used anywhere that a double-T deck member would be specified. In fact, the wise specifier will allow the option of bidding the single T as an alternative for double-T construction, thereby increasing competition with potential resulting economy.

**Figure 31.2** FAB parking structure, Los Angeles, California. Spread single T's span 63 ft to provide clear-span parking in this seven-level structure. Hollow-core planks span transversely between the T's. Temporary shoring provided efficient use of the composite T section.

In the erection stage, additional bracing is required for single T's. The flanges are usually shored at installation to prevent overturning prior to pouring cast-in-place concrete topping. Often flange weld plates are provided so that the most recently placed unit can be immediately welded off to the adjacent structure, thereby rendering the T stable. When using single T's in a spread-apart design supporting hollow-core planks, the T flange requires temporary support, usually consisting of a series of pipe shores supporting a continuous header beneath the flange tip. Also, the plank should be supported on continuous hardboard or Korolath bearing strips placed back 2 in from the edge of the flange to prevent localized spalling and bond failures at the flange tip.

Shown in Fig. 31.3 is the "standard" single-T section made by many prestressers. Note that the flange thickness, slope, and chamfer dimensions may vary slightly with individual manufacturers. It is also wise to check the availability of the deeper sections in the preliminary stages of project development.

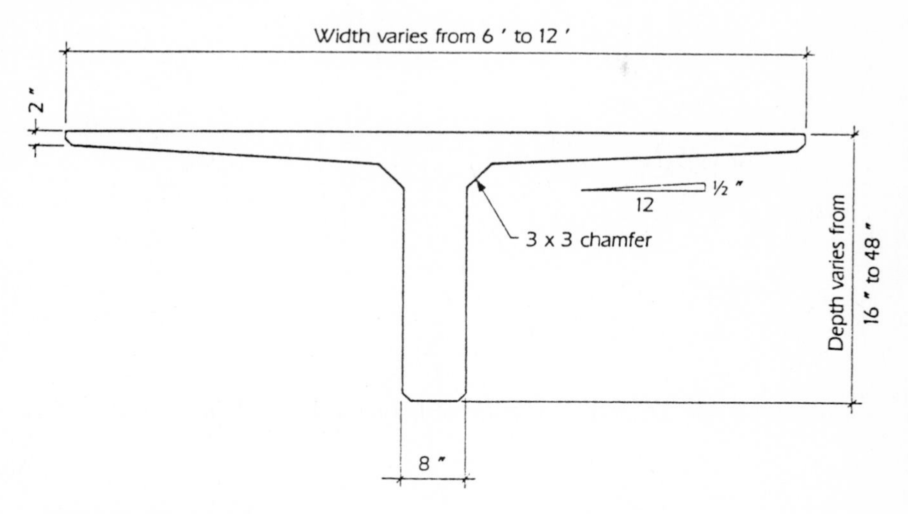


**Figure 31.3** Single-T section.

Preliminary design charts for 8-ft-wide and 10-ft-wide single T's are shown in Figs. 31.4 and 31.5 and an example of the use of single T's is shown in Fig. 31.6.

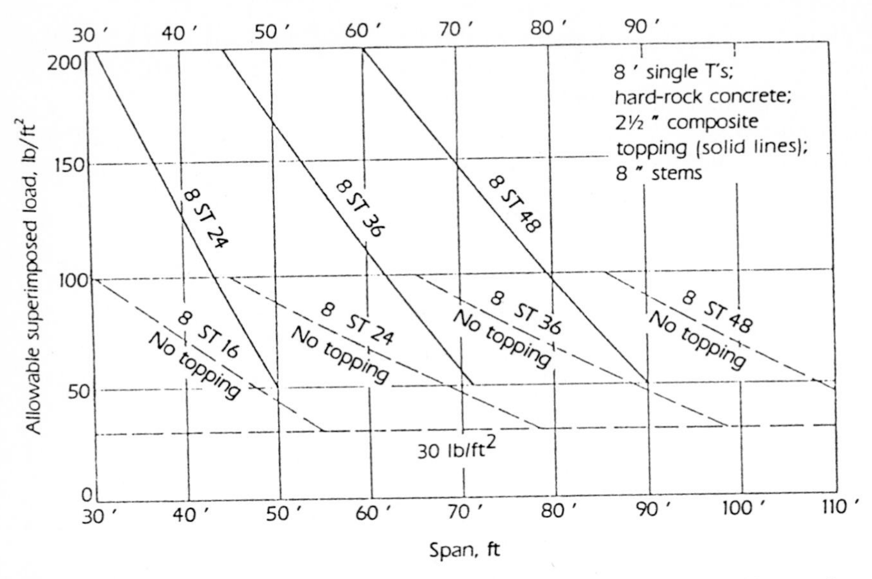


**Figure 31.4** 8-ft-wide single-T load tables. 8ST24 means 8-ft-wide single T 24 in deep.

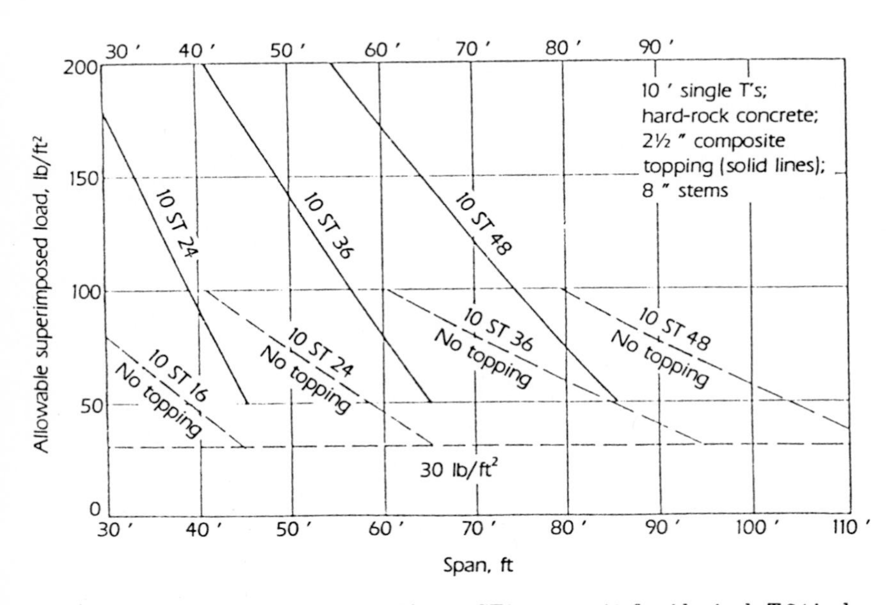


**Figure 31.5** 10-ft-wide single-T load tables. 10ST24 means 10-ft-wide single T 24 in deep.

**Figure 31.6** Ambassador College physical education facility, Pasadena, California. Single T's form both the walls and the roof of this multibuilding complex, providing beauty as well as diversity of function.

## Stadium Slab Units

In the early days of the industry, the use of prestressed concrete elements to provide an economical solution to stadium construction was obvious, and that advantage has not changed. The early prestressers made a shape that is generally not available today, namely, the 4-ft-wide double T. This once-standard unit was modified slightly to provide a slab unit varying from 10 to 16 in in depth in a 3-ft width. This shallow-slab solution consisted of a continuous wood seat unit placed directly on the top horizontal edge of the prestressed unit. Single T's were also modified to achieve a similar solution, with the overhanging flange constituting the seat.

Later, more sophisticated seating requirements for auditoriums coupled with higher-angle seating to accommodate semienclosed or fully enclosed facilities dictated the selection of a deeper unit. The dimensions commonly selected for slab units now reflect this philosophy, coupled with dimensions which are multiples of optimum tread and riser dimensions. For large projects, casting repetition justifies the initial expense in setting up for multiple slab units three wide. Figures 31.7, 31.9, and 31.10 demonstrate the three most common design solutions for stadium slab units, and Figs. 31.8 and 31.11 show examples of stadium slab use.

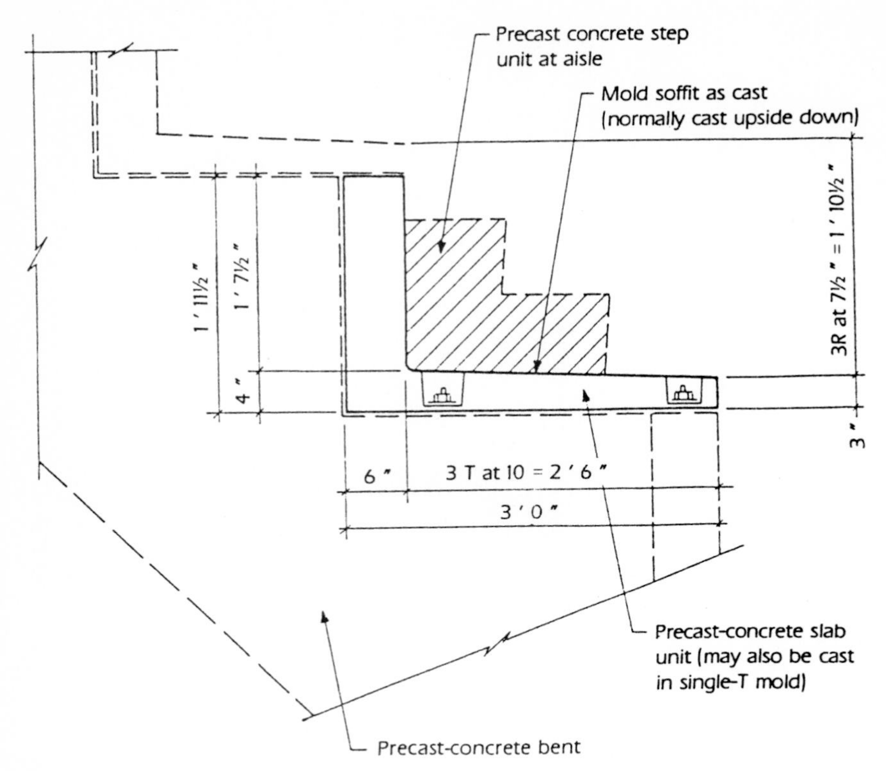


**Figure 31.7** Stadium slab unit.

**Figure 31.8** The Forum, Inglewood, California. These single-riser units were cast inverted in a single-T mold and rotated 180° prior to erection. The sawtooth cast-in-place concrete bents could have been economically made in precast concrete also.

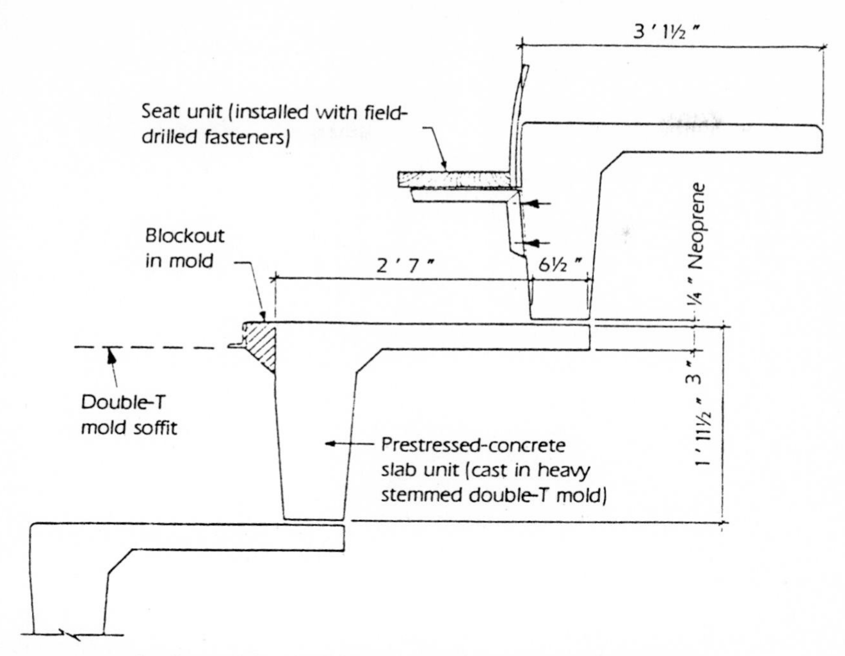


**Figure 31.9** Stadium slab unit fabricated in standard double-T mold.

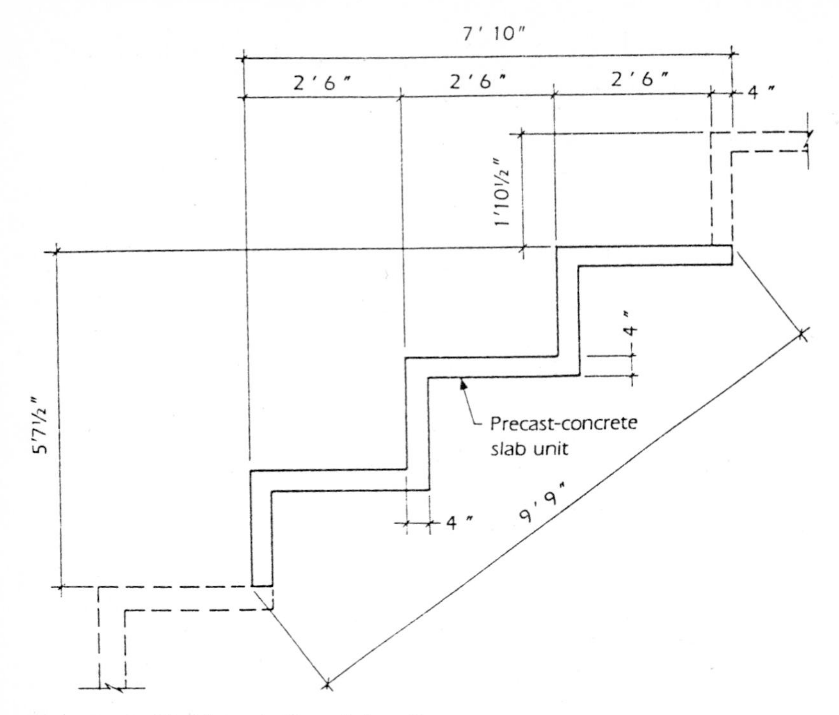


**Figure 31.10** Multiple-stadium-slab unit.

**Figure 31.11** Candlestick Park, South San Francisco, California. By looking closely at the underside of these stadium slabs one can see evidence of hand finishing, indicating that the slabs were cast inverted. The smooth mold finish on the exposed surface has optimum weathering characteristics and appearance.

# Part 4

# Related Topics

Figure 32.1 California First Bank Building, San Francisco, California. Careful consideration of desired finishes and required tolerances in drafting the specifications for this project, along with consultation with plant-cast precast concrete manufacturing personnel in preliminary project planning, helped to make this project a success. A total of 110,000 $ft^2$ of architectural precast concrete facade in 1576 pieces clad this 324-ft-high building. The use of white cement and natural granite fine and coarse aggregates cast under factory-controlled conditions made possible the uniform off-white color desired by the architect. Plant casting also made possible the close tolerances specified to assure that the complex interlocking column cover and spandrel units fit together.

Chapter

# 32

# Specifications

Specifications are a very important part of contract documents along with the drawings and the subcontract agreement between the general contractor and the precast concrete manufacturer. The purpose of the specifications is to define the extent of the precast concrete contractor's responsibilities and to give clearly defined demarcations between the contractor's work and that to be performed by other trades. The specifications outline the work to be performed, the quality standards that are to be met, the materials to be used, methods of application, and the responsibilities for design, quality control, fabrication, handling in the plant, transportation, erection, cleaning, and satisfaction of contract requirements for final payment.

The specifications are written for each specific project and are tailor-made for the scope and complexity of the precast and prestressed concrete components required on a specific project. Because of this, the specification writer must be thoroughly familiar with precast concrete materials, manufacturing processes, and erection and connection practices. Often the problems that occur on a project are a result of poorly written specifications that do not clearly define the scope of the work and the extent of responsibilities of the various subcontractors that interface with the work of the precaster.

Guide specifications are available from the Prestressed Concrete Institute or the Construction Specifications Institute to assist the designer or specification writer in preparing a detailed specification for a specific project. When defining specific requirements in the specification, the desired results, or performance criteria, should be given without detailing the manufacturing or construction procedure. In this way, the end result is correctly assigned prime importance rather than procedures which may vary with various precast concrete manufacturers.

Some of the items of principal importance in a properly written specification are:

1. *Quality control.* In-plant quality control should be performed by the manufacturer's in-house quality control department, as certified by either the Prestressed Concrete Institute's Plant Certification Program or the International Conference of Building Officials (ICBO) certification. In the absence of effective in-plant quality control, the specifications should automatically provide for an outside testing agency to perform this function if owner-provided testing is not called for in Division 1 of the specifications. (See Figs. 32.2 and 32.4.)

2. *Mock-up.* The correct way to satisfy finish requirements for architectural precast concrete projects is to provide for a mock-up, usually consisting of the first piece cast on a project, which the architect subsequently approves to set the standard for the project. Many misunderstandings have been created by basing the required finish on a 12-in by 12-in sample. (See Fig. 32.3.)

3. *Handling.* Handling information should be shown on the precaster's drawings. Handling points for stripping, storage, and shipping should be shown on the shop drawings. Erection pick points and final lifting procedures should be shown on the erection drawings. It is not sufficient merely to state that the precaster is responsible for the handling of the units in such a manner as to preclude cracking; this information should be indicated on the drawings for ready reference by shop, field, and inspection personnel.

4. *Bracing plan.* When the precast or prestressed components are not permanently tied into the structure prior to releasing from erection equipment, temporary bracing is required to maintain stability to resist construction, wind, or seismic forces. The method by which this bracing is to be performed should be shown on the bracing plan, based upon an analysis by a licensed civil or structural engineer. Once again, it is not sufficient to state that this is the precaster's responsibility; the above documentation is essential to assure the safety of other trades and personnel working on the site.

5. *Water-cement ratio.* A water-cement ratio of 0.40 is recommended for both architectural precast concrete and prestressed concrete projects to assure sufficient durability to stand the test of time. In addition, minimum cement content should be specified when there is evidence that the members will be exposed to corrosive atmospheres or seawater.

6. *Erection access.* The general contractor's responsibility to provide wide, *firmly compacted* access roads and sufficient access all around and/or into the structure as required should be clearly defined. Heavy loads and long reaches with heavy precast elements require firm, solid footing for crane outrigger supports.

7. *Camber.* The general contractor and interfacing subcontractors should be aware of *camber* and *differential camber* in estimating quantities of topping pours, setting screeds, correctly attaching partitions, etc.

8. *Tolerances.* Realistic tolerances for both production and other structural elements in the field should be spelled out to permit efficient installation of precast members. These required tolerances also affect detailing of materials that connect to the precast concrete, such as windows and curtain wall systems.

9. *Patching.* In the normal course of handling and erection, spalls and chipping do occur, and patching is done in the field to correct these conditions. Proper bonding agents and admixtures should be indicated in the specifications to be used for this patching.

10. *Prepayment in the yard.* Precast concrete manufacturers spend large amounts of money for material and labor in producing and stockpiling prefabricated elements while work goes on at the site. If the members were site-cast, progress payments would be made; so payment for in-plant value is justified.

**Figure 32.2** Quality control laboratory, prestressed concrete plant. Here we see concrete cylinders being broken early in the morning to ascertain if adequate release strength has been attained to strip yesterday's prestress bed pours. A precast plant quality control inspector performs the test while the plant superintendent looks on.

Figure 32.3 Hastings College of Law, San Francisco, California. The panel exterior finish is as the architect saw it in his dream because he had the wisdom to provide for a mock-up in the specifications for his project. Below we see the mock-up panel for the cladding in the fabricator's yard, approved by the architect, and setting the standard for production on the project. This unit was subsequently installed on the building.

## Specifications for Architectural Precast Concrete (Plant-Cast), CSI Designation 03450

### This document

This document provides a basis for specifying in-plant fabrication and field erection of architectural precast concrete with a variety of textures and finishes. It does not cover field-fabricated precast concrete panels and precast structural concrete, nor does it include dampproofing, special coatings applied to the panels, caulking around the panels, or loose attaching hardware.

### Drawings and specifications

**Drawings.** The architect's or engineer's plans will show locations and necessary sections and dimensions to define the size and shape of the architectural precast concrete. Indicate the location of joints, both functional and aesthetic, and illustrate details between units. When more than one type of panel material or finish is used, indicate the location of each type on the drawings. Illustrate the details of corners of the structure and interfacing with other materials. Whether or not sizes and locations of steel reinforcement and details and locations of typical and special connection items and inserts are shown may be determined by local practices. If reinforcement and connections are not detailed, identify the requirements for design and indicate load support points and space allowed for connections.

**Specifications.** Describe the type and quality of the materials incorporated into the units, the design strength of the concrete, the finishes, and the tolerances for casting and erection. The methods and techniques required to achieve similar results will vary with individual precasters. Specifying the results desired without specifically defining manufacturing procedures will ensure a concise and accurate interpretation and in turn encourage the best competitive bidding.

**Coordination.** The responsibility for supplying items to be placed on or in the structure in order to receive the precast concrete units depends on the type of structure and varies with local practice. Clearly specify responsibility for supply and installation of hardware. When the building frame is structural steel, erection hardware is normally supplied and installed as part of the structural steel. When the building frame is cast-in-place concrete, hardware, if not predesigned or shown on drawings, is normally supplied by the precast manufacturer and placed by the general contractor to a hardware layout prepared by the precast supplier. Assurance that type and quantity of hardware items required to be cast into the precast units for other trades are specified and not duplicated is of greater importance than the supplier. Specialty items, however, should be supplied from the trade requiring them. Verify that materials specified in the section on flashings are galvanically compatible with reglets or counterflashing receivers installed under this section. Check that concrete coatings, adhesives, and sealants specified in other sections are compatible with each other and with the form release agents or surfaces to which they are to be applied.

Sample or guide specifications for architectural precast concrete are available from the Prestressed Concrete Institute in Chicago or the Construction Specifications Institute in Washington, D.C.

**Figure 32.4** In-plant quality control, structural precast plant. Strict quality control is essential to the production of prestressed concrete products. This company's in-house quality control program is periodically reviewed by the PCI Plant Certification Program. Here we see the quality control inspector rodding concrete cylinders according to ASTM C-31.

## Specifications for Structural Precast Prestressed Concrete (Plant-Cast) CSI Designation 03420

### This document

This document provides a basis for specifying in-plant fabrication and field erection of structural precast and prestressed concrete units, such as double T's, beams, hollow-core plank, flat slabs, columns, prestressed structural wall elements, bearing piles, and sheet piling. It does not cover architectural precast concrete panels, site-cast precast and posttensioned concrete, or cast-in-place concrete topping and closure pours.

### Drawings and specifications

**Drawings.** Indicate the number and size of precast prestressed sections on a layout plan. Show the magnitude and distribution of imposed loads. Indicate required reinforcing, prestressing, and connection material. Indicate the location and size of openings and supports and the position of inserts, anchors, and connections. Include setting plans with details showing the connection of precast prestressed sections to adjacent construction such as beam to column, beam to girder, and column base. Illustrate related items such as bearing pads, grout, cast-in-place concrete, and weld plates. For complicated or unusual situations with temporarily unstable elements provide an erection drawing indicating the bracing method to be employed.

**Specifications.** State the required qualifications of the precast prestressed section manufacturer and erector. Define the materials to be used including concrete, tendons, and bearing pads. Require submission of shop drawings and erection procedures. List the allowable tolerances for fabrication and erection. State fire ratings if required. Define in-plant quality assurance measures to be used.

**Coordination.** Assure that specifications for framing and supporting members contain tolerances compatible with precast prestressed sections. When the building foundations of an interfacing structure contain cast-in hardware to facilitate connections to precast elements, provide hardware setting drawings to the general contractor if these are predesigned or shown on the contract drawings.

Sample or guide specifications for structural precast concrete are available from the Prestressed Concrete Institute in Chicago or the Construction Specifications Institute in Washington, D.C.

**Figure 33.1** Kodak Center, San Jose, California. Metal-stud-stiffened spandrels, column covers, and entry beam wraps demonstrate the versatility of this type of GFRC panel, featuring dark-red granite inset pieces contrasting with the coral finish of the panel face mix.

Chapter

# 33

# Glass-Fiber Reinforced Concrete

Glass-fiber reinforced concrete is a material consisting of a sand-and-cement slurry interspersed with alkaline-resistant glass fibers to give added tensile and flexural strength to the resultant cured material. In cladding panel applications, the usage discussed in this chapter, the material is hand-sprayed into the molds in several layers to build up the required thickness.

Fibers have been used for thousands of years to reinforce wet-mixed matrices. Straw was added to clay in ancient times to improve its tensile strength. Other examples such as horsehair and plaster, asbestos and grout, and, in more recent times, plain glass fibers added to plastic (FRP) and steel fibers in concrete come to mind. Experiments in the Soviet Union in 1941 to use plain glass fibers demonstrated that the plain E glass fibers were eroded in a matter of weeks by the alkalines generated by the portland cement hydration reactions. A major breakthrough occurred in 1967, when the Building Research Establishment in England developed a process for making alkaline-resistant glass fibers by the addition of zirconia to the fiber-manufacturing process.

Pilkington Brothers, Ltd., in England refined and patented a commercially feasible process for manufacturing alkaline-resistant fibers called AR glass fibers. Today, both Pilkington and Nippon Electric Glass of Japan sell AR glass for the manufacture of GFRC products. Additionally, there has been developed an alternative process using a polymer latex emulsion called Forton, which when used in the sand-cement-water slurry in a concentration of 10 to 15 percent by volume permits plain E glass fibers to be used and not be degraded by the alkalines generated by the portland cement matrix.

GFRC is used internationally to provide a variety of products. In the United States it has been used primarily to make architectural cladding panels for buildings. In this application, projects are in service dating from 1968 in Great Britain and from 1974 in the United States.

The advantages of GFRC used to produce cladding panels are the following:

- Economy
- Light weight (8 to 25 lb/ft$^2$)
- Savings in supporting structure and foundations
- High strength; impact resistance
- Fire resistance
- Impermeability to moisture
- Ease of installation
- Low maintenance
- Excellent appearance; natural concrete finishes used on exposed surfaces
- High plant-manufactured quality

GFRC also possesses some unusual characteristics which make it unlike plain reinforced concrete:

1. *Higher shrinkage.* This is due to the higher ratio of cement to sand in the mix.

2. *Moisture movement (reversible shrinkage).* An expansion or contraction in volume when the final ambient relative humidity differs from that present during initial curing. The use of a polymer latex emulsion in the mix will decrease the amount of this reversible movement and provide more resistance to shrinkage cracking.

3. *Time-dependent strength.* Over a long period of time, there is a significant reduction of ultimate strength, on the order of 25 to 50 percent, in the initial, or 28-day test, value. This is probably due to the fact that some alkaline attack on the fibers may continue to occur over the lifetime of the material. For this reason, the ultimate strength used in designs is approximately 25 percent of the initial, or 28-day, value ascertained from actual testing. This assures that service load stresses will always be well below the actual ultimate flexural strength of the composite. (See Fig. 33.2 for a graphical representation of this strength

degradation phenomenon.)

Typical GFRC mix proportions are as follows:

100 lb of Type III portland cement (white or gray) *
100 lb 0 mesh sand
23 lb (3 gal) of water
12 lb of AR glass fibers (5 percent by weight)
10 lb of Forton polymer
10 oz of superplasticizer
2 oz of retarder (when ambient temperature is above 80°F)

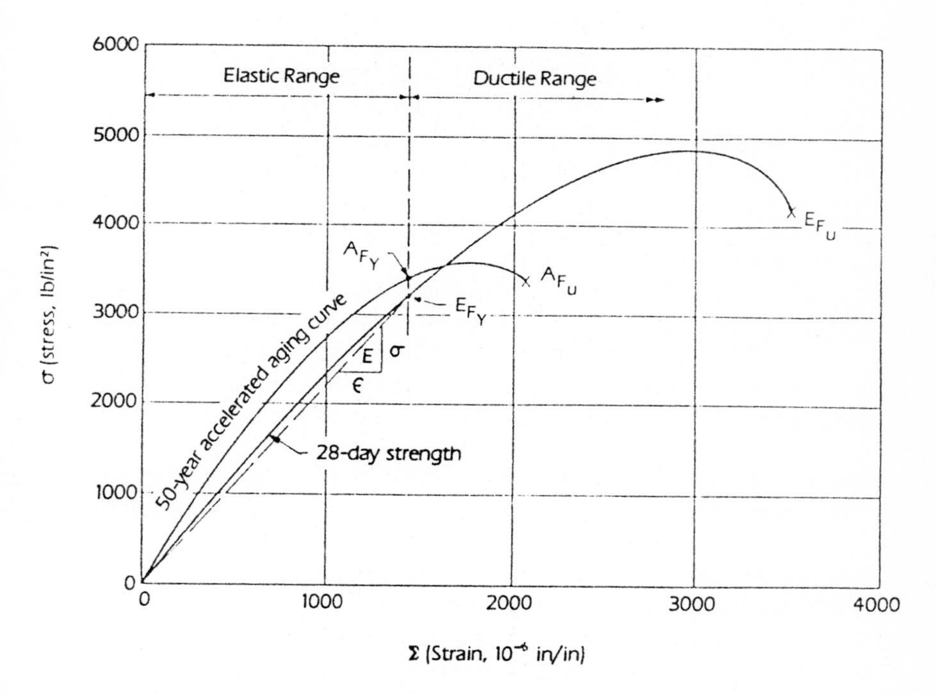


**Figure 33.2** GFRC degradation (for demonstration purposes only).

The resultant GFRC cladding skin of approximately ½- in thickness from a three-pass layered spray-up process will possess the following physical properties:

| | |
|---|---|
| Weight | 110–125 lb/ft³ (pure GFRC) |
| Early flexural yield (EFY) | 1000–1800 lb/in² † |
| Early flexural ultimate (EFU) | 2000–4500 lb/in² † |
| Aged flexural yield (AFY) | 1000–1700 lb/in² † |

* The first backup layer behind white cement face mixes will use white cement.
† Tensile value is one-half of this.

| | |
|---|---|
| Aged flexural ultimate (AFU) | 1500–2400 lb/in$^2$ * |
| Compressive strength (edgewise) | 7000 lb/in$^2$ |
| Shear strength (interlaminar) | 500 lb/in$^2$ |
| Shear strength (in plane) | 1000 lb/in$^2$ |
| Punching shear strength | 4000 lb/in$^2$ |
| Modulus of elasticity (E) | 1.0–3.0 $\times 10^6$ lb/in$^2$ |
| Coefficient of thermal expansion | 4–7 $\times 10^6$ in/(in · °F) |

### Manufacturing Procedure

The GFRC mix presented above is a structural backup for a sprayed-on face mix consisting of 100 lb of cement, 75 lb of $^1/_4$-in stone, and 75 lb of sand. This face mix is deposited in the molds by spraying with a grout pump and hose with a plaster nozzle attachment. The importance of face mix with small coarse aggregate will be discussed in the section "Architectural Finishes." The GFRC material is deposited by hand-spraying the sand-cement slurry with a slurry spray gun which is fed by a pump. Attached to the slurry spray gun is an air-powered chopper gun which chops a continuous AR glass roving into 1$^1/_2$-in-long pieces and deposits them simultaneously with the sand-cement slurry. Three passes of slurry and glass fibers are deposited and rolled out to a thickness of $^1/_8$ in per layer. The resulting thickness of face mix and backup GFRC is approximately $^3/_4$ in. If the face mix contains white cement, then the first backup pass of GFRC will contain white cement also, since gray cement could bleed through the face mix. After overnight curing, the panel is removed from the mold for finishing, similarly to the procedure used for architectural precast concrete panels, so that a daily manufacturing cycle is maintained.

Throughout the manufacturing process, routine quality control measures are used to assure proper spray-up rates of slurry and fiber, panel skin thicknesses, and slurry slump consistencies. Sample coupons are sprayed up to perform fiber content checks; cured coupons are used to determine bulk density, absorption, and flexural tensile and shear strengths.

### Types of GFRC Cladding Panels

Two types of panel systems are fabricated by using GFRC: the integral-rib and the metal-stud-stiffened systems. The integral-rib system uses hidden formers, usually Styrofoam battens, to spray GFRC material around to develop stiffening sections sufficient to resist lateral, longitudinal, and vertical loads without additional support being required from

* Tensile value is one-half of this.

secondary building frame members. An example design of an integral-rib panel can be found in App. B of the *Recommended Practice for Glass Fiber Reinforced Concrete Panels,* published by the Prestressed Concrete Institute in Chicago. Note that panel sizes and connection configurations are limited in the integral-rib system owing to the time-dependent characteristics of relatively high shrinkage and reversible moisture movement peculiar to GFRC which have been discussed previously.

The metal-stud-stiffened system consists of a flat-sheet exterior skin which is anchored to a metal frame by bent rods located at approximately 2-ft centers in each direction. The bent rods, or flex anchors, as they are referred to in the *PCI Recommended Practice*, enable gravity, wind, and seismic loads to be resisted elastically by various arrangements of these rods, yet permit volumetric changes in the skin to be accommodated by small localized deformations of these flex anchors. This enables the skin to be effectively "isolated" from the metal stud frame, which can then be connected to the building frame by using connection techniques similar to those employed for traditional hard precast concrete panels. Large panel configurations may be achieved by using this system, their size being limited only by handling and trucking requirements.

When panel lengths exceed 20 ft, a real joint is provided in the skin as a control joint to allow for skin movement. This joint is subsequently caulked. The steel stud frame is designed to span the full length of the panel.

The advantages of the metal stud system over the integral-rib system are:

- The isolated skin eliminates problems caused by higher shrinkage, moisture movements, and thermal differentials.
- The interior cavity of the metal frame panel is compatible with traditional methods of insulating, fireproofing, and achieving interior finishes.
- Connector problems are eliminated.
- Larger panel sizes are possible.
- The "embrittlement," or time-dependent strength degradation phenomenon, is less of a concern with closely spaced flex anchors.

## Architectural Finishes

The use of face mixes consisting of sand with small coarse aggregate in the proportions given previously is highly recommended in order to minimize color variation problems, which are typical with higher-cement-content matrices. Smooth finishes in either white or gray cement are to be avoided at all costs, since even with uniform production and finishing techniques these surfaces will exhibit very high degrees of color variation between individual panel elements. The sand-stone texture in sandblasted or retarded finishes minimizes the visual impact of the cement matrix and the inherent color variation characteristics unique to the thin GFRC section. These stone-sand finishes also help prevent the "telegraphing" of a shaded effect through the skin at locations of GFRC ribs or flex anchor boss attachments to the GFRC skin. (See Fig. 33.3.)

A wealth of information and detail is given in the *Recommended Practice* available from the PCI. The reader interested in further information on GFRC should obtain this document. Since no design examples for metal-stud frame panels are presented in the PCI document, an example design of a GFRC metal-stud-stiffened spandrel is given here. (See also Fig. 33.4.)

**Figure 33.3** Bishop Ranch No. 8, San Ramon, California. GFRC elements clad this prestigious office complex, achieving uniformity of finish due to the selection of no. 8 mesh basalt chips mixed with an equal weight of dolomite sand and white cement for the face mix used. (See inset for a close-up of the finish.)

**Figure 33.4** Lincoln Center, Campbell, California. Metal-stud-stiffened spandrels and integral-rib column covers clad this three-story office building. The following design example is based upon the actual spandrel design used for this project.

## Design Example: Metal-Stud-Stiffened GFRC Panel

Design a metal-stud-stiffened GFRC panel for the building elevation shown in Fig. 33.5. Use Uniform Building Code seismic Zone 4 design criteria. The 9-ft-1-in-deep roof parapet will be the critical panel. (See Fig. 33.6.)

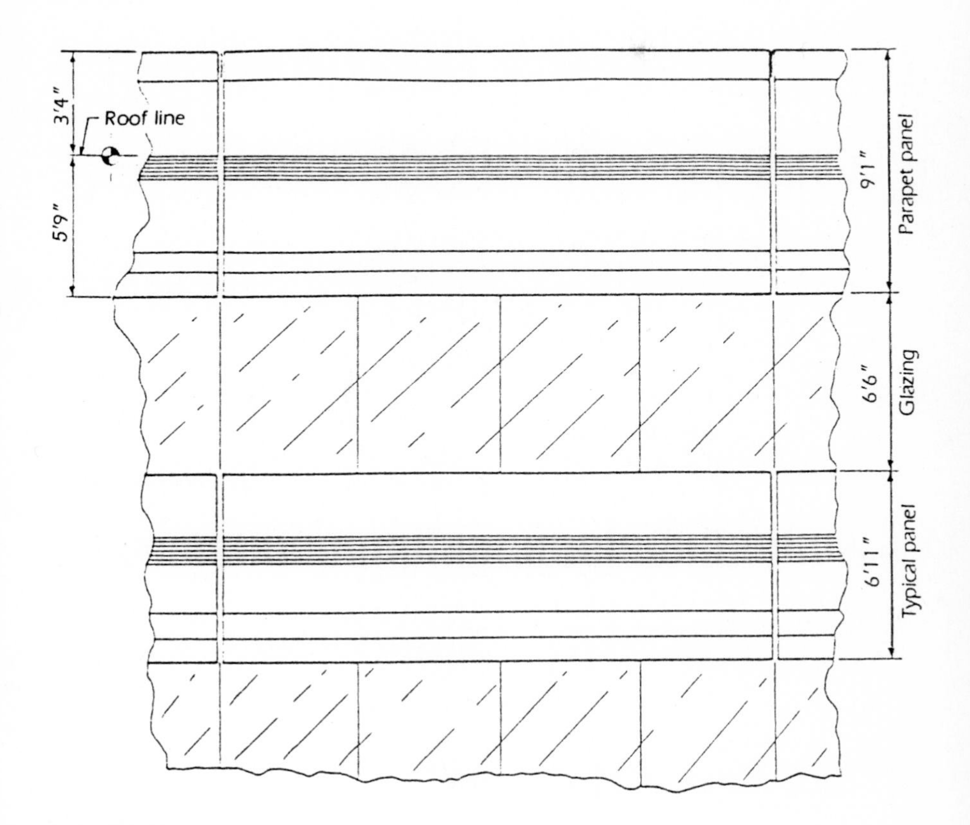


**Figure 33.5** Exterior elevation.

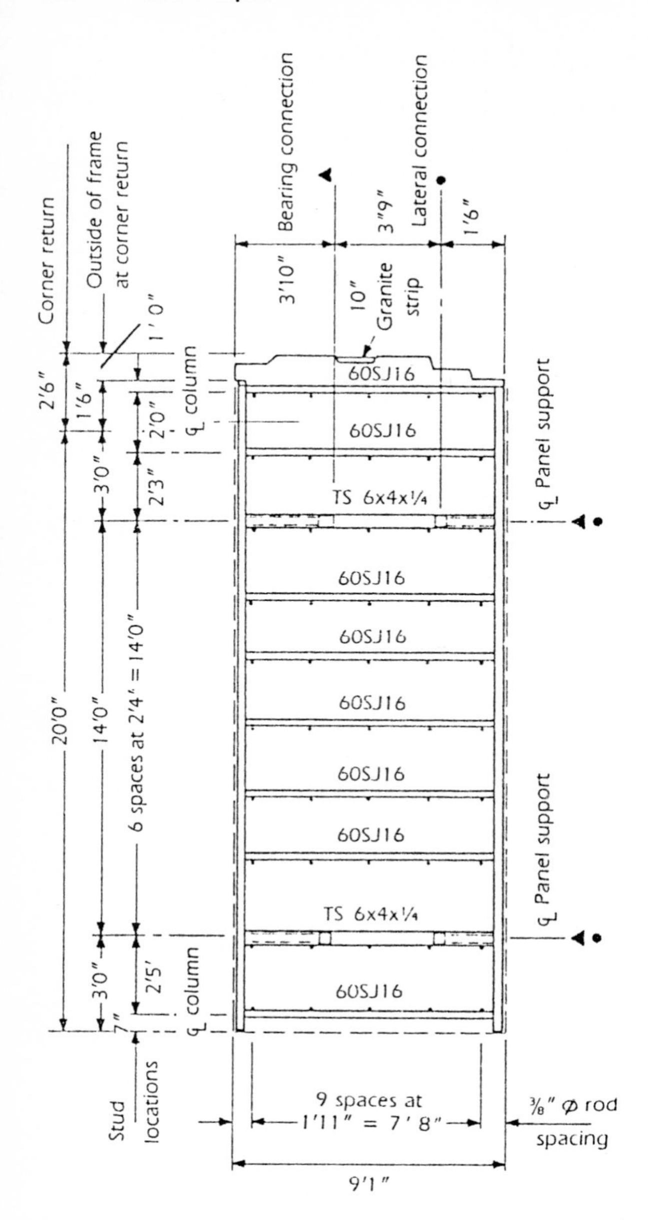


**Figure 33.6** GFRC metal stud panel arrangement.

**Frame Weight**

| | Pounds |
|---|---|
| Two TS 6 × 3 × 3/16 × 22.5 ft at 10.7 lb/ft = | 482 |
| Two TS 6 × 4 × 1/4 × 8.6 ft at 12.0 lb/ft = | 206 |
| Eight steel studs 16 × 8.6 ft at 2.1 lb/ft = | 145 |
| Sixty hooks at 1.5 lb each = | 90 |
| Connection material = | 50 |
| | 973 (43 lb/ft) |

**Summary**

| | lb/ft |
|---|---|
| Frame | 43 |
| GRFC, 0.75"/12 × 130 × 11.9 | 97 |
| Granite | 7 |
| Sheetrock, 9.08 × 2.5 | 23 |
| | 170 lb/ft* |

*170 lb/ft × 22.5 ft = 3825 lb (19 lb/ft²).

**GFRC skin for 28-in span**

Check GFRC skin for 28-in span, simply supported (see Prestressed Concrete Institute *Recommended Practice for GFRC*, page 21).

$$f_u = \phi s f'_u$$

where $\phi = 0.67$ and $s = 1.0$.

$$f'_u = \frac{f_{yr}(1 - 2.539V)}{0.9} \leq 1300 \text{ lb/in}^2$$

$$= \frac{1000(1 - 2.539 \times 0.10)}{0.9} = 829 \text{ lb/in}^2$$

$$f_u = 0.67 \times 1.0 \times 829 = 555 \text{ lb/in}^2$$

**Ultimate load to skin.** Wind governs.

$$UW = 26 \times 1.3 = 34 \text{ lb/ft}^2$$

$$UM = 1/8 \times 34 \times \left(\frac{28}{12}\right)^2 \times 12 = 278 \text{ in}\cdot\text{lb}$$

For ½-in-thick GFRC,

$$S = 1/6 \times 12 \times 0.5^2 = 0.5 \text{ in}^3$$

$$f_{u,\text{actual}} = \frac{UM}{S} = \frac{278}{0.5} = 556 \text{ lb/in}^2 \cong 555 \text{ lb/in}^2 \quad \text{(OK)}$$

**Steel stud elements**

Design steel stud elements (see *AISI Cold Formed Steel Design Manual,* 1977 edition).

$$M_u = 1/8 \times 26 \times 2.33 \times 8.5^2 = 548 \text{ lb}\cdot\text{ft}$$

Try 6-in-deep by 16-gauge cold-formed steel studs with stiffened flanges (unpunched). From the manufacturer's literature:

$$A = 0.51 \text{ in}^2 \quad \text{Weight} = 2.1 \text{ lb/ft} \quad F_y = 42 \text{ ksi}$$

$$S_x = 0.99 \text{ in}^3 \quad I_y = 0.195 \text{ in}^4 \quad F_b = 28 \text{ ksi}$$

Determine $F_b$ for $L_u = 8.0$ ft (with fully supported compression flange). See AISI specifications, Sec. 3.3 and Design Chart V-3.3 (A).

$$\frac{L^2 S_{xc}}{dI_{yc}} = \frac{(8.5 \times 12)^2 (0.99)}{(6)(0.098)} = 17{,}520 \text{ lb/in}^2 > 0.36\pi^2 E/F_y \quad \text{(OK)}$$

$$> 1.8\pi^2 E/F_y \quad \text{(no good)}$$

$$\therefore F_b = 0.6\pi^2 E C_b \frac{dI_{yc}}{L^2 S_{xc}} \qquad \text{where } I_{yc} = 1/2 I_y = 0.098$$

$$C_b = 1.0$$

$$= 0.6\pi^2(29{,}500)(1.0)\frac{(6)(0.098)}{(8.5 \times 12)^2(0.99)} = 9.97 \text{ ksi}$$

$$f_b = \frac{548 \times 12}{1000 \times 0.99} = 6.64 \text{ ksi} < F_b \quad \text{(OK)}$$

Use 6-in-deep, 16-gauge unpunched galvanized steel studs.

**Top and bottom frame members**

Design top and bottom frame members. Use steel tube sections; one-half of the total panel vertical and lateral loading is carried by each horizontal tube to vertical tube supports.

**Gravity loading.** See Fig. 33.7.

$$\text{Panel weight} = 1/2 \times 0.170 = 0.085^{k}/\text{ft}$$

$$\text{Glazing} \frac{0.006 \times 6.5}{2} = 0.020$$

$$\text{Total Weight} = 0.105^{k}/\text{ft}$$

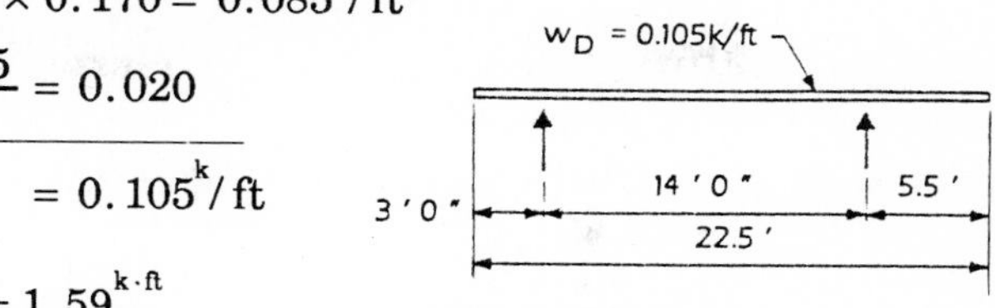


**Figure 33.7** Gravity loading.

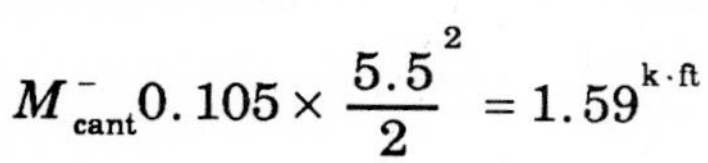

$$M^{-}_{cant} 0.105 \times \frac{5.5^{2}}{2} = 1.59^{k\cdot ft}$$

$$\Delta_{cant} = \frac{wL^{4}}{8EI} = \frac{0.105 \times 5.5^{4} \times 1728}{8 \times 29{,}500 I_{y}} = \frac{0.70}{I_{y}}$$

$$M^{+}_{int} \cong 1/2 \times 0.105 \times 14^{2} = 1.72^{k\cdot ft}$$

$$\Delta_{int} \cong 0.004 \frac{wL^{4}}{EI} = \frac{0.004 \times 0.105 \times 14^{4} \times 1728}{29{,}500 I_{y}} = \frac{0.95}{I_{y}}$$

Deflection governs design. (The authors recommend limiting vertical deflection to $L/720$.)

$$\therefore \Delta_{v,max} = \frac{14 \times 12}{720} = 0.23 \text{ in } (1/4 \text{ in})$$

$$I_{y} \text{ required} = \frac{0.95}{0.23} = 4.13 \text{ in}^{4}$$

Try TS 6 x 3 x 3/16:

$$I_{y} = 4.83 \text{ in}^{4} \quad I_{x} = 14.3 \text{ in}^{4}$$

$$S_{y} = 3.22 \text{ in}^{3} \quad S_{x} = 4.76 \text{ in}^{3}$$

$$f_{b} = \frac{1.72 \times 12}{3.22} = 6.41^{k}/\text{in}^{2}$$

$$< F_{b} = 0.66 \times 42 = 28 \text{ ksi (OK)}$$

**Lateral loading.** Maximum wind loading is 26 lb/ft².

$$w = 1/2 \times 15.5 \text{ ft} \times 0.026 = 0.20^{k}/\text{ft}$$

(The authors recommend limiting horizontal deflection to $L/360$.)

$$\Delta_{int} = 0.004 \frac{wL^{4}}{EI} = \frac{0.004 \times 0.20 \times 14^{4} \times 1728}{29{,}500 I_{x}} = \frac{1.80}{I_{x}}$$

$$\Delta_{H,\max} = \frac{14 \times 12}{360} = 0.47 \text{ in}$$

$$I_x \text{ required} = \frac{0.80}{0.47} = 3.83 \text{ in}^4 \quad (\text{OK})$$

**Combined stresses.** Check combined stresses on tube.

$$f_b = \frac{M_x}{S_x} + \frac{M_y}{S_y} = \frac{1/12 \times 0.2 \times 14^2 \times 12}{4.76}$$

$$+ 6.41 = 14.65 \text{ ksi} < 1.33F_b \quad (\text{OK})$$

**Rod connectors**

Design rod connectors: GFRC skin to frame.

Check 12-in cantilevered skin.

$$UM = 34 \times 1 \times 6 = 204^{\text{lb}\cdot\text{in}}$$

$$f_u = \frac{204}{0.5} = 406 \text{ lb/in}^2 < 555 \text{ lb/in}^2 \quad (\text{OK})$$

$$\Delta = \frac{(26)(1^4)(1728)}{(8)(1.5 \times 10^6)(0.125)} = 0.03 \text{ in}$$

$$< \frac{L}{360} \quad (\text{OK})$$

See Figs. 33.8 and 33.9.

**Rod sizes**

**Rod size for wind.** Check rod size required for wind at 26 lb/ft$^2$.

$$P_C = 2.0 \text{ ft} \times 2.5 \text{ ft} \times 26 = 130 \text{ lb} \quad L_u = 13 \text{ in}$$

Using 3/8-in $\phi$ smooth rod,

$$I = \frac{\pi d^4}{64} = \frac{\pi \times 0.375^4}{64} = 0.001 \text{ in}^4$$

$$A = \frac{\pi d^2}{4} = \frac{\pi \times 0.375^2}{4} = 0.11 \text{ in}^2$$

$$r = \sqrt{\frac{I}{A}} = \sqrt{\frac{0.001}{0.11}} = 0.095 \text{ in}$$

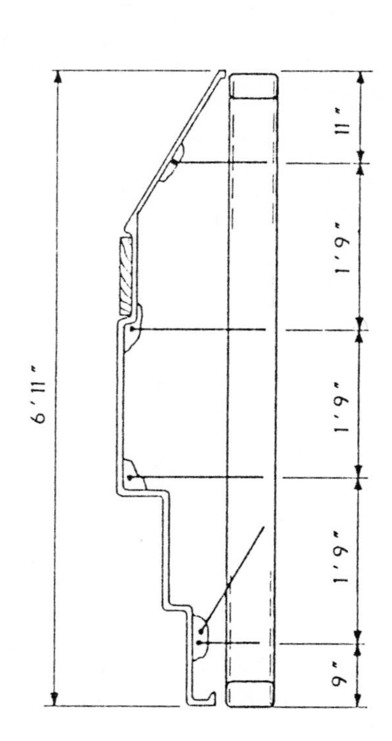


Figure 33.8 Typical floor panel.

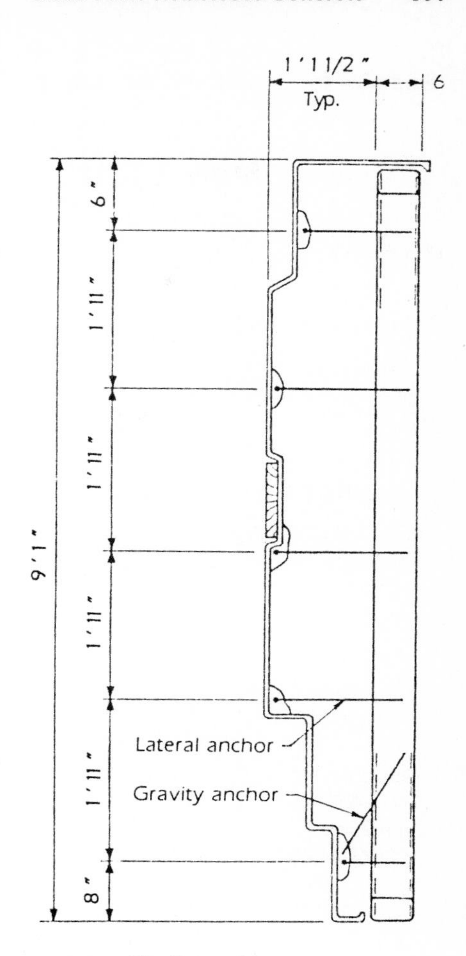


Figure 33.9 Roof panel.

$$\frac{KL_u}{r} = \frac{13}{0.095} = 137$$

$$F_a = 1.33 \times 7.96 \times 10.6 \text{ ksi}$$

$$f_a = \frac{0.130^{\text{k}}}{0.11} = 1.18 \text{ ksi} < F_a \quad (\text{OK})$$

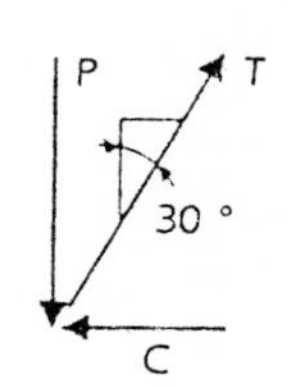

**Rod size for gravity anchor.** Check rod size for gravity anchor.

$$P = 104 \text{ lb/ft} \times 2.5 \text{ ft} = 260 \text{ lb}$$

$$\dot{T} = \frac{260}{0.866} = 300 \text{ lb} \quad C = 150 \text{ lb}$$

Use ³/₈-in $\phi$ rod; $f_T = 2.8$ ksi. (OK)

**Longitudinal anchors designed for 1.2-*g* seismic.** See Fig. 33.10.

Using one set of rods at each vertical tube support,

$$P_E = \frac{1.2 \times 104 \text{ lb/ft} \times 22.5 \text{ ft}}{2 \times 5} = 280 \text{ lb/rod}$$

Use ³/₈-in rod.

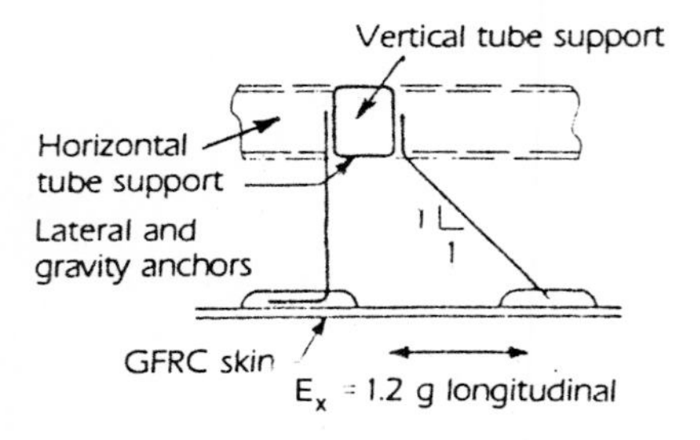


**Figure 33.10** Longitudinal anchor detail.

**Capacity of bosses connecting rods to GFRC skin.** See *PCI Recommended Practice for GFRC*, page 23.

Only $P_c$ = 1000 lb in each three principal directions according to the manufacturer's test data.

**Weld capacity: ³/₈-in $\phi$ rod to stud.** Use ¹/₈-in groove weld.

( ¹/₁₆-in effective throat)

$$\left(P_{\text{effective}} = 0.93 \times 2 = 1.86^{\text{k/rod}}\right)$$

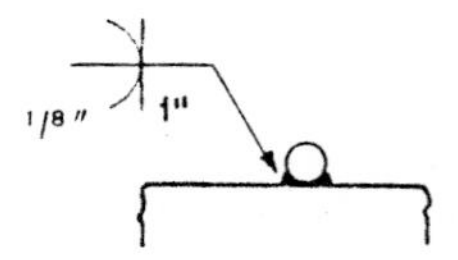

**Seismic analysis**

See Fig. 33.11.

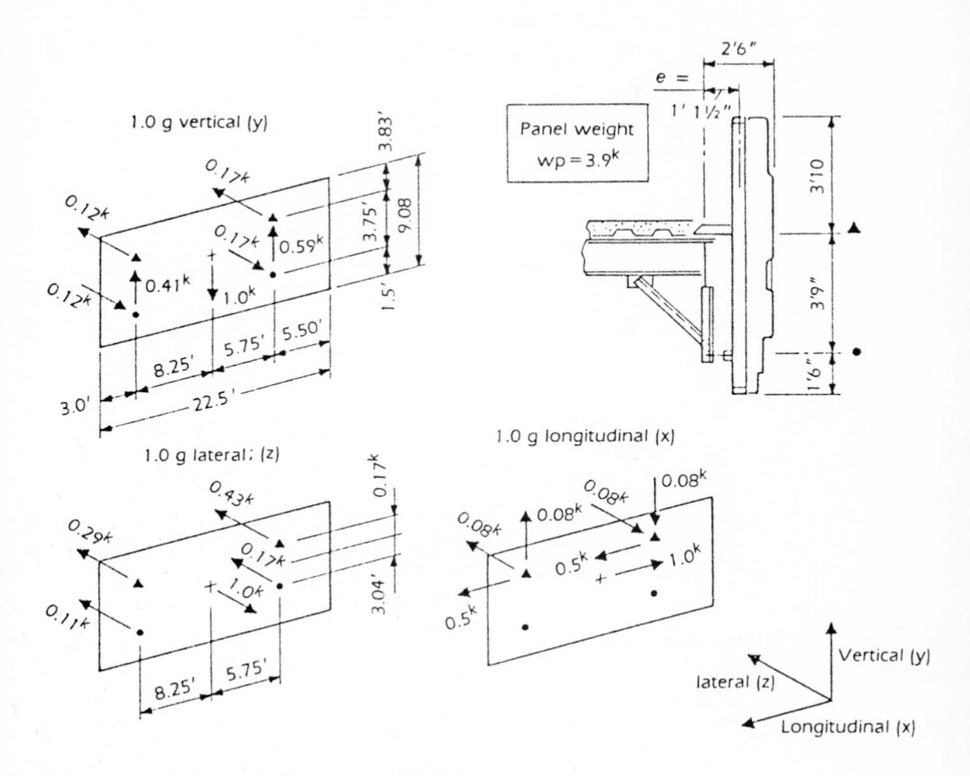


**Figure 33.11** Seismic analysis: loading conditions.

**Bottom lateral connection.** Design bottom lateral connection.

$$D + 1.2E_z = (3.9 \times 0.17) + (1.2 \times 3.9 \times 0.17) = 1.46^{\mathrm{k}}$$

or $$D + \text{wind} = 0.66 + [0.026 \times (9.08 + 3.25) \times 22.5 \times 0.17]$$

$$= 1.89^{\mathrm{k}} \text{ (wind governs)}$$

$$f_a = \frac{1.89^{\mathrm{k}}}{0.31} = 6.1 \text{ ksi (OK)}$$

Use 3/4-in $\phi$ coil rod ($A_{net} = 0.31\ \text{in}^2$).

$$\frac{KL}{r} = \frac{10}{0.16} = 63 \quad F_a \cong 1.33 \times 21.85 = 29.1 \text{ ksi}$$

Connect to ³⁄₄-in-deep coil nut welded to vertical tube support in shop.

**Bearing connection.** Design bearing connection.
Use cantilevered tube design. (Analyze for $D + 0.5\,E_{\text{vert}}$.) See Fig. 33.12.

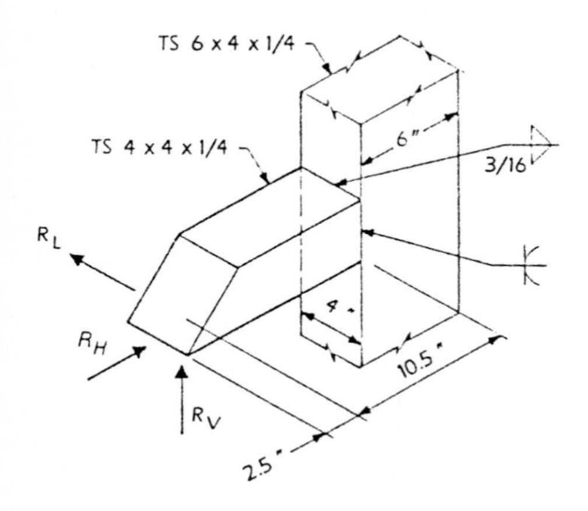


**Figure 33.12** Bearing connection.

$$R_V = 1.5 \times 0.59 \times 3.9 = 3.45^{\text{k}}$$

$$R_H = 1.5 \times 0.17 \times 3.9 = 1.00^{\text{k}}$$

$$M_V = 3.45 \times 10.5 = 36.2^{\text{k}\cdot\text{in}}$$

$$M_H = 1.00 \times 10.5 = 10.5^{\text{k}\cdot\text{in}}$$

Try TS 4 × 4 × 1/4 .

$$f_b = \frac{M_x + M_y}{S}$$

$$= \frac{36.2 + 10.5}{4.11} = 11.4 \text{ ksi} \quad (\text{OK})$$

Analyze for $D + 0.4\,g_{\text{long.}}$

$$R_V = (0.59 + 0.4 \times 0.08)(3.9) = 2.4^{k}$$

$$R_H = (0.17 + 0.4 \times 0.08)(3.9) = 0.8^{k}$$

$$R_L = 0.4 \times 0.5 \times 3.9 = 0.8^{k}$$

$$\frac{M_V + M_L}{S} + \frac{R_H}{A} = \frac{(2.4 + 0.8)(10.5)}{4.11} + \frac{0.8}{3.59}$$

$$= 8.18 + 0.22 = 8.4 \text{ ksi (OK)}$$

Load to weld of TS $4 \times 4$ to vertical TS $6 \times 4$ $(D + 1.2\, g_{long})$.
Weld properties :

$$S = 16 + \frac{16}{3} = 21.3^{2}$$

$$I_p = \frac{(2 \times 4)^3}{6} = 85.3 \text{ in}^3$$

4″

4′

$$R_V = (0.59 + 1.2 \times 0.08)(3.9) = 2.7^{k}$$

$$R_H = (0.17 + 1.2 \times 0.08)(3.9) = 1.0^{k}$$

$$R_L = 1.2 \times 0.5 \times 3.9 = 2.3^{k}$$

Due to shear :

$$f_{Hv} = \frac{1.0}{16} = 0.06^{k}/\text{in}$$

$$f_{Lv} = \frac{2.3}{16} = 0.14^{k}/\text{in}$$

$$f_{Vv} = \frac{2.7}{16} = 0.17^{k}/\text{in}$$

Due to bending :

$$f_{Hb} = \frac{1.0 \times 2}{21.3} = 0.10^{k}/\text{in}$$

$$f_{Lb} = \frac{2.3 \times 10.5}{21.3} = 1.13^{k}/\text{in}$$

$$f_{Vb} = \frac{2.7 \times 10.5}{21.3} = 1.33^{k}/\text{in}$$

Due to torsion :

$$f_{LT} = \frac{R_L ed}{I_p} = \frac{2.3 \times 2.0 \times 10.5}{85.3} = 0.57^k / \text{in}$$

$$f_r = \sqrt{(f_{Hv} + f_{Hb} + f_{Lb} + f_{Vb})^2 + (f_{Lv} + f_{LT})^2 + (f_{Vv} + f_{LT})^2}$$

$$= \sqrt{(0.06 + 0.10 + 1.13 + 1.33)^2 + (0.14 + 0.57)^2 + (0.17 + 0.57)^2}$$

$$= 2.81^k / \text{in}$$

Using 3/16-in welds,

$$F_r = 0.93 \times 3 \times 1.33 = 3.71^k / \text{in} \quad \text{(OK)}$$

**Welding outrigger to spandrel.** Design field weld of TS 4×4 outrigger to wide-flange spandrel. (See Fig. 33.13.)

Use L 3×3×1/4×3 in long, each side. Weld to TS 4× 4.

$$f_V = \frac{2.7}{2 \times 3} = 0.45^k / \text{in}$$

$$f_H = \frac{1.0}{2 \times 3} = 0.17^k / \text{in}$$

$$f_L = \frac{2.3}{2 \times 3} = 0.38^k / \text{in}$$

$$\Sigma f_r = 1.00^k / \text{in}$$

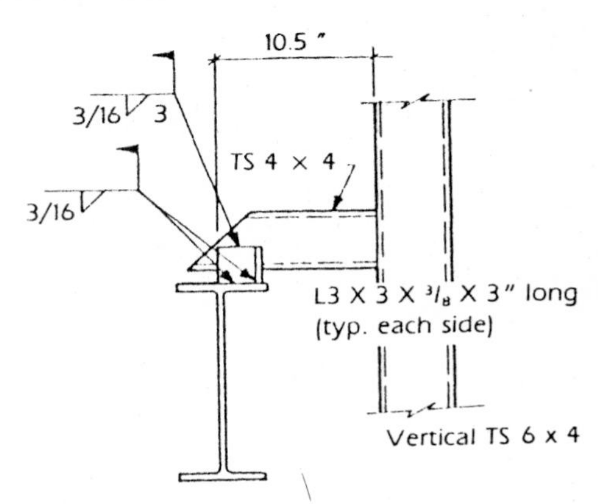


**Figure 33.13** Welding tube support to spandrel.

Use 3/16-in fillet weld ($F_r = 3.71^k$/in).

Weld to wide-flange spandrel.

Section modulus of weld group (see Fig. 33.14):

$$S = \frac{4bd + d^2}{6} = \frac{(4 \times 2.6) + 2.6^2}{6}$$

$$= 5.6 \text{ in}^2$$

$$f_{Hb} = \frac{1.0 \times 2}{2 \times 5.6} = 0.18^k / \text{in}$$

$$f_{Hv} = \frac{1.0}{4 \times 2.6} = 0.10^{k}/\text{in}$$

$$f_{Lb} = \frac{2.3 \times 2}{2 \times 5.6} = 0.41^{k}/\text{in}$$

$$f_{Lv} = \frac{2.3}{4 \times 2.6} = 0.22^{k}/\text{in}$$

$$f_{Vv} = \frac{2.7}{4 \times 2.6} = 0.26^{k}/\text{in}$$

$$f_r = \sqrt{(f_{Hb} + f_{Lb} + f_{Vv})^2 + f_{Hv}^2 + f_{Lv}^2}$$

$$= \sqrt{(0.18 + 0.41 + 0.26)^2 + 0.10^2 + 0.22^2} = 0.95^{k}/\text{in}$$

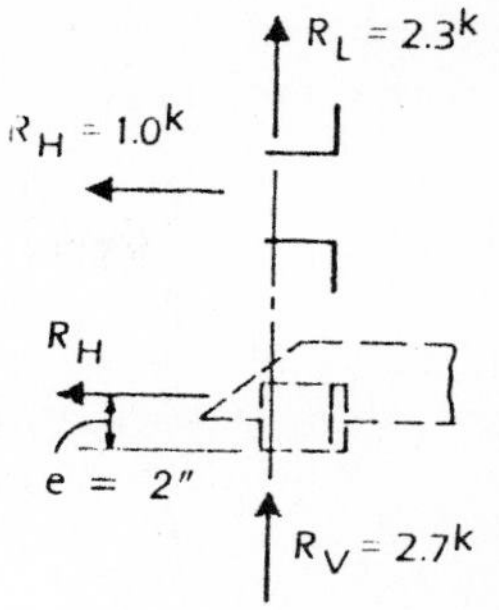

**Figure 33.14** Section modulus of weld group.

Use $^3/_{16}$-in fillet welds.

**Torsion.** Check vertical tube support for torsion.

$$T = 0.8^{k} \times 15\ \text{in} = 12.0^{k \cdot \text{in}} \qquad D + 0.4E_L = 0.8^{k}$$

$$f_{vT} = \frac{T}{2At} = \frac{12.0}{2 \times 6 \times 4 \times 0.25} = 1.00\ \text{ksi}\ \ (\text{OK})$$

Angle of twist $\theta$ (see Fig. 33.15):

$$q_{vT} = \frac{T}{2A} = \frac{12.0}{2 \times 6 \times 4} = 0.25^{k}/\text{in}$$

$$\theta = \frac{L_q}{2AG} \times \Sigma\frac{\Delta_s}{t} = \frac{63 \times 0.25}{2 \times 6 \times 4 \times 11{,}200} \times \frac{20}{1/4}$$

$$= 0.0023\ \text{rad}\ (0.13°)(\text{small})$$

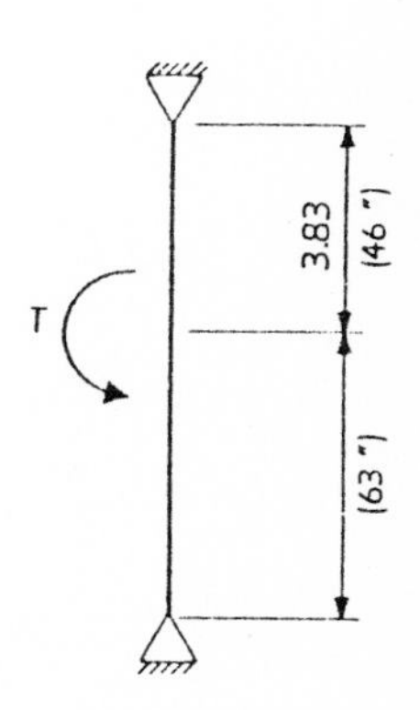


**Figure 33.15** Angle of twist.

**Combined stresses.** Check combined stresses on TS 6× 4 ×1/4 vertical frame member.

$$D + 0.5E_v = 3.45^k \quad P_A = 3.45^k$$

$$M_V = 3.45 \times 15 = 52^{k \cdot in}$$

$$M_H = 1.0 \times 15 = 15^{k \cdot in}$$

$$f_{bx} = \frac{52}{7.36} = 7.1^k/\text{in}^2 \quad f_{by} = \frac{15}{5.87} = 2.6^k/\text{in}^2$$

$$f_a = \frac{3.45}{4.59} = 0.8^k/\text{in}^2 < 0.15F_a$$

$$f_b = 9.7^k/\text{in}^2 < F_b \quad (\text{OK})$$

**Connection summary**

Connection summary and details are shown in Fig. 33.16.

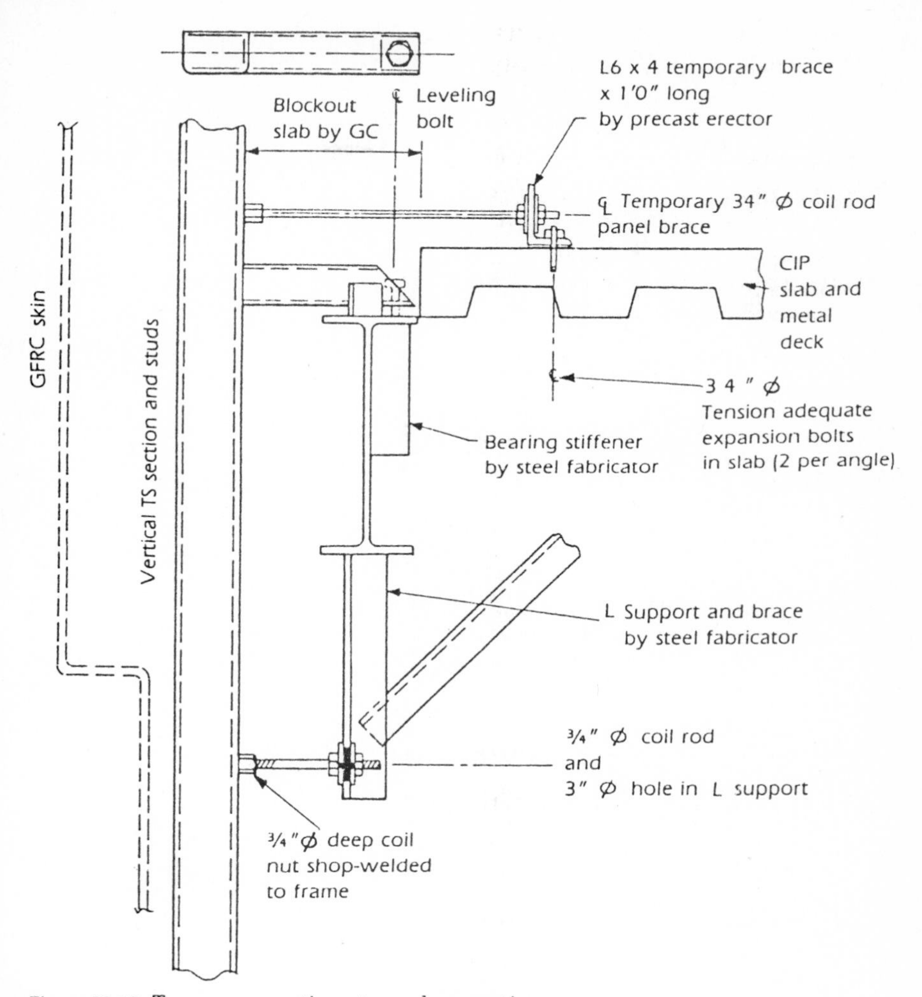


**Figure 33.16** Transverse section at panel connections.

Chapter

# 34

# Metric Design

The United States is one of the few countries still using the British foot-pound system of measurements. The majority of other countries use a metric system. This metric system may be in International System (SI) units or may be in kilogram-force, kilogram-force-per-square-millimeter units, depending on the country. For this chapter the SI units will be used.

Canada not too long ago used the foot-pound British system of units (imperial system), and pretensioning beds accordingly were in foot-inch measurements. Self-stressing forms are expensive, and therefore many of those beds are still in use; a soft conversion is used. The newer forms are all in meter-millimeter measurements. Strand, reinforcing steel, and structural steel manufactured outside the United States are all in metric units. When these are used in the United States, a soft conversion is made. Strand, reinforcing steel, mesh, and structural steel made in the United States are in British (imperial) units; when these are used outside the United States, a soft conversion is made to metric units. [A soft conversion converts exact metric measurements into slightly smaller imperial nomenclature. The reverse is true when converting products produced in British (imperial) units to metric.] There is considerable international trade in strand, reinforcing steel, and structural steel. There is also trade across borders for finished products, as between Canada and the United States and between Mexico and the United States. For finished products problems could arise because of the different measurement systems, which must be checked.

Students of architecture or engineering must become familiar with metric units, especially if they become involved in international projects. To design in metric one must think in metric. To use conversion factors will lead to error and confusion. To specify 31.63-MPa concrete in a country using the metric system is just as wrong as to specify 4.268-ksi concrete in the United States.

When working in another country, you must use the material standards and minimum live loads of that country. Awareness of conversion factors is necessary for a relationship between the two systems, but you cannot design by using conversion factors.

The SI metric system and the British system are compared in Table 34.1.

**TABLE 34.1 Conversion of Metric and U.S. Units**

| Quantity | Metric units to U.S. units | U.S. units to metric units |
|---|---|---|
| Length | 1 meter (m) = 3.28084 feet (ft) | 1 ft = 0.3048 m |
| | 1 centimeter (cm) = 0.3937 inch (in) | 1 in = 2.54 cm |
| | 1 millimeter (mm) = 0.0937 in | 1 in = 25.4 mm |
| Area | 1 square meter ($m^2$) = 10.7643 square feet ($ft^2$) | 1 $ft^2$ = 0.0929 $m^2$ |
| | 1 square millimeter ($mm^2$) = 0.00155 square inch ($in^2$) | 1 $in^2$ = 645.2 $mm^2$ |
| Volume | 1 cubic meter ($m^3$) = 35.3107 $ft^3$ | 1 $ft^3$ = 0.020832 $m^3$ |
| Gravity | 981 $m/s^2$ (g) = 32.2 $ft/s^2$ (386.4 $in/s^2$) | |
| Mass | 1 kilogram (kg) = 2.20462 pounds (lb) | 1 lb = 0.45359 kg |
| | 1 $kg/m^3$ = 0.0624 $lb/ft^3$ | 1 $lb/ft^3$ = 16.02 $kg/m^3$ |
| Force | 1 newton (N; kg $m/s^2$) = 0.2248 lb | 1 lb = 4.448 N |
| | 1 kilonewton (kN) = 0.2248 kip | 1 kip = 4.448 N |
| | 1 newton per meter (N/m) = 0.06852 lb/ft | 1 lb/ft = 14.5939 N/m |
| | 1 kilonewton per meter (kN/m) = 0.06852 k/ft | 1 k/ft = 14.5939 kN/m |
| Stress or pressure | 1 $N/m^2$ = 0.0208855 $lb/ft^2$ | 1 $lb/ft^2$ = 47.88 $N/m^2$ |
| | 1 $kN/m^2$ = 20.92575 $lb/ft^3$ | 1 $lb/ft^2$ = 0.04788 $kN/m^2$ |
| | 1 pascal (Pa; $N/m^2$) = 0.000145 $lb/in^2$ | 1 $lb/in^2$ = 6894.757 Pa |
| | 1 kilopascal (kPa) = 0.14504 $lb/in^2$ | 1 $lb/in^2$ = 6.89476 kPa |
| | 1 megapascal (MPa) = 0.14504 $kip/in^2$ | 1 $kip/in^2$ = 6.89476 MPa |
| Other units | | |
| Frequency | 1 hertz (Hz) = 1 cycle per second | |
| Moment of inertia (I) | 1 $mm^4$ = $2.4 \times 10^{-6} in^4$ | 1 $in^4$ = 416,231 $mm^4$ |
| Section modulus | 1 $mm^3$ = $6.1 \times 10^{-5} in^3$ | 1 $in^3$ = 16,387 $mm^3$ |

## Live Loads

Each country through its building codes specifies minimum live loads for public safety. These values are not always the same from country to country. A few of the values used are shown in Table 34.2. (1 lb/ft² = 0.04788 kN/m².)

**TABLE 34.2 Live Loads**

| Occupancy | Minimum live load, kN/m² | Occupancy | Minimum live load, kN/m² |
|---|---|---|---|
| Assembly | | Residential | |
| Fixed seats | 2.4 | Private apartments | 1.9 |
| Movable seats | 4.8 | Dwellings | 1.9 |
| Dining rooms and restaurants | 4.8 | Hotel guest rooms | 1.9 |
| Garages (passenger cars only) | 2.4 | Public areas | 4.8 |
| Libraries | | Schools | |
| Reading rooms | 2.9 | Classrooms | 2.4 |
| Stack areas | 7.2 | Corridors | 4.8 |
| Manufacturing | 6.0 | Stores, retail and wholesale | 4.8 |
| Office buildings | | Theaters | 4.8 |
| First floor and basement | 4.8 | | |
| Upper floors | 2.4 | | |
| Corridors | 4.8 | | |

## Dead Loads

Dead loads are the force (weight) of the building materials and any permanent attached equipment. For office buildings a load is included for the force (weight) of any movable partitions, usually 1.0 kN/m². The force of a few common building materials is shown in Table 34.3.

**TABLE 34.3 Dead Loads**

| Material | Force, kN/m² |
|---|---|
| Normal concrete, mass density 2300 kg/m³ (10 mm thick) | 0.24 |
| Medium-density concrete (sand-lightweight), mass density 1900 kg/m³ (10 mm thick) | 0.19 |
| Hardwood flooring (22 mm thick) | 0.19 |
| Softwood flooring (20 mm thick) | 0.13 |
| Suspended-ceiling system | 0.10 |
| Acoustical tile | 0.05 |
| 20-mm plaster on metal lath | 0.40 |
| 20-mm plaster on concrete | 0.26 |
| 12.7-mm gypsum board | 0.10 |
| Five-ply built-up roofing and gravel | 0.31 |
| Rigid insulation (50 mm thick) | 0.04 |
| Walls | |
| Steel stud with 15.9-mm gypsum on each side | 0.28 |
| Steel stud with 20-mm plaster on each side | 0.86 |
| Brick (200 mm thick) | 3.77 |
| Glass with frame or storefront glazing | 0.38 |

## Concrete, Prestressed Strand, and Reinforcing

The modulus of elasticity of concrete in the metric system is $E_c = 0.043\, w_c^{1.5} f_c'^{\,0.5}$; $f'_c$ is in megapascals, and $w_c$ is mass density in kilograms per cubic meter. Force/acceleration of gravity = mass density, or mass density $\times g$ = force. The concrete strengths used for prestressed concrete are 35, 40, 45, 50 and 55 MPa (usually 35 or 40 MPa). For nonprestressed concrete or topping the concrete strengths used are 20, 25, 30, and 35 MPa (usually 25 MPa). The mass density for normal-density concrete is 2300 kg/m$^3$; for medium-density concrete, 1900 kg/m$^3$.

The prestressing strand is $f_{pu}$ = 1860 MPa with $E_s$ = 190,000 MPa. The strand used today is primarily low-relaxation (LR) seven-wire strand. Strand size and area are as shown in Table 34.4.

**TABLE 34.4 1860-MPa Seven-Wire Strand**

| Size designation | Diameter, mm | Area, mm² |
|---|---|---|
| 9 | 9.53 | 55 |
| 11 | 11.13 | 74 |
| 13 | 12.7 | 99 |
| 15 | 15.24 | 140 |

Mild reinforcing steel used with prestressed members may be either grade 300-MPa or grade 400-MPa. Grade 400-MPa is usually employed. Welded wire fabric is grade 400-MPa. Bar sizes and areas are shown in Table 34.5.

**TABLE 34.5 Reinforcing Steel Bars**

| Size designation | Diameter, mm | Area, mm² | Perimeter, mm |
|---|---|---|---|
| 10 | 11.3 | 100 | 36 |
| 15 | 16.0 | 200 | 50 |
| 20 | 19.5 | 300 | 61 |
| 25 | 25.2 | 500 | 79 |
| 30 | 29.9 | 700 | 94 |
| 35 | 35.7 | 1000 | 112 |
| 45 | 43.7 | 1500 | 137 |
| 55 | 56.4 | 2500 | 177 |

Welded wire fabric nomenclature is as follows: longitudinal wire spacing × transverse wire spacing–longitudinal wire size × transverse wire size. Example: 152×305–MW 25.8× MW 18.7. Wire size nomenclature is the area of the wire in square millimeters (see Table 34.6). The usual center-to center spacings are 51, 76, 102, 152, 203, 254, and 305 millimeters.

**TABLE 34.6 Wire Fabric ***

| | | $A_s$ = mm²/m Spacing, mm | | | | |
|---|---|---|---|---|---|---|
| Wire size, mm | Wire area, mm² | 102 | 152 | 203 | 254 | 305 |
| 9.1 | 9.1 | 89.5 | 59.8 | 44.3 | 36 | 30 |
| 11.1 | 11.1 | 109 | 73 | 54 | 44 | 36.4 |
| 13.3 | 13.3 | 130 | 87.5 | 64.7 | 52.6 | 43.6 |
| 18.7 | 18.7 | 184 | 123 | 91 | 74 | 61.4 |
| 25.8 | 25.8 | 254 | 169 | 127 | 102 | 84.7 |
| 34.9 | 34.9 | 343.4 | 229.5 | 169.8 | 138 | 114.6 |
| 47.6 | 47.6 | 468 | 313 | 231 | 188 | 156 |

* Other wire sizes and spacings are available. See manufacturers' brochures.

## Allowable Prestressing Steel Stresses: Working Stress Method

The building codes of the various countries differ, and so do the allowable stresses. Many countries pattern their concrete codes on the ACI 318 building code of the American Concrete Institute. The following is in accord with ACI 318-83.

Allowable tendon stress after anchoring prior to losses = $0.74f_{pu}$ for pretensioning. The posttensioning allowable stress is $0.7f_{pu}$($f_{pu}$ = 1860 MPa). The initial loss at detensioning is approximately 10 percent. The loss under service is approximately 14 to 18 percent for LR strand; 15 percent is often used. For stress-relieved strand service loss is approximately 18 to 22 percent. In practice, the computed loss, for which there are several recommended methods, is often used.

## Allowable Concrete Stresses

Allowable concrete stresses also vary from country to country. The following is relevant for those countries which pattern their codes on the ACI 318 code.

At detensioning (transfer):

$$\text{Compression} = 0.6 f'_{ci}$$

$$\text{Tension at ends of member} = 0.5 f'^{0.5}_{ci}$$

$$\text{Tension elsewhere} = 0.25 f'^{0.5}_{ci}$$

At service:

$$\text{Compression} = 0.45 f'_c$$

$$\text{Tension in precompressed tension zone} = 0.5 f'^{0.5}_c$$

$$\text{or if bilinear analysis is used} = 1.0 f'^{0.5}_c$$

The ACI 318 load factors used for ultimate strength are 1.4 for dead load, 1.7 for live load, 1.3 for wind, and 1.43 for seismic loads. The resistance (phi) factors are 0.9 for flexure and 0.85 for shear. The resistance combinations considered are (1) $1.4DL + 1.7LL$, (2) $0.9DL + 1.43E$, and (3) $0.75\,(1.4DL + 1.7LL + 1.87E)$. For the Canadian code the load factors are 1.25 for dead load, 1.5 for live load, and 1.5 for seismic loads. The load factors are 0.6 for concrete, 0.85 for reinforcing steel, and 0.9 for prestressing steel.

Shear is designed by using the ultimate strength method. There is also an ultimate strength check required for flexure.

For shear the usual allowable stress for the concrete is either

$$\phi_c V_c = 0.6 \times 0.2 f'^{0.5}_c\, b_w\, d\, \lambda^*$$

or, if effective prestressed force exceeds 40 percent of the tension strength of the flexural reinforcement, then

$$\phi_c V_c = \phi_c \left[ 0.06 f'^{0.5}_c + 6(V_u d_p / M_u) \right] b_w\, d\, \lambda = < 0.4 f'^{0.5}_c\, b_w\, d$$

$$\phi_c V_{c,\max} = \phi_c \times 0.4 \lambda f'^{0.5}_c \left[ (1 + f_{pc}) / \left( 0.4 \lambda\, \phi_c f'^{0.5}_c \right) \right] b_w\, d_p + \phi_p V_p$$

The following design examples are in SI metric units and are based upon the current Canadian building code, except for the double-T example, which is based upon the ACI code. Comparing the double-T example with the inverted-T example shows the code change. Because building codes are updated periodically, the designer must consult the code governing the area where the project is to be constructed.

## Example Problem 1: Solid Cladding Panel (SI Units)

### Non-load-bearing spandrel panel

Design a typical solid concrete non-load-bearing spandrel panel. The

*$\lambda$= 1.0 for normal-weight concrete, 0.85 for sand-lightweight concrete, and 0.75 for all-lightweight concrete.

panel will be for an office building. The concrete is normal-weight with a density of 2400 kg/m³ and a strength at stripping of 30 MPa and at erection and service of 40 MPa. Precast concrete spandrel panels may be used with precast concrete building systems, cast-in-place concrete frame systems, or steel frame systems. See Fig. 34.1.

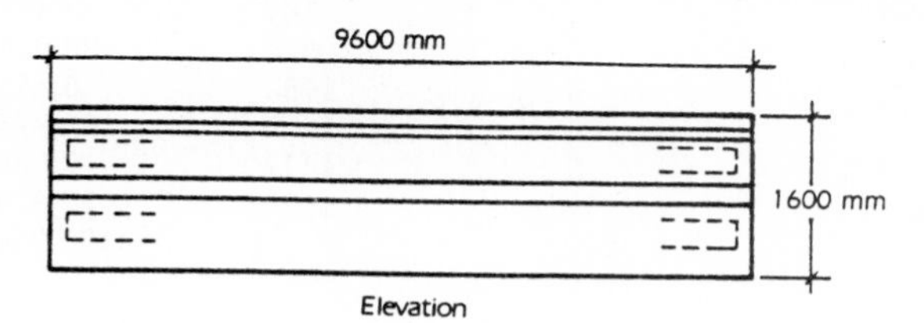


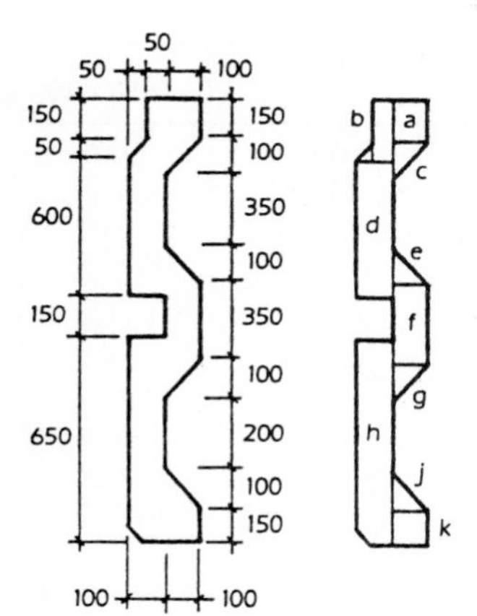


Compute Y from right

| Part | Size | Area | y | ay |
|---|---|---|---|---|
| a | 100 × 150 | 15 * 10³ | 75 | 1125* 10³ |
| b | 50 × 200 | 10 | 100 | 1000 |
| c | 100 × 100/2 | 5 | 183 | 915 |
| d | 100 × 600 | 60 | 500 | 30000 |
| e | 100 × 100/2 | 5 | 667 | 3335 |
| f | 100 × 350 | 35 | 875 | 30625 |
| g | 100 × 100/2 | 5 | 1083 | 5415 |
| h | 100 × 650 | 65 | 1275 | 82875 |
| j | 100 × 100/2 | 5 | 1417 | 7085 |
| k | 100 × 150 | 15 | 1525 | 22875 |
| | | 220* 10³ | | 185250* 10³ |

$Y_{top}$ = 185250/220 = 842 mm
$Y_{bot}$ = 1600 − 842 = 758 mm

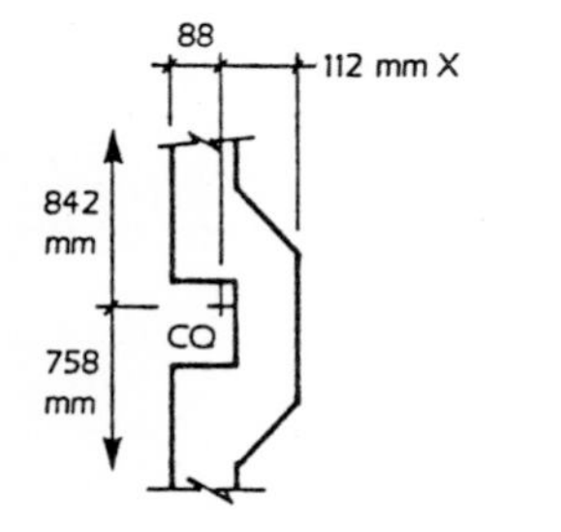


Compute X from right

| Part | Size | Area | x | ax |
|---|---|---|---|---|
| a | 100 × 150 | 15 * 10³ | 50 | 750 * 10³ |
| b | 50 × 200 | 10 | 125 | 1250 |
| c | 100 × 100/2 | 5 | 67 | 335 |
| d | 100 × 600 | 60 | 150 | 9000 |
| e | 100 × 100/2 | 5 | 67 | 335 |
| f | 100 × 350 | 35 | 50 | 1750 |
| g | 100 × 100/2 | 5 | 67 | 335 |
| h | 100 × 650 | 65 | 150 | 9750 |
| j | 100 × 100/2 | 5 | 67 | 335 |
| k | 100 × 150 | 15 | 50 | 750 |
| | | 220* 10³ | | 24590* 10³ |

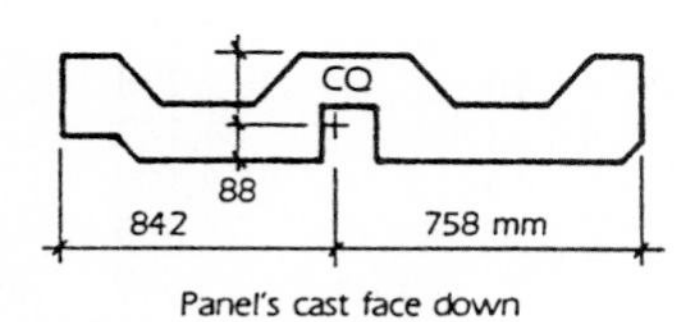


$X_t$ = ax/a = 24590/220 = 112 mm
$X_b$ = 200 − 112 = 88 mm

**Figure 34.1** Spandrel panel properties. Compute $\overline{Y}$ from the top. $Y_{top}$ = 185,250/20 = 842 mm; $Y_{bottom}$ = 1600 - 842 = 758 mm. (Legend continued on page 578.)

| Part | Size | Area, × $10^3$ | $y$ | ay, × $10^3$ |
|---|---|---|---|---|
| $a$ | 100 ×150 | 15 | 75 | 1.125 |
| $b$ | 50 × 200 | 10 | 100 | 1,000 |
| $c$ | 100 × 100/2 | 5 | 183 | 915 |
| $d$ | 100 × 600 | 60 | 500 | 30,000 |
| $e$ | 100 × 100/2 | 5 | 667 | 3,335 |
| $f$ | 100 × 350 | 35 | 875 | 30,625 |
| $g$ | 100 × 100/2 | 5 | 1083 | 5,415 |
| $h$ | 100 × 650 | 65 | 1275 | 82,875 |
| $j$ | 100 × 100/2 | 5 | 1417 | 7,085 |
| $k$ | 100 × 150 | 15 | 1525 | 22,875 |
| | | 220 | | 185,250 |

Compute $X$ from the right. $X_t = ax/a = 24{,}590/220 = 112$ mm; $X_b = 200 - 112 = 88$ mm.

| Part | Size | Area, × $10^3$ | $x$ | $ax$, × $10^3$ |
|---|---|---|---|---|
| $a$ | 100 × 150 | 15 | 50 | 750 |
| $b$ | 50 × 200 | 10 | 125 | 1,250 |
| $c$ | 100 × 100/2 | 5 | 67 | 335 |
| $d$ | 100 × 600 | 60 | 150 | 9,000 |
| $e$ | 100 × 100/2 | 5 | 67 | 335 |
| $f$ | 100 × 350 | 35 | 50 | 1,750 |
| $g$ | 100 × 100/2 | 5 | 67 | 335 |
| $h$ | 100 × 650 | 65 | 150 | 9,750 |
| $j$ | 100 × 100/2 | 5 | 67 | 335 |
| $k$ | 100 × 150 | 15 | 50 | 750 |
| | | 220 | | 24,950 |

Compute moment of inertia about the $X$ axis since the panel is flat for stripping and handling. $I_o$ for a rectangular section is $bd^3/12$; for a triangle, $bd^3/36$. $I = I_o + ad_i^2$. See Table 34.7.

**TABLE 34.7 Moment of Inertia and Section Modulus**

| Part | $b$ | $d$ | a, × $10^3$ | $d_i$ | $ad_i^2$, × $10^3$ | $I_o$, × $10^3$ |
|---|---|---|---|---|---|---|
| $a$ | 150 | 300 | 15 | 62 | 57,825 | 12,500 |
| $b$ | 200 | 50/2 | 10 | 13 | 1,690 | 2,083 |
| $c$ | 100 | 100 | 5 | 45 | 10,125 | 2,778 |
| $d$ | 600 | 100 | 60 | 38 | 86,640 | 50,000 |
| $e$ | 100 | 100 | 5 | 45 | 10,125 | 2,778 |
| $f$ | 350 | 100 | 35 | 62 | 134,540 | 29,167 |
| $g$ | 100 | 100 | 5 | 45 | 10,125 | 2,778 |
| $h$ | 650 | 100 | 65 | 38 | 93,860 | 54,167 |
| $j$ | 100 | 100 | 5 | 45 | 10,125 | 2,778 |
| $k$ | 150 | 100 | 15 | 62 | 57,660 | 12,500 |
| | | | | | 472,715 | 171,529 |

$I = 472{,}715 + 171{,}529 = 644{,}244 \times 10^3$. $S_{top} = I/X_t = 644{,}244/112 = 5752.18 \times 10^3$. $S_{bottom} = I/X_b = 644{,}244/88 = 7320.96 \times 10^3$.

The dead load per meter due to the panel mass is 220,000 mm²/10⁶ × 2400 kg/m³ × 9.81 m/s²/1000 = 5.18 kN/m.

### Stripping

See Fig. 34.2.

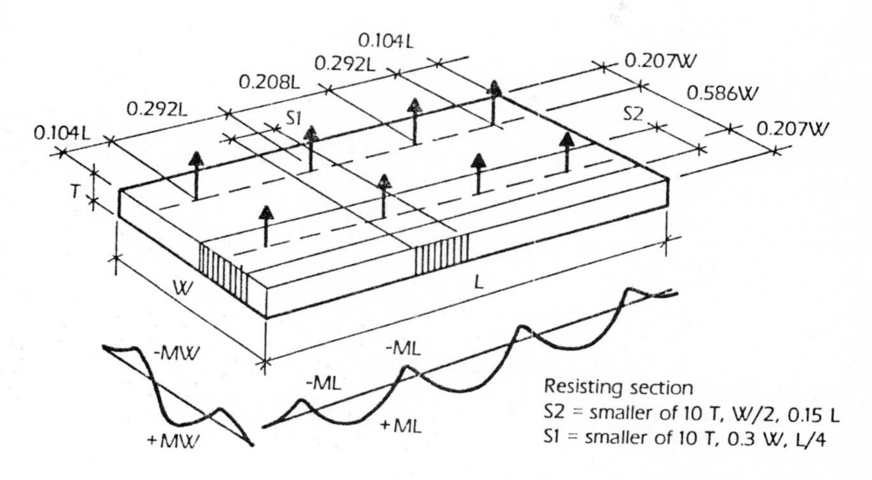


**Figure 34.2** Pickup points for stripping.

**Moment in the *L* direction.** $ML = 0.0054/2 \times 5.18$ kN/m $\times 9.6$ mm² = 1.29 kN · m. The resisting section is the smallest of 10 × 100 mm = 1000 mm, 1600 mm/2 = 800 mm, or 0.15 × 9600 = 1440. Use 800 mm (half the panel width); therefore, use half of the section modulus of the panel or twice *ML* and the full section modulus. Since the panel is not flat, for this example use twice the moment and the full section modulus.

$$ML = 0.0054 \times 5.18 \text{ kN/m} \times 9.6 \text{m}^2 = 2.58 \text{ kN}\cdot\text{m}$$

$$f_{\text{ten}} = M/S = 2.58 \times 10^6 / 5752 \times 10^3 = 0.45 \text{ MPa}$$

$$f_{\text{allowable}} = 0.6\gamma f_{ci}'^{0.5} / \text{factor of safety} = 0.6 \times 1.0 \times (30 \text{ MPa})^{0.5} / 2.0$$
$$= 1.64 \text{ MPa} > 0.45 \text{ MPa (OK)}$$

NOTE: Some designers will multiply the moment by a load factor from 1.4 to 2.0 and not use the factor of safety as in the above equation while others will write the equation as above. The end result is the same. The factor of safety, or multiplier, is used for suction during stripping and impact. For sculptured panels 2.0 is commonly used; for plain panels, 1.5.

**Moment in the *W* direction.** See Fig. 34.3. For the resisting section a minimum of $10T = 10 \times 100 = 1000$ mm, $0.3W = 0.3 \times 1600 = 4800$ mm, and $L/4 = 9600/4 = 2400$ mm. Therefore, use 480 mm.

$$\text{Section modulus} = bd^2/6 = 480 \times 100^2/6 = 800 \times 10^3$$

$$f_{ten} = M/S = 0.542 \times 10^6/(800 \times 10^3) = 0.678 \text{ MPa} < 1.64 \text{ MPa (OK)}$$

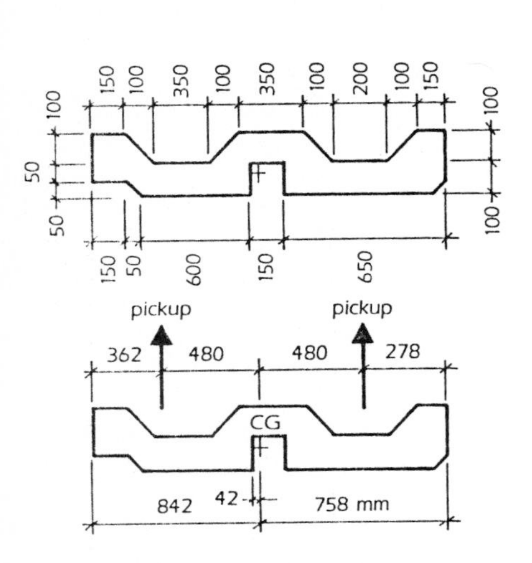


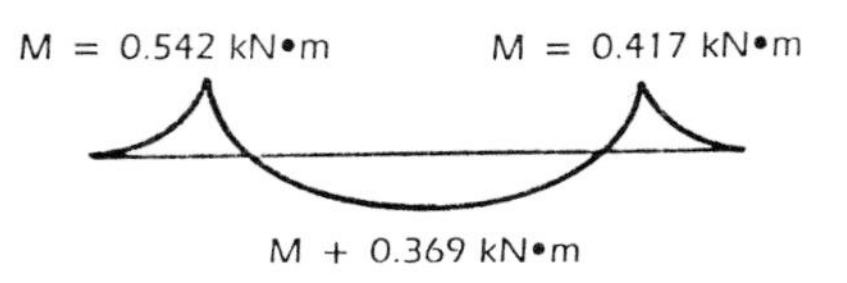


Since the shape is complex, use volume X lever arm, then convert to load moment.

Example for the negative moment:

$M_{neg}$ 1 = (150x150x287 + 212x100x106 + 100x50x179) x (9.6 m/4) x (2400 kg/m$^3$ x 9.81) $10^{12}$

= 0.542 kN • m moment

Same process was used for M+ and M- at the other end.

**Figure 34.3 Moments in the wide direction.**

The fiber stress at the extreme fiber is less than the fiber stress at rupture divided by the factor of safety in each direction. Therefore, the section will not crack during stripping. Since the panel is a noncracked section, reinforcing steel is not necessary. However, minimum wall steel is provided for shrinkage and thermal effects. The Uniform Building Code and the ACI codes require a minimum of 0.002 times the cross-sectional area. In the $L$ direction this would be

$$\text{Number of 10M bars} = 0.002 \times 220{,}000 \text{ mm}^2/100 \text{ mm}^2/\text{no. 10M bar}$$
$$= \text{minimum of 5 bars}$$

In the $W$ direction, with a maximum bar spacing of 5 times the thickness = 5000 mm, use

$$A_s = 0.002 \times 500 \text{ mm} \times 100 \text{ mm} = 100 \text{ mm}^2 \text{ or no. 10M bars at 500-mm spacing}$$

Some designers will place reinforcing in the section to save the panel if somehow it should crack. $A_s = M/f_s jd$. Let $j = 0.9$. Let $f_s = 0.5\,f_y$ for 300-MPa steel or $0.4f_y$ for 400-MPa steel. Let $d = 150$ mm. Compute for the whole section; hence use 2 × moment:

$$A_s = 2 \times 1.29 \text{ kN·m} \times 10^6/(0.5 \times 300 \text{ MPa} \times 0.9 \times 150 \text{ mm})$$
$$= 127 \text{ mm}^2, \text{or 2 bars, 10M}$$

Owing to the shape of the panel, bar placement will often determine the quanitity of bars. In this case there are 10 no. 10M bars parallel, which is greater than the minimum of 5 bars. (See Fig. 34.4.)

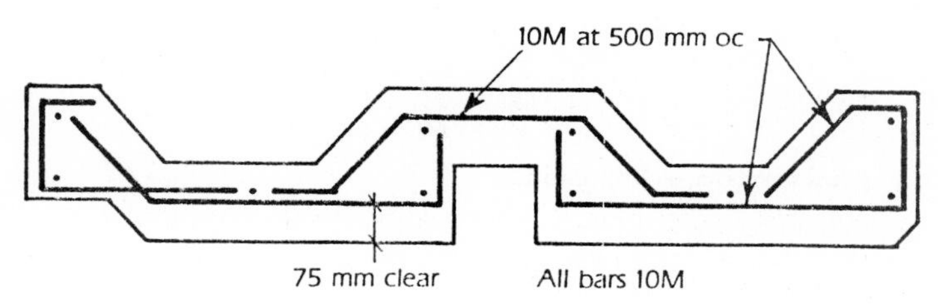


**Figure 34.4** Steel placement.

### Handling and erection

See Fig. 34.5. Moments in the $L$ direction are the same as for stripping. The $W$ direction is treated as a unit strip since the one edge is continuously supported during the rotation process. This moment is as follows:

Using the volume lever arm and converting to kilonewton meters as before,

$$MW^+ \text{ at center of gravity} = (150 \times 150 \times 767 + 212 \times 100 \times 586 + 100 \times 50 \times 659 + 480 \times 100 \times 240 + 100 \times 50 \times 175 + 100 \times 100 \times 90 - 110{,}000 \times 842)$$
$$(2400 \text{ kg/m}^3 \times 9.81 \text{m/s}^2/10^{12})$$
$$= -1.09 \text{ kN·m/m of length}$$

$$S = bt^2/6 = 1000 \times 100^2/6 = 1667 \times 10^3$$
$$f_{\text{ten}} = M/S = 1.09 \times 10^6/1667 \times 10^3$$
$$= 0.653 \text{ MPa} < f_{r,\text{allowable}} \text{ of } 1.64 \text{ MPa (OK)}$$

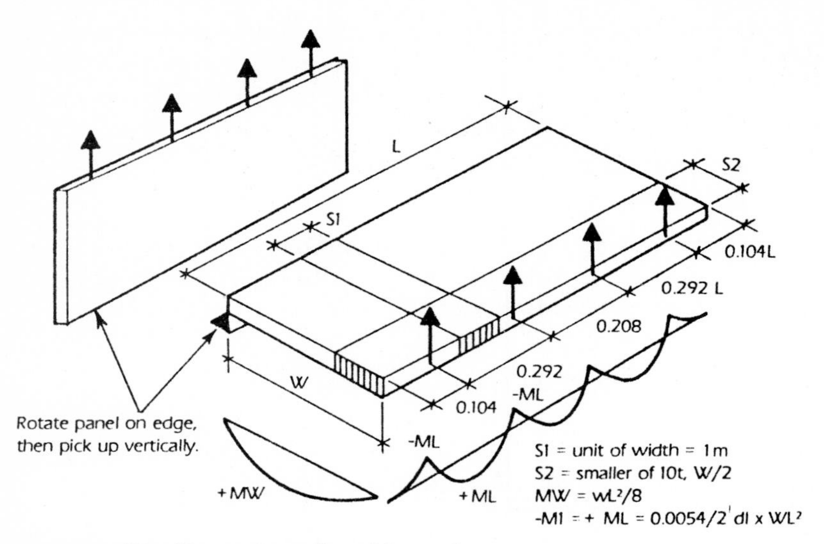


**Figure 34.5** Handling and erection pickup points.

## Shipping

The panel is shipped vertically because of two-point support during shipping. See Fig. 34.6.

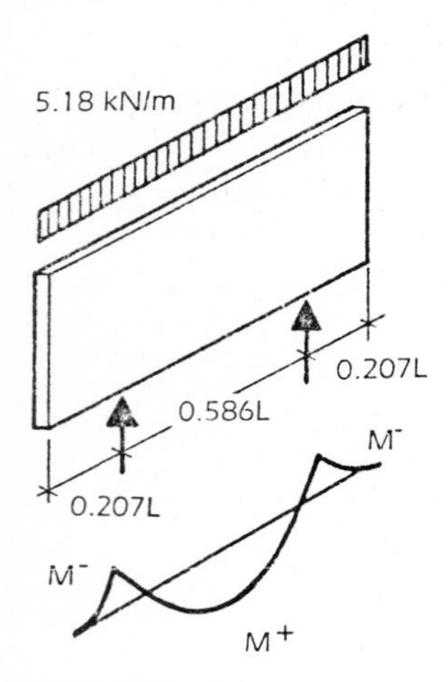


**Figure 34.6** Shipping pickup points. $^{-}M = {}^{+}M = 0.0215wL^2 = 0.0215 \times 5.18 \times 9.62^2 = 10.264$ kN · m. $I = ad^2 + I_o = 49{,}380 \times 10^6$. $S_{top} = I/Y_{top} = 49{,}380/842 = 58.65 \times 10^6$. $S_{bottom} = I/Y_{bottom} = 49{,}380/758 = 65.15 \times 166$. $f_{ten} = M = 10.264/58.65 = 0.175$ **MPa < 1.64 MPa (therefore** OK, uncracked).

**Moment of Inertia Vertical Position**

| Part | Area, × $10^3$ | $d_i$, mm | $ad_i^2$, × $10^6$ | $I_o$, × $10^6$ |
|---|---|---|---|---|
| *a* | 15 | 767 | 8,824 | 28.125 |
| *b* | 10 | 742 | 5,505 | 33.33 |
| *c* | 5 | 659 | 2,171 | 2.78 |
| *d* | 60 | 342 | 7,018 | 1800.0 |
| *e* | 5 | 175 | 153 | 2.78 |
| *f* | 35 | 33 | 38 | 357.0 |
| *g* | 5 | 241 | 290 | 2.78 |
| *h* | 65 | 433 | 12,186 | 2288.0 |
| *j* | 5 | 575 | 1,653 | 2.78 |
| *k* | 15 | 683 | 6,997 | 28.13 |
| | | | 44,835 | 4545.705 |

With an impact factor of 1.5 to 2.0, say, 2.0, then $M = 2.0 \times 10.264 =$ 20.528 kN·m. Check minimum steel requirement if panel should crack.

$$A_s = M/f_s jd = 20.528 \text{ kN·m} \times 10^6/(0.5 \times 300 \text{ MPa} \times 0.9 \times 1600 \text{ mm}) = 95 \text{ mm}^2$$

Two no. 10M bars as recommended above $= 2 \times 100 \text{ mm}^2 = 200 \text{ mm}^2$, which is greater than 95 and therefore OK.

### Service loads

The service loads are gravity, wind, and seismic loads both out of plane and in plane. When designing concrete connections, factored loads generally are used. The steel elements are designed with both the working stress method and the ultimate strength method.

In the United States load factors, resistance (phi) factors, and seismic factors are not the same as in other countries. The metric system is not currently being used in the United States, and the designer must therefore design in accordance with the building code applicable where the project is to be built. Since the following depends on the metric system, Canadian factors will be used.

In the United States the load factors are 1.4 for dead load, 1.7 for live load, 1.3 for wind, and 1.43 for seismic loads. The resistance (phi) factors are 0.9 for flexure and 0.85 for shear. Wind values are subject to local wind velocities, building height, and exposure. The seismic equation is

$$F_p = ZIC_pW_p$$

where $Z$ = the seismic zone and is 0.4 for Zone 4, $I$ is the building's importance factor, usually 1.0 for nonessential buildings, $C_p$ = the factor for the wall panel and is 0.75, and $W_p$ is the panel weight. That is, $F_p = 0.4 \times 1.0 \times 0.75W_p = 0.3W_p$ or 30 percent of gravity. For connection design the force $F_p$ is multiplied by 1.33 for the body of the connector and by 4.0 for the fastener of the connector.

In Canada the load factors are 1.2 for dead load, 1.5 for live load, and 1.5 for seismic and wind loads. The resistance (phi) factors are 0.6 for concrete, 0.85 for reinforcing steel, 0.9 for prestress tendons, and 0.9 for structural steel. Conditions affecting the wind force are similar to those in the United States, i.e., local velocity, height, and exposure. For this example a wind force of 1.2 kN/m² will be used. The wall panel seismic equation is

$$V_p = vS_pW_p$$

where the zone value $v$ is 0.4 for seismic Zone 6 and is a velocity ratio. This value is obtained either from a map of the zones or from a map of the peak horizontal ground velocities. The highest intensity contour is 0.32. The Vancouver Island coastal region extending up to Alaska is in Zone 6 and contour 0.32, as indicated on the maps. The $S_p$ factor is 0.9 for wall panel design and 11.0 for connector design. $W_p$ is the dead load of the panel.

For this panel the dead load is 5.18 kN/m $\times$ 9.6$M$ = 49.728 kN,say,50 kN. The wind load is 1.2 kN/ m² $\times$ 1.6 m $\times$ 9.6 m = 18.432 kN, the panel seismic force = $0.4 \times 0.9 \times 49.728$ = 17.90 kN, and the connection seismic force = $0.4 \times 11.0 \times 49.728$ = 218.8 kN.

Canadian seismic requirements are greater than current Uniform Building Code provisions for comparable seismic risk zones.

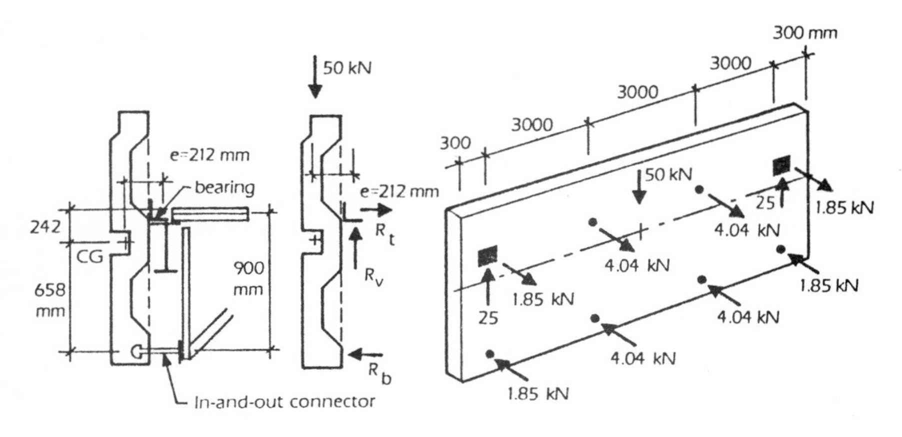


**Figure 34.7** Gravity loads.

## Gravity loads

Gravity loads are shown in Fig. 34.7. The lateral reactions were obtained from moment distribution by using a unit load and then multiplying by the $R_t$ and $R_b$ values.

## Seismic loads

Seismic connection loads out of plane are shown in Fig. 34.8. The resultant = 218.8 kN. Seismic connection loads in plane are shown in Fig. 34.9.

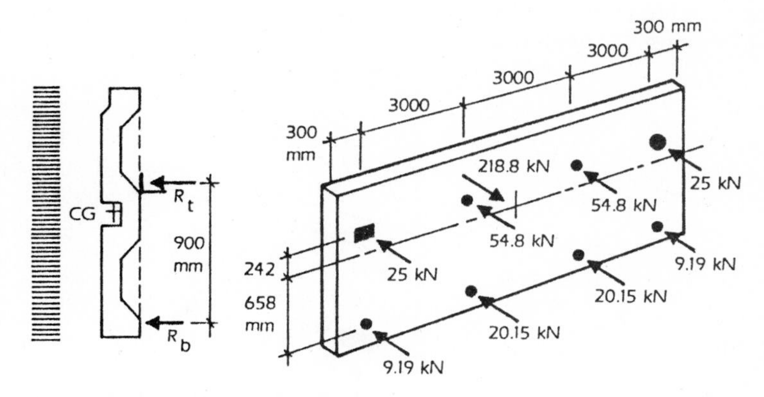


**Figure 34.8** Seismic loads out of plane.

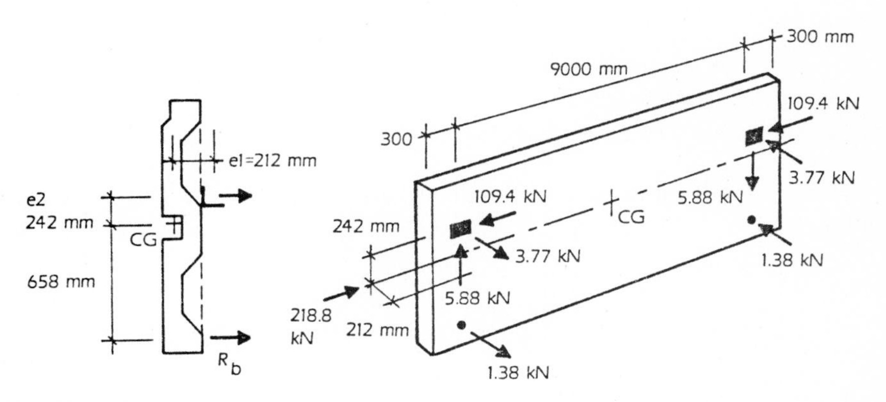


**Figure 34.9** Seismic loads in plane.

Gravity loads and seismic loads out of plane are shown to the left in Fig. 34.10. Gravity loads and seismic loads in plane are shown to the right.

**Connectors**

With load factors of 1.2 for dead load and 1.5 for seismic load the factored design forces on the connectors are as follows :

Upper in-and-out connector = $1.2\times 4.04$ kN$^+$ − $1.5\times 54.8$ kN
= 87 kN for tension, 77.38 kN for compression

Lower in-and-out connector = $-1.2\times 4.04$ kN$^-$ + $1.5\times 20.15$ kN
= 35 kN for compression, 25.38 kN for tension

Vertical-bearing connector = $1.2\times 25$ kN + $1.5\times 5.88$ kN = 32.82 kN
Horizontal-bearing connector = $1.5\times 109.4$ = 164.1 kN
In-and-out bearing connector = $1.2\times 1.85$ + - 3.72 kN
= 7.85 kN for tension

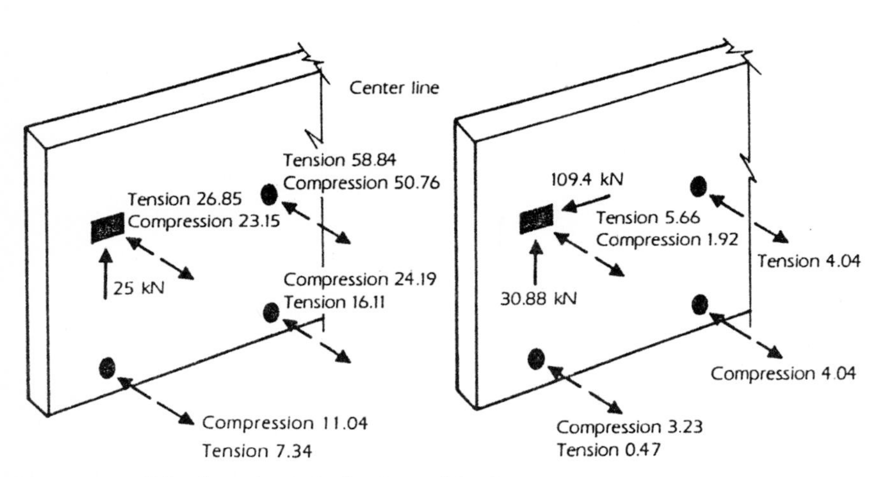


**Figure 34.10** Seismic and gravity loads combined.

Gravity loads and seismic loads in plane govern for the bearing connection.

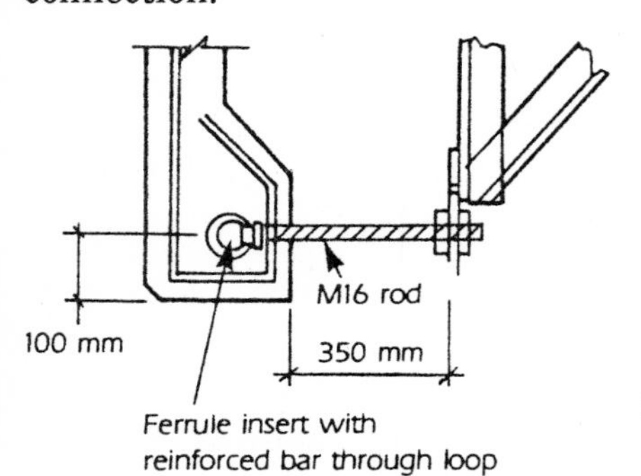


**Figure 34.11** Bottom in-and-out connection.

A bottom in-and-out connection is shown in Fig. 34.11. The factored force for compression is 35 kN and for tension 25.38 kN. Try an M16 bolt A-307M or A-36M threaded rod. The area is 201 mm²; $r$ = 4 mm; $kL/r$ = 350/4 = 87.5. By using the *AISC Load and Resistance Factor Design Manual* and converting to metric units,

$$kL/r = 1.0 \times 14/0.64 = 21{,}875, \text{ say, } 22$$

$$\phi_s f_{cr} = 29.83 \text{ ksi (by using table, } \phi = 0.85)$$

$$29.83 \times 6.895 = 205 \text{ MPa}$$

$$205 \text{ MPa} \times 201 \text{ mm}^2/1000 = 41.2 \text{ kN} > 35 \text{ kN (OK)}$$

Check tension:

$$\text{Allowable tension factored} = \phi_s \times 0.75F_u \times \text{area} = 0.9 \times 0.75 \times 400 \text{ MPa} \times 201 \text{ mm}^2/1000 = 54.27 \text{ kN} > 25.38 \text{ kN (OK)}$$

Check the manufacturer's brochure. Connector's factored capacity > 25.38 in tension; this is OK.

A top in-and-out connection is shown in Fig. 34.12. The tension is 87 kN and the compression 77.35 kN. Try threaded bar M22. The area is 380 mm²; $kL/r$ = 1.0 × 22/5.5 = 31.8. By using the *AISC Load and Resistance Factor Design Manual* and converting to metric units,

$$kL/r = 1.0 \times 7/0.88 = 7.95$$

$$\phi F_c = 30.5 \text{ ksi} \times 6.895 = 210.3 \text{ MPa}$$

$$210.3 \text{ MPa} \times 380 \text{ mm}^2/1000 = 79.9 \text{ kN} > 77.35 \text{ kN (OK)}$$

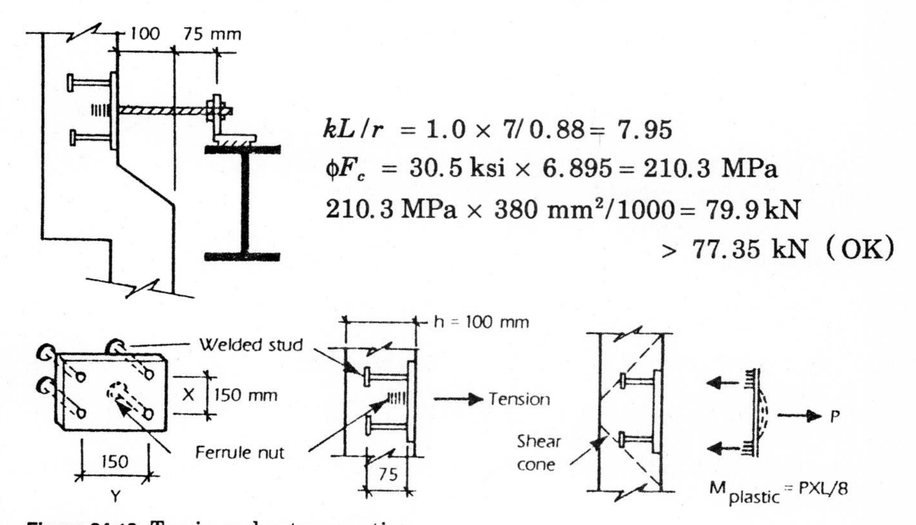


**Figure 34.12** Top in-and-out connection.

Check tension:

$$0.9 \times 0.75 \times 400 \text{ MPa} \times 380 \text{ mm}^2/1000 = 102.6 \text{ kN} > 87 \text{ kN (OK)}$$

Assume four welded studs $12.7 \times 75$ mm; then check if this will work. From the manufacturing data, $L_e = 67.2$ mm, $dh = 25$ mm, area = 121.67 mm², $f_y = 360$ MPa, and $\phi_s = 0.9$. Use the smallest of the following four conditions.

$$1.\ P_c = \phi_c \times 0.33 f_c'^{0.5} \left[2h(X + Y + 4L_e - 4h) + 4h^2\right]$$
$$= 0.6 \times 0.33 \times 40\ \text{MPa}^{0.5}[2 \times 100(150 + 150 + 4 \times 67.2$$
$$- 4 \times 100) + 4 \times 100^2]/1000$$
$$= 92.36\ \text{kN}$$

$$2.\ P_c = \phi_c \times 0.33 f_c'^{0.5} \left[XY + 2L_e(X + Y) + 4L_e^2\right]$$
$$= 0.6 \times 0.33 \times 40^{0.5}[150 \times 150 + 2 \times 67.2(150 + 150)$$
$$+ 4 \times 67.2^2]/1000$$
$$= 101.28\ \text{kN}$$

$$3.\ P_c = \phi_c \times 0.33 f_c'^{0.5} \times 2^{0.5} L_e \pi (L_e + dh) \times \text{number of studs}$$
$$= 0.6 \times 0.33 \times 40^{0.5} \times 2^{0.5} \times 67.2 \times 3.1416(67.2 + 25)/$$
$$1000 \times 4 = 137.8\ \text{kN}$$

$$4.\ P_s = \phi_s f_y \times \text{area of stud} \times \text{number of studs}$$
$$= 0.85 \times 360\ \text{MPa} \times 121.67\ \text{mm}^2/1000$$
$$\times 4\ \text{studs} = 148.9\ \text{kN}$$

Use the smallest, which is 92.36 kN. Since this is greater than 87 kN, it is OK. Use four studs.

**Thickness of plate.** Determine thickness of plate.

$$\text{Moment}_{\text{plastic}} = PL/8 = 87\ \text{kN} \times 150\ \text{mm}/(8 \times 1000) = 1.6\ \text{kN}\cdot\text{m}$$

$$t_p = (4M_p/b\,\phi_s f_y)^{0.5} = [4 \times 1.6\ \text{kN} \times 10^6/$$
$$(200\ \text{mm} \times 0.9 \times 345\ \text{MPa})]^{0.5} = 10.15\ \text{mm}$$

Use plate 12 mm thick, 200 mm by 200 mm, 345-MPa steel (50 ksi) to match studs.

**Punching shear.** Check punching shear due to compression.

$$P_{\text{compression}} \leqq \phi_c \times 0.33 f_c'^{0.5} h b_o$$
$$b_o = 2(X + h + Y + h)$$

$P_{compression} = 0.6 \times 0.33 \times 40\text{MPa}^{0.5} \times 100 \times 2(200 + 100 + 200 + 100)/$

$1000 = 150 \text{ kN} > 77.35 \text{ (OK)}$

**Clip angle.** Design clip angle (see Fig. 34.13).

$P = 87\text{kN}$. Try angle 100 mm × 100 mm × 16 mm × 150 mm long.
Steel $f_y = 345 \text{ MPa}$

$k = 27 \text{ mm} \quad a = 62 - 27 = 35 \text{ mm}$

$M = 87 \text{ kN} \cdot \text{m} \times 35 \text{ mm}/1000 = 3.045 \text{ kN} \cdot \text{m}$

$t = (4M/b\,\phi_s f_y)^{0.5}$

$t = [4 \times 3.045 \text{ kN} \cdot \text{m} \times 10^6/(150 \text{ mm} \times 0.9$
$\times 345 \text{ MPa})]^{0.5} = 16 \text{ mm}$

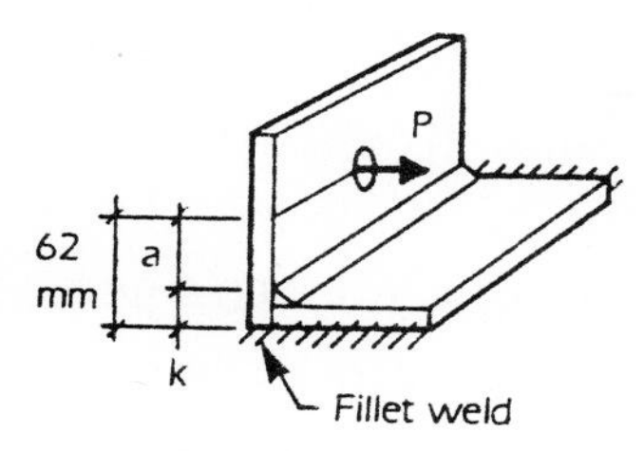


**Figure 34.13** Clip angle design.

For welding, use 480xx electrode.

Allowable stress $= \phi_w \times 0.6 f_{w,xx} \times 0.707$

$= 0.75 \times 0.6 \times 480 \text{ MPa} \times 0.707 = 152.7 \text{ MPa}$

$f_v = 87 \text{ kN}/2 \text{ welds } 100 \text{ mm long} = 0.435 \text{ kN/mm}$

$I = 1 \times 100 \text{ mm}^3/12 \times 2 = 166{,}667 \text{ mm}^2$

$C = 50 \text{ mm}$

$f_m = M \times C/1 = 87 \text{ kN} \times 62 \text{ mm} \times 50 \text{ mm}/166{,}667 \text{ m}^3 = 1.63 \text{ kN/mm}$

$R = (f_v^2 + f_m^2)^{0.5} = (0.435^2 + 1.63^2)^{0.5} = 1.687 \text{ kN/mm}$

$1.687 \text{kN/mm} \times 1000/152.7 \text{ MPa} = 11\text{-mm weld size}$

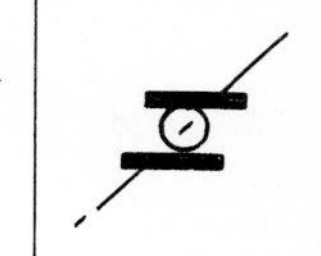

Use 12-mm weld 100 mm long on two edges.

**Bearing connection.** Assume six studs size 22×150 mm long. Then check to see if the connection is adequate.

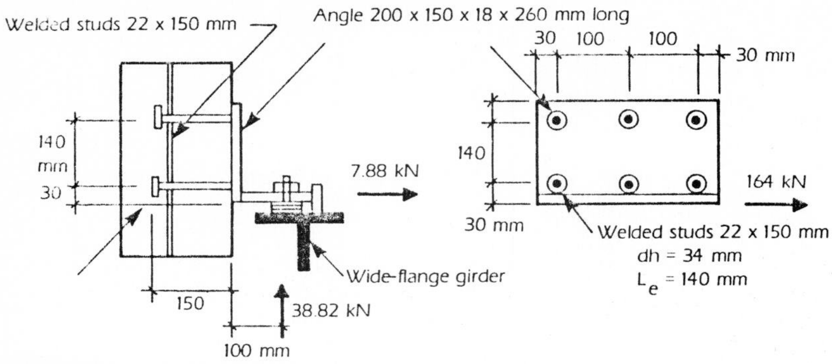


**Figure 34.14** Bearing connection.

Welded stud capacities:

dh

$L_e$

$$\phi_s P_s = \phi_s f_y \times \text{area of stud}$$

$$= 0.85 \times 360 \text{ MPa} \times 380 \text{m}^2/1000 = 116.28 \text{ kN}$$

$$\phi_c P_c = \phi_c \times 0.33 f_c'^{0.5} \times 2^{0.5} L_e \pi (L_e + dh)$$

$$= 0.6 \times 0.33 \times 40 \text{ MPa}^{0.5} \times 2^{0.5} \times 140 \text{ mm} \times 3.1416$$

$$(140 + 34)/1000 = 135.5 \text{ kN}$$

$$\phi_s V_s = 0.83 \phi_s f_y A_s = 0.83 \times 0.85 \times 360 \text{ MPa} \times 380 \text{ mm}^2/1000 = 96.5 \text{ kN}$$

$$\phi_c V_c = \phi_c m_u A_s f_y = 0.6 \times 1.0 \times 380 \text{mm}^2 \times 360 \text{ MPa}/1000 = 82 \text{kN}$$

$$\phi_c V \leqq \phi_c P_c$$

Check angle thickness for bending:

$$M = 38.82 \text{ kN} \times 100 \text{ mm}/1000 = 3.882 \text{ kN} \cdot \text{m}$$

$$t = (4M \times 1000/b_s f_y)^{0.5}$$

$$= \left[4 \times 3.882 \times 10^6/(260 \times 0.9 \times 345 \text{ MPa})\right]^{0.5} = 13.87 \text{ mm} < 18 \text{ mm} \quad \text{(OK)}$$

Check studs in shear and in tension (see Fig. 34.15).

100 mm
100
70
70
100 mm
164 kN
100 mm

Shear = 164 kN/6 studs = 27.33 kN
Tension = 164x100/(200x2) = 41.0 kN

Moment = 164 kN x 100 mm
Let R = 1.0
f = MxRxd1/(X²+ Y²)

**Figure 34.15** Stud forces.

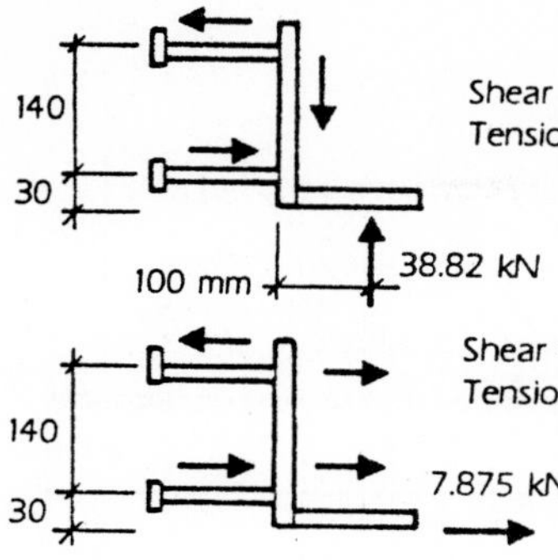

140
30
100 mm
Shear = 38.82 kN / 6 studs = 6.47 kN vertical
Tension = 38.82 kN x 100 mm/(140x3) = 9.24 kN
38.82 kN

140
30
Shear = 0.0
Tension = 7.875 kN/6 + 7.875 x 170/(140x3) = 4.5 kN
7.875 kN

**Figure 34.15** (*Continued*)

$$\text{Shear vertical} = 164 \times 100 \times 1.0 \times 100/\left(6 \times 70^2 + 4 \times 100^2\right)$$
$$= 23.63 \text{ kN}$$

$$\text{Shear horizontal} = 164 \times 70 \times 1.0 \times 70/\left(6 \times 70^2 + 4 \times 100^2\right)$$
$$= 16.54 \text{ kN}$$

$$P_f = \text{sum of tension} = 9.24 + 4.5 + 41 = 54.74 \text{ kN}$$

$$V_f = \text{resultant of shears} = \left[(6.47 + 23.62)^2 + (27.33 + 16.54)^2\right]^{0.5}$$
$$= 53.42 \text{ kN}$$

The following are interaction equations for tension and shear:

For steel, $(P_f/\phi_s P_s)^2 + (V_f/\phi_s V_s)^2 =< 1.0$

$$= (54.75/116.28)^2 + (54.46/96.5)^2$$
$$= 0.53 < 1.0 \text{ (OK)}$$

For concrete, $(P_f \ \phi P_c)^{1.33} + (V_f / \phi_c V_c)^{1.33} =< 1.0$

$$= (54.74/135.5)^{1.33} + (53.42/82)^{1.33}$$
$$= 0.86 < 1.0 \text{ (OK)}$$

NOTE: In the 1988 Uniform Building Code the concrete equation is

$$1/\phi_c\left[P_f/P_c^{\ 2} + V_f/V_c^{\ 2}\right] =< 1.0$$

After the panel has been leveled with shims, then the angle is welded to the wide flange girder with the use of steel bars as fillers. Where the framing system is concrete, then weld plates are cast into the concrete girders to receive the panel clip angles. When the weight is high, then bracketed angles or steel tubes are used. See Fig. 34.16.

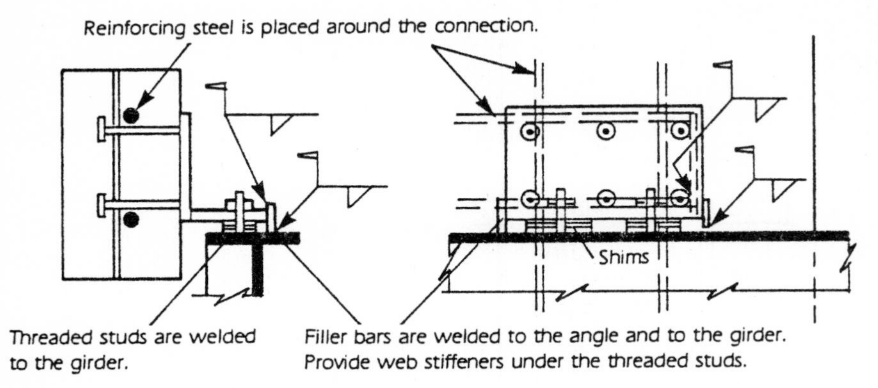


**Figure 34.16** Final bottom bearing connection.

## Example Problem 2: Window Panel (SI Units)

The design of a window panel is shown in Fig. 34.17.

$f_{ten} = \lambda 0.6 f_c'^{0.5} / FS$

FS= 1.5 to 2.0 for stripping, 1.2 to 1.5 for shipping and erection. Use 2.0.

$\lambda$= 1.0 normal-weight concrete, 0.85 for sand-lightweight concrete.

$f_c'$= 25 MPa at stripping, 35 MPa at erection.

$f_{ten,allowable} = 0.85 \times 0.6 \times 25 \text{ MPa}^{0.5}/2.0$

$= 1.275$

$f = M/S$

$F = M^+/S_b = 2.89 \text{ kN} \cdot \text{m} \times 10^6/13.968 \times 10^6$

$= 0.207 \text{ MPa} < 1.275 \text{ MPa}$ (therefore OK)

$f = M^-/S_t = 5.304 \text{ kN} \cdot \text{m} \times 10^6/15.96 \times 10^6$

$= 0.322 \text{ MPa} < 1.275 \text{ MPa}$ ( therefore OK)

Section is uncracked during stripping and erection.

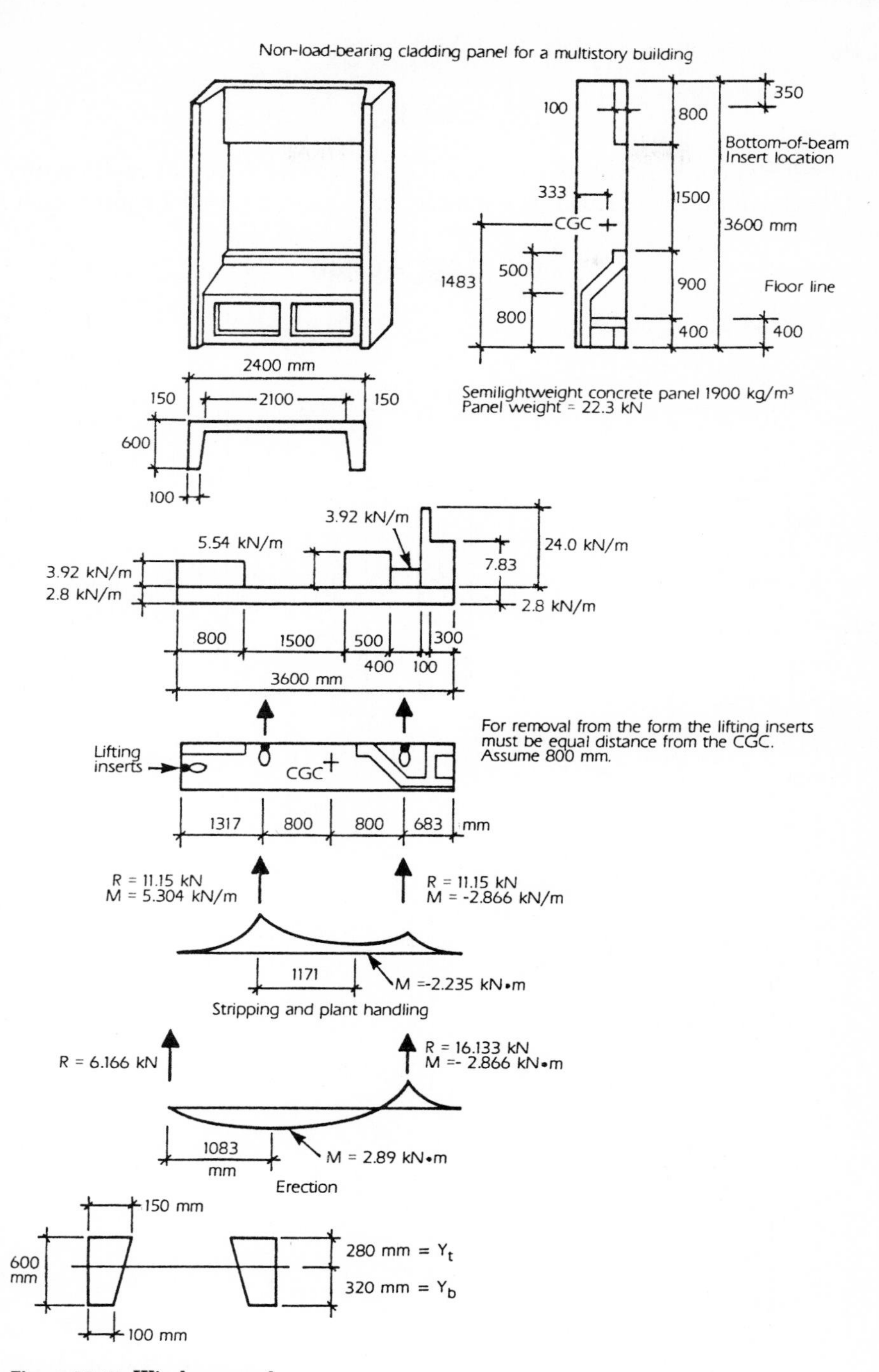


**Figure 34.17** Window panel.

**Lateral seismic force**

$$V_p = vS_pW_p$$

where $v$ = seismic zone velocity (use 0.4); $S_p$ = seismic panel coefficient, 0.9 for the panel and 11.0 for the connection; and $W_p$ = panel weight = 22.3 kN.

$$V_p = 0.4 \times 0.9W_p = 0.36W_p$$

Use Fig. 34.18.

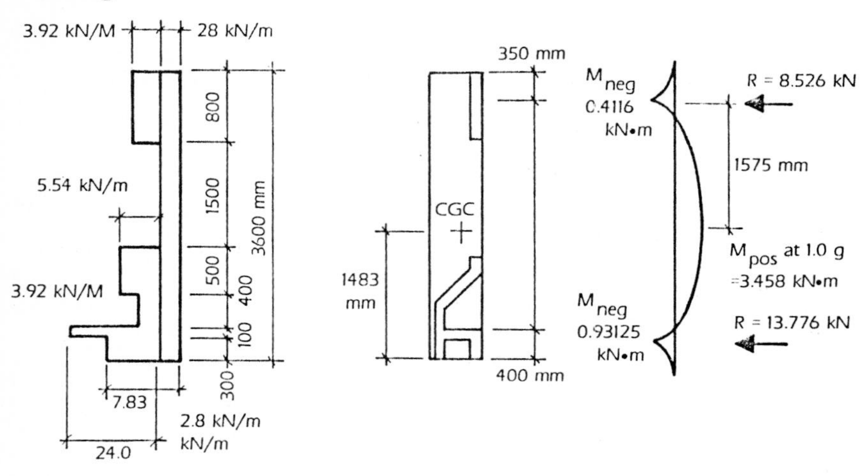


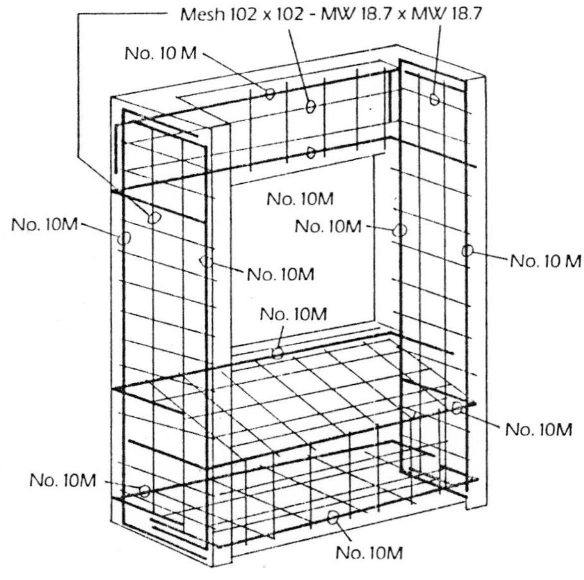


**Figure 34.18** Seismic force out of plane and reinforcing.

$M^+ = 0.36 \times 3.458\ \text{kN}\cdot\text{m} = 1.2449\ \text{kN}\cdot\text{m}$

Rib section modulus top $= 15.96 \times 10^6$

Rib section modulus bottom $= 13.968 \times 10^6$

Concrete strength at erection $=$ 35 MPa

$f = M/S = 1.2449\ \text{kN}\cdot\text{m} \times 10^6/13.968 \times 10^6 = 0.089\ \text{MPa}$

$f_{\text{ten, allowable}} = \lambda \times 0.6 f_c'^{0.5}$

$f_{\text{ten}} = 0.85 \times 0.6 \times 35^{0.5} = 3.017\ \text{MPa}$

$3.017 > 0.089$; therefore, a noncracked section

Reinforcing steel is not required for the seismic force, but steel is provided for temperature change and shrinkage and for tying the panel together. The building codes do not cover reinforcing steel for precast panels except to refer to cast-in-place concrete.

The temperature steel requirement is $0.0018bt$. For vertical wall steel the minimum is 0.0012; for horizontal, 0.0020. The Prestressed Concrete Institute recommends a minimum of $0.001bt$.

$$A_s = 0.0018 \times 102\ \text{mm} \times 100\ \text{mm} = 18.36\ \text{mm}^2$$

Use mesh measuring 102×102–MW 18.7×MW 18.7 if ribs were cracked; moment = 1.25 kN·m. Use moment and concrete strength from stripping.

$M = 5.304\ \text{kN}$ $\qquad f_c' = 25\ \text{MPa}$

$b = 100\ \text{mm}$ $\qquad d = 600\ \text{mm} - 40\ \text{mm} = 560\ \text{mm}$

$$a = d - \left[- 2M_u/(\phi_c \times 0.85 f_c' b) + d^2)\right]^{0.5}$$

$$a = 560 - \left[- 2 \times 5.3 \times 10^6/(0.6 \times 0.85 \times 25\ \text{MPa} \times 100\ \text{mm}) + 560^2\right]^{0.5}$$

$a = 7.47\ \text{mm}$

$A_s = \phi_c \times 0.85 f_c' ba/\phi_s f_y$

$A_s = 0.6 \times 0.85 \times 25\ \text{MPa} \times 100\ \text{mm} \times 7.47\ \text{mm}/(0.85 \times 400\ \text{MPa})$

$A_s = 28\ \text{mm}^2$

Use no. 10M area = 100 mm², top and bottom. Place no. 10M dowels through all insert loops and adjacent to studs.

### Connection design

Various connection forces are shown in Fig. 34.19.

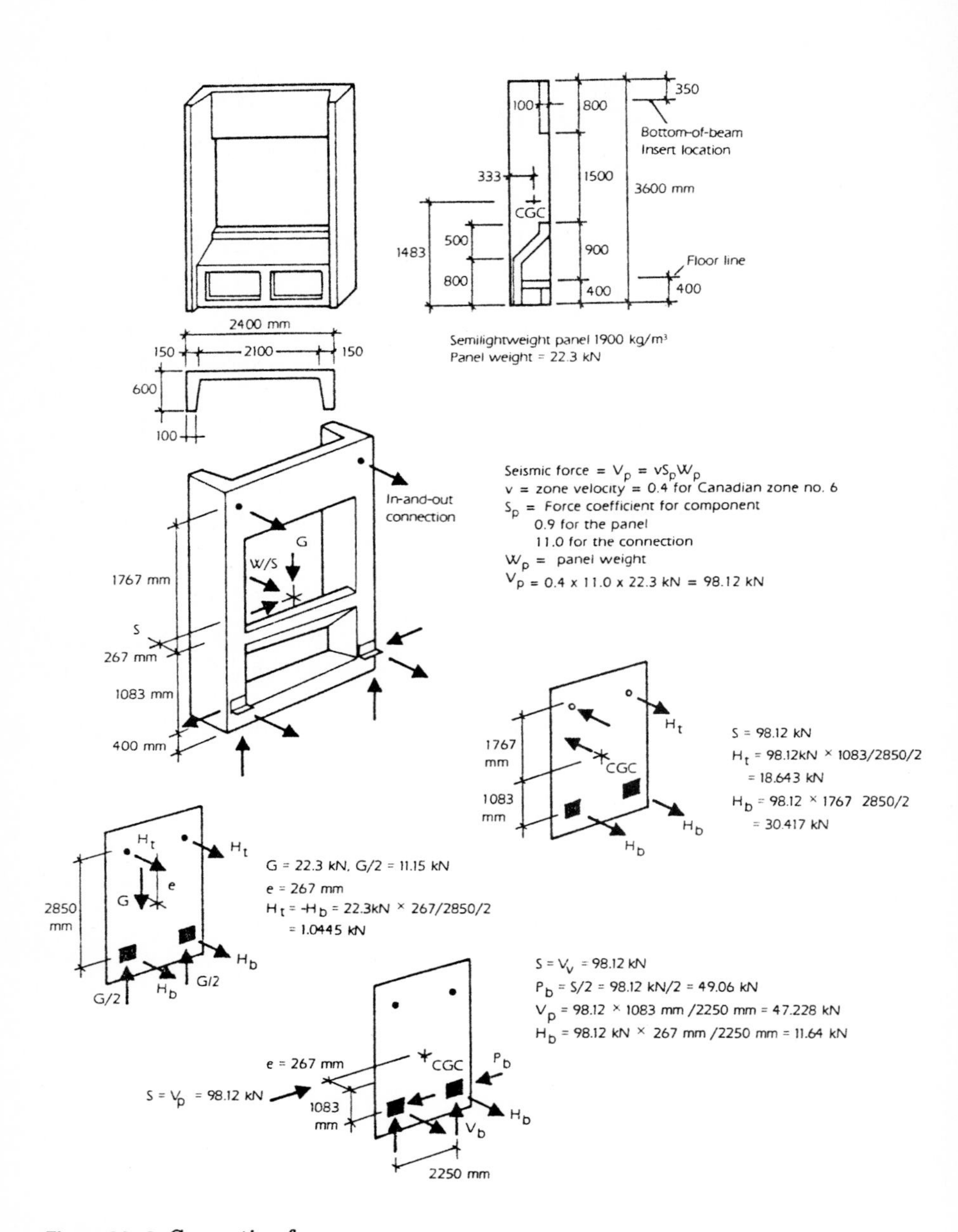


**Figure 34.19** Connection forces.

Concrete and the concrete inserts are designed by the ultimate strength method. The steel components may be designed either by the working stress method or by the load and resistance factor design method. Since the load and resistance factor design method for the steel components is compatible with the concrete design, it is used for this example. Canadian load factors are 1.25 for dead load and 1.5 for live load and for seismic or wind loads. The resistance (phi) factor for concrete is 0.6, for reinforcing steel 0.85, and for structural steel and fasteners 0.9. The Canadian code seismic equation for components is $V_p = vS_pW_p$. For this example the value of $v$ is 0.4, which is for the critical seismic area of British Columbia. The value of $S_p$ for the panel design is 0.9 and for the connections 11.0. In comparison, the equation in the Uniform Building Code (1988 edition) is $F_p = ZIC_pW_p$, in which $Z$ is the seismic zone factor, which for California is 0.4; $I$ is the importance factor, which is 1.0 for an office building; and $C_p$ is 0.75 for the panel. For the body of the connector this is multiplied by 1.33, which gives 1.0; for the fasteners the value of $C_p$ is multiplied by 4, which gives 3.0. According to the recommended provisions of the National Earthquake Hazards Reduction Program of the U.S. Federal Emergency Management Agency (Building Seismic Safety Council), the equation is $F_p = A_vC_pPW_p$, where $A_v$ is 0.4 for the areas of greatest seismic risk; $P$ is the occupancy factor, which is 1.0 for an office-type occupancy; and $C_p$ is 0.9 for the panel, 3.0 for the body of the connector, and 6.0 for the fasteners. The $U$ factor for seismic loads is 1.4 in the United States and 1.5 in Canada. The Canadian requirements are thus more stringent than those of the United States.

**Top connection.** See Fig. 34.20.

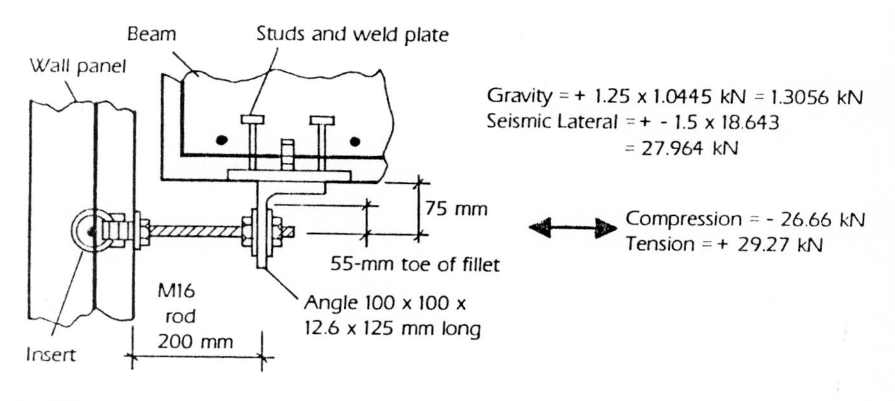


**Figure 34.20** Top connection.

Clip angle design:

$$t = (4M_u / \phi_s f_y b)^{0.5} = [(4 \times 2927\,\text{kN} \times 1000)/(0.9 \times 360\ \text{MPa} \times 125\ \text{mm})]^{0.5} = 12.5\ \text{mm}$$

Use an angle 100× 100×12.6 mm long.

Threaded rod design: For M16 threaded rod of A-307 steel the allowable ultimate tension is 55 kN, which is greater than 29.27 kN and therefore OK.

$$0.9 \times 0.75 \times 400\ \text{MPa} \times 201\ \text{mm}^2 = 54.27\ \text{kN}$$

where 0.9 = $\phi_s$ and 0.75 = tension factor for threaded rods.

$k = 1.0$ $\quad$ $r$ = diameter/4 = 16 mm/4 = 4 mm

$L = 200$ mm $\quad$ $kL/r = 1{,}0 \times 200/4 = 50$

By using the *AISC Load and Resistance Factor Design Manual*, page 6-124, $f_a$ = 26.83 ksi. In metric units, 26.83 ksi ×6.895 = 185 MPa.

201 mm²×185 MPa/1000 = 37 kN > 27.96 kN (OK)

Use M16 threaded rod.

Beam insert: See Fig. 34.21.

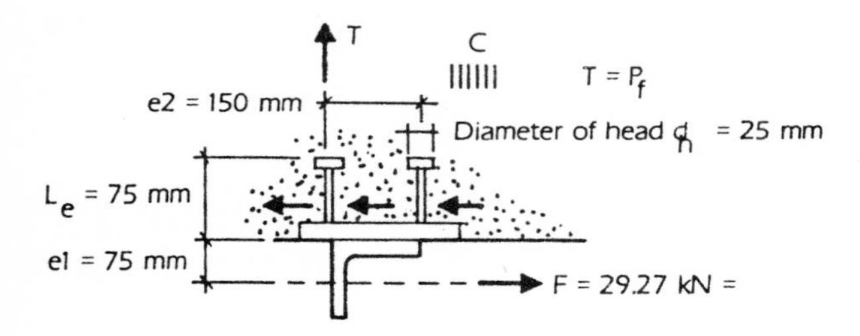


**Figure 34.21** Beam insert.

Normal-weight concrete, 35 MPa at erection

Shear = 29.27 kN

Stud tension = $T \times e_1/e_2$ $\quad$ $e_2$ = stud spacing

$T$ = 29.25 kN × 57 mm/150 mm = 14.635 kN

**Headed Stud Values ($\phi_s$ = 0.85; $f_y$ = 360 MPa)**

| Stud diameter, mm | 6.3 | 9.5 | 12.7 | 15.9 | 19.0 | 22.2 |
|---|---|---|---|---|---|---|
| $\phi_s V_s$, kN | 8.0 | 18.0 | 32.2 | 50.3 | 68.7 | 98.5 |
| $\phi_s P_s$, kN | 9.7 | 21.9 | 38.7 | 60.5 | 87.2 | 118.6 |

Try 12.7-mm studs 75 mm long, $\phi_s P_s = 38.7$ kN,$\phi_s V_s = 32.2$ kN.

$$\phi_c P_c = \phi_c \lambda \times 0.33 f_c'^{0.5} A_o$$

where $A_o = 2^{0.5} L_e \pi(L_e + dh)$.

$$\phi_c P_c = 0.6 \times 1.0 \times 0.33 \times 35 \text{ MPa}^{0.5} \times 2.05 \times 75 \times 3.1416 \times (75 + 25)/1000 = 20 \text{ kN/ stud}$$

$$\phi_c V_c = \phi_s m_u A_s f_s = 0.9 \times 1.0 \times 2 \text{ studs} \times 126 \text{ mm}^2 \times 0.9 \times 400 \text{ MPa}/1000 = 81.14 \text{ kN}$$

Combine tension and shear equations:
For concrete,

$$(P_f / \phi_c P_c)^{4/3} + (V_f / \phi_c V_c)^{4/3} \leqq 1.0$$

$$(14.635/20)^{4/3} + (29.27/81.14)^{4/3} = 0.916 < 1.0 \text{ (OK)}$$

For studs,

$$(P_f / \phi_s P_s)^2 + (V_f / \phi_s V_s)^2 < 1.0$$

$$(14.635/20)^2 + [(29.27/(2 \times 32.2)]^2 = 0.353 \leqq 1.0 \text{ (OK)}$$

Weld clip angle to plate: see Fig. 34.22.

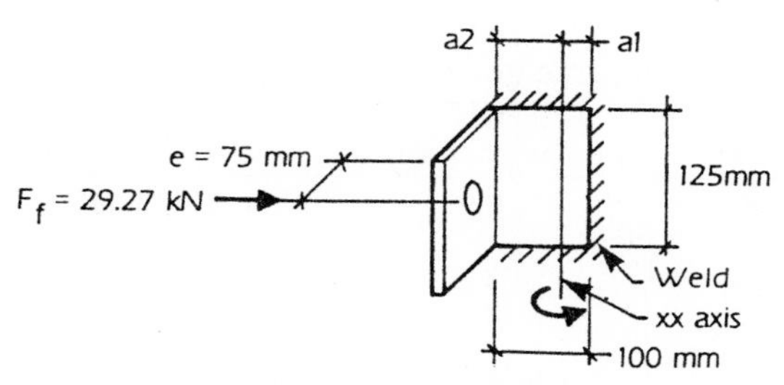


**Figure 34.22** Weld clip angle to plate.

$$a_1 = (2 \times 100 \times 100/2 + 125 \times 0.0)/(2 \times 100 + 125) = 30.77 \text{ mm}$$

$$a_2 = 100 - 30.77 = 69.23 \text{ mm}$$

$$I_{xx} = 2 \times 69.23^3/3 + 2 \times 30.77^3/3 + 125 \times 30.77^3 = 359{,}934 \text{ mm}^4$$

$$f_t = M \times c / I_{xx} = 29.27 \text{ kN} \times 75 \text{ mm} \times 69.23/359{,}934 = 0.4222 \text{ kN/ mm}$$

$$f_v = 29.27 \text{kN}/325 \text{ mm} = 0.090 \text{ kN/ mm}$$

Allowable for the weld using 480xx electrode $\Rightarrow \phi_w \times 0.6 \times f_u$ weld $\times 0.707$. fillet coefficient = $0.75 \times 0.6 \times 480$ MPa $\times 0.707$ = 152.7 MPa.

$$f_r = \left(f_t^2 + f_v^2\right)^{0.5} = 0.432 \text{ kN/mm}$$

The minimum weld size = 0.432 kN/mm ×1000/152.7 MPa = 3.0 mm. Use a 7-mm fillet weld. The minimum weld size for the 12.6-mm angle = 5 mm.

**Bottom connection.** See Fig. 34.23.

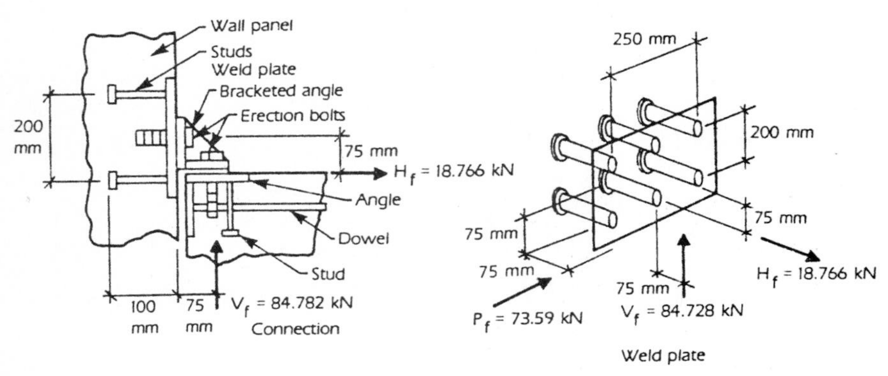


**Figure 34.23** Bottom connection.

Gravity plus seismic longitudinal force governs.

| | Vertical, kN | In-plane horizontal | In-and-out, kN |
|---|---|---|---|
| Gravity × 1.25 = | 13.940 | 0.0 | 1.3056 |
| Seismic in-plane × 1.5 = | 70.842 | 73.59 | 17.4600 |
| Total | 84.782 | 73.59 | 18.7656 |

Shear:

| | Vertical, kN | Horizontal, kN |
|---|---|---|
| $V_f$ /6 studs = 84.782 kN/6 = | 14.13 | |
| $P_f$ 6 studs = 73.59 kN/6 = | | 12.265 |
| $P_f\, ed\, /d^2$ | | |
| Vertical = 73.59 kN × 75 mm × 125mm / $\left(6 \times 100^2 + 2 \times 0.0^2 + 4 \times 125^2\right)$ = | 5.632 | |
| Horizontal = 73.59 kN × 75 mm × 100 mm / $\left(6 \times 100^2 + 2 \times 0.0^2 + 4 \times 125^2\right)$ = | | 4.5055 |
| | 19.762 | 16.7705 |

$$V_f \text{ resultant} = (19.762^2 + 16.7705^2)^{0.5} = 26.05 \text{ kN}$$

Tension :

$$V_f e/(3 \text{ studs} \times 200\,\text{mm})$$

$$= 83.782 \text{ kN} \times 75 \text{ mm}/(3 \times 200 \text{ mm}) = 10.598 \text{ kN}$$

$$P_f e/(2 \text{ studs} \times 250\,\text{mm})$$

$$= 73.59 \text{ kN} \times 75 \text{ mm}/(2 \times 250 \text{ mm}) = 11.0385 \text{ kN}$$

$$H_f e/(3 \text{ studs} \times 200\,\text{mm})$$

$$= 18.766 \text{ kN} \times 75/\text{mm}/(3 \times 200 \text{ mm}) = 2.346 \text{ kN}$$

$$H_f \text{ 6 studs} = 18.766/6 = 3.128 \text{ kN}$$

$$\text{Total tension} = 27.1105 \text{ kN}$$

Try 15.9-mm studs × 100-mm length.

Area = 198 $\text{mm}^2$

$\phi_s V_s = 50.3$ kN $\quad \phi_s P_s = 60.5$ kN (from stud table)

$$\phi_c P_c = \phi_c \lambda \times 0.33 f_c'^{0.5} A_o$$

where $A_o$ for a single cone $= 20.5\pi L_e (L_e + db)$.

$$\phi_c P_c = 0.6 \times 0.85 \times 0.33 \times 35 \text{ MPa}^{0.5} \times 2^{0.5}$$
$$\times 3.1416 \times 100(100 + 25) = 55.28 \text{ kN}$$

$$\phi_c V_c = \text{smaller of } \phi_c P_c \text{ versus } \phi_c \mu_c A_s \phi_s f_{us}$$
$$= 55.28 \text{ kN versus } 0.6 \times 1.0 \times 198\text{mm}^2 \times 0.9 \times 400 \text{ MPa}$$
$$= 55.28 \text{ kN versus } 42.768 \text{ kN}$$

Use 42.768 kN.

Combine tension and shear equations:

For studs,

$$(P_f/\phi_s P_s)^2 + (V_f/\phi_s V_s)^2 = 1.0$$

$$(27.11/60.5)^2 + (26.05/50.3)^2 = 0.46 \text{ (OK)}$$

For concrete,

$$(P_f/\phi_c P_c)^{1.333} + (V_f/\phi_s V_s)^{1.333} < 1.0$$

$$(27.11/64.04)^{1.333} + (26.05/42.77)^{1.333} = 0.829 < 1.0 \text{ (OK)}$$

Note that the value of $\phi_c P_c$ for a single cone has been used because the tension is not uniform.

Weld angle to weld plate: see Fig. 34.24.

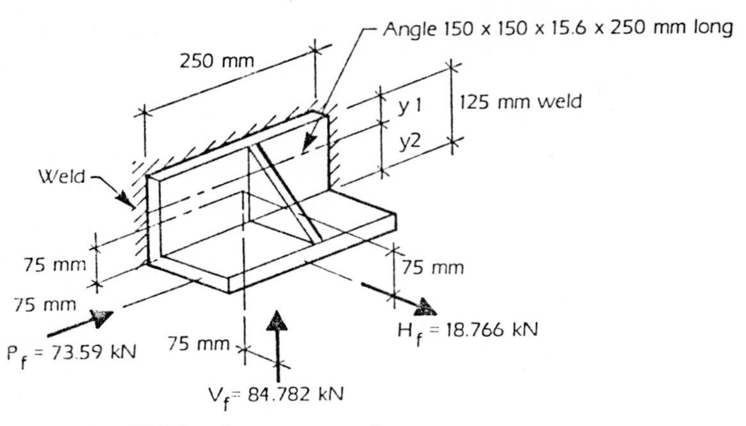


**Figure 34.24** Welding: bottom connection.

$$y_1 = 2 \times 125/2 + 250 \times 0.0/(2 \times 125 + 250) = 31.25\,\text{mm}$$

$$y_2 = 125 - 31.25 = 93.75\,\text{mm}$$

$$I_{yy} = 2 \times 93.75^3/3 + 2 \times 31.25^3/3 + 250 \times 31.25^2 = 813{,}802\,\text{mm}^4$$

$$I_{xx} = 2 \times 125 \times 125^2 + 250^3/12 = 5{,}208{,}333$$

$$J = I_{xx} + I_{yy} = 6{,}022{,}135$$

| | Vertical, kN/mm | In-plane, kN/mm | Out-of-plane, kN/mm |
|---|---|---|---|
| Verticalforce | | | |
| 84.728 kN/500 mm = | 0.170 | | |
| 85.728 × 75 × 93.75/813,802 = | | | 0.732 |
| Horizontal force | | | |
| 18.766 kN/500 mm = | | | 0.038 |
| 18.766 × 75 × 93.75/813.802 = | | | 0.160 |
| In-plane force | | | |
| 73.59 kN/500 mm = | | 0.147 | |
| 73.59 × 75 × 125 mm/5,208,333 = | | | 0.1325 |
| 73.59 × 75 × 125 mm/6,022,135 = | 0.115 | | |
| 73.59 × 75 × 125 mm/6,022,135 = | | 0.086 | |
| Total | 0.285 | 0.233 | 1.845 |

$$F_f \text{ weld} = \left(0.285^2 + 0.233^2 + 1.845^2\right)^{0.5} = 1.881\ \text{kN/mm}$$

For weld use 480xx electrode allowable = $\phi_w \times 0.6 f_w \times 0.707$ = $0.75 \times 0.6 \times 480$ MPa $\times 0.707$ = 152.7 MPa

$$\text{Weld size} = 1.881\ \text{kN/mm} \times 1000/152.7\,\text{MPa} = 12.6\ \text{mm}$$

Use 12.6-mm weld.

Note that if the angle were not bracketed, then the minimum angle thickness would be

$$t = (4M/\phi_s f_y b)^{0.5} = [4(84.728\text{ kN} + 18.766\text{ kN})$$

$$\times 75\text{ mm}/(0.9 \times 400\text{ MPa} \times 250\text{mm})]^{0.5} = 18.85\text{ mm}$$

Use angle 18.77 mm thick.

After leveling provide end closure plates and weld angle to the floor weld plates.

## Example Problem 3: Solid Slab Composite (SI Units)

Solid precast concrete slabs are commonly used for floor and roof systems in buildings, load-bearing and non-load-bearing wall panels, spandrel panels, and bridge decks. The panels are prestressed in one direction and designed to resist bending in that direction only. They are generally used as simple beams but can also be designed and used as continuous beams subject to shipping and handling length restrictions. For floor systems the panels may be noncomposite if there is no topping or composite if there is a topping. For roof systems the panels are usually noncomposite. For wall systems the panels are noncomposite and may be either single-thickness or sandwiched. Topping thickness is usually governed by depth requirements for floor electrical outlet and junction boxes. If the floor system is not electrified, the topping is usually 50 mm thick. A topping also serves as a diaphragm to transfer and resist lateral forces, for which some building codes require a minimum thickness of 60 mm.

When designing any prestressed concrete system, the designer must consult the building code governing the location where the project is to be constructed. The load factors (*U* values) and the resistance (phi) values vary with the country. For the United States the *U* values are 1.4*DL* and 1.7*LL*; the phi values are 0.9 for flexure and 0.85 for shear. For Canada the load factors are 1.25*DL* and 1.5*LL*; the phi factors are 0.6 for concrete, 0.85 for reinforcing steel, and 0.9 for prestressing strand.

As an example, take the solid slab in Fig. 34.25:

Live load $= 2.4\text{kN/m}^2$

Superimposed dead load $= 1.0\text{ kN/m}^2$

Unit force $= 100\text{ mm} \times 1.0\text{m}^2 \times 1850\text{ kg/m}^3$

$$\times 9.81\text{ m/s}^2/10^6 = 1.81\text{ kN/m}^2$$

Topping $= 60 \times 1.0\text{ m}^2 \times 2400\text{kg/m}^3$

$$\times 9.81\text{ m/s}^2/10^6 = 1.4\text{ kN/m}^2$$

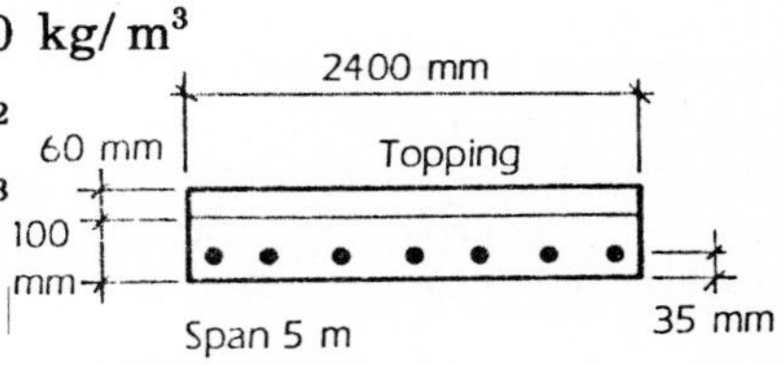

2400 mm
60 mm
Topping
100 mm
35 mm
Span 5 m

**Figure 34.25** Solid slab.

Note that the superimposed dead load is for movable partitions and mechanical items.

Normal density topping = 2400 kg/m$^3$ $f'_{ct}$ = 30 MPa

Sand-lightweight-density unit = 1850 kg/m$^3$

$f'_c$ = 35 MPa $f'_{ci}$ = 25 MPa

For no. 13 low-relaxation strand, $f_{pu}$ = 1860 MPa, $E_{ps}$ = 190,000 MPa, and area = 99 mm$^2$. The loss of prestress initially is 10 percent; in service, 18 percent.

**Section properties**

**Noncomposite.** Noncomposite properties are as follows:

$$\text{Area} = 100 \times 2400 = 240{,}000 \text{ mm}^2$$

$$\text{Moment of inertia} = 2400 \times 100^3/12 = 200 \times 10^6$$

$$\text{Top and bottom section modulus} = I/C = 200 \times 10^6/50 \text{ mm} = 4000 \times 10^3$$

**Composite.** See Fig. 34.26.

$$E_{ct} = 0.043 w_c^{1.5} f'^{0.5}_{ci} = 0.043 \times 2400^{1.5} \times 30^{0.5} = 26{,}791$$

$$E_c = 0.043 w_c^{1.5} f'^{0.5}_c = 0.043 \times 1850^{1.5} \times 35^{0.5} = 20{,}242$$

$$\text{Effective topping width} = 2400 \text{ mm} \times E_{ct}/E_c = 2400 \times 26{,}791/20{,}242 = 3280 \text{ mm}$$

$$Y_{comp} = \left(240{,}000 \text{ mm}^2 \times 50 \text{ mm} + 60 \text{ mm} \times 3280 \text{ mm} \times 130 \text{ mm}\right) (240{,}000 + 60 \times 3280) = 86 \text{ mm}$$

$$I_{o,\,topping} = bd^3/12 = 3280 \times 60^3/12 = 59 \times 10^6$$

$$\text{Area of topping} = 3280 \times 60 = 196{,}800 \text{ mm}^2$$

**Composite I Data**

| Part | Area | $d$ | $Ad^2$ | | $I_o$ |
|---|---|---|---|---|---|
| Unit | 240,000 | (86-50) | $311 \times 10^6$ | | $200 \times 10^6$ |
| Topping | 196,800 | (160-30-86) | $381 \times 10^6$ | | $59 \times 10^6$ |
| | | | $692 \times 10^6$ | + | $259 \times 10^6 = 951 \times 10^6$ |

Topping 60 mm
14 mm
100 mm
74 mm
NA
86 mm
Span 5 m

**Figure 34.26** Composite properties. $I_{composite} = 951 \times 10^6$. $S_{ct} = 951 \times 10^6/74 = 12,851 \times 10^3$. $S_{cc} = 951 \times 10^6/14 = 67,928 \times 10^3$. $S_{bc} = 951 \times 106/86 = 11,058 \times 10^3$.

**Moments for simple span = $wL^2/8$**

$$\text{Unit force} = 1.81\ \text{kN/m}^2 \times 2.4\ \text{m} \times 5^2/8 = 13,575\ \text{kN}\cdot\text{m}$$

$$\text{Topping} = 1.4\ \text{kN/m}^2 \times 2.4\ \text{m} \times 5^2/8 = 10.5\ \text{kN}\cdot\text{m}$$

$$\text{Superimposed dead load} = 1.0\ \text{kN/m}^2 \times 2.4\ \text{m} \times 5^2/8 = 7.5\ \text{kN}\cdot\text{m}$$

$$\text{Live load} = 2.4\ \text{kN/m}^2 \times 2.4\ \text{m} \times 5^2/8 = 18.5\ \text{kN}\cdot\text{m}$$

**Allowable stress**

Concrete at transfer:

| | |
|---|---|
| Compression, $0.6f'_{ci} = 0.6 \times 25\text{MPa} =$ | 15.0 MPa |
| Midspan tension, $0.25f'^{0.5}_{ci} = 0.25 \times 25^{0.5} =$ | 1.25 MPa |

Concrete at service:

| | |
|---|---|
| Compression, $0.45f'_c = 0.45 \times 35\ \text{MPa} =$ | 15.75 MPa |
| Tension, $0.5f'^{0.5}_c$ H $= 0.5 \times 35^{0.5} =$ | 2.59 MPa |
| If corrosive environment, $0.25f'^{0.5}_c$ H $=$ | 1.259 MPa |

Topping:

| | |
|---|---|
| Compression, $0.45f'_{ct} = 0.45 \times 30\ \text{MPa}$ | 13.5 MPa |

Low-relaxation strand, $0.74f_{pu}$ (100 percent −loss percent) 100× area/ strand

| | |
|---|---|
| At transfer, $0.74 \times 1860(100 - 10)/100 \times 99\ \text{mm}^2 =$ | 122.6 kN/strand |
| At service, $0.74 \times 1860(100 - 18)/100 \times 99\ \text{mm}^2 =$ | 111.73 kN/strand |

**Required prestress**

$$F_{p,\min} = \frac{\dfrac{M_{DL} + M_{top}}{S_b} + \dfrac{M_{DLS} + M_{LL}}{S_{bc} - f_{ten,allowable}}}{1/A + e/S_b}$$

$$F_{p,\min} = [(13.575 + 10.5) \times 1000/4000 + (7.5 + 18.5)$$
$$\times 1000/11,058 - 2.959\ \text{MPa}]/(1000/240,000 + 15/4000) = 683\ \text{kN}$$

Number of strands = 683 kN/111.73 kN/strand = 6.11 strands
Use seven strands. For placement, see Fig. 34.27.

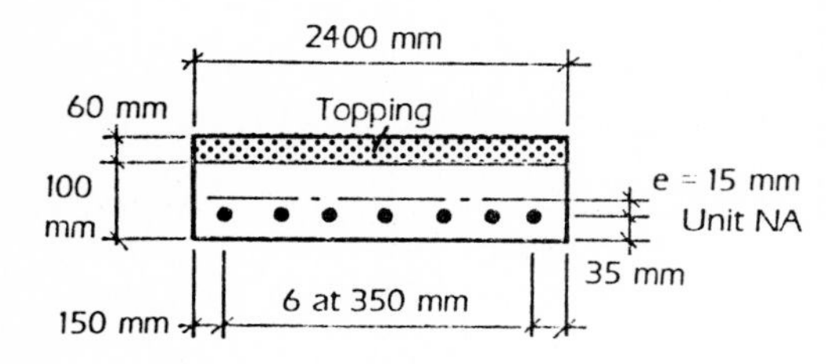


**Figure 34.27** Strand placement.

**Stress check**

At detensioning:

At ends: $-F_{ps}/A^{+} - F_{ps}e/S = -7 \times 122.6 \times 1000/240{,}000^{+}$

$-7 \times 122.6 \times 1000 \times 15/(4000 \times 1000)$

$=$ (top) $-$ 0.357 MPa < 15 MPa (OK)

$=$ (bottom) $-$ 6.794 MPa < 15 MPa (OK)

At midspan:

$-F_{ps}/A^{+} - F_{ps}e/S^{-} + M_{unit}/S = -7 \times 122.6 \times 1000/240{,}000^{+}$

$-7 \times 122.6 \times 1000 \times 15/(4000 \times 1000)^{-}$

$-+ 13.575 \times 1000/4000$

$=$ (top) $-$ 3.75 MPa < 15 MPa (OK)

$=$ (bottom) $-$ 3.40 MPa < 15 MPa (OK)

At erection at midspan, topping application, assume the unit is 30 days old:

$-F_{ps}/A^{+} - F_{ps}e/S^{-} + (M_{unit} + M_{top})/S = -7 \times 111.73$

$\times 1000/240{,}000^{+}$

$-7 \times 111.73 \times 1000 \times 15/(4000 \times 1000)^{-} + (13.575 + 10.5)$

$\times 10^{6}/(4000 \times 10^{3})$

$=$ (top) $-$ 6 .3446 MPa < 15.75 MPa (OK)

$=$ (bottom) $-$ 0.1729 MPa < 15.75 MPa (OK)

At service:

At top of topping:

$-(M_{DLS} + M_{LL})/S_{ct} - (7.5 + 18.5) \times 10^{6}/(12{,}851 \times 10^{3})$

$=$ 2.023 MPa < 13.5 MPa (OK)

At top and bottom of precast unit:

$$-F_{ps}/A^{+} - F_{ps}e\ /S^{-} + (M_{DL} + M_{top})/S^{-} + (M_{DLS} + M_{LL})/S_c$$

$$= -7 \times 111.73 \times 1000/240{,}000^{+} - 7 \times 111.73 \times 1000 \times 15/(4000 \times 1000)^{-}$$

$$+ (13.575 + 10.5) \times 10^6/(4000 \times 10^3)^{-} + (7.5 + 18.5) \times 10^6/$$

$$(67{,}928 \times 10^3) = (\text{top}) - 6.727 \text{ MPa} < 15.75 \text{ MPa (OK)}$$

$$(11{,}058 \times 10^3) = (\text{bottom}) + 2.178 \text{ MPa} < 2.959 \text{ MPa (OK)}$$

### Ultimate strength

Check ultimate strength. For this example use the Canadian load factors and phi factors.

$$M_u = 1.25(13.575 + 10.5 + 7.5) + 1.5 \times 18.5 = 67.218 \text{ kN} \cdot \text{m}$$

$$pp = A_{ps}/bd_p = 7 \times 99/[2400 \times (160 - 35)] = 0.00231$$

$$f_{ps} = f_{pu}\{1 - G_M / \beta[pp \times f_{pu}/f'_c + d/d_p(w - w')]\}$$

where $G_m = 0.28$ for low-relaxation strand
$\beta = 0.81$ for 35-MPa concrete
$d = 0$
$w = 0$
$w' = 0$

$$f_{ps} = 1860[1 - 0.28/0.81(0.00231 \times 1860/35)] = 1781 \text{ MPa}$$

$$a = \phi_s/\phi_c(A_s f_y + A_{ps} f_{ps})/(0.85 f'_c b)$$

$$= 0.9/0.6 \times 7 \times 99 \times 1781/(0.85 \times 30 \times 2400) = 21.35 \text{ mm}$$

$$M_u = \phi_s A_{ps} f_{ps}(d_p - a/2) = 0.9 \times 7 \times 99 \times 1781/10^6$$

$$(125 \text{ mm} - 21.35 \text{ mm}/2) = 127 \text{ kN} \cdot \text{m} > 67.218 \text{ kN} \cdot \text{m (OK)}$$

Check upper limit, $w_p = 0.36/\beta$.

$$w_p = A_{ps}/(bd)f_{ps}/f'_c = 7 \times 99/(2400 \times 125) \times 1860/35 = 0.12276$$

$$0.36 \times 0.81 = 0.2916 < 0.12276 \text{ (OK)}$$

Check lower limit, $M_u = 1.2M_{\text{cracked}}$.

$$M_{\text{cracked}} = 1.2S_{cb}\left(f_{ps}/A + F_{ps}e\ /S_b + 0.6f'^{\,0.5}_c\right)$$

$$= 1.2 \times 11{,}058 \times 10^3/10^6(7 \times 11.73 \times 1000/240{,}000 + 7 \times 11.73)$$

$$\times 15 \times 1000/(4000 \times 1000) + 0.5 \times 35^{0.5})$$
$$= 121.41 \text{ kN} \cdot \text{m} < 127 \text{kN} \cdot \text{m} \quad \text{(OK)}$$

**Shear**

Check shear:

$$V_u = \phi V_c$$
$$V_u = wL/2 = [1.25 \times 2.4\text{m}(1.8 + 1.4 + 1.0\text{kN/m}^2)$$
$$+ 1.5 \times 2.4\text{m} \times 2.4\text{kN/m}^2] \times 5\text{m}/2 = 53.1\text{ kN}$$
$$\phi_c V_c = \phi_c \times 0.17 f_c'^{0.5} bd_p = 0.6 \times 0.17 \times 37^{0.5} \times 2400 \times 125/1000$$
$$= 181 \text{ kN} > 53.1 \quad \text{(OK)}$$

Check horizontal shear:

$$V_h \geq, \text{ then the smaller of}, C \text{ versus } T$$
$$T = A_{ps} f_{ps} + A_s h_y$$

Since $A_s = 0.0$, then

$$T = A_{ps} f_{ps} = 7 \times 99 \times 1781/1000 = 1234 \text{ kN}$$
$$C = 0.85 f_c' A_{ct} = 0.85 \times 35 \text{ MPa} \times 2400 \text{ mm} \times 60 \text{ mm}/1000$$
$$= 4284 \text{ kN}$$
$$V_h = 0.42\, bv\ Lvh$$
$$= 0.42 \text{ MPa} \times 2400 \text{ mm} \times 5\text{m}/2 \times 1000 \text{ mm/m}/1000$$
$$= 2520 \text{ kN}$$

where 0.55/2 is the average shear stress allowable.

Since 2580 kN is greater than 1234 kN, shear ties are not required. However, the surface must be roughened.

**Temperature and shrinkage steel**

Consult the governing code for this requirement. The American Concrete Institute code and the Uniform Building Code require 0.0018*bt*. The PCI and the CPCI recommend a minimum of 0.001 *bt*. The following example uses 0.0018 *bt*.

For the topping, 0.0018× 60 mm ×1000 mm = 108 mm²/m. Use mesh 152× 152 – MW 18.7 × 18.7 area/meter = 123 mm².

For the unit in the transverse direction, 0.0018×100×1000 = 180 mm²/m. The maximum spacing $5t$ = 5 × 100 = 500 mm. Use no. 10M reinforcing bars at 500 mm on center = 200 mm²/m.

**Deflection**

By using the PCI and CPCI coefficients for the following,

$$E_{ci} = 0.043 \times 1850^{1.5} \times 25^{0.5} = 17{,}107$$

$$E_c = 0.043 \times 1850^{1.5} \times 35^{0.5} = 20{,}242$$

At detensioning:

Camber $= F_{pi}eL^2/8E_{ci}I = 7 \times 122.6 \times 15 \text{ mm} \times 5^2 \times 10^9/$
$(8 \times 17{,}107 \times 200 \times 10^6) = 11.7578 \text{ mm up}$

Deflection $= 5wL^4/384E_{ci}I = 5 \times 2.4 \times 1.81 \text{ kN/m}^2 \times 5^4$
$\times 10^{12}/(384 \times 17{,}107 \times 200 \times 10^6) = 10.3325 \text{ mm down}$

Net $= 1.425$ mm up

At erection after casting the topping:

Topping $= 5wL^4/384E_cI = 5 \times 2.4 \text{ m} \times 1.4 \text{ kN/m}^2 \times 5^4 \times 10^{12}$
$(384 \times 20{,}242 \times 200 \times 10^6) = 6.754 \text{ mm down}$

Camber $= 1.83 \times 11.758 = 21.164$ mm up

Deflection unit $= 1.85 \times 10.3325 = 19.115$ mm down

Net $= 4.7$ mm down

At service after 5 years:

Camber $= 2.2 \times 11.758 = 25.867$ mm up

Deflection unit $= 2.4 \times 10.3325 = 24.798$ mm down

Deflection topping $= 2.3 \times 6.754 = 15.535$ mm down

50 percent of superimposed dead load (probable partitions)

$= 5 \times 2.4 \times 1.0 \times 5^4 \times 10^{12}/(384 \times 20{,}242 \times 951 \times 10^6)$

$= 0.507 \text{ mm}(1 + \text{creep factor of } 2) = 1.5 \text{ mm down}$

Net $= 15.96$ mm down

Live load deflection $= 5wL^4/384E_cI_c$

$= 5 \times 2.4 \text{ m} \times 2.4 \text{ kN/m}^2 \times 5 \text{ m}^4 \times 10^{12}/$
$(384 \times 20{,}242 \times 951 \times 10^6)$
$= 2.435 \text{ mm down}$

Maximum live-load deflection $= L/360 = 5 \text{ m} \times 1000 \text{ mm/m}/360$
$= 13.9 \text{mm} > 2.435 \text{ mm}$ (OK)

Maximum dead-load creep deflection (elements attached but not likely to be damaged)

$= L/240 = 5 \text{ m} \times 1000 \text{ mm/m}/240 = 20.83 \text{ mm} > (15.96 - 4.7)$ (OK)

Note that for this case a percentage of the live load was not considered because of the probable partitions.

Maximum dead-load creep deflection (elements attached but likely to be damaged) = $L/480 = 5$ m 1000 mm/m 480 = 10.416 mm < (15.96 - 4.7)

Therefore, all elements attached likely to be damaged owing to large deflections must be designed to allow for creep deflection, i.e., clearance of separation.

### Handling

Check handling (see Fig. 34.28).

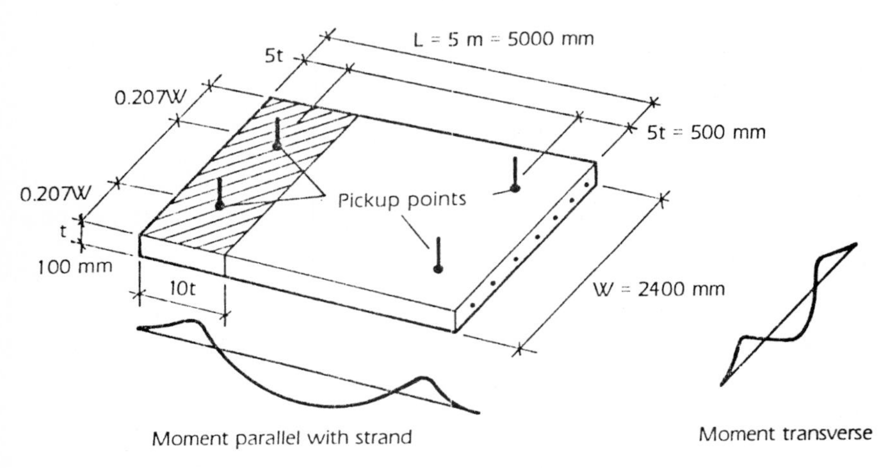


**Figure 34.28** Handling.

**Parallel with the strand.** Design as a 1-m strip:

$$M^- = wL^2/2 = 1.8\ \text{kN/m}^2 \times 1\ \text{m} \times 0.5\ \text{m}^2/2 = 0.225\ \text{kN·m}$$

$$M^+ = wL^2/8 - M^- = 1.8\ \text{kN/m}^2 \times 1.0\ \text{m} \times 4\text{m}^2/8 - 0.225$$

$$= 3.375\ \text{kN·m}$$

$$f = M/S$$

$$S = bd^2/6 = 1000 \times 100^2/6 = 1666.7 \times 10^3$$

$$f = 3.37\ \text{kN·m} \times 1000/6 = 1666.7 \times 10^3$$

$$f = 2.025\text{MPa}$$

$$f_{\text{allowable}} = \left(F_p/A + \lambda \times 0.5 f_c'^{0.5}\right)/2$$

where $\lambda = 0.85$ for sand-lightweight concrete.

$$f_{\text{allowable}} = (7 \times 122.6 \times 1000/240{,}000 + 0.85 \times 0.5 \times 25^{0.5})/2.0$$
$$= 2.85 \text{ MPa}$$

Since 2.85 MPa is greater than 2.025 MPa, it is OK to pick up without cracking.

**Transverse to the strand.** Design for the theoretical internal beam, with a width of $10t = 10 \times 100 = 1000$ m (1 m).

$$M^- = M^+ = 1.8 \text{kN/m}^2 \times 5 \text{m}/2\,(0.207 \times 2.4)^2/2 = 0.555 \text{ kN}\cdot\text{m}$$

$$S = bd^2/6 = 10 \times 100 \times 100^2/6 = 1666.7 \times 10^3$$

$$f = M/S = 0.555 \text{ kN}\cdot\text{m} \times 10^6/1666.7 \times 10^3 = 0.333 \text{ MPa}$$

$$f_{\text{allowable}} = \lambda \times 0.5 f_c'^{0.5}/\text{FS} = 0.85 \times 0.5 \times 25^{0.5}/2 = 1.0625 \text{ MPa}$$
$$> 0.333 \text{ MPa (OK)}$$

FS = factor of safety

## Example Problem 4: Double-T Composite Flexure Design

### Design conditions

| | |
|---|---|
| Span of floor member | 15 m |
| Live load | 2.4 kN/m$^2$ |
| Superimposed dead load | 1.2 kN/m$^2$ |
| Normal-weight concrete | |
| Mass density | 2300 kg/m$^3$ |
| $f_c'$ | 35 MPa |
| $f_{ci}'$ | 25 MPa |
| Normal-weight concrete topping | |
| Mass density | 2300 kg/m$^3$ |
| $f_c'$ | 25 MPa |
| Low-relaxation strand, 1860 MPa, no. 11 | |
| Diameter | 11.13 mm |
| Area | 74 mm$^2$ |

The initial loss is 10 percent; assume a service loss of 18 percent. Verify the loss when the number of strands and hence the prestressing force are known.

Assume member size; determine shape properties or use data from the manufacturer. See Fig. 34.29.

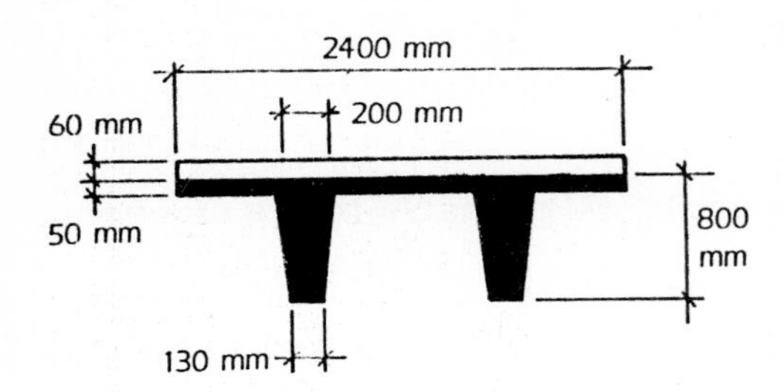


**Figure 34.29** Double-T properties.

| Noncomposite (area = 367,500 mm$^2$) | Composite |
|---|---|
| $Y_t$ = 276 mm | $Y_{ct}$ = 260 mm |
| $Y_b$ = 523 mm | $Y_{cc}$ = 200 mm |
| $I = 22{,}725 \times 10^6$ | $Y_{cb}$ = 600 mm |
| $S_t = 82{,}180 \times 10^3$ | $I_c = 31{,}351 \times 10^6$ |
| $S_b = 43{,}413 \times 10^3$ | $S_{ct} = 120{,}448 \times 10^3$ |
| Force = 3.45 kN/m$^2$ | $S_{cc} = 15{,}653 \times 10^3$ |
| Force = 8.28 kN.m | $S_{cb} = 52{,}277 \times 10^3$ |

**Allowable stresses**

Determine allowable stresses:

$$f_{psi}/\text{strand} = 0.74 \times 1860 \times 0.925 \times 74/1000 = 91.67 \text{ kN/strand}$$

$$f_{ps}/\text{strand} = 0.74 \times 1860 \times 0.82 \times 74/1000 = 83.52 \text{ kN/strand}$$

Allowable concrete stresses at transfer:

$$\text{Compression} = 0.60 f'_{ci} = 0.60 \times 25 \text{ MPa} = 15 \text{ MPa}$$

$$\text{Tension at ends} = 0.5 f'^{0.5}_{ci} = 0.5 \times 25^{0.5} = 2.5 \text{ MPa}$$

$$\text{Tension elsewhere} = 0.25 f'^{0.5}_{ci} = 0.25 \times 25^{0.5} \text{ MPa}$$

Allowable concrete stresses at service:

$$\text{Compression} = 0.45 f'_c = 0.45 \times 35 = 15.75 \text{ MPa}$$

$$\text{Tension} = 0.5 f'^{0.5}_c = 0.5 \times 35^{0.5} = 2.959 \text{ MPa}$$

If bilinear analysis is employed, then use $1.0 f_c'^{0.5} = 1.0 \times 35^{0.5}$ = 5.916 MPa.

**$e_c$ optimum.** Determine $e_c$ optimum. Since the number of strands is not known, an assumption is made for the preliminary design, which may later be adjusted. As an example (see Fig. 34.30),

$e_c = Y_b$ − clearance − stirrup
strand diameter − spacing −
strand diameter/2$e_c$ = 523 mm −
40 − 10 − 12 − 50 − 8 = 403 mm

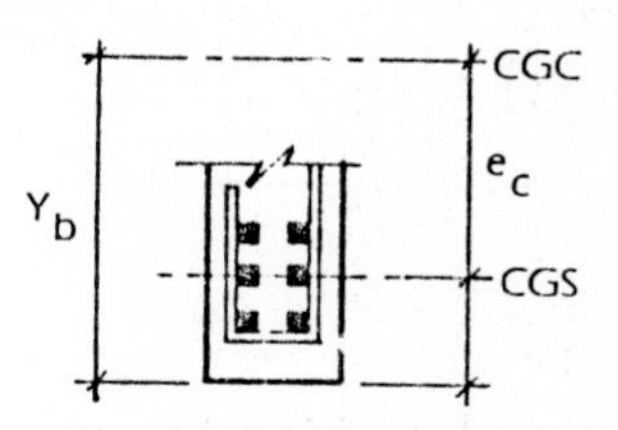


**Figure 34.30** $e_c$ optimum.

Let $e_c$ = 403 mm for this example.

**Moments**

Determine moments.

Unit dead load = 2300 kg/m³ × 9.81 m/s² × 367,500 mm²/10⁶ mm²/m²/1000 m·kg/s² = 8.28 kN/m

(8.28/2400 mm × 1000 mm/m = 3.45 kN/m²)

Topping load = 2300 kg/m³ × 9.81 m/s² × 60 mm × 2400 mm/10⁶ mm²/m²/1000 m·kg/s² = 3.25 kN/m

Moment live load = $wL^2/8$ = 2.4 m · 2.4 kN/m² × 15 m²/8 = 162 kN · m

Moment dead load = 8.28 kN/m × 15 m²/8 = 232.875

Moment superimposed dead load = 1.2 kN/m² × 2.4 m × 15 m²/8 = 81 kN · m

Moment topping dead load = 3.25 kN/m × 15 m²/8 = 91.38 kN · m

**Number of strands for service conditions**

Determine minimum number of strands for service conditions.

Force prestressing:

$$F_{ps,\ \text{minimum}} = [(M_{DL} + M_{TDL})1000/S_b + (M_{SDL} + M_{LL})\,1000/S_{cb} - f_{\text{ten, allowable}}]/(1000/A_c + e_c/S_b)$$

$$F_{ps,\ \text{minimum}} = [(232.8 + 91.38) \times 1000/43{,}413 + (81 + 162) \times 1000/52{,}277 - 0.424]/(1000/367{,}500 + 400/43{,}413) = 763\ \text{kN}$$

Number of strands required = 763 kN/83.51 kN/strand = 9.14 strands

Use 10 strands since this is a double T and stems should be balanced.

Let $e_c$ = 403 mm since stirrups may be placed in the required clearance.

Let $e_c$ = 0.0. For strand location see Fig. 34.31.

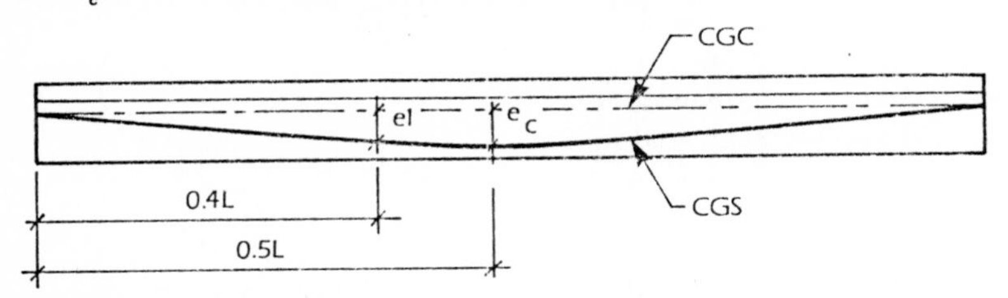


**Figure 34.31** Preliminary strand location. Try a single harp. If the single harp will not work at 0.4*L*, use a double harp.

**Stresses**

**Transfer.** Check stresses at transfer.

$$F_{psi} = 10 \text{ strands} \times 91.67 \text{ kN/strand} = 916.7 \text{ kN prestressing force}$$

At ends:

$$-F_{psi}/A_c = f_{top} = -916.7 \text{ kN} \times 1000/367{,}500 \text{ mm}^2$$

$$= -2.459 \text{ MPa} < -15 \text{ MPa (OK)}$$

$$-F_{psi}/A_c = f_{bottom} = -916.7 \text{kN} \times 1000/367{,}500 \text{ mm}^2$$

$$= -2.459 \text{ MPa} < -15 \text{ MPa (OK)}$$

At midspan:

$$-F_{psi}/A_c + F_{psi}e_c/S_t - M_{DL}/S_t = f_{top}$$

$$-916.7 \times 1000/367{,}500 + 916.7 \times 403/82{,}181 - 232.8$$

$$\times 1000/82{,}181 = -0.83 \text{MPa} < -15 \text{MPa (OK)}$$

$$-F_{psi}/A_c - F_{psi}E_c/S_b + M_{DL}/S_b = f_{bottom}$$

$$-916.7 \times 1000/367{,}500 + 916.7 \times 403/43{,}413 + 232.8$$

$$\times 1000/43{,}413 = -5.64 \text{ MPa} < -15 \text{ MPa (OK)}$$

**Service.** Check stresses at service.

$F_{ps}$ = 10 strands × 83.51 kN/strand = 835.1-kN prestressing force

At ends:

$$F_{ps}/A_c = f_{top} = f_{bottom} = -835.1 \text{ kN} \times 1000/367{,}500 \text{ mm}^2$$

$$= -2.27 \text{ MPa} < -15.75 \text{ MPa (OK)}$$

At midspan at top of topping:

$$-(M_{SDL} + M_{LL})/S_{ct} = -(81 + 162) \times 1000/120{,}448$$
$$= -2.017 \text{ MPa} < -11.25 \text{ MPa (OK)}$$

At midspan at top of unit (topping intersection):

$$-F_{ps}/A_c + F_{ps}e_c/S_t - (M_{DL} + M_{TDL})/S_t - (M_{SDL} + M_{LL})S_{cc} = f_{cc}$$
$$-835.1 \times 1000/367{,}500 + 835.1 \times 403/82{,}180 - (232.87 + 91.3)$$
$$\times 1000/82{,}180 - (162 + 81) \times 1000/156{,}530$$
$$= -3.67 \text{MPa} < -15.75 \text{ MPa (OK)}$$

At midspan at bottom of unit:

$$-F_{ps}/A_c - f_{ps}e_c/S_t + (M_{DL} + M_{TDL})/S_b + (M_{SDL} + M_{LL})/S_{cb} = f_{cb}$$
$$-835.1 \times 1000/367{,}500 - 835.1 \times 403/43{,}413 + (232.87 + 91.3)$$
$$\times 1000/43{,}413 + (162 + 81) \times 1000/52{,}277$$
$$= 2.09 \text{ MPa} < 2.958 \text{ MPa (OK)}$$

At $0.4L$,

$e1 = 0.8\, e_c = 0.8 \times 403 = 322$ Moment $= 0.96 \times \text{moment}_{max}$

$$-F_{ps}/A_c + F_{ps}e_i/S_t - 0.96(M_{DL} + M_{TDL})/S_t - 0.96\,(M_{SDL} + M_{LL})/$$
$$S_{cc} = f_{cc} - 835.1 \times 1000/367{,}500 + 835.1 \times 403 \times 0.8/82{,}180 - 0.96$$
$$\times (232.87 + 91.3) \times 1000/82{,}180 - 0.96 \times (162 + 81) \times 1000/156{,}530$$
$$= -4.25 \text{MPa} < -15.75 \text{ MPa (OK)}$$
$$-F_{ps}/A_c - F_{ps}e\,1/S_b + 0.96(M_{DL} + M_{TDL})/S_b + (M_{SDL} + M_{LL}/S_{cb}$$
$$= f_{bottom} - 835.1 \times 1000/367{,}500 - 835.1 \times 403 \times 0.8/43{,}413$$
$$+ (232.87 + 91.3) \times 0.96 \times 1000/43{,}413 + (162 + 81) \times 0.96$$
$$\times 1000/52{,}277 = 3.158 \text{ MPa} > 2.958 \text{ MPa}$$

Too high. The solution is to double-harp or to increase the number of strands. Since double harping is less expensive than adding two strands, use double harping at $0.4L$. Final strand location is shown in Fig. 34.32.

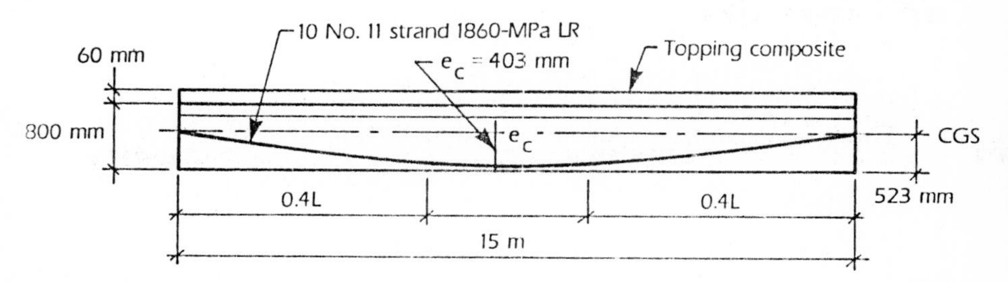


**Figure 34.32** Final strand location.

**Ultimate strength**

Check the design for ultimate strength.

$$pp = A_{ps}/b\,d_p$$

where $A_{ps} = 10\times 74 = 740$

$b = 2400$ mm

$d_p = e_c + Y_t + t_t = 403 + 276 + 60 = 739$

$$pp = 740/(2400\times 739) = 0.000417$$

$G_p = 0.28$ for low-relaxation (LR) strand if $f_{py}/f_{pu} = 0.9$ (LR strand $f_{py} = 0.9\,f_{pu}$).

β for 35-MPa concrete = 0.81. Since there is no mild steel or compression steel in the design thus far, $w$ and $w' = 0$.

$$f_{ps} = f_{pu}\{1 - G_p/\beta[pp \times f_{pu}/f'_c + d/d_p(w - w')]\}$$

$$f_{ps} = 1860\{1 - 0.28/0.81[(0.000417\times 1860/35 + 0/739(0-0))]\}$$

$$= 1845 \text{ MPa}$$

$$a = (A_{ps}f_{ps} + A_s f_y)/0.85f'_c b = (10\times 74 \text{ mm}^2 \times 1845 \text{ MPa} + 0\times 0)/(0.85\times 35\text{MPa}\times 2400 \text{ mm}) = 19.12 \text{ mm} < 60 \text{ mm}$$

Therefore, $a$ is within the flange, and the design is rectangular.

$$\phi M_n = \phi[A_{ps}f_{ps}(d_p - a/2) + A_s f_y(d - a/2)] = 0.9 \times [740\times 1845(739 - 19.12/2)]/10^6 = 896 \text{ kN·m allowable}$$

$$M_{u,\text{actual}} = 1.4(232.875 + 91.38 + 81) + 1.7(162) = 842.76 < 896 \text{ (OK)}$$

If $M_{u,\text{ actual}}$ had been greater than $\phi M_n$, mild steel would have been added or the number of prestressing strands increased.

Check upper limit of prestressing and reinforcing steel:

$$w_p \text{ or } w_p + d/d_p(w - w') \leqq 0.36\beta$$

$$w_p = A_{ps}/(bd_p)\times f'_{ps}/f'_c$$

$$w = A_s/(bd)\times f'_y/f'_c$$

There is no compression steel. Therefore, $w' = 0.0$; $\beta = 0.81$.

$$10\times 74/(2400\times 739)\times 1845/35 = 0.022 < 0.36\times 0.81 \text{ (OK)}$$

Check lower limit, i.e., 1.2 ×moment crack:

$$\phi M_n = 1.2 \text{ moment crack or } \phi M_n > 2M_u$$

$$1.2M_{\text{cracked}} = 1.2S_{cb}\left(F_{ps}/A_c + F_{ps}e_c/S_b + 0.6f_c'^{0.5}\right)$$

$$= 1.2 \times 52{,}277/10^3(835.2 \times 1000/367{,}500 + 835.2$$

$$\times 402/43{,}413 + 0.6 \times 35^2)$$

$$851.6 \text{ kN} \cdot \text{m} < 896 \text{ kN} \cdot \text{m} \quad \text{(OK)}$$

If the member is not composite, use 1.2 $S_b$ instead of 1.2 $S_{cb}$.

Figure 34.33 presents the computer program printout for the double-T composite flexure design.

```
        DESIGN  CONDITIONS

   Roof member, span =  15    metres
   Live load in kN/sq.m =    2.4
   Superimposed dead load in kN/sq.m =   1.2
   Concrete type is NW
Concrete mass density in kg/cub.m =   2300
   Concrete strength f'c =   35  MPa, and f'ci=  25    MPa
 Topping concrete type is NW
Topping concrete mass density in kg/cub.m = 2300
Topping concrete strength in MPa =   25
 Topping thickness in mm=   60
Topping load in kN/sq. m = 1.35378
   Low Relaxation strand, Fu=   1860    MPa
   Strand size = 11.13 mm, and strand area is   74     sq.mm
   Initial loss of prestress =   10    %
   Service loss of prestress =   18    %

        SECTION    PROPERTIES

   Flange width =   2500    mm and thickness =   50    mm
   Member depth = 800    mm
   Stem thickness at bottom =   130   and at the top =   200   mm
   Cross sectional area =   367500 sq. mm
   Moment of Inertia =   22725 *10^6
   Y top =   276    and Y bottom =   523   mm
   Sectiom modulus top = 82180 section modulus bottom = 43413 *10^3
   Ratio   V/S =   48
   Member weight =   20.7    kN/m or   8.28  kN/sq.m
   Moment of Inertia composite= 31351 *10^6
   YCT= 260 YCC= 200 YCB= 600   mm
Section mod. comp. SCT= 120448 SCC 156530 SCB= 52277 *10^3
   Specified eccentricity at ends = 0 mm, mid span =   403 mm

        ALLOWABLE   STRESS

   Strand allowable force at detensioning =   91.66 kN/strand
                                 at service = 83.51 kN/strand
   At transfer allowable concrete stress in compression=   15   MPa
                                            in tension at ends = 2.5   MPa
                                    in tension at mid span = 1.25  MPa
   Allowable concrete stress at service in compression =   15.75  MPa
                                                 in tension =   2.958  MPa
                              Bi-linear tension stress =   5.916  MPa

          MOMENTS

   Moment in kN.m are Live Load = 168.75
                              Dead Load = 582.1874
                              Topping= 95.18766
                  Superimposed Dead Load = 84.375

          SOLUTION

  Required min. force prestressed at service =   1456.76   kN
   Total number of strands = 18
  Number of strans required per stem =   9
  Force prestressed at service per strand =   83.51   kN
   Total force prestressed at service =   1503.359   kN
  MUST DOUBLE HARP
     STRESS AT TRANSFER
        At ends at top stress = -4.49   MPa
        At ends at bottom stress = -4.49  MPa
        At mid span at top stress = -3.483   MPa
        At mid span at bottom stress = -6.397   MPa
     STRESS AT SERVICE
        At ends at top, stress = -4.091   MPa
        At ends at bottom, stress = -4.091 MPa
        At mid span at top of topping, stress = 2.101   MPa
        At mid span at flange top, stress = -6.579   MPa
        At mid span at bottom, stress=   2.398 MPa
        At .4L at top of topping, stress = 2.017 MPa
        At .4L at flange top, stress = -6.184 MPa
        At .4L at bottom, stress =   1.58 MPa

          ULTIMATE   STRENGTH   CHECK

  A RECTANGULAR BEAM SOLUTION
    RHO P =   7.846834E-04    Gamma P=   .28    fps=   1833.188   MPa
     Mn=   1764.41 kN.m   phi* Mn =   1587.969
     Mu=   1353.32 kN.m
  Member is OK
  prestressing steel is less than maximum therefore OK
     Moment cracked =   937.54 kN.m
 phi*Mn > 1.2 Mcr. ( 1125.048   ) therefore OK
  If 1.2 *Mcr > phi Mn then check to see if phi Mn > 2* Mu, if so then
 Scbe for composite members Scb is used to compute Moment Cracked ie 1.
  for non composite members Sb is used.
```

**Figure 34.33** Computer program printout for double-T composite flexure design. The author assumes no liability for the use of this program.

Figure 34.34 presents the computer program listing for the double-T composite flexure design.

```
LPRINT
LPRINT TAB(10);"DESIGN for FLEXURE - DOUBLE TEE COMPOSITE SI units":LPRINT
LPRINT "Note: the author assumes no liability relative to the use of this program"
LPRINT
LPRINT STRING$(60,42)
LPRINT CHR$(27);"n":LPRINT CHR$(14):LPRINT TAB(5);"DESIGN CONDITIONS"
LPRINT CHR$(15)
INPUT " If a floor member type F, if a roof member type R ",A$
INPUT " Input span in metres = ",L
IF A$="F" THEN LPRINT TAB(4);" Floor member,  span = ";L;" metres."
IF A$="R" THEN LPRINT TAB(4);"Roof member, span = ";L;" metres"
INPUT " Input Live Load in kN/sq.m = ",LL
LPRINT TAB(4);"Live load in kN/sq m = ",LL
INPUT "Input superimposed dead load in kN/sq m= ",SDL
LPRINT TAB(4);"Superimposed dead load in kN/sq.m = ";SDL
INPUT " Input type of concrete ie light wt = "; TC$
LPRINT TAB(4);"Concrete type is ";TC$
INPUT " Concrete mass density in kg/cub.m = ",WT
LPRINT "Concrete mass density in kg/cub.m = "; WT
INPUT "Input concrete strength f'c and f'ci in MPa = ",FPC , FPCI
LPRINT TAB(4);"Concrete strength f'c = ";FPC;" MPa, and f'ci = ";FPCI;" MPa"
INPUT " Input topping concrete type = ",TT$
LPRINT " Topping concrete type is ";TT$
INPUT "Topping concrete mass density in kg/cub  m=",TMS
LPRINT "Topping concrete mass density in kg/cub m =";TMS
INPUT " Topping concrete strength in MPa =",FCT
LPRINT "Topping concrete strength in MPa = ";FCT
INPUT "Input topping thickness in mm = ",TMM
LPRINT " Topping thickness in mm= ";TMM
TF=TMS*9.81*TMM/10^6
LPRINT "Topping load in kN/sq. m =";TF
INPUT "Input strand type ie LR for low relaxation or SR for stress relieved =",ST$
INPUT "Input strand ultimate strength Fu in MPa = ",FU
IF ST$="LR" THEN LPRINT TAB(4);"Low Relaxation strand, Fu= ";FU;" MPa"
IF ST$ = "SR" THEN LPRINT TAB(4);" Stress relieved strand. Fu= ";FU;" MPa"
INPUT " Input strand size in mm, and area in sq. mm. = ",SS$, SA
LPRINT TAB(4);"Strand size = ";SS$;" mm, and strand area is ",SA;" sq.mm"
INPUT " Input initial loss of prestress in percent =",IL
LPRINT TAB(4);"Initial loss of prestress = ",IL;" %"
INPUT " Input service loss of prestress in percent = ",SL
LPRINT TAB(4);"Service loss of prestress = ";SL;" %"
LPRINT CHR$(27);"n": LPRINT CHR$(14);TAB(5);"SECTION  PROPERTIES"
LPRINT CHR$(15)
INPUT "Input flange width and thickness in mm ",FW,FT
LPRINT TAB(4);"Flange width = ";FW;" mm and thickness = ";FT," mm"
INPUT " Input member depth in mm = ",D
LPRINT TAB(4);"Member depth =",D;" mm"
INPUT " Input stem width at the bottom then at the top in mm =",WB,WT
LPRINT TAB(4);"Stem thickness at bottom = ";WB;" and at the top = ";WT; " mm"
INPUT " Input member cross sectional area in sq. mm =",AC
LPRINT TAB(4);"Cross sectional area = ";AC;"sq. mm"
INPUT " Input Moment of Inertia *10^6= ",I
LPRINT TAB(4);"Moment of Inertia = ";I;"*10^6"
INPUT " Input Y top and Y bottom in mm = ",YT, YB
LPRINT TAB(4);"Y top = ",YT;" and Y bottom = ",YB, " mm "
INPUT "Input section modulus top and bottom *10^3= ",ST,SB
LPRINT TAB(4);"Sectiom modulus top =";ST;"section modulus bottom =";SB;"*10^3"
INPUT " Input the ratio Y/S = ",YS
LPRINT TAB(4);"Ratio  Y/S = ";YS
INPUT "Input member load in kN/sq m = ",DLLF
DL = DLLF*FW/1000
LPRINT TAB(4);"Member weight = ";DL;" kN/m or ";DLLF;" kN/sq.m"
INPUT " Moment of inertia composite *10^6=",IC
LPRINT TAB(4);"Moment of inertia composite=";IC;"*10^6"
INPUT "YCT,YCC,YCB in mm=",YCT,YCC,YCB
LPRINT TAB(4);"YCT=";YCT;"YCC=";YCC;"YCB=";YCB;" mm"
INPUT " SCT,SCC,SCB  *10^3",SCT,SCC,SCB
LPRINT "Section mod. comp. SCT=",SCT;"SCC",SCC;"SCB=";SCB;"*10^3"
INPUT " Input eccentricity at end EE in mm and EC at mid span",EE,EC
IF EC= 0! THEN EC= YB-40-10-12-50-8
LPRINT TAB(4);"Specified eccentricity at ends =";EE;"mm, mid span = ";EC,"mm"
LPRINT CHR$(27);"n": LPRINT CHR$(14);TAB(5);"ALLOWABLE STRESS"
LPRINT CHR$(15)
FPSI= .74*FU*SA*(100-IL)/100/1000 : FPSS = .74*FU*SA*(100-SL)/100/1000
LPRINT TAB(4);"Strand allowable force at detensioning = ";INT(FPSI*100)/100;"kN/strand"
LPRINT TAB(30);"at service =",INT(FPSS*100)/100;"kN/strand"
FCTC = .6* FPCI : FCTTE=.5*FPCI^.5:FCTTM= 25*FPCI^.5
LPRINT TAB(4);"At transfer allowable concrete stress in compression= ";INT(FCTC*1000)
/1000;" MPa"
LPRINT TAB(37);"in tension at ends =";INT(FCTTE*1000)/1000;" MPa"
LPRINT TAB(33);"in tension at mid span =";INT(FCTTM*1000)/1000;" MPa"
```

**Figure 34.34** Computer program listing for double-T composite flexure design.

```
FCSC= .45*FPC : FCST= .5*FPC^.5 :FCSTB = 11*FPC^.5
LPRINT TAB(4);"Allowable concrete stress at service in compression = ";INT(FCSC*1000)
  /1000;" MPa"
LPRINT TAB(42);"in tension = ";INT(FCST*1000)/1000;" MPa"
LPRINT TAB(28);"Bi-linear tension stress = ";INT(FCSTB*1000)/1000; " MPa"
LPRINT CHR$(27);"n": LPRINT CHR$(14);TAB(5);" MOMENTS"
LPRINT CHR$(15)
MLL = FW/1000*LL*L^2/8 : MDL=FW/1000*DLLF*L^2/8:MSDL =FW/1000*SDL*L^2/8
MTDL=TF*FW/1000*L^2/8
LPRINT TAB(4);"Moment in kN.m are Live Load =";MLL
LPRINT TAB(28);"Dead Load =";MDL
LPRINT TAB(28);"Topping=",MTDL
LPRINT TAB(15);"Superimposed Dead Load =";MSDL
LPRINT CHR$(27);"n":LPRINT CHR$(14);TAB(5);"SOLUTION"
LPRINT CHR$(15)
FPSM=((MDL+MTDL)*1000/SB+(MSDL+MLL)*1000/SCB-FCST)/(1000/AC+EC/SB)
LPRINT TAB(2);"Required min. force prestressed at service = ";INT(FPSM*100)/100;" kN"
NS=FPSM/FPSS : IF NS> INT(NS) THEN NS = INT(NS)+1
IF NS/2> INT(NS/2) THEN NS= 2*(INT(NS/2)+1)
5
FPS=NS*FPSS
LPRINT TAB(2);" Total number of strands =";NS
LPRINT TAB(2);"Number of strans required per stem = ",NS/2
LPRINT TAB(2);"Force prestressed at service per strand = ";INT(FPSS*100)/100;" kN"
LPRINT TAB(2);" Total force prestressed at service = ";INT(FPS*1000)/1000;" kN"
REM CHECK STRESS AT TRANSFER
FPI=NS*FPSI
FCIET = -FPI*1000/AC+FPI*EE/ST : IF FCIET>FCTTE THEN LPRINT"At transfer tension stress
  at end at top exceeds allowable"
FCIEB = - FPI*1000/AC-FPI*EE/SB : IF FCIEB <-FCTC THEN LPRINT"At transfer compression
  stress at bottom exceeds allowable"
FCICT= -FPI*1000/AC+FPI*EC/ST-MDL*1000/ST : IF FCICT <- FCTC OR FCICT > FCTTM
  THEN LPRINT "At transfer stress at mid span at top exceeds allowable"
FCICB = -FPI*1000/AC-FPI*EC/SB + MDL*1000/SB : IF FCICB > FCTTE OR FCICB < -FCTC
  THEN LPRINT "At transfer stress at mid span bottom exceeds allowable"
REM CHECK STRESS AT SERVICE
FCSET = -FPS*1000/AC+FPS*EE/ST : IF FCSET>FCSC THEN LPRINT " At service stress at end
  top exceeds allowable"
FCSEB = -FPS*1000/AC-FPS*EE/SB : IF FCSEB <-FCSC THEN LPRINT " At service stress at
  end bottom exceeds allowable"
FCSMTC=(MSDL+MLL)*1000/SCT : IF FCSMTC >.45*FCT THEN LPRINT " At service stress at
  the top of the topping exceeds allowable"
FCSMT= -FPS*1000/AC+FPS*EC/ST-(MDL+MTDL)*1000/ST-(MSDL+MLL)*1000/SCC:IF
FCSMT<-FCSC THEN LPRINT " At service stress at top mid span exceeds allowable"
FCSMB=-FPS*1000/AC-FPS*EC/SB+(MDL+MTDL)*1000/SB +(MSDL+MLL)*1000/SCB:IF
FCSMB >FCST THEN LPRINT " At service stress at bottom mid span exceeds allowable"
E1= EE+ .3*(EC-EE)
10
FCSQT=-FPS*1000/AC+FPS*E1/ST-(MDL+MTDL)*1000*.96/ST-(MSDL+MLL)*1000*.96
  /SCC
FCSQB=-FPS*1000/AC-FPS*E1/SB+(MDL+MTDL)*1000*.96/SB+(MSDL+MLL)*1000*.9
  6/SCB
FCSQC=(MSDL+MLL)*1000*.96/SCT
IF FCSQT <- FCSC OR FCSQB > FCST THEN LPRINT " MUST DOUBLE HARP":E1=EC:GOTO 10
LPRINT TAB(4);" STRESS AT TRANSFER"
LPRINT TAB(8);"At ends at top stress = ";INT(FCIET*1000)/1000;" MPa"
LPRINT TAB(8);"At ends at bottom stress = ";INT(FCIEB*1000)/1000;" MPa"
LPRINT TAB(8);"At mid span at top stress = ";INT(FCICT*1000)/1000;" MPa"
LPRINT TAB(8);"At mid span at bottom stress = ";INT(FCICB*1000)/1000;" MPa"
LPRINT TAB(4);" STRESS AT SERVICE"
LPRINT TAB(8);"At ends at top, stress = ";INT(FCSET*1000)/1000;" MPa"
LPRINT TAB(8);"At ends at bottom, stress = ",INT(FCSEB *1000)/1000; "MPa"
LPRINT TAB(8);"At mid span at top of topping, stress =",INT(FCSMTC*1000)/1000," MPa"
LPRINT TAB(8);"At mid span at flange top, stress = ",INT(FCSMT *1000)/1000;" MPa"
LPRINT TAB(8);"At mid span at bottom, stress= ";INT(FCSMB*1000)/1000;"MPa"
LPRINT TAB(8);"At .4L at top of topping, stress =";INT(FCSQC*1000)/1000;"MPa"
LPRINT TAB(8);"At .4L at flange top, stress = "; INT(FCSQT*1000)/1000;"MPa"
LPRINT TAB(8);"At .4L at bottom, stress = ";INT(FCSQB*1000)/1000;"MPa"
REM ULTIMATE STRENGTH CHECK
LPRINT CHR$(27);"n": LPRINT CHR$(14);TAB(5);"ULTIMATE STRENGTH CHECK"
LPRINT CHR$(15)
PP=NS*SA/FW/(EC+YT): IF ST$="LR" THEN GP=.28 ELSE GP=.4
IF FPC=30 THEN BETA=.85
IF FPC>30 THEN BETA = .85-.008*(FPC-30)
IF BETA < .65 THEN BETA =.65
UFPS=FU*(1-GP/BETA*(PP*FU/FPC))
A=(NS*SA*UFPS)/(.85*FPC*FW): IF A=<FT THEN LPRINT " A RECTANGULAR BEAM SOLUTION"
IF A> FT THEN LPRINT " A TEE BEAM SOLUTION HENCE NOT SOLVED BY THIS PROGRAM"
MN=NS*SA*UFPS*((YT+EC+TMM)-A/2)/10^6 : MNPHI=MN*.9
MU= 1.7*MLL+1.4*(MDL+MSDL+MTDL)
```

**Figure 34.34** *(Continued)*

```
LPRINT TAB(4);"RHO P = ";PP;" Gamma P= ";GP;" fps= ";UFPS;" MPa"
LPRINT TAB(4);" Mn= ";INT (MN *100)/100;"kN.m phi* Mn = "; INT (MN*100)/100*.9
LPRINT TAB(4);" Mu= "; INT(MU*100)/100;"kN.m"
IF MU=< MNPHI THEN LPRINT " Member is OK "
IF MU > MNPHI THEN LPRINT "Must add mild steel or increase prestressing"
REM CHECK UPPER LIMIT OF PRESTRESSING
WP=NS*SA/FW/(YT+EC)*UFPS/FPC
IF WP=< .36 * BETA THEN LPRINT " prestressing steel is less than maximum therefore OK"
REM CHECK LOWER LIMIT OF PRESTRESSING IE MOMENT CRACKED
IF SCB= 0! THEN SCB=SB
MCR=SB/10^3*(FPS*1000/AC+FPS*EC/SB+.6*1!*FPC^.5)
LPRINT TAB(4);" Moment cracked = ";INT (MCR*100)/100; "kN.m"
IF MNPHI > 1.2*MCR THEN LPRINT "phi*Mn > 1.2 Mcr. (";1.2*INT(MCR*100)/100;" )
   therefore OK"
LPRINT " If 1.2 *Mcr > phi Mn then check to see if phi Mn > 2* Mu, if so then OK"
LPRINT "Note for composite members Scb is used to compute Moment Cracked ie 1.2*Scb";
LPRINT " for non composite members Sb is used."
LPRINT : LPRINT STRING$(60,42)
LPRINT : LPRINT TAB(5);" END OF RUN "
END
```

**Figure 34.34** *(Continued)*

## Example Problem 5: Design for Shear, End Bearing, Camber, and Deflection; Double-T Composite (SI Units)

### Shear design

Shear must be considered for prestressed concrete as it is for reinforced concrete. For beams shear reinforcing must be used when $V_u > V_c/2$. For slabs, hollow planks, and wall panels shear must be considered when $V_u > V_c$, with these members being designed so that $V_u$ is always less than $\phi V_c$. For beams $V_{u,\,maximum} = 5\phi V_c$. If $V_u$ is greater than $5\phi V_c$, then the web thickness must be increased. If $V_u$ is greater than $\phi V_c / 2$ and $V_u$ is less than $\phi V_c$, then minimum shear reinforcing is required.

$$A_{v,\,minimum} = A_{ps}/80 \times f_{pu}/f_y \times S/d\,(d/b_w)^{0.5}$$

where $A_v$ = area of shear reinforcing
$A_{ps}$ = area of prestressing strand
$f_{pu}$ = ultimate strength of strand
$f_y$ = yield strength of shear reinforcing
$S$ = spacing of shear reinforcing ($S_{maximum}$ = 0.75 member depth, or $3b_w$.)
$b_w$ = average width of web
$d$ = larger of 0.8× member depth or effective depth of prestressing strand ($d_p$)

When $V_u$ is greater than $\phi V_c$, then the shear reinforcing must be designed:

$$V_s = (V_u - \phi V_c)/\phi$$

$$A_v = V_s S/f_y d \quad \text{or} \quad A_v = (V_u - \phi V_c) \times S/\phi f_y d$$

$$\text{or} \quad S = A_v f_y \times d/V_s \quad \text{or} \quad S = A_v f_y d\,\phi/(V_u - \phi V_c)$$

where $V_s$ = shear resisted by shear reinforcing
$V_u$ = ultimate shear force on member at point under consideration
$\phi$ = 0.85 for shear

The value of $V_c$ is shown in Fig. 34.35. From the curve of $\phi V_c$, $V_u$ and $\phi V_c/2$ are plotted. When the $V_u$ curve is above the $\phi V_c$ curve, shear reinforcing must be designed. When $V_u$ is between the $\phi V_c$ curve and the $\phi V_c/2$ curve, then minimum shear reinforcing is required. When $V_u$ is below the $\phi V_c/2$ curve, shear reinforcing is not required.

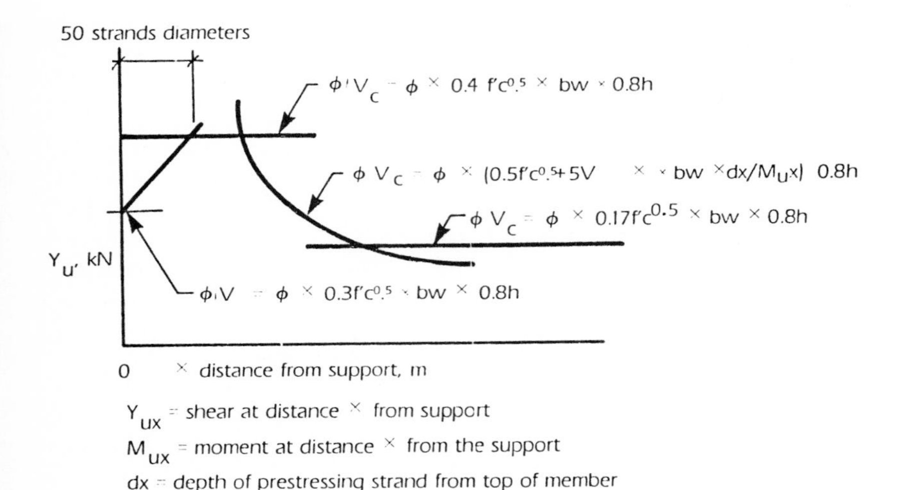


**Figure 34.35** Allowable shear diagram.

In the term $V_{ux}d_{px}/M_{ux}$, $d_p$ is the true distance from the extreme fiber in compression of the concrete to the centroid of the prestressing steel, and the term is taken to be equal to or less than 1.0. For simple spans with a uniform load, which is the usual case for pretension members, the shear at distance $X$ from the end of a member, $V_{ux} = w_u(L/2 - X)$. The term $V_u d_p / M_u$ becomes $d_p w_u(L/2 - X)/(w_u L/2 - w_u X^2/2)$, which reduces to $d_p/X(L - 2X)/(L - X)$. Since $L$ and $X$ are in meters and $d_p$ is in millimeters, you must divide by 1000. Since the term $\left(0.5f_c'^{0.5} + 5V_{ux}d_{px}/M_{ux}\right) \times \phi / b_w$

$\times 0.8h$ is in megapascals, you must divide by 1000 to convert to kilonewtons.

The following format may be used for a solution: If $w_u = 29.25$ kN/m and $L = 15$ m, then

50 strand diameters = 50× 11.13 mm(no.11 strand) = 556 mm

The distance $X$ from the end of the member to midspan is as follows in 0.5-m steps:

| $X$, m | $V_u = w_u(L/2 - X)$, kN | $X$, m | $V_u = w_u(L/2 - X)$, kN |
|---|---|---|---|
| 0.5 | 104.8 | 2.5 | 74.8 |
| 1.0 | 97.8 | 3.0 | 67.3 |
| 1.5 | 89.8 | 3.5 | 59.9 |
| 2.0 | 82.3 | | |

If double harping is used and $e = 0.0$, $e_c = 403$ mm, $Y_t = 276$ mm, $T_t = 60$ mm, and $b_w = 165$ mm, then

$$d_p \text{ (mm)}/1000 \text{ mm/m } X \text{ (m)} \times (L - 2X)\ (L - X) = 1.0$$

| $X$,m | $e$, mm $(e_c \cdot e_e)X/0.5L$ or $0.4L$ | $d$,m $e + y_t + T_t$ | kN $\left[0.5f_c'^{0.5} + 5d_p/X\ (L - 2X)/(L - X)\right] \times \phi b_w \times 0.8h/1000$ | $\phi V_c$, kN |
|---|---|---|---|---|
| 0.5 | 33 | 369 | 372.3 | 186.15 |
| 1.0 | 67 | 403 | 209.0 | 104.5 |
| 1.5 | 100 | 436 | 153.1 | 76.5 |
| 2.0 | 134 | 470 | 124.4 | 62.2 |
| 2.5 | 167 | 503 | 106.1 | 53.05 |
| 3.0 | 201 | 537 | 97.0 | 48.5 |
| 3.5 | 235 | 571 | 97.0 | 48.5 |

$$\begin{aligned}\phi V_c \text{ upper limit} &= \phi \times 0.4 f_c'^{0.5} \times b_w \times 0.8h \\ &= 0.85 \times 0.4 \times 35^{0.5} \times 165 \times 0.8(800 + 60)/1000 \\ &= 228.34 \text{ kN}\end{aligned}$$

$$\begin{aligned}\phi V_c \text{ lower limit} &= \phi \times 0.17 f_c'^{0.5} \times b_w \times 0.8h \\ &= 0.85 \times 0.17 \times 35^{0.5} \times 165 \times 0.8(800 + 60)/1000 \\ &= 97.0 \text{ kN}\end{aligned}$$

$\phi V_c$ need not be lower than the lower limit, and it cannot exceed the upper limit beyond X ≧ 50 strand diameters.

For this example $V_u < \phi V_c$ ; hence shear reinforcing need not be designed. However, $V_u > \phi V_c / 2$; hence minimum shear reinforcing is required. Maximum spacing is the smaller of $0.75h$ or $3b_w$:

$$0.75 \times 860 \text{ mm} = 645 \text{ mm} \quad \text{or} \quad 3 \times 165 = 495 \text{ mm}$$

Hence the maximum spacing is 495 mm.

$$A_{v,\text{minimum}} = A_{ps} / 80 \times f_{pu} / f_y \times S_d (d / b_w)^{0.5}$$

$$= 10 \times 74/80 \times 1860/400 \times 495/(0.8 \times 860)(0.8 \times 860/165)^{0.5}$$

$$= 63 \text{ mm}^2$$

For double-T and T beams with a narrow stem, mesh is often used. To save time some precasters use prefabricated shear reinforcing such as Dur-O-WAL prefabricated web reinforcing. The first wire or bar is placed at a minimum distance $S/2$ from the ends. Usually special bearing cages will take care of the end condition. The maximum yield stress that can be used for shear reinforcing is 400 MPa.

For mesh at a spacing 305 mm the wire size would be MW 47.6.

$$63 \text{ mm}^2/495\text{-mm spacing} = \text{wire size/spacing of } 305 \text{ mm}$$

The wire size required is 38.8 mm². Such a size could be used, but it is not common. A no. 10 reinforcing bar has an area of 100 mm². While this is greater than 63 mm² and could be used, it would not be economical to do so.

### Horizontal shear design

For a composite member horizontal shear occurs at the interface of the topping and the precast unit. If the surface is roughened to a minimum amplitude of 6 mm, the allowable horizontal shear resistance is 0.55 MPa. If the shear stress is greater than this, shear ties must be used. With $V_u h$ of 0.55 MPa at the end of the member to 0.0 at distance $Lvh$, the average value is 0.28 MPa.

$$V_{uh} = V_u / \phi b_v d \qquad V_u = V_{uh} \phi b_v d \leqq \phi \times 0.55 \text{ MPa} \times b_v\, d$$

If the allowable average value of $\phi \times 0.28 \text{ MPa} \times b_v Lvh$ is greater than the smaller of the compression force in the topping or the tension force in the strand plus any tension mild steel, then the roughened surface is sufficient without shear ties. If shear ties are required, then the minimum area is

$$A_{sv} = 0.35 b_v Lvh / f_y$$

$$A_{sv} = F_h / \phi \mu f_y$$

$$F_{h,\text{maximum}} = 0.25 f'_c b_v Lvh$$

If ties are used, the maximum spacing is 4 times the topping thickness. Horizontal shear is diagramed in Fig. 34.36.

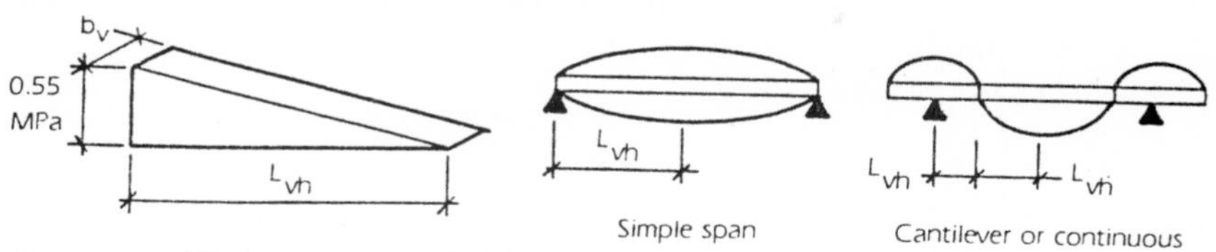


**Figure 34.36** Horizontal shear diagram.

$\phi \times 0.28\ \text{MPa} \times b_v\ Lvh = 0.85 \times 0.28\ \text{MPa} \times 2400\ \text{mm} \times 15\ \text{m} \times 1000/\text{m}/2/1000 = 4284\ \text{kN}$

Maximum compression $= \phi \times 0.85 f'_c b_v T_t = 0.9 \times 0.85 \times 20\ \text{MPa} \times 2400 \times 60/1000 = 2203\ \text{kN}$

Maximum tension $= \phi A_{ps} f_{ps} = 0.9 \times 10 \times 74\text{mm}^2 \times 1845\ \text{MPA}/1000 = 1228.8\ \text{kN}$

Since 1228.8 is less than 4284, shear ties are not required.

### End-bearing design

The bearing ends of precast pretensioned members rest on bearing pads and are designed to resist shear cracking failure, tension due to creep and other types of tension forces, and bearing on the pad. If the tension is zero or unknown, the connection is designed for a minimum tension force of 20 percent of the end reaction ($0.2V_u$). For shear cracking the potential crack pattern is determined, and by using the shear friction method the necessary reinforcing steel is computed. This reinforcing is added to that required to resist the tension forces. As an example,

$V_{u,LL} = 1.7 \times 2.4\ \text{kN/m}^2 \times 2400\ \text{mm}/1000 \times 15\ \text{m}/2/2$ stems $=$ 43.47 kN

$V_{u,DL} = 1.4 \times 3.45\ \text{kN/m}^2 \times 2400\ \text{mm}/1000 \times 15\text{m}/2/2$ stems $=$ 36.72 kN

$V_{u,TDL} = 1.4 \times 1.35\ \text{kN/m}^2 \times 2400\ \text{mm}/1000 \times 15\ \text{m}/2/2$ stems $=$ 17.01 kN

$V_{u,SDL} = 1.4 \times 1.2\ \text{kN/m}^2 \times 2400\ \text{mm}/1000 \times 15\ \text{m}/2/2$ stems $=$ 15.12 kN

$V_{u,\text{total}} =$ 112.32 kN

The bearing-pad length is $V_u$/stem width × bearing value of the pad. The bearing-pad width is approximately 20°mm wider than the stem of the double T to allow for placement tolerance. By assuming an allowable bearing value for the pad of 7 MPa and a stem width of 130 mm, the pad length is 112.32 kN/ (130 mm ×7.0 MPa) ×1000 = 123.4 mm.Use a pad 150 mm wide by 150 mm long. The 150-mm length will permit a placement or slip tolerance of approximately 25 mm. Design for a 125-mm bearing length. End-bearing forces are shown in Fig. 34.37.

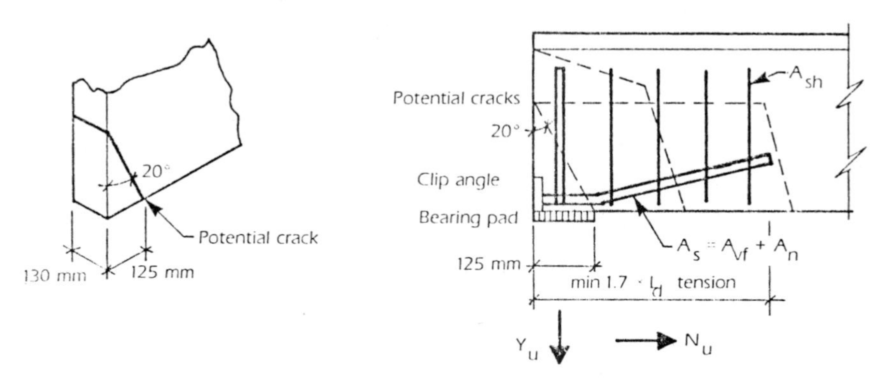


**Figure 34.37** End-bearing forces.

The area of the potential crack = 130 mm ×125 mm/sin 20° = 47,511 mm$^2$. The Uniform Building Code requires that $V_c$ be the smaller of $0.2f'_c A_{\text{cracked}}$ or 5.5 MPa × $A_{\text{cracked}}$. The Prestressed Concrete Institute and the Canadian Prestressed Concrete Institute recommend the smaller of $0.3f'_c A_{\text{cracked}}$ or 6.9 MPa ×$A_{\text{cracked}}$.

By using the Uniform Building Code,

$$\phi V_c = \text{smaller of } \phi \times 0.2f'_c A_{\text{cracked}}$$

$$= 0.85 \times 0.2 \times 35 \text{ MPa} \times 47{,}511/1000 = 282.7 \text{ kN}$$

or

$$\phi \times 5.5 \text{ MPa} \times A_{\text{cracked}}$$

$$= 0.85 \times 5.5 \text{ MPa} \times 47{,}511/1000 = 222 \text{ kN}$$

Since 222 kN is greater than 112.32, this is OK.

The reinforcing steel

$$A_s = A_{vf} + A_n = V_u / \phi f_y \mu + 0.2V_u / \phi f_y$$

where $A_{vf}$ = shear friction steel

$A_n$ = tension steel

$\phi$ = 0.85 for shear

$\mu$ = 1.4 for monolithically placed concrete

$$A_s = 112.323\ \text{kN}/(0.85 \times 400\ \text{MPa} \times 1.4) \times 1000$$
$$+\ 0.2 \times 112.32\ \text{kN}/(0.85 \times 400\ \text{MPa}) \times 1000 = 302\ \text{mm}^2$$

Use one no. 20M bar (300 mm²). Since these bars will be welded, specify ASTM A-706 weldable reinforcing steel, $f_y = 400$ MPa. Length development tension is the larger of $0.019 A_s f_y / f_c'^{0.5}$ or 0.058 bar diameter$\times f_y$ or 300 mm.

$$L_{d,\text{ten}} = 0.019 \times 300\ \text{mm}^2 \times 400\ \text{MPa}/35^{0.5} = 385\ \text{mm}$$
$$L_{d,\text{ten}} = 0.058 \times 19.5\ \text{mm} \times 400 = 452\ \text{mm (the largest value)}$$
$$L_{d,\text{ten}} = 300\ \text{mm}$$
$$1.7 L_{d,\text{ten}} = 1.7 \times 0.452 = 768\ \text{mm}$$

The optimal value is $2L_{d,\text{ten}} = 2\times452$ mm = 904 mm. Use the 800-mm length.

The vertical bar at the end is designed for shear friction:

$$A_{vf} = V_u / \phi f_y\, \mu = 112.32\ \text{kN}/(0.85 \times 400\ \text{MPa} \times 1.4) \times 1000 = 236\ \text{mm}^2$$

Use either one no. 20M bar or two no. 15M bars. The area of a 20M bar is 300 mm², and that of a15M bar is 200 mm².

The $A_{sh}$ steel is not required by either the Uniform Building Code or by the American Concrete Institute model code. The PCI and the CPCI do recommend this steel owing to possible horizontal cracking around the $A_s$ steel. If we assume the $A_{sh}$ steel is 300 MPa,

$$A_{sh} = (A_{vf} + A_n) \times f_y / \mu f_y \text{ hoop}$$
$$= 300\ \text{mm}^2 \times 400\ \text{MPa}/(1.4 \times 300\ \text{MPa}) = 286\ \text{mm}^2$$

Use two no. 10M U stirrups equally spaced along the horizontal bar; $2\times2\times100$ mm² = 400 mm². Since 400 is greater than 286 mm, this is OK.

**Typical end assemblies.** Typical end-bearing assemblies are shown in Fig. 34.38.

Weld size = bar diameter/3; 19.5 mm/3 = 6.5 mm. Use 7-mm size. Fillet-weld (flare-weld) both sides of the bar. For the vertical bar weld all around with a 6-mm fillet weld. The minimum plate size = weld size + 1.5 mm = 6 + 1.5 = 7.5. Use an 8-mm plate or clip angle.

The length of the weld is designed to develop the full bar strength. Bar strength = 300 mm²$\times$400 MPa/1000 = 120 kN.

According to the Uniform Building Code, the fillet weld strength = 0.3 FeXX$\times$0.707, where FeXX is the ultimate strength of the welding electrode. With a 480xx-MPa electrode this is 0.3$\times$480 MPa$\times$0.707/

1000 = 0.1 kN/mm²; for a 7-mm weld, 0.7 kN/mm. For ultimate strength design the PCI recommends a $U$ factor of 1.67 and a phi factor of 1.0.

The 1986 edition of the *AISC Load and Resistance Factor Design Manual*, Chap. J2.3, gives 0.6 FeXX (where $\phi$ = 0.75)×0.707; however, the load factors are 1.2 for dead load, 1.6 for floor live load, and 0.5 for roof live load.

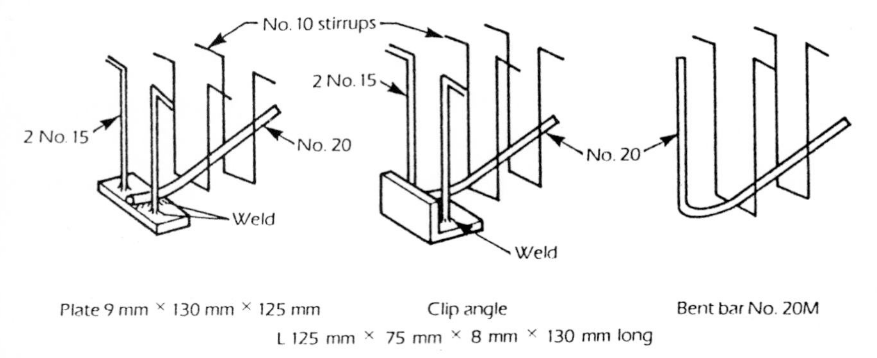


**Figure 34.38** Typical end-bearing assemblies.

By using the Uniform Building Code, 0.3×480 MPa×0.707/1000 = 0.1 kN/mm². For a 7-mm weld size, this is

$$0.1\,\text{kN/mm}^2 \times 7\,\text{mm} = 0.7\,\text{kN/mm weld length}$$

$$120\ \text{kN}/0.7\ \text{kN/mm}/2\ \text{sides} = 85.\underline{7}\ \text{mm weld length}$$

Use a 7-mm weld 90 mm long. If the 1.67 factor is used, the weld would be 85.7/1.67 = 51.3 mm; or use a minimum 60-mm weld length.

## Flange design

When designing the flange of a double T (see Fig. 34.39) that will become composite, the designer must consider a construction live load along with the weight of the topping and the double-T flange. The construction live load will vary with the local building code and is subject to the placement of the topping. If the member is a roof member and the roof live load is small, the flange is usually designed for a larger live load for the time when a construction worker is standing on the edge of the double-T flange. Hence good judgment is required when selecting a design live load. For this composite member the critical load occurs during construction.

$$\text{Flange force} = 50\,\text{mm} \times 1000\ \text{mm} \times 2300\,\text{kg/m}^3 \times 9.81\,\text{m/s}^2/$$

$$1000\,\text{kg/s}^2/10^6 = 1.128\ \text{kN/m}^2$$

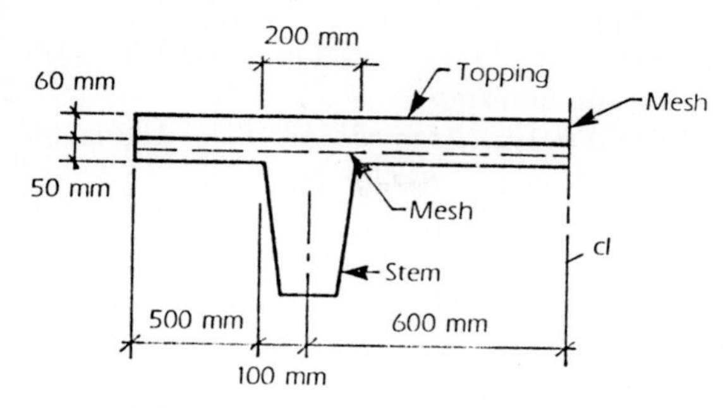


**Figure 34.39** Flange section. Construction live load = 2.4 kN/m²; topping dead load = 1.35 kN/m²; top flange = 1.128 kN/m²; unit $f_c$ = 35 MPa; service live load = 2.4 kN/m²; superimposed dead load = 1.2 kN/m²; topping $f'_c$ = 25 MPa.

During construction,

| | |
|---|---|
| Moment construction live load $= 1.7 \times 2.4\ \text{kN/m}^2 \times 1\text{m} \times 0.5\text{m}^2/2 =$ | $0.51\ \text{kN}\cdot\text{m}$ |
| Moment topping dead load $= 1.4 \times 1.35\ \text{kN/m}^2 \times 1\text{m} \times 0.5\ \text{m}^2/2 =$ | $0.236\ \text{kN}\cdot\text{m}$ |
| Moment flange dead load $= 1.4 \times 1.28\ \text{kN/m}^2 \times 1\text{m} \times 0.5\ \text{m}^2/2 =$ | $0.197\ \text{kN}\cdot\text{m}$ |
| Total moment = | $0.943\ \text{kN}\cdot\text{m}$ |

Place mesh in the center of the flange for both negative and positive movement; hence $d$ = 25 mm.

$$a = d - \left[-2M_u/(\phi \times 0.85 f'_c b) + d^2\right]^{0.5}$$

$$a = 25 - (-2 \times 0.943\,\text{kN}\cdot\text{m} \times 1000\,\text{mm/m} \times 1000/$$

$$\left[0.9 \times 0.85 \times 35\,\text{MPa} \times 1000\ \text{mm}) + 25^2\right]^{0.5}$$

$$a = 1.45\,\text{mm}$$

$$A_s/m = 0.85 f'_c ba/f_y = 0.85 \times 35\ \text{MPa} \times 1000\ \text{mm} \times$$

$$1.45\ \text{mm}/400\,\text{MPa} = 107.84\ \text{mm}^2$$

Minimum temperature steel = $0.0018bT_f = 0.0018 \times$ $1000\ \text{mm} \times 50\ \text{mm} = 90\ \text{mm}^2$

Structural steel will govern.

Maximum spacing = $3T_f = 3 \times 50\ \text{mm} = 150\ \text{mm}$

Temperature steel spacing = $5t_f = 5 \times 50\ \text{mm} = 250\ \text{mm}$

The temperature steel requirement parallel to member length is 90 $mm^2$. Use MW 16.0 wire (area, 106 $mm^2/m$).

Use mesh 152×152(MW 18.7×MW 16). Area of MW 18.7 = 123 $m^2/m$ > 107 $mm^2/m$.

## Completed design

The completed design of the composite double T is shown in Fig. 34.40.

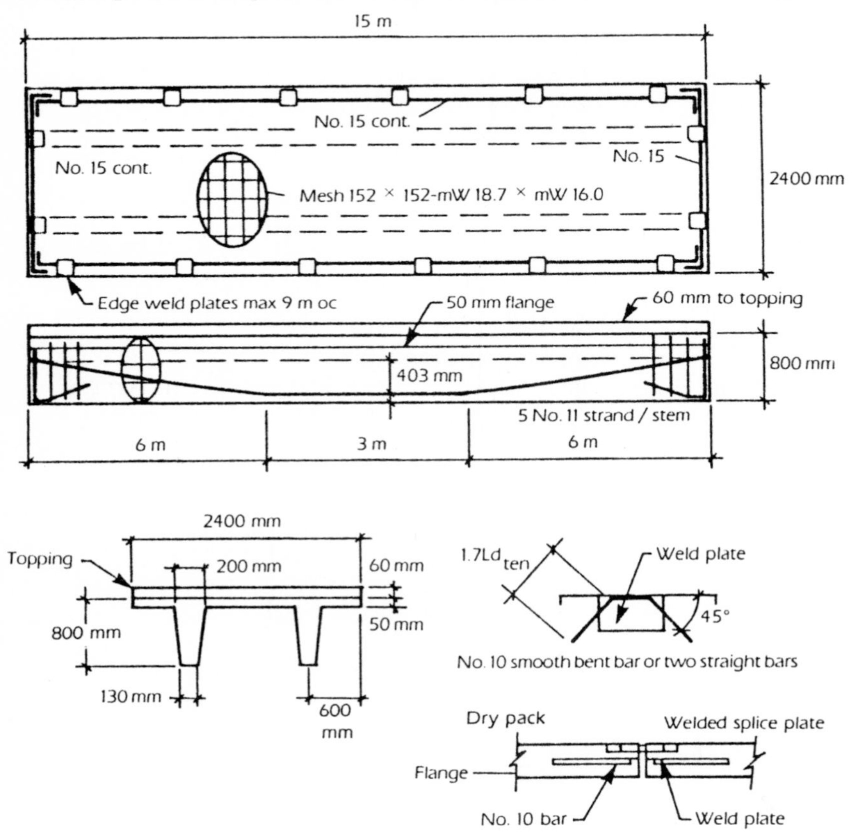


**Figure 34.40** Completed design.

## Deflection

Deflection is an important consideration in the design of any floor or roof system. In prestressed concrete, unlike nonprestressed concrete, deflection can be controlled by controlling the camber due to prestressing. While this is frequently done in posttensioned slabs, it is not a prime consideration in the design of pretensioned precast systems. Camber in precast pretensioned systems is the result of the placement of the strands to resist the design moment and the concrete stresses. If the camber is excessive or needs to be increased, then the strand profile and number of strands can be altered subject to the design concrete stresses. With partial prestressing, the use of mild steel can give additional control of deflection.

Owing to the variations of the design mix, curing, storage, relative humidity, time of detensioning, time of loading, creep phenomena, and strand relaxation, it is impossible to determine the exact deflection which will occur. Laboratory-determined factors based upon a time-step method with controlled humidity, while suited for the laboratory, are often not suited for actual conditions. Therefore, the designer should not specify the camber or the deflection.

The various building codes place a limit on deflection, and the designer should consult the code applicable to the project location. The Uniform Building Code, 1988 edition, in Sec. 2307 and Tables 23-D and 23-E, gives the following limits:

For roof members supporting a plaster ceiling and floor members: live-load deflection span length/360 and live load + $K$ times dead-load deflection span length/240. $K$ is a creep factor, and $K = \left[2 - 1.2\left(A'_s / A_s\right)\right] = 0.6$. However, Sec. 2609 of the Uniform Building Code gives the following value of $\lambda$ to be used for $K$: $\lambda = T / (1 + 50p')$ where $p' = A'_s / bd$, and $T$ for sustained loads = 2.0 for 5 years, 1.4 for 12 months, 1.2 for 6 months, and 1.0 for 3 months.

Some codes such as the New Zealand and Canadian national codes recommend the following:

| Member | Type of load | Recommendation |
|---|---|---|
| Roofs without attached partition systems | Immediate live load | $L/180$ |
| Floors without attached partitions | Immediate live load | $L/360$ |
| Roofs or floors with attached partitions not likely to be damaged by deflection | Creep × dead load + percent live load | $L/240$ |
| Roofs or floors with attached partitions likely to be damaged by deflection | Creep × dead load + percent live load | $L/480$ |

Camber first occurs at detensioning of the strand (transfer of the prestressing force to the concrete). As the member gains strength with time and as loss of prestress occurs with time, the magnitude of camber will change. At detensioning the member normally is a noncracked section, and the elastic concept of deflection therefore is used. (See Fig. 34.41.)

If the member is designed as an uncracked section, the elastic equations for deflection are used (see Fig. 34.42). Here the moment of inertia

$I$ is the gross moment of inertia. $E$ is the modulus of elasticity of the concrete at the time load being considered. The modulus of elasticity of the concrete $= E_c = 0.043 w_c^{1.5} f_c'^{0.5}$.

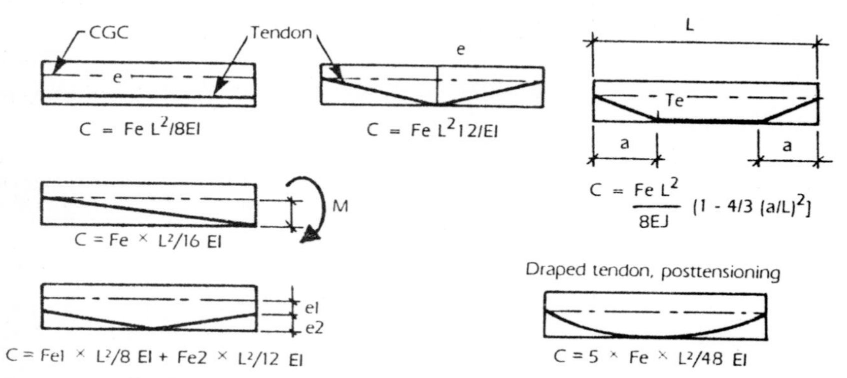


**Figure 34.41** Camber equations. $C$ = camber; $F$ = prestressing force.

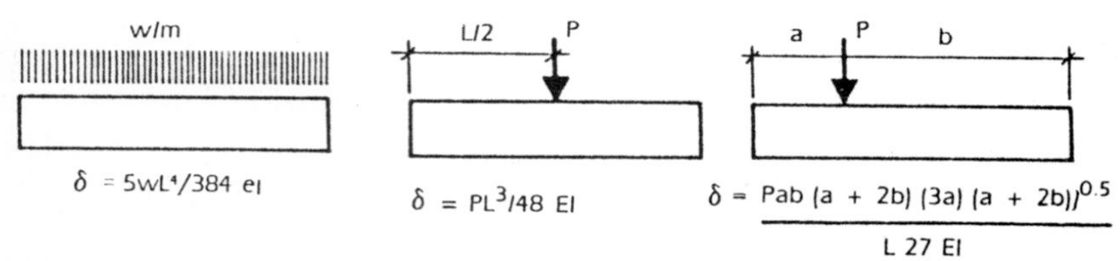


**Figure 34.42** Deflection equations.

If the member is a cracked section under nonfactored loads, then either a bilinear analysis or the $I$ effective value is used. With bilinear analysis,

$$\Delta = L^2 / 48E \left[ 5M_{\text{cracked}} / I_g + 5(M - M_{\text{cracked}}) / I_{\text{cracked}} \right]$$

$$M_{\text{cracked}} = S_b (F / A + Fe / S_b + f_r)$$

where $f_r = 0.6\lambda \times f_c'^{0.5}$

$\lambda$ = 1.0 for normal-weight concrete and 0.85 for sand-lightweight concrete

The mass of normal-density concrete $w_c$ = 2300 to 2400 kg/m³, and for sand-lightweight concrete is 1800 to 2000 kg/m³.

If the $I$ effective method is used (see Fig. 34.43),

$$\Delta = 5wL^4 / 384 E_c I_{\text{eff}}$$

where $I_{\text{eff}} = (M_{\text{cracked}} / M_a)^3 \times I_g + \left[1 - (M_{\text{cracked}} / M_a)^3 \times I_{\text{cracked}}\right]$

[For prestressed concrete $M_{\text{cracked}} / M_a$ is expressed as $1 - (f_t - f_r)/f_{LL}$].

$f_t$ = calculated stress in member under load in tension

$f_r$ = modulus of rupture

$f_{LL}$ = stress due to live load in tension

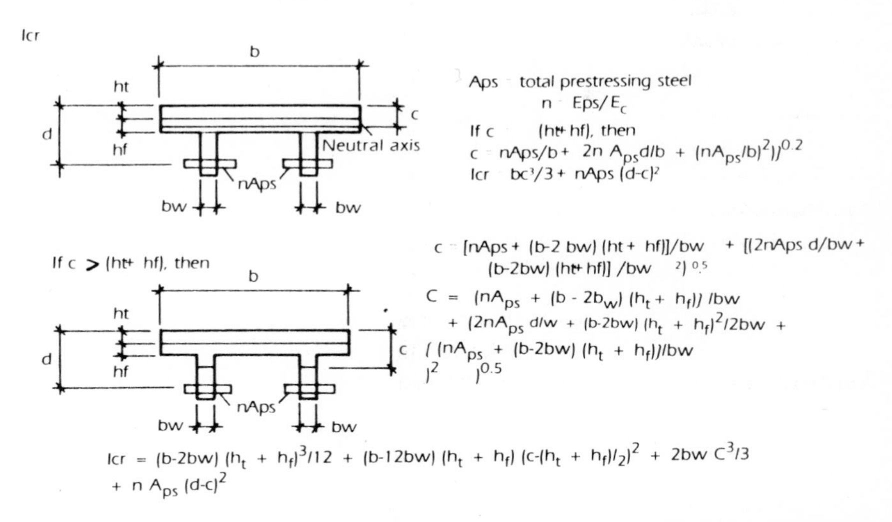


**Figure 34.43** Equation for $I_c$. If there is only one stem, use $b_w$ instead of $2b_w$. For double T's, $c$ is usually in the flange or the topping.

**Long-term camber or deflection.** In lieu of a time-step method of computing long-term deflection both the Prestressed Concrete Institute and the Canadian Prestressed Concrete Institute recommend the simplified method shown in Table 34.8.

**Table 34.8 Seven PCI and CPCI Deflection Coefficients**

| | Multipliers, $C$ | |
|---|---|---|
| | Non-composite | Composite topping |
| At erection | | |
| 1. Camber: apply to elastic camber computed at release | 1.8 | 1.8 |
| 2. Deflection: apply to elastic deflection, member density, at release | 1.85 | 1.85 |
| Final | | |
| 3. Camber: apply to elastic camber at release | 2.45 | 2.20 |
| 4. Deflection: apply to elastic deflection, member density, at release | 2.7 | 2.4 |
| 5. Deflection: superimposed dead load, using $E_c$ at 28 days | 3.0 | 3.0 |
| 6. Deflection due to topping density, using $E_c$ at 28 days | | 3.0 |
| 7. Deflection using $E_c$ at 28 days | 1.0 | 1.0 |

NOTE: If a percentage of live load is considered as a long-time loading, then treat as if it were a superimposed dead load. If compression steel is used, then modify the multipliers by $C_{\text{modified}} = \left(C + A'_s / A_{ps}\right) / \left(1 + A'_s / A_{ps}\right)$.

**Example:** An example double T is shown in Fig. 34.44.

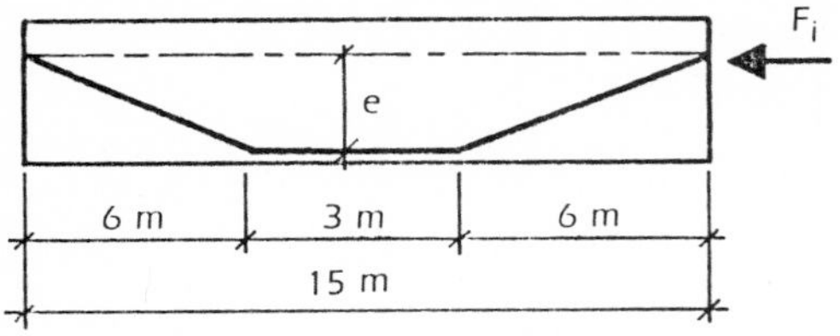

$F_i$

$e$

6 m 3 m 6 m

15 m

$e_{ci} = 0.043 \times 2300^{1.5} \times 25^{0.5} = 23{,}700$ MPa

$e_c = 0.043 \times 2300^{1.5} \times 35^{0.5} = 28{,}700$ MPa

**Figure 34.44** Example double T. $f'_c = 35$ MPa; $f'_{ci} = 25$ MPa. $I_{g,\text{noncomposite}} = 22{,}725 \times 10^6$; $I_{g,\text{composite}} = 31{,}351 \times 10^6$. $F_{pi}$ = 10 strands × 91.67 kN/strand = 916.7 kN. Unit dead load = 8.28 kN/m; topping = 3.25 kN/m; superimposed dead load = 2.88 kN/m; live load = 5.76 kN/m.

$$\text{Camber} = F_i e L^2 \left[1 - 4/3(a/L)^2\right] / 8E_{ci} I_g = 916.7 \times 403 \times 15^2 \times 10^9$$
$$\left[1 - 4/3(6/15)^2\right] / \left(8 \times 23{,}700 \times 22{,}725 \times 10^6\right)$$
$$= 15.17\,\text{mm}$$

$$\text{Deflection unit load} = 5wL^4/384E_{ci}I_g \text{ noncomposite}$$
$$= 5 \times 8.28 \times 15^4 \times 10^{12} / \left(384 \times 23{,}700 \times 22{,}725 \times 10^6\right)$$
$$= 10.13\,\text{mm}$$

$$\text{Deflection topping} = 5wL^4/384E_{ci}I_g \text{ noncomposite}$$
$$= 5 \times 3.25 \times 15^4 \times 10^{12} / \left(384 \times 28{,}000 \times 22{,}725 \times 10^6\right)$$
$$= 3.37\,\text{mm}$$

$$\text{Deflection superimposed dead load} = 5wL^4/384E_c I_g \text{ composite}$$
$$= 5 \times 2.88 \times 15^4 \times 10^{12} / \left(384 \times 28{,}000 \times 31{,}351 \times 10^6\right)$$
$$= 2.16\,\text{mm}$$

$$\text{Deflection live load} = 5wL^4/384E_c I_g \text{ composite}$$
$$= 5 \times 5.76 \times 15^4 \times 10^{12} \left(384 \times 28{,}000 \times 31{,}351 \times 10^6\right)$$
$$= 4.32\,\text{mm}$$

**Summary.** Camber and deflection data may be summarized as follows :

| | Multiplier | Movement |
|---|---|---|
| At erection | | |
| Camber, 15.17 mm | 1.8 | 27.31 mm up |
| Deflection unit dead load, 10.13 mm | 1.85 | 18.74 mm down |
| Net | | 8.57 mm up |
| Final (after 2 years) | | |
| Camber, 15.17 mm | 2.2 | 33.37 mm up |
| Deflection unit dead load, 10.13 mm | 2.4 | 24.31 mm down |
| Deflection topping, 3.37 mm | 2.3 | 7.75 mm down |
| Deflection superimposed dead load, 2.16 mm | 3.0 | 6.48 mm down |
| Net | | 5.16 mm down |
| Deflection live load, 4.32 mm | 1.0 | 4.32 mm down |
| Net | | 9.48 mm down |

## Example Problem 6: Inverted-TGirder (SI Units)

Inverted-T and L-shaped girders are used to support other precast prestressed members such as double T's, hollow planks, and solid slabs, which rest on the flange of the inverted-T or L girder. This allows for the reduction of space required for the floor-framing system. Rectangular-shaped and I-shaped girders are used when hollow planks or solid slabs rest on top of the girders and are designed similarly to inverted-T girders. When a topping is used, the girders are usually designed as composite members.

Precast prestressed concrete manufacturers have standardized shapes because of the cost of steel forms, and they provide load-span tables as a guide to the designer. However, nonstandard shapes can be produced with wood and fiberglass forms. The width of the ledge is determined by the required bearing length of the supported double T or hollow plank, while allowing for erection tolerance.

When designing any prestressed concrete system, the designer must consult the building code governing the location where the project is to be constructed. The load factors ($U$ values), resistance factors (phi values), material strengths, allowable stresses, and equations vary with the country. For example, in the United States the standard reinforcing steel used is $f_y$ 60 ksi, which equates to 413.7 MPa. In Canada the standard is 400 MPa, and in New Zealand it is 380 MPa. However, the general theory is similar in the various countries.

The example inverted-T design is for an office building (see Fig. 34.45).

| Concrete | Density, kg/m$^3$ | $f'_c$, MPa | $f'_{ci}$, MPa |
|---|---|---|---|
| Inverted T | 1800 | 40 | 30 |
| Double T | 1800 | 40 | |
| Topping | 2400 | 30 | |

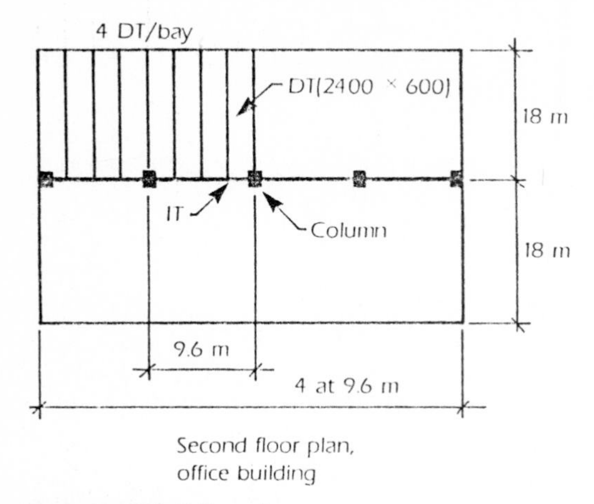


**Figure 34.45** Building plan.

**Loads.** The inverted-T tributary area = 9.6 × 18 m = 172.8 m². Use a 40 percent reduction. The live load for an office building is 2.4 kN/m².

Live load:

$$0.6 \times 2.4 \text{ kN/m}^2 \times 18\text{m} = \qquad 26 \text{ kN/m}$$

Dead loads:

$$\text{Inverted T} = \qquad 6.84 \text{ kN/m}$$
$$\text{Double T} = 2.0 \text{ kN/m}^2 \times 18\text{m} = 36 \text{ kN/m}$$
$$\text{Topping} = 1.4 \text{ kN/m}^2 \times 18\text{m} = 25.2 \text{ kN/m}$$

**Properties**

**Noncomposite properties.** See Fig. 34.46.

| | |
|---|---|
| Area | 372,500 mm² |
| Moment of inertia | 23,000 × 10⁶ |
| $Y_t$ | 480.37 mm |
| $Y_b$ | 369.63 mm |
| $S_t$ | $47.88 \times 10^6$ |
| $S_b$ | $62.22 \times 10^6$ |
| Unit density | 1,800 kg/m³ |
| $w$ | 6.58 kN/m |
| Topping density | 2,400 kg/m³ |
| $w$ topping | 1.4 kN/m² |

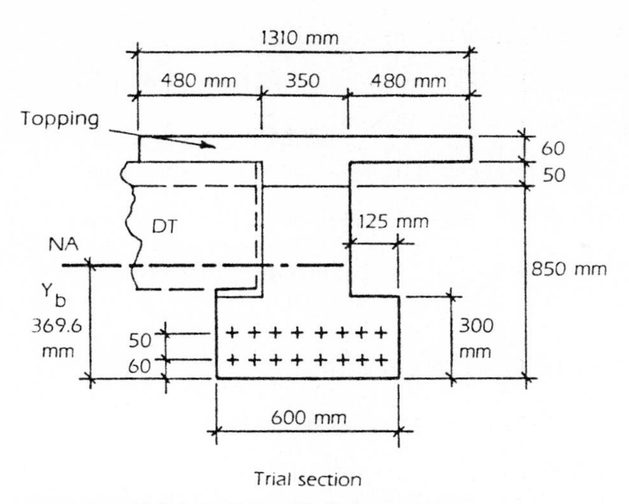


**Figure 34.46** Trial section: physical properties.

$$\text{Area} = 300\times 600\times 550\times 350 \text{ m} = 372{,}500\text{m}^2$$

$$\text{Unit force/ m} = 372{,}500 \text{ mm}^2 \times 1800 \text{ kg/ m}^3 \times 9.81\text{m / s}^2/10^9 = 6.58 \text{ kN/ m}$$

$$\text{Topping force/ m}^2 = 60 \text{ mm} \times 1 \text{ m}^2 \times 2400 \text{kg/ m}^3 \times 9.81/10^6 - 1.4 \text{ kN/ m}^2$$

$$Y_b = \left(250\times 300^2/2 + 350\times 850^2/2\right)/372{,}500 = 369.63 \text{ mm}$$

$$Y_t = 850 - 369.63 = 480.37 \text{ mm}$$

$$I = 350(850 - 369.63)^3/3 + 350\times 369.63^3/3 + 250\times 300 (369.63 - 150)^2 + 250\times 300^3/12 = 23{,}000\times 10^6$$

$$S_t = I/Y_t = 23{,}000/480.37 = 47.88\times 10^6$$

$$S_b = I/Y_b = 23{,}000/369.63 = 62.22\times 10^6$$

**Composite properties.** See Fig. 34.37. Since the topping is not of the same modulus of elasticity as the unit, the effective width of the topping must be determined:

$$E_{\text{unit}} = 0.043w^{1.5}f_c'^{0.5} = 0.043\times 1800^{1.5} \times 40^{0.5} = 20{,}768 \text{MPa}$$

$$E_{\text{topping}} = 0.043w^{1.5}f_{ct}^{0.5} = 0.043\times 2400^{1.5} \times 30^{0.5} = 27{,}691 \text{ MPa}$$

Effective topping width= 1310 mm × 27,691/ 20,768 = 1746 mm

Effective stem width= 350 mm × 27,691/ 20,768 = 466 mm

$$Y_{ct} = [1746\ \text{mm} \times 60\ \text{mm}^2/2 + 50\ \text{mm} \times 466 \times 85\ \text{mm} + 372{,}500\ (480.37 + 110)]/(1746 \times 60 + 466 \times 50 + 372{,}500) = 449.57\ \text{mm}$$

$$Y_{cb} = 960\ \text{mm} - 449.57\ \text{mm} = 510.43\ \text{mm}$$

$$\text{Distance between neutral axes} = 510.43 - 369.63 = 140.8\ \text{mm}$$

$$I_c = 23{,}000 \times 10^6 + 372{,}500 \times 140.8^2 + 1746 \times 60^3/12 + 1746 \times 60\ (449.57 - 30)^2 + 466 \times 50^3/12 + 466 \times 50\ (449.57 - 60 - 25)^2$$

$$= 51{,}959 \times 10^6$$

$$S_{ct} = I_c/C = 51{,}959/449.57 = 115.477 \times 10^6$$

$$S_{cc} = I_c/C = 51{,}959/(449.56 - 110) = 153.017 \times 10^6$$

$$S_{cb} = I_c/C = 51{,}959/510.43 = 101.79 \times 10^6$$

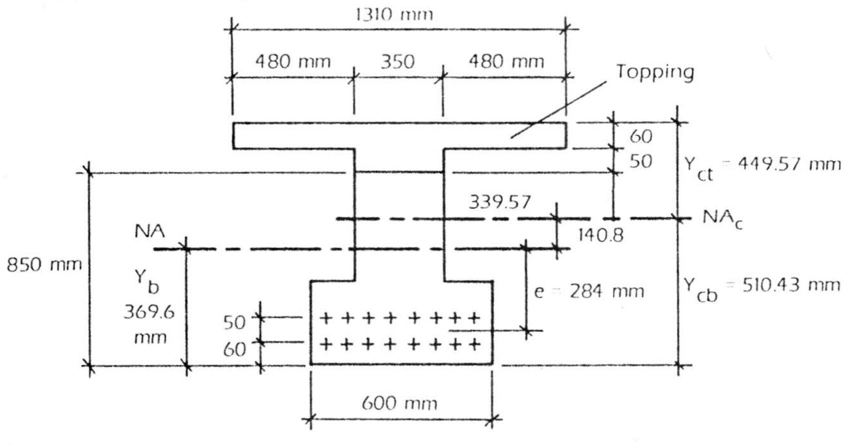


**Figure 34.47** Composite properties.

### Allowable stress

**Strand.** The no. 13 low-relaxation strand, 1860 MPa and area of 99 mm$^2$, has an initial loss of 10 percent and a service loss of 16 percent.

At detensioning:

$$0.74 \times 1860\ \text{MPa} \times 0.9 \times 99/1000 = 122.637\ \text{kN/strand}$$

At service:

$$0.74 \times 1860\ \text{MPa} \times 0.84 \times 99/1000 = 114.46\ \text{kN/strand}$$

**Concrete.** Stress is as follows.

Inverted T at detensioning:

$$\text{Compression} = 0.6 f'_{ci} = 0.6 \times 30\ \text{MPa} = 18\ \text{MPa}$$

$$\text{Tension at ends} = \lambda 0.5 f_{ci}^{\prime 0.5} = 0.85 \times 0.5 \times 30^{0.5} = 2.328\,\text{MPa}$$

$$\text{Tension at midspan} = \lambda 0.5 f_{ci}^{\prime 0.5} = 0.85 \times 0.25 \times 30^{0.5} = 1.164\ \text{MPa}$$

Inverted T at service:

$$\text{Compression} = 0.45 f_c' = 0.45 \times 40\,\text{MPa} = 18\ \text{MPa}$$

$$\text{Tension} = \lambda\, 0.5_c^{0.5} = 0.85 \times 0.5 \times 40^{0.5} = 2.69\ \text{MPa}$$

Topping:

$$\text{Compression} = 0.45 f_{ct}' = 0.45 \times 30 = 13.5\,\text{MPa}$$

**Moments:**

| | $M$, kN·m | $U$ factor | $M_u$, kN·m |
|---|---|---|---|
| $M_{LL} = wL^2/8 = 26\,\text{kN/m} \times 9.6\ \text{m}^2/8 =$ | 299.52 | 1.5 | 449.28 |
| $M_{DL}$, inverted T $= wL^2/8 = 6.58\ \text{kN/m} \times 9.6\ \text{m}^2/8 =$ | 75.80 | 1.25 | 94.75 |
| $M_{DL}$, double T $= wL^2/8 = 36\,\text{kN/m} \times 9.6\ \text{m}^2/8 =$ | 414.72 | 1.25 | 518.40 |
| $M_{DL}$, topping $= wL^2/8 = 25.2\ \text{kN/m} \times 9.6\,\text{m}^2/8 =$ | 290.30 | 1.25 | 362.88 |
| | | | 1425.31 |

**Prestressing**

Determine prestressing force.

$$F_{ps}\ \text{required} = \left[\left(M_{\text{inverted T}} + M_{\text{double T}} + M_{\text{topping}}\right)/S_b + M_{LL}/S_{cb} - f_{\text{ten}}\right]/$$
$$(1/A + e/S_b) = [(75.8 + 414.72 + 290.3)$$
$$\times 10^6/\left(62.22 \times 10^6\right) + 299.52 \times 10^6/\left(101.79 \times 10^6\right)$$
$$- 2.69\,\text{MPa}]/\left(1000/372{,}500 + 284 \times 1000/62.66 \times 10^6\right)$$
$$= 1766.016\ \text{kN}$$

No. of strands $= f_{ps}$/force per strand = 1766.016/114.46 = 15.43
Use 16 strands. Place the strands in two rows of 8 strands each.

**Possible strand placement**

The concept is to reduce tension in the precompressed zone of the section. Of the three methods shown in Fig. 34.48, method A is the most common if tension exists at midspan prior to becoming composite. Method B is the most common if there is allowable tension and there is no objection to debonding some of the strand. Method C is used where there is allowable tension in the precompressed zone and where debonding is not desirable. For this example method A will be used.

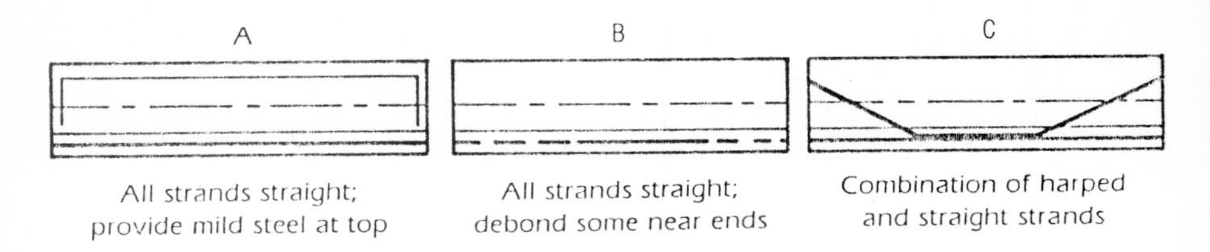


**Figure 34.48** Possible strand location.

### Stress check

At detensioning at midspan:

$$-F_{pi}/A + -F_{pi}e/S_{t,b} - +M_{it}/S_{t,b} = f_{t,b}$$

$$\text{At top} = -16\times 122.637\times 1000/372{,}500 + 16\times 122.637\times 1000 \times 284/(47.88\times 10^6) - 75.8\times 10^6/(47.88\times 10^6) = 4.78 \text{ MPa}$$

Since this is greater than 1.164 MPa, provide mild steel at top.

$$\text{At bottom} = -16\times 122.637\times 1000/372{,}500 + 16\times 122.637\times 1000\times 284/62.22\times 10^6 - 75.8\times 10^6/(62.22\times 10^6) - 13 \text{ MPa} < -18 \text{ MPa (OK)}$$

At detensioning at 50 strand diameters from end where strand has full capacity:

$$M = 6.58 \text{ kN/m} \times 9.6\text{ m}/2 \times 650\text{ mm}/1000 \text{ mm/m} - 6.58 \times 0.65 \text{ m}^2/2 = 19.14 \text{ kN}\cdot\text{m}$$

$$\text{At top} = -16\times 122.637\times 1000/372{,}500 - 16\times 122.637\times 1000\times 284/(47.88\times 10^6) - 19.139\times 10^6/(47.88\times 10^6) = 5.97 \text{ MPa}$$

Since this is greater than 2.328 MPa, provide mild steel at top.

$$\text{At bottom} = -16\times 122.637\times 1000/372{,}500 - 16\times 122.637\times 1000 \times 284/(62.22\times 10^6) + 19.139\times 10^6/(62.22\times 10^6) = -13.91 \text{ MPa} < -18 \text{ MPa (OK)}$$

Tension steel requirements are diagramed in Fig. 34.49.

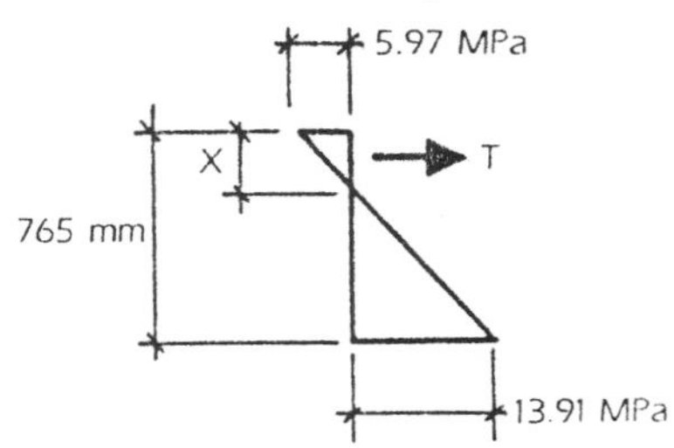


**Figure 34.49** Tension steel requirements.

$$\frac{765}{5.97+13.91}=\frac{X}{5.97} \quad X = 229.7 \text{ mm}$$

$$T = 5.97 \text{ MPa} \times 229.7 \times 350 \text{ mm}/1000 = 480 \text{ kN}$$

Allowable steel stress $= 0.6f_y =< 200$ MPa

$0.6 \times 400 \text{ MPa} = 240 \text{ MPa}$

Since this is greater than 200 MPa, use 200 MPa.

$$480 \text{ kN} \times 1000/200 \text{ MPa} = 2400 \text{ mm}^2$$

Use two 30M and two 25M bars; $A_s = 2400 \text{ mm}^2$.

Place reinforcing in the top of the girder to resist tension in the precompression zone. Run reinforcing the full length of the girder.

**Erection.** Check stress at erection.

At 50 strand diameters:

$$\text{Load/meter} = 6.58 \text{ kN/m} + 36 \text{ kN/m} + 25.2 \text{ kN/m} = 67.78 \text{ kN/m}$$

$$M = 67.78 \text{kN/m} \times 9.6 \text{m}/2 \times 0.65 \text{m} - 67.78 \times 0.65^2/2$$
$$= 197.16 \text{kN} \cdot \text{m}$$

$$\text{At top} = -F_{ps}/A + F_{ps}e/S_t - M/S_t$$
$$= -16 \times 114.46 \times 1000/372{,}500 + 16 \times 114.46 \times 1000 \times 284/(47.88 \times 10^6) - 197.16/47.88 = 1.828 \text{MPa} < 2.69 \text{MPa (OK)}$$

At midspan:

$$\text{At top} = -F_{ps}/A + F_{ps}e/S_t + (M_{\text{inverted T}} + M_{\text{double T}} + M_{\text{topping}})/S_t$$
$$= -16 \times 114.46 \times 1000/372{,}500 + 16 \times 114.46 \times 100 \times 284/(47.88 \times 10^6) - (75.8 + 414.72 + 290.3)/47.88$$
$$= -10.36 \text{ MPa} < -18 \text{ MPa (OK)}$$

$$\text{At bottom} = -F_{ps}/A - F_{ps}e/S_t + (M_{\text{inverted T}} + M_{\text{double T}} + M_{\text{topping}})/S_t$$
$$= -16 \times 114.46 \times 1000/372{,}500 - 16 \times 114.46 \times 100 \times 284/(47.88 \times 10^6) + (75.8 + 414.72 + 290.3) \times 10^6/(47.88 \times 10^6) = -726 \text{ MPa} < 18.0 \text{ MPa (OK)}$$

**Service at midspan.** Check stress at service at midspan.

$$\text{At top of inverted T} = -F_{ps}/A + F_{ps}e/S_t - (M_{\text{inverted T}} + M_{\text{double T}} + M_{\text{topping}})/S_t - M_{LL}/S_{cc}$$

$$= -\ 10.36\ \text{MPa} - 299.52 \times 10^6/(153.017 \times 10^6)$$
$$= 12.317\ \text{MPa} < 18.0\ \text{MPa}\ \ (\text{OK})$$

At bottom of inverted T $= -F_{ps}/A - F_{ps}e/S_t +$

$$(M_{\text{inverted T}} + M_{\text{double T}} + M_{\text{topping}})/S_t - M_{LL}/S_{cb}$$
$$= -\ 0.726 + 299.52 \times 10^6/(101.79 \times 10^6)$$
$$= 2.21\ \text{MPa} < 2.69\ \text{MPa}\ \ (\text{OK})$$

At top of topping $= -M_{LL}/S_{ct} = 299.52 \times 10^6/$

$$(115.577 \times 10^6) = 2.59\ \text{MPa} < 13.5\ \text{MPa}\ \ (\text{OK})$$

**Ultimate strength**

Canadian load factors and resistance factors will be used to check ultimate strength. Strain compatibility will be used to determine stress in the strand (see Fig. 34.50). Stress in the double-T flange will be neglected.

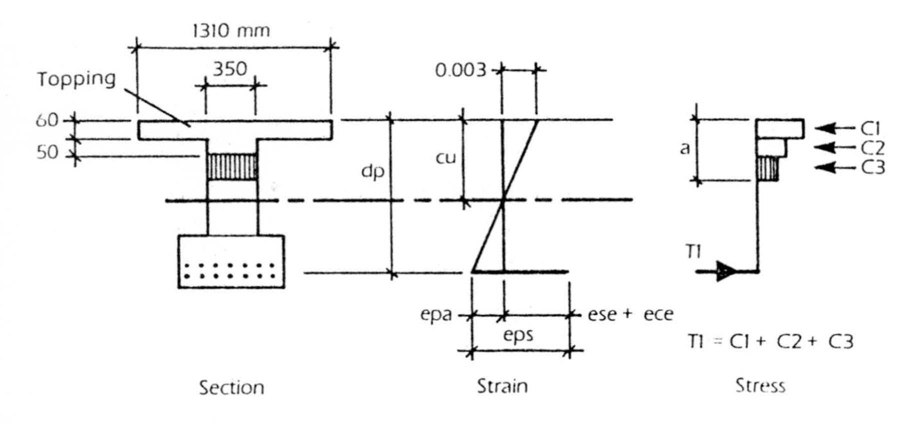


**Figure 34.50** Strain compatibility.

The unknowns are $f_{ps}$, $e_{pa}$, $c_u$, and $a; a = \beta c_u$. The stress-strain curve is used to determine $f_{ps}$. An iteration method is employed. Assume a value for $c_u$ and solve for $e_{pa}$, $f_{ps}$, $T1$, and $C3$. If $T1$ does not equal $C1 + C2 + C3$, then assume a new value for $c_u$ until within 1 percent of the correct result.

One method of arriving at the first trial value of $c_u$ is to assume that $f_{ps} = f_{pu}$, solve for $c_u$, then test for this value of $c_u$ using the true value of $f_{ps}$ obtained from the stress-strain curve for the prestressing strand using $e_{pa}$.

$$C1 = \phi_c \times 0.85 f_c' bt / 1000$$
$$= 0.6 \times 0.85 \times 30 \text{ MPa} \times 350 \text{ mm} \times 60 \text{ mm}/1000 = 1202.8 \text{ kN}$$
$$C2 = 0.6 \times 0.85 \times 30 \text{ MPa} \times 350 \text{ mm} \times 50 \text{ mm}/1000 = 267.75 \text{ kN}$$
$$C3 = 0.6 \times 0.85 \times 40 \text{ MPa} \times 350 \text{ mm} \times (0.77c_u - 110)/1000$$
$$= 5.4978c_u - 785.4 \text{ kN}$$

For a trial assume $f_{ps} = f_{pu}$.

$$T1 = \phi_p A_{ps} f_{se} = 0.9 \times 16 \times 99 \times 1860/1000 = 2651 \text{ kN}$$
$$d_p = 875 \text{ mm}$$
$$c_u / d_p = (T1 - C1 - C2 + 785.4)/(5.4978 \times 875)$$
$$= (2651 - 1202.58 - 267.75 + 785.4)/(5.4978 \times 875) = 0.409$$
$$k_p = 3(1 - f_{py}/f_{pu}) = 3(1 - 0.9 \times 1860/1860)$$
$$= 0.3 \text{ for low-relaxation strand} (0.45 \text{ for standard strand})$$
$$f_{ps} = f_{pu}(1 - k_p c_u / d_p) = 1860(1 - 0.3 \times 0.409) = 1632$$

Let $f_{ps}$ = 1632; then,

$$T1 = 0.9 \times 16 \times 99 \times 1632/1000 = 2326 \text{ kN}$$
$$c_u = (T1 - C1 - C2 + 785.4)/5.4978$$
$$= (2326.6 - 1202.58 - 267.75 + 785.4)/5.4978 = 298.6 \text{ mm}$$

Round off this value of 298.6 mm and use as the first trial value for $c_u$. For the first trial let $c_u$ = 300 mm.

$$f_{se} = 0.74 f_{pu} (100 \text{ percent} - \text{loss in percent})/1000$$
$$= 0.74 \times 1860 \times 0.84 = 1156 \text{ MPa}$$
$$E_c = 0.0043 w^{1.5} f_c'^{0.5} = 0.043 \times 1800^{1.5} \times 40^{0.5} = 20{,}768 \text{ MPa}$$
$$e_{se} = f_{se}/E_c = 1156 \text{ MPa}/20{,}768 \text{ MPa} = 0.006085$$
$$f_{ce} = f_{ps}(1/a + e^2/1) = 16 \times 99 \times 1156 \text{ MPa} \left(1/327{,}500 + 284^2\right)/\left(23{,}000 \times 10^6\right) = 11.337 \text{ MPa}$$
$$e_{ce} = f_{ce}/E_c = 11.337 \text{ MPa}/20{,}768 \text{ MPa} = 0.0005458$$
$$e_{se} + e_{ce} = 0.006085 + 0.0005458 = 0.006308$$
$$e_{pa} = 0.003 d_p / c_u - 0.003 = 0.003 \times 875/300 - 0.003 = 0.00575$$
$$e_{ps} = e_{pa} + e_{se} + e_{ce} = 0.00575 + 0.006308 = 0.01184$$

With $e_{ps}$ known, find $f_{ps}$ from the stress-strain curve or use the following approximate equation:

$$f_{ps} = 1848 - 0.517/(e_{ps} - 0.0065) = 1848 - 0.517/(0.01184 - 0.0065)$$
$$= 1760 \text{ MPa}$$
$$T1 = \phi_p A_{ps} f_{ps} = 0.9 \times 16 \times 99 \times 1760 \text{ MPa}/1000 = 2509 \text{ kN}$$
$$a_3 = \beta c_u - a_1 - a_2 = 0.77 \times 300 - 60 - 50 = 121 \text{ mm}$$
$$C3 = \phi_c \times 0.85 f'_c b a_3 = 0.6 \times 0.85 \times 40 \text{ MPa} \times 350 \text{ mm} \times 121 \text{ mm}/1000$$
$$= 863.94 \text{ kN}$$
$$C1 + C2 + C3 = 1202.58 + 267.75 + 863.94 = 2334.27 \text{ kN} \neq 2509 \text{ kN}$$

Therefore, increase $c_u$. Try 328 mm.

$$e_{ps} = 0.003 \times 875/328 - 0.003 + 0.006638 = 0.01164$$
$$f_{ps} = 1848 - 0.517/(0.01164 - 0.0065) = 1747.4$$
$$C3 = 0.9 \times 16 \times 99 \times 1747.4/1000 = 2491.146$$
$$a = 0.77 \times 328 - 110 = 142.56$$
$$T1 = 0.6 \times 0.85 \times 40 \text{ MPa} \times 350(0.77 \times 328 - 110)/1000 = 1017.88$$
$$C1 + C2 + C3 = 1202.58 + 267.75 + 1017.88 = 2488.2$$

This is close to 2491.146 and therefore OK.

$$\text{Moment resisting } (M_r) = 1202.58(328 - 30) + 267.75(328 - 60 - 25)$$
$$+ 1017.88(328 - 60 - 50 - 142.56/2) +$$
$$2491.146(875 - 328)/1000 = 1935.45 \text{ kN} \cdot \text{m}$$
$$M_u = 1425.31 \text{ kN} \cdot \text{m} < 1935.45 \text{ kN} \cdot \text{m} \quad \text{(OK)}$$

#### Horizontal shear

The unfactored horizontal shear force is the smaller of $T$ or $C$. Since the compression block extends into the inverted-T girder, then $C$ is smaller than $T$, as indicated above.

$$C = 0.85 f'_{c,\text{topping}} A_{\text{topping concrete}} = 0.85 \times 30 \text{ MPa}$$
$$(1310 \times 60 + 350 \times 50)/1000 = 2450 \text{ kN}$$
$$f_h = C = 2450 \text{ kN}$$
$$b_v = 350 \text{ mm} \qquad Lvh = 9600 \text{ mm}/2 = 4800 \text{ mm}$$

If the surface is intentionally roughened,

$$f_{h,\max} = 0.42 b_v$$
$$= 0.42 \text{ MPa} \times 350 \text{ mm} \times 4800 \text{ mm}/1000 = 705.6 \text{ kN}$$

Since this is less than 2450 kN, therefore minimum ties must be used with the surfaces intentionally roughened.

$$f_{h,max} = 1.8b_v Lvh$$

$$= 1.8 \times 350 \times 4800/1000 = 3024 \text{ kN}$$

Since 3024 kN > 2450 kN, minimum ties are OK.

Provide minimum ties:

$$A_{cs} = 0.35b_v Lvh/f_y = 0.35 \text{ MPa} \times 350 \text{ mm}$$

$$\times 4800 \text{ mm}/300 \text{ MPa} = 1960 \text{ mm}^2$$

Use no. 10M ties, $f_y$ 300 MPa. The maximum spacing is 600 mm, or 4 times the minimum dimension of the section at the connection.

$$4 \times 110 = 440 \text{ mm} < 600 \text{ mm}$$

Therefore, the $s$ maximum = 440 mm.

$$Lvh/s \times \text{hoop area} = A_{cs}$$

Rewriting the equation and solving for $s$,

$$s = Lvh \quad \text{hoop area}/A_{cs} = 4800 \times 2 \times 100/1960 = 49 \text{ mm} > 490 \text{ mm}$$

Therefore, space 10M hoop ties at 440 mm on center.

Note than if $F_h$ is greater than 3024 kN, the shear friction method must be used. As an example,

$$F_{fh} = \text{factored force } C = 2450\,\phi_c = 2450 \times 0.6 = 1470 \text{ kN}$$

$$A_{cs} = F_{fh}/\phi_s \mu f_y$$

where $\mu$ = 0.9 for intentionally roughened surface

$$\phi_s = 0.85$$

$$A_{cs} = 1470 \text{ kN} \times 1000/(0.85 \times 0.9 \times 300 \text{ kN}) = 6405 \text{ mm}^2$$

**Shear design**

Ties required because of horizontal shear are no. 10M hoops at 440 mm center. If vertical shear requirements are greater, the tie spacing would be reduced.

| | kN/m | $U$ factor | kN/m |
|---|---|---|---|
| Live load = | 26.0 | × 1.5 = | 39.0 |
| Dead load | | | |
| Inverted T = | 6.58 | × 1.25 = | 8.225 |
| Double T = | 36.0 | × 1.25 = | 45.0 |
| Topping = | 25.2 | × 1.25 = | 31.5 |
| | | | 123.725 |

$V_f$ at support $= 123.725$ kN/m $\times 9.6$ m $/2 = 593.88$ kN

$V_f$ at $h/2$ from support $= 593.88$ kN $- 123.725 \times 0.960/2 = 534.5$ kN

$V_f$ at 50 strand diameters $= 593.88 - 123.725 \times 0.650 = 513.45$ kN

$V_{fx} = 593.88 - 123.725X \qquad M_{fx} = 593.88X - X^2/2$

$d_p = 0.85\,\text{m} \qquad \gamma = 0.85 \qquad d = d_p$

$$V_{cx} = \left(0.6\gamma f_c'^{0.5} + 6d_pV_{fx}\right)\phi_c b_w d/1000$$

If $d_pV_{fx}/M_{fx} > 1.0$, then $d_pV_{fx}/M_{fx} = 1.0$.

| $X$, m | $V_{fx}$, kN | $M_{fx}$, kN/m | $d_pV_{fx}/M_{fx}$ | $V_{cx}$, kN |
|---|---|---|---|---|
| 0.5 | 532.0 | 281.5 | 1.0 | 1161.7 |
| 1.0 | 470.2 | 532.0 | 0.773 | 911.7 |
| 1.5 | 408.3 | 751.6 | 0.475 | 583.2 |
| 2.0 | 346.4 | 940.3 | 0.3223 | 414.6 |
| 2.5 | 284.6 | 1098.0 | 0.2267 | 309.6 |
| 3.0 | 222.7 | 1224.9 | 0.1591 | 234.6 |
| 3.5 | 160.8 | 1320.7 | 0.1066 | 176.6 (lower limit = 197.6) |
| 4.0 | 99.0 | 1385.7 | 0.0625 | 128.1 (lower limit = 197.6) |
| 4.5 | 37.1 | 1419.7 | 0.0229 | 84.4 (lower limit = 197.6) |

$V_c$ at end $= 0.4\gamma\,\phi_c f_c'^{0.5} b_w d_p$

$= 0.4 \times 0.85 \times 0.6 \times 40^{0.5} \times 350 \times 875/1000 = 395.12$ kN

$$V_{cw,\max} = 0.4\gamma\,\phi_c f_c'^{0.5} b_w d_p \left\{\left[1 + f_{pc}/\left(0.4\phi_c \gamma f_c'\right)\right]^{0.5}\right\}$$

where $f_{pc} = f_{ps}/A$.

$V_{cw,\max} = 0.4 \times 0.85 \times 0.6 \times 40^{0.5} \times 350 \times 875$

$\{[1 + (16 \times 114.46 \times 1000/372{,}500)/(0.4 \times 0.6 \times 0.6 \times 0.85 \times 40)]^{0.5}\}$

$= 500.18$ kN

$V_c$ lower limit $= 0.2\gamma\,\phi_c f_c'^{0.5} b_w d_p$

$= 0.2 \times 0.85 \times 0.6 \times 40^{0.5} \times 350 \times 875/1000 = 197.56$ kN

A shear diagram is presented in Fig. 34.51.

For maximum shear spacing,

$s_{\max} = 0.75h$, not 600 mm

$s_{\max} = 0.75h = 0.75 \times 960$ mm $= 720$ mm $> 600$ mm

Therefore, 600 mm governs.

By using no. 10M $f_y$ 300-MPa stirrups,

$$s_{max} = 80A_v f_y d_p (b_w d_p)^{0.5} / A_{ps} f_{pu}$$
$$= 80 \times 200\text{mm}^2 \times 300\text{MPa} \times 875\text{mm}(350/875)^{0.5} /$$
$$(16.99\text{mm}^2 \times 1860\text{MPa}) = 570\text{mm}$$

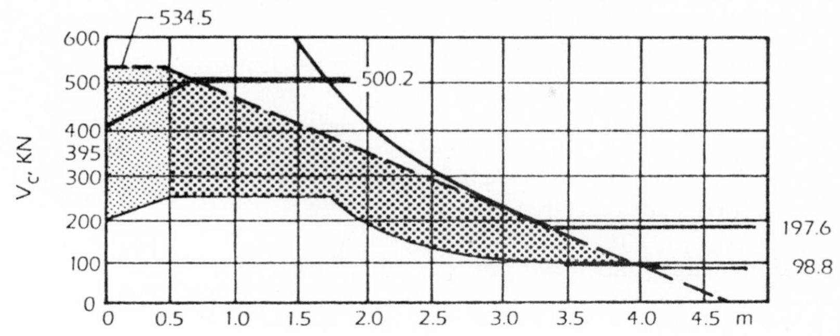


**Figure 34.51** Shear diagram.

The required horizontal shear tie spacing of 440 mm governs for the area of minimum stirrups.

$$V_s = V_f - V_c = 534.5\,\text{kN} - 395.12\ \text{kN} = 139.38\ \text{kN}$$
$$V_s = A_v f_y d\ \phi_s / s$$
$$\text{Spacing required} = A_v f_y d\ \phi_s / V_s$$
$$= 200\ \text{mm}^2 \times 300\ \text{MPa} \times 875\ \text{mm} \times 0.85/$$
$$(139.38\ \text{kN} \times 1000)$$
$$= 320\ \text{mm}$$

Place two No. 10M U stirrups from each end at 320 mm on center, then at 440 mm on center for the rest of the girder. Bend the top of the stirrups in the topping area to form hoops and to serve as the horizontal shear ties. The first stirrups are at 100 mm from each end.

### Camber and deflection

PCI multipliers are used for the following camber and deflection calculations.

$$E_{ci} = 0.043w^{1.5} f'^{0.5}_{ci} = 0.043 \times 1800^{1.5} \times 30^{0.5} = 17{,}986$$
$$E_c = 0.043w^{1.5} f'^{0.5}_c = 0.043 \times 1800^{1.5} \times 40^{0.5} = 20{,}768$$

Noncomposite moment of inertia = $23{,}000 \times 10^6$

Composite moment of inertia = $51{,}959 \times 10^6$

$F_{psi}$ = 16 strands × 122.637 kN/strand = 1962.192 kN

Strand eccentricity = 284 mm

$$\text{Camber} = F_{psi} eL^2 / 8E_{ci} I$$

$$= 1962.192 \text{ kN} \times 284 \text{ mm} \times 9.6 \text{ m}^2 \times 10^9 / (8 \times 17{,}986 \text{ MPa} \times 23{,}000 \times 10^6)$$

$$= 15.51 \text{ mm up}$$

$$\text{Deflection girder load} = 5w/\text{m} \times L^4 / 384E_{ci} I$$

$$= 5 \times 6.85 \text{ kN/m} \times 9.6 \text{ m}^4 \times 10^{12} / (384 \times 17{,}986 \text{ MPa} \times 23{,}000 \times 10^6) = 1.83 \text{ mm down}$$

$$\text{Deflection double-T load} = 5w / \times L^4 / 384E_c I$$

$$= 5 \times 36 \text{ kN/m} \times 9.6 \text{ m}^4 \times 10^{12} / (384 \times 20{,}786 \text{ MPa} \times 23{,}000 \times 10^6) = 8.33 \text{ mm down}$$

$$\text{Deflection topping load} = 5w/m \times L^4 / 384E_c I$$

$$= 5 \times 25.2 \text{ kN/m} \times 9.6 \text{ m}^4 \times 10^{12} / (384 \times 20{,}786 \text{ MPa} \times 23{,}000 \times 10^6) = 5.83 \text{ mm down}$$

$$\text{Deflection live load} = 5w/m \times L^4 / 384E_c I_c$$

$$= 5 \times 26 \text{ kN/m} \times 9.6 \text{ m}^4 \times 10^{12} / (384 \times 20{,}786 \text{ MPa} \times 51{,}959 \times 10^6) = 2.9$$

Camber and deflection data may be summarized as follows, using PCI multipliers:

| | Multiplier | Position |
|---|---|---|
| At detensioning | | |
| Camber, 15,51 mm | 1.0 | 15.51 mm up |
| Deflection inverted T, 1.83 mm | 1.0 | 1.83 mm down |
| Net | | 13.68 mm up |
| At erection | | |
| Camber, 15.51 mm | 1.8 | 27.92 mm up |
| Deflection inverted T, 1.83 mm | 1.85 | 3.38 mm down |
| Deflection double T, 8.33 mm | 1.0 | 8.33 mm down |
| Deflection topping, 5.83 mm | 1.0 | 5.83 mm down |
| Net | | 10.38 mm up |

| | Multiplier | Position |
|---|---|---|
| At service (after 2 years) | | |
| Camber, 15.51 mm | 2.2 | 34.12 mm up |
| Deflection inverted T, 1.83 mm | 2.4 | 4.39 mm down |
| Deflection double T, 8.33 mm | 2.3 | 19.16 mm down |
| Deflection topping, 5.83 mm | 2.3 | 13.41 mm down |
| Net | | 2.84 mm down |
| Deflection live load, 2.9 mm | 1.0 | 2.90 mm down |
| Net | | 5.78 mm down |

**Ledge design (see Fig. 34.52)**

| | kN | $U$ factor | kN |
|---|---|---|---|
| Double T = 2.0 kN/m² ×18 m/2×2.4 m/2 stems = | 21.6 | × 1.25 = | 29.00 |
| Topping = 1.4 kN/m² ×18 m/2×2.4 m/2stems = | 15.2 | × 1.25 = | 18.90 |
| Live load = 0.6 ×2.4 kN/m² ×18/2 ×2.4 m/ | 15.5 | × 1.50 = | 23.33 |
| 2 stems = | 52.3 | | 69.23 |

The double-T stem width = 130 mm, and the bearing-pad strength is 7 MPa.

$$69.23 \text{ kN} \times 1000/(130 \times 7) = 76 \text{ mm}$$

Use a pad length of 75 mm. The ledge length is

$$l_p = 25\text{-mm tolerance} + 75 \text{ mm} + 25 \text{ mm edge distance} = 125 \text{ mm}$$

Note that this computation usually precedes selecting the inverted-T dimensions. In addition, the pad bearing capacity is listed as unfactored values. Hence, 52.3 kN/1000/(130×7 MPa) = 57.5 mm $N_u$ if zero is taken as 0.2 $V_f$ as a minimum due to creep and temperature change.

$$N_u = 0.2 \times 69.23 \text{ kN} = 13.84$$

$$a = 2/3 \times l_p + 20 \text{ mm} = 2/3 \times 125 + 20 = 104 \text{ mm}$$

$$A_s = A_m + A_n = 1/f_y \left\{\left[V_f a + N_f (h - d)\right]/\phi_c d + N_f/\phi_s\right\}$$

$$A_s = 1000/300 \text{ MPa}\left[(69.23 \text{ kN} \times 104 \text{ mm} + 13.84 \text{ kN} \times 20 \text{ mm})/(0.6 \times 280 \text{ mm}) + 13.84/0.85\right] = 202 \text{ mm}^2$$

$$A_{s,\min} = 1.4(b + h)d/f_y = 1.4(130 + 300) \times 280/300 = 560 \text{ mm}$$

If $A_s$ is smaller than $A_{s,\min}$, then use the smaller of $1.33 \times A_s$ or $A_{s,\min}$.

$$1.33 \times 2.02 = 269 \text{ mm}$$

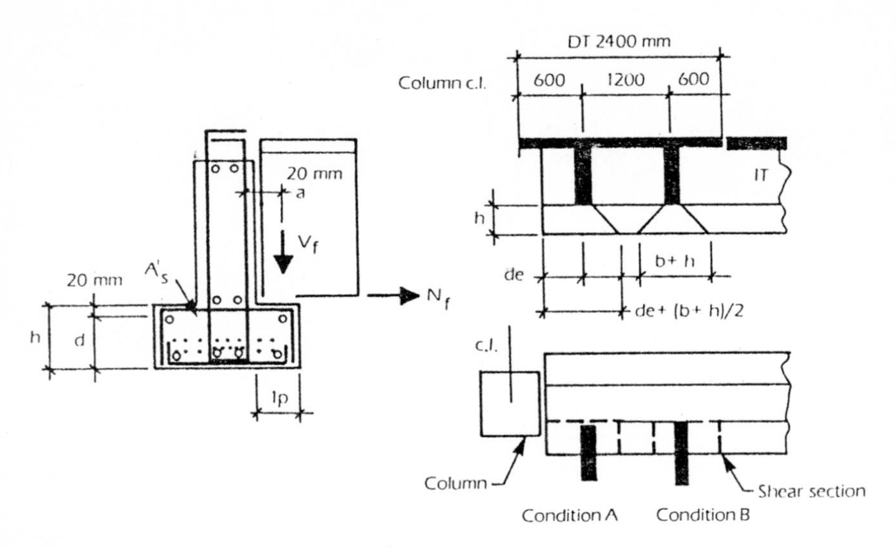


**Figure 34.52** Ledge design. $h$ = 300 mm; $d$ = 280 mm; $l_p$ = 125 mm.

Use $A_s$ of 300 mm. Use no. 10M bars:

$$300 \text{ mm}^2/100 \text{ mm}^2 = 3 \text{ bars}$$

Space bars within the smaller of 1200 mm or 6$h$.

$$6 \times 300 \text{ mm} = 1800 \text{ mm}$$

Therefore, use 1200 mm.

$$1200 \text{ mm}/3 \text{ bars} = 400 \text{ mm on center}$$

Maximum spacing = $h$ = 300 mm. Therefore use 10M bars at 300 mm on center.

In condition $B$ in Fig. 34.52,

$$V_f = V_r$$

$$V_r = 0.25\phi_c \gamma f_c'^{0.5} h\,[2l_p + (b + h)) = 0.25 \times 0.6 \times 0.85 \times 40^{0.5}$$
$$\times 300\,[2 \times 125 + (130 + 130)]/1000 = 164.5 \text{ kN}$$
$$> 69.23 \text{ kN (OK)}$$

In condition $A$,

$$V_f < V_r$$

$$V_r = 0.166\phi_c \gamma f_c'^{0.5} h\,[l_p + (b + h)/2 + d_e]$$

Assume that the column is 400 mm square; then $d_e$ = 600 mm - 400 mm/2 = 400 mm.

$$V_r = 0.166 \times 0.6 \times 0.85 \times 40^{0.5} \times 300\,[125 + (130 + 300)/$$
$$2 + 400]/1000 = 118.8 \text{ kN} > 69.23 \text{ kN (OK)}$$

Place two additional bars at the ends. Parallel ties are to be no. 10M at the edge of the ledge and as shown to tie the hoops and $A_s$ bars together. Check vertical steel:

$$A_{sv} = V_f / \phi_s f_y = 69.23 \text{ kN} \times 1000/(0.85 \times 300) = 271.5 \text{mm}^2$$

Shear reinforcing = 10M at 440 mm on center.

$$1200 \text{ mm}/440 = 2.72 \text{ bars}$$

$$2.72 \times 100 = 272 \text{ mm}^2$$

Since 272 mm² is greater than 271.5 mm², shear reinforcing is OK; otherwise, shear reinforcing spacing would have to be decreased.

### Example Problem 7: Prestressed Columns (SI Units)

There is both an advantage and a disadvantage in prestressing precast concrete columns. The disadvantage is the reduction of the axial compresssive force that can be applied to the column. The advantage is the increase in moment-resisting capacity. Precast columns are cast in the horizontal position, are handled and shipped in the horizontal position, and on the project site are raised to a vertical position and set in place. In the horizontal position during handling induced tensile stresses are high. Therefore, prestressing is very helpful. In precast concrete multistory buildings the girder forces are applied to the columns with the use of corbels, thus inducing moments to the column. The corbels are cast with the columns and may be on one or more sides. See Fig. 34.53.

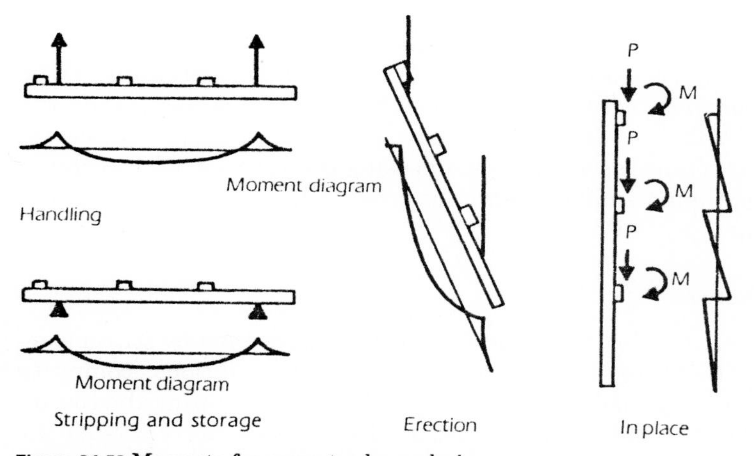


**Figure 34.53** Moments for precast column design.

### In-place forces on a precast column

To determine the moment on the column, the column and corbel size is assumed, hence *e*. $M = Pe$. See Fig. 34.54.

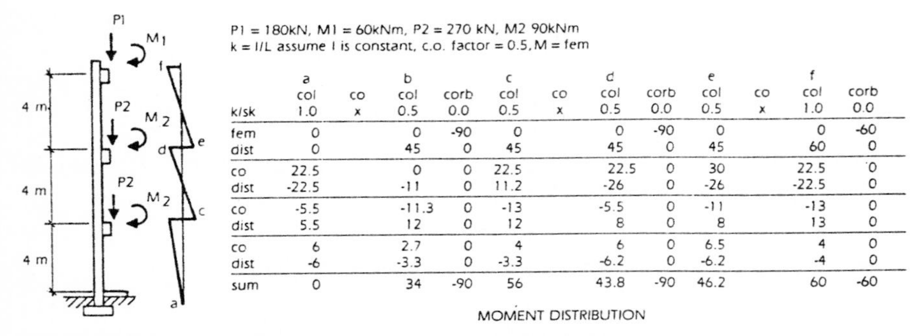


| | a col | co | b col | corb | c col | co | d col | corb | e col | co | f col | corb |
|---|---|---|---|---|---|---|---|---|---|---|---|---|
| k/sk | 1.0 | x | 0.5 | 0.0 | 0.5 | x | 0.5 | 0.0 | 0.5 | x | 1.0 | 0.0 |
| fem | 0 | | 0 | -90 | 0 | | 0 | -90 | 0 | | 0 | -60 |
| dist | 0 | | 45 | 0 | 45 | | 45 | 0 | 45 | | 60 | 0 |
| co | 22.5 | | 0 | 0 | 22.5 | | 22.5 | 0 | 30 | | 22.5 | 0 |
| dist | -22.5 | | -11 | 0 | 11.2 | | -26 | 0 | -26 | | -22.5 | 0 |
| co | -5.5 | | -11.3 | 0 | -13 | | -5.5 | 0 | -11 | | -13 | 0 |
| dist | 5.5 | | 12 | 0 | 12 | | 8 | 0 | 8 | | 13 | 0 |
| co | 6 | | 2.7 | 0 | 4 | | 6 | 0 | 6.5 | | 4 | 0 |
| dist | -6 | | -3.3 | 0 | -3.3 | | -6.2 | 0 | -6.2 | | -4 | 0 |
| sum | 0 | | 34 | -90 | 56 | | 43.8 | -90 | 46.2 | | 60 | -60 |

**Figure 34.54** Moment distribution for column design. $P_1 = 180$ kN; $M_1 = 60$ kN m; $P_2 = 270$ kN; $M_2 = 90$ kN· m; $k = I/L$. Assume $I$ is constant. CO factor = 0.5; $M$ = fixed-end moment (FEM).

**Moment Distribution**

| | *a* Column | CO | *b* Column | Corbel | *c* Column | CO | *d* Column | Corbel | *e* Column | CO | *f* Column | Corbel |
|---|---|---|---|---|---|---|---|---|---|---|---|---|
| *k/sk* | 1.0 | X | 0.5 | 0.0 | 0.5 | X | 0.5 | 0.0 | 0.5 | X | 0.5 | 0.0 |
| FEM | 0 | | 0 | -90 | 0 | | 0 | -90 | 0 | | 0 | -60 |
| Distribution | 0 | | 45 | 0 | 45 | | 45 | 0 | 45 | | 60 | 0 |
| CO | 22.5 | | 0 | 0 | 22.5 | | 22.5 | 0 | 30 | | 22.5 | 0 |
| Distribution | -22.5 | | -11 | 0 | -11.2 | | -26 | 0 | -26 | | -22.5 | 0 |
| CO | -5.5 | | -11.3 | 0 | -13 | | -5.5 | 0 | -11 | | -13 | 0 |
| Distribution | 5.5 | | 12 | 0 | 12 | | 8 | 0 | 8 | | 13 | 0 |
| CO | 6 | | 2.7 | 0 | 4 | | 6 | 0 | 6.5 | | 4 | 0 |
| Distribution | -6 | | -3.3 | 0 | -3.3 | | -6.2 | 0 | -6.2 | | -4 | 0 |
| Total | 0 | | 34 | -90 | 56 | | 43.8 | -90 | 46.2 | | 60 | -60 |

A column size and the amount of prestressing are assumed; then the interaction curve is plotted for that column size and prestressing. If the load and moment conditions fit within the interaction curve, then the column is satisfactory. If any load and moment set fall outside the curve, then a larger column or/and the prestressing is increased. If the column is excessive, then a smaller column is tried. The minimum amount of prestressing is 1.5 MPa.

### Interaction curve

The coordinates of the interaction curve are determined by strain compatibility. Owing to the repetitious use of the equations, computer programs are employed for the solution. When determining the coordinates of the curve, the designer must use the stress-strain curve for the strand to be used. This is necessary because there is no plastic range for

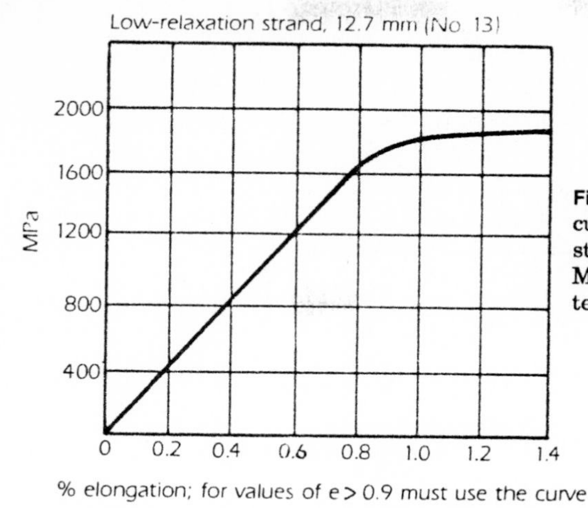


**Figure 34.55** Typical stress-strain curve for low-relaxation 1860-MPa strand. $E_s$ = 190,000 MPa; $f_{pu}$ = 1860 MPa; $f_{py}$ = 1810 MPa, 1 percent extension.

| $e$ | MPa |
|---|---|
| 0.8 | 1540 |
| 0.9 | 1710 |
| 1.0 | 1810 |
| 1.1 | 1825 |
| 1.2 | 1840 |
| 1.3 | 1850 |
| 1.4 | 1860 |

high-strength strand or steel, and so there is no true yield point. The yield point is determined by the 0.2 percent offset method or by the 1 percent extension method. Copies of the stress-strain curve are obtained from the strand suppliers or from the manufacturer.

A typical stress-strain curve for 1860-MPa strand is shown in Fig. 34.55.

The area of the strand is as follows:

| | |
|---|---|
| No. 9 diameter = 9.53 mm | Area = 55 $mm^2$ |
| No. 11 diameter = 11.13 mm | Area = 96 $mm^2$ |
| No. 13 diameter = 12.7 mm | Area = 99 $mm^2$ |
| No.15 diameter = 15.24 mm | Area = 140 $mm^2$ |

Initial loss is usually assumed to be 7½ to 10 percent. Service loss ranges from 14 to 20 percent.

$$f_{psi} = 0.74_{pu}(100 - \text{loss in percent})/100$$

Load and resistance factors used in design vary with the country where the project is located. For the United States load factors (*U* values ) are 1.7 for live load and 1.4 for dead load. In Canada load factors are 1.25 for dead load and 1.5 for live load. Resistance (phi) factors in the United States are 0.9 for flexure, 0.85 for shear, 0.75 for spiral columns, and 0.7 for tied columns. In Canada resistance factors are 0.6 for concrete, 0.85 for reinforcing steel, and 0.9 for prestressing steel. The designer is advised to consult the code governing the location of the project.

The beta factors used for metric design are as follows:

| $f'_c$, MPa | beta | $f'_c$, MPa | beta |
|---|---|---|---|
| < 30 | 0.85 | 45 | 0.73 |
| 35 | 0.81 | 50 | 0.69 |
| 40 | 0.77 | >55 | 0.65 |

The equations for the interaction curve are shown in Fig. 34.56. When using these equations, proper resistance factors should be employed.

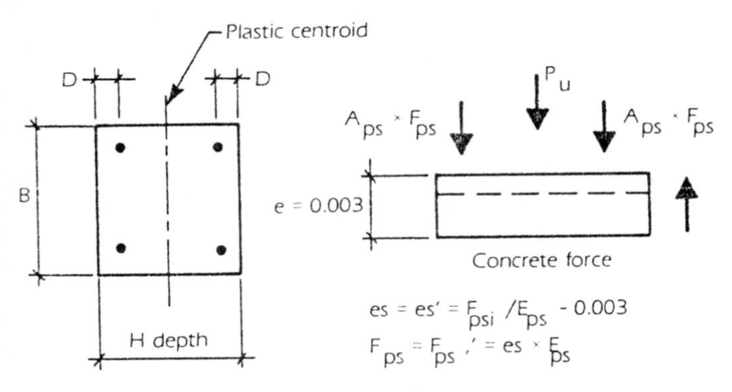


**Figure 34.56** Column forces for $P_o$ condition.

The following data are related to the resistance factors used in Canada and must be altered for use in other countries.

In the equation shown $f_{psi}$ is the stress in the strand less the service

$$P_o = \phi_c \times 0.85 f'_c \left(A_g - A_{ps} - A'_{ps}\right) - \phi_p \left(A'_{ps} f'_{ps} + A_{ps} f_{ps}\right)$$

where $\phi_c = 0.6$ and $\phi_c = 0.9$.

In the United States the equation is

$$P_o = \phi\left[\,0.85 f'_c \left(A_g - A_{ps} - A'_{ps}\right) - \left(A_{ps} f'_{ps} + A'_{ps} f'_{ps}\right)\right]$$

where $\phi$ = 0.7 for tied columns and 0.75 for spiral columns. $P_o$ is often referred to as $\phi P_n$.

$P_o$ is the maximum axial force that the column can support without considering any eccentric condition. However, there is no such condition where the load is perfectly symmetrical. Therefore, the codes set a maximum value to take care of minor eccentricities.

$P_{u,\max} = 0.8P_o$ for tied columns and $0.85P_o$ for spiral columns in both Canada and the United States.

For prestressed concrete there is no $P_f$ balance or $M_f$ balance since these values do not exist. $P_f$ is the same as $P_u$, and $M_f$ is the same as $M_u$. In Canada the term $P_f$ is used; in the United States, the term is $P_u$.

The coordinates of the points of the interaction curve are computed on the basis of strain compatibility. Here the values of $C$ are inserted in the equation for each point. Owing to the repetition, a loop (For-Next) statement is used in the computer program. The values of $C$ range from a high equal to the column dimension divided by $\beta$ to a low of $D'$. The step is a negative step in values of 10 mm or larger. The reason for setting an upper limit for $C$ is that the concrete stress block does not exceed the area of the column ($a = < H$). The upper value of $C$ is set with a factor of 10 so that values are in units of 10 mm. (See Fig. 34.57.)

$$e_{\text{base}} = f_{psi}/E_{ps} = 0.74 \times 1860(100 - 16)/100/190{,}000 = 0.00608$$

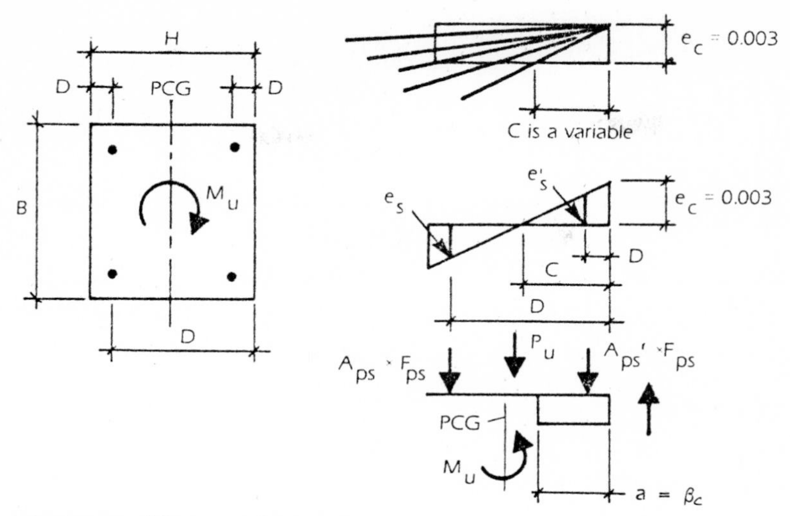


**Figure 34.57** Column with moment.

$$e_s = e_{\text{base}} - 0.003(C - D)/C$$

$f'_s = e_s E_{ps}$ if $e_s < 0.009$. If $e_s > 0.009$, then the stress-strain curve must be used to determine $f'_s$. When writing the computer program, the stress-strain curve is modeled for strains greater than 0.009.

$$P_f = \phi_c \times 0.85 f'_c b\, \beta C - \phi_p A'_{ps} f'_{ps} - \phi_p A_{ps} f_{ps}$$

where $\phi_c = 0.6$ and $\phi_p = 0.9$.

In the United States,

$$P_u = \phi\left(0.85 f'_c B\, \beta C - A'_{ps} f'_{ps} - A_{ps} f_{ps}\right)$$

where $\phi = 0.7$ for tied columns and 0.75 for spiral columns.

$$M_f = \phi_c \times 0.85 f'_c B\, \beta C\ (H/2 - \beta C/2)$$

$$- \phi_p A'_{ps} f'_{ps} (H/2 - D') + \phi_p A_{ps} f_{ps} (H/2 - D')$$

where $\phi_c = 0.6$ and $\phi_p = 0.9$

In the United States,

$$M_u = [\phi\, 0.85 f'_c B\, \beta C\ (H/2 - \beta C/2)$$

$$- A'_{ps} f'_{ps} (H/2 - D') + A_{ps} f_{ps} (H/2 - D')]$$

where $\phi = 0.7$ from $P_u = P_o$ to $P_u = 0.1\,P_o$. Below $0.1P_o\,\phi$ can be increased from 0.7 to 0.9 when $P_u = 0.0$. That is, if $P_u < 0.1P_o$ then $\phi = 0.9 - 0.2(P_u/0.1P_o)$.

Note that the revised $\phi$ value is used for moment $M_u$ but not for $P_u$. An example column is shown in Fig. 34.58.

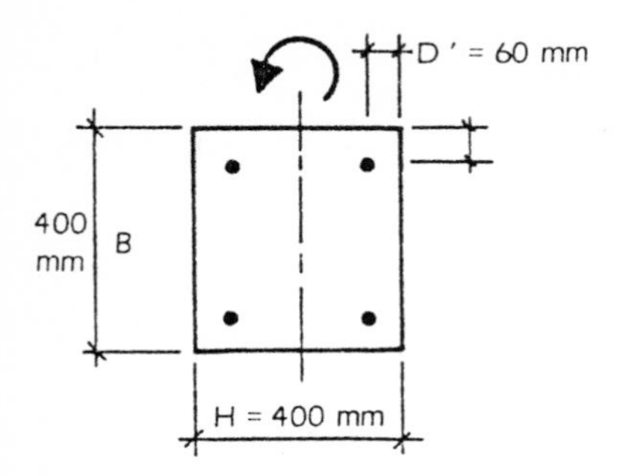


**Figure 34.58** Example column.

Use four no. 13 low-relaxation strands, 1860 MPa; $E_{ps}$ = 190,000 MPa; strand area = 99 mm$^2$.

Assume an 18 percent loss of prestress (14 to 20 percent).

$f_{se}$ = 0.74 ×1860(100 − 18 percent)/ 100 = 1128.6MPa

Use 1120 MPa.

Concrete = 40 MPa = β 0.77
Density = 2400 kg/m$^3$

Minimum prestress = 1.5 MPa $\quad 1.5 \times 400^2/1000 = 240$ kN

$$4 \text{ strands} \times 99\ \text{mm}^2 \times 1120\ \text{MPa}/1000$$

$$= 443.5\ \text{kN} > 240\ \text{kN minimum (OK)}$$

Note that the United States load factors are 1.4 dead load and 1.7 live load and $\phi = 0.7$. The Canadian load factors are 1.25 dead load and 1.5 live load and $\phi = 0.6$ for concrete, 0.9 for strand, and 0.85 for reinforced steel.

$$e_{\text{base}} = f_{se}/E_{ps} = 1120/190{,}000 = 0.00589\text{mm/mm}$$

$$e_s = e_{\text{base}} - 0.003 = 0.00589\,\text{mm} - 0.003 = 0.00289$$

$$f_{ps} = e_s E_{ps} = 0.00289 \times 190{,}000 = 550\ \text{MPa}$$

$$P_o - \phi_c \times 0.85 f_c' \left(A_g - A'_{ps} - A_{ps}\right) - \phi_p \left(A'_{ps} f'_{ps} + A_{ps} f_{ps}\right)$$

For Canada,

$$P_o = 0.6 \times 0.85 \times 40\left(400^2 - 2 \times 99 - 2 \times 99\right)$$

$$- 0.9 \times 2 \times 99 \times 550 - 0.9 \times 2 \times 99 \times 550]\ /1000 = 3060\ \text{kN}$$

$$P_{f,\max} = 0.8 \times 3060 = 2448\ \text{kN}$$

For the United States,

$$P_o = 0.7/1000\left[0.85 \times 40\left(400^2 - 2 \times 99 - 2 \times 99\right)\right.$$

$$\left. - 2 \times 99 \times 550 - 2 \times 99 \times 550\right] = 3646\ \text{kN}$$

$$P_{u,\max} = 0.8 P_o \times 0.8 \times 3646\ \text{kN} = 2916.8\ \text{kN}$$

To plot the curve a computer program is used. The following is the solution for one point on the curve (see Fig. 34.59):

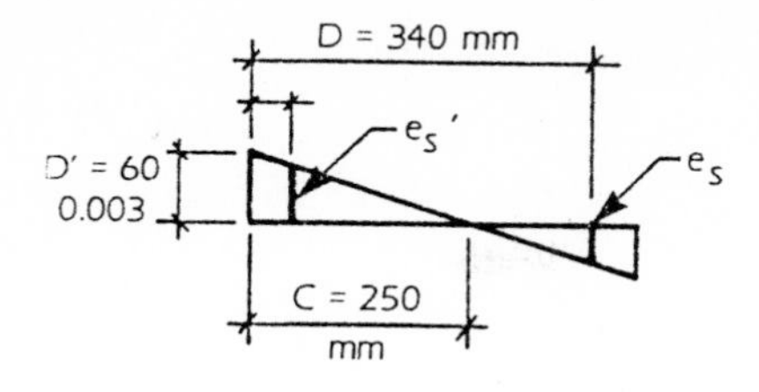


**Figure 34.59** Column strain diagram.

For $P_f$ and $M_f$ at $C = 250$ mm, $D' = 60$ mm and $D = 400 - 60 = 340$ mm.

$$e'_s = e_{\text{base}} - 0.003(C - D')/C$$

$$f'_{ps} = e'_s E_{ps}$$

$$e_s = e_{\text{base}} - 0.003(C - D)/C$$

If $e_s < 0.009$, then $f_{ps} = e_s E_{ps}$.

If $e_s > 0.009$, then $f_{ps}$ is obtained from the stress-strain curve.

$$e'_s = 0.00589 - 0.003(250 - 60)/250 = 0.00361$$

$$f'_{ps} = 0.00361 \times 190{,}000 = 686 \text{ MPa}$$

$$e_s = 0.00589 - 0.003(250 - 340)/250 = 0.00697$$

$$f_{ps} = 0.00697 \times 190{,}000 = 1324 \text{ MPa}$$

$$P_f = \phi_c \times 0.85 f'_c B\ \beta C - \phi_p \left(A'_{ps} f'_{ps} - A_{ps} f_{ps}\right)$$

$$M_f = \phi_c \times 0.85 f'_c B\ \beta C\ (H/2 - \beta C/2) - \phi_p A'_{ps} f'_{ps} (H/2 - D') + \phi_p A_{ps} f_{ps} (H/2 - D')$$

For Canada,

$$P_u = 0.6 \times 0.85 \times 40 \times 400 \times 0.77 \times 250 - 0.9 \times 2 \times 99 \times 686 - 0.9 \times 2 \times 99 \times 1324)/1000 = 1212.6 \text{ kN}$$

$$M_u = [0.6 \times 0.85 \times 40 \times 400 \times 0.77 \times 250(400/2 - 0.77 \times 250/2) - 0.9 \times 2 \times 99 \times 686(400/2 - 60) + 0.9 \times 2 \times 99\ 1324(400/2 - 60)]/1000/1000 = 178.9 \text{ kN} \cdot \text{m}$$

For the United States,

$$P_u = 0.7/1000(0.85 \times 40 \times 400 \times 0.77 \times 250 - 2 \times 99 \times 686 - 2 \times 99 \times 686 - 2 \times 99 \times 1324) = 1554 \text{ kN}$$

$$M_u = 0.7/1000[0.85 \times 40 \times 400 \times 0.77 \times 250(400/2 - 0.77 \times 250/2) - 2 \times 99 \times 686(400/2 - 60) + 2 \times 99 \times 1324(400/2 - 60)]/1000 = 202 \text{ kN} \cdot \text{m}$$

Figure 34.60 shows the column interaction curve.

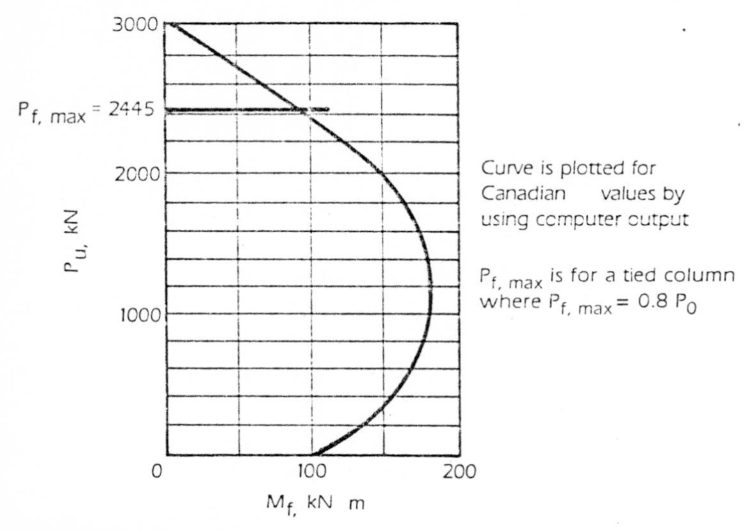


**Figure 34.60** Column interaction curve.

precast pretensioned systems is the result of the placement of the strands to resist the design moment and the concrete stresses. If the camber is excessive or needs to be increased, then the strand profile and number of strands can be altered subject to the design concrete stresses. With partial prestressing, the use of mild steel can give additional control of deflection.

Owing to the variations of the design mix, curing, storage, relative humidity, time of detensioning, time of loading, creep phenomena, and strand relaxation, it is impossible to determine the exact deflection which will occur. Laboratory-determined factors based upon a time-step method with controlled humidity, while suited for the laboratory, are often not suited for actual conditions. Therefore, the designer should not specify the camber or the deflection.

The various building codes place a limit on deflection, and the designer should consult the code applicable to the project location. The Uniform Building Code, 1988 edition, in Sec. 2307 and Tables 23-D and 23-E, gives the following limits:

For roof members supporting a plaster ceiling and floor members: live-load deflection span length/360 and live load + $K$ times dead-load deflection span length/240. $K$ is a creep factor, and $K = \lfloor 2 - 1.2(A'_s / A_s) \rfloor = 0.6$. However, Sec. 2609 of the UniformBuilding Code gives the following value of $\lambda$ to be used for $K$: $\lambda = T / (1 + 50p')$ where $p' = A'_s / bd$, and $T$ for sustained loads = 2.0 for 5 years, 1.4 for 12 months, 1.2 for 6 months, and 1.0 for 3 months.

Some codes such as the New Zealand and Canadian national codes recommend the following:

| Member | Type of load | Recommendation |
|---|---|---|
| Roofs without attached partition systems | Immediate live load | *L*/180 |
| Floors without attached partitions | Immediate live load | *L*/360 |
| Roofs or floors with attached partitions not likely to be damaged by deflection | Creep × dead load + percent live load | *L*/240 |
| Roofs or floors with attached partitions likely to be damaged by deflection | Creep × dead load + percent live load | *L*/480 |

Camber first occurs at detensioning of the strand (transfer of the prestressing force to the concrete). As the member gains strength with time and as loss of prestress occurs with time, the magnitude of camber will change. At detensioning the member normally is a noncracked section, and the elastic concept of deflection therefore is used. (See Fig. 34.41.)

If the member is designed as an uncracked section, the elastic equations for deflection are used (see Fig. 34.42). Here the moment of inertia

$I$ is the gross moment of inertia. $E$ is the modulus of elasticity of the concrete at the time load being considered. The modulus of elasticity of the concrete $= E_c = 0.043 w_c^{1.5} f_c'^{0.5}$.

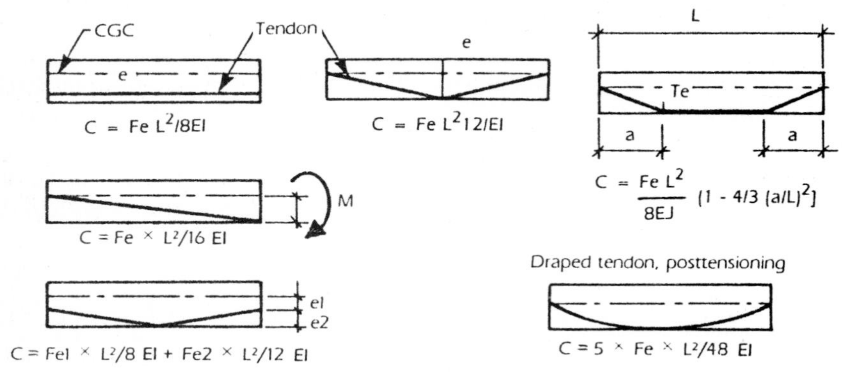


**Figure 34.41** Camber equations. $C$ = camber; $F$ = prestressing force.

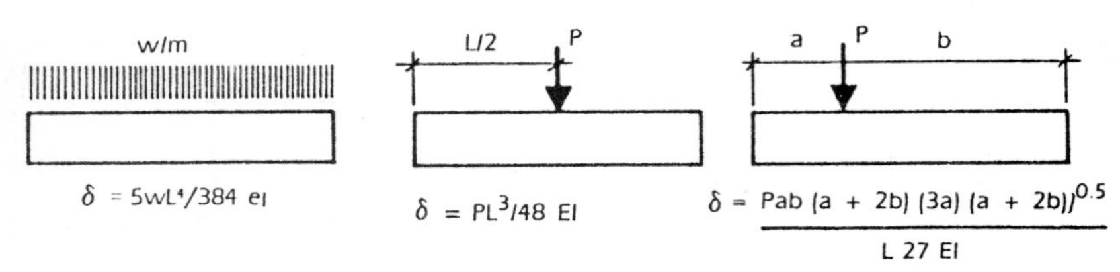


**Figure 34.42** Deflection equations.

If the member is a cracked section under nonfactored loads, then either a bilinear analysis or the $I$ effective value is used. With bilinear analysis,

$$\Delta = L^2 / 48E \left[ 5M_{\text{cracked}} / I_g + 5(M - M_{\text{cracked}}) / I_{\text{cracked}} \right]$$

$$M_{\text{cracked}} = S_b (F/A + Fe/S_b + f_r)$$

where $f_r = 0.6\lambda \times f_c'^{0.5}$

$\lambda$ = 1.0 for normal-weight concrete and 0.85 for sand-lightweight concrete

The mass of normal-density concrete $w_c$ = 2300 to 2400 kg/m³, and for sand-lightweight concrete is 1800 to 2000 kg/m³.

If the $I$ effective method is used (see Fig. 34.43),

$$\Delta = 5wL^4 / 384E_c I_{\text{eff}}$$

where $I_{\text{eff}} = (M_{\text{cracked}}/M_a)^3 \times I_g + \left[1 - (M_{\text{cracked}}/M_a)^3 \times I_{\text{cracked}}\right]$

$\left[\text{For prestressed concrete } M_{\text{cracked}}/M_a \text{ is expressed as } 1 - (f_t - f_r)/f_{LL}\right]$.

Figure 34.61 presents the computer program printout for prestressed column.

```
DESIGN    INPUT

COLUMN DEPTH IN mm =     400
COLUMN WIDTH IN mm =     400
CONCRETE STRENGTH IN MPa =   40
BETA =   .77
DISTANCE FROM EDGE OF COLUMN TO STRAND =   60     mm
NUMBER OF STRAND =   4
AREA PER STRAND =   99    sq.mm
TYPE OF STRAND = LR
STRAND STRENGTH IN MPa =   1860
STRAND MODULUS OF ELASTICITY IN MPa =   190000
% OF LOSS OF PRESTRESS =   18
PU REQUIRED =   1200   kN
MU REQUIRED =   160   kN.m

      DESIGN      OUTPUT

phi values used are those for  CANADA
    PO=   3056.8   kN        P-MAX. =   2445.4     kN
   VALUES OF C IN mm ,PU IN kN, AND MU  IN kN.m
```

| VALUES OF C | PU | MU | e=MU/PU | es-tension | ec-conc. tensic |
|---|---|---|---|---|---|
| 510 | 2925.6 | 19.5 | 0 | 4.940253E-03 | -6.470588E-04 |
| 500 | 2861.2 | 31.5 | .01 | 4.980253E-03 | -.0006 |
| 490 | 2796.7 | 43 | .01 | 5.021886E-03 | -5.510204E-04 |
| 480 | 2732.1 | 54.1 | .01 | 5.065253E-03 | -.0005 |
| 470 | 2667.5 | 64.7 | .02 | 5.110465E-03 | -4.468085E-04 |
| 460 | 2602.8 | 74.8 | .02 | 5.157644E-03 | -3.913044E-04 |
| 450 | 2538 | 84.4 | .03 | 5.20692E-03 | -3.333334E-04 |
| 440 | 2473.1 | 93.6 | .03 | 5.258434E-03 | -2.727273E-04 |
| 430 | 2408.1 | 102.3 | .04 | 5.312346E-03 | -2.093023E-04 |
| 420 | 2343.1 | 110.5 | .04 | 5.368824E-03 | -1.428571E-04 |
| 410 | 2277.9 | 118.2 | .05 | 5.428058E-03 | -7.317073E-05 |
| 400 | 2212.6 | 125.5 | .05 | 5.490253E-03 | 0 |
| 390 | 2147.1 | 132.3 | .06 | 5.555638E-03 | 7.692308E-05 |
| 380 | 2081.5 | 138.6 | .06 | 5.624463E-03 | 1.578947E-04 |
| 370 | 2015.8 | 144.5 | .07 | 5.69701E-03 | 2.432433E-04 |
| 360 | 1949.9 | 149.9 | .07 | 5.773586E-03 | 3.333334E-04 |
| 350 | 1883.9 | 154.8 | .08 | 5.854539E-03 | 4.285715E-04 |
| 340 | 1817.6 | 159.3 | .08 | 5.940253E-03 | 5.294118E-04 |
| 330 | 1751.2 | 163.3 | .09 | 6.031162E-03 | 6.363636E-04 |
| 320 | 1684.5 | 166.8 | .09 | 6.127753E-03 | .00075 |
| 310 | 1617.6 | 169.9 | .1 | 6.230575E-03 | 8.709677E-04 |
| 300 | 1550.4 | 172.5 | .11 | 6.340253E-03 | .001 |
| 290 | 1482.9 | 174.7 | .11 | 6.45/495E-03 | 1.13/931E-03 |
| 280 | 1415 | 176.4 | .12 | 6.58311E-03 | 1.285714E-03 |
| 270 | 1346.8 | 177.6 | .13 | 6.718031E-03 | 1.444445E-03 |
| 260 | 1278.2 | 178.5 | .13 | 6.86333E-03 | 1.615385E-03 |
| 250 | 1209.1 | 178.8 | .14 | 7.020253E-03 | .0018 |
| 240 | 1139.5 | 173.8 | .15 | 7.190253E-03 | .002 |
| 230 | 1069.3 | 178.3 | .16 | 7.375035E-03 | 2.217391E-03 |
| 220 | 998.5 | 177.4 | .17 | 7.576616E-03 | 2.454546E-03 |
| 210 | 926.8 | 176.1 | .19 | 7.797396E-03 | 2.714286E-03 |
| 200 | 850.9 | 174.9 | .2 | 8.040253E-03 | .003 |
| 190 | 778.4 | 172.7 | .22 | 8.308674E-03 | 3.315789E-03 |
| 180 | 704.7 | 170.1 | .24 | 8.60692E-03 | 3.666667E-03 |
| 170 | 629.8 | 167.1 | .26 | 8.940253E-03 | 4.058823E-03 |
| 160 | 557.3 | 163.3 | .29 | 9.315253E-03 | .0045 |
| 150 | 484.3 | 158.9 | .32 | 9.740253E-03 | .005 |
| 140 | 414.7 | 153.5 | .37 | 1.022597E-02 | 5.571429E-03 |
| 130 | 346.1 | 147.3 | .42 | 1.078641E-02 | 6.230769E-03 |
| 120 | 277.2 | 140.6 | .5 | 1.144025E-02 | .007 |
| 110 | 207.9 | 133.2 | .64 | 1.221298E-02 | 7.909091E-03 |
| 100 | 137.9 | 125.1 | .9 | 1.314025E-02 | 9.000001E-03 |
| 90 | 66.7 | 116.4 | 1.74 | 1.427359E-02 | 1.033333E-02 |
| 80 | -4.6 | 106.8 | -23.22 | 1.569025E-02 | .012 |
| es exceeds 0.016 mm/mm strain | | | | | |
| 70 | -78.3 | 96.3 | -1.23 | 1.751168E-02 | 1.414286E-02 |
| es exceeds 0.016 mm/mm strain | | | | | |
| 60 | -155.6 | 84.9 | -.55 | 1.994025E-02 | .017 |

**Figure 34.61** Computer program printout for interaction curve data for prestressed column.

```
PU REQUIRED =   1200     kN.+++ MU REQUIRED =   160   kN.m
FOR PU REQUIRED MU ALLOWABLE = 178.8   kN.m
    PU ALLOWABLE = 1209.1   kN,    MU/PU =   .1478786
 C=    250     STRAIN IN THE COMPRESSION STEEL =    3.660253E-03     mm/mm
```

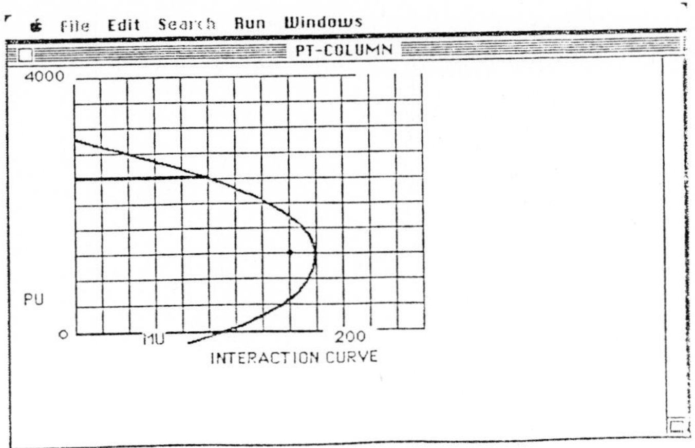


**Figure 34.61** *(Continued)*

Figure 34.62 presents the computer program listing for prestressed column.

```
LPRINT TAB(11);"PRE TENSION COLUMN ++ metric ++
LPRINT TAB(10);"INTERACTION CURVE DATA" : LPRINT
LPRINT TAB(5);STRING$(40,42 )
DIM P(100),M(100)
100 INPUT " COLUMN DEPTH IN mm= ";L
INPUT " COLUMN WIDTH IN mm = ";T
INPUT " DISTANCE TO STRAND FROM EDGE OF COLUMN IN mm = ";D1
INPUT " CONCRETE STRENGTH IN MPa AND BETA = ";FC,BE
INPUT " NUMBER OF TOTAL STRANDS AND AREA PER STRAND IN sq. mm= ";N,SA
AP=N/2*SA
INPUT " STRAND FPU IN MPa AND MOD. OF ELASTICITY IN MPa= ";FU,ES
132 INPUT " IF STRAND IS STRESS RELIEVE TYPE SR, IF LOW RELAX TYPE LR ? ";TS$
IF TS$="SR" THEN FI=.7 * FU : GOTO 138
IF TS$ = "LR" THEN FI=.74*FU : GOTO 138
GOTO 132
138 INPUT " SERVICE LOSS OF PRESTRESS IN PERCENT = ";PL
INPUT " INPUT YOUR REQUIRED PU IN kN = ";PR
INPUT "YOUR REQUIRED MOMENT IN kN.m = ";MR
INPUT " are the phi values for the U.S.A. or Canada, if U.S. type US in Canada type CA =";C$
REM COMPUTE PO AND P MAX
EB=(FI*(100-PL)/100)/ES
E1=EB-.003 :F1=E1*ES :F2=F1
IF C$="US" THEN PO=.7*(.85*FC*(L*T-2*AP)-AP*(F1+F2))/1000
IF C$="CA" THEN PO=(.85*FC*.6*(L*T-2*AP)-.9*AP*(F1+F2))/1000
PM = .8*PO
LPRINT CHR$(27);"n":LPRINT CHR$(14)
LPRINT " DESIGN INPUT" : LPRINT CHR$(15) : LPRINT
LPRINT " COLUMN DEPTH IN mm = ";L
LPRINT " COLUMN WIDTH IN mm = ";T
LPRINT " CONCRETE STRENGTH IN MPa = ";FC
LPRINT " BETA = " ; BE
LPRINT " DISTANCE FROM EDGE OF COLUMN TO STRAND = ";D1; " mm"
LPRINT " NUMBER OF STRAND = ";N
LPRINT " AREA PER STRAND = " ; SA; " sq.mm"
LPRINT " TYPE OF STRAND = ";TS$
LPRINT " STRAND STRENGTH IN MPa = ";FU
LPRINT " STRAND MODULUS OF ELASTICITY IN MPa = ";ES
LPRINT " % OF LOSS OF PRESTRESS = "; PL
LPRINT " PU REQUIRED = ";PR;" kN"
LPRINT " MU REQUIRED = ";MR;" kN.m"
LPRINT CHR$(27);"n":LPRINT CHR$(14)
LPRINT "   DESIGN  OUTPUT" : LPRINT CHR$(15) : LPRINT
IF C$="US" THEN D$=" U.S.A." ELSE D$=" CANADA"
LPRINT " phi values used are those for ";D$
LPRINT TAB(5); " PO= "; INT(PO*10)/10;" kN    P-MAX. = "; INT(PM * 10 )/10; "  kN"
LPRINT TAB(5);"VALUES OF C IN mm ,PU IN kN, AND MU  IN kN.m "
LPRINT
LPRINT TAB(1);"VALUES OF C";TAB(14);"  U";TAB(24);"MU";TAB(30);"e=MU/PU"
  ;TAB(41);"es-tension";TAB(56);"ec-conc. tension
LPRINT TAB(4);STRING$(60,45) : LPRINT
H=INT (L/BE/10) *10: D=L-D1
Z=0! : MX=0!
FOR C=H TO D1 STEP - 10
E1=EB-.003*(C-D1)/C : F1=E1*ES
E2= EB-.003*(C-D)/C :F2= E2*ES
IF E2> .008 AND E2< .009 THEN F2=1540+170*(E2-.008)/.001
IF E2>.009 AND E2<.01 THEN F2=1710+ 100*(E2-.009 )/.001
IF E2>.01 AND E2< .011 THEN F2=1800+25*(E2-.01)/.001
IF E2>.011 AND E2< .012 THEN F2=1825+15*(E2-.011)/.001
IF E2>.012 AND E2<.013 THEN F2=1840 + 10*(E2-.012)/.001
IF E2>.013 AND E2<.014 THEN F2=1850+10*(E2-.013)/.001
IF E2 > .014 THEN F2= 1860
IF E2 > .016 THEN LPRINT "es exceeds 0.016 mm/mm strain"
IF C$="US" THEN PU=.7*(.85*FC*T*BE*C-AP*F1-AP*F2)/1000
IF C$="CA" THEN PU=(.6*.85*FC*T*BE*C-.9*AP*F1 - .9*AP*F2)/1000
IF C$="US" THEN
```

**Figure 34.62** Computer program listing for prestressed column.

```
MU=.7/1000*(.85*FC*T*BE*C*(L/2-BE*C/2)-AP*F1*(L/2-D1)+AP*F2*(L/2-D1))
  /1000
IF C$="US" AND  PU < .1*PO THEN MU=MU/.7*(.9-.2*PU/(.1*PO))
IF C$="CA" THEN
MU=(.6*.85*FC*T*BE*C*(L/2-BE*C/2)-.9*AP*F1*(L/2-D1)+.9*AP*F2*(L/2-D1))
  /1000/1000
ET = .003*(L-C)/C
PU= INT (PU*10)/10 : MU = INT (MU*10)/10
IF PU> PR THEN MA=MU : CD=C : ED=MU/PU : ECD=E1 : PA=PU
LPRINT TAB(4);C;TAB(11);PU;TAB(22);MU;TAB(32);INT(MU/PU*100)/100;  TAB(41)
  ;E2;TAB(56);ET
P(Z)=PU : M(Z)=MU : Z=Z+1: IF MX> MU THEN MX=MU
NEXT C
LPRINT TAB(4); STRING$(60,45) : LPRINT
LPRINT "PU REQUIRED = ";PR;" kN.+++ MU REQUIRED = ";MR;" kN.m"
LPRINT "FOR PU REQUIRED MU ALLOWABLE =";MA;" kN.m"
LPRINT "    PU ALLOWABLE =";PA;" kN,  MU/PU = "; ED
LPRINT " C=  "; CD; "  STRAIN IN THE COMPRESSION STEEL = "; ECD; "  mm/mm"
IF MA< MR THEN LPRINT "Column will not work, increase the number of strands and run again "
IF PR > PM THEN LPRINT " Column will not work, increase the column size and run again"
LPRINT
REM GRAPHIC ROUTINE
IF PO =< 2000 THEN SP=10 GOTO 600
IF PO =< 4000 THEN SP=20: GOTO 600
IF PO =< 6000 THEN SP=30 .GOTO 600
IF PO =<8000 THEN SP=40 .GOTO 600
IF PO > 8000 THEN SP=50
600
IF MX =< 260 THEN SM=1  GOTO 700
IF MX =< 520 THEN SM=2 .GOTO 700
IF MX =< 780 THEN SM=3 :GOTO 700
IF MX =< 1040 THEN SM=4 :GOTO 700
IF MX > 1040 THEN SM=5
700
CLS
LINE (50,10)-(50,210)
LINE (50,210)-(260,211),,BF
LINE (50,10)-(260,10)
FOR J=1 TO 13 : LINE (50+J*20,10)-(50+J*20,210) : NEXT J
CIRCLE (40,210),3
PRINT PTAB (4);200*SP
FOR K=1 TO 10 : PRINT : NEXT K
PRINT PTAB(10);"PU"
PRINT : PRINT
PRINT PTAB(100);"MU"; PTAB(230);200*SM
PRINT PTAB(150);"INTERACTION CURVE"
CALL PENSIZE (2,2)
CALL MOVETO ( 50,210-INT (PO/SP))
FOR I=0 TO Z-1
CALL LINETO (50+INT(M(I)/SM),210-INT(P(I)/SP))
NEXT I
CIRCLE (50+INT(MR/SM),210-INT(PR/SP)),2
LINE (50,210-INT(PM/SP))-(150,210-INT(PM/SP)+1),,BF
LCOPY
PRINT " END OF THE RUN, IF YOU WISH TO RERUN TYPE YES ELSE TYPE NO"
INPUT " YOUR REPLY ";REP$
IF REP$="YES" THEN GOTO 100
END
```

**Figure 34.62** *(Continued)*

## Example Problem 8: Corbel Design (SI Units)

Corbels are designed by using the shear friction concept when $A/D \leqq 1.0$ and $N_u \leqq V_u$. This is the usual case for corbels placed on columns and wall panels. The designer must review the building code applicable in the project location. In the United States the $U$ values are 1.4 dead load and 1.7 live load, and the phi value for shear is 0.85. In Canada the $U$ or load factor values are 1.25 dead load and 1.5 live load, and the phi values are 0.6 for concrete and 0.85 for reinforcing steel. The mu($\mu$) value for monolithically cast concrete is 1.4 in the United States and 1.25 in Canada. See Fig. 34.63.

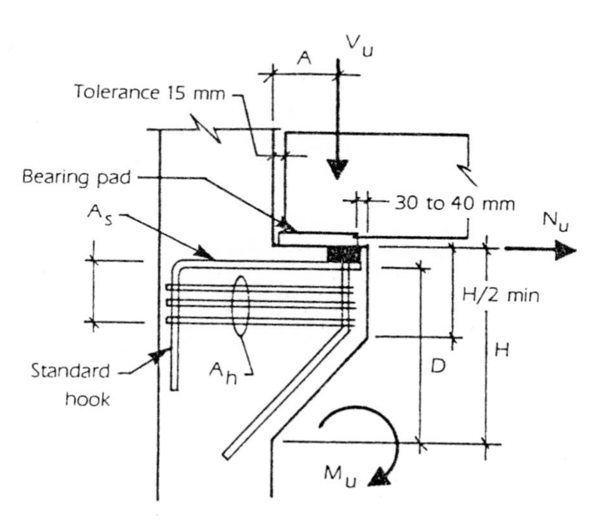


**Figure 34.63** Corbel. It must be designed to resist shear $V_u$, moment $V_uA + N_u(H - D)$, tension $N_u$. Minimum tension = $0.2V_u$; $A_{vf} = V_u/\phi f_y \mu$; $A_n = N_u/\phi f_y$. $\phi$ varies with the code used.

The length of the bearing pad $L_{bp} = V_u$/(allowable unit bearing for the pad material × the smaller of the corbel width or the beam width). Some designers use the corbel width less 25 to 30 mm tolerance versus the beam width. This allows for tolerance in case the beam slips to one side.

$$A = 2/3\,L_{bp} + 30 \text{ mm to allow for } 2\times \text{ tolerance of } 15 \text{ mm}$$

Other designers will use $A$ = 3/4 length of corbel, thus allowing for tolerance.

The minimum length of the corbel is $L_{bp}$ + 30 mm seating tolerance (2 × tolerance) + 25 to 40 mm for the distance from the end of the bearing pad to the end of the corbel. This will reduce the probability of a tension failure at the end of the corbel.

For the shear friction equation $A_{vf} = V_u / \phi_s f_y \mu$, $\mu = 1.4$ in the United States for the corbel placed monolithically with the column. In Canada the value used for $\mu$ is 1.25. For shear of the concrete $V_u < \phi V_c$, $\phi_c = \phi_c \times$ the smaller of 5.5 MPa or $0.2 f_c' \times$ the width of the corbel $\times D$. For sand-lightweight concrete, multiply $\phi_c V_c$ by 0.85. $A_s$ is the smaller of

$$A_s = A_m + A_n = [V_u A + N_u (H - D)] / \phi_c f_y D + N_u / \phi_s f_y$$

or $$A_s = 2/3 A_{vf} + A_n = 2/3 V_u / \phi_s f_y \mu + N_u / \phi_s f_y$$

If the corbel is sloped, then $D$ is to the interface of the corbel to the column. This is true because the compression strut is at an angle.

$$A_{h,\min} = 0.5(A_s - A_n)$$

The bearing area must not extend beyond the end of $A_s$ or beyond the interior face of the transverse bar or the edge of a transverse plate if used instead of a transverse bar (see Fig. 34.64).

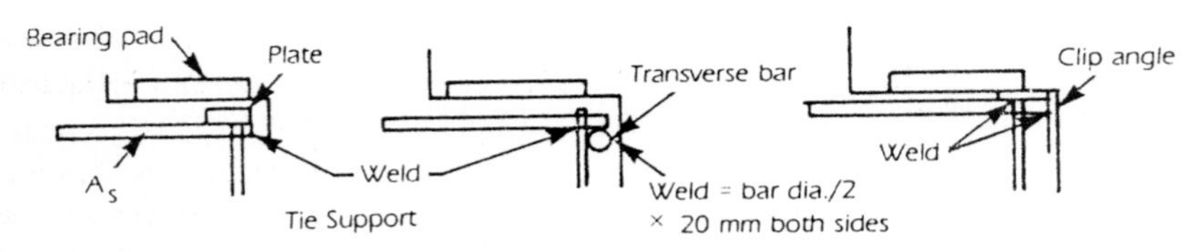


**Figure 34.64** Corbel end condition.

An example corbel design is shown in Fig. 34.65.

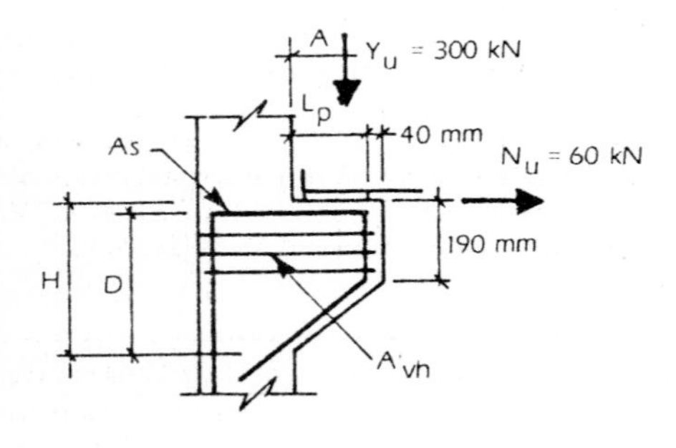


**Figure 34.65** Corbel design example.

The column measures 400 by 400 mm; $f_c' = 40$ MPa. The beam width is 350 mm. For the bearing pad, assume 4 MPa (60 Durometer pad).

$$\frac{300\,\text{kN} \times 1000}{4\,\text{MPa} \times 350\,\text{mm}} = 214\,\text{mm}$$

say, 220 mm

$$2\times \text{tolerance of 15 mm} = \underline{30 \text{ mm}}$$
$$L_p = 250 \text{ mm}$$
$$\text{Edge distance} = \underline{40 \text{ mm}}$$
$$\text{Corbel length} = 290 \text{ mm}$$

Use A-706 reinforcing steel if welded, $f_y$= 400 MPa. For hoops, $f_y$ = 300 MPa.

$A$ is the smaller of 3/4 $L_p$ or 2/3 pad length + 2 × tolerance.

$A = 3/4 \times 250 = 187.5$ mm or $A = 2/3 \times 220 + 30 = 176.7$ mm

Use $A = 190$ mm.

For a grade 60 Durometer pad with a shape factor of 3.38, the allowable compressive stress is 5 MPa. Since this is greater than MPa, it is OK.

Assume the use of Canadian phi values for this example:

$$D_{\min} = V_u \times 1000/(B \times 5.5 \text{MPa} \times \phi_c) = 300 \text{ MPa} \times 1000/$$
$$(400 \text{ mm} \times 5.5 \text{MPa} \times 0.6) = 227 \text{ mm}$$
$$A/D = < 1.0$$

Since 190/227 is less than 1.0, it is OK. Use $D$ = 350 mm and $H$ = 370 mm. Use of $D$ of 350 mm will reduce steel requirements.

$A_s$ shall be the larger of the following :

$$A_s = A_f + A_n = 1/f_y\,[V_u A + N_u (H - D)/(\phi_c \times D) + N_u / \phi_s]$$
$$= 100/400 \text{ MPa } [300 \text{ kN} \times 190 \text{ mm} + 60 \text{ kN}(370 \text{ mm} - 350 \text{ mm})/$$
$$(0.6 \times 350 \text{ mm}) + 60 \text{ kN}/0.85] = 869 \text{ mm}^2$$
$$A_s = 1/\phi_s f_y \times (2/3 V_u / \mu + N_u)$$
$$= 1000/(0.85 \times 400) \times (2/3 \times 300 \text{ kN}/1.25 + 60) = 647 \text{ mm}^2$$
$$A_{s,\min} = 0.04 f_c'/f_y bd = 0.04 \times 40/(400 \times 400 \text{ mm} \times 350 \text{ mm}) = 560 \text{ mm}^2$$
$$A_s \text{ required} = 869 \text{ mm}^2$$

Use three no. 20M bars:

$$3 \times 300 \text{ mm}^2 = 900 \text{ mm}^2 > 869 \text{ mm}^2 \quad \text{(OK)}$$

Or use two no. 25M bars :

$$2 \times 500 \text{ mm}^2 = 1000 \text{ mm}^2$$
$$A_{vh} = 0.5(A_s - A_n) = 0.5$$
$$[869 \text{ mm}^2 - 60 \text{ kN}/1000/(0.85 \times 400 \text{ MPa})] = 347 \text{ mm}^2$$

If the hoop steel is of grade 300 MPa,

$$A_{vh} = 400 \text{ MPa}/300 \text{ MPa} \times 347 = 463 \text{ mm}^2 \text{ required}$$

Use three no. 10M hoops.

Area of steel = 3 hoops × 2 bars per hoop × 100 mm$^2$ = 600 mm$^2$

Since 600 mm$^2$ > 463 mm$^2$, it is OK.

The final corbel design is shown in Fig. 34.66.

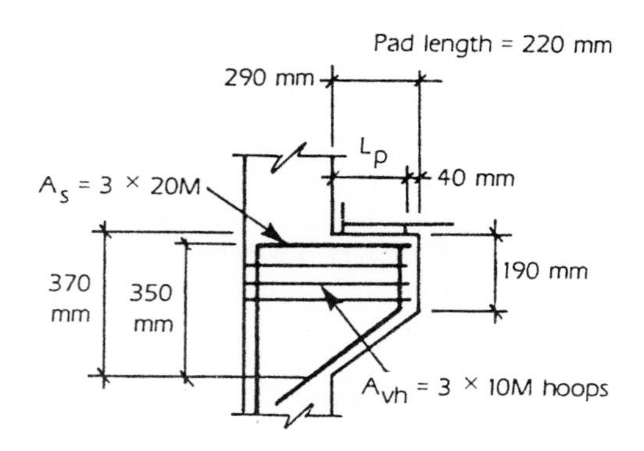


**Figure 34.66** Final corbel design.

Chapter

# 35

# Seismic Design in Metric Units

## Introduction

Building code requirements for seismic design vary from country to country and, in some countries, from region to region or from city to city. For example, in the United States there is the *NEHRP Recommended Provisions for the Development of Seismic Regulations for New Buildings* by the Building Seismic Safety Council (BSSC) and the Federal Emergency Management Agency (FEMA) of the United States government. In the western and midwestern parts of the country the Uniform Building Code is used; in the southern part of the country, the Southern Building Code, and in the northeastern part of the country, the National Building Code. However, many large cities have their own building codes, among them Los Angeles, San Francisco, Chicago, and New York. The National Building Code of Canada also differs from the Uniform Building Code and the BSSC recommendations. The designer therefore must consult the building code governing the location where the project is to be constructed.

No building code will guarantee that damage will not occur during a major or moderate earthquake. To design a building to resist total damage would render the cost of construction so high that no one could afford to own, lease, or operate the building (nuclear reactors are an exception). The intent of building codes is to recognize that damage will occur but that, if the building is subject to a major earthquake, collapse will be prevented and occupants can leave the building without serious injury. Building codes are based upon probability studies of the magnitude and frequency of earthquakes in the various regions of a country. They are updated periodically to reflect current concepts and lessons learned from recent earthquakes. Some building codes require that buildings over 10 stories in height be provided with not less than three approved accelerographs for earthquake recordings: one placed at the top of the building, one at midheight, and one in the basement. Traces from these accelerographs are used to study the response of the structure to the earthquake and to gain new knowledge for future design. Some cities, for new high-rise construction, require that a dynamic test be performed on the completed building by using an eccentric-weight vibration generator. The purpose is to compare the natural frequency of the structure with that of the design. This will give a correlation between the design methodology and the completed structure.

The design of the building is the responsibility of the architect and the structural engineer working with the owner. The building code is only a minimum standard to safeguard life, health, and public safety.

## Nature of Earthquakes

It is important that the designer understand the nature and cause of earthquakes as well as normal building response to seismic waves. To understand the cause of earthquakes a brief introduction to plate tectonics is necessary.

The crust of the earth is composed of plates which are in constant motion and are constantly changing. As one plate separates from another, rifts are formed and magma rises to form a new or wider sea floor. Usually occurring in the ocean areas where the plates are thin, this development is called spreading. In continental plates also plate movement causes rifts where the earth crust thins as it stretches and large blocks of the surface drop, as in New Mexico owing to the Rio Grande Rift. At the rift near Borah Peak in Idaho on October 8, 1983, the gound surface fell approximately 3 m (10 ft), producing an earthquake of a magnitude of 7.3.

As the older, heavier ocean plate dives under the lighter continental plate, back in the interior of the earth trenches are formed, the lower plate melts, and the heat generated melts through the overlying plate and volcanoes are formed, or existing volcanoes become active. The uplift also lifts the mountains. As one plate slides beneath the other, because of the marine peaks and irregular surface the plate at some point tends to hang up and bend. As the stress created becomes excessive, the earth snaps and jumps at that point, creating a fault, shock waves, and an earthquake. This is called subduction (see Fig.

35.1). Examples are the Mexico City earthquake of 1985, magnitude 8.1; the Alaska earthquake of 1964, magnitude 8.4; the Alaska earthquake of November 1987, magnitude 6.5; and the Queen Charlotte earthquake in British Columbia in 1949, magnitude 8.0.

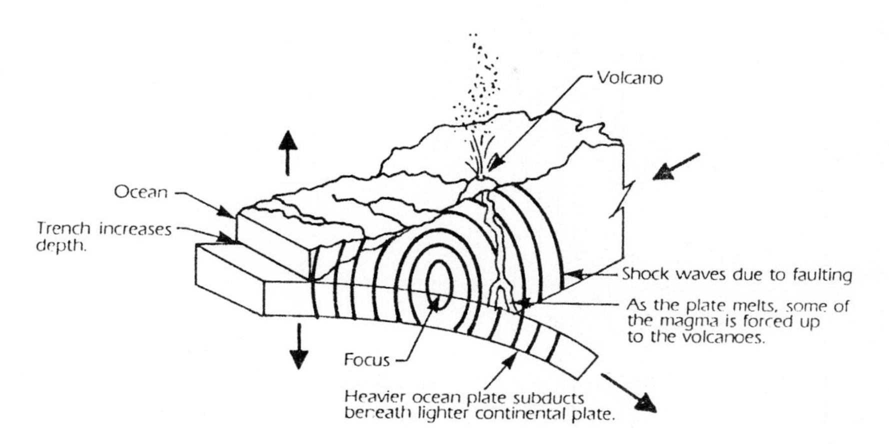


**Figure 35.1** Subduction.

As the plates move horizontally relative to each other, a grinding action causes faulting or slippage. Sections tend to lock up owing to the irregular fractured surface and the compressive stress between the two plates. When the stress becomes excessive, faulting occurs, resulting in an earthquake (see Fig. 35.2). For example, the Pacific side of the San Andreas fault near its center is moving north at approximately 3.6 cm a year relative to the North American plate. The sections where little movement occurs are locked up and eventually will give and jump to relieve the built-up stress. The edge between two plates consists of more than one fault plane running both parallel and transversely. Examples of earthquakes resulting from fault movement are the San Francisco earthquake of 1906 on the San Andreas fault, estimated magnitude 8.3; the El Centro earthquake in the Imperial Valley in 1940, magnitude 7.1; the El Centro earthquake of 1979, magnitude 6.6; and the Parkfield earthquake of 1966, magnitude 6.6. Earthquakes on adjacent faults include the Inglewood fault →, closed up both sides, Long Beach earthquake of 1933, magnitude 6.3; the San Fernando earthquake of 1971, magnitude 6.4; the Coalinga earthquake of 1983, magnitude 6.5; the Morgan Hill earthquake of 1984, magnitude 6.2; and the Whittier earthquake of 1987, magnitude 6.1.

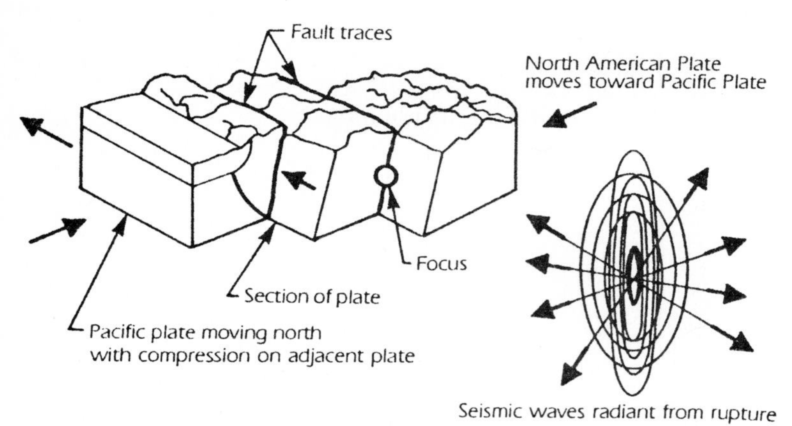


**Figure 35.2** Faulting between plates.

As two plates come together but are too light to subduct, collision occurs and mountains are pushed up. As the compressive stresses build, they are released in jerks as the mountains rise.

Spreading occurs most often under the ocean where the crust is thin (see Fig. 35.3). Magma rises, and new ocean floor is created, as well as ridges and underwater volcanoes. As the volcanoes rise to the surface, islands are formed. The rising of the domes in jerks and the eruptions cause seismic waves. Spreading can also occur within continental plates.

As the plates move, one against the other, seamounts and other chunks of crust are broken off. These are called terranes, and they move like debris in a stream. They fuse to overlying plates in a jarring collision. Alaska and many other parts of western North America were formed in this fashion.

As the plates move relative to each other, strains are developed in the earth until rupture occurs. When rupture occurs, energy waves are produced, causing earthquakes. (See Fig. 35.4.)

The waves thus produced fall into two categories: body waves and surface waves. The body waves are of two types (1) P waves or compression waves which travel at a higher velocity than (2) S waves. The S, or shear, waves move transversely. The P waves pass through the molten core of the earth and can be recorded by seismographs on opposite sides of the globe. The S waves rebound off the molten core of the earth. Both P waves and S waves radiate out from the point of rupture. Owing to the variation of the mass of the crust, they are bent during travel. As the S waves reach the surface, surface waves are formed. These surface waves are classed as Love waves, which have a horizontal motion, and Rayleigh

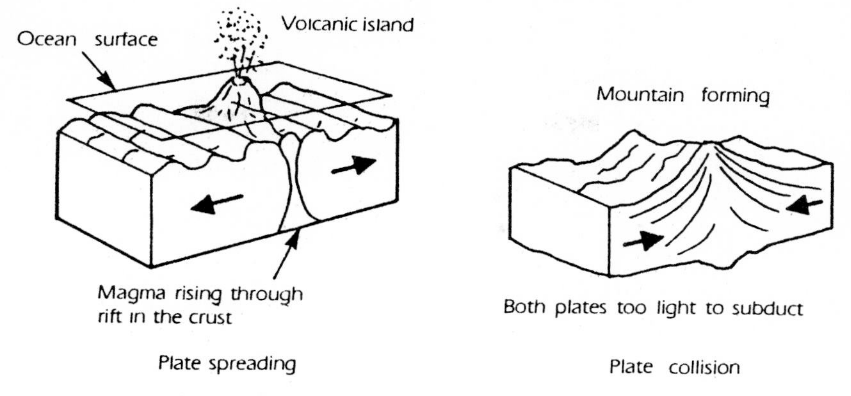


**Figure 35.3** Plate spreading and collision.

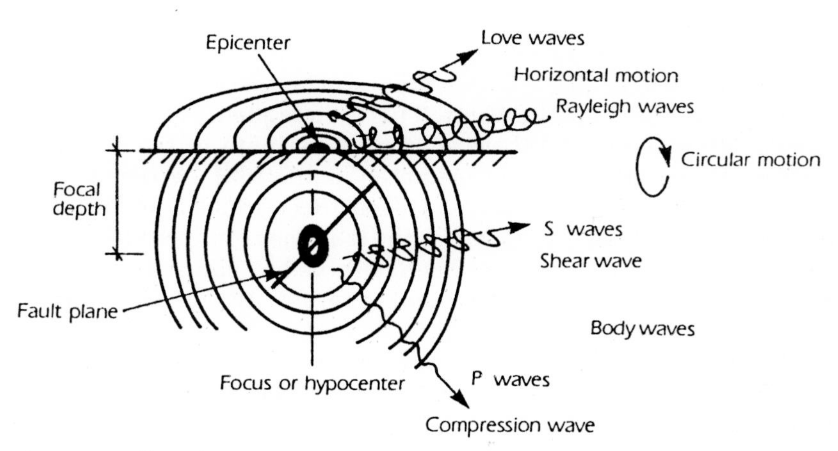


**Figure 35.4** Seismic waves.

waves, which have an orbital motion. Surface waves have a lower frequency than body waves and cause the greatest damage to buildings. The amplitude of the low-frequency surface waves decays less with distance than that of the higher-frequency waves relative to distance from the fault. The frequency and velocity of the waves also change as the waves pass through different densities and types of soil.

The velocity of P waves ranges from 5 to 7 km/s and that of S waves from 3 to 4 km/s. S waves travel at roughly half the speed of P waves. P waves reach the recording instrument first and trip the instrument. From the time interval between the P and S waves the distance to the focus can be determined. By triangulation the location of the seismic event can also be determined. (See Fig. 35.5.)

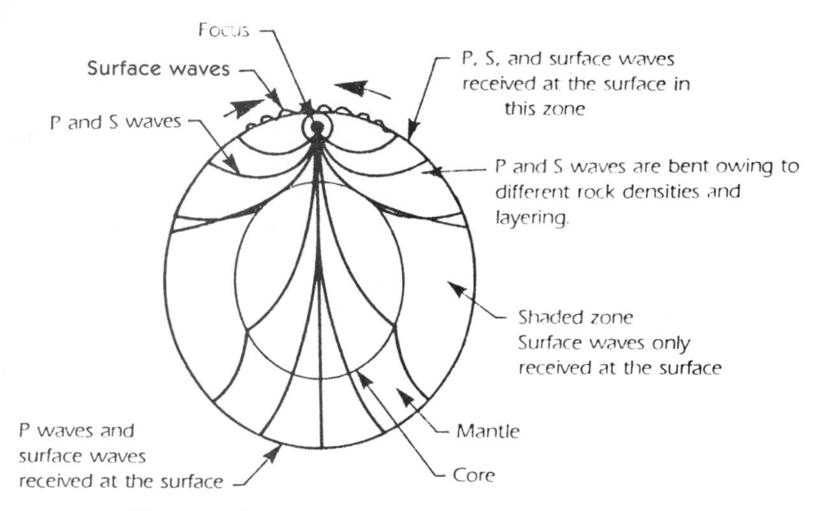


**Figure 35.5** Wave travel.

Seismic waves resemble the waves that form on the surface of a pond when a rock is dropped into the water. There is a sequence of waves with decreasing amplitudes, and the waves also dissipate in travel. The shapes are similar and are approximately sinusoidal. Whereas the water of the pond is of one density, the earth's surface is of variable densities and variable soil types; hence the true wave patterns are more complex. The wave patterns have been observed as sinusoidal, but to date there has been no real measurement of amplitude. Visual observations have ranged from 0.3 to 1.2 m (1 to 4 ft). The wavelength is a product the period and the velocity. For a period of 0.2 second and a wave velocity of 1 km/s the wavelength would be 100 m, or approximately 328 ft. This, however, is not valid since period and velocity are directly related to both the soil conditions at the surface and the wave intensity in the rock substrata. When the wave passes into soil of different densities and types, it is amplified with an increased period and wavelength and decreased velocity and acceleration. An example is the Mexico City earthquake of 1985, where the period of the lake bed area changed from 0.2 to 2 seconds and the acceleration was reduced from 0.4 to 0.2 $g$. When the soil period reached the period of the taller buildings, resonance set in and there was major damage. (See Fig. 35.6.)

In summary, earthquakes can be caused by (1) volcanic activity, (2) subduction between two plates, (3) faulting or slippage between adjacent plates such as the San Andreas fault or parallel or transverse faults, (4) spreading between plates called rifts, (5) collision between plates called thrusts, and (6) blast due to explosions. In the areas of faulting or slippage there can also be compression thrusts and spreading since the fault is not along a straight line.

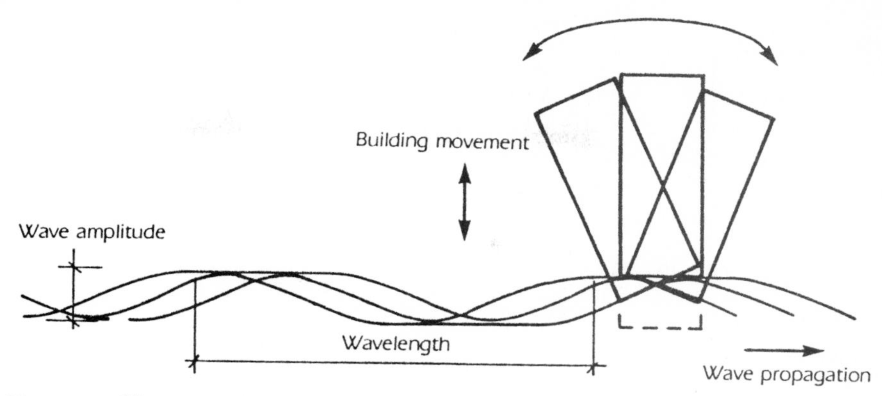


**Figure 35.6** Wave propagation.

## Measurement Scales

The two scales currently used to measure earthquakes are the Richter Scale, which is a measurement of magnitude; and the Modified Mercalli Scale, which measures observed intensity.

The Richter Scale is the common logarithm of the recorded maximum amplitude in micrometers taken at a distance of 100 km from the epicenter of an earthquake that would be measured on an Anderson and Wood torsion seismograph. Hence a value of 7 is 10 times as great as a value of 6 on the scale. The energy at the hypocenter (focus) in ergs = log $w = 11.8 + 1.5M$, where $M$ = the Richter magnitude and $w$ = the energy released in ergs.

From the accelograph of a recorded earthquake, with the time of arrival and the amplitude of the acceleration known, both the distance to the hypocenter and the Richter magnitude can be determined by using alignment charts (see Fig. 35.7).

The Richter Scale was developed in 1935 by C. F. Richter. Prior to that time earthquakes had been measured by visual observation by using either the Rossi-Forel Scale or the Mercalli Scale. The Mercalli Scale was modified to better represent the type of construction in the United States. The Modified Mercalli Scale measures the observed intensity of the earthquake as shown in Table 35.1.

It is difficult to compare the Richter Scale and the Mercalli Scale since the Richter Scale relates to measured acceleration as recorded at a distance from the epicenter, whereas the Mercalli Scale measures only observed damage. Ground or soil conditions and the construction of the buildings are also factors that must be considered when assessing damage.

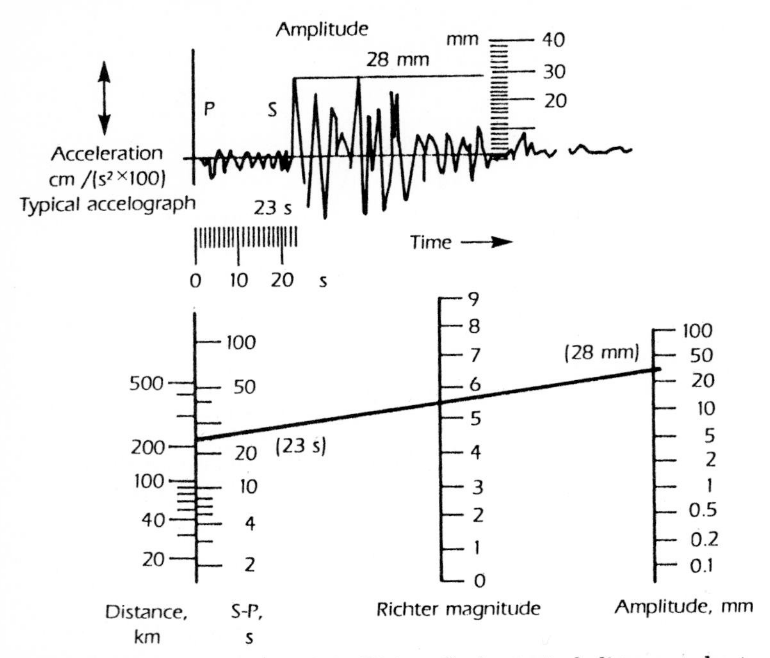


**Figure 35.7** Acceleration and the Richter Scale: typical alignment chart.

## Seismic Recordings

The strong-motion accelerographs most commonly used consist of three accelerometers and three strip chart recorders. One accelerometer measures acceleration in a longitudinal direction, one in a transverse direction, and one in the vertical direction. The instrument is usually oriented in the north-south direction, thus giving readings in the north-south, east-west, and vertical directions. When P waves strike the instrument, it is triggered and recording of acceleration/ acceleration of gravity versus time in seconds begins.

There are many different types of accelerographs. The AR-240 records on 12-in paper with a scale of 7.6 cm/*g*. The SMA-1 records on 70-mm film at a scale of 1.5 to 2.0 cm/*g*. Another type records on 35-mm film. Recordings from the San Fernando earthquake of 1971 were available from over 200 different accelerographs. The majority of these were in buildings, thus giving a good record of their response. An example of an accelerograph trace is shown in Fig. 35.8.

**TABLE 35.1 Modified Mercalli Intensity Scale**

| Scale value | Description (condensed) |
|---|---|
| I | Felt only in especially favorable circumstances. |
| II | Felt by persons at rest or on upper floors. |
| III | Felt indoors. Hanging objects swing and vibration is like the passing of a light truck. Duration is estimated. |
| IV | Hanging objects swing, vibration is like the passing of a heavy truck, standing cars rock, and windows, dishes, and doors rattle. |
| V | Felt outdoors; direction is estimated. Liquids are disturbed and small unstable objects are displaced or upset. Doors swing open or close. Pendulum clocks stop, start, or change their rate. |
| VI | Felt by all. Many people are frightened and run outdoors. Windows, dishes, and glassware are broken, books fall off shelves, and furniture is moved. Plaster is cracked and trees shake visibly. |
| VII | Difficult to stand; noticed by drivers. Hanging objects quiver, and masonry cracks. Weak chimneys break at the roofline. Plaster, roof tiles, cornices, and unbraced parapets fall. Waves form on water. Large bells ring. |
| VIII | Steering of cars is difficult; damage to masonry. Twisting and fall of chimneys and stacks, monuments, towers, and elevated tanks occur. Branches are broken on trees. There is a change in the flow of spring water. Cracks appear in wet ground. |
| IX | General panic; masonry walls are seriously damaged. There is general damage to foundations, and frames are racked. Underground pipes are broken; cracks appear in the ground. In alluviated areas sand and mud are ejected, sand craters form, and liquefaction occurs. |
| X | Most nonreinforced and framed structures are destroyed. Some well-built wood structures are destroyed. There is serious damage to dams, dikes, and embankments. Rails are bent slightly. |
| XI | Rails are bent greatly; underground pipes are completely out of service. |
| XII | Damage is nearly total. Large rock masses are displaced. Objects are thrown into the air. The line of sight and level are distorted. |

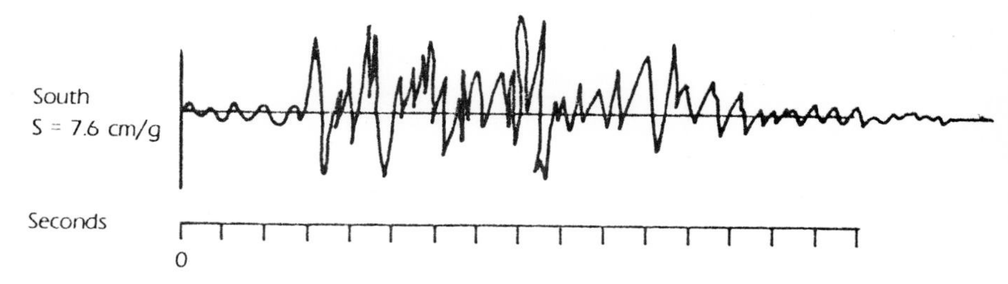


**Figure 35.8** Typical acceleration trace.

After an earthquake the recordings are reproduced and then digitalized. The data are integrated to obtain the velocity, then integrated a second time to obtain the displacement. This information is made available in both graphical representation (see Fig. 35.9) and digital format. Copies are usually sent to the building owner and the structural engineer of record and are made available to architects, structural engineers, and researchers. Clues relative to major structural damage in high-rise buildings could be indicated since most large cities in active seismic regions require accelerographs to be located in the basement, at midheight, and on top of buildings 10 stories or greater in height.

## Response Spectrum

A response spectrum is an envelope of the maximum response of single-degree-of-freedom oscillators subject to a specified seismic disturbance and plotted as a function of maximum acceleration versus natural period of the oscillator or of velocity versus period. Different sets of curves are plotted with different percentages of critical damping: 0, 2, 5, and 10 percent. They are also plotted for different ground conditions. The set of curves may indicate acceleration versus period or velocity versus period or, more commonly, have a tripartite form which plots acceleration, velocity, and displacement relative to period or frequency. The plot may show peaks and valleys or a normalized smooth line. For design purposes smooth-line response spectra are used. (See Fig. 35.10.)

The effective peak acceleration (EPA) is the normalized value of a smooth tripartite or linear curve for average peak accelerations between 0.5 second and 1 second for 5 percent damping. The effective peak velocity (EPV) is the normalized value for a period range of about 1 second from about 0.5 to 1.5 seconds. These values for probable earthquakes during a 50-year life are plotted on a map of the country or state and are used in the static equivalent design methods outlined in some building codes. Examples of EPA and EPV are shown in Figs. 35.11 and 35.12.

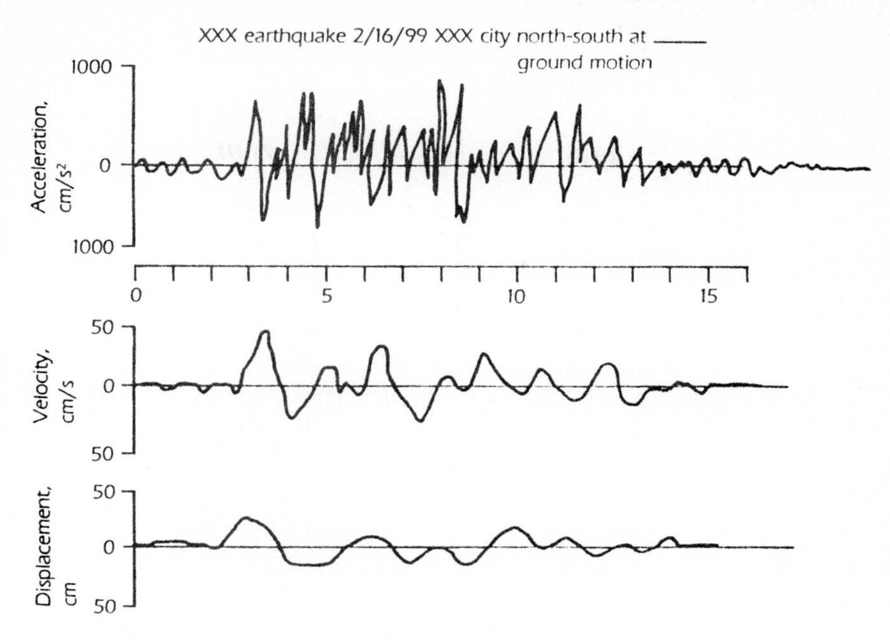


**Figure 35.9** Acceleration versus velocity and displacement.

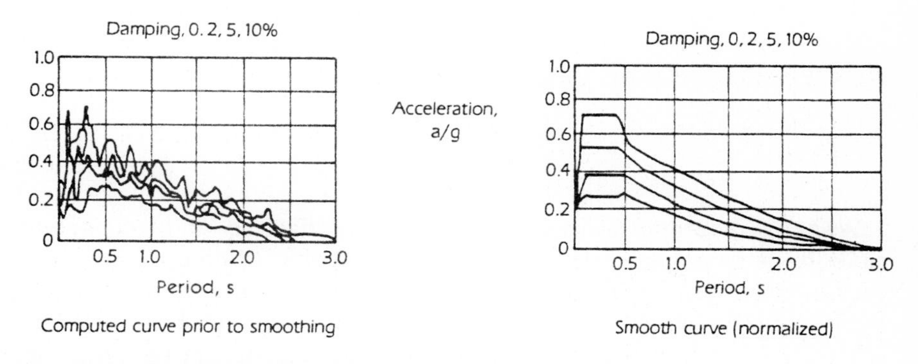


**Figure 35.10** Typical acceleration response spectrum.

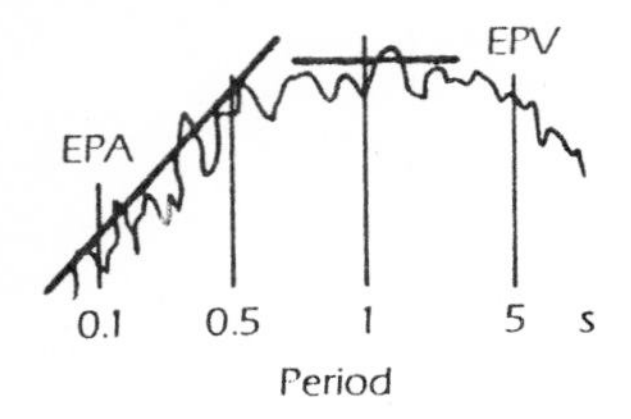


**Figure 35.11** Effective peak acceleration.

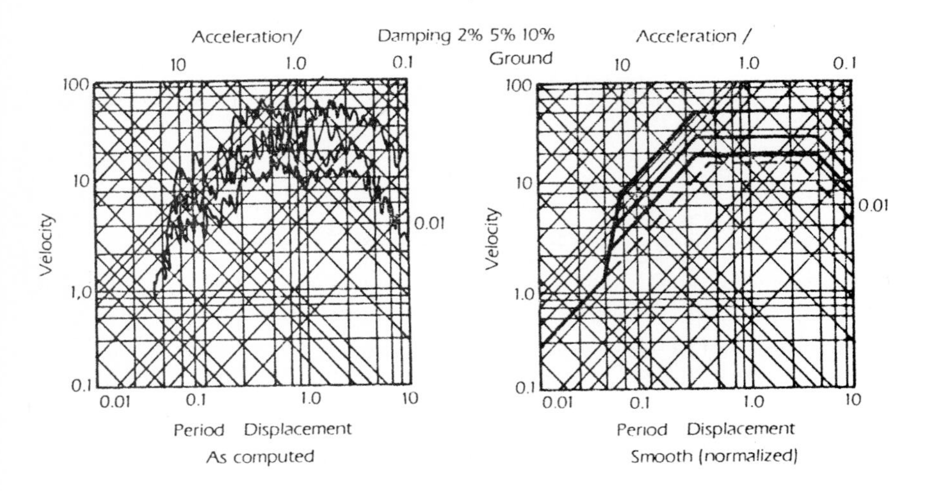


**Figure 35.12** Typical tripartite response spectrum.

When a dynamic analysis is required or desired, a geotechnical consultant is engaged to prepare both probable (damage-level) and credible (collapse-level) response spectra for the specified site, considering the faults which may produce earthquakes. For example, the probable response spectrum may be for a 50-year design life, with an estimated occurrence of 50 to 60 percent for an earthquake of a magnitude of 6.3 at a distance from 0.5 to 2 mi from either the Newport-Inglewood fault or the Santa Monica – Hollywood fault. The credible response spectrum may be for a 50-year life for an earthquake of a magnitude of 8.3 occurring on the San Andreas fault a distance of 45 mi, with a probability of less than 10 percent. These response spectra will take into consideration the soil conditions and the depth of the various layers of soil over bedrock. Therefore, soil investigations with drilling logs are necessary.

## Damping

Critical damping is the minimum value of damping that will allow a displacement oscillator to return to its initial position without oscillating. The damping ratio is the ratio of the damping coefficient to the critical damping coefficient for a single-degree-of-freedom oscillator. It is a fraction of critical damping and is expressed in a percentage. It is difficult to determine the damping ratio for a building owing to the elastic and inelastic energy absorption of the structural system during an earthquake. Nonstructural elements also have an effect upon damping. For buildings, damping is usually estimated at 5 percent.

Plate boundaries and seismic and volcanic activity in the world are shown in Figs. 35.13 and 35.14, and effective peak ground acceleration and velocity contours for the United States and Canada are shown in Figs. 35.15 and 35.16.

## Building Response

As the seismic waves pass under a building, the earth moves in a complex sine-wave pattern. The foundation, being embedded, moves with the earth. The superstructure tends to remain in its original position, then tries to catch up with the ground or foundation movement. This lag between the superstructure and the foundation causes distortions and develops forces on the superstructure (see Fig. 35.17). The force is mass × acceleration. Seismic waves can come from any direction with unknown magnitude, frequency, wavelength, amplitude, and duration. If the period of ground motion is the same as the period of the building in any of its modes of vibration, then resonance can occur with increasing damage to the building. If the building is nonsymmetrical at any or all of its floor levels, then torsion will occur. (See Fig. 35.18.) Since no building is 100 percent symmetrical, torsion will occur to some degree. In tall buildings it is possible to have torsion in more than one mode, in which one floor may try to rotate in one direction and another floor in the opposite direction. The torsional modes may be in phase or out of phase with planar modes; this can be readily observed with model testing. The building may also be subject to aftershocks as well as foreshocks. The period of the building can also change after a primary shock if inelasticity (yielding) occurs. The damage caused by aftershocks can be added to the damage caused by the primary shock. Aftershocks may occur within hours, days, or weeks of the main shock.

Torsion occurs when the center of mass is not at the same point as the center of rigidity. Since the seismic wave can come in any direction, rotation can also be in either direction. As the building oscillates back and forth, rotation will also occur in alternating directions. (See Fig. 35.19.)

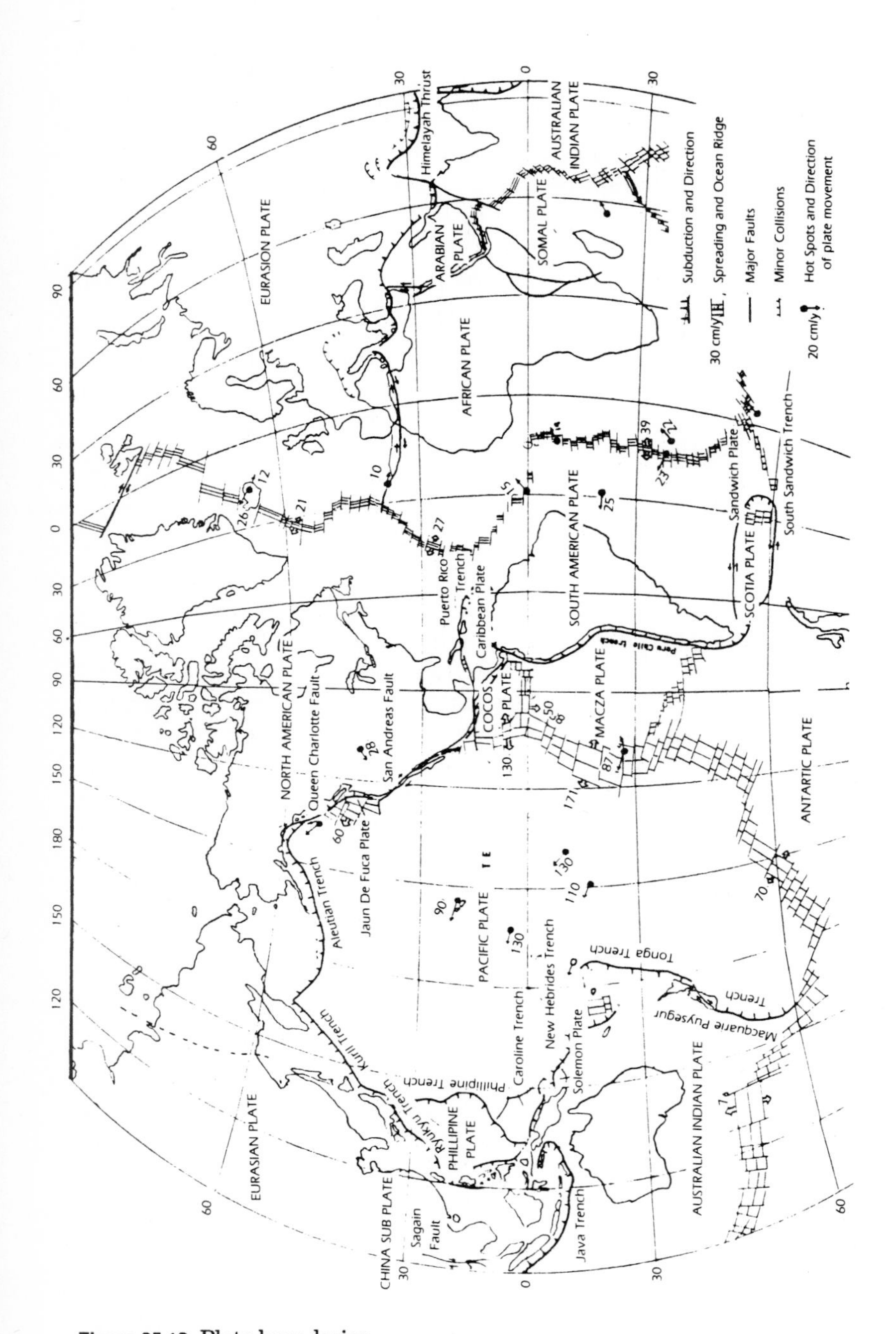


**Figure 35.13** Plate boundaries.

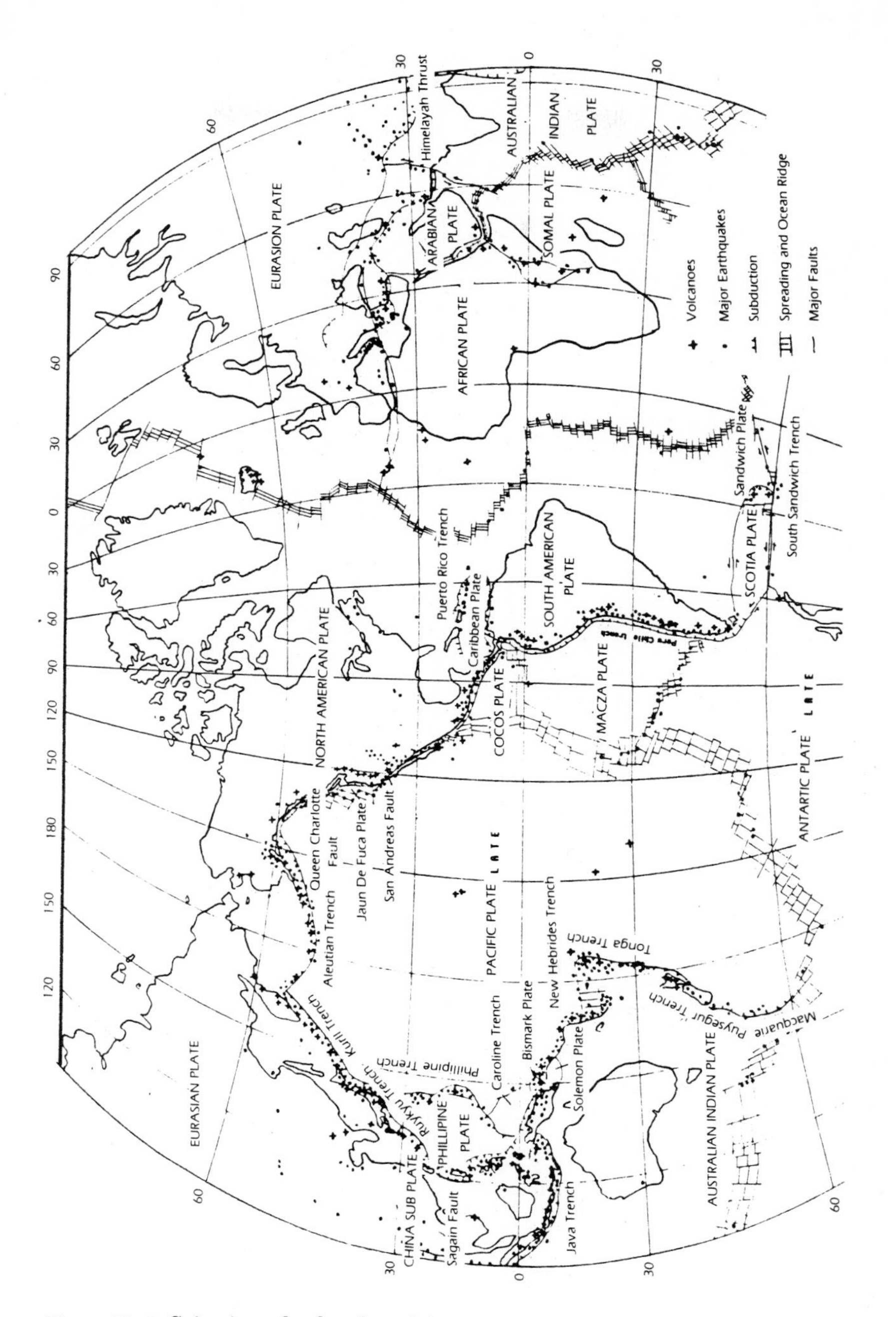


**Figure 35.14** Seismic and volcanic sctivity.

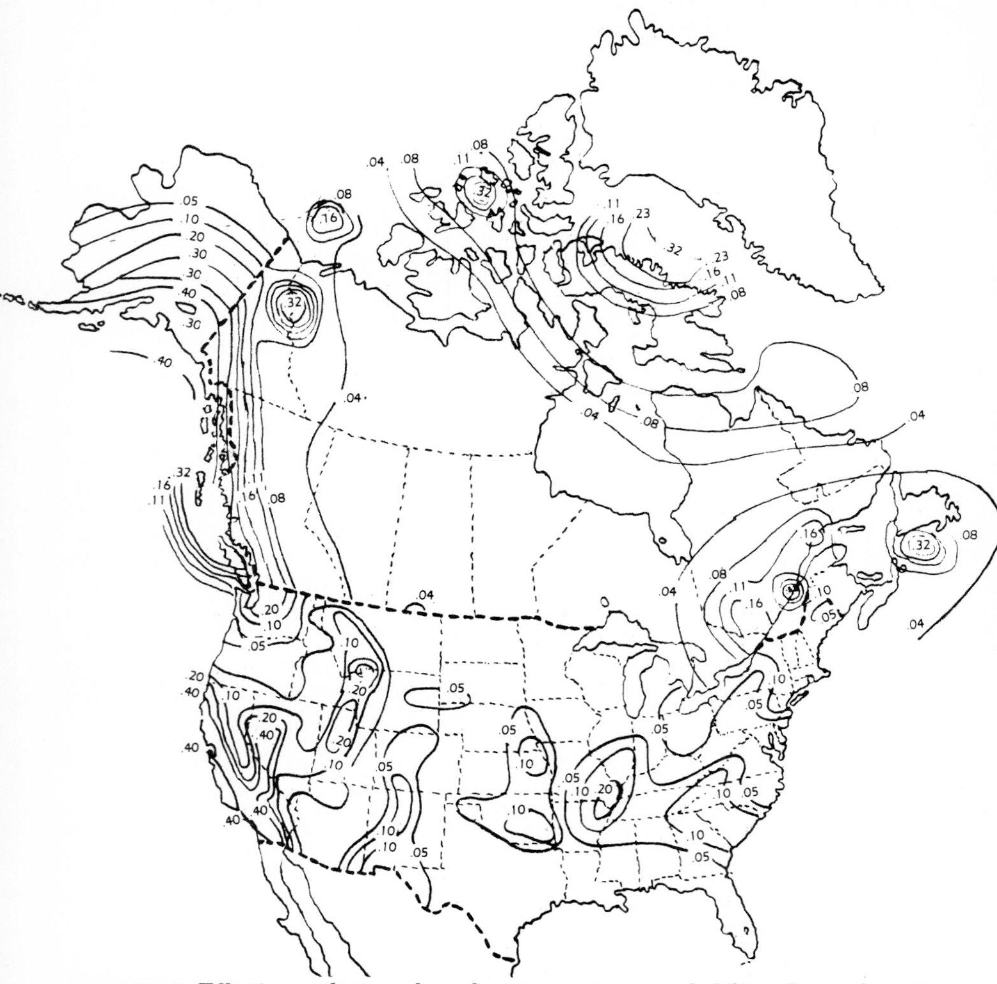


**Figure 35.15** Effective peak ground acceleration contours: probability of exceeding 10 percent in 50 years. The map is approximate and serves only to indicate the regions of seismic probability. The designer must consult the local building code for the correct values of acceleration where his or her building is to be constructed. United States contour lines: 0.05, 0.10, 0.20, 0.40; Canadian contour lines: 0.04, 0.08, 0.11, 0.16, 0.23, 0.32.

**Canadian Zones**

| *a/g* | Zone |
|---|---|
| 0. 00 - | 0 |
| 0. 04 - | 1 |
| 0. 08 - | 2 |
| 0. 11 - | 3 |
| 0. 16 - | 4 |
| 0. 23 - | 5 |
| 0. 32 - | 6 |

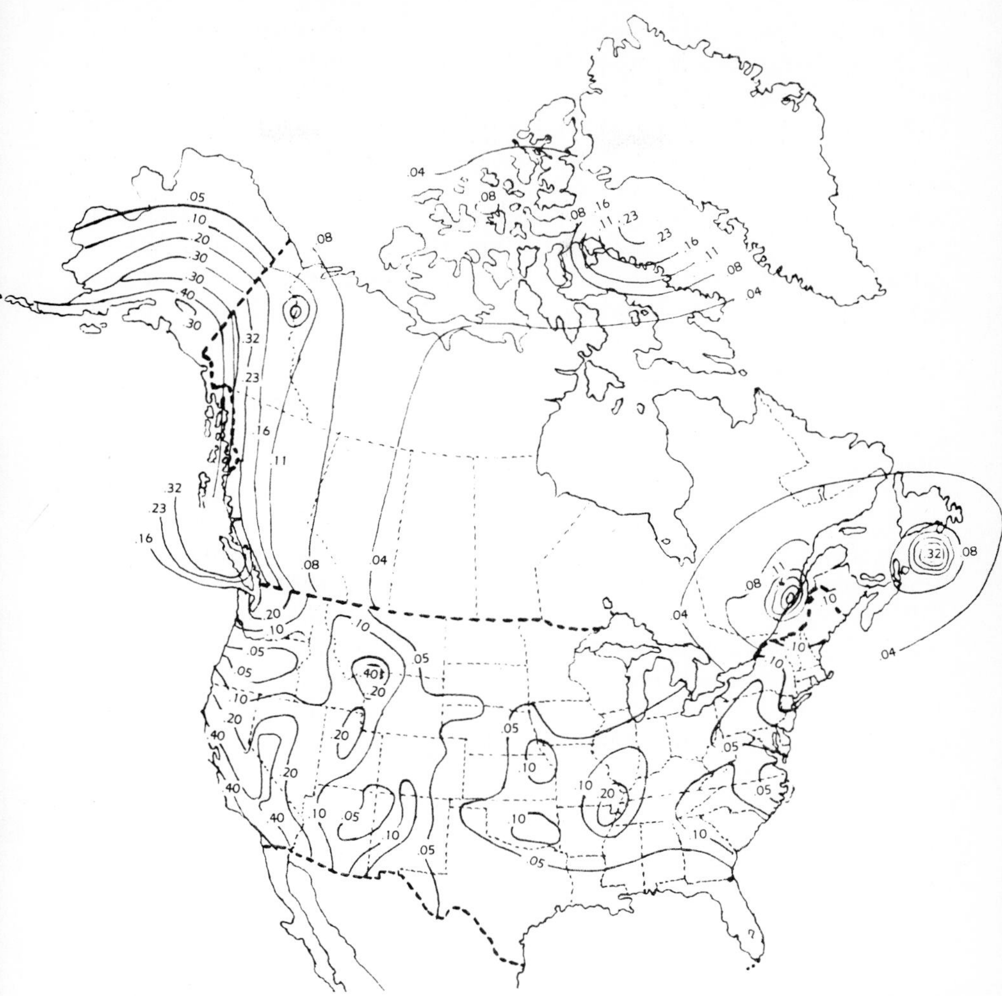


**Figure 35.16** Effective peak ground velocity contours: probability of exceeding 10 percent in 50 years. The map is approximate and serves only to indicate the regions of seismic probability. The designer must consult the local building code for the correct values of velocity where his or her building is to be constructed. United States contour lines: 0.05, 0.10, 0.20, 0.40; Canadian contour lines: 0.04, 0.08, 0.11, 0.16, 0.23, 0.32. $Z_v$ = zone for velocity.

**Canadian Zones**

| m/s | V | $Z_v$ |
|---|---|---|
| 0.00 - | 0.00 - | 0 |
| 0.04 - | 0.05 - | 1 |
| 0.08 - | 0.10 - | 2 |
| 0.11 - | 0.15 - | 3 |
| 0.16 - | 0.20 - | 4 |
| 0.23 - | 0.30 - | 5 |
| 0.32 - | 0.40 - | 6 |

If the building is larger than the seismic wavelength, the walls or columns (if a frame system) will rotate and cause tension or compression on the diaphragm (see Fig. 35.20). Hence the wall-to-diaphragm connection is critical. In a frame system additional moment will be produced in the column-girder connection. The wave amplitude is not very large, ranging from approximately 25 cm to 1 m. There has been no real measurement. Integration of vertical acceleration data gives about 30 cm up and about 15 cm down for the San Fernando earthquake. Where surface faulting occurs, this distance can be much greater. If surface faulting occurs under a building, considerable damage will occur. In the San Fernando earthquake surface faulting up to 2 m vertical and 1.5 m lateral occurred. In the Mexico earthquake of 1985 near the coast 1 m vertical and 2.5 m lateral occurred. Surface faulting did occur under some buildings in the San Fernando earthquake but of a smaller magnitude. These buildings suffered major damage.

For multistory buildings the amplitude of acceleration recorded for the top floor is greater than that for the basement. This is logical since the top floor has farther to move than the ground floor (a whiplike action). Since force is mass × acceleration, the force occurring at the upper floor will be greater than that for the lower floors and will decrease with the decreasing height of the building. This is reflected in the building codes when the static equivalent method is used for design.

The types of forces which develop on a shear wall building during an earthquake are shown in Figs. 35.21 through 35.24.

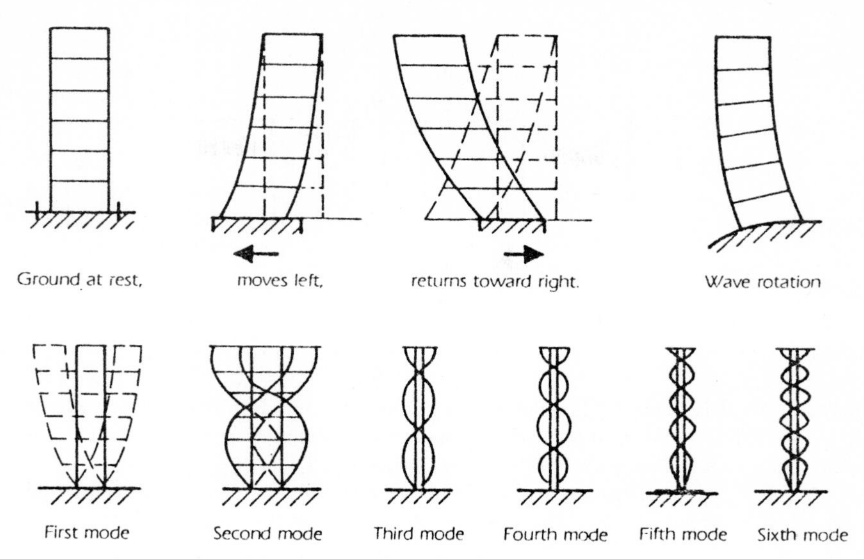


**Figure 35.17** Ground motion and lateral displacement.

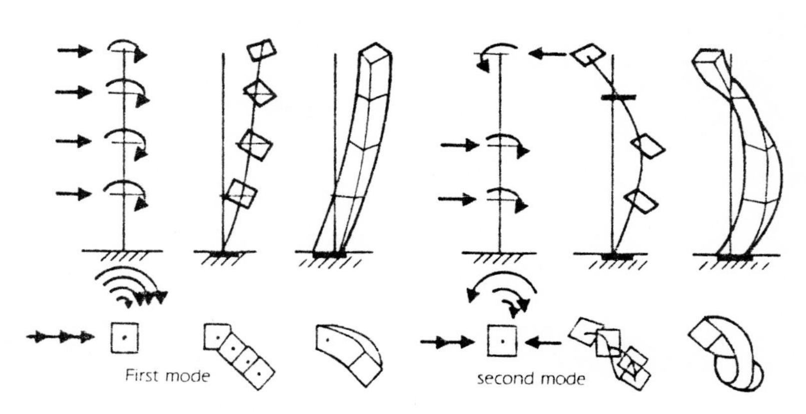


**Figure 35.18** Torsion and lateral displacement.

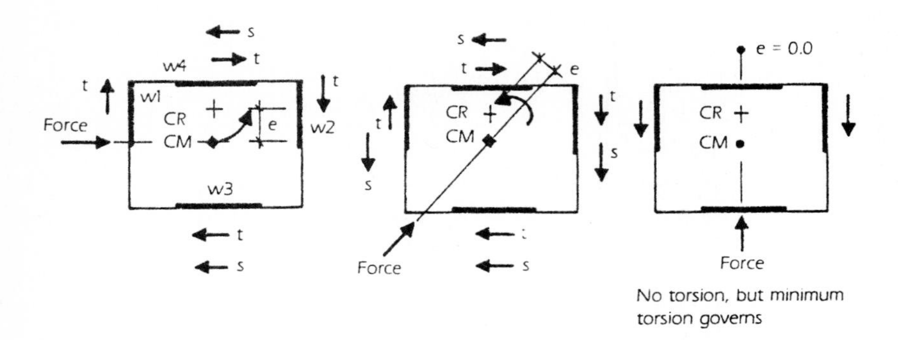


**Figure 35.19** Torsional forces. Moment torsion = force $\times e = M_T$. Minimum torsion = force $\times e_{min}$; $e_{min} = 0.05 \times$building dimension perpendicular to direction of the force. $d$ = distance from wall to center of rigidity; $r$ = wall rigidity. $T = M_T r_{wall} d_{wall} / \Sigma r d^2$. $s$ = force $r_{wall} / \Sigma r_{parallel\ walls}$. If wall is at an angle to the direction of force, use $r$ as computed for the parallel direction where the moment of inertia is about an axis normal to the direction of the force.

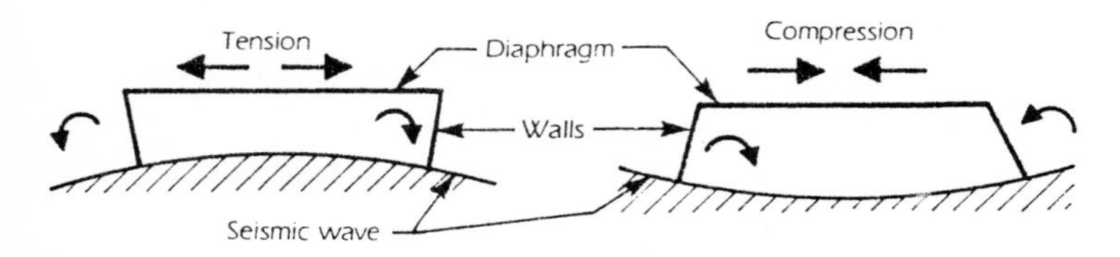


**Figure 35.20** Wave effects on a building.

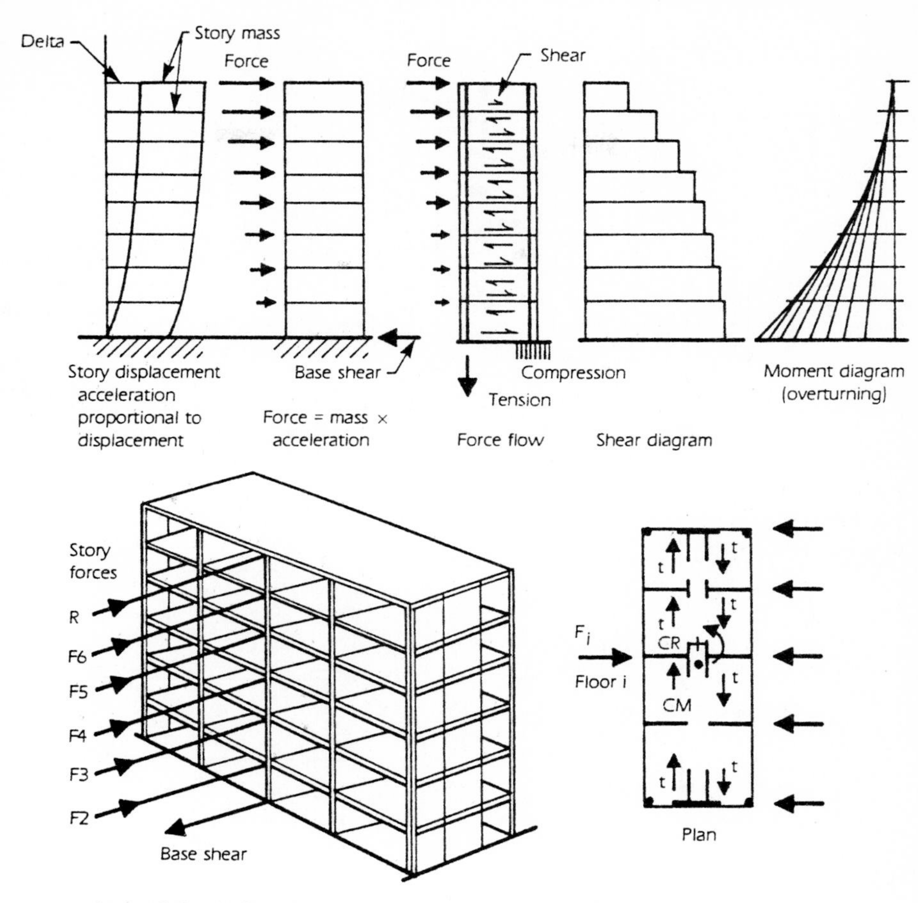


**Figure 35.21** Seismic forces on a shear wall apartment building.

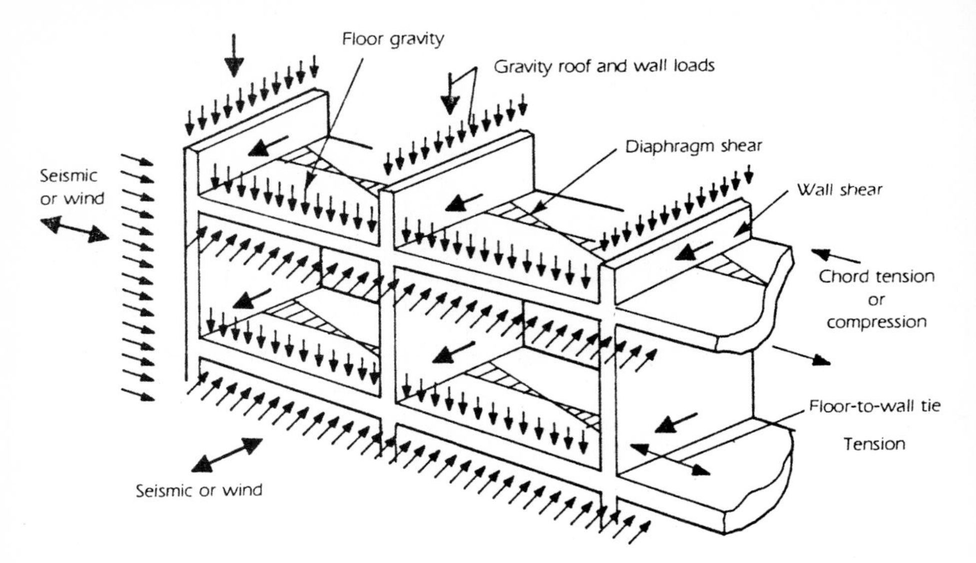


**Figure 35.22** Forces relative to floor-wall intersection.

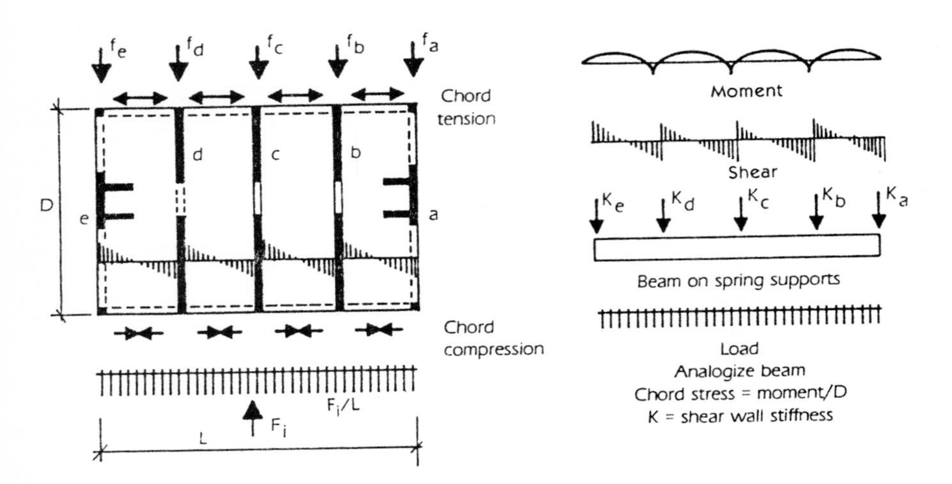


**Figure 35.23** Diaphragm forces.

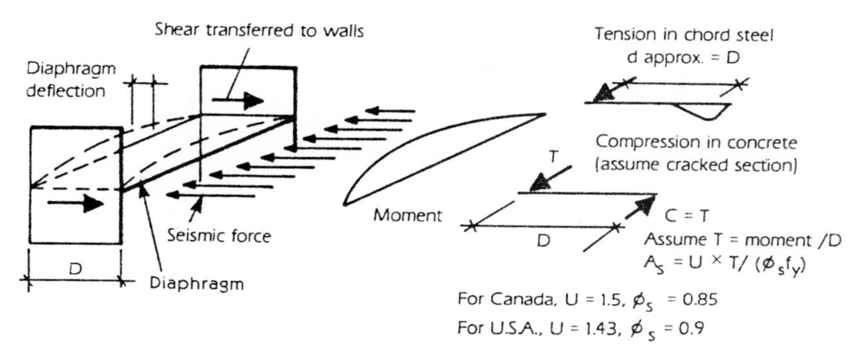


**Figure 35.24** Chord forces. The chord steel may be placed in the wall or, for the precast concrete system, in the floor system topping adjacent to the wall.

Horizontal roof truss systems can be used where large skylight areas occur. The members may be concrete or steel. If the building system is concrete, the truss members will also be concrete. (See Fig. 35.25.)

In an earthquake motion is transferred from the ground up into the building, and the forces generated by this motion must flow back to the ground. Therefore, the load path must be as direct as possible. Once the forces from the diaphragms enter the walls, they should remain in the walls until the footing is reached. If discontinuity occurs, an adequate transfer system must be created to transfer the force to another lateral-load-resisting element. In an earthquake the forces will try to reach the ground by the most direct route.

## Moment Frame and Braced Frame Systems

Moment frame systems for lateral forces can be used with cast-in-place concrete, precast concrete, and steel frame systems. The braced frame system can be used with cast-in-place concrete but is most common with steel frame systems.Precast concrete is seldom used for moment frame systems. (See Figs. 35.26, 35. 27, and 35.28.)

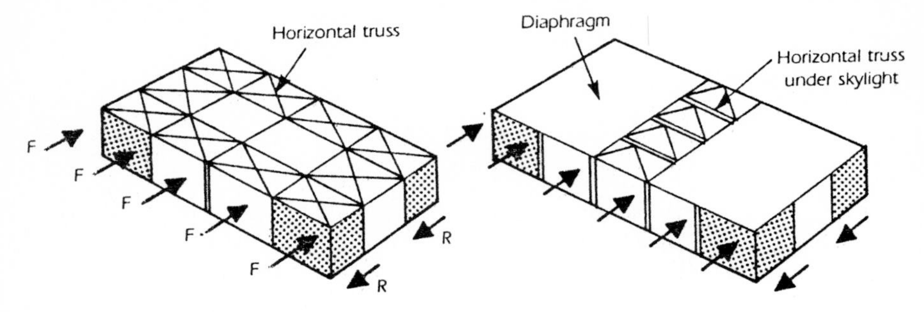


**Figure 35.25** Horizontal truss systems for lateral forces.

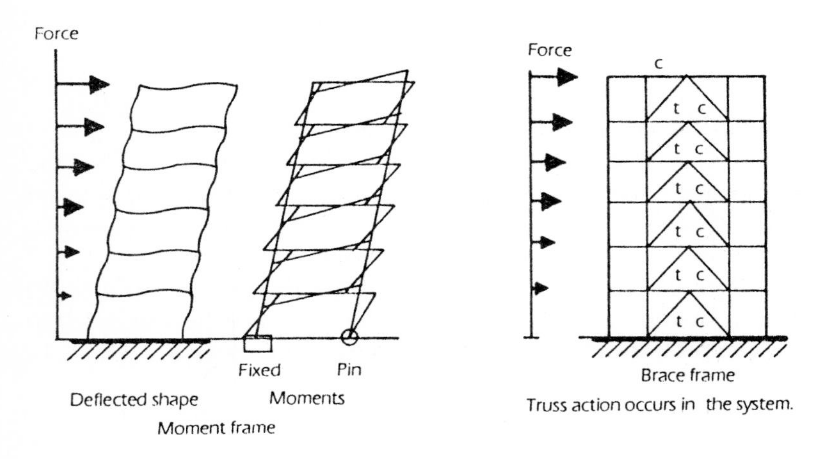


**Figure 35.26** Moment frame and braced frame response.

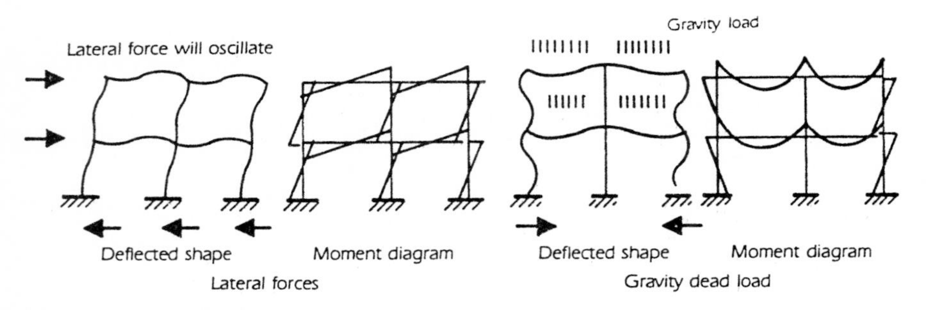


**Figure 35.27** Moment frame response.

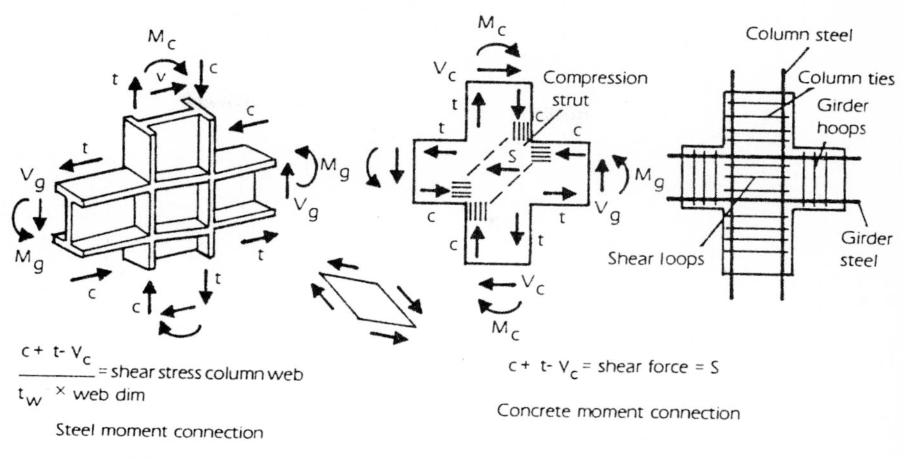


**Figure 35.28** Moment connections.

For moment frames seismic energy is dissipated in the flexing of column-girder connections, in the elastic range for small earthquakes and in the inelastic range for large earthquakes. In design the concept is to have a strong column and a weak girder; that is, the column is to be in the elastic range and the girder in the inelastic range. The sum of column moment capacity must be 1.2 times the sum of girder moment capacity at the connection in order to account for flexural overstrength capacity of the girder reinforcing. (See Fig. 35.29.)

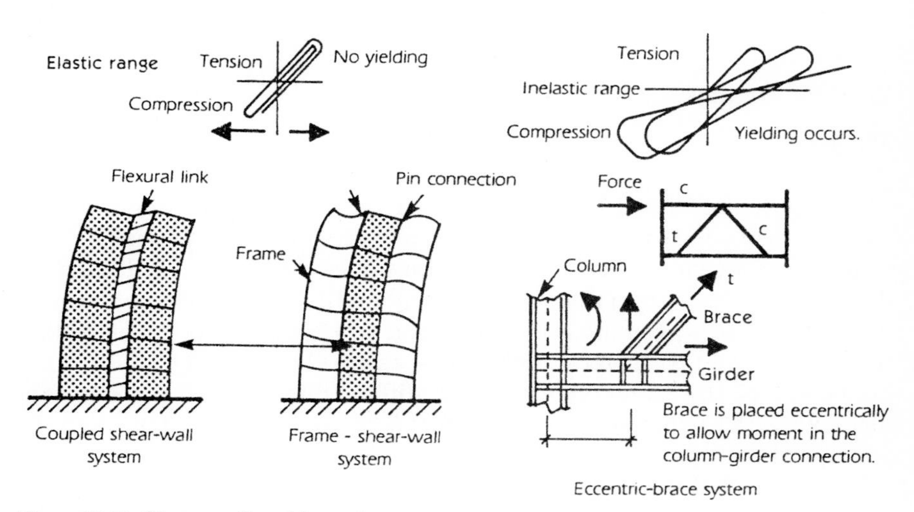


**Figure 35.29** Shear wall and braced systems.

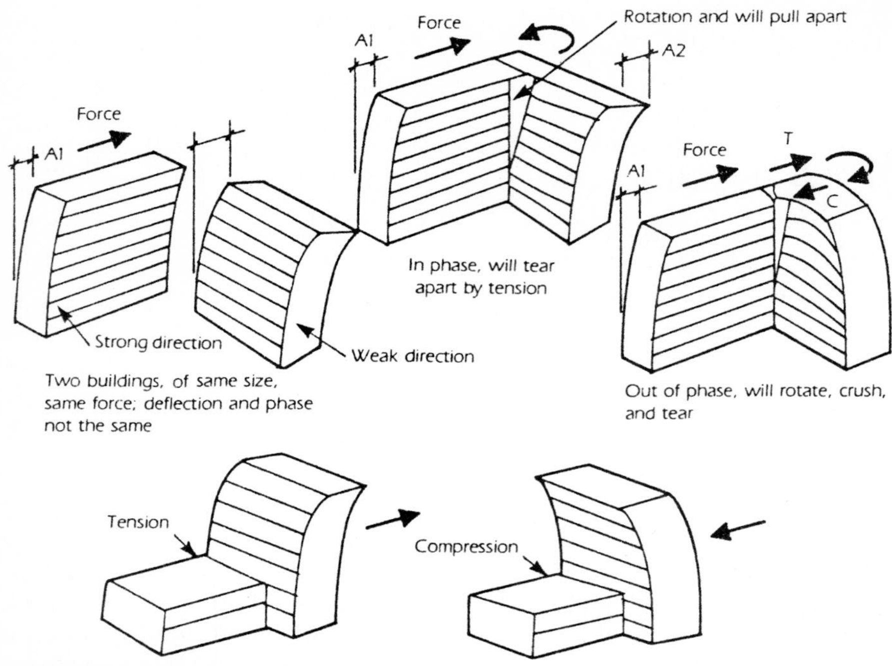


**Figure 35.30** Configuration effects.

## Building Shapes

The configuration of a building is an important consideration in seismic design. Displacement is dependent on both the magnitude of the earthquake and the stiffness of the structure. If the building is not a symmetrical cube or cylinder, rotation will occur because of wings. The stiffness of the two wings will not be the same; one element will tend to deflect differently than the other and will oscillate in a different phase. To overcome this problem, separation and slip joints should be provided between the wings. Several problem areas are shown in Fig. 35.30. The deflection of the taller element will differ from that of the shorter element, and the phase also will differ. The elements will tend to pull apart and then push together, causing both tearing and crushing.

Separation between adjacent buildings is very important. If buildings are too close together, hammering (battering) can occur. Since the period of earthquakes is unknown, the range of displacements for the natural frequency of each mode of vibration must be checked. The Mexico City earthquake of 1985 had a period of 2 seconds in parts of the city owing to soil conditions. This caused increased displacements and hence a greater possibility of hammering between adjacent buildings. Considerable damage during this earthquake was due to hammering. (See Fig. 35.31.)

For shear wall buildings in which the roof or floor levels are not the same, one level or diaphragm will distort differently than the other, which will cause additional out-of-plane forces on intersecting walls. These forces must be considered in the design of the walls. It may be necessary to provide slip joints if the diaphragm deflection and hence the force are excessive or to design the buildings as two separate buildings.

A "soft story" occurs when the stiffness of a story is not the same as that of the story above or the story below. It can also occur where shear walls are discontinuous or the load path is interrupted. This usually occurs at the first floor, where a greater number of openings is needed owing to the occupancy requirements of the building, and also where greater story height is required. However, it can occur in high-rise buildings at equipment floors. The effects of a soft story can be offset by stiffening the column or support system at that floor. The columns of a soft story are subjected to large shear and bending stresses, which often lead to failure. (See Fig. 35.32.)

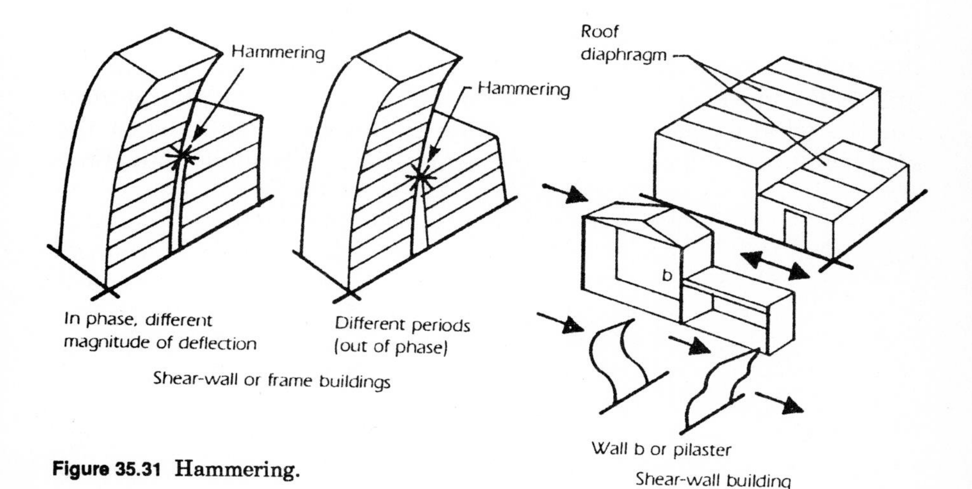


**Figure 35.31** Hammering.

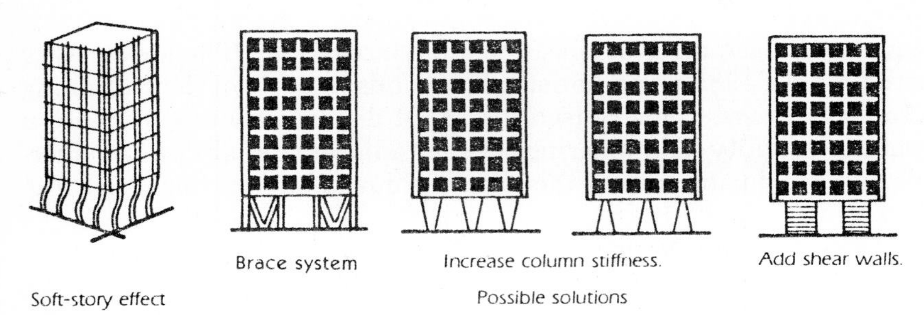


**Figure 35.32** Soft story.

## Liquefaction

Liquefaction can occur with sandy and cohesionless soils and with a few types of clay soils which are highly saturated (thixotropic). As granular soils are vibrated, the pore pressure which supports the particles escapes to the surface and the particles settle into a more compact mass. As the water blows to the surface, finer sand particles also flow up, creating sand boils. Buildings located on soil when liquefaction occurs will settle vertically, and if the settling is uneven, they will tilt and even turn over. This occurred in the Niigata, Japan, earthquake of 1964 and in the Alaska earthquake of 1964. If liquefaction occurs with sloping ground, slides can also occur, as in the Alaska earthquake. In the San Fernando earthquake of 1971 liquefaction occurred in some areas, but little building damage was associated with it. (See Fig. 35.33.)

## Response of Equipment and Furnishings

The stiffer the structure, the greater the shock or force on the buildng but the smaller the force on electrical and mechanical equipment and furnishings. The more flexible the building structural system, the greater the movement and the smaller the shock or force on the building but the greater the force on building equipment. The force on the equipment is the mass of the equipment × the acceleration of the floor

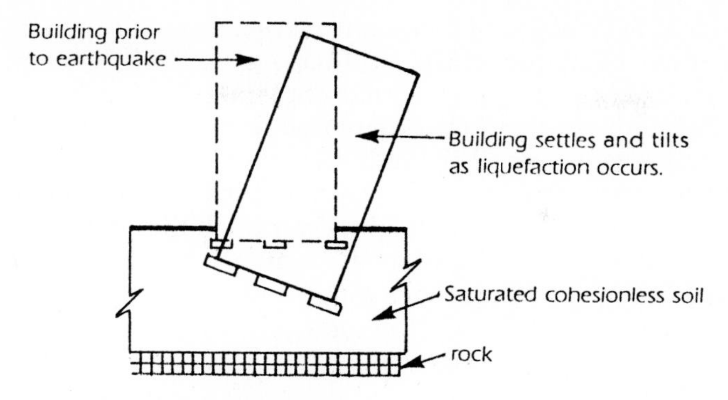


**Figure 35.33** Effects of liquefaction.

or roof system to which the equipment is attached. If the equipment is pin-connected, it will oscillate to its own period. Most building codes will give a factor × the weight of the equipment. The factor is the assumed acceleration divided by the acceleration of gravity. (See Fig. 35.34.)

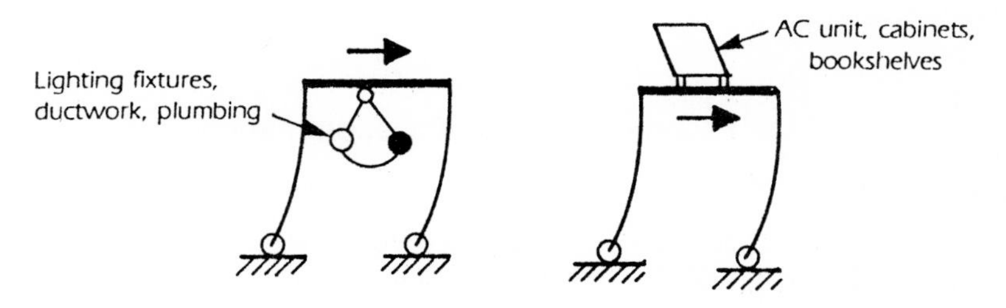


**Figure 35.34** Building equipment response.

## Building Codes

The response of a building subjected to seismic motion depends upon its structural properties, including mass, stiffness, structural system, strength of the materials, damping, period of the building, load path, and characteristics of ground motion including period, intensity, soil type, and direction. Building codes usually include a static equivalent method for design which takes these factors into account. If the building is of an irregular shape or exceeds a specified height, then the building official or the code may require a dynamic analysis in addition to the static equivalent method. Since building codes differ and also are updated periodically, the designer must consult the code governing the location where the project is to be constructed.

It is impossible to cover all the various building codes or the overlapping of codes in any one document. Here is presented a brief summary of the requirements of three codes or recommendations. These are the 1988 Uniform Building Code (UBC) *Seismic Design Requirements*, the National Earthquake Hazards Reduction Program (NEHRP) *Recommended Provisions for the Development of Seismic Regulations for New Buildings* by the Building Seismic Safety Council (BSSC) for the Federal Emergency Management Agency (FEMA) of the United States government, and the National Building Code of Canada. The reader is directed to these codes for specific information.

## Brief Summary of the Proposed 1988 UBC Recommendations

### Static lateral force method

In this method the seismic response of the structure is considered as if a static force is applied at each of the floor levels and at the roof level. This seismic force may come in any direction but may be considered to come from each of the principal axis directions.

Regular structures under 240 ft in height may be designed by the static lateral force method or by both the dynamic method and the static lateral force method. Irregular structures may be designed by the static lateral force method if (1) in seismic Zone 1, (2) in seismic Zone 2 if of special or standard occupancy, and (3) not more than 50 ft or five stories in height.

### Dynamic analysis

A dynamic lateral force procedure is required for (1) all buildings over 240 ft in height, (2) structures over five stories or 50 ft in height in seismic Zones 3 and 4 not having the same structural system throughout, and (3) structures over two stories or 30 ft in height if the weaker story has less than 65 percent of the strength of the story above it.

With the dynamic method, modal analysis using response spectra for the specific site or those given in the code for the soil classification can be employed. A ground motion time-history method for approved past earthquake records may also be used in conjunction with a modal analysis employing the response spectrum for the specific site.

### Static lateral force procedure

The base shear equation is

$$V = ZICW/R_w \quad C = 1.25S/T^{2/3}$$

where $Z$ = zone coefficient related to ground velocity (The zone is obtained from a zone map of the United States. For Zone 1, $Z = 0.1$; for Zone 2, $Z = 0.2$; for Zone 3, $Z = 0.3$; and for Zone 4, $Z = 0.4$. Note that this is similar to the BSSC method, in which the values are the effective peak acceleration.)

$I$ = importance factor (Important or essential occupancy = 1.2;
special and standard occupancy = 1.0.)

$C$ = coefficient relative to soil interaction and period of the building

$W$ = dead load of the building plus 10 lb/ft$^2$ for movable parti tions (For storage buildings add 25 percent of live load.)

$R_w$ = coefficient relative to system response

$S$ = site coefficient for soil conditions

$T$ = fundamental period of vibration = 1/frequency of first mode

$C$ need not exceed 2.75.

$$T = C_t h_n^{3/4}$$

where $C_t$ = 0.035 for steel moment frame, 0.03 for concrete frame and steel eccentric-braced frame, and 0.02 or $0.2/A_c^2$ for all other buildings

$h_n$ = height of structure above base $n$, ft

$$A_c = \Sigma D_e A_e / D_m$$

where $A_c$ = combined effective area of shear wall

$D_e$ = length of shear wall, ft

$A_e$ = effective horizontal cross-sectional area of shear wall, ft²

$D_m$ = length of longest shear wall, ft

The period can be obtained by using Rayleigh's equation.

$$T = 2\pi\left( \Sigma w_i \Delta_i^2 / g \Sigma f_i \Delta_i \right)^{0.5}$$

where $w_i$ = story weight at story $i$

$f_i$ = story force (any story force obtained from distribution rela tive to equation in subsection "Vertical distribution of sei smic force")

$\pi$ = 3.1416

$g$ = acceleration of gravity = 384 in/s², or 9.81 m/s²

When using the Rayleigh equation the value of T shall not be less than 80 percent of that obtained from the equation above for the period.

$R_w$ values are shown in Table 35.2 and $S$ factors in Table 35.3.

TABLE 35.2 $R_w$ Values

| Basic structural system | Lateral-load-resisting system | $R_w$ | Height limit, ft |
|---|---|---|---|
| A. Bearing wall system | 1. Light frame walls with shear panels | | |
| | *a*. Plywood walls | 8 | 65 |
| | *b*. All other light frame walls | 6 | 65 |
| | 2. Shear walls | | |
| | *a*. Concrete | 6 | 160 |
| | *b*. Masonry | 6 | 120 |
| | 3. Light steel frame with tension bracing | 4 | 65 |
| | 4. Braced frame with brace carrying gravity loads | | |
| | *a*. Steel | 8 | 160 |
| | *b*. Concrete (not permitted in Zones 3 and 4) | 4 | |
| | *c*. Heavy timber | 4 | 65 |
| B. Building frame system | 1. Steel eccentric-braced frame | 10 | 240 |
| | 2. Light frame wall with shear panel | | |
| | *a*. Plywood | 9 | 65 |
| | *b*. All other light frame walls | 7 | 65 |
| | 3. Shear walls | | |
| | *a*. Concrete | 8 | 240 |

**TABLE 35.2 (continued) $R_w$ Values**

| Basic structural system | Lateral-load-resisting system | | |
|---|---|---|---|
| | *b*. Masonry | 8 | 160 |
| | 4. Concentric-braced frames | | |
| | *a*. Steel | 8 | 160 |
| | *b*. Concrete (not permitted in Zones 3 and 4) | 8 | |
| C. Moment-resisting frame systems | 1. Special moment-resisting frames | | |
| | *a*. Steel | 12 | No limit |
| | *b*. Concrete | 12 | No limit |
| | 2. Concrete intermediate moment space frame (not permitted in Zones 3 and 4), exceptions | 7 | |
| | 3. Ordinary moment-resisting space frames | | |
| | *a*. Steel | 6 | 160 |
| | *b*. Concrete (not permitted in Zones 3 and 4 | 5 | |
| D. Dual systems | 1. Shear walls | | |
| | *a*. Concrete with special moment-resisting frame | 12 | No limit |
| | *b*. Concrete with intermediate moment frame | 9 | 160 |
| | *c*. Masonry with special moment-resisting frame | 8 | 160 |
| | 2. Steel with eccentric-braced frame and special moment-resisting space frame | 12 | No limit |
| | 3. Concentric-braced frame | | |
| | *a*. Steel with steel special moment-resisting frame | 10 | No limit |
| | *b*. Concrete with concrete special resistance frame (not permitted in Zones 3 and 4) | 9 | |
| | *c*. Concrete with intermediate moment-resisting space frame | 6 | |

**TABLE 35.3 Site Coefficients: S factor Due to Soil Conditions**

| Soil type | Description | S factor |
|---|---|---|
| *S*1 | Rocklike material characterized by a shear wave velocity greater than 2500 ft/s or by other suitable means of classification; or by a stiff or dense soil condition where the soil depth is less than 200 ft | 1.0 |
| *S*2 | Soil material with a dense or stiff soil condition where the soil depth exceeds 200 ft | 1.2 |
| *S*3 | Soil profile 40 ft or more in depth and containing more than 20 ft of soft to medium stiff clay but not more than 40 ft of soft clay | 1.5 |
| *S*4 | Soil profile containing more than 40 ft of soft clay | 2.0 |

The soil profile is established by geotechnical data for the specific building site. If the properties are unknown, use soil type *S*3 or *S*4 subject to approval by the building official.

### Vertical distribution of seismic force

$$V = F_t + \Sigma F_i \text{ from } i = 1 \text{ to } n$$

where $n$ = number of top floor
$F_t$ = concentrated force to be applied at top in addition to $F_n$ (need not exceed $0.25V$ and is 0 when $T < 0.7$ s)

$$F_t = 0.07TV$$

$$F_x = [(V - F_t)(w_x h_x)] / \Sigma w_i h_i$$

where $F_x$ = force to be applied at level $x$. At each level designated as $x$ the force $F_x$ shall be applied over the area of the floor in accordance with the mass distribution of that floor.

The design story shears $V_x$ at any floor are the sum of $F_x + F_t$ above that floor. $V_x$ shall be distributed to the various elements of the vertical-lateral-force-resistance system relative to the rigidities of the element, considering also the rigidity of the diaphragm.

### Horizontal torsional moments

Provision shall be made for additional shears resulting from horizontal torsion where the diaphragms are not flexible. (A flexible diaphragm is one in which diaphragm deflection is twice that of the average story drift i.e., wood.) The torsional moment is the story force × the distance between the center of mass and the center of rigidity. The minimum torsion moment is 5 percent × story mass × the building dimension perpendicular to the direction of force under consideration.

### Overturning

Every structure shall be designed to resist moment overturning caused by seismic forces. For shear wall buildings the magnitude of the seismic force is determined by the rigidity of the element. The overturning effects on the element shall be carried down to the foundation. In computing the overturning the gravity dead load shall be multiplied by 0.85 when combined with seismic forces. (See Fig. 35.35.)

### Story drift

Story drift shall not exceed $0.03/R_w$ or $0.004 \times$ the story height.

### Lateral forces on elements of the structure and nonstructural components

Parts and portions of the structure, permanent nonstructural components and their attachments, and attachments for equipment supported

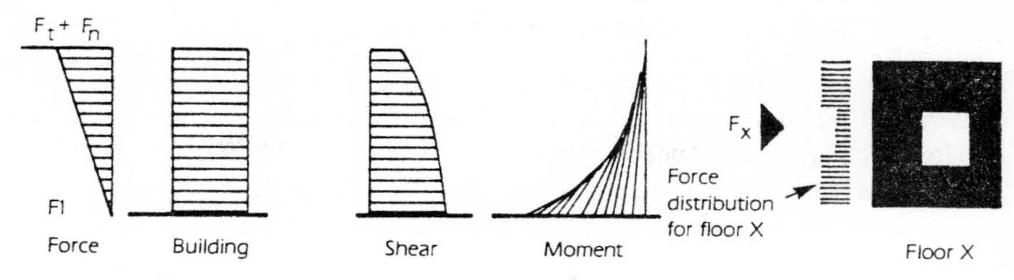


**Figure 35.35** Overturning.

by the structure must be designed to resist lateral forces. The minimum design force is

$$F_p = ZIC_pW_p$$

where $Z$ = value given under "Static lateral force procedure"
$I$ = value given under "Static lateral force procedure"
$W_p$ = weight of element
$C_p$ = coefficient shown in Table 35.4

**TABLE 35.4 Horizontal Force Factors, $C_p$**

| Element | $C_p$ |
|---|---|
| Part or portion of structure | |
| 1. Walls | |
| *a*. Unbraced cantilevered parapets | 2.0 |
| *b*. Other exterior walls aboveground | 0.75 |
| *c*. Interior bearing and nonbearing walls and partitions | 0.75 |
| *d*. Masonry or concrete fences over 6 ft high | 0.75 |
| 2. Penthouses | |
| 3. Connections for prefabricated structural elements other than walls | 0.75 |
| Nonstructural components | |
| 1. Exterior and interior ornamentation and appendages | 2.0 |
| 2. Chimneys, stacks, truss towers, and tanks on legs | |
| *a*. Supported as unbraced cantilever above the roof | 2.0 |
| *b*. All others | 0.75 |
| 3. Signs and billboards | 2.0 |
| 4. Mechanical, plumbing, and electrical equipment and machinery | 0.75 |
| 5. Tanks plus contents | |
| 6. Storage racks | 0.75 |
| 7. Anchorage for permanent floor-supported cabinets and bookstacks more than 5 ft in height | 0.75 |
| 8. Anchorage for suspended ceilings and light fixtures | 0.75 |

Exterior nonstructural wall panels or elements attached to the exterior are to be designed for the force $F_p$ and the following considerations. Connections and panel joints shall allow for a relative movement between stories of $3(R_w/8) \times$ the calculated elastic story drift with a minimum of ½ in. The body of the connector is to be designed for $1.33F_p$. The connector fastener (welds, bolts) is to be designed for $4F_p$.

## BSSC (NEHRP) Federal Recommendations

The Building Seismic Safety Council was created to provide a national recommendation for seismic design of new buildings. Also known as the National Earthquake Hazards Reduction Program (NEHRP), this recommendation sets forth provisions for the development of seismic regulations for new buildings. The provisions were prepared for vote by the BSSC in 1984 after a series of trial designs had been conducted throughout the United States in which many structural engineering firms were given contracts to prepare trial designs using both the BSSC recommendations and local codes. The engineers were also asked to give recommendations regarding approval and suggestions for improvement. The recommendations were updated in 1986 (document FEMA 96/February 1986).

The BSSC recommendations cover both an equivalent lateral force method and a dynamic modal analysis method. A brief summary of the equivalent lateral force procedure is as follows:

### Seismic base shear

The equation for seismic base shear in the direction under consideration is

$$V = C_s W$$

where $V$ = base shear
$C_s$ = seismic coefficient
$W$ = total gravity weight of building including permanent partitions and equipment (For storage and warehouse buildings include 25 percent of live load.)

### Seismic coefficient

$$C_s = 1.2 A_v S / R T^{2/3}$$

where $A_v$ = coefficient of effective peak velocity–related acceleration obtained from a seismic map of the United States
$S$ = coefficient for soil profile
$R$ = response modification factor relative to building system (see Table 35.7)
$T$ = fundamental period of building

$C_s$ need not be greater than the value determined by the following equations. If the period of the building is not calculated, the value of $C_s$ is

$$C_s = 2.5 A_a / R$$

where $A_a$ = effective peak acceleration obtained from a map of the United States. For soil profile type $S3$ in areas where $A_a \geq 0.3$, $C_s$ shall be determined from the equation

$$C_s = 2A_a/R$$

**Period determination**

The fundamental period of the building can be determined on the basis of the properties of the structural system by using established methods of mechanical analysis, that is, Rayleigh's equation, eigenvalue analysis, etc. However, the fundamental period $T$ shall not exceed $C_a T_a$. $C_a$ is given in Table 35.5.

$$T_a = C_t h_n^{3/4}$$

where $C_t = 0.035$ for steel frame buildings and 0.03 for concrete frames
$h_n$ = building height above base

For all other buildings $T_a = 0.05h_n/L^{0.5}$, where $L$ = overall length of the building at the base in the direction under consideration.

**TABLE 35.5 Coefficient for Upper Limit on Calculated Period**

| $A_v$ | $C_a$ | $A_v$ | $C_a$ |
|---|---|---|---|
| 0.4 | 1.2 | 0.15 | 1.5 |
| 0.3 | 1.3 | 0.1 and | 1.7 |
| 0.2 | 1.4 | 0.05 | |

**Vertical distribution of seismic forces**

The seismic shear force of any level $x = F_x$ where

$$F_x = C_{vx} V$$

where $C_{vx} = w_x h_x^k / \Sigma w_i h_i^k$

$k$ = 1.0 for buildings with period of 0.5 s or less and 2.0 for buildings with a period of 2.5 s, linear interpolated between $k$ = 1.0 and $k$ = 2.0

$w_i$, $w_x$ = $w$ assigned to level $i$ or $x$

$h_i$, $h_x$ = height above level $i$ or $x$

**Horizontal shear distribution and torsion**

The seismic shear force at any level is determined by

$$V_x = \Sigma F_i$$

where $i = x$ to $n$. The torsional moment $M_t$ = building mass × distance from the center of mass to the center of rigidity plus the torsional moment $M_{t,\text{ accidental}}$ caused by an assumed displacement of the building mass each way from its actual location by a distance of 5 percent of the building dimension perpendicular to the direction of the applied force.

### Overturning

At any level the increment of overturning moment in the story under consideration shall be distributed to the various walls or frames in proportion to their rigidity in the same way as for horizontal shear.

$$M_x = k\, \Sigma F_i (h_i - h_x)$$

where $i = x$ to $n$

$k$ =1.0 for top 10 stories and 0.8 for the twentieth story from top and below, linear interpolated between 10 and 20 stories for $k$

For foundation overturning use $k$ = 0.75 for all building heights.

### Story drift

Story drift is the difference between $\Delta_x$ at the top and bottom of a story. It shall be computed by

$$\alpha_x = C_d\, \alpha_{xe}$$

where $C_d$ = deflection amplification from Table 35.7

$\alpha_x$ = deflection at level $x$

$\alpha_{xe}$ = deflection determined by elastic analysis

Soil profile coefficients are shown in Table 35.6 and $R$ values in Table 35.7.

**TABLE 35.6 Soil Profile Coefficients***

| Soil type | Description | Coefficient |
|---|---|---|
| $S1$ | Rock of any characteristic, either shalelike or crystalline in nature; shear wave velocity greater than 2500 ft/s or stiff soil where soil depth is less than 200 ft and the soil types overlying rock are stable deposits of sand, gravels, or stiff clays. | 1.0 |
| $S2$ | Deep cohesionless or stiff clay conditions, including sites where soil depths exceed 200 ft and the soil types overlying rock are stable deposits of sands, gravels, or stiff clays. | 1.2 |
| $S3$ | Profile with soft to medium stiff clays and sands, characterized by 30 ft or more of soft to medium clays without intervening layers of sand or other cohesionless soils. | 1.5 |

* If the soil profile is unknown, use $S2$ or $S3$, whichever gives the higher value of the seismic coefficient $C_s$.

**TABLE 35.7 *R* Values : Response Modification Coefficients (Condensed)**

| Type of structural system | Vertical resisting system | $R$ | $C_d$ |
|---|---|---|---|
| Bearing-wall building with shear wall or frame for lateral force resistance | Light frame shear panels | 6.5 | 4 |
| | Concrete shear walls | 4.5 | 4 |
| | Reinforced-masonry shear walls | 3.5 | 3 |
| | Braced frames | 4.0 | 3.5 |
| | Unreinforced masonry | 1.25 | 1.25 |
| Building frame for vertical loads and shear wall or brace for lateral force resistance | Light frame with shear panels | 7 | 4.5 |
| | Concrete shear walls | 5.5 | 5 |
| | Reinforced-masonry shear walls | 4.5 | 4 |
| | Braced frames | 5 | 4.5 |
| | Unreinforced-masonry shear walls | 1.5 | 1.5 |
| Moment-resisting frame for vertical and lateral forces | Special moment frame: steel | 8 | 5.5 |
| | Special frame: reinforced concrete | 8 | 5.5 |
| | Ordinary frame: steel | 4.5 | 4 |
| | Ordinary frame: reinforced concrete | 2 | 2 |
| | Intermediate frame of concrete | 4 | 3.5 |
| Dual system, frame 100 percent of vertical and 25 percent of lateral forces; combined frame, shear walls, braced system 100 percent of lateral forces | Reinforced-concrete shear walls | 8 | 6.5 |
| | Reinforced masonry | 6.5 | 5.5 |
| | Wood-sheathed shear walls | 8 | 5 |
| | Braced frame | 6 | 5 |
| Intermediate moment frame of concrete or steel in combination with above | Reinforced-concrete shear walls | 6 | 5 |
| | Reinforced-masonry shear walls | 6 | 5 |
| | Wood shear panels | 7 | 4.5 |
| | Braced frames | 5 | 4.5 |
| Inverted-pendulum structures | Special steel moment frame | 2.5 | 2.5 |
| | Special concrete moment frame | 2.5 | 2.5 |
| | Ordinary steel moment frame | 1.25 | 1.25 |

**TABLE 35.8 Seismic Coefficients and Performance Levels (Condensed)**

| | | Hazard Level | | |
|---|---|---|---|---|
| Architectural components | $C_c$ factor | III | II | I |
| Appendages | 0.9 | *S* | *G* | *L* |
| Exterior and nonbearing walls | 3.0 | *S* | *G* | *L* |
| Wall attachments | 6.0 | | | |
| Connector fasteners | 3.0 | *G* | *G* | *L* |
| Veneer attachments | 0.6 | *G* | *G* | Not required |
| Roofing units | | | | |
| Partitions | | | | |
| Stairs and shafts | 1.5 | *S* | *G* | *G* |
| Elevator shafts | 1.5 | *S* | *L* | *L* |
| Vertical shafts | 0.9 | *S* | *L* | *L* |
| Horizontal exits including ceilings | 0.9 | *S* | *S* | *G* |
| Public corridors | 0.9 | *S* | *G* | *L* |
| Private corridors | 0.6 | *S* | *L* | Not required |
| Ceilings | | | | |

**TABLE 35.8 Seismic Coefficients and Performance Levels (Condensed)(*continued*)**

| Architectural components | $C_c$ factor | Hazard Level III | II | I |
|---|---|---|---|---|
| Fire-rated | 0.9 | *S* | *G* | *G* |
| Non-fire-rated | 0.6 | *G* | *G* | *L* |
| Architectural equipment (cabinets, etc.) | 0.9 | *S* | *G* | *L* |

### Architectural systems and components and their attachments

These systems, components, and attachments must be designed to resist a minimum seismic force of

$$F_p = A_v C_c P W_c$$

where $A_v$ = effective peak velocity obtained from the seismic map of the United States
$C_c$ = seismic coefficient for component obtained from Table 35.8
$P$ = performance factor as given below
$W_c$ = weight of component

The seismic hazard groups are as follows:

III Facilities necessary for postearthquake recovery
II Buildings with a large number of occupants in which mobility is impaired
I All other buildings

Performance characteristic levels, obtained from the $C_c$ column in Table 35.8 relative to the hazard group, are as follows:

| | | *P* |
|---|---|---|
| *S* | Superior | 1.5 |
| *G* | Good | 1.0 |
| *L* | Low | 0.5 |

### National Building Code of Canada: Summary of Seismic Design Requirements

Information from the National Building Code of Canada is included since Canada uses the metric system whereas the United States is currently designing with the old British system of units. The following is condensed from the code; hence the designer should refer directly to the latest version of the Canadian code and the building code requirements of the province where the building is to be constructed.

The code recommends a dynamic analysis for tall or unusual buildings with irregular layouts, large setbacks, or a torsion eccentric distance greater than 25 percent of the building width, and critical industrial buildings. If the period of the building is greater than 1 second, it is usually beneficial to have a dynamic analysis. The dynamic analysis required is a modal analysis using a response spectrum for the site with a probable earthquake exceeding 10 percent in 50 years. If the dynamic analysis determines a base shear less than that determined by a static equivalent method of the code, the lower value can be used, except that the lower value should not be less than 90 percent of the value determined by the static equivalent method and that the story forces, shears, and overturning moment be proportioned accordingly.

#### Static equivalent method

A brief summary of the static equivalent method outlined in the code is as follows:

$$\text{Base shear } V = vSKIFW$$

where $v$ = zone velocity ratio (effective peak velocity) as indicated on the zone map of Canada

$S$ = seismic response factor based on period of the building and the ratio $Z_a/Z_v$ for the site

$Z_a$ = acceleration zone as indicated on the zone map

$Z_v$ = velocity zone as indicated on the zone map

$K$ = coefficient relative to structural system

$I$ = importance factor

$F$ = foundation factor

$W$ = dead load of building plus 25 percent of snow load and, if a storage building, 60 percent of storage load

The period $T$ is 1/fundamental frequency of the first mode of vibration. It may be determined from an eigenvalue solution, by using the Rayleigh method, by any number of available computer programs, or by the equation

$$T = 0.09\, h_n / d_s^{0.5}$$

where $h_n$ = height of building,

$d_s$ = dimension of building in direction parallel to applied force, m

The base shear is distributed over the height of the building by

$$F_x = \frac{(V - F_t)\, W_x h_x}{\Sigma w_i h_i}$$

where $F_t$ = 0.0 if $h_{n/}d_s$ < 3.0, but otherwise = $0.004V(h_n/d_{s)}^2$, al though it need not exceed 0.15$V$

$i$ = 1 to $n$

$W_x$ = dead load of floor $x$

$W_i$ = dead load of floor $i$

$h_x$ = height to floor $x$

$h_i$ = height to floor $i$

Seismic response factors are shown in Table 35.9, $K$ values in Table 35.10, importance factors in Table 35.11, and foundation factors in Table 35.12.

**TABLE 35.9 Seismic Response Factor S**

| Period $T$, s | $Z_a/Z_v$ | $S$ |
|---|---|---|
| ≤ 0.25 s | > 1.0 | 0.62 |
| | 1.0 | 0.44 |
| | < 1.0 | 0.31 |
| > 0.25 s but ≤ 0.5 s | > 1.0 | 0.62 - 1.23($T$ - 0.25) |
| | 1.0 | 0.44 = 0.50($T$ - 0.25) |
| | < 1.0 | 0.31 |
| > 0.50 s | All values | $0.22/T^{0.5}$ |

**TABLE 35.10 $K$ Values (Condensed)**

| Case | Resisting elements | Value of $K$ |
|---|---|---|
| 1. | Ductile moment-resisting space frame | 0.7 |
| 2. | Dual system: moment frame to resist vertical and 25 percent of lateral; ductile flexible walls and frame to resist 100 percent of lateral | 0.7 |
| 3. | Dual system: moment frame to resist vertical and 25 percent of lateral; shear walls or steel bracing acting independently to resist 100 percent of lateral | 0.8 |
| 4. | Buildings with ductile flexible walls and buildings with ductile frame systems not covered in cases 1, 2, 3, or 5 | 1.0 |
| 5. | Dual system with ductile moment space frame resisting vertical forces and 25 percent of lateral forces; masonry infill wall system to resist 100 percent of lateral forces | 1.3 |
| 6. | Buildings other than cases 1 ,2, 3, 4, or 5 with reinforced concrete, structural steel, or reinfo rced-masonry shear walls | 1.3 |
| 7. | Buildings of unreinforced masonry | 2.0 |
| 8. | Elevated tanks plus contents on legs not supported by the building | 3.0 |

**TABLE 35.11 Importance Factors**

| Type of occupancy | I |
|---|---|
| Postdisaster buildings and schools | 1.3 |
| Other buildings | 1.0 |

**TABLE 35.12 Foundation Factors**

| Type and depth of soil measured from foundation level | $F$ |
|---|---|
| Rock, dense coarse-grain soils, stiff and hard fine-grain soils, compact coarse-grain and stiff fine-grain soils with a depth of 0 to 15 m | 1.0 |
| Compact coarse-grain, firm, and stiff fine-grain soils with a depth greater than 15 m; loose coarse-grain and soft fine-grain soils from 0 to 15 m | 1.3 |
| Loose coarse-grain and soft fine-grain soils with a depth greater than 15 m | 1.5 |

### Parts of the building and their anchorage

Parts of the building, including non-load-bearing elements, and their anchorage must be designed for a lateral force of

$$V_p = vSW_p$$

where $V$ = velocity from the zone map of Canada
$S_p$ = horizontal force factor for element (see Table 35.13)
$W_p$ = weight or dead load of element

**TABLE 35.13 Values of $S_p$**

| Category | Part or portion of building | Direction of force | Value of $S_p$ |
|---|---|---|---|
| 1. | All exterior and interior walls | Normal to surface | 0.9 |
| 2. | Cantilever parapets and cantilever walls except retaining walls | Normal to surface | 4.4 |
| 3. | Exterior and interior ornamentation | Any direction | 4.4 |
| 4. | Machinery, fixtures, and equipment rigidly connected to the building | Any direction | 0.9 |
| 5. | Towers, chimneys, stacks, and penthouses | Any direction | 1.3 |
| 6. | Tanks plus contents resting on ground within the building | Any direction | 0.9 |
| 7. | Floors and roof acting as diaphragms | Any direction | 0.45 |
| 8. | Connections for exterior and interior walls and elements except those forming part of the structural system | Any direction | 11.0 |

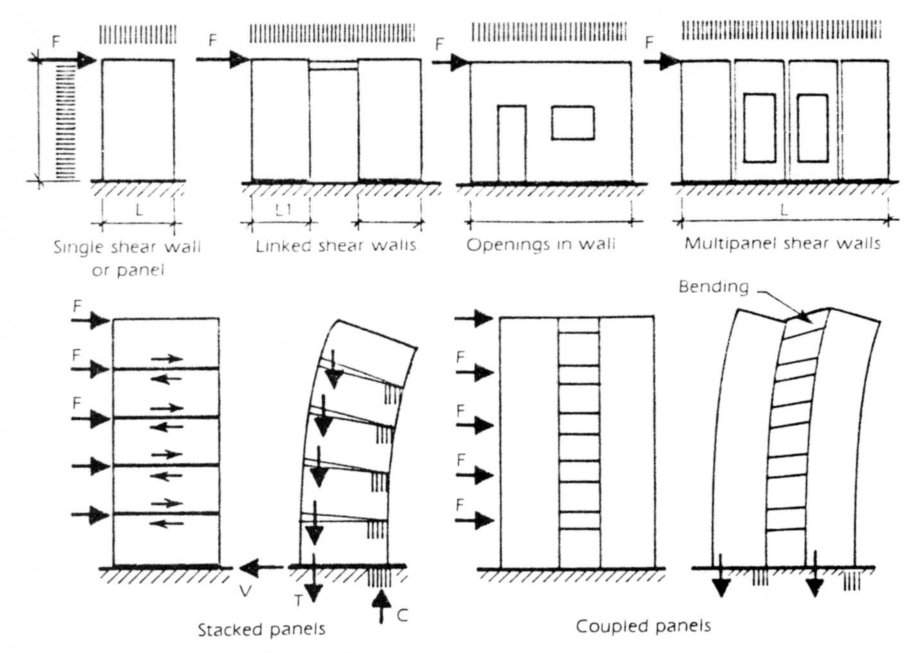


**Figure 35.36** Forces on shear walls.

## Shear Walls: In-Plane Loading

A few of the various types of concrete shear walls are shown in Fig. 35.36.

Piers within a wall which has openings are treated the same as walls fixed at top and bottom. If the wall is rigidly connected to the floor system or the beam system, then it is considered fixed at top and bottom. For a one-story building with its roof system either of wood or of steel, the top is assumed to be pinned; hence the panel is assumed to be fixed at the bottom only. Rigidity is computed for each wall of each floor level. For a multistory precast wall panel where the floor system does not restrain the wall panel against rotation, the panel is assumed to be a cantilever and hence fixed at the bottom only. This is not exact since the topping of the floor system is doweled to the wall panel; however, the topping is not as thick as the wall, and the pinned assumption is normally used.

### Single shear wall or single panel

Shear wall rigidity is shown in Fig. 35.37.

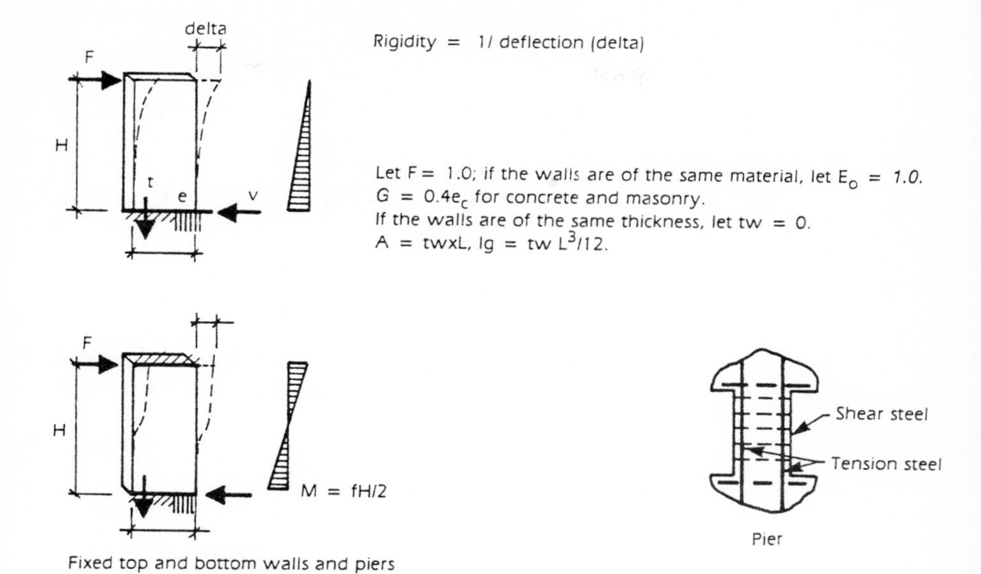


**Figure 35.37** Shear wall rigidity.

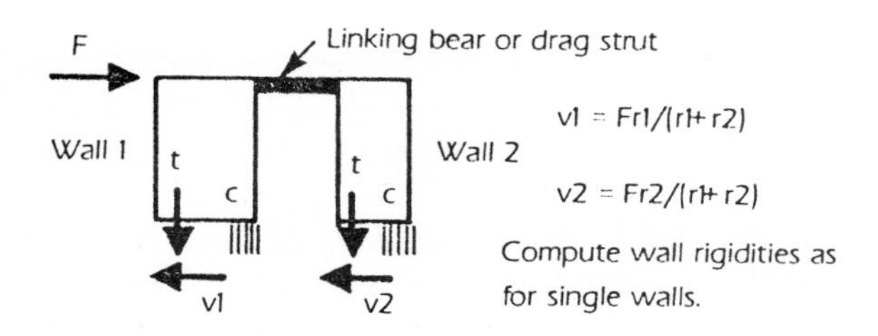


**Figure 35.38** Linked shear walls.

### Linked shear walls

Linked or coupled shear walls may be either pinned at the top or fixed at the top subject to the floor- or roof-to-wall connection. The force carried by the beam (drag strut) is distributed to each shear wall according to the rigidity of that wall. (See Fig. 35.38.)

### Walls with openings

For walls with openings three different methods may be used, subject to the degree of accuracy desired. (See Fig. 35.39.)

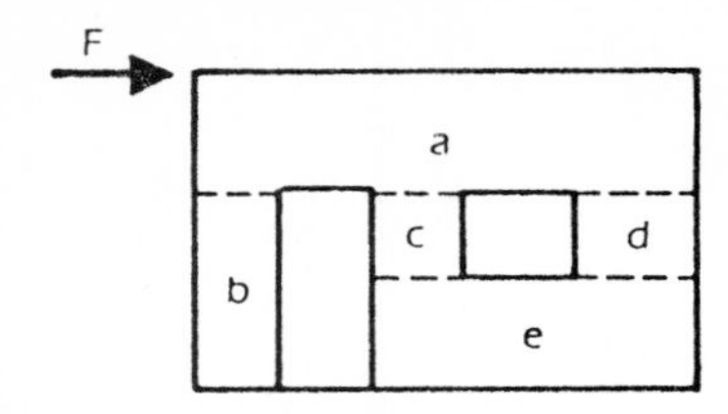


**Figure 35.39** Shear walls with openings.

**Method 1.** Neglect the wall above and below the piers.

$$R_{\text{wall}} = R_b + R_c + R_d$$

Shear force on pier $c$:

$$V_c = FR_c/(R_b + R_c + R_d)$$

Compute $R$ as for a single panel, fixed top and bottom.

**Method 2.** This method considers all parts of the wall. If the wall is fixed at the top, this method is good, but if the wall is pinned at the top, then use method 3.

$$R_{\text{wall}} = 1/\Delta_{\text{wall}}$$

$$\Delta_{\text{wall}} = \Delta_a + \cfrac{1}{\cfrac{1}{\Delta_b} + \cfrac{1}{\Delta_c + \cfrac{1}{\Delta_c + \cfrac{1}{\text{etc.}}}}}$$

Compute $\Delta$ as for a single wall, fixed or pinned as applicable.

**Method 3.** This method is most accurate when the top of the wall is pinned.

1. Compute the wall as if solid and as a cantilever pin at the top = $+\Delta\ (1)$
2. Compute parts $b$, $c$, $d$, and $e$ as if solid and fixed top and bottom = $-\Delta\ (2)$
3. Compute $\Delta_c$, then $R_c$.
4. Compute $\Delta_d$, then $R_d$.
5. Add $R_c$ and $R_d$, then determine $\Delta_{cd}$ $[1/(R_c + R_d)]$.
6. Compute $\Delta_e$, then add to $\Delta_{cd}$.
7. Determine $R_{cde} = 1/\Delta_{e+c+d}$.
8. Determine $\Delta_{ecd} = 1/R_{cde}$ = $+\ (8)$

$$\overline{\Delta_{\text{wall}}}$$

$$R_{\text{wall}} = 1/\Delta_{\text{wall}}$$

After the force on the wall has been obtained, the force to each pier is determined by

$$V_b = F_{\text{wall}} R_b / (R_b + R_{cde})$$
$$V_c = (F_{\text{wall}} - V_b) \times R_c / (R_c + R_d)$$

In method 2,

$$R_{cde} = 1/\{1/\Delta_e + [1/(1/\Delta_c + 1/\Delta_d)]\}$$

**Multipanel shear wall**

See Fig. 35.40. The rigidity of the wall depends on the case for which the wall is designed. For the semirigid ductile connection rigidity may be computed for both the elastic and the plastic ranges.

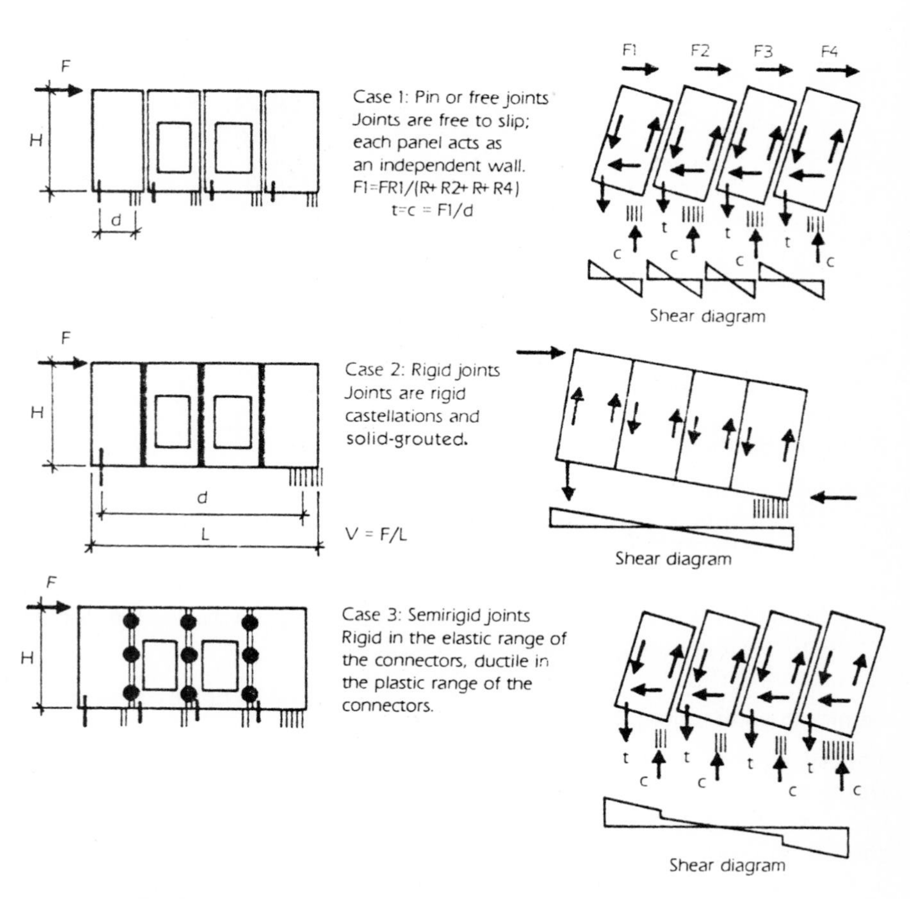


**Figure 35.40** Multipanel shear walls.

### Free or pinned joints

Free or pinned joints are free to move in the vertical direction and offer no resistance in that direction. A pinned joint may be designed to offer resistance in a horizontal direction. For a pinned joint one side of the connecting strap may be welded with the other side bolted, or both sides may be bolted. One side of the strap should be slotted to allow for vertical movement. The bolts may be screwed into ferrule nuts welded to the embedded plate, or a threaded stud may be welded to the plate. The connection is on the inside surface of the panels and would be recessed and dry-packed for protection. (See Fig. 35.41.)

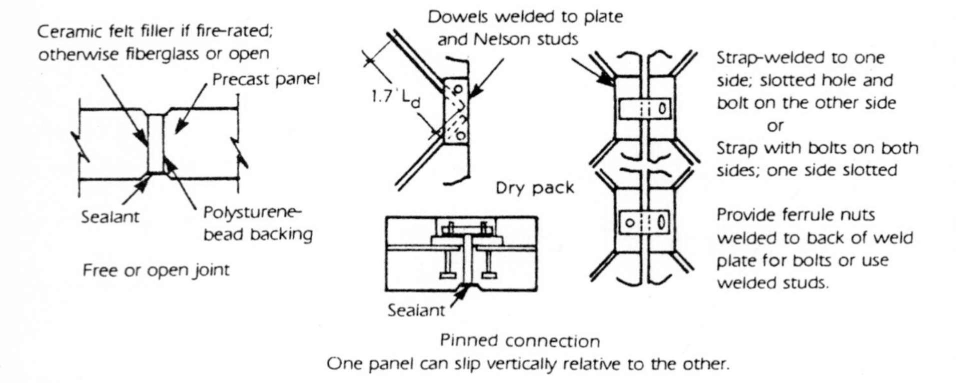


**Figure 35.41** Pinned connections.

### Rigid connections

The most common type of rigid joint is the castellated joint. For this joint the full capacity of the shear wall in shear is obtained. The joint can be evaluated either by shear friction or by truss analogy. When a rigid joint is used, the wall is treated as one wall and not as separate panels. If the panels used to make up the wall contain openings, the wall is treated as one wall with openings. (See Fig. 35.42.)

### Semirigid connections (ductile connections)

This type of connection can be brittle or ductile according to the welding, shape of the connecting plate, and class of reinforcing bars used. If the bars are ASTM A-706 (weldable reinforcing steel), the weld on the bars should have an elastic strength exceeding that of the plate connector bodies. The design concept is that the connecting plate should go into the plastic range prior to the weld plate and the welded reinforcing bars. The connection is designed to be in the elastic range for the probable earthquake but to go into the plastic range for the critical earthquake. To achieve this goal the connecting plate can be punched or shaped to reduce the section. Several possibilities are shown in Fig. 35.43.

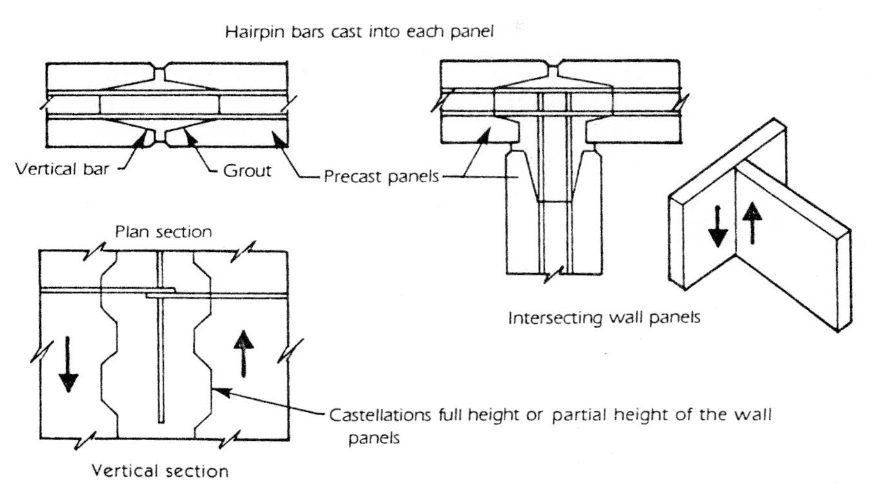


**Figure 35.42** Rigid connections.

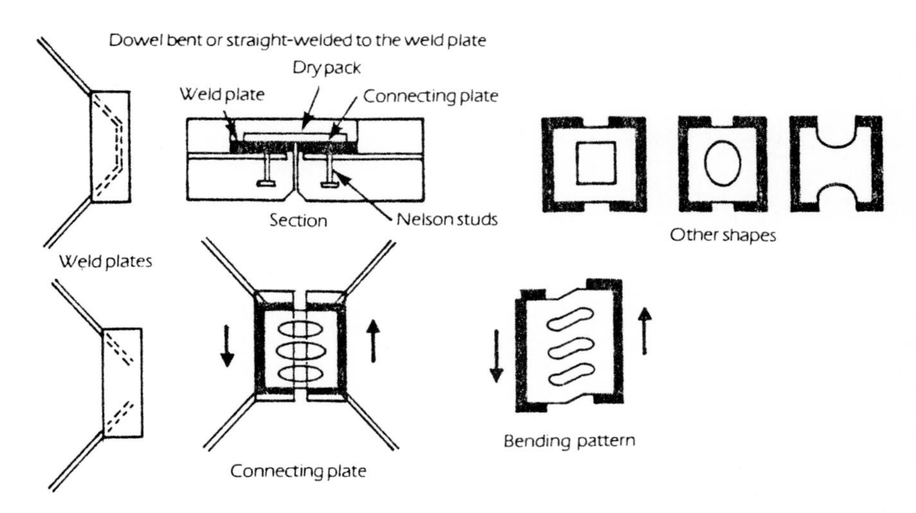


**Figure 35.43** Semirigid connections.

## Reinforced-concrete stacked panels

See Fig. 35.44.

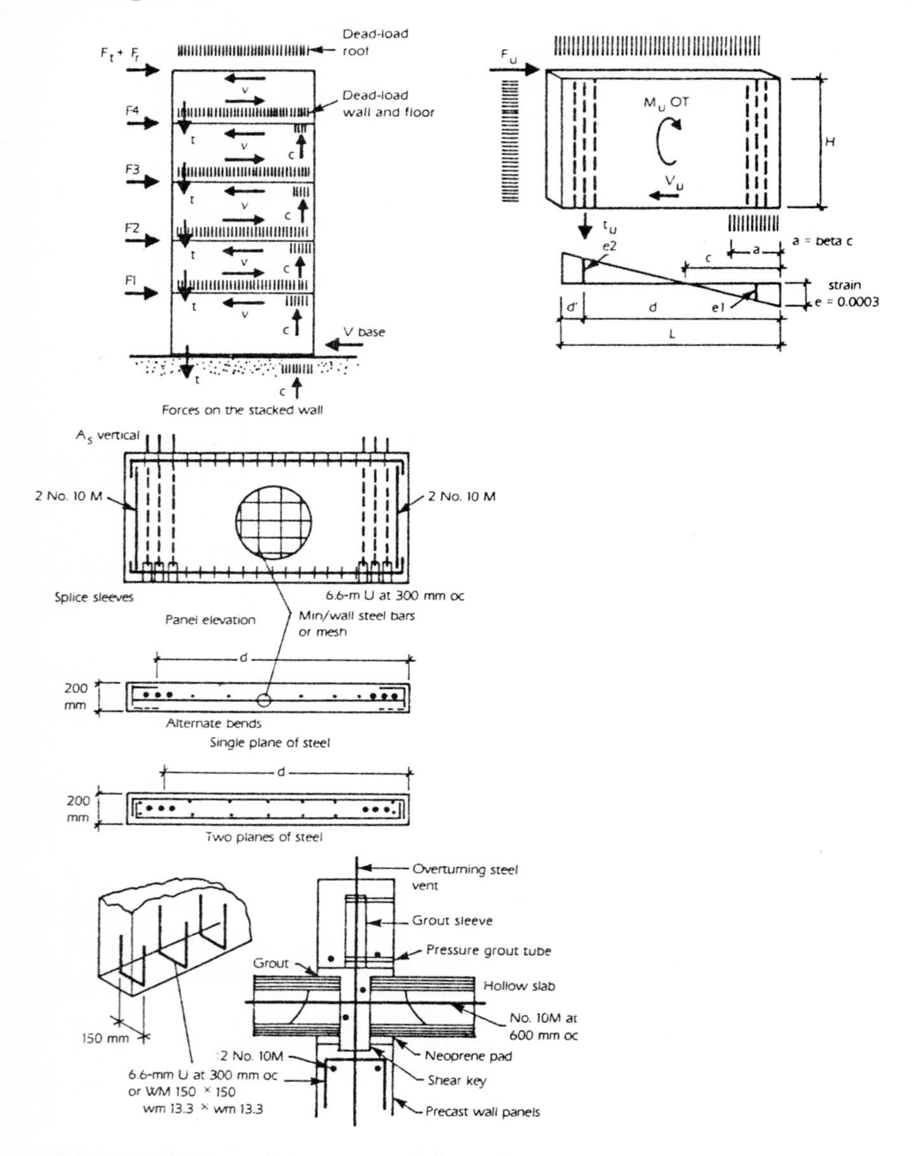


**Figure 35.44** Reinforced-concrete stacked panels.

Minimum-wall steel:

If $V_u < \phi V_c/2$, vertical steel = $0.0012Lt_w$ and horizontal steel = $0.0020Ht_w$.

If $V_u > \phi V_c/2$, vertical steel and horizontal steel = $0.0025bt_w$.
If $V_u > \phi V_c$, design the steel.

$$A_{s,\text{ horizontal}} = (V_u - \phi V_c)S / (\phi_s F_y \times 0.8H)$$

$$\rho_h = A_{s,h} / (t_w \times 0.8H)$$

$$\rho_v = 0.0025 + 0.5 \times (2.5 - H/L) \times (\rho_h - 0.0025)$$

$$\phi_c V_c = \phi_c \times 0.2 f_c'^{0.5} t_w \times 0.8L$$

$$S_{\max} \text{ horizontal steel} = L/5,\ 3t_w,\ 450 \text{ mm}$$

$$S_{\max} \text{ vertical steel} = L/3,\ 3t_w,\ 450 \text{ mm}$$

Flexural wall steel, $A_{s,\text{ vertical}}$:
Assume steel; then run an interaction curve program in plane bending and compression.

$$P_o = \phi_c \times 0.85 f_c' (A_g - A_{s1} - A_{s2}) + \phi_s F_y (A_{s1} A_{s2})$$

$$P_{u,\max} = 0.8 P_o$$

Then loop for $c = L$ to $L/5$ step = 100 mm.

$$e_1 = 0.003(c - d_1)/c \qquad f_{s1} = e_1 E_s \leqq F_y$$

$$e_2 = 0.003(d - c)/c \qquad f_{s2} = e_2 E_s \leqq F_y$$

If $e_2$ is negative, then $f_{s2} \leqq F_y$.

$$P_u = \phi_c \times 0.85 f_c' t_w \beta c + \phi_s A_c' f^{s1} - \phi_s A_s f_{s2}$$

$$M_u = \phi_c \times 0.85 f_c' t_w \beta c (L/2 - \beta c/2) + \phi_s (A_s f_{s1} + A_s f_{s2})(L/2 - d_1)$$

The design condition must fall within the interaction curve.
Horizontal shear wall floor wall joint:
Without special connectors,

$$V_u \leqq \phi V_{ch}$$

$$\phi_c V_{ch} = \phi_c K (0.85 f_c' \beta c t_w)$$

$K$ = friction factor and ranges from 0.3 to 0.8; 0.35 and 0.4 are common in the United States, whereas 0.5 to 0.7 are common in Europe, and are not given in building codes.

If $V_u > \phi_c V_c$, then provide shear connectors or dowels designed by the shear friction method.

## Posttensioned stacked panels

See Fig. 35.45.

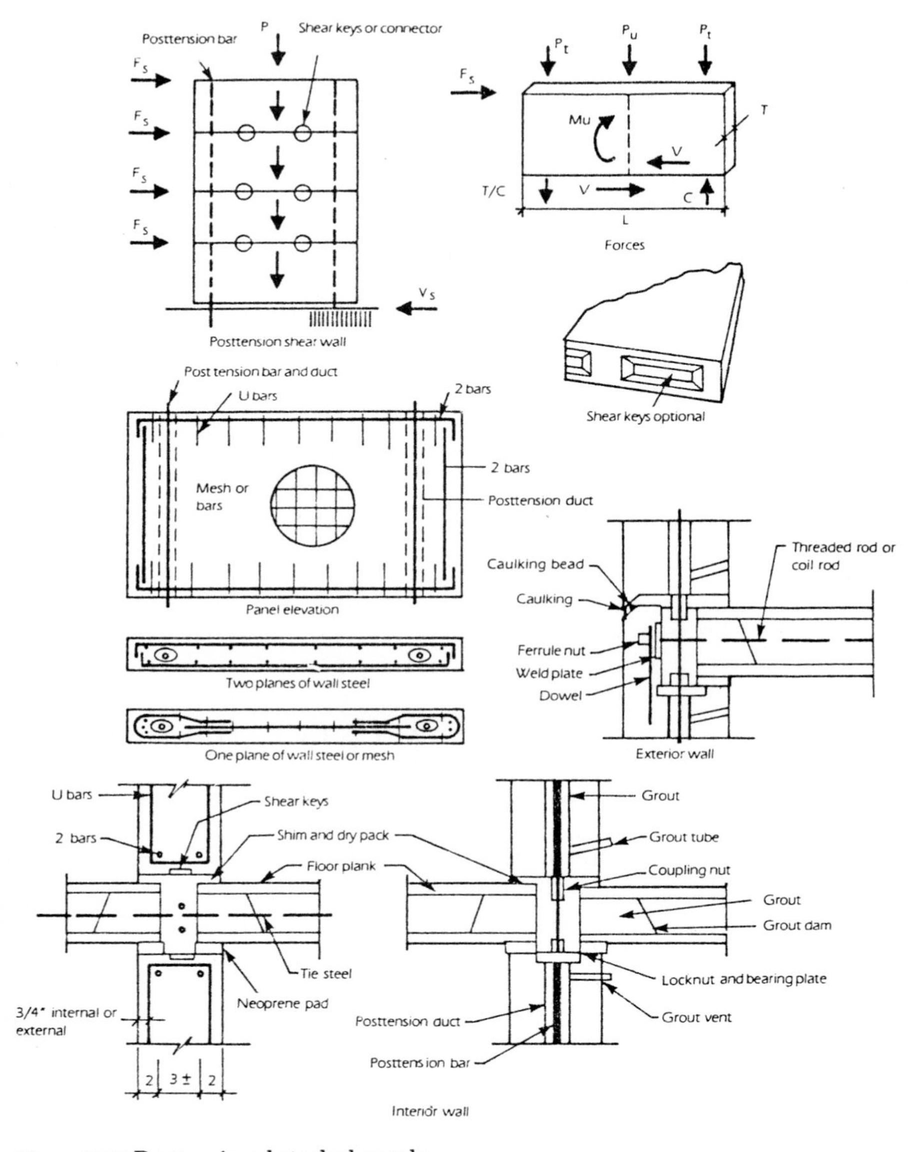


**Figure 35.45** Posttensioned stacked panels.

## Design of Posttensioned Walls: In-Plane Bending

Walls subjected to in-plane bending are similar to columns since they are both subjected to axial and flexural stresses. Therefore, interaction curves or computer programs are used. The first step is to assume the amount of prestressing steel, then to use the curve or program to check the allowable axial load and moment combination with the design axial load and moment. If the applied loads and moments fit within the curve, the design is acceptable. For posttensioned wall systems, tension stresses in the concrete are avoided or kept within the noncracked range so as to reduce the possibility of corrosion of the tendon and to avoid the reduction of horizontal shear resistance if the shear friction method is used.

The Canadian load factors are 1.5 for wind or seismic loads and 0.85 for dead loads resisting overturning. Resistance factors are 0.6 for concrete and 0.9 for the prestressing tendons. For this type of construction grouted Dywidag threaded bars are used for the prestressing.

**TABLE 35.14 Physical Properties of Dywidag Threaded Bars**

| Bar diameter, mm | Area, mm² | Sheathing inside diameter, mm | $f_{pu}$, MPa |
|---|---|---|---|
| 15 | 177 | 19 | 1080 |
| 26 | 548 | 32 | 1030 |
| 32 | 806 | 38 | 1030 |
| 36 | 1018 | 44 | 1030 |

The modulus of elasticity = 190,000 MPa. When the strain exceeds 0.0042 mm/mm, the stress-strain curve must be used to determine the stress in the tendon.

The loss of prestress due to steel relaxation and concrete creep is approximately 10 percent. There is no shrinkage because the panels are usually older than 30 days when they are posttensioned. Friction and curvature are nil because of the short story height and the straight rod and duct. The anchor nut is torqued; hence anchor loss is small. Elastic shortening is considered in the equations.

Allowable stress in the tendon is

$$f_{ps} = 0.7\, f_{pu}\ (100\ \text{percent} - \text{loss in percent})/100$$

For preliminary bar selection,

$$A_{ps,\text{total}} \geq (P_u/A + {}^{-}M_{ot}/S)\, A/f_{ps}$$

$$A_{ps} = A'_{ps} = A_{ps,\text{total}}/2$$

$$f_{ps} = 0.7\ (100\ \text{percent}-\text{loss in percent})/100$$

At $P_o$,

$$e_s = e'_s = f_{ps}/E_{ps} - 0.003$$

$$A_g = t_w L_w$$

$$f_{ps} = f'_{ps} = e_s E_{ps}$$

$$P_o = \phi_c \times 0.85 f'_c (A_g - A_{ps} - A'_{ps}) - \phi_p A_{ps} f_{ps} - \phi_p A'_{ps} f'_{ps}$$

$$P_{u,\max} = 0.8\, P_o$$

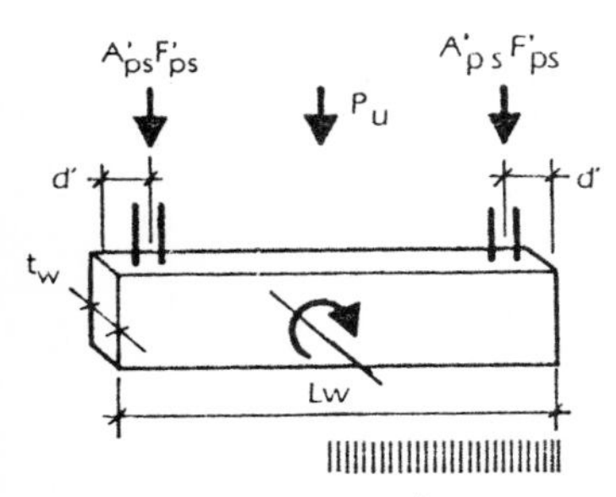


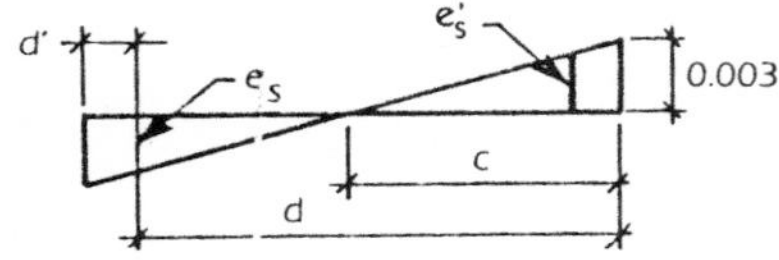


**Figure 35.46** Theory of posttensioned wall interaction equations.

$P_{bal}$ and $M_{bal}$ do not exist for high-strength steels.

For $e > e$ proportional limit, the stress-strain curve must be used for values of $f_{ps}$.

$$e_{ps} = f_{ps}/E_{ps} \qquad e_s = e_{ps} - 0.003(c - d)/c \qquad f_{ps} = e_s E_{ps}$$

$$e'_s = e_{ps} - 0.003(c - d')/c \quad f'_{ps} = e'_s E_{ps}$$

$$P_u = \phi_c \times 0.85 f'_c t_w \beta c - \phi_p (A_{ps} f_{ps} + A'_{ps} f'_{ps})$$

$$M_u = \phi_c \times 0.85 f'_c t_w \beta c\ (L_w/2 - \beta c/2) - + \phi_p A_{ps} f_{ps} (L_w/2 - d')$$

$$+ \ \phi_p A'_{ps} f'_{ps} (L/2 - d')$$

Write a computer program for the above with a loop for $c = L_w$ to $L_w/4$ step-150 mm. Use the following or similar statements to model the stress-strain curve:

If $e_{ps} > 0.0060$, then $f_{ps} = 1000$ MPa: goto ______.
If $e_{ps} > 0.0055$, then $f_{ps} = 995$ MPa: goto ______.

If $e_{ps} > 0.0050$, then $f_{ps} = 971$ MPa + $(e_{ps} - 0.0050) / 0.0005 \times 24$: goto ________.

If $e_{ps} > 0.0045$, then $f_{ps} = 932$ MPa + $(e_{ps} - 0.0045) / 0.0005 \times 39$: goto ________.

### Example Problem

The example building is shown in Fig. 35.47. For this example problem the Canadian code will be used because Canada employs the metric system whereas the United States is currently using the foot-pound system of units. The reader should review the code governing the location where the project is to be built and design accordingly. While the methods of the various codes differ and the magnitude of the seismic force to be considered also differs, the concepts of the force path and the way in which the forces are resisted are the same.

For this example a hypothetical site in Parksville, British Columbia, was selected. The site is in the Canadian seismic Zone VI, with a velocity ratio of 0.4 and an acceleration ratio of 0.4. The ratio of acceleration to velocity is 1.0. The soil condition is compacted coarse-grain layered with firm fine-grain soils with a total depth of 18 m. The soil factor $F$ is 1.3. The occupancy is an office-type usage with an importance factor $I$ of 1.0. The framing or resistance system is shear walls without a moment frame; hence the $K$ factor is 1.3. Since the building is only two stories high, the code equation for the period $T$ will be used: $T = 0.9h_n d_s^{0.5}$; $h_n$ is 7.2 m, and the base dimension in the direction of the seismic force is 38.4 m (the base dimension is the same in both directions for this building). $T = 0.09 \times 7.2/38.4^{0.5} = 0.10$s. By using the code table for a period of 0.1 s and the acceleration-velocity ratio of 1.0, the value of the seismic response factor is 0.44. The base shear equation is $V = vSKIFW$.

$$V_{\text{base}} = 0.4 \times 0.44 \times 1.3 \times 1.0 \times 1.3W = 0.3W$$

The dead load on the system is as follows:

| | |
|---|---|
| Roof system including double T's, 50-mm lightweight concrete, roofing, 25 percent of snow load, and inverted-T girder and mechanical-electrical systems | 5.6 kN/m² |
| Floor system including double T's, inverted-T girders, 60-mm topping, ceiling system, mechanical-electrical systems, and movable partitions | 6.4 kN/m² |
| Walls: 150-mm-thick precast concrete panels | 3.6 kN/m² |
| Glass window areas | 0.4 kN/m² |

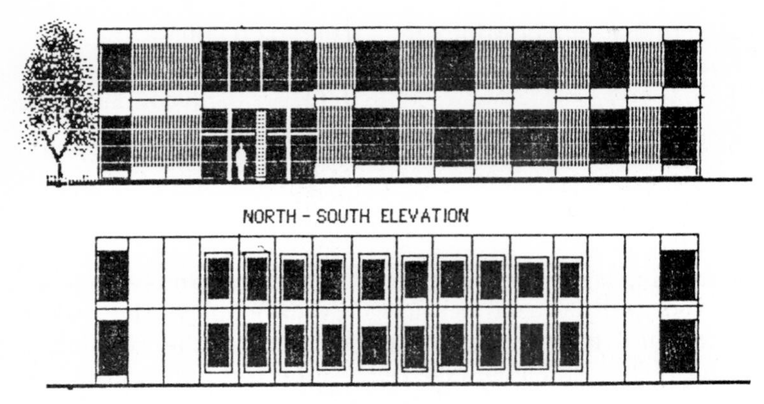


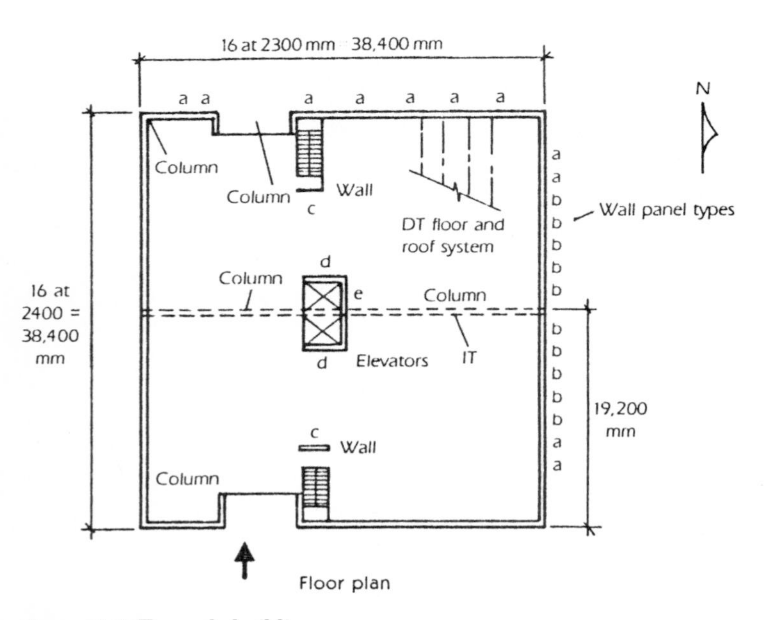


**Figure 35.47** Example building.

To compute the load attributed to each level use the floor or roof load plus one-half of the wall weight above and below that level. (See Fig. 35.48.)

Roof zone weight = 38.4 m²×5.6 kN/m² + (3.6 m/2×24 panels)×2.4 m × 3.6 kN/m²+ 3.6 m/²× 50 percent glass × 20 panels × 2.4 m × 3.6 kN/m² + 3.6/ m /2× 12 panels × 2.4 m × 0.4kN/m² + 3.6 m/2× 50 percent glass × 20 panels × 2.4 m × 0.4 kN/m² = 8257 kN + 567 kN = 8824 kN

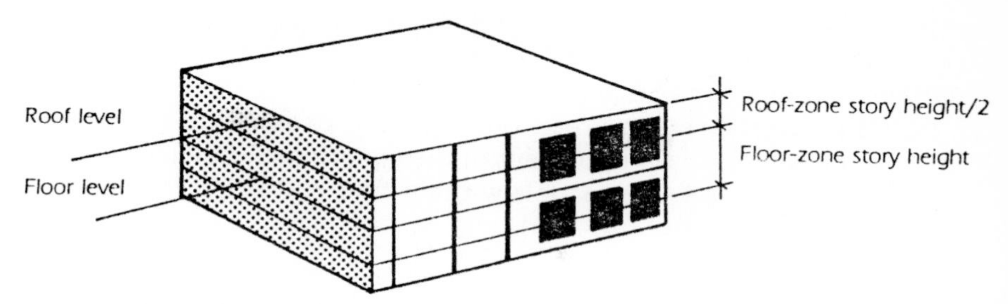


**Figure 35.48** Determination of mass weights.

$$\text{Floor zone weight} = 38.4\ \text{m}^2 \times 6.5\ \text{kN/m}^2 + 2 \times 567\ \text{kN} = 9585\ \text{kN} + 1134\ \text{kN} = 10{,}720\ \text{kN}$$

Building period < 0.27 s $\therefore F_t = 0.0\ (h_n/d_s < 3.0)$

$F_x = (V - F_t) W_x h_x / \Sigma W_i h_i$

$F_x$ = seismic force at level $x$

| Item | $W$, kN | $h_x$, m | $W_i h_i$ | $W_i h_i / W_i h_i \times$ | | $V$ | = | $F_x$, kN |
|---|---|---|---|---|---|---|---|---|
| Roof | 8,824 | 7.2 | 63,533 | 0.62 | × | 5863 | | 3625 |
| Floor | 10,720 | 3.6 | 38,592 | 0.38 | × | 5863 | | 2228 |
| Total | 19,544 | | 102,125 | 1.00 | | | | 5863 |

$$\text{Base shear} = 0.3W = 0.3 \times 19{,}544 = 5863\ \text{kN}$$

To determine the force on a wall or a panel or a pier the relative rigidity of that wall, panel, or pier must be determined along with the center of mass and center of rigidity. The building is symmetrical; hence the center of mass is at the center of the building. Owing to the interior shear walls, the center of rigidity is not at the center of the building.

The following assumptions are made:

1. The walls between the first and second floors are the same as the walls between the second floor and the roof; hence their relative rigidities are the same.
2. Treat the first-floor – second-floor walls as cantilevers, and use the same relative rigidities as for the second-floor – roof walls.
3. Neglect the brace effect of the stairways.
4. The connection of an *a* panel to an *a* panel is to be solid-grouted with castellations; hence this is a rigid connection.
5. The connection of an *a* panel to a *b* panel and the connection of a *b* panel to a *b* panel are to be pinned. This is not exact if the weld plate

connectors are designed for full elastic load conditions. For this example the connectors will be designed for full elastic loads as an exercise but will be treated as pinned in the analysis.

6. Since there are sunscreen fins on the $b$ panels, the moment-of-inertia and area terms will be kept in the equations for deflection; that is, $\Delta = 4h^3/12I + 3h/A$. For panels without fins, $\Delta = 4h^3/t_wL^3 = 3H/t_wL$.

7. Relative rigidity = $1/\Delta$.

**Panel deflection and rigidity.** Panel properties are shown in Fig. 35.49.

Panel $a$ $\Delta = 4h^3/t_wL3 + 3b/t_wL$. Thickness of wall panel is 150 mm, $h$ = story height, and $L$ = panel length.

$$4 \times 3.6\ \text{m}^3/(0.15 \times 2.4\ \text{m}^3) + 3 \times 3.6\ \text{m}/(0.15 \times 2.4\ \text{m}) = \Delta = 120$$
$$R = 0.00833$$

Panel $aa$ (two $a$ panels rigidly tied together):

$$4 \times 3.6\ \text{m}^3/(0.15 \times 4.8\ \text{m}^3) + 3 \times 3.6\text{m}/(0.15 \times 4.8\ \text{m}) = \Delta = 26.25$$
$$R = 0.038095$$

Panel $c$ = panel $a$.

Panel $d$ (3 m long):

$$4 \times 3.6\ \text{m}^3/(0.15 \times 3\ \text{m}^3) + 3 \times 3.6\ \text{m}/(0.15 \times 3\ \text{m}) = \Delta = 70.08$$
$$R = 0.01427$$

Panel $e$ = panel $aa$.

Panel $b$. Use the subtraction method for this panel since it is cantilevered. For a cantilevered solid panel without openings, $\Delta$ = 120.00. Subtract the lower part as if it were fixed:

$$2.6^3/\left(0.15 \times 2.4^3 + 3 \times 2.6\right)(0.15 \times 2.4) = -30.1427$$

Compute $\Delta$ for the piers.

$$\Delta \text{ for panel } b = 120.00 - 30.1427 + 288.0 + 6.542 = 384.4$$
$$R = 1/384.4 = 0.0025$$

Sum of panels of east-west wall $= R = 2 \times 0.038095 + 10 \times 0.0026 = 0.10219$

Sum of panels of north-south wall $= R = 0.038095 + 5 \times 0.00833 = 0.0797$

Sum of all walls in east-west direction $= 2 \times 0.0797 + 2 \times 0.00833 + 2 \times 0.1427 = 0.2046$

Sum of all walls in north-south direction $= 2 \times 0.10219 + 4 \times 0.0083 + 0.038095 = 0.27567$

In the north-south direction the center of rigidity is on the centerline of the building because of symmetry. In the east-west direction the center of rigidity is not on the centerline and so must be computed.

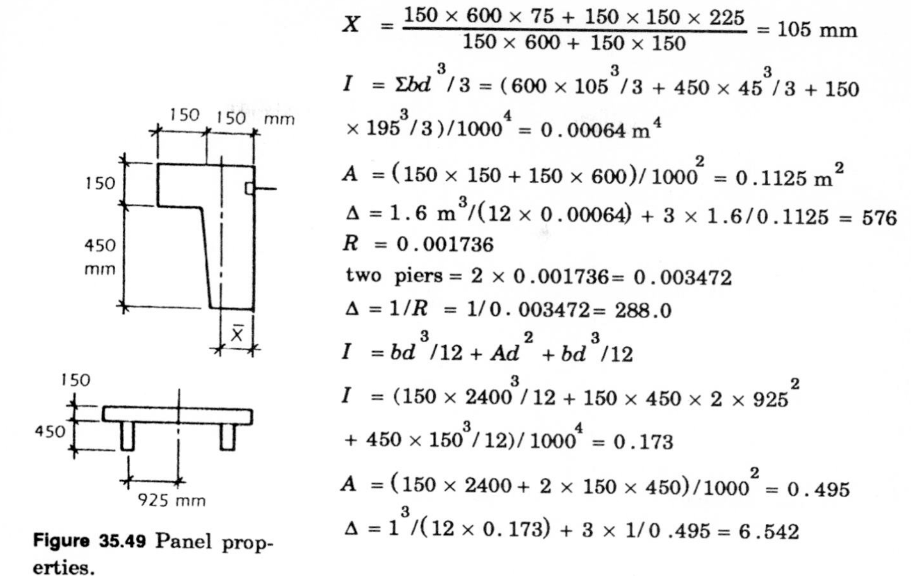


$$X = \frac{150 \times 600 \times 75 + 150 \times 150 \times 225}{150 \times 600 + 150 \times 150} = 105 \text{ mm}$$

$$I = \Sigma bd^3/3 = (600 \times 105^3/3 + 450 \times 45^3/3 + 150 \times 195^3/3)/1000^4 = 0.00064 \text{ m}^4$$

$$A = (150 \times 150 + 150 \times 600)/1000^2 = 0.1125 \text{ m}^2$$

$$\Delta = 1.6 \text{ m}^3/(12 \times 0.00064) + 3 \times 1.6/0.1125 = 576$$

$$R = 0.001736$$

$$\text{two piers} = 2 \times 0.001736 = 0.003472$$

$$\Delta = 1/R = 1/0.003472 = 288.0$$

$$I = bd^3/12 + Ad^2 + bd^3/12$$

$$I = (150 \times 2400^3/12 + 150 \times 450 \times 2 \times 925^2 + 450 \times 150^3/12)/1000^4 = 0.173$$

$$A = (150 \times 2400 + 2 \times 150 \times 450)/1000^2 = 0.495$$

$$\Delta = 1^3/(12 \times 0.173) + 3 \times 1/0.495 = 6.542$$

**Figure 35.49** Panel properties.

Taking moments from the left (west) of all walls running north and south and dividing by the sum of the walls running north and south give the location of the center of rigidity.

$$X = (0 \times 0.10219 + 38.4 \text{ m} \times 0.10219 + 19.2 \text{ m} \times 0.038095 + 2 \times 7.2 \text{ m} \times 0.00833 + 2 \times 14.4 \text{ m} \times 0.00833)/0.27567 = 18.18 \text{ m}$$

$$e = 19.2 - 18.18 = 1.02 \text{ m}$$

$$e_{\min} = 0.05 \times 38.4 = 1.92 \text{ m (accidental torsion)}$$

Note that some codes will use the larger of these two values, while others will add the two values. For this example the sum will be used when computing the torsional moment.

The equation for the force on a wall is

$$F_{\text{wall1}} = F_x R_1/\text{sum of } R \text{ parallel walls} + F_x e R_1 d_1/\Sigma Rd^2 \text{ for all walls}$$

If the wall is made up of panels, then $F_p = F_{\text{wall}} R_{p1}/R_{\text{wall}}$.
A structural framing plan is shown in Fig. 35.50.

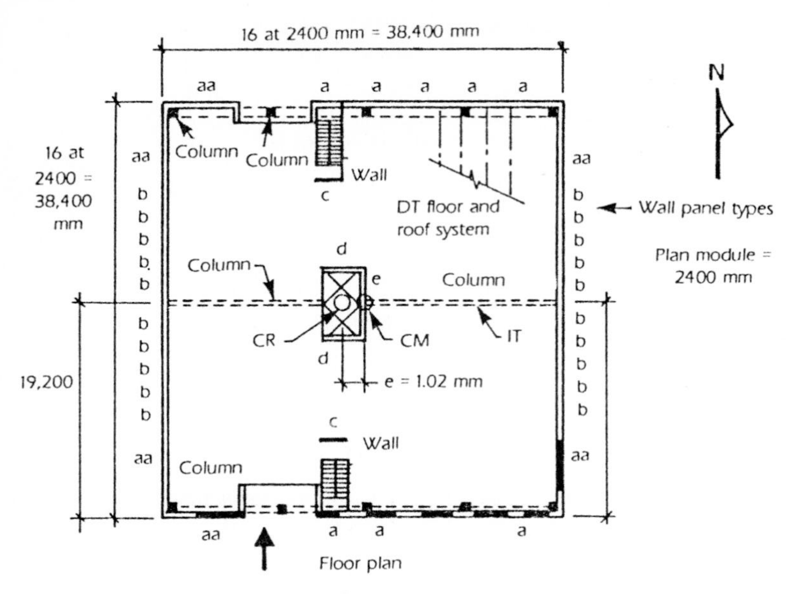


**Figure 35.50** Structural framing plan.

**Seismic forces.** The seismic force in the east-west direction is as follows:

$$F_{x,\text{roof}} = 3635\,\text{kN} \qquad F_{x,\text{floor}} = 2228\,\text{kN}$$

$$V_{\text{base}} = 5863\,\text{kN} \qquad e = 1.92\ \text{m}\,(e_{\min})$$

| Wall | $R$ | $d$ | $Rd$ | $Rd^2$ | $\text{Force}_{\text{roof}}$, kN | $\text{Force}_{\text{roof + floor}}$, kN |
|---|---|---|---|---|---|---|
| North | 0.0797 | 19.2 | 1.53 | 29.38 | 1492.8 | 2407.8 |
| South | 0.0797 | 19.2 | 1.53 | 29.38 | 1492.8 | 2407.8 |
| *c* | 0.00833 | 12.0 | 0.10 | 1.2 | 153.0 | 246.8 |
| *c* | 0.00833 | 12.0 | 0.10 | 1.2 | 153.0 | 246.8 |
| *d* | 0.01427 | 2.4 | 0.034 | 0.082 | 255.2 | 411.7 |
| *c* | 0.01427 | 2.4 | 0.034 | 0.082 | 255.2 | 411.7 |
| | 0.2046 | | | 61.324 | | |

The seismic force in the north-south direction is as follows:

$$e = 2.94 \text{ m}$$

| Wall | $R$ | $d$ | $Rd$ | $Rd^2$ | $\text{Force}_{roof}$, kN | $\text{Force}_{roof + floor}$, kN |
|---|---|---|---|---|---|---|
| East | 0.10219 | 19.2 | 1.962 | 37.67 | 1497.8 | 2415.9 |
| West | 0.10219 | 19.2 | 1.962 | 37.67 | 1497.8 | 2415.9 |
| $a1$ | 0.00833 | 10.95 | 0.091 | 0.999 | 116.8 | 188.4 |
| $a1$ | 0.00833 | 10.95 | 0.091 | 0.999 | 116.8 | 188.4 |
| $a2$ | 0.00833 | 3.75 | 0.031 | 0.117 | 112.2 | 180.9 |
| $a2$ | 0.00833 | 3.75 | 0.031 | 0.117 | 112.2 | 180.9 |
| $e$ | 0.03809 | 1.02 | 0.0388 | 0.0396 | 505.0 | 814.6 |
| | 0.27579 | | | 77.6116 | | |

The sum of $Rd^2 = 61.324 + 77.6116 = 138.9356$ for all walls.

**Panel forces.** South wall panel forces are as follows:

$$F_p = F_{wall} R_p / R_{wall}$$

Panel *aa* at roof:

$$F_p = 1492.8 \text{ kN} \times 0.038095/0.0797 = 713.53 \text{ kN}$$

Panel *aa* at second floor:

$$F_p = 2407.8 \text{ kN} \times 0.038095/0.0797 = 1150.9 \text{ kN}$$

(applied at floor = 437.39 kN)

Panel *a* at roof:

$$F_p = 1492.8 \text{ kN} \quad 0.00833/0.0797 = 156.0 \text{ kN}$$

Panel *a* at second floor:

$$F_p = 2407.8 \text{ kN} \times 0.00833/0.0797 = 251.6 \text{ kN}$$

(applied at floor = 95.65 kN)

East wall panel forces are as follows:

Panel *aa* at roof:

$$F_p = 1497.8 \text{ kN} \quad 0.038095/0.10219 = 558.36 \text{ kN}$$

Panel *aa* at second floor:

$$F_p = 2415.9 \text{ kN} \times 0.038095/0.10219 = 900.6 \text{ kN}$$

(applied at floor = 342.25 kN)

Panel *b* at at roof:

$$F_p = 1497.8 \text{ kN} \times 0.0025/0.10219 = 36.64 \text{ kN}$$

Panel $b$ at second floor:

$$F_p = 2415.9 \text{ kN} \times 0.0025/0.10219 = 59.10 \text{ kN}$$

$$(\text{applied at floor} = 22.46 \text{ kN})$$

Design panel $aa$ for 713.53 kN applied at roof and 437.39 kN applied at second floor with a base shear of 1150.9 kN.

### Element and connection design

The concrete elements and connections are designed by the load factor and resistance method (ultimate strength), and load factors must therefore be applied to the forces. The load factors ($U$ values) and resistance factors ($\phi$ values) vary with the building code used. For the Canadian code load factors are 1.5 for seismic loads, 1.5 for live loads, 1.25 for dead loads, and 0.85 for a dead load resisting overturning. The $\phi$ factors are 0.6 for concrete, 0.85 for reinforcing steel, and 0.9 for prestressing tendons or strand. For the United States the load factors ($U$ values) are 1.43 for seismic loads, 1.4 for dead loads, 1.7 for live loads, and 0.9 for a dead load resisting overturning. The $\phi$ factors are 0.9 for flexure, 0.85 for shear, and 0.7 for compression. For this example Canadian load and resistance factors will be used since the solution is in metric units.

### Diaphragm shears

Diaphragm shears are shown in Fig. 35.51. Because of the larger eccentric moment in the north-south direction, diaphragm shears will be determined for that direction.

Assume that the torsion in the diaphragm will be resisted by the two end walls in lieu of being resisted on all four sides. This is conservative, but the difference is small. For the roof slab the shear equals the reaction $= 1.5 \times 3636 \text{ kN}/2 + 1.5 \times 3635 \text{ kN} \times 2.92 \text{ m}/38.4 \text{ m} = 3140 \text{ kN}$. [If four sides were used, then for this building the shear would be $1.5 \times 3635 \text{ kN}/2 + 1.5 \times 3635 \times 2.94/(4 \times 19.2) = 2935 \text{ kN}$.] Unit shear per meter $= 3140.4/38.4 \text{ m} = 81.78 \text{ kN/m}$. If a concrete topping of 50 mm of 30-MPa concrete were used, the shear strength of the topping would be $\phi_c V_c$.

$$\phi_c V_c = \phi_c \times 2 f_c'^{0.5} t_t b = 0.6 \times 2 \times 30 \text{ MPa}^{0.5}$$

$$\times 50 \text{ mm} \times 1000 \text{ mm}/1000 = 328.6 \text{ kN/m}$$

which is greater then 81.78 kN/m. Therefore, the topping is adequate, and shear connectors are not required between the double T's. However, they are used to tie the double T's together. In this case connectors at 2400 mm on center usually are employed.

If a topping was not used, the connectors must transfer the shear. The resistance of the connectors is shown in Fig. 35.52.

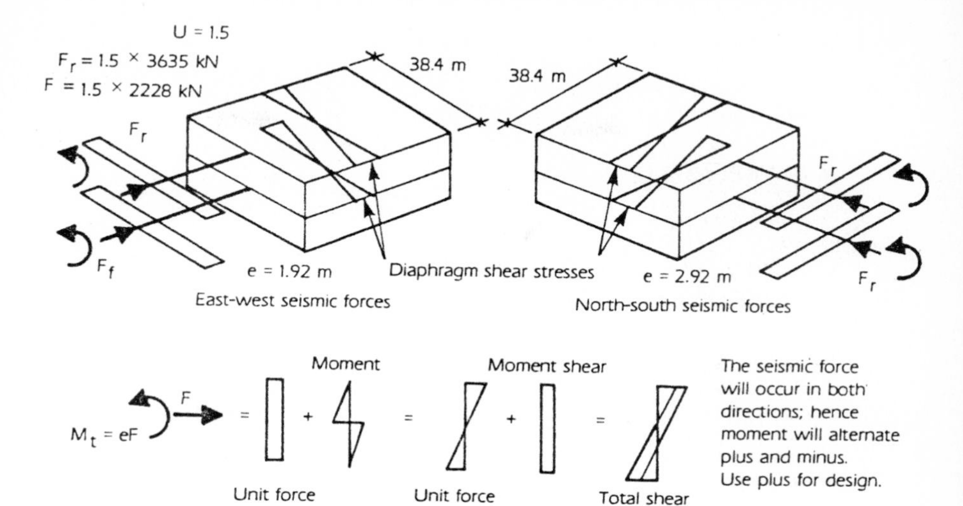


**Figure 35.51** Diaphragm shears.

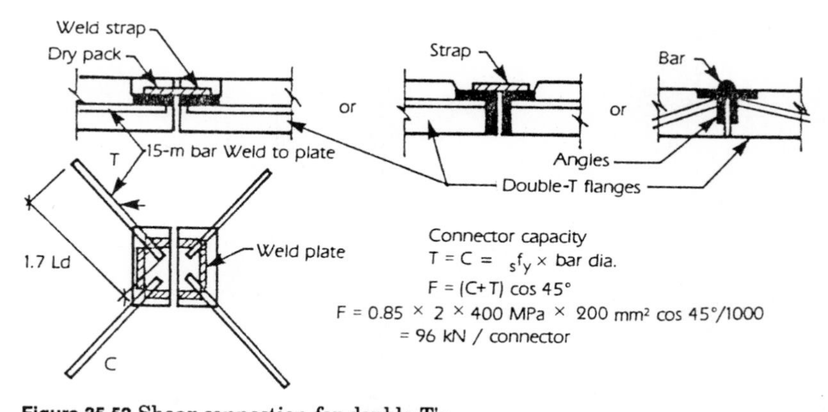


**Figure 35.52** Shear connection for double T's.

The shear force at one double T from the edge is 3140 kN/(1.5× 3635/2)×38.4/2 = 22.11 m to zero shear. Shear force = 3140 kN ×(22.11 m−2.4 m)/22.11= 2800 kN. The number of connectors per double T = 2800 kN/96 kN/connector = 30 connectors. Connector spacing = 38,400 mm/30 = 1152 mm; place at 1150 mm on center all double-T edges. At a distance of two double T's from the edge the shear force = 3140 kN ×( 22.11 m = 4.8 m)/22.11 m = 2458 kN. The number of connectors would be reduced to 2458 kN/96 kN per connector = 26 connectors for a spacing of 38,400 mm/26 m= 1470 mm on center.

The diaphragm at the second-floor level would be designed in the same way.

**Connection double T over inverted-T girder**

See Fig. 35.53. Horizontal shear at a point = $VQ/I = V6S_{min} \times (B - S_{min})/B^3$. At this connection $S_{min} = B/2$. Therefore,

$$v = V6B/2 \times (B - B/2)/B^3$$
$$v = V \times 1.5/B = 3140\text{ kN} \times 1.5/38.4\text{ m} = 122.65\text{ kN/m}$$

By using the shear friction concept with fresh concrete cast against hardened concrete, $\mu = 0.9$.

$$A_{vf} = (\mu\phi_s f_y) = 122.65\text{ kN}/(0.9 \times 0.85 \times 400\text{ MPa}) \times 1000 = 400\text{ mm}^2/\text{m}$$

Use 15M bar at 500 mm on center (area = 200 mm²/bar).

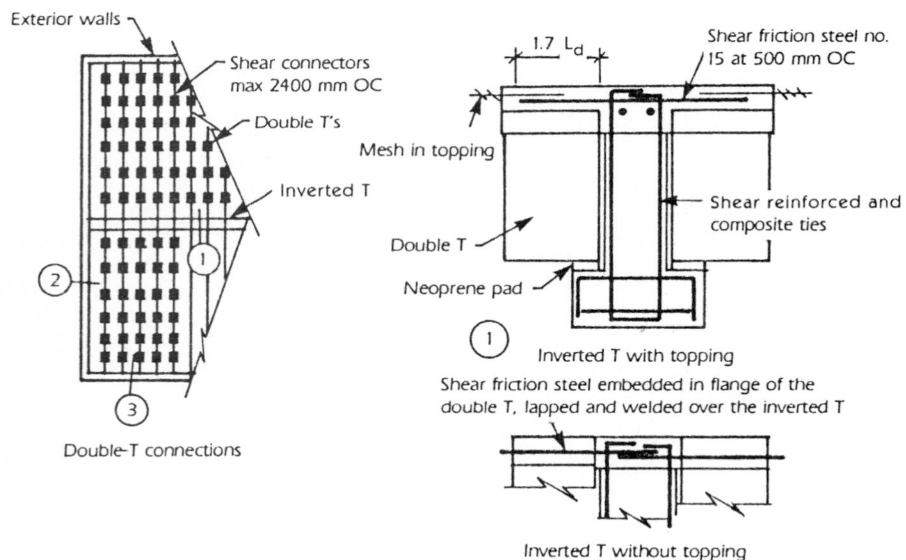


**Figure 35.53** Double-T and inverted-T connections.

If a topping is used, the bar is placed in the topping. If there is no topping, the bar is placed in the ends of the double T and lapped over the inverted T, welded or weld plates are placed in the ends of the double T's, and strap is placed across the inverted T and welded to the weld plates.

**Chord stress and chord steel design**

See Fig. 35.54. Note that the center of mass is at the center of the building. Neglect the stair and elevator openings.

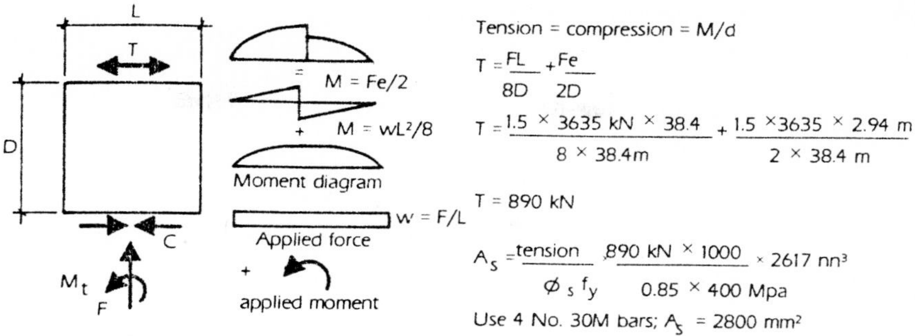


**Figure 35.54** Chord steel design.

### Wall-panel-to-diaphragm connections

The connection from the roof or floor diaphragm to the precast shear wall panels is designed by the shear friction method when a topping or a pour strip is used. If there is no topping or pour strip, the connection is designed to transfer the shear stresses through a weld plate system. (See Fig. 35.55.)

In addition to designing the connection for shear, the out-of-plane tension compression seismic force must be considered. Thus the area of steel required for shear friction is added to the area of steel required for tension out of plane for selecting the final bar or coil rod size.

For shear friction the equation for the reinforcing across the joint is

$$A_{vf} = V_u / \mu \phi_s f_y$$

If the surface of the wall panel in this area is intentionally roughened, the $\mu$ value is 0.9, and if the surface at the connection is not intentionally roughened, the value of $\mu$ = 0.5, for fresh concrete cast against hardened concrete ( $\phi_s$ = 0.85). In addition, the concrete shear stress cannot exceed

$$V \leq \phi_c V_c$$

$$\phi_c V_c = \phi_c \times 2.5 f_c' \times \text{area of contact surface}$$

$$= < \phi_c \times 6.9 \text{ MPa} A_c \ (A_c = \text{area of the contact surface})$$

$$\phi_c = 0.6$$

For out-of-plane tension steel (see Fig. 35.56),

$$A_{\text{ten}} = F_t U / \phi_s f_y$$

where $U$ = load factor of 1.5 and $\phi_s$ = 0.85.

$F_t = vS_pW_p$

$v$ = 0.4 for this seismic zone

$S_p$ = component factor of 11.0 for connections

$W_p$ = weight of panel contributing to this force

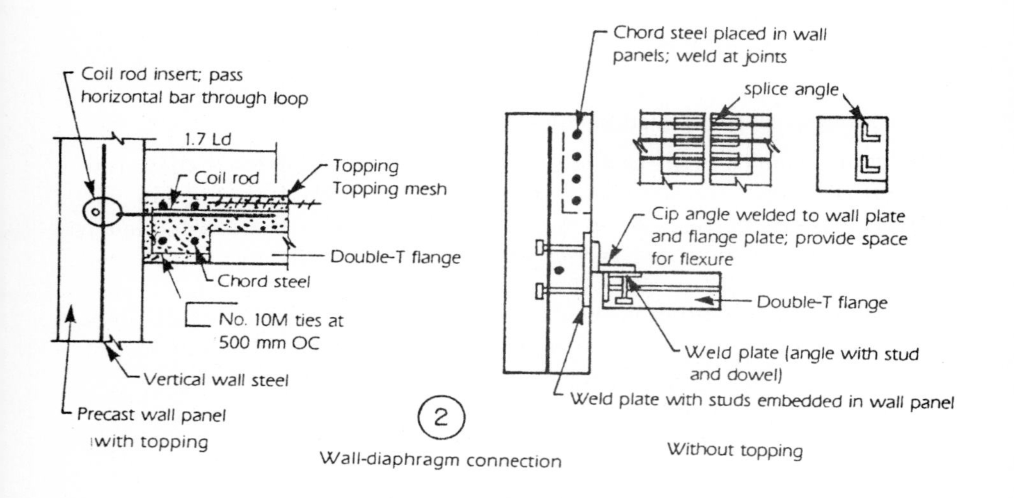


**Figure 35.55** Diaphragm-to-wall connections.

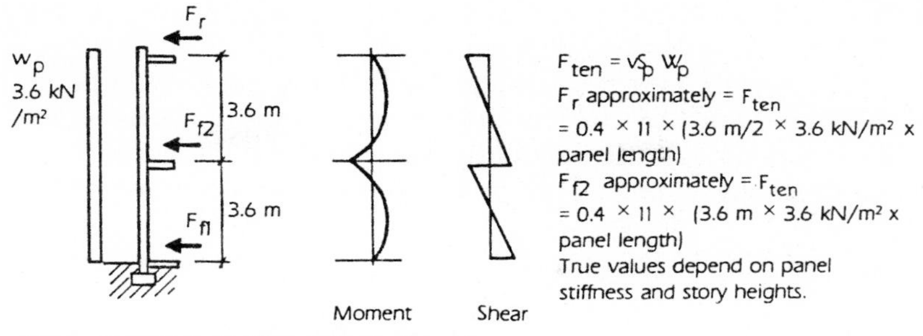


**Figure 35.56** Out-of-plane seismic forces.

If all the wall panels have the same rigidity or if the panels are rigidly tied together to form one wall, the connectors would be evenly distributed along the wall. If the panels are of different rigidities and are pin-connected together, the connectors are designed for each panel separately. The L-shaped girder on chord steel encasement acts as a collector or drag strut to collect the diaphragm shears, which are then transferred to the wall panels according to their rigidities.

Take as an example the north-south seismic direction panel *aa*:

$$\Delta_{vf} = UV / \mu\phi_s f_y$$

$$A_{vf} = 1.5 \times 558.36\,\text{kN} \times 1000/(0.5 \times 0.85 \times 400\,\text{MPa}) = 4927\,\text{mm}^2$$

$$\Delta_{ten} = F / \phi_s f_y$$

$$\Delta_{ten} = 0.4 \times 11 \times (3.6\text{m}/2 \times$$

$$3.6\,\text{kN/m}^2 \times 4.8\,\text{m}) \times 1000/(0.85 \times 400\,\text{MPa}) = \quad 402\,\text{mm}^2$$

$$\text{Total required area} = \quad 5329\,\text{mm}^2/\text{panel}$$

Try using 24M coil rod; area = 450 mm²:

$$5329/450 = 12 \text{ coil rods required for panel } aa$$

Since this panel is composed of two separate panels, use six coil rods for each panel.

| Coil rod size, mm | Approximate area, mm$^2$ |
|---|---|
| 12 | 110 |
| 24 | 450 |
| 36 | 1000 |
| 48 | 1800 |

$$f_y = 400\,\text{MPa}$$

Note that the designer must contact the manufacturer for sizes, areas, and strength of coil rods and coil rod inserts.

If the panel-to-slab interface is roughened, the μ value could be increased to 0.9 instead of 0.5 with a reduction in the number of connectors.

Some designers do not recommend the use of coil rod inserts because of the unknown effects under cyclic load and possible loosening at the insert thread.

**North - South Direction**

| Panel | $V$ = seismic force | Panel length, m | $A_{vf}$ | $A_{ten}$ | $A_{s,total}$ | Required connectors |
|---|---|---|---|---|---|---|
| *aa* | 1.5 × 558.3 kN | 4.8 | 926 | 402 | 5328 | 12 24M coil rods |
| *b* | 1.5 × 36.64 kN | 2.4 | 323 | 201 | 524 | 2 24M coil rods |

NOTE: For this example the panels are assumed to be pin-jointed.

**East-West Direction**

| Panel | $V$ = seismic force | Panel length, m | $A_{vf}$ | $A_{ten}$ | $A_s$ | Required connectors |
|---|---|---|---|---|---|---|
| *aa* | 1.5 × 558.3 kN | 4.8 | 6296 | 402 | 6698 | 16 24M coil rods* |
| *a* | 1.5 × 36.64 kN | 2.4 | 1376 | 201 | 1577 | 4 24M coil rods |

NOTE: These are separate panels.
* If the interface is roughened, the connectors could be reduced to 8.

Check allowable concrete shear stress for panel *aa*, maximum condition:

$$vV_u = 1.5 \times 713.53 \text{ kN} = 1070 \text{ kN}$$

$$\phi_c V_c = 0.6 \times 0.25 \times 30 \text{ MPa} \times 100 \text{ mm} \times 4800 \text{ mm}/1000 = 2160 \text{ kN}$$

$$\phi_c V_c = 0.6 \times 6.5 \text{MPa} \times 100 \text{ mm} \times 4800 \text{ mm}/1000$$

$$= 1872 \text{ kN} > 1070 \text{ kN} \quad (\text{OK})$$

### Panel design: in-plane seismic loading

For this example the panel *aa* south wall is illustrated, since this panel has the maximum lateral load (see Fig. 35.57). There is no vertical load on the panel other than its own weight because an L-shaped girder and columns will resist the vertical gravity loads. If there were a vertical load, then a combined axial load and moment interaction curve computer program would be used to solve the problem.

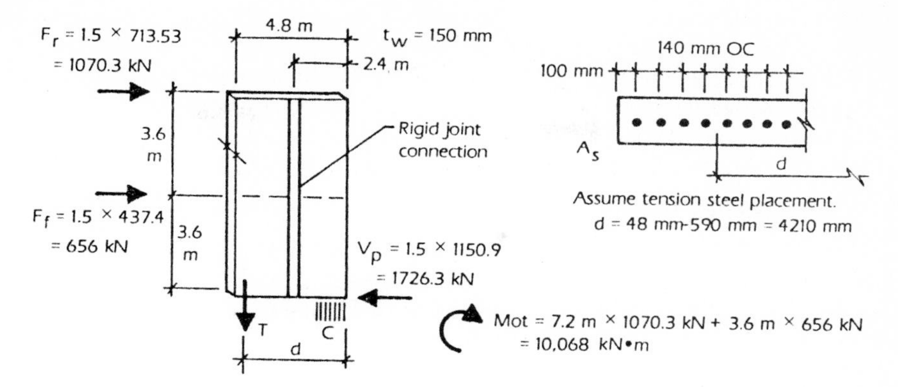


**Figure 35.57** In-plane seismic forces.

For Fig. 35.57,

$$a = d - \left[-2M_u/(\phi_c 0.85 f'_c t_w) + d^2\right]^{0.5}$$

$$a = 4210\text{ mm} - [-2 \times 10{,}068\text{ kN}\cdot\text{m} \times 1000\text{ mm/m} \times 1000/(0.6 \times 0.85 \times 35\text{ MPa} \times 150\text{ mm}) + 4210^2]^{0.5}$$

$$a = 1015.7\text{ mm}$$

$$A_s\text{ required} = \phi_c 0.85 f'_c t_w a/\phi_s f_y = 0.6 \times 35\text{ MPa} \times 150\text{ mm} \times 1015.7/(0.85 \times 400\text{ MPa}) = 7998\text{ mm}^2$$

Use eight 35M bars; $As = 8000\text{ mm}^2$.

**Sliding.** Check sliding at the base.
Sliding friction resistance:

$$\text{Coefficient of sliding friction} = 0.35 = K_{sf}$$

$$C = 0.85 f'_c \phi_c t_w a$$

$$\text{Force resistance} = K_{sf} C = 0.35 \times 0.85 \times 35\text{ MPa} \times 0.6 \times 150\text{ mm} \times 959\text{ mm} = 898.7\text{ kN}$$

Since 898.7 kN < 1726.3 kN, use shear friction method.

$$A_{vf} = V_p/\phi_s \mu f_y = 1726.3\text{ kN} \times 1000/(0.85 \times 0.5 \times 400\text{ MPa}) = 10{,}154\text{ mm}^2$$

$$\text{Eight 35M bars at each end} = 8 \times 2 \times 1000\text{ mm}^2 = 16{,}000\text{ mm}^2 > 10{,}154\text{ mm}^2\ (\text{OK})$$

**Shear.** Check shear.

$$V_p < \phi_c V_c/2$$

$$V_p = 1726.3\text{ kN}$$

$$\phi_c V_c/2 = \phi_c \times 2 f_c'^{0.5} t_w \times \text{length of panel}/2$$

$$= 0.6 \times 2 \times 35\text{ MPa}^{0.5} \times 150\text{ mm} \times 4800\text{ mm}/1000/2 = 2555.7\text{ kN}$$

Since 1726.3 kN < 2555.7 k/N, special shear reinforcing is not required. Use minimum wall steel. If $V_p > \phi_c V_c/2$, then provide minimum shear steel of $0.0025 t_w b$ in each direction ($b$ = unit of measure). If $V_p > \phi_c V_c$, the shear reinforcing must be designed, where

$$A_{s,\text{horizontal}} = (V_u - \phi_c V_c) \times \text{spacing}/\phi_s f_y d$$

$$\rho_h = A_{s,\text{horizontal}}/t_w b$$

$$A_{s,\text{vertical}} = t_w b \times \left[0.0025 + 0.5(2.5 - h_w/L_w)(\rho_h - 0.0025)\right]$$

Since only minimum wall is required, use the following (ACI requirement): horizontal steel, $0.002 t_w b$; and vertical steel, $0.0012 t_w b$.

$$\text{Horizontal steel} = 0.002 \times 150\text{ mm} \times 72\text{ m} \times 1000\text{ mm/m}$$
$$= 2160\text{ mm}^2$$

Use 10M bars.

$$2160\text{ mm}^2/100\text{ mm}^2\text{ per bar} = 22\text{ bars minimum}$$

$$7200\text{ mm}/22\text{ bars} = 327\text{mm on center}$$

Use 10M bars at 300 mm on center.

$$\text{Vertical steel} = 0.0012 t_w b = 0.0012 \times 150\text{ mm}$$
$$\times 4.8\text{ m} \times 1000/\text{ mm/m} = 864\text{ mm}^2$$
$$864\text{ mm}^2/100\text{ mm}^2/\text{bar} = 9\text{ bars}$$

Try 10 bars (5 for each section).

$$2400\text{ mm}/5 = 480\text{ mm on center}$$

Use 10M bars at 450 mm on center.

**Out-of-plane bending.** Check out-of -plane bending. See Fig. 35.58.

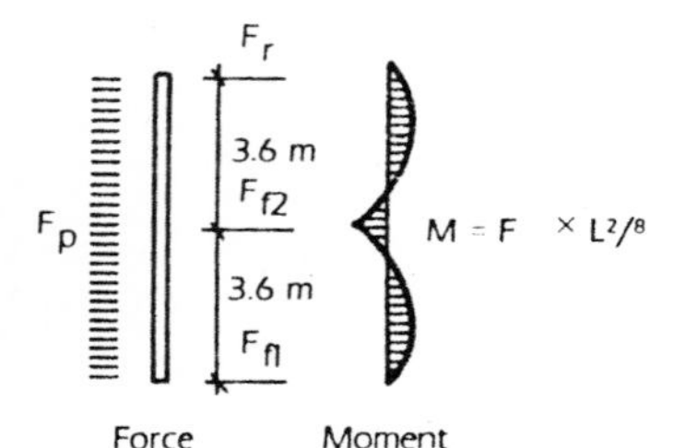


**Figure 35.58** Out-of-plane seismic forces.

For Fig. 35.58,

$$F_p = vS_pW_p$$

where $v$ = 0.4 for this building location
$S_p$ = 0.9 for exterior walls
$W_p$ = panel weight = 3.6 kN/m²
$F_p$ = $0.4 \times 0.9 \times 3.6$ kN/m² = 1.3 kN/m²
$M_u$ = $1.5 \times 1.3$ kN/m² $\times 1$ m $\times 3.6$ m²/8 = 3.16 kN m
$d$ = $t_w/2$ = 150 mm/2 = 75 mm (steel in center of wall)
$a$ = $d - [-2M_u/(\phi_c 0.85 f'_c B_u) + d^2]^{0.5}$
$a$ = 75 mm $-[-2 \times 3.16$ kN $\cdot$ m $\times 1000$ mm/m $\times 1000/$ $(0.6 \times 0.85 \times 35$ MPa $\times 1000)$ + 75 mm²$]^{0.5}$
$a$ = 2.4 mm
$A_s$ = $\phi_c\, 0.85 f'_c Ba/\phi_s f_y$ = $0.6 \times 0.85 \times 35$ MPa $\times 1000$ mm $\times$ 2.4 mm/(0.85 $\times$ 400 MPa)
= 126 mm²/m of wall length

Use minimum-wall steel 10M at 450 mm on center = 222 mm²/m. (OK)
See Fig. 35.59 for wall panel *aa* as designed.

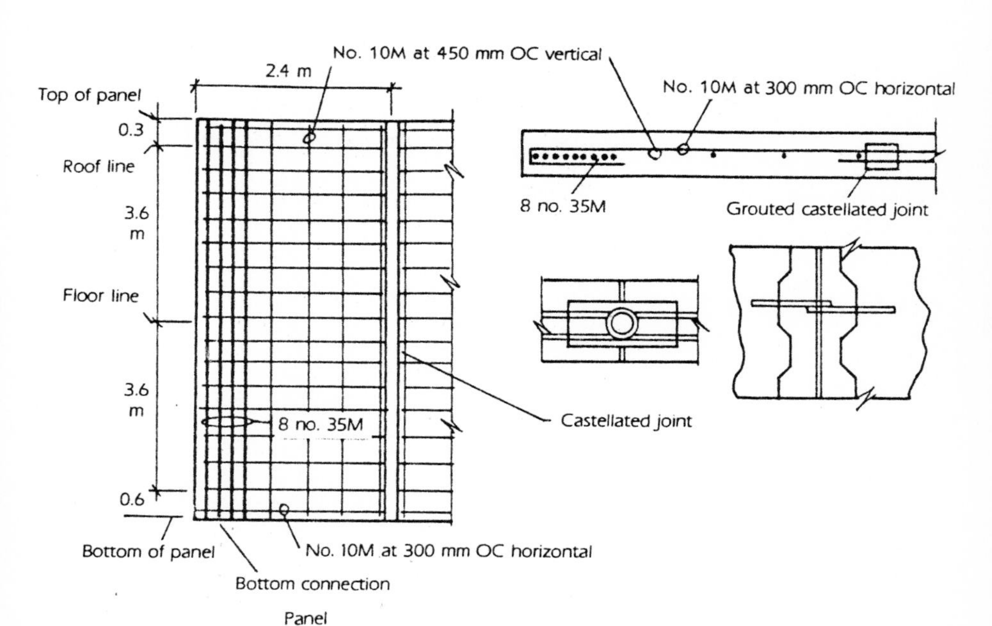


**Figure 35.59** Wall panel as designed.

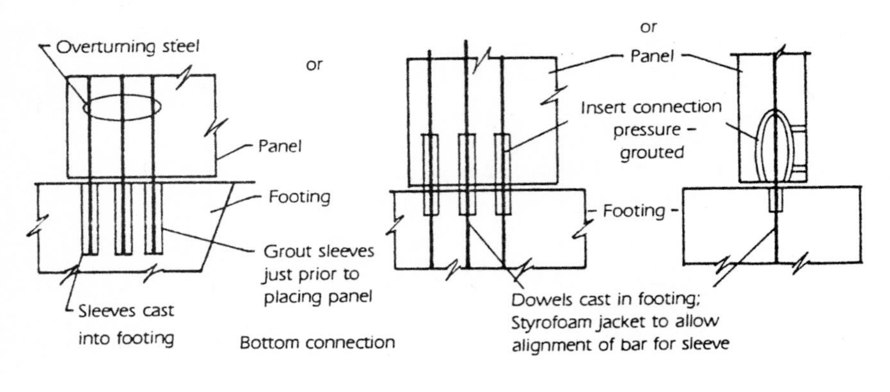


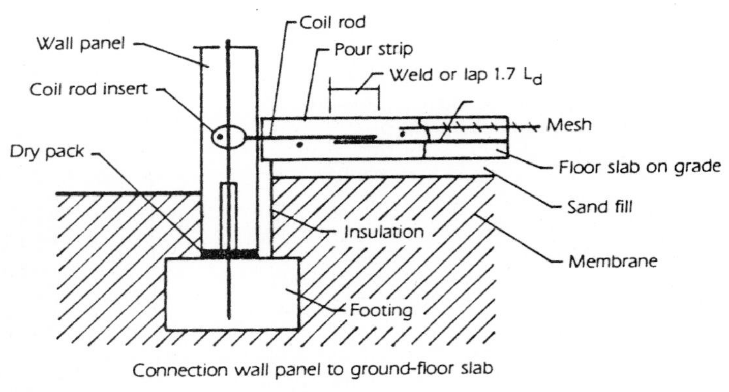


**Figure 35.59 Wall panel *aa* as designed. (*continued*)**

Chapter

# 36

# Computer Programs for Analysis and Design

## Introduction

As the cost of personal computers is continuously being reduced and their capacities increased, nearly every architectural and engineering student today either owns a personal computer or has access to a personal computer through his or her university personal computer laboratories or design studios. Most architectural and engineering offices have personal computers. Some of the offices also have minicomputers with terminals at each work station or a network system tieing their personal computers in with a minicomputer system or tieing their personal computers together.

The current standards for personal computers comprise both 16-bit and 32-bit processors and from 1 to 8 megabytes of random-access memory. Hard disk drives are available with from 20 to 80 or more megabytes of storage, either built in or external. Subject to the application software, computers have graphic, analytical, and word-processing capacities. Architectural and engineering offices are using computers for design, delineation, analysis, computations, working drawings, estimating, specifications, reports, accounting, and general administrative activities. Many different software products are available, and their number is ever increasing.

The programs in this chapter were written on the Macintosh Plus using Microsoft BASIC. The programs will also work on the Macintosh II. The Macintosh II with the 32-bit processor would be preferred. The BASIC language is the language most commonly used with personal computers; it is taught in elementary schools, junior high and high schools, and universities. Each language has its own advantages and disadvantages. The authors prefer BASIC over other programming languages.

The following programs, which are in United States units, can quickly be modified to reflect local building code differences and desired formatting. Reserve words vary from application software to software and must therefore be checked if one enters the program using a different software or computer. The symbols (variable names) used in the programs are not the same symbols used in hand calculations because of the problem of reserve words and noncomputerized symbols. Many of the programs subject to modifications and code update have been used in the classroom since 1976. See Chaps. 34 and 35 for metric computer programs and design examples.

The authors and the publisher assume no responsibility or liability for the use of the following programs. Before using any program the user must first become knowledgeable in the theory and hand solutions, then test the program and revise it as necessary to meet local building code requirements, project requirements, and his or her own professional approach. The following programs are intended as a base upon which the user can construct his or her own programs.

## Program 1: Loss of Prestress

Loss of prestress is the reduction of tension stress in the strand, which in turn reduces the total prestressing force. It is due to the following:

| Type of loss | Pretensioning | Posttensioning |
|---|---|---|
| 1. Anchor seating | nil | X |
| 2. Elastic shortening | X | X |
| 3. Creep in the concrete | X | X |
| 4. Shrinkage of the concrete | X | X |
| 5. Relaxation of the strand | X | X |
| 6. Friction loss due to curvature and wobble | ... | X |

Owing to the inherent properties of concrete, it is almost impossible to derive the exact magnitude of the loss of prestress. The variables are time, exact value of the modulus of elasticity of concrete (since an empirical equation is used, whereas the curve has a continuous degree of change of curvature), cement content of the mix, ratio of fine aggregate to coarse aggregate, humidity during curing and in use, density due to the use of a water-reducing admixture, air content, age and method of curing, slump, change of concrete strength in time, volume-to-surface ratio, magnitude of dead load, and possible presence of nonprestressed reinforcement. The losses after transfer are interrelated as to time. Laboratory research has been conducted and papers have been written in an attempt to arrive at a finer method for design, and papers will continue to be written in the future.

In 1958 the ACI-ASCE Committee 423 recommended lump-sum estimated losses of 35 ksi (241 N/mm$^2$) for pretensioning beams and 25 ksi (172 N/mm$^2$) for posttensioning.* These values appeared in the 1963 American Concrete Institute Building Code 318 and in the American Association of State Highway and Transportation Officials Standard Specifications for Highway Bridges until 1975.

In 1975 the PCI Committee on Prestress Losses recommended both a simplified method and a general method for computing losses.* The simplified method appeared in the second edition (1978) of the *PCI Design Handbook* with a recommendation for use of the general method. However, in the third edition (1985) the ACI-ASCE Committee 423 method of 1979† appears without reference to the previous PCI Committee's method. In the 1982 edition of the Canadian *CPCI Design Handbook* the PCI Committee's simplified method appears with reference to the general method. However, in the second edition (1987) the PCI simplified method still appears, but for a detailed method the method proposed by Neville, Dilger, and Brooks (1983)‡ is recommended with examples.

The reader is referred to the third edition of the *PCI Design Handbook: Precast and Prestressed Concrete*, published by the Prestressed Concrete Institute, and to the third edition of the *Metric Design Manual: Precast and Prestressed Concrete,* published by the Canadian Prestressed Concrete Institute.

For preliminary design many designers use for initial loss at transfer 10 percent for stress-relieved strand and 7.5 to 10 percent for low-relaxation strand; for service conditions, 20 to 22 percent for stress-relieved strand and 12 to 20 percent for low-relaxation strand. When the magnitude of prestressing is small or the dead loads are light, these values will be too high. The loss of prestress has no effect upon ultimate-strength design unless the final stress is less than $0.5f_{pu}$. However, the loss of prestress will affect service conditions such as camber, deflection, cracking, and service stresses. To convert from a loss in ksi to a percentage of loss, divide by $0.74f_{pu}$ for pretensioning or $0.70f_{pu}$ for posttensioning.

---

*PCI Committee on Prestress Losses, "Recommendations for Estimating Prestress Losses," *PCI Journal*, July–August 1975.

†Paul Zia, H. K. Preston, N. L. Scott, and E. B. Workman, ACI-ASCE Committee 423: Prestress Concrete, "Estimating Prestress Losses," *Concrete International*, vol. 1, no. 6, June 1979; also Notes on ACI 318-83 Building Code Requirements for Reinforced Concrete with Design Applications by the Portland Cement Association.

‡A. M. Neville, W. H. Dilger, and J. J. Brooks, *Creep of Plain and Neutral Concrete*, Construction Press, London and New York, 1985.

### PCI simplified method

For normal-weight concrete:

Stress-relieved strand: total loss = $33.0 + 13.8f_{cir} - 4.5f_{cds}$
Low-relaxation strand: total loss = $19.8 + 16.3f_{cir} - 5.4f_{cds}$

For sand-lightweight concrete:

Stress-relieved strand: total loss = $31.2 + 16.8f_{cir} - 3.8f_{cds}$
Low-relaxation strand: total loss = $17.5 + 20.4f_{cir} - 4.8f_{cds}$

where $f_{cir} = f_{pst}/A_c + f_{pst}e^2/I - M_d e/I$

$f_{cds}$ = moment service dead load superimposed $e/I$ ($f_{cir}$ and $f_{cds}$ are stress in concrete at strand location)
$f_{pst}$ = prestress force after transfer after initial loss of 10 or 7.5 percent (elastic shortening and shrinkage up to transfer)

### ACI-PCA method

This follows the 1985 PCI recommendations and the Portland Cement Association notes on the 1983 ACI code.

$$\text{Total loss} = \text{ES} + \text{CR} + \text{SH} + \text{RE}$$

where ES = elastic shortening = $K_{es}E_s/E_{ci}\,f_{cir}$
$K_{es}$ = 1.0 for pretensioning and 0.5 to 1.0 for posttensioning; i.e.,$(N-1)/2N$, and $N$ = number of tendons in a beam stressed independently
$f_{cir}$ = $K_{cir}(f_{pa}/A_c + f_{pa}e^2/I) - M_d e/I$
$f_{pa}$ = initial force in strand after anchoring prior to losses
$A_c$ = area of concrete section
$E_s$ = modulus of elasticity of strand
$E_{ci}$ = modulus of elasticity of concrete at detensioning
$K_{cir}$ = 0.9 for pretensioning(100 percent- initial loss in percent) 100 and 1.0 for posttensioning
$e$ = strand eccentricity from $CG_s$ to $CG_c$
$I$ = moment of inertia
CR = creep of concrete under dead load = $K_{cr}E_s/E_c(f_{cir} - f_{cds})$
$K_{cr}$ = 2.0 for normal-weight concrete and 1.6 for sand-light weight concrete (creep factor for concrete, assuming noncompression steel)

$f_{cds}$ $= M_{ds}e/I$ ($M_{ds}$ is moment due to all superimposed dead loads)
SH = shrinkage of concrete = $(8.2 \times 10^{-6})\ E_s K_{sh}(1-0.06V/S)$ $(100-\text{RH})$
$K_{sh}$ = 1.0 for pretensioning; for posttensioning use table value related to time between end of moist curing and posttensioning:

| Days | 1 | 3 | 5 | 7 | 10 | 20 | 30 | 60 |
|---|---|---|---|---|---|---|---|---|
| $K_{sh}$ | 0.92 | 0.85 | 0.80 | 0.77 | 0.73 | 0.64 | 0.58 | 0.45 |

$V/S$ = volume-to-surface ratio for member
RH =relative humidity at project site
RE = strand relaxation = $[K_{re} - J(\text{SH} + \text{CR} + \text{ES})]C$

The values of $K_{re}$, $J$, and $C$ are taken from tables. $K_{re}$ and $J$ are relative to the grade of the strand and whether it is stress-relieved or low-relaxation. The value $C$ is from a table relating to $f_{pi}/A_{ps}$ and the type of strand. For 270-ksi strand, low-relaxation, $K_{re} = 5$ ksi, $J = 0.04$, and for $f_{py}/f_{pu}$ of 0.74, $C = 0.95$. For 270-ksi strand, stress-relieved, $K_{re} = 20$ ksi, $J = 0.15$, and for $f_{pi}/f_{pu}$ of 0.74, $C = 1.36$. For other strand strengths or $f_{pi}/f_{pu}$ ratios, refer to the 1985 PCI recommendations and PCA notes on the 1983 ACI code.

By using this method the maximum prestress loss values which need not be exceeded are:

| Strand type | Normal-weight concrete | Sand-lightweight concrete |
|---|---|---|
| Stress-relieved | 50 ksi | 55 ksi |
| Low-relaxation | 40 ksi | 45 ksi |

The recommended minimum value is 30 ksi, although some designers use 25 ksi for low-relaxation strand.

### AASHTO 1983 Design Code with interim update to 1985

The American Association of State Highway and Transportation Officials code gives both lump-sum values and a detailed method. The lump-sum values are:

| | 4-ksi concrete | 5-ksi concrete |
|---|---|---|
| Pretensioning | .... | 45 ksi maximum |
| Posttensioning strand | 32 ksi | 33 ksi maximum |
| Posttensioning bars | 22 ksi | 23 ksi maximum |

The detailed method is as follows:

$$\text{Total loss} = \text{ES} + \text{CR} + \text{SH} + \text{RE}$$

| Item | Pretension | Posttension |
|---|---|---|
| ES (elastic shortening): $f_{cir} = (f_{pi}/A + f_p e^2/I_g) - M_d e/I_g$ | $E_s/E_c f_{cir}$ | $0.5E_s/E_c f_{cir}$ |
| SH shrinkage, lb/in² | 17,000-100RH | 0.8(17,000 - 150RH |
| CR (concrete creep): $f_{cd} = M_{dis} e/I_g$ | $12f_{cir} - 7f_{cd}$ | $12f_{cir} - 7f_{cd}$ |
| RE (relaxation of prestressing steel, psi | | |
| SR (stress-relieved strand | 20,000 - 0.4ES -0.2(SH + CR) | 20,000-0.3FR-0.4ES -0.2(SH + CR) |
| LR (low-relaxation strand) | 5000 - 0.1ES -0.05(SH + CR) | 5000-0.07FR-0.1ES -0.05(SH +CR) |

NOTE: FR = friction and wobble loss.

### PTI 1985 recommendation

The Post-Tensioning Institute (PTI) recommends the following for lump-sum losses :

| | Slab | Beams and joists |
|---|---|---|
| SR strand | 30 ksi | 35 ksi |
| LR strand | 15 ksi | 20 ksi |
| Bars | 20 ksi | |

The Post-Tensioning Institute does not make a recommendation for a detailed method, which is left to the discretion of the designer.

### Ontario Highway Bridge Design Code, 1983

Recommendations are in megapascals: 1 ksi = 6.9 MPa.

Loss of prestress at transfer = ANC + FR + $REL_1$ + ES + = $\Delta f_{ps1}$

where ANC = anchor loss = seating slip, inches/tendon length, inches $\times E_s$

FR = friction loss due to curvature and wobble (posttensioning)

ES = elastic shortening for pretensioning = $E_s/E_{ci} f_{cir}$; for posttensioning = $(N - 1)/2N \times E_s/E_{ci} f_{cir}$

$f_{cir} = f_{pst}/A_c + f_{pst}e^2/I - M_d e/I$

$N$ = number of tendons in a beam stressed independently

$REL_1 = \log 24t/10[(f_{si}/f_{py}) - 0.55] f_{si}$ for stress-relieved strand

$= \log 24t/45[(f_{si}/f_{py}) - 0.55] f_{si}$ for low-relaxation strand

Loss of prestress after transfer = CR + SH + $REL_2$ = $\Delta_{fps2}$

where CR = creep = $(1.37 - 0.77 \times 0.01H2) K_{cr} E_s/E_c (f_{cir} - f_{cds})$

$F_{cds} = M_{ds} e/I$

$K_{cr}$ = 2.0 for pretensioning and 1.6 for posttensioning

$H$ = relative humidity (see map of area)

SH = shrinkage = 117−1.05$H$ for pretensioned members of normal-density concrete; 94−0.85$H$ for posttensioned members

$REL_2 = [(f_{st}/f_{pu}) - 0.52]\,[0.42 - (CR + SH)/1.25 f_{pu}]\, f_{pu} = >0.01 f_{pu}$ for stress-relieved strand

$= [(f_{st}/f_{pu}) - 0.55]\,[0.34 - (CR + SH)/1.25 f_{pu}][f_{pu}/3 = >0.002 f_{pu}$ for low-relaxation strand

$f_{pu}$ = 1860 MPa

LR strand = $0.75 f_{pu}$ SR strand = $0.7 f_{pu}$

### PCI detailed method: a time-step method

This method is based on the *PCI Design Handbook* of 1978 and the *CPCI Design Handbook* of 1982. The number of time steps is up to the designer.

### CPCI detailed method, 1987

For this detailed method the designer is referred to the *Metric Design Manual: Precast and Prestressed Concrete*, published by the Canadian Prestressed Concrete Institute.

### Example

The following two computer runs are for the same member with the same loads. One run uses the ACI-ASCE method, and the other uses the PCI detailed method. A third comparison uses the PCI simplified method. The attempt is not to discredit any of the methods but to illustrate each one. If a member has light dead loads, the loss of prestress is small. For the PCI detailed method the initial loss input has an effect on the final results. (See Figs. 36.1 and 36.2.)

## Program 2: Pretensioned Concrete Wall Panels; Single Plane of Prestressing

The empirical equation for wall design as shown in both the Uniform Building Code and the American Concrete Institute Recommended Building Code ACI-318 cannot be used when designing prestressed concrete walls. The $P$-$\Delta$ method could be used if it were modified for prestressing. For design, the interaction curve method is most commonly used.

The interaction equations for walls are similar to those for columns. Walls usually have a single plane of prestressing where the wall thickness is less than 8 in. For wall thicknesses from 8 to 10 in either one or two planes are used. For wall thicknesses greater than 10 in two planes of prestressing are used. If two planes of prestressing are used

```
INPUT DATA

Live Load in lb./sq.ft = 16
Dead Load Superimposed and topping wt. in lb / sq.ft.= 12
Weight of the prestressed unit in lb/lin.ft.= 540
Span of the unit in ft. = 60
WIDTH of the unit in ft.= 8
Unit cros sectional area in sq. in = 535
Moment of Inertia = 44458
Eccentric distance e from cgc to cgs in inches = 17.4
surface perimeter /section area V/S = 1.737 inch
Strand strength fpu = 270 Es = 28000 ksi
Number of strands = 6 Area per strand = .153
Area of prestress steel total = .918 sq.in.
Type of strand is Low Relaxation ,fps= 0.74*fpu = 199.8 ksi
 System is pretensioned
Type of concrete is Normal Wt.
 Ec in ksi = 4074.281 Eci in ksi = 3644.147
Relative Humidity = 75 percent


SOLUTION

 PCI SIMPLIFIED METHOD - PRETENSIONED
    LR Strand NW Concrete , fcir= .3312454 fcds= .2028917
TOTAL LOSS = 24.1 KSI
     Percent of loss = 12

------------------------------

 ACI-ASCE METHOD

FCIR= .2914479 FCDS = .2028917
KRE= 5 J= .04 C= .95 KES= 1 KCR= 2 KSH= 1
elastic shorenting ES= 2.239355 creep CR= 1.217183
shrinkage sh= 5.141777 relaxation RE= 4.423264 ksi
 Total loss in ksi = 13.02
 Percent of loss = 6.5

*********************************************
 END OF PROGRAM
```

**Figure 36.1** Computer printout: loss of prestress; ACI-ASCE method.

and the distance from the strand to the face of the wall is the same for both sides of the wall (which is the usual case), then the prestressed column interaction curve computer program is used. In this case the column width is the unit wall length. If there is only one plane of prestressing, the following program can be used.

The average prestressed force on the concrete section ($f_{pc}$) should be greater than 225 psi. If so, the minimum wall reinforcing steel can be eliminated. However, many designers will still use mild steel transverse to the prestressing strand.

The width of the wall used for design is the unit of measurement. In the United States this is the foot (12 in); under the metric system this is the meter. Where there is a concentrated load, the unit length is usually 4 times the wall thickness plus the bearing width of the concentrated load, i.e., $b + 4t$.

```
LPRINT : LPRINT TAB(60);" Page  - 6" : LPRINT : LPRINT
LPRINT TAB(6);STRING$(55,42): LPRINT
LPRINT TAB(15);"LOSS OF PRESTRESSED ACI-ASCE METHOD":LPRINT
LPRINT TAB(6); STRING$(55,42)
LPRINT CHR$(27);"n" :LPRINT CHR$(14) : LPRINT TAB(5)" INPUT  DATA "
LPRINT CHR$(15): LPRINT CHR$(27);"N"
INPUT " Input Live Load in lb./sq.ft.",LL : LPRINT TAB(6);"Live Load in lb./sq.ft =";LL
INPUT " Input dead Load Superimposed and topping wt in lb/ sq.ft. ",DLS
LPRINT TAB(6);"Dead Load Superimposed and topping wt. in lb / sq.ft.=";DLS
INPUT "Input weight of the prestressed unit in lb/ lin ft. ",DL : LPRINT TAB(6)"Weight of the
    prestressed unit in lb/lin.ft.=";DL
INPUT " Input unit span in ft = ",L : LPRINT TAB(6);"Span of the unit in ft. =";L
INPUT "Input width of the unit in ft.=",B : LPRINT TAB(6);"WIDTH of the unit in ft.=";B
INPUT "input cross sectional area of the unit in sq. in.=",AC : LPRINT TAB(6);"Unit cros
    sectional area in sq. in =";AC
INPUT "Input moment of Inertia of the unit=",I:LPRINT TAB(6);"Moment of Inertia = ";I
INPUT " Input eccentric distance e from cgc to cgs in inches=",E: LPRINT TAB(6);"Eccentric
    distance e from cgc to cgs  in inches =";E
INPUT " Input V/S , (surface perimeter / area) =",VS : LPRINT TAB(6);"surface perimeter
    /section area V/S = ";VS;" inch"
INPUT "Input strand fpu in ksi and Modulus of Elasticity in ksi= ",FPU, ES : LPRINT
TAB(6);"Strand strength fpu = ";FPU;" Es = ";ES;" ksi"
INPUT " Input number of strands and area of one strand in sq. in. = ",NS,SAS
APS =NS*SAS
LPRINT TAB(6);"Number of strands = ";NS;" Area per strand =";SAS
LPRINT TAB(6);"Area of prestress steel total =";APS;" sq.in."
10
INPUT " Type LR if strand is Low Relax or SR if strand is Stress Releived =",T$
IF T$ = "LR" THEN LPRINT TAB(6);"Type of strand is Low Relaxation ,fps= 0.74*fpu
    =";.74*FPU;" ksi": GOTO 20
IF T$ ="SR" THEN LPRINT TAB(6);"Type of strand is Stress Releived, fps=0.70*fpu =";.7*FPU
    ;" ksi": GOTO 20
PRINT " Try again ":GOTO 10
20
IF T$="LR" THEN PI =.74*FPU*APS
IF T$="SR" THEN PI = .7 *FPU*APS
30
INPUT " If the member is pretensioned type PRET, if post tension typePOST your reply= ",P$
IF P$="PRET" THEN LPRINT TAB(6);" System is pretensioned": GOTO 40
IF P$="POST" THEN LPRINT TAB(6);" System is post tensioned" :GOTO 40
PRINT " Try again" : GOTO 30
40
INPUT " Input strength of concrete f'ci in ksi at detensioning = ",FCI
INPUT "Input strength of concrete at 28 days f'c in ksi =",FC
50
INPUT "Type NW for normal wt concrete or type SLW FOR sand light wt. concrete, your
    reply=",W$
IF W$="NW" THEN LPRINT TAB(6);"Type of concrete is Normal Wt." : GOTO 60
IF W$="SLW" THEN LPRINT TAB(6);"Type of concrete is Sand Light Wt.: GOTO 60
PRINT "TRY AGAIN " : GOTO 50
60
INPUT " Input concrete wt, in lb/ft.cub. = ", WC
ECI=33*WC^1.5*(FCI*1000)^.5/1000 : EC=33*WC^1.5*(FC*1000)^.5/1000
LPRINT TAB(6);" Ec in ksi = ";EC; " Eci in ksi = ";ECI
INPUT " Input relative humidity at project location in percent ",RH :LPRINT TAB(6);"Relative
    Humidity =";RH;" percent"
LPRINT CHR$(27);"n" : LPRINT CHR$(14): LPRINT TAB(5);" SOLUTION" :LPRINT CHR$(15) :
LPRINT CHR$(27);"N"
MDL = DL *L^2/8 *12/1000: MDLS= DLS*B*L^2/8*12/1000
IF T$="LR" THEN C1=.925 ELSE C1=.9
FCIR=PI*C1/AC+PI*C1*E^2/I - MDL*E/I : FCDS=MDLS*E/I
```

**Figure 36.2** Computer program listing: loss of prestress; ACI-ASCE method.

```
IF T$="LR" AND W$="NW" THEN TL=19.8+16.3*FCIR-5.4*FCDS
IF T$="SR" AND W$ = "NW" THEN TL=33+13.8*FCIR-4.5*FCDS
IF T$="LR" AND W$="SLW" THEN TL = 17.5+20.4*FCIR-4.8*FCDS
IF T$="SR" AND W$="SLW" THEN TL= 31.2+16.8*FCIR-3.8*FCDS
LPRINT TAB(6)" PCI SIMPLIFIED METHOD - PRETENSIONED"
LPRINT TAB(10) ; T$;" Strand ";W$;" Concrete , fcir=";FCIR;"fcds= ";FCDS
LPRINT TAB(6);"TOTAL LOSS =";INT(TL*100)/100; " KSI"
LPRINT TAB(10);" Percent of loss = ";INT(TL/(PI/APS)*100)
LPRINT : LPRINT TAB(6);STRING$(30,45): LPRINT
LPRINT TAB(6)" ACI-ASCE METHOD" : LPRINT
KES=1!
IF T$="LR" THEN C=.95 ELSE C=1!
IF FPU=270 AND T$="LR" THEN KRE=5 : J=.04
IF FPU =270 AND T$="SR" THEN KRE=20 : J=.15
IF FPU = 250 AND T$="SR" THEN KRE=18.5 : J=.14
IF P$="PRET" THEN KCIR=.9 :KCR=2!: KSH=1!
IF P$="POST" THEN KCIR=1!: KCR=1.6 : GOSUB 200
IF T$="LR" THEN PI=.74*FPU*APS
IF T$="SR" THEN PI=.7*FPU*APS
FCIR=KCIR*(PI/AC+PI*E^2/I)-MDL*E/I
FCDS=MDLS*E/I
ESS=KES*ES/ECI*FCIR
CR=KCR*ES/EC*(FCIR-FCDS)
SH=8.2*10^-6*KSH*ES*(1-.06*VS)*(100-RH)
RE=(KRE-J*(SH+CR+ESS))*C
TLACI=ESS+CR+SH+RE : TLACI = INT(TLACI*100)/100
LPRINT TAB(6);"FCIR= ";FCIR;" FCDS = ";FCDS
LPRINT TAB(6);"KRE=";KRE;" J= ";J;" C= ";C; "KES=";KES;"KCR=";KCR;" KSH=";KSH
LPRINT TAB(6);"elastic shorenting ES= ";ESS;"creep CR=";CR
LPRINT TAB(6);"shrinkage sh=";SH;"relaxation RE=";RE;" ksi"
LPRINT TAB(6);" Total loss in ksi = ";TLACI
LPRINT TAB(6)" Percent of loss = ";INT(TLACI/(PI/APS)*1000)/10
LPRINT : LPRINT TAB(6);STRING$(50,42) : LPRINT TAB(6);" END OF PROGRAM"
END
200
INPUT "Input age at post tensioning in days =",TD
IF TD=<1 THEN KSH=.92 : GOTO 300
IF TD=<3 THEN KSH=.85 : GOTO 300
IF TD=< 5 THEN KSH=.8 : GOTO 300
IF TD=< 7 THEN KSH=.77 : GOTO 300
IF TD =< 10 THEN KSH=.73 : GOTO 300
IF TD =< 20 THEN KSH=.64 :GOTO 300
IF TD=< 30 THEN KSH=.58 : GOTO 300
IF TD > 30 THEN KSH=.6 : GOTO 300
300
RETURN
```

**Figure 36.2** (*Continued*)

Precast wall panels if placed vertically are usually pretensioned. If the wall system is a vertical stacked panel system, as for multistory housing-type buildings, the panels are usually posttensioned. If the wall panels are placed horizontally, they are usually not prestressed or posttensioned. The following computer program is designed for panels placed vertically and precast pretensioned. These panels may be more than one story in height and contain either corbels or ledges to receive the vertical loads. Moments on the panel are caused by the loads being placed eccentrically plus moments due to wind force and suction or seismic forces in opposite directions. The larger of the wind or the seismic force governs.

Strand is used for precast pretensioned panels. The diameter may be $^3/_8$, $^7/_{16}$, or $^1/_2$ in. The strand may be 250 or 270 ksi and may be either stress-relieved or low-relaxation. Today low-relaxation strand is common.

For 270-ksi strand, $E_{ps}$ = 28,000 ksi.

| Diameter, in | Area, in² |
|---|---|
| $^3/_8$ | 0.085 |
| $^7/_{16}$ | 0.015 |
| $^1/_2$ | 0.153 |

To determine the magnitude of stress beyond the proportional limit the stress-strain curve must be used. The following program models the stress-strain curve for low-relaxation strand. The user should obtain stress-strain curves from the strand manufacturer and then readjust the modeling accordingly. (See Figs. 36.3 through 36.6.)

The computer program will compute the values of $P_o$ and $P_{u,\max}$ and the values of $P_u$ and $M_u$ for the variable $C_u$. The interaction curve can be constructed from these values, or the values from the listing can be used. The program will also draw the curve and locate the given required $P_u$ and $M_u$ values. The required $P_u$-$M_u$ value usually is located in the lower portion of the interaction curve. For wall panels the moment is zero at the base, and since the required $P_u$ value is small, reinforcing steel usually need not be added within the strand development length of 50 bar diameters. However, this should be checked. When entering the required $M_u$ value, slenderness must be considered. Slenderness will require that the moment be magnified.

## Program 3: Double-T Properties

Double-T sections are often designed as composite members. The thickness of the topping ranges from 2 to 3 in and either is determined by the space requirements for floor electrical outlet and distribution boxes or is a minimum of 2½ in to develop the diaphragm. The topping

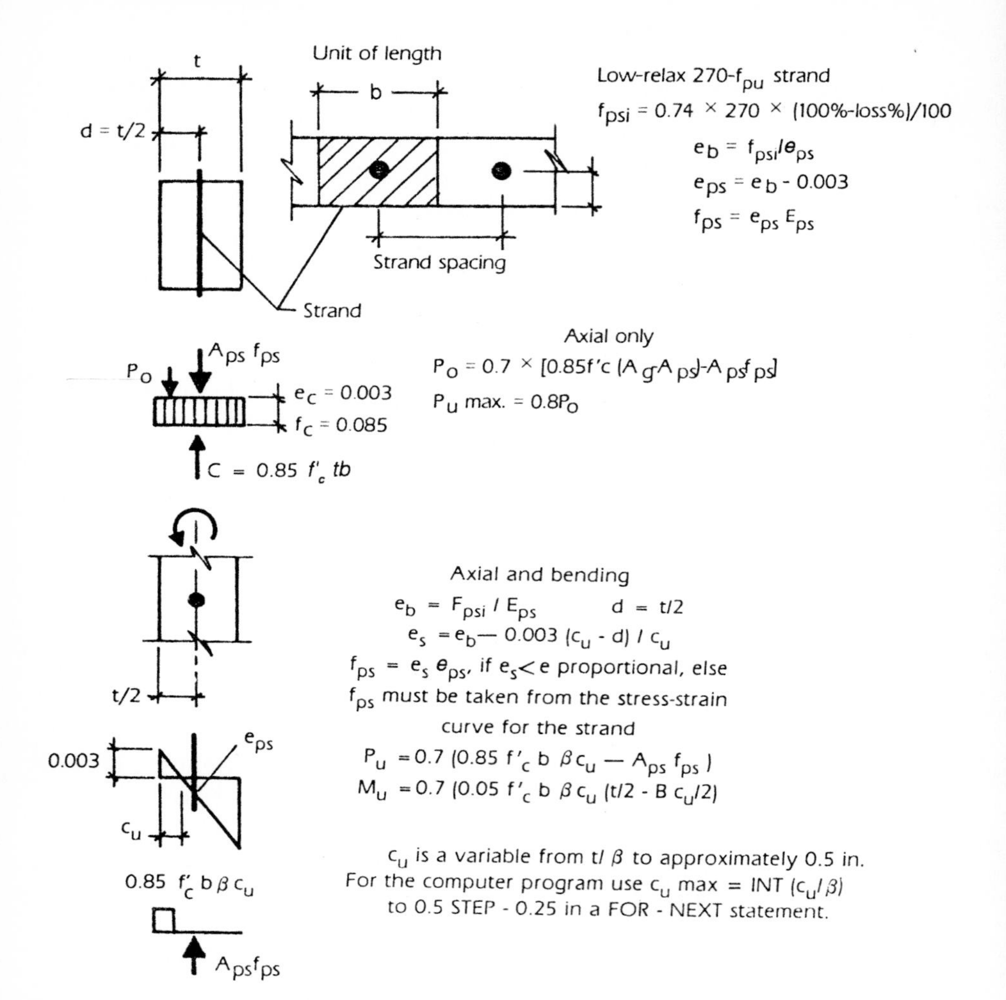


**Figure 36.3** Theory: single plane of prestressing for wall panels.

also serves as a leveling element and as a diaphragm to transfer the lateral forces, due to wind or earthquake, to the wall or frame system. The strength of the concrete used for the topping is usually lower than that used for the prestressed concrete unit. Also the topping may be normal-weight, whereas the prestressed unit may be sand-lightweight concrete. If the modulus of elasticity of the topping is not the same as that of the unit, the effective topping width is used in computing the composite moment of inertia. The modulus of elasticity is based upon both the unit weight or mass of the concrete and its strength.

When using the program, a picture of a composite double T appears

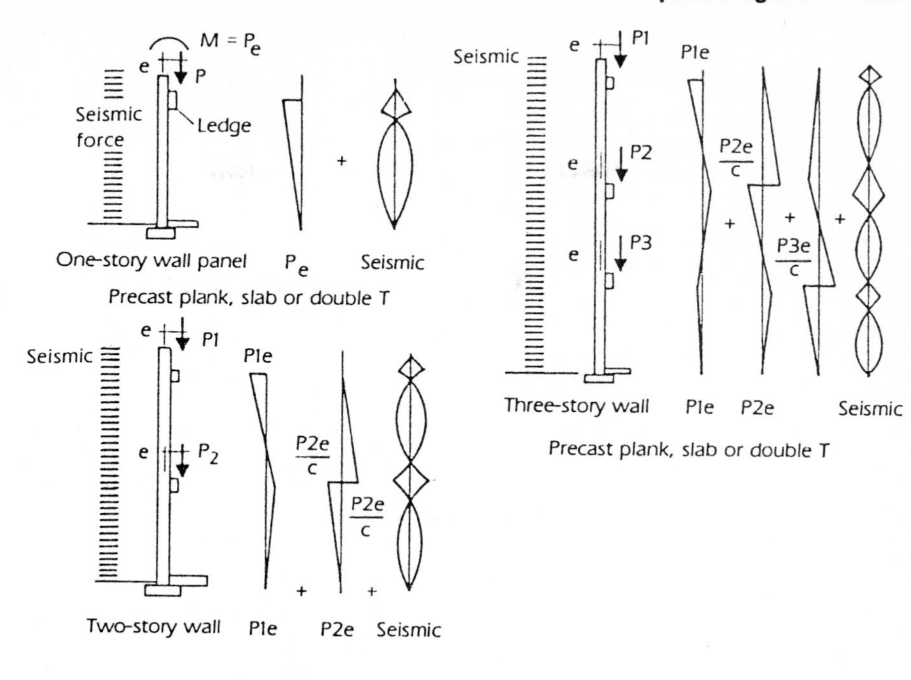


**Figure 36.4** Moments on wall panels. $K = I/L$; $I$ is assumed to be constant; cof = 1/2.

**Example of Moment Distribution for Solution**

| | Column | cof | Column | Beam | Column | cof | Column | Beam |
|---|---|---|---|---|---|---|---|---|
| k/ Σk | 1.0 | × | 0.5 | 0.0 | 0.5 | × | 1.0 | 0.0 |
| FEM * | 0.0 | | 0.0 | +80 | 0.0 | | 0.0 | +40.0 |
| Distribution | 0.0 | | -40.0 | 0 | -40.0 | | -40.0 | 0.0 |
| Carryover | -20.0 | | 0.0 | 0 | -20.0 | | -20.0 | 0.0 |
| Distribution | +20.0 | | +10.0 | 0 | +10.0 | | +20.0 | 0.0 |
| Carryover | +5.0 | | +10.0 | 0 | +10.0 | | +5.0 | 0.0 |
| Distribution | -5.0 | | -10.0 | 0 | -10.0 | | -5.0 | 0.0 |
| Carryover | -5.0 | | -2.5 | 0 | -2.5 | | -5.0 | 0.0 |
| Distribution | +5.0 | | +2.5 | 0 | +2.5 | | +5.0 | 0.0 |
| Total | 0.0 | | -30.0 | +80 | -50.0 | | +40.0 | +40.0 |

* FEM = fixed-end moment.

## DESIGN INPUT

```
WALL DEPTH IN INCHES =   6
UNIT WALL WIDTH IN INCHES =  12 INCHES
CONCRETE STRENGTH IN KSI =  5
BETA =  .8
DISTANCE FROM EDGE OF WALL TO STRAND =  3
SPACING OF STRAND =  12
AREA PER STRAND =  .115
TYPE OF STRAND = LR
STRAND STRENGTH IN KSI =  270
STRAND MODULUS OF ELASTICITY IN KSI =  28000
% OF LOSS OF PRESTRESS =  18
Pu REQUIRED =  50  kips per ft.
Mu REQUIRED = 120  inch kips 0  0
```

## DESIGN OUTPUT

```
 PO=   207.4   KIPS      P-MAX. =  165.9     KIPS
VALUES OF PU AND MU FOR VALUES OF C IN KIPS AND INCH KIPS
```

| VALUES OF C | PU | MU | Ecc | Es-tension | Ec- tension |
|---|---|---|---|---|---|
| 7 | 190.5 | 39.9 | .2 | .004137 | 0 |
| 6.75 | 183.3 | 57.8 | .31 | 4.184619E-03 | 0 |
| 6.5 | 176 | 74.2 | .42 | 4.235901E-03 | 0 |
| 6.25 | 168.8 | 89.2 | .52 | 4.291286E-03 | 0 |
| 6 | 161.5 | 102.8 | .63 | 4.351286E-03 | 0 |
| 5.75 | 154.2 | 114.9 | .74 | 4.416503E-03 | 1.304348E-04 |
| 5.5 | 146.9 | 125.6 | .85 | 4.487649E-03 | 2.727273E-04 |
| 5.25 | 139.6 | 134.9 | .96 | 4.565571E-03 | 4.285714E-04 |
| 5 | 132.3 | 142.8 | 1.07 | 4.651286E-03 | .0006 |
| 4.75 | 124.9 | 149.2 | 1.19 | 4.746023E-03 | 7.894737E-04 |
| 4.5 | 117.5 | 154.2 | 1.31 | 4.851285E-03 | 9.999999E-04 |
| 4.25 | 110.1 | 157.7 | 1.43 | 4.968932E-03 | 1.235294E-03 |
| 4 | 102.7 | 159.9 | 1.55 | 5.101285E-03 | .0015 |
| 3.75 | 95.2 | 160.6 | 1.68 | 5.251286E-03 | .0018 |
| 3.5 | 87.7 | 159.9 | 1.82 | 5.422714E-03 | 2.142857E-03 |
| 3.25 | 80.1 | 157.7 | 1.96 | 5.620516E-03 | 2.538461E-03 |
| 3 | 72.4 | 154.2 | 2.12 | 5.851286E-03 | .003 |
| 2.75 | 64.7 | 149.2 | 2.3 | 6.124013E-03 | 3.545455E-03 |
| 2.5 | 56.8 | 142.8 | 2.51 | 6.451285E-03 | .0042 |
| 2.25 | 48.8 | 134.9 | 2.76 | 6.851286E-03 | .005 |
| 2 | 40.5 | 125.6 | 3.1 | 7.351286E-03 | .006 |
| 1.75 | 31.9 | 114.9 | 3.6 | 7.994143E-03 | 7.285714E-03 |
| 1.5 | 23.4 | 102.8 | 4.39 | 8.851285E-03 | 9.000001E-03 |
| 1.25 | 14.8 | 96.4 | 6.51 | 1.005129E-02 | .0114 |
| 1 | 6.9 | 88.3 | 12.79 | 1.185129E-02 | .015 |
| .75 | -.4 | 74.6 | -186.5 | 1.485129E-02 | .021 |
| .5 | -7.5 | 55.5 | -7.4 | 2.085129E-02 | .033 |

```
Pu REQUIRED = 50   kips per ft
Mu REQUIRED =  120  inch kips  per ft.or  10  ft kips per ft.
FOR PU REQUIRED MU ALLOWABLE =  142.8  inch kips'
    PU ALLOWABLE = 56.8  kips, MU/PU =  2.514085
```

**Figure 36.5** Computer printout: prestressed wall out of plane; intersection curve data.

END OF PROGRAM.

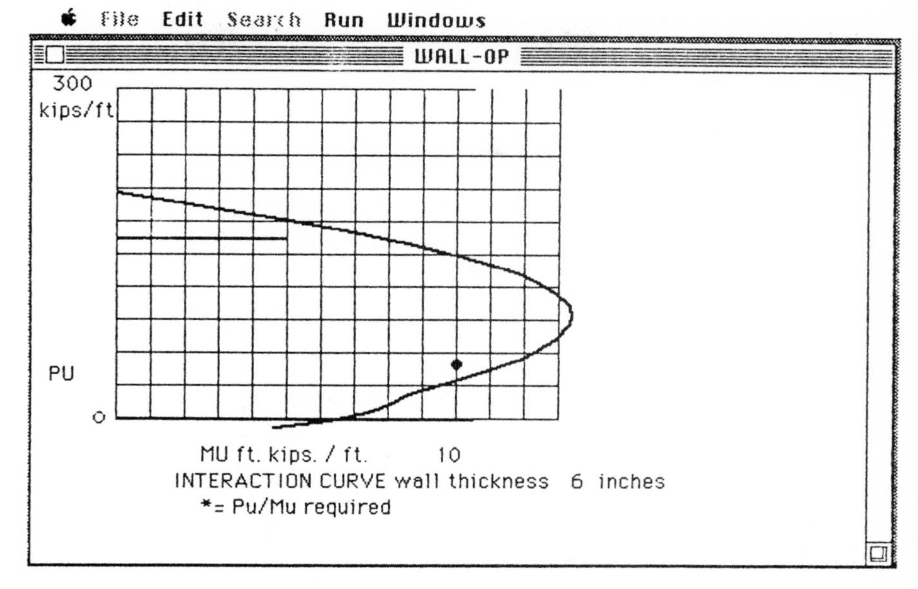


**Figure 36.5** *(Continued)*

and questions are asked regarding the dimensions and concrete weight and strength. The program then computes the physical properties, redraws the picture, and adds the physical properties. The picture is then sent to the printer to provide a printed output. (See Figs. 36.7 and 36.8.)

## Program 4: Noncomposite Double-T Design (United States System)

Double T's are used for both roof-framing systems and floor-framing systems. For roof-framing systems they are usually noncomposite, while for floor-framing systems they are usually composite. Because of the high cost of steel forms standard shapes are used. However, some nonstandard shapes can be obtained by placing blockouts in the standard forms. The designer must consult the regional precast prestressed concrete manufacturer regarding the standard sizes available. The precaster will also furnish data on the physical properties and suggested load-span tables. The variables in design are the loads, the number and size of the strand, the type and strength of the concrete, and mild steel (if used).

The program is for a noncomposite double-T design, using the United States (foot-pound) system. This was formerly known as the British

```
LPRINT : LPRINT TAB(10);STRING$(50,42):LPRINT
LPRINT TAB(20);"PRESTRESSED WALL OUT OF PLANE "
LPRINT TAB(20);"INTERACTION CURVE DATA": LPRINT
LPRINT TAB(10);STRING$(50,42)
DIM P(800),M(80)
100 INPUT " WALLTHICKNESS IN INCHES = ";L
PRINT" UNIT OF WIDTH = 12 INCHES "
INPUT " DISTANCE TO STRAND FROM EDGE OF WALL = ";D
INPUT " CONCRETE STRENGTH IN KSI AND BETA = ";FC,BE
INPUT "AREA PER STRAND AND STRAND SPACING";SA,SP
AP=SA*12/SP
INPUT " STRAND FPU AND MOD. OF ELASTICITY = ";FU,ES
132 INPUT " IF STRAND IS STRESS RELIEVE TYPE SR, IF LOW RELAX TYPE LR ? ";TS$
IF TS$="SR" THEN FI=.7 * FU : GOTO 138
IF TS$ = "LR" THEN FI=.74*FU : GOTO 138
GOTO 132
138 INPUT " SERVICE LOSS OF PRESTRESS IN PERCENT = ";PL
INPUT "Input your required Pu in kips per ft. =",PR
INPUT "Input you required Mu in inch kips per ft.=",MR
REM COMPUTE PO AND P MAX
EB=(FI*(100-PL)/100)/ES
E1=EB-.003 :F1=E1*ES
PO=.7*(.85*FC*(L*12-AP)-AP*F1)
PM = .8*PO
LPRINT CHR$(27);"n":LPRINT CHR$(14):
LPRINT TAB(6);" DESIGN  INPUT" : LPRINT CHR$(15) : LPRINT CHR$(27);"N"
LPRINT TAB(6);" WALL DEPTH IN INCHES =  ";L
LPRINT TAB(6);" UNIT WALL WIDTH IN INCHES =  12 INCHES"
LPRINT TAB(6);" CONCRETE STRENGTH IN KSI = ";FC
LPRINT TAB(6);" BETA = " ; BE
LPRINT TAB(6);" DISTANCE FROM EDGE OF WALL TO STRAND = ";D
LPRINT TAB(6);" SPACING OF STRAND = ";SP
LPRINT TAB(6);" AREA PER STRAND = " ; SA
LPRINT TAB(6);" TYPE OF STRAND = ";TS$
LPRINT TAB(6);" STRAND STRENGTH IN KSI = ";FU
LPRINT TAB(6);" STRAND MODULUS OF ELASTICITY IN KSI = ";ES
LPRINT TAB(6);" % OF LOSS OF PRESTRESS = "; PL
LPRINT TAB(6);" Pu REQUIRED = ";PR;" kips per ft."
LPRINT TAB(6);" Mu REQUIRED =";MR;" inch kips" per ft.
LPRINT CHR$(27);"n":LPRINT CHR$(14) :
LPRINT TAB(6);"   DESIGN  OUTPUT" : LPRINT CHR$(15) : LPRINT  CHR$(27);"N"
LPRINT TAB(6); " PO=  "; INT(PO*10)/10;"  KIPS    P-MAX. = "; INT(PM * 10 )/10; "KIPS"
LPRINT TAB(6);"VALUES OF PU AND MU FOR VALUES OF C IN KIPS AND INCH KIPS "
LPRINT
LPRINT TAB(6);"VALUES OF C";TAB(20);"PU";TAB(30);"MU";TAB(38); "Ecc";TAB(46);
     "Es-tension";TAB(63);"Ec- tension
LPRINT TAB(6);STRING$(60,45)
H=INT (L/BE)
Z=0! : MX=0!
FOR C=H TO .5 STEP - .25
E2= EB-.003*(C-D)/C :F2= E2*ES
IF E2> .008 AND E2< .009 THEN F2=222+22*(E2-.008)/.001
IF E2>.009 AND E2<.01 THEN F2=244 + 14*(E2-.009 )/.001
IF E2>.01 AND E2< .011 THEN F2=258+8*(E2-.01)/.001
IF E2>.011 AND E2< .012 THEN F2=266+3*(E2-.011)/.001
IF E2 > .012 THEN F2= 270
PU=.7*(.85*FC*12*BE*C-AP*F2)
MU=.7*(.85*FC*12*BE*C*(L/2-BE*C/2)-AP*F2*(L/2-D))
IF PU < .1*PO THEN MU=MU/.7*(.9-.2*PU/(.1*PO))
ET = .003*(L-C)/C
```

**Figure 36.6** Computer program listing: prestressed concrete wall panel out of plane; single plane of prestressing.

```
IF C>L THEN ET=0!
PU= INT (PU*10)/10 : MU = INT (MU*10)/10
IF PU>PR THEN MA=MU  : ED=MU/PU :  PA=PU
LPRINT TAB(8);C;TAB(18);PU;TAB(28);MU;TAB(37);INT(MU/PU*100)/100;
TAB(44);E2;TAB(61);ET
P(Z)=PU : M(Z)=MU : Z=Z+1 : IF MX > MU THEN MX=MU
NEXT C
LPRINT TAB(6);STRING$(60,45): LPRINT: LPRINT
LPRINT TAB(6);" Pu REQUIRED =";PR;"  kips per ft"
LPRINT TAB(6);" Mu REQUIRED = "MR;" inch kips  per ft.or ";MR/12;" ft kips per ft."
LPRINT TAB(6);" FOR PU REQUIRED MU ALLOWABLE = ";MA;" inch kips'
LPRINT TAB(6);"     PU ALLOWABLE =";PA;" kips, MU/PU = ";ED
IF MA<MR THEN LPRINT "Wall will not work, increase wall thickness or decrease strand spacing
   and run again"
IF PR>PM THEN LPRINT "Wall will not work, increase wall thickness and run again"
LPRINT CHR$(27);"n":LPRINT CHR$(14) :LPRINT TAB(2);" END OF PROGRAM, HAVE A COKE"
LPRINT CHR$(15) : LPRINT CHR$(27);"E"
REM GRAPHIC ROUTINE
IF PO =< 100 THEN SP=.5 : GOTO 600
IF PO =< 200 THEN SP = 1! : GOTO 600
IF PO=< 300 THEN SP=1.5 : GOTO 600
IF PO=< 400 THEN SP=2! : GOTO 600
IF PO=< 500 THEN SP=2.5 : GOTO 600
IF PO=< 600 THEN SP = 3! : GOTO 600
IF PO> 600 THEN GOTO 1000
600
IF MX/12 =< 13 THEN SM=.05 : GOTO 700
IF MX/12=<26 THEN SM=.1 : GOTO 700
IF MX/12=<39 THEN SM=.15 : GOTO 700
IF MX/12 =< 52 THEN SM=.2 : GOTO 700
IF MX/12 =<65 THEN SM=.25 : GOTO 700
IF MX/12 =< 78 THEN SM=.3 : GOTO 700
IF MX/12 > 78 THEN GOTO 1000
700
CLS
LINE (50,10)-(50,210) : LINE (50,210)-(260,211),, BF
LINE (50,10)-(260,10)
FOR I=1 TO 10 : LINE(50,10+I*20)-(310,10+I*20) : NEXT I
FOR J=1 TO 13 : LINE (50+J*20,10)-(50+J*20,210) : NEXT J
CIRCLE (40,210),3
PRINT PTAB(4);200*SP
PRINT PTAB(4);"kips/ft"
FOR K=1 TO 9 : PRINT : NEXT K
PRINT PTAB(10);"PU"
PRINT : PRINT
PRINT PTAB(100);"MU ft. kips. / ft." ; PTAB(230);200*SM
PRINT PTAB(80);" INTERACTION CURVE wall thickness ";L;"inches"
PRINT PTAB (100);"*= Pu/Mu required"
CALL PENSIZE (2,2)
CALL MOVETO (50,210-INT (PO/SP))
FOR I=0 TO Z-1 : CALL LINETO (50+INT(M(I)/12/SM),210-INT(P(I)/SP)) : NEXT I
FOR A=1 TO 3
CIRCLE ( 50+INT(MR/12/SM),210-INT(PR/SP)),A  : NEXT A
LINE (50,210-INT (PM/SP))-(150,210-INT(PM/SP)+1),,BF
LCOPY
1000
PRINT " END OF THE RUN, IF YOU WISH TO RERUN TYPE YES ELSE TYPE NO"
INPUT " YOUR REPLY ";REP$
IF REP$="YES" THEN GOTO 100
END
```

**Figure 36.6** *(Continued)*

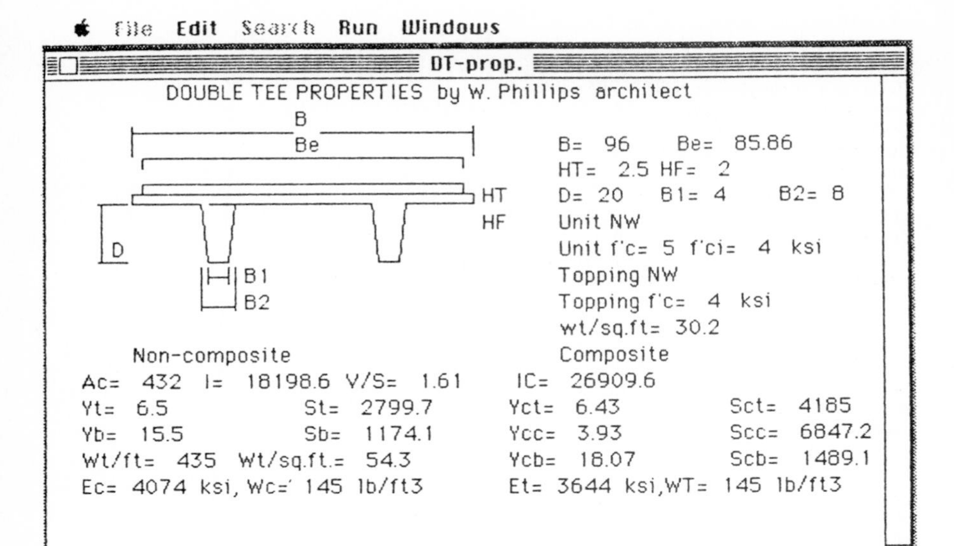


**Figure 36.7** Computer program printout: double-T properties.

system; however, the British now use a metric system. Load and resistance factors are those of the Uniform Building Code and the ACI recommended building code. Included with the program are a flowchart (Fig. 36.9), a computer run (Figs. 36.10 and 36.11), and a hand solution of the same problem.

### Hand solution for noncomposite prestressed precast double T

**Design conditions.** The roof member has a span of 60 ft, and the superimposed dead load is 12 lb/ft². The live load is 16 lb/ft². Normal-weight concrete = 145 lb/ft³.

$$f'_c = 5\,\text{ksi} \quad f'_{ci} = 4\,\text{ksi}$$

Use ½-in-diameter, 270-ksi, low-relaxation strand with an initial loss of 7½ percent. Assume a service loss of 18 percent; then verify the loss when the number of strands and hence the prestressing force are known.

**Member size.** See Fig. 36.12. Determine shape properties or use data from the manufacturer.

$$\text{Area} = 535\ \text{in}^2 \qquad V/S = 1.737 \qquad \text{Weight/ft} = 540\ \text{lb}$$

$$Y_t = 9.59\ \text{in} \qquad I = 44.458 \qquad \text{Weight/ft}^2 = 67.5\ \text{lb}$$

$$Y_b = 20.41\ \text{in} \qquad S_b = 2178$$

$$S_t = 4635$$

```
REM Sectional properties of a Double Tee
GOSUB 1000
GOSUB 1300
REM Input dimensions of the section
INPUT "Input width of the unit in inches =",B
INPUT "Input depth of the flange in inches = ",HF
INPUT " Input width of stem at bottom in inches=",B1
INPUT " Input stem width at bottom of flange in inches=",B2
INPUT " Input stem depth in inches= ",D
REM Calculate non composite properties
AC=B*HF+2*(B1+B2)/2 * D : SA=2*B+2*HF+4*D-2*B2+2*B1
YS=AC/SA : YS= INT (YS*100)/100
YB=(HF*B*(D+HF/2)+2*B2*D^2/2-4*(B2-B1)/2*D/2*D/3)/AC
YB=INT (YB*100)/100 : YT=D+HF-YB
B3=YB/D*(B2-B1)/2
I=B/12*HF^3+B*HF*(YT-HF/2)^2+2/3*B1*YB^3+4*B3/12*YB^3+4*B3/3*(D-YB)^3+
    4*((B2-B1)/2-B3)/4*(D-YB)^3+2/3*B1*(D-YB)^3
I=INT (I*10)/10 : ST=INT (I/YT*10)/10 : SB=INT(I/YB*10)/10
INPUT " Input type of concrete for the DT =",U$
INPUT " Input weight of the concrete for the DT in lb/ft.cub. =",WC
INPUT " Strength of the DT concrete f'c, in ksi =",FC
INPUT "Detensioning strength of the concrete f'ci, in ksi =",FIC
EC= 33*WC^1.5*(FC*1000)^.5 / 1000 : EC= INT (EC)
UW= INT ( AC/144*WC*10)/10
INPUT " If the section is composite type Y for yes else type N for no =",A$
IF A$ = "N" THEN HT=0 : TFC=0:WT=0: T$ = " NO TOPPING" :ET=0:BT=):YCB=0:YCT=0
    :YCC=0:IC=0:B4=0:SCT=0:SCB=0:SCC=0: GOTO 700
INPUT " Input topping thickness in inches =",HT
INPUT "Input type of concrete for the topping =",T$
INPUT "Input weight of the topping concrete in lb/ft.cub.=",WT
INPUT "Input strength of the topping concrete in ksi =",TFC
ET=33*WT^1.5*(TFC*1000)^.5/1000 : ET = INT (ET)
BT= B*ET/EC : BT = INT(BT*100)/100
YCB=(YB*AC+BT*HT*(D+HF+HT/2))/(AC+BT*HT) : YCB=INT(YCB*100)/100
YCT= D+HF+HT-YCB :YCC=YCT-HT
B4= YCB/D*(B2-B1)/2
IC=BT/12*HT^3+BT*HT*(YCT-HT/2)^2+B/12*HF^3+B*HF*(YCT-HT-HF/2)^2+2*B1/3*
YCB^3+2*B1/3*(D-YCB)^3+4*B4/12*YCB^3+4*B4/3*(D-YCB)^3+4*((B2-B1)/2-B4)
    /4*(D-YCB)^3
IC=INT (IC*10)/10 : SCT=INT (10*IC/YCT)/10 :SCC=INT(10*IC/YCC)/10 :
SCB=INT(10*IC/YCB)/10
700 CLS
GOSUB 1000
GOSUB 1500
GOTO 1800
1000 CLS
LINE (56,64)-(244,70),,B :LINE (50,70)-(250,76),,B :LINE (90,76)-(94,111)
LINE(94,111)-(106,111):LINE (106,111)-(110,76):LINE (190,76)-(194,111)
LINE (194,111)-(206,111):LINE(206,111)-(210,76) : LINE (91,76)-(109,73),30
LINE (191,76)-(209,76),30 : LINE (50,32)-(250,32): LINE (56,48)-(244,48)
LINE (50,28)-(50,46) : LINE(250,28)-(250,46):LINE (56,48)-(56,54)
LINE (244,48)-(244,54) : LINE (94,120)-(106,120):LINE(90,138)-(110,138)
LINE (94,114)-(94,124): LINE (106,114)-(106,124) : LINE (90,114)-(90,140)
LINE (110,114)-(110,140) : LINE (30,76)-(47,76) : LINE(30,111)-(47,111)
LINE ( 32,76)-(32,111)
RETURN
1300 REM Label the shape
PRINT
PRINT PTAB (145);"B" : PRINT PTAB(145);"Be" : PRINT PTAB(255);"HT"
PRINT PTAB(275);"HF" :PRINT PTAB(37);"D" :PRINT PTAB(115);"B1"
PRINT PTAB(115);"B2"
```

**Figure 36.8** Computer program listing: double-T properties.

```
RETURN
1500
REM Print final text
PRINT PTAB(70);"DOUBLE TEE PROPERTIES by W. Phillips architect"
PRINT PTAB(145);"B"
PRINT PTAB(145);"Be"; PTAB(300);"B= ";B;PTAB(370);"Be= ";BT
PRINT PTAB(300);"HT= ";HT;PTAB(360);"HF= ";HF
PRINT PTAB(255);"HT";PTAB(300);"D=";D;PTAB(360);"B1=";B1;PTAB(430);"B2=";B2
PRINT PTAB(255);"HF";PTAB(300);"Unit ";U$
PRINT PTAB(38);"D";PTAB(300);"Unit f'c=";FC;"f'ci= ";FIC;" ksi"
PRINT PTAB(115);"B1";PTAB(300);"Topping ";T$
PRINT PTAB(115);"B2";PTAB(300);"Topping f'c= ";TFC;" ksi"
PRINT PTAB(300);"wt/sq.ft=";INT(10*WT/12*HT)/10
PRINT PTAB(50);"Non-composite";PTAB(300);"Composite"
PRINT PTAB(20);"Ac= ";AC;" I= ";I;"Y/S= ";YS;PTAB(270);" IC= ";IC
PRINT PTAB(20);"Yt= ";YT;PTAB(150);"St= ";ST;PTAB(270);"Yct= ";YCT;PTAB(400);"Sct=
    ";SCT
PRINT PTAB(20);"Yb= ";YB;PTAB(150);"Sb= ";SB;PTAB(270);"Ycc= ";YCC;PTAB(400)
    ;"Scc= ";SCC
PRINT PTAB(20);"Wt/ft= ";UW;" Wt/sq.ft.= ";INT(10*UW/8)/10;PTAB(270);"Ycb= ";YCB;
PTAB(400);"Scb= ";SCB
PRINT PTAB(20);"Ec=";EC;"ksi,
Wc=";WC;"lb/ft3";PTAB(270);"Et=";ET;"ksi,WT=";WT;"lb/ft3"
RETURN
1800 LCOPY
END
```

**Figure 36.8** *(Continued)*

**Allowable stresses.** Determine allowable stresses.

$$f_{psi}/\text{strand} = 0.74 \times 270 \times 0.925 \times 0.153 = 28.27\,\text{kips/strand}$$

$$f_{psi}/\text{strand} = 0.74 \times 270 \times 0.82 \times 0.153 = 25.0\,\text{kips/strand}$$

Allowable concrete stresses at transfer:

$$\text{Compression} = 0.60 f'_{ci} = 0.60 \times 4\,\text{ksi} = 2.4\,\text{ksi}$$

$$\text{Tension at ends} = 6 f'^{\,0.5}_{ci} = 6 \times 4000^{0.5}/1000 = 0.379\ \text{ksi}$$

$$\text{Tension elsewhere} = 3 f'^{\,0.5}_{ci} = 3 \times 4000^{0.5}/1000 = 0.190\ \text{ksi}$$

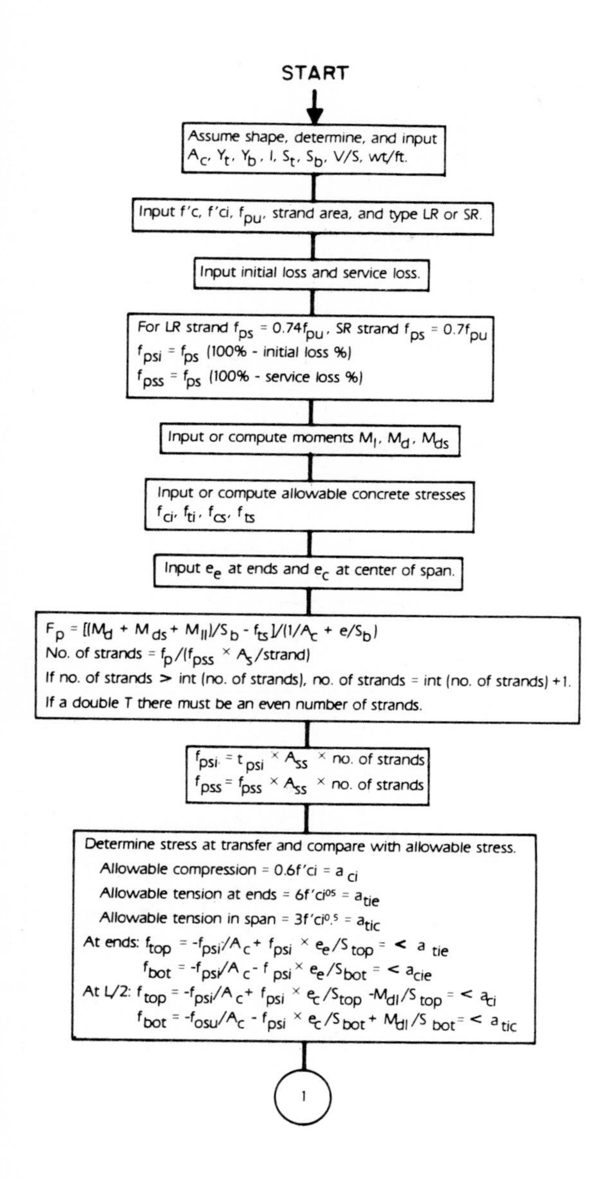


**Figure 36.9** Flowchart: noncomposite double T in flexure.

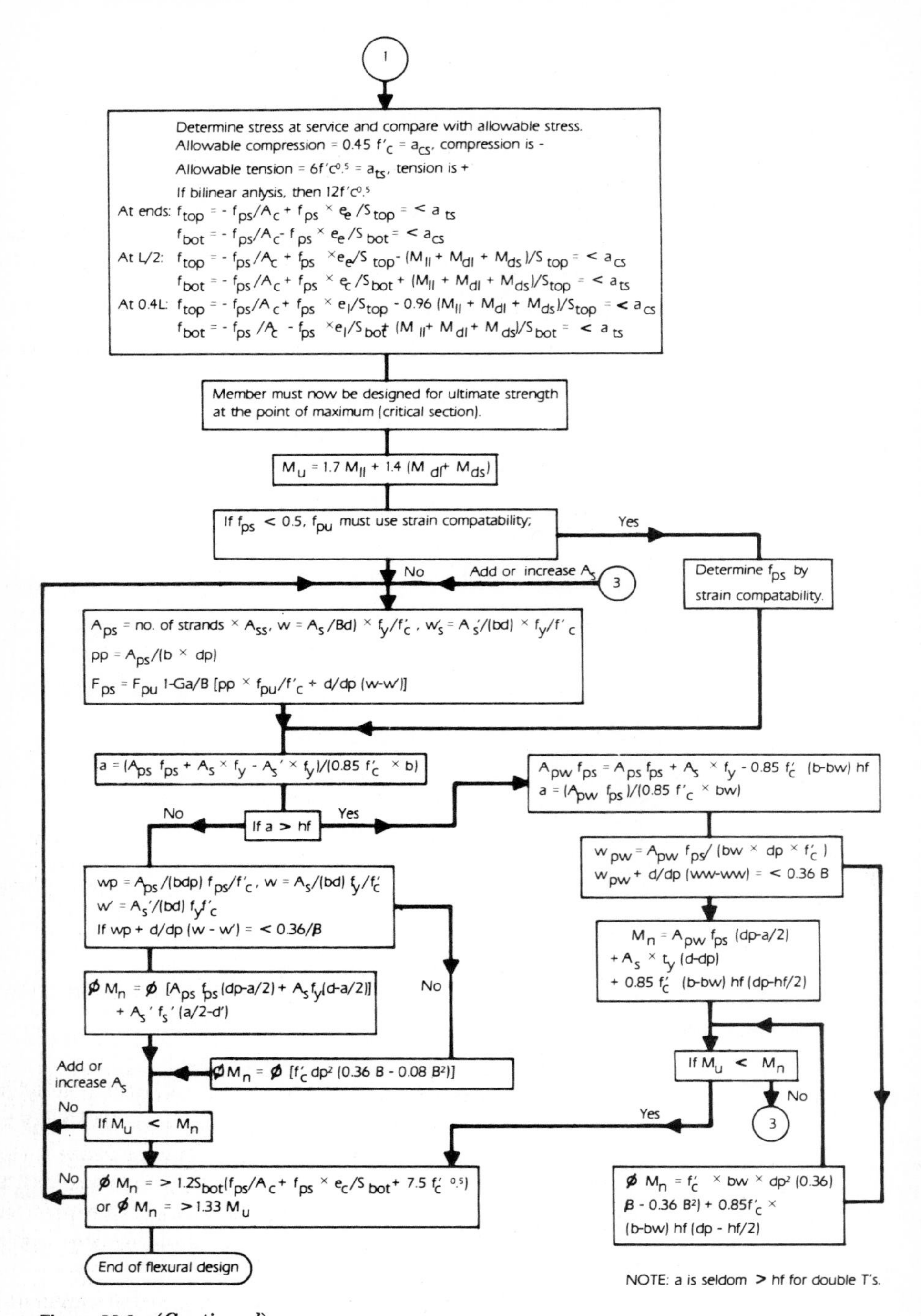


**Figure 36.9** *(Continued)*

```
DESIGN CONDITIONS

Roof member, span =  60   ft.
Live load in lb./ sq. ft. =   16
Superimposed dead load in lb./sq.ft =  12
Concrete type is NORMAL WEIGHT
Concrete strength f'c =  5  ksi, and f'ci=  4   ksi
Low Relaxation strand, Fu=  270   ksi
Strand size = 1/2"  strand area is  .153   sq.in.
Initial loss of prestress =  7.5   %
Service loss of prestress =  18   %

SECTION  PROPERTIES

Flange width =  96   inches and thickness =  2   inches
Member depth = 30   inches
Stem thickness at bottom =  4.25  and at the top =  8  inches
Cross sectional area =  535   square inches
Moment of Inertia =  44458
Y top =  9.59   and Y bottom =  20.41   inches
Sectiom modulus top =  4635   section modulus bottom = 2178
Ratio  V/S =  1.737
Member weight =  540   lb./ lin ft. or  67.5  lb / sq. ft
Specified eccentricity at ends = 0 inches, mid span =  17.41 inches

ALLOWABLE  STRESS

Strand allowable force at detensioning =  28.27  kips
                        at service = 25.06  kips
At transfer allowable concrete stress in compression=  2.4  ksi
                                in tension at ends = .379  ksi
                            in tension at mid span = .189  ksi
Allowable concrete stress at service in compression =  2.25   ksi
                                     in tension =  .424  ksi
                         Bi-linear tension stress =  .848  ksi

MOMENTS

Moment in inch kips are Live Load = 691.2
                        Dead Load = 2916
         Superimposed Dead Load = 518.4

SOLUTION

Required min. force prestressed at service =  149.04  kips
Number of strans required per stem =  3  Total number of strands = 6
Force prestressed at service per strand =  25.06  kips
 Total force prestressed at service =  150.401  KIPS
 MUST DOUBLE HARP try e @ 0.4L =e center
   STRESS AT TRANSFER
    At ends at top stress = -.318   ksi
    At ends at bottom stress = -.318   ksi
    At mid span at top stress = -.309   ksi
    At mid span at bottom stress = -.335   ksi
   STRESS AT SERVICE
    At ends at top, stress = -.282   ksi
    At ends at bottom, stress = -.282   ksi
    At mid span at top, stress = -.607   ksi
    At mid span at bottom, stress=  .41   ksi
  e at 0.4L =  17.41
    At .4L at top, stress = -.571   ksi
    At .4L at bottom, stress =  .335   ksi
```

**Figure 36.10** Computer program printout: noncomposite double T in flexure.

ULTIMATE STRENGTH CHECK

```
     A RECTANGULAR BEAM SOLUTION
    RHO P =  3.541667E-04    Gamma P=  .28   fps=  268.1927   ksi
    Mn=  6573.14    Inch kips  phi* Mn =  5915.826
    Mu=  5983.2    Inch kips
Mu < phi*Mn, MUST ADD MILD STEEL
     prestressing steel is less than maximum therefore OK
     Moment cracked =  4385.83    inch kips
    phi*Mn > 1.2 Mcr. ( 5262.996  ) therefore OK
     If 1.2 *Mcr > phi Mn then check to see if phi Mn > 2* Mu, if so then OK

    ************************************************************
     END OF PROGRAM
```

**Figure 36.10** *(Continued)*

Allowable concrete stresses at service:

$$\text{Compression} = 0.45 f'_c = 0.45 \times 5\,\text{ksi} = 2.250\ \text{ksi}$$

$$\text{Tension} = 6 \times f_c'^{\,0.5} = 6 \times 5000^{0.5}/1000 = 0.424\ \text{ksi}$$

If bilinear analysis is employed, use $12 f_c'^{\,0.5}$.

Determine $e_c$ optimum. See Fig. 36.13.

$$e_c = Y_b - \text{clearance} - \text{stirrup} - \text{strand diameter} - \text{spacing}/2$$

$$e_c = 20.41\ \text{in} = 1\tfrac{1}{2}\ \text{in} - \tfrac{1}{4}\ \text{in} - \tfrac{1}{2} - 2\tfrac{1}{2}\ \text{in}$$

$$= 17.16\ \text{in}$$

Let $e_c = 17$ in.

**Moments.** Determine moments.

Moment for dead load

$$= M_d = wl^2/8 = 0.540\ \text{kips} \times 60^2 \times 12\,\text{ft/ft}/8 = 2916\ \text{in} \cdot \text{kips}$$

Moment for dead superimposed load

$$= M_{ds} = 8\,\text{ft} \times 0.012\ \text{ksf} \times 60^2 \times 12\,\text{ft/ft}/8 = 518.4\ \text{in} \cdot \text{kips}$$

Moment for live load

$$= M_{ll} = 8\,\text{ft} \times 0.016\ \text{ksf} \times 60^2 \times 12\ \text{ft/ft}/8 = 691.2\ \text{in} \cdot \text{kips}$$

```
LPRINT : LPRINT TAB(60);"Page - 25" : LPRINT : LPRINT
LPRINT TAB(6) STRING$(60,42) : LPRINT
LPRINT TAB(16);"DOUBLE TEE, PRESTRESSED PRECAST - non composite-":LPRINT
LPRINT TAB(6);STRING$(60,42)
LPRINT CHR$(27);"n" : LPRINT CHR$(14) : LPRINT TAB(5);"DESIGN CONDITIONS"
LPRINT CHR$(15) : LPRINT CHR$(27);"E"
INPUT " If a floor member type F, if a roof member type R " , A$
INPUT " Input span in feet = ",L
IF A$="F" THEN LPRINT TAB(6);" Floor member, span = ";L;" ft."
IF A$="R" THEN LPRINT TAB(6);"Roof member, span = ";L;" ft."
INPUT " Input Live Load in lbs/ sq. ft. = ",LL
LPRINT TAB(6);"Live load in lb./ sq. ft. = ";LL
INPUT "Input superimposed dead load in lb./sq.ft= ",SDL
LPRINT TAB(6);"Superimposed dead load in lb./sq.ft = ";SDL
INPUT " Input type of concrete ie light wt = "; TC$
LPRINT TAB(6);"Concrete type is ";TC$
INPUT "Input concrete strength f'c and f'ci in ksi = ",FPC , FPCI
LPRINT TAB(6);"Concrete strength f'c = ";FPC;" ksi, and f'ci= ";FPCI; " ksi"
INPUT "Input strand type ie LR for low relaxation or SR for stress relieved =",ST$
INPUT "Input strand ultimate strength Fu in ksi = ",FU
IF ST$="LR" THEN LPRINT TAB(6);"Low Relaxation strand, Fu= ";FU;" ksi"
IF ST$ = "SR" THEN LPRINT TAB(6);" Stress relieved strand. Fu= ";FU;" ksi"
INPUT " Input strand size and area in sq. in. = ",SS$, SA
LPRINT TAB(6);"Strand size = ";SS$;" strand area is ";SA;" sq.in."
INPUT " Input initial loss of prestress in percent =",IL
LPRINT TAB(6);"Initial loss of prestress = ";IL; " %"
INPUT " Input service loss of prestress in percent = ",SL
LPRINT TAB(6);"Service loss of prestress = ";SL;" %"
LPRINT CHR$(27);"n": LPRINT CHR$(14);TAB(5);"SECTION PROPERTIES"
LPRINT CHR$(15): LPRINT CHR$(27);"E"
INPUT "Input flange width and thickness in inches ",FW,FT
LPRINT TAB(6);"Flange width = ";FW; " inches and thickness = ";FT;" inches"
INPUT " Input member depth in inches = ",D
LPRINT TAB(6);"Member depth =";D;" inches"
INPUT " Input stem width at the bottom then at the top in inches =",WB,WT
LPRINT TAB(6);"Stem thickness at bottom = ";WB; " and at the top = ";WT; " inches"
INPUT " Input member cross sectional area in sq. inches =",AC
LPRINT TAB(6);"Cross sectional area = ";AC;" square inches"
INPUT " Input Moment of Inertia = ",I
LPRINT TAB(6);"Moment of Inertia = ";I
INPUT " Input Y top and Y bottom in inches = ",YT, YB
LPRINT TAB(6);"Y top = ";YT;" and Y bottom = ";YB; " inches "
INPUT "Input section modulus top and bottom = ",ST,SB
LPRINT TAB(6);"Sectiom modulus top = ";ST;" section modulus bottom =";SB
INPUT " Input the ratio V/S = ",VS
LPRINT TAB(6);"Ratio V/S = ";VS
INPUT "Input member weight in lb./ lin. ft = ",DLLF
DL = DLLF /FW*12
LPRINT TAB(6);"Member weight = ";DLLF;" lb./ lin ft. or ";DL;" lb / sq. ft"
INPUT " Input eccentricity at end EE in inches and EC at mid span",EE,EC
IF EC= 0! THEN EC= YB-1.5-.375-.5-1
LPRINT TAB(6);"Specified eccentricity at ends =";EE;"inches, mid span = ";EC;"inches"
LPRINT CHR$(27);"n": LPRINT CHR$(14);TAB(5);"ALLOWABLE STRESS"
LPRINT CHR$(15):LPRINT CHR$(27);"E"
FPSI= .74*FU*SA*(100-IL)/100 : FPSS = .74*FU*SA*(100-SL)/100
LPRINT TAB(6);"Strand allowable force at detensioning = ";INT(FPSI*100)/100;" kips"
LPRINT TAB(30);"at service =";INT(FPSS*100)/100;" kips"
FCTC = .6* FPCI : FCTTE=6*(FPCI*1000)^.5/1000:FCTTM=3*(FPCI*1000)^.5/1000
LPRINT TAB(6);"At transfer allowable concrete stress in compression= ";INT(FCTC*1000)
   /1000;" ksi"
```

**Figure 36.11** Computer program listing: noncomposite double T in flexure.

```
LPRINT TAB(37);"in tension at ends =";INT(FCTTE*1000)/1000;" ksi"
LPRINT TAB(33);"in tension at mid span =";INT(FCTTM*1000)/1000;" ksi"
FCSC= .45*FPC : FCST= 6*(FPC*1000)^.5/1000 :FCSTB = 12*(FPC*1000)^.5/1000
LPRINT TAB(6);"Allowable concrete stress at serrvice in compression =";INT (FCSC *1000)/
   1000;" ksi"
LPRINT TAB(42);"in tension = ";INT(FCST*1000)/1000;" ksi"
LPRINT TAB(28);"Bi-linear tension stress = ";INT(FCSTB*1000)/1000; " ksi"
LPRINT CHR$(27);"n": LPRINT CHR$(14);TAB(5);" MOMENTS"
LPRINT CHR$(15):LPRINT CHR$(27);"E"
MLL = FW/12*LL*L^2*12/8/1000 : MDL=FW/12*DL*L^2*12/8/1000:MSDL =FW/12*
   SDL*L^2*12/8/1000
LPRINT TAB(6);"Moment in inch kips are Live Load =";MLL
LPRINT TAB(28);"Dead Load =";MDL
LPRINT TAB(15);"Superimposed Dead Load =";MSDL
LPRINT : LPRINT STRING$(60,45)
LPRINT CHR$(12)
LPRINT: LPRINT TAB(60);"Page - 26 " : LPRINT :LPRINT
LPRINT CHR$(27);"n":LPRINT CHR$(14);TAB(5);"SOLUTION"
LPRINT CHR$(15):LPRINT CHR$(27);"E"
FPSM=((MLL+MDL+MSDL)/SB-FCST)/(1/AC+EC/SB)
LPRINT TAB(6);"Required min. force prestressed at service = ";INT(FPSM*100)/100;" kips"
NS=FPSM/FPSS : IF NS> INT(NS)THEN NS=INT(NS)+1
IF NS/2> INT(NS/2) THEN NS= 2*(INT(NS/2)+1)
COUNT = 1
5
FPS=NS*FPSS
LPRINT TAB(6);"Number of strans required per stem = ";NS/2;" Total number of strands
   =";NS
LPRINT TAB(6);"Force prestressed at service per strand = ";INT(FPSS*100)/100;" kips"
LPRINT TAB(6);" Total force prestressed at service = ";INT(FPS*1000)/1000;" KIPS"
REM CHECK STRESS AT TRANSFER
FPI=NS*FPSI
FCIET = -FPI/AC+FPI*EE/ST : IF FCIET>FCTTE THEN LPRINT"At transfer tension stress at end
   at top exceeds allowable"
FCIEB = - FPI/AC-FPI*EE/SB : IF FCIEB <-FCTC THEN LPRINT"At transfer compression stress
   at bottom exceeds allowable"
FCICT= -FPI/AC+FPI*EC/ST-MDL/ST : IF FCICT <- FCTC OR FCICT > FCTTM THEN LPRINT "At
   transfer stress at mid span at top exceeds allowable"
FCICB = -FPI/AC-FPI*EC/SB + MDL/SB : IF FCICB > FCTTE OR FCICB < -FCTC THEN LPRINT "
   At transfer stress at mid span bottom exceeds allowable"
FCSET = -FPS/AC+FPS*EE/ST : IF FCSET>FCSC THEN LPRINT " At service stress at end top
   exceeds allowable"
FCSEB = -FPS/AC-FPS*EE/SB : IF FCSEB <-FCSC THEN LPRINT " At service stress at end
   bottom exceeds allowable"
FCSMT= -FPS/AC+FPS*EC/ST-(MLL+MDL+MSDL)/ST:IF FCSMT<-FCSC THEN LPRINT " At
   service stress at top mid span exceeds allowable"
FCSMB=-FPS/AC-FPS*EC/SB+(MLL+MDL+MSDL)/SB :IF FCSMB >FCST THEN LPRINT " At
   service stress at bottom mid span exceeds allowable"
E1= EE+ .8*(EC-EE)
10
IF COUNT =3 THEN LPRINT " INCREASE NUMBER OF STRANDS BY 2": NS=NS+2:GOTO 5
FCSQT= -FPS/AC+FPS*E1/ST-.96*(MLL+MDL+MSDL)/ST
FCSQB=-FPS/AC-FPS*E1/SB+.96*(MLL+MDL+MSDL)/SB
IF FCSQT <- FCSC OR FCSQB > FCST THEN LPRINT TAB(6);" MUST DOUBLE HARP try e @ 0.4L
   =e center":E1=EC:COUNT=COUNT + 1 : GOTO 10
LPRINT TAB(8);" STRESS AT TRANSFER"
LPRINT TAB(10);"At ends at top stress = ";INT(FCIET*1000)/1000;" ksi"
LPRINT TAB(10);"At ends at bottom stress = ";INT(FCIEB*1000)/1000;" ksi"
LPRINT TAB(10);"At mid span at top stress = ";INT(FCICT*1000)/1000;" ksi"
LPRINT TAB(10);"At mid span at bottom stress = ";INT(FCICB*1000)/1000;" ksi"
LPRINT TAB(8);" STRESS AT SERVICE"
```

**Figure 36.11** *(Continued)*

```
LPRINT TAB(10);"At ends at top, stress = ";INT(FCSET*1000)/1000;" ksi"
LPRINT TAB(10);"At ends at bottom, stress = ";INT(FCSEB *1000)/1000; " ksi"
LPRINT TAB(10);"At mid span at top, stress = ";INT(FCSMT *1000)/1000;" ksi"
LPRINT TAB(10);"At mid span at bottom, stress= ";INT(FCSMB*1000)/1000;" ksi"
LPRINT TAB(8);"e at 0.4L = ";E1
LPRINT TAB(10);"At .4L at top, stress = "; INT(FCSQT*1000)/1000;" ksi"
LPRINT TAB(10);"At .4L at bottom, stress = ";INT(FCSQB*1000)/1000;" ksi"
REM ULTIMATE STRENGTH CHECK
LPRINT CHR$(27);"n": LPRINT CHR$(14);TAB(5);"ULTIMATE STRENGTH CHECK"
LPRINT CHR$(15):LPRINT CHR$(27);"E"
PP=NS*SA/FW/(EC+YT): IF ST$="LR" THEN GP=.28 ELSE GP=.4
IF FPC=4 THEN BETA=.85
IF FPC>4 THEN BETA = .85-.05*(FPC-4)
IF BETA < .65 THEN BETA =.65
UFPS=FU*(1-GP/BETA*(PP*FU/FPC))
A=(NS*SA*UFPS)/(.85*FPC*FW) : IF A=<FT THEN LPRINT TAB(6)" A RECTANGULAR BEAM
   SOLUTION "
IF A> FT THEN LPRINT TAB(6)" A TEE BEAM SOLUTION HENCE NOT SOLVED BY THIS PROGRAM"
MN=NS*SA*UFPS*((YT+EC)-A/2) : MNPHI=MN*.9
MU= 1.7*MLL+1.4*(MDL+MSDL)
LPRINT TAB(6);"RHO P = ";PP;" Gamma P= ";GP;" fps= ";UFPS;" ksi
LPRINT TAB(6);"Mn= ";INT (MN *100)/100;" Inch kips phi* Mn = "; INT (MN*100)/
   100*.9
LPRINT TAB(6);"Mu= "; INT(MU*100)/100;" Inch kips"
IF MU=< MNPHI THEN LPRINT TAB(6)" Member is OK " ELSE LPRINT " Mu < phi*Mn, MUST
   ADD MILD STEEL"
REM CHECK UPPER LIMIT OF PRESTRESSING
WP=NS*SA/FW/(YT+EC)*UFPS/FPC
IF WP=< .36 * BETA THEN LPRINT TAB(6);" prestressing steel is less than maximum therefore
   OK"
REM CHECK LOWER LIMIT OF PRESTRESSING IE MOMENT CRACKED
MCR=SB*(FPS/AC+FPS*EC/SB+7.5*(FPC*1000)^.5/1000)
LPRINT TAB(6);" Moment cracked = ";INT (MCR*100)/100; " inch kips"
IF MNPHI > 1.2*MCR THEN LPRINT TAB(6)"phi*Mn > 1.2 Mcr. (";1.2*INT(MCR*100)/
   100;" ) therefore OK"
LPRINT TAB(6);" If 1.2 *Mcr > phi Mn then check to see if phi Mn > 2* Mu, if so then OK"
LPRINT : LPRINT TAB(6);STRING$(60,42)
LPRINT TAB(6);" END OF PROGRAM"
END
```

**Figure 36.11** *(Continued)*

**Number of strands.** Determine minimum number of strands for service conditions.

Force prestressing:

$$f_{ps,\ \min} = (\Sigma\ moments\ /S_b - f_{ten,\ \text{allowable}})/(1/A_c + e_c/S_b)$$

$$= [(2916 + 518.5 + 691.2)/2178 - 0.424]/(1/535 + 17/2178)$$

$$= 151.96\ kips$$

The number of strands required = 151.96/25.0 = 6.078 strand, or 8 strands, 4 strands in each leg of the double tee. Try increasing $e_c$ to 17.41 in, since stirrups may be placed in the required clearance. Then, $f_{ps,\min}$ = 149.06 kips, and the number of strands is 6. Use six strands with $e_c$ = 17.41 in.

See Fig. 36.14 for a trial strand profile.

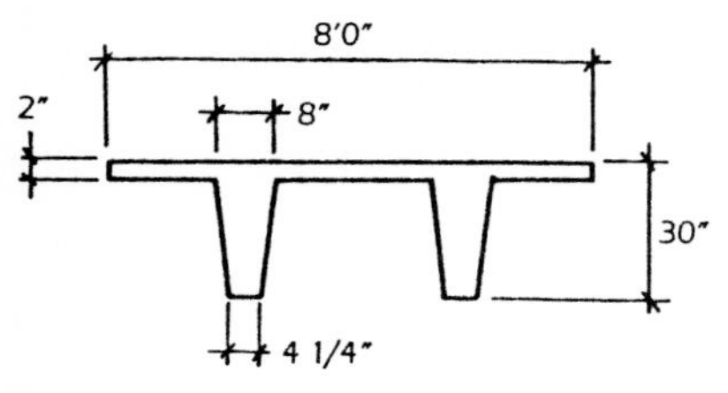


Figure 36.12 Shape properties of a double T.

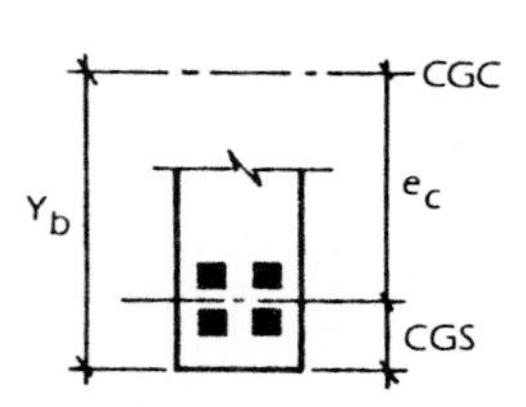


Figure 36.13 Determination of $e_c$ optimum for a double T.

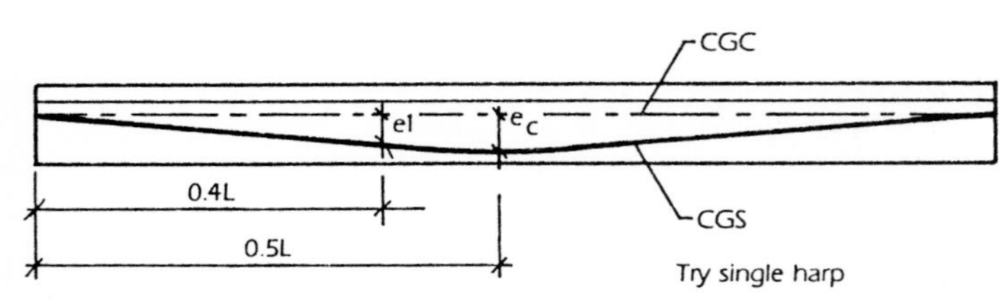


Figure 36.14 Trial strand profile.

**Stresses at transfer.** Check stresses at transfer.

$$f_{psi} = 6 \times 0.153 \times 0.74 \times 270 \times 0.925 = 169.67 \text{ kips}$$

At end:

$$-f_{psi}/A_c = f_{top} = 169.67/535 = -\ 0.317 < -\ 2.4 \ \text{(OK)}$$

$$-f_{psi}/A_c = f_{bottom} = -\ 169.67/535 = -\ 0.317 < -2.4 \ \text{(OK)}$$

At midspan:

$$-f_{psi}/A_c + f_{ps}\, e_c/S_t - M_d/S_t = f_{top}$$

$$-\ 169.67/S\ 535 + 169.67 \times 17.41/4635 - 2916/4635$$

$$= -\ 0.308 < -\ 2.4 \ \text{(OK)}$$

$$-f_{psi}/A_c - f_{ps}\, e_c/S_b + M_d/S_b = f_{bottom}$$

$$-\ 169.67/535 - 169.67 \times 17.41/2178 + 2916/2178$$

$$= -\ 0334 < -\ 2.4 \ \text{(OK)}$$

**Stresses at servcice. Check stresses at service.**

$$f_{ps} = 6 \times 0.153 \times 0.74 \times 270 \times 0.82 = 150.4 \text{ kips}$$

At end:

$$-f_{ps}/A_c = f_{top} = f_{bottom} = -\ 150.4/535 = -\ 0.281 < 2.25 \ \text{(OK)}$$

At midspan:

$$-f_{ps}/A_c + f_{ps}e_c/S_t - (M_d + M_{ds} + M_{ll})/S_t = f_{top}$$

$$-150.4/535 + 150.4 \times 17.41/4635 - (2916 + 518.4 + 691.2)/4635$$

$$= -0.606 < -2.25 \text{ (OK)}$$

$$-f_{ps}/A_c - f_{ps}e_c/S_b + (M_d + M_{ds} + M_{ll})/S_b = f_{bottom}$$

$$-150.4/535 + 150.4 \times 17.41/2178 + (2916 + 518.4 + 691.2)/2178$$

$$= -0.410 < 0.424 \text{ (OK)}$$

At 0.4$L$,

$$e_1 = 0.8e_c = 0.8 \times 17.41 = 13.928 \text{ in} \quad \text{Moment} = 0.96M_{max}$$

$$-f_{ps}/A_c + f_{ps}e_1/S_t - 0.96(M_d + M_{ds} + M_{ll})/S_t = f_{top}$$

$$-150.4/535 + 150.4 \times 13.298/4635 - 0.96(2916 + 518.4 + 691.2)/4635$$

$$= -0.684 < -2.25 \text{ (OK)}$$

$$-f_{ps}/A_c - f_{ps}e_1/S_b + 0.96(M_d + M_{ds} + M_{ll})/S_t = f_{bottom}$$

$$-150.4/535 - 150.4 \times 13.928/2178 + 0.96(2916 + 518.4 + 691.2)/2178$$

$$= 0.576 > 0.424$$

Too high. The solution is either to double-harp or to increase the number of strands. Use double harping at 0.4$L$, which is less expensive than two additional strands.

With double harping let $e$ at 0.4$L$ = $e$ at center = 17.41 in. At 0.4$L$ $e_1$ = 17.41 in.

$$-f_{ps}/A_c + f_{ps}e_1/S_t - 0.96(M_d + M_{ds} + M_{ll})/S_t = f_{top} - 150.4/535$$

$$+ 150.4 \times 17.41/4635 - 0.96(2916 + 518.4 + 691.2)/4635$$

$$= -0.571 < -2.25 \text{ (OK)}$$

$$-f_{ps}/A_c - f_{ps}e_1/S_t + 0.96(M_d + M_{ds} + M_{ll})/S_t = f_{bottom} - 150.4/535$$

$$-150.4 \times 17.41/2178 + 0.96(2916 + 518.4 + 691.2)/2178$$

$$= 0.335 < 0.424 \text{ (OK)}$$

See Fig. 36.15 for the final strand profile.

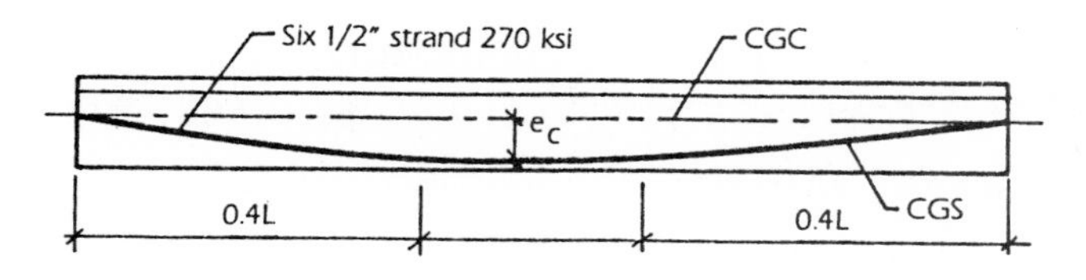


**Figure 36.15** Final strand profile.

**Ultimate-strength design.** Check ultimate strength.

$$P_p = A_{ps}/bd_p$$

where $A_{ps} = 6 \times 0.153 = 0.918$

$b = 96$ in

$d_p = e_c + Y_t = 17.41 + 9.59 = 27$ in

$P_p = 0.918/(96 \times 27) = 0.000354$

$G_p = 0.28$ for low-relaxation strand if $f_{py} = f_{pu} = 0.9$ (for low-relaxation strand $f_{py} = 0.9\,f_{pu}$). For 5-ksi concrete $\beta = 0.8$. Since there is no mild steel or compression steel in the design thus far, $w$ and $w' = 0$.

$$f_{ps} = f_{pu}\left\{1 - G_p/\beta\left[P_p f_{pu}/f'_c + d/d_p(w - w')\right]\right\}$$

$$f_{ps} = 270\{1 - 0.28/0.8\,[0.000354 \times 270/5 + 0/27(0 - 0)]\}$$

$$= 268.19 \text{ ksi}$$

$$a = (A_{ps} f_{ps} + A_s f_y)/0.85 f'_c b$$

$$= (0.918 \times 268.19 + 0 \times 0)/(0.85 \times 5 \times 96) = 0.603 < 2 \text{ in}$$

Therefore, $a$ is within the flange, and the design is rectangular.

$$\phi M_n = \phi\left[A_{ps} f_{ps}(d - a/2) + A_s f_y(d - a/2)\right]$$

$$= 0.9[0.918 \times 268.19\,(27 - 0.603/2)]$$

$$= 5915.816 \text{ in} \cdot \text{kips (allowable)}$$

$$M_{u,\text{act}} = 1.4\,(2916 + 518.4) + 1.7(691.2) = 5983.2 \text{ in} \cdot \text{kips}$$

Since this is greater than $\phi M_n$, either mild steel must be provided or the prestressing must be increased to eight strands. The least expensive option at this point is to increase the number of strands to eight and to single-harp. This decision is based upon the cost of the harping hardware and additional labor versus the cost of the additional strand. A current estimate should be made in each case. For this example add mild steel and double-harp. Try one no. 6 at $d = 28$ in.

$$f_{ps} = 270\{1 - 0.28/0.8\,[0.00354 \times 270/5 + 28/27$$

$$(0.44/96/28 \times 60/5 - 0)]\} = 268.0$$

$$a = (0.918 \times 268 + 0.44 \times 60)/(0.85 \times 5 \times 96) = 0.6677$$

$$M_n = 0.9[0.918 \times 268(27 - 0.6677/2) + 0.44 \times 60\,(28 - 0.6677/2)]$$

$$= 6561.8 \text{ in} \cdot \text{kips } M_u \text{ (OK)}$$

**Prestressing and reinforcing steel.** Check upper limit.

$$w_p \text{ or } w_p + d/d_p(w - w') \leq 0.36\beta$$

where $w_p = A_{ps} / bd_p \times f_{ps} f'_c$

$w = A_s / bd \times f_y f'_c$

$w' = 0$ since there is no compression steel

$\beta = 0.8$

$$0.918/(96 \times 27) \times 268/5 + 28.27[0.44/(96 \times 28)60/5]$$

$$= 0.02 < 0.36 \times 0.8 \text{ (OK)}$$

Check lower limit; that is, 1.2 moment$_{crack}$.

$$\phi M_n = 1.2 M_{cr} \quad \text{or} \quad \phi M_n > 2 M_u$$

$$1.2 M_{cr} = 1.2 S_b \left( f_{ps} / A_c + f_{ps} e_c / S_b + 7.5 f_c'^{0.5} / 1000 \right)$$

$$= 1.2 \times 2178 (150.4/535 + 150.4 \times 17.41/2178 + 7.5 \times 5000^{0.5}/1000)$$

$$= 5262.9 \text{ kips} < 6561.8 \text{ kips (OK)}$$

## Program 5: Shear Design Double T

Like reinforced-concrete members, prestressed concrete members must be designed for shear. For beams, when $V_u > \phi V_c / 2$, and for slabs and walls, when $V_u > \phi V_c$, the web thickness of the member must be increased $V_u > \phi \times 10 f_c'^{0.5}$. For beams, when $V_u > \phi V_c / 2$ and $V_u \leq \phi V_c$, minumum shear reinforcing is required. The minimum shear reinforcing requirement is

$$A_{v,\ min} = A_{ps} / 80 f_{pu} / f_y \ s / d \ (d / b_w)^{0.5}$$

When $V_u > \phi V_c$, shear reinforcing must be designed. For the design of the shear reinforcing,

$$V_s = (V_u - \phi V_c)/\phi \quad \text{and} \quad V_s = A_v f_y d / s$$

The second equation may be written as

$$A_v = V_s s / f_y d \quad \text{or} \quad A_v = (V_u - \phi V_c) \times s / (\phi f_y d)$$

The term $d$ is the larger of $0.8h$ (height of prestressed member) or distance from the extreme compression fiber to the centroid of the prestressing steel. Near the ends of the member, $d$ usually is smaller than $0.8h$; therefore, the term $0.8h$ is used in the following equations instead of $d$.

The value of $V_c$ is shown in Fig. 36.16. $V_u$ is plotted on the curve along with $\phi V_c / 2$. When $V_u$ is above $\phi V_c / 2$ and below $\phi V_c$, minimum shear reinforcing is required. When $V_u$ is above $\phi V_c$, shear reinforcing must be designed. In the term $V_u d_p / M_u$, $d_p$ is the true distance from the extreme fiber in compression of the concrete to the centroid of the prestressing steel. The term is taken as equal to or less than 1.0.

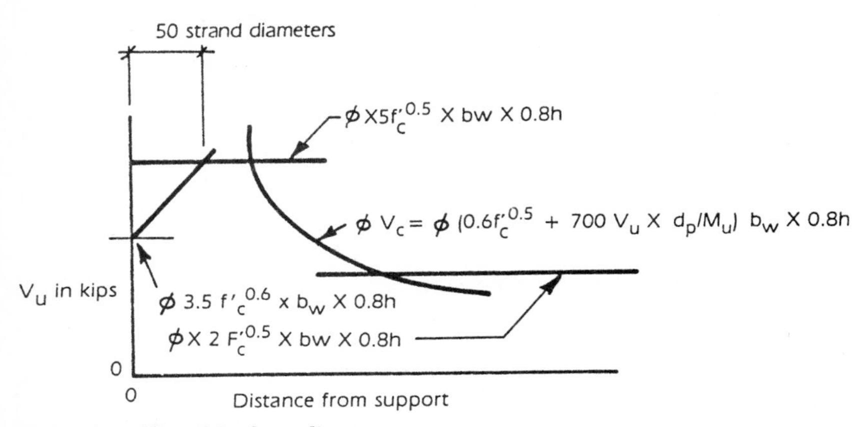


**Figure 36.16** Allowable shear diagram.

For simple spans with a uniform load (which is the usual case for pretensioned precast concrete members), the shear at the distance $x$ from the end of a member is $V_{ux} = w_u(L/2 - x)$, and the moment is $M_{ux} = w_u L/2 \times x \times w_u x^2.2$. The term $V_u d_p/M_u$ becomes

$$dx \; [(L - 2x)/(L - x)]$$

which reduces to

$$d_p w_u (L/2 - x)/(w_u L/2 - w_u x^2/2)$$

For a solution of the above equations a computer program is convenient (see Figs. 36.17 and 36.18). The selection of the type of shear reinforcing, that is, reinforcing bars of welded wire fabric, is left to the designer.

For T beams with a narrow stem, mesh is often used. To save time some precasters use prewelded shear reinforcing such as Dur-O-WAL prefabricated web reinforcing. The maximum spacing in web reinforcing is 3/4$h$ or 24 in on center, with the first wire or bar at a minimum distance of $s$/2 from the ends. Usually special bearing cages will take care of the end condition. The maximum yield stress that can be used for shear reinforcing is 60 ksi.

## Program 6: Prestressed Concrete Column Interaction Diagram

The design solution for prestressed concrete columns is given in the form of an interaction diagram. The coordinates of the interaction curve are determined from strain compatibility equations. Because of the repetitious use of the equations, computer programs are ideal for the solution. When determining the coordinates of the curve the designer must work with the stress-strain curve for the strand to be used. This is so because

INPUT

```
Member span in ft.=  60
Member width in ft.=  8
Member depth in inches = 30
Ave. stem width in inches =  6.125
Number of stems =  2
Y top =  9.59  inches
ee at end = 0  ec at center =  17.41  inches
MEMBER IS DOUBLE HARPED AT 0.4 L
Live load in lb/sq.ft/=  16
Superimposed dead LOAD in lb/sq.ft. = 12
Unit weight in lb./lin ft.=  410
Concrete strength f'c in ksi =  5
Normal weight concrete
phi factor for concrete=  .8
U factors for live load =  1.7  U dead load =  1.4
```

SHEAR DESIGN DATA PER STEM

```
phi Vc at ends per stem= phi*3.5*gamma*f'c'0.5*bw*0.8h = 29 kips
phi Vc at 50 strand dia. per stem=phi*5.0*gamma*f'c*bw*0.8h= 41 kips
phi Vc min. per stem = phi * 2.0 * gamma*f'c'0.5*bw*0.8H = 16 kips
phi Vc min /2 =  8  kips

PROVIDE MIN. SHEAR REINFORCING WHEN Vu> phiVc/2 and when Vu =< phiVc
DESIGN SHEAR REINFORCING WHERE Vu > phi Vc
```

TABLE FOR COMPUTING phi Vc

phi Vc = phi*gamma*(0.6*f'c'0.5+700*dp*Vu/Mu)*bw*0.8h

| distance ft. from support | Vu | Mu | dp*Vu/Mu | Vc | Vc/2 | Vu > Vc/2 lower limit |
|---|---|---|---|---|---|---|
| 2 | 12.9 | 322 | .44 | 41.5 | 20.7 | |
| 3 | 12.5 | 475 | .3 | 30.4 | 15.2 | |
| 4 | 12 | 622 | .24 | 24.8 | 12.4 | |
| 5 | 11.5 | 763 | .2 | 21.4 | 10.7 | yes |
| 6 | 11.1 | 900 | .17 | 19.1 | 9.5 | yes |
| 7 | 10.6 | 1030 | .15 | 17.4 | 8.7 | yes |
| 8 | 10.1 | 1155 | .13 | 16.1 | 8 | yes |
| 9 | 9.7 | 1275 | .12 | 15.1 | 7.5 | yes |
| 10 | 9.2 | 1389 | .11 | 14.2 | 7.1 | yes |
| 11 | 8.7 | 1497 | .1 | 13.4 | 6.7 | yes |
| 12 | 8.3 | 1600 | .09 | 12.8 | 6.4 | yes |
| 13 | 7.8 | 1697 | .08 | 12.2 | 6.1 | |
| 14 | 7.4 | 1789 | .08 | 11.7 | 5.8 | |
| 15 | 6.9 | 1875 | .07 | 11.2 | 5.6 | |
| 16 | 6.4 | 1955 | .07 | 10.7 | 5.3 | |
| 17 | 6 | 2030 | .06 | 10.3 | 5.1 | |
| 18 | 5.5 | 2100 | .05 | 9.9 | 4.9 | |
| 19 | 5 | 2164 | .05 | 9.5 | 4.7 | |
| 20 | 4.6 | 2222 | .05 | 9.1 | 4.5 | |
| 21 | 4.1 | 2275 | .04 | 8.7 | 4.3 | |
| 22 | 3.7 | 2322 | .04 | 8.3 | 4.1 | |
| 23 | 3.2 | 2364 | .03 | 7.9 | 3.9 | |
| 24 | 2.7 | 2400 | .03 | 7.5 | 3.7 | |
| 25 | 2.3 | 2430 | .02 | 7.1 | 3.5 | |
| 26 | 1.8 | 2455 | .02 | 6.7 | 3.3 | |
| 27 | 1.3 | 2475 | .01 | 6.3 | 3.1 | |
| 28 | .9 | 2489 | .01 | 5.9 | 2.9 | |
| 29 | .4 | 2497 | 0 | 5.4 | 2.7 | |
| 30 | 0 | 2500 | 0 | 4.9 | 2.4 | |

```
Vs max=  0  which occurs at  0  ft from the end
Vu max per stem, at end of the beam to h = 12732.5
Vc max at 50 strand dia from end = 41
Max shear reinforcing spacing, smaller of 24 in. vs 0.75*h = 22.5
Av min at S max = Aps/80*fpu/fy*S/0.8h*(0.8h/bw)'0.5 = .09  sq. in.
If Vu > Vc then must design the shear reinforcing
Av= (Vu-phi*Vc)*S/(phi*fy*.8h)

  ***************************************************
                END  OF  PROGRAM
```

**Figure 36.17** Computer program printout: shear design prestressed concrete double T.

```
LPRINT : LPRINT TAB(60);" Page  - 35" : LPRINT : LPRINT
LPRINT TAB(6);STRING$(50,42) : LPRINT
LPRINT TAB(14);"SHEAR DESIGN PRESTRESSED CONCRETE - double tee": LPRINT
LPRINT TAB(6); STRING$(50,42)
LPRINT CHR$(27);"n" : LPRINT CHR$(14) : LPRINT TAB(10);"INPUT " : LPRINT CHR$(15)
LPRINT CHR$(27);"E"
INPUT " input member span in ft = ",L : LPRINT TAB(6);" Member span in ft.= ";L
INPUT "input member width in ft= ",B : LPRINT TAB(6);" Member width in ft.= ";B
INPUT "input member depth in inches = ",H : LPRINT TAB(6);" Member depth in inches =";H
INPUT "input ave. stem width in inches = ",BW : LPRINT TAB(6);" Ave. stem width in inches
   = ";BW
INPUT "input no. of stems = ",NS : LPRINT TAB(6);" Number of stems = ";NS
INPUT "input Y top in inches = ",YT : LPRINT TAB(6);" Y top = ";YT;" inches"
INPUT " input ee at ends and ec at center in inches = ",EE,EC : LPRINT TAB(6);" ee at end =";ee;"
    ec at center = ";EC; " inches"
INPUT " if member is double harped at 0.4 L type Y else type N = ",H$
IF H$="Y" THEN LPRINT TAB(6);" MEMBER IS DOUBLE HARPED AT 0.4 L" ELSE LPRINT "
MEMBER IS SINGLE HARPED"
INPUT "input live load in lb/sq.ft. unfactored = ",LL : LPRINT TAB(6);" Live load in lb/sq.ft/
    = ";LL
INPUT "input superimposed dead load in lb/sq.ft. unfactored =",DLS : LPRINT TAB(6);"
     Superimposed dead LOAD in lb/sq.ft. =";DLS
INPUT "input unit weigth in lb/lin ft = ",DL : LPRINT TAB(6);" Unit weight in lb./lin ft.= ";DL
INPUT " input concrete strength f'c in ksi = ",FC : LPRINT TAB(6);" Concrete strength f'c in
    ksi = ";FC
5
INPUT " if concrete is sand light wt. type SLW else type NW = ",T$
IF T$="SLW" THEN LPRINT TAB(6);" Sand light wt concrete " : GA=.85 : GOTO 10
IF T$ ="NW" THEN LPRINT TAB(6);" Normal weight concrete " : GA=1! : GOTO 10
PRINT " YOU TYPED THE WRONG ANSWER, TRY AGAIN " : GOTO 5
10
INPUT " input phi factor for concrete = ",PHIC : LPRINT TAB(6);" phi factor for concrete
     = ";PHIC
INPUT "input U factors for LL, and DL = ",ULL,UDL : LPRINT TAB(6);"U factors for live load
     = ";ULL;"-U dead load = ";UDL
LPRINT CHR$(27);"n" :LPRINT CHR$(14):LPRINT TAB(5);"SHEAR DESIGN DATA PER STEM " :
LPRINT CHR$(15)
LPRINT CHR$(27);"E"
V1=INT(PHIC*3.5*GA*(FC*1000)^.5/1000*BW*.8*H)
LPRINT TAB(6);" phi Vc at ends per stem= phi*3.5*gamma*f'c^0.5*bw*0.8h =";V1;"kips-
V2= INT(PHIC*5!*GA*(FC*1000)^.5/1000*BW*.8*H)
LPRINT TAB(6);" phi Vc at 50 strand dia. per stem=phi*5.0*gamma*f'c*bw*0.8h=";V2;"kips"
V3= INT(PHIC*2!*GA*(FC*1000)^.5/1000*BW*.8*H)
LPRINT TAB(6);" phi Vc min. per stem = phi * 2.0 * gamma*f'c^0.5*bw*0.8H =";V3;"kips"
LPRINT TAB(6);" phi Vc min /2 = "; V3/2; " kips"
LPRINT
LPRINT  TAB(6);" PROVIDE MIN. SHEAR REINFORCING WHEN Vu> phiVc/2 and when Vu
     =< phiVc"
LPRINT TAB(6);" DESIGN SHEAR REINFORCING WHERE Vu > phi Vc "
LPRINT : LPRINT CHR$(12)
LPRINT : LPRINT TAB(60);" Page  - 36" : LPRINT : LPRINT
LPRINT TAB(20);"TABLE  FOR  COMPUTING phi Vc"
LPRINT TAB(6);" phi Vc = phi*gamma*(0.6*f'c^0.5+ 700*dp*Vu/Mu)*bw*0.8h"
LPRINT TAB(6)"distance ft.";TAB(20);"Vu";TAB(25);"Mu";TAB(30);"dp*Vu/Mu"; TAB(42);
     "Vc";TAB(48);"Vc/2";TAB(55);"Vu > Vc/2"
LPRINT TAB(6)"from support"; TAB(55);"lower limit"
LPRINT TAB(6)STRING$(60,45)
VM=0! : XM=0!
WU=B*(UDL*DLS+ULL*LL)+UDL*DL
FOR X=2 TO INT(L/2) STEP 1
VU=WU*(L/2-X)/1000/NS: MU = (WU*L/2*X-WU*X^2/2)/1000/NS*12
```

**Figure 36.18** Computer program listing: shear design prestressed concrete double T.

```
IF H$="Y" THEN DP=((EC-EE)*X/(.4*L))+YT+EE ELSE DP=((EC-EE)*X/(.5*L))+YT+EE
F1= DP*VU/MU : IF F1 > 1! THEN F1=1!
VC=PHIC*GA*(.6*(FC*1000)^.5 + 700*F1)*BW*.8*H/1000
VC2=VC/2
IF VU > VC2 AND VU > V3/2 THEN A$="yes" ELSE A$=" "
IF VU>VC THEN VM=VU-VC : XM=X
VU=INT(VU*10)/10 : MU= INT (MU): F1 = INT (F1*100)/100 : VC= INT(VC*10)/10 :
VC2=INT(VC2*10)/10
LPRINT TAB(6);X;TAB(18);VU;TAB(25);MU;TAB(33);F1;TAB(40); VC;TAB(48); VC2;
    TAB(55);A$
NEXT X
LPRINT TAB(6);STRING$(60,45)
LPRINT TAB(6);" Vs max= ";VM; " which occurs at ";XM;" ft from the end"
LPRINT TAB(6);" Vu max per stem, at end of the beam to h =" WU*(L/2-H/12)/NS
LPRINT TAB(6);" Vc max at 50 strand dia from end =";V2
SP = 24 : IF SP > .75 * H THEN SP = .75*H
LPRINT TAB(6);" Max shear reinforcing spacing, smaller of 24 in. vs 0.75*h =";SP
INPUT " input Aps in sq. in and fpu in ksi = " ,APS, FPU
INPUT "input fy for the shear reinforcing in ksi =",FY
AVM = APS/80*FPU/FY*SP/(.8*H)*(.8*H/BW)^.5
LPRINT TAB(6);" Av min at S max = Aps/80*fpu/fy*S/0.8h*(0.8h/bw) ^0.5=";
    INT(AVM*100) /100;" sq. in."
LPRINT TAB(6);" If Vu > Vc then must design the shear reinforcing"
LPRINT TAB(6);" Av= (Vu-phi*Vc)*S/(phi*fy*.8h)"
LPRINT : LPRINT TAB(6);STRING$(50,42)
LPRINT : LPRINT TAB(10);"END OF PROGRAM"        END
```

**Figure 36.18** *(Continued)*

there is no plastic range for high-strength steel and therefore no true yield point. As a result, there is no balance point for a prestressed concrete interaction diagram. For strains beyond the proportional limit the value of the stress can only be determined from the stress-strain curve. The curve for each grade of steel must therefore be modeled in separate computer programs. The yield point is determined by the 0.2 percent offset method or by the 1 percent extension method. Copies of stress-strain curves for the specified grade can be obtained from the strand manufacturer or suppliers.

A typical stress-strain curve for $f_{pu}$-270-ksi low-relaxation ½ -in strand is shown in Fig. 36.19.

The area of the strand is

| Diameter, in | Area, in² |
|---|---|
| 3/8 | 0.085 |
| 7/16 | 0.115 |
| 1/2 | 0.153 |
| 0.6 | 0.217 |

Initial loss is usually assumed to be 7½ percent. Service loss ranges from 14 to 20 percent.

$$f_{psi} = 0.74 f_{pu}(100 - \text{loss})/100$$

The equation for $P_o$ is shown in Fig. 36.20.

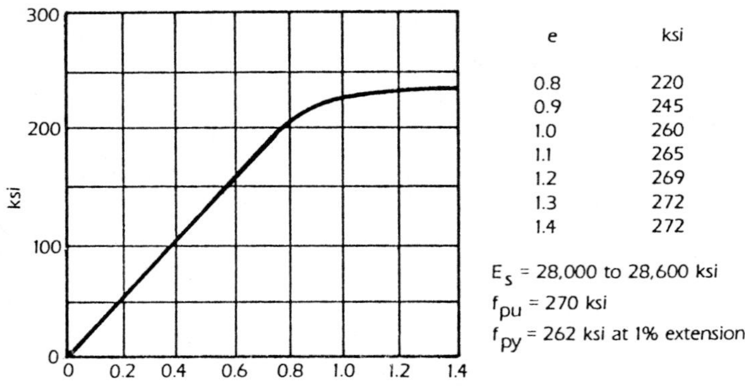


**Figure 36.19** Typical stress-strain curve: 270-ksi strand.

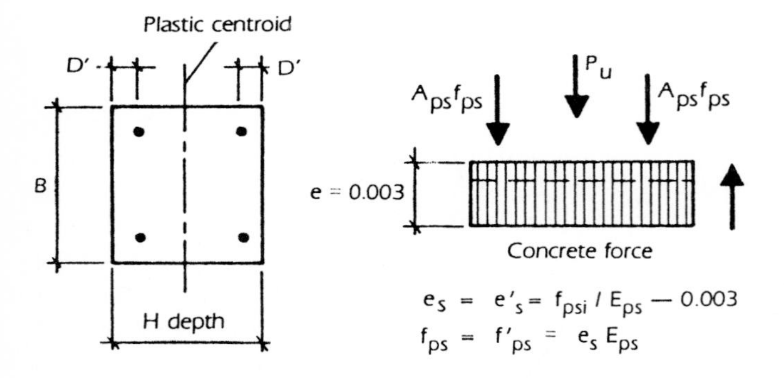


**Figure 36.20** Equation for $P_o$.

In this equation, $f_{psi}$ is the stress in the strand less the service loss.

$$P_o = \phi\left[0.85 f'_c\left(A_g - A_{ps} - A'_{ps}\right) - A'_{ps} f'_{ps} - A_{ps} f_{ps}\right]$$

$P_o$ is the maximum axial force which the column can support without considering any eccentric condition. However, there is no condition where the load is perfectly symmetrical, and the code therefore sets a maximum value to take care of minor eccentricities: $P_{u,\max} = 0.8P_o$ for tied columns and $0.85P_o$ for spiral columns. Since there is no moment balance or $P_u$ balance, these values do not exist.

The coordinates of the points on the interaction curve are computed on the basis of strain compatibility. Here the values of $C$ are inserted in the equations to solve for each point. Because of the repetition a loop (For-

Next) statement is used The values of $C$ range from a high limit of $H/\beta$ to a low limit$D'$ or larger. This step is a negative step of values of 1 in, 1/2 in or 1/4 in, according to the accuracy desired for the curve. The reason the high limit is set at $C/\beta$ is that the $a$ dimension of the stress block cannot exceed the column depth. (See Fig. 36.21.)

If $e_s > 0.009$, the stress-strain curve data must be used to find $f_{ps}$. To model the stress-strain curve "if" statements such as "If $e_s > 0.0012$, then $f_{ps} = 269 + [e_s - 0.0012(272 - 269)]$, " are used.

$$e'_s = e_{\text{base}} - 0.003(C - D')/C$$

$$f'_{ps} = e'_s E_{ps}$$

$e$ tension of concrete at extreme tension fiber $= 0.003(H - C)/C$

$$P_u = \phi(0.85f'_c B \beta C - A'_{ps} f'_{ps} - A_{ps} f_{ps})$$

where $\phi$ is 0.7 for compression.

$$M_u = \phi[0.85f'_c B \beta C (H/2 - \beta C/2) - A'_{ps} f'_{ps} (h/2 - D') + A_{ps} f_{ps} (h/2 - D')]$$

If $P_u < 0.10P_o$, for the moment calculation increase $\phi$ from 0.7 toward 0.9 linearly from $0.10P_o$ to $0.0P_o$. That is, if $P_u < 0.1P_o$,

$$M_u = M_u/0.7\{0.9 - 0.2[P_u/(0.1P_o)]\}$$

The interaction curve is plotted by using the coordinates $P_u$, $M_u$ for each value of $C$.

If the design condition of the $P_u$, $M_u$ (corrected for slenderness) coordinate fits within the curve, the column is adequate. If the coordinate falls outside the curve, either the column must be increased in size or the number of strands increased.

Computer program printout and program listing for the pretension concrete column interaction curve data are shown in Figs. 36.22 and 36.23.

### Program 7: Corbel Design (United States System)

Corbels are designed by using the shear friction concept when $A/D$ is equal to or less than 1.0 and $N_u < V_u$. This is the usual case for corbels placed on columns and wall panels. (See Fig. 36.24.)

Corbels must be designed to resist:
Shear $V_u$
Moment $V_u A + N_u(H - D)$
Tension $N_u$

Minimum tension $= 0.2N_u$
$\phi$ for all occasions $= 0.85$
$A_{vf} = V_u(\phi f_y \mu)$
$A_n = N_u(\phi f_y)$

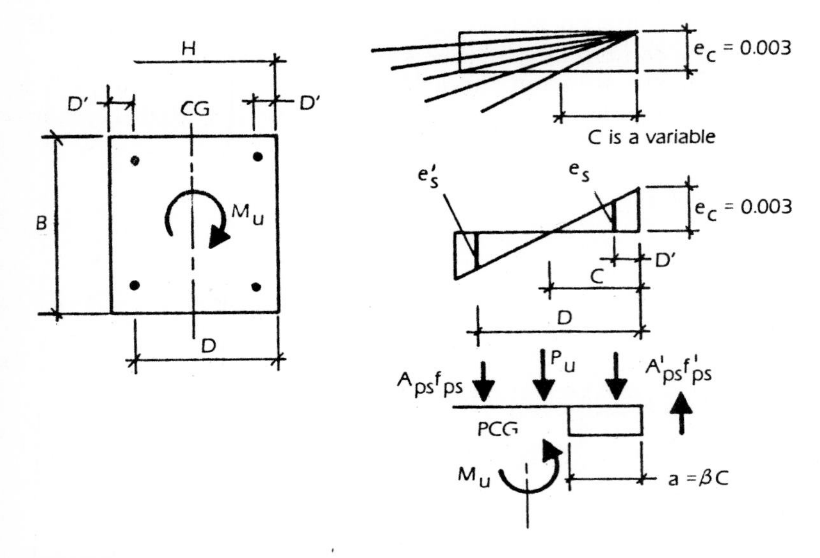


**Figure 36.21** Stress block with variable $C$.

$$e_{\text{base}} = f_{psi}/E_s = 0.74 \times 270(100 - 16)/100/\ 28{,}000 = 0.00599$$

$$e_s = e_{\text{base}} - 0.003(C - D)/C$$

$$f_{ps} = e_s E_s \text{ if } e_s < 0.009$$

## DESIGN INPUT

COLUMN DEPTH IN INCHES = 12
COLUMN WIDTH IN INCHES = 12
CONCRETE STRENGTH IN KSI = 5
BETA = .8
DISTANCE FROM EDGE OF COLUMN TO STRAND = 2.5 INCHES
NUMBER OF STRAND = 4
AREA PER STRAND = .153 SQ.IN.
TYPE OF STRAND = LR
STRAND STRENGTH IN KSI = 270
STRAND MODULUS OF ELASTICITY IN KSI = 28000
% OF LOSS OF PRESTRESS = 16
PU REQUIRED = 200 kips
MU REQUIRED = 320 inch kips

## DESIGN OUTPUT

PO= 390.6 KIPS P-MAX. = 312.5 KIPS
VALUES OF PU AND MU FOR VALUES OF C IN KIPS AND INCH KIPS

| C VALUES | PU | MU | e=MU/PU | es-tension | ec-conc. tension |
|---|---|---|---|---|---|
| 15 | 378 | 29.3 | .07 | .004894 | -.0006 |
| 14.5 | 363.3 | 113.2 | .31 | 4.959518E-03 | -5.172414E-04 |
| 14 | 348.5 | 191.4 | .54 | 5.029715E-03 | -4.285714E-04 |
| 13.5 | 333.6 | 263.9 | .79 | 5.105112E-03 | -3.333333E-04 |
| 13 | 318.7 | 330.9 | 1.03 | 5.186308E-03 | -2.307692E-04 |
| 12.5 | 303.8 | 392.2 | 1.29 | 5.274001E-03 | -.00012 |
| 12 | 288.8 | 447.9 | 1.55 | 5.369001E-03 | 0 |
| 11.5 | 273.7 | 498.1 | 1.81 | 5.472261E-03 | 1.304348E-04 |
| 11 | 258.6 | 542.7 | 2.09 | 5.58491E-03 | 2.727273E-04 |
| 10.5 | 243.4 | 581.7 | 2.38 | 5.708286E-03 | 4.285714E-04 |
| 10 | 228 | 615.2 | 2.69 | .005844 | .0006 |
| 9.5 | 212.6 | 643.3 | 3.02 | .005994 | 7.894737E-04 |
| 9 | 197.1 | 665.8 | 3.37 | 6.160667E-03 | 9.999999E-04 |
| 8.5 | 181.4 | 683 | 3.76 | 6.346941E-03 | 1.235294E-03 |
| 8 | 165.5 | 694.8 | 4.19 | 6.556501E-03 | .0015 |
| 7.5 | 149.4 | 701.3 | 4.69 | 6.794001E-03 | .0018 |
| 7 | 133.1 | 702.7 | 5.27 | 7.065429E-03 | 2.142857E-03 |
| 6.5 | 116.5 | 698.9 | 5.99 | 7.378616E-03 | 2.538461E-03 |
| 6 | 99.4 | 690.3 | 6.94 | .007744 | .003 |
| 5.5 | 82.5 | 674.7 | 8.17 | 8.175818E-03 | 3.545455E-03 |
| 5 | 65 | 654.7 | 10.07 | 8.694001E-03 | .0042 |
| 4.5 | 47.3 | 628.3 | 13.28 | 9.327333E-03 | .005 |
| 4 | 29.5 | 635.8 | 21.55 | .010119 | .006 |
| 3.5 | 12 | 660.5 | 55.04 | 1.113686E-02 | 7.285714E-03 |
| 3 | -5.2 | 659.3 | -126.79 | .012494 | 9.000001E-03 |
| es exceeds 0.014 IN/IN strain | | | | | |
| 2.5 | -22.4 | 628.4 | -28.06 | .014394 | .0114 |

PU REQUIRED = 200 KIPS.
MU REQUIRED = 320 INCH KIPS, OR 26.66667 FT. KIPS
FOR PU REQUIRED MU ALLOWABLE = 643.3 INCH KIPS
PU ALLOWABLE = 212.6 KIPS, MU/PU = 3.02587
C= 9.5 STRAIN IN THE COMPRESSION STEEL = 3.783474E-03 IN/IN

**Figure 36.22** Computer program printout: pretensioned concrete column interaction curve data.

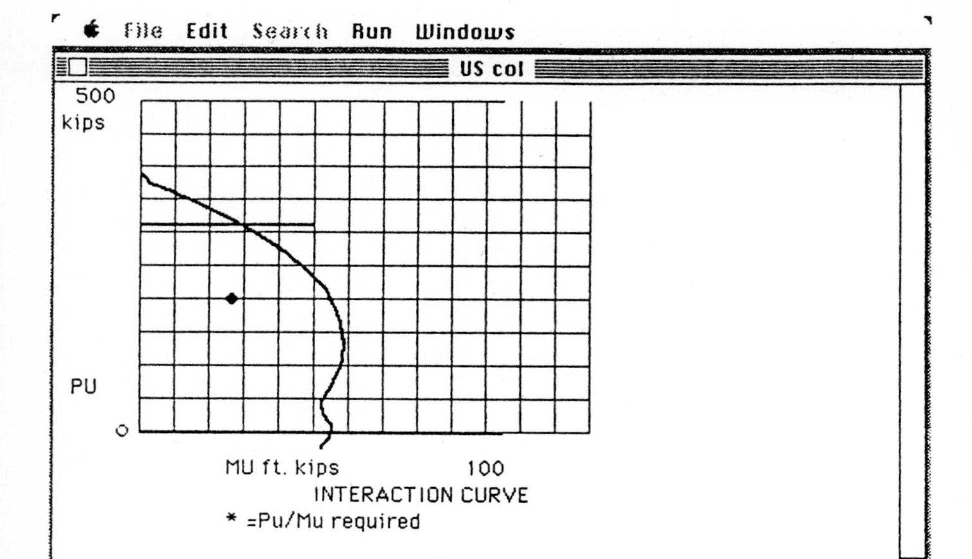


**Figure 36.22** (*Continued*)

```
LPRINT : LPRINT TAB(60);" Page  - 41": LPRINT : LPRINT
LPRINT : LPRINT TAB(6);STRING$(40,42):LPRINT
LPRINT TAB(11);"PRETENSIONED  COLUMN "
LPRINT TAB(10);"INTERACTION CURVE DATA" :LPRINT
LPRINT TAB(6);STRING$(40,42 )
DIM P(100), M(100)
100 INPUT " COLUMN DEPTH IN INCHES = ";L
INPUT " COLUMN WIDTH IN INCHES = ";T
INPUT " DISTANCE TO STRAND FROM EDGE OF COLUMN IN INCHES = ";D1
INPUT " CONCRETE STRENGTH IN KSI AND BETA = ";FC,BE
INPUT " NUMBER OF TOTAL STRANDS AND AREA PER STRAND in sq. in.= ";N,SA
AP=N/2*SA
INPUT " STRAND FPU IN KSI AND MOD. OF ELASTICITY in ksi= ";FPU,ES
132 INPUT " IF STRAND IS STRESS RELIEVE TYPE SR, IF LOW RELAX TYPE LR ? ";TS$
IF TS$="SR" THEN FI=.7 * FPU : GOTO 138
IF TS$ = "LR" THEN FI=.74*FPU : GOTO 138
GOTO 132
138 INPUT " SERVICE LOSS OF PRESTRESS IN PERCENT = ";PL
INPUT " INPUT YOUR REQUIRED PU IN KIPS = ";PR
INPUT "YOUR REQUIRED MOMENT IN IN. KIPS = ";MR
REM COMPUTE PO AND P MAX
EB=(FI*(100-PL)/100)/ES
E1=EB-.003 :F1=E1*ES :F2=F1
PO=.7*(.85*FC*(L*T-2*AP)-AP*(F1+F2))
PM = .8*PO
LPRINT CHR$(27);"n":LPRINT CHR$(14)
LPRINT TAB(4)" DESIGN  INPUT" : LPRINT CHR$(15) : LPRINT CHR$(27);"E"
LPRINT TAB(6);" COLUMN DEPTH IN INCHES =  ";L
LPRINT TAB(6);" COLUMN WIDTH IN INCHES =  ";T
LPRINT TAB(6);" CONCRETE STRENGTH IN KSI = ";FC
LPRINT TAB(6);" BETA = " ; BE
LPRINT TAB(6);" DISTANCE FROM EDGE OF COLUMN TO STRAND = ";D1; "  INCHES"
LPRINT TAB(6);" NUMBER OF STRAND = ";N
LPRINT TAB(6);" AREA PER STRAND = " ; SA; "  SQ.IN."
LPRINT TAB(6);" TYPE OF STRAND = ";TS$
LPRINT TAB(6);" STRAND STRENGTH IN KSI = ";FPU
LPRINT TAB(6);" STRAND MODULUS OF ELASTICITY IN KSI = ";ES
LPRINT TAB(6);" % OF LOSS OF PRESTRESS = "; PL
LPRINT TAB(6);" PU REQUIRED = ";PR;"kips
LPRINT TAB(6);" MU REQUIRED = ";MR;" inch kips
LPRINT CHR$(27);"n":LPRINT CHR$(14)
LPRINT TAB(4)"DESIGN   OUTPUT" : LPRINT CHR$(15) : LPRINT CHR$(27);"E"
LPRINT TAB(5); " PO=  "; INT(PO*10)/10;"  KIPS     P-MAX. = "; INT(PM * 10 )/10; " KIPS"
LPRINT TAB(5);"VALUES OF PU AND MU FOR VALUES OF C IN KIPS AND INCH KIPS "
LPRINT TAB([illegible]);"C
VALUES";TAB(17);"PU";TAB(24);"MU";TAB(30);"e=MU/PU";TAB(41);"es-tension";
    TAB(56);"ec-conc. tension
LPRINT TAB(4);STRING$(65,45) : LPRINT
H=INT (L/BE) : D=L-D1
Z=0 : MX=0
FOR C=H TO D1 STEP - .5
E1=EB-.003*(C-D1)/C : F1=E1*ES
E2= EB-.003*(C-D)/C :F2= E2*ES
IF E2> .008 AND E2< .009 THEN F2=222+22*(E2-.008)/.001
IF E2>.009 AND E2<.01 THEN F2=244 +  14*(E2-.009 )/.001
IF E2>.01 AND E2< .011 THEN F2=258+8*(E2-.01)/.001
IF E2>.011 AND E2< .012 THEN F2=266+3*(E2-.011)/.001
IF E2 > .012 THEN F2= 270
IF E2 > .014 THEN LPRINT "es exceeds 0.014 IN/IN strain"
```

**Figure 36.23** Computer program listing : pretensioned concrete column interaction curve data.

```
MU=.7*(.85*FC*T*BE*C*(L/2-BE*C/2)-AP*F1*(L/2-D1)+AP*F2*(L/2-D1))
IF PU < .1*PO THEN MU=MU/.7*(.9-.2*PU/(.1*PO))
ET = .003*(L-C)/C
PU= INT (PU*10)/10 : MU = INT (MU*10)/10
IF PU> PR THEN MA=MU : CD=C : ED=MU/PU : ECD=E1 : PA=PU
LPRINT TAB(7);C;TAB(15);PU;TAB(22);MU;TAB(30);INT(MU/PU*100)/100; TAB(40);
   E2;TAB(56);ET
P(Z)=PU : M(Z)=MU : Z=Z+1 : IF MX>MU THEN MX=MU
NEXT C
LPRINT TAB(4); STRING$(60,45)
LPRINT CHR$(12) : LPRINT : LPRINT TAB(60);" Page  - 42" : LPRINT :LPRINT
LPRINT TAB(6);"PU REQUIRED = ";PR;"  KIPS."
LPRINT TAB(6);"MU REQUIRED = ";MR;" INCH KIPS, OR ";MR/12;" FT. KIPS"
LPRINT TAB(6);"FOR PU REQUIRED MU ALLOWABLE =";MA;" INCH KIPS"
LPRINT TAB(6);"    PU ALLOWABLE =";PA;" KIPS,  MU/PU = "; ED
LPRINT TAB(6);" C=  "; CD; "   STRAIN IN THE COMPRESSION STEEL = "; ECD; "  IN/IN "
IF MA< MR THEN LPRINT TAB(6);" Column will not work, increase the number of strands and
    run again "
IF PR > PM THEN LPRINT TAB(6);" Column will not work, increase the column size and run
    again"
LPRINT
REM GRAPHIC ROUTINE
IF PO =< 500 THEN SP=2.5 : GOTO 600
IF PO =< 1000 THEN SP=5 : GOTO 600
IF PO=<1500 THEN SP=7.5 : GOTO 600
IF PO=< 2000 THEN SP=10 : GOTO 600
IF PO =< 2500 THEN SP=12.5 : GOTO 600
IF PO > 2500 THEN GOTO 1000
600
IF MX/12=< 130 THEN SM=.5 : GOTO 700
IF MX/12=< 260 THEN SM=1 : GOTO 700
IF MX/12=<390 THEN SM=1.5: GOTO 700
IF MX/12 =< 520 THEN SM=2 : GOTO 700
IF MX/12 =< 650 THEN SM=2.5 : GOTO 700
CLS
LINE (50,10)-(50,210) : LINE (50,210)-(260,211),,BF
LINE (50,10)-(260,10)
FOR I=1 TO 10 : LINE (50,10+I*20)-(310,10+I*20) : NEXT I
FOR J=1 TO 13 : LINE (50+J*20,10)-(50+J*20,210) : NEXT J
CIRCLE ( 40,210),3
PRINT PTAB(4); 200*SP
PRINT PTAB(4);"kips"
FOR K=1 TO 9 : PRINT : NEXT K
PRINT PTAB(10);"PU"
PRINT : PRINT
PRINT PTAB(100);"MU ft. kips"; PTAB(230);200*SM
PRINT PTAB(150);"INTERACTION CURVE"
PRINT PTAB (100) "* =Pu/Mu required"
CALL PENSIZE (2,2)
CALL MOVETO ( 50,210-INT (PO/SP))
FOR I=0 TO Z-1 : CALL LINETO (50+INT(M(I)/12/SM),210-INT(P(I)/SP)) : NEXT I
FOR A =1 TO 3
CIRCLE ( 50+INT(MR/12/SM), 210-INT(PR/SP)),A : NEXT A
LINE (50,210-INT(PM/SP))-(150,210-INT(PM/SP)+1),,BF
LCOPY
1000
PRINT " END OF THE RUN, IF YOU WISH TO RERUN TYPE YES ELSE TYPE NO"
INPUT " YOUR REPLY ";REP$
IF REP$="YES" THEN GOTO 100
END
```

**Figure 36.23** *(Continued)*

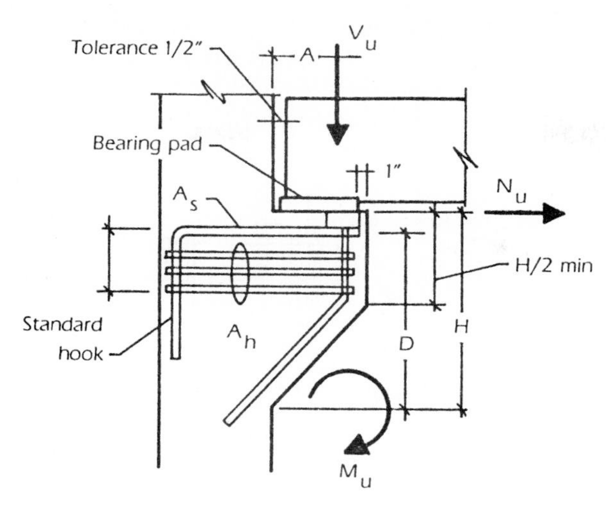


**Figure 36.24** Corbel.

The length of the bearing pad $L_{bp}$ is $V_u$/ (allowable unit bearing for pad material× smaller of corbel width or beam width). Some designers will use the corbel width less 1-in tolerance versus the beam width. This will allow for tolerance in case the beam slips to one side.

$$A = 2/3L_{bp} + 1 \text{ in to allow for } 2 \times 1/2\text{-in tolerance}$$

Some designers use $A = 3/4 \times$ length of corbel, thus allowing for erection and creep. The minimum length of the corbel is $L_{bp}$ + 1-in seating tolerance (2 × tolerance) + 1 to 1 1/2 in for distance from the end of the bearing pad to the end of the corbel. This will reduce the probability of a tension failure at the end of the corbel.

For the shear friction $A_{vf} = V_u / \phi f_y \mu$, $\mu = 1.4$ for the corbel placed monolithically with the column. For shear of the concrete $V_u < V_c$, $\phi V_c = \phi \times$ the smaller of 800 lb/in² or $0.2f'_c \times$ the width of the corbel $\times D$. For sand-lightweight concrete multiply $\phi V_c$ by 0.85.

$A_s$ is the larger of

$$A_s = A_m + A_n = [V_u A + N_u (H - D)]/\phi f_y D + N_u / \phi f_y$$

$$= (V_u A + N_u H)/\phi f_y D$$

or $$A_s = 2/3A_{vf} + A_n = 2/3V_u/\phi f_y \mu + N_u / \phi f_y$$

$$A_{s,\min} = 0.04 f'_c / f_y B_{\text{corbel}} D$$

If the corbel is sloped, then $D$ is to the interface of the corbel to the column. This is due to the fact that the compression strut is at an angle.

$$A_h = 0.5(A_s - A_n)$$

The bearing area (Fig. 36.25) must not extend beyond the end of $A_s$ or the interior face of the transverse bar or the edge of a transverse plate (if used). All welded reinforcing bars are ASTM A-706 steel.
Computer program printout and program listing for the corbel design are shown in Figs. 36.26 and 36.27.

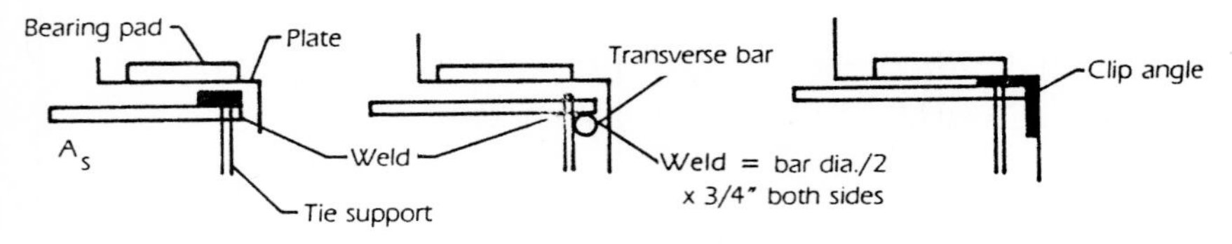


**Figure 36.25** Corbel bearing area.

```
DESIGN    INPUT

Column width = 12 inches
Column depth = 12 inches
Concrete strength f'c = 5 ksi
Normal weight concrete
Corbel steel strength fy = 40  ksi
Bearing pad strength =  1  ksi
Beam width = 14  INCHES
Vu=  80  kips
Nu =  16 kips
a/d ratio = .6

DESIGN    SOLUTION

Corbel height =  11.25  INCHES
effective depth d =  10  INCHES
Ratio A/D = .6
Beam width =  14 Column width= 12 Corbel width= 12 inches
Bearing pad width = 12  Bearing pad length =  7  inches
Vu=  80  Nu= 16  kips, a=  6
As=  1.94  sq.in., use 2-#9)
Ah=  .73  sq.in. , use 4-#3 HOOPS
Ah min =  .59  sq. in.

END OF RUN
```

**Figure 36.26** Computer program printout: corbel design.

```
LPRINT : LPRINT TAB(60);"Page - 46" : LPRINT : LPRINT
LPRINT TAB(10);STRING$(50,42) : LPRINT
LPRINT TAB(20);" CORBEL DESIGN - U.S.A. system" : LPRINT
LPRINT TAB(10);STRING$(50,42)
LPRINT CHR$(27);"n" : LPRINT CHR$(14) : LPRINT TAB(6);"DESIGN INPUT "
LPRINT CHR$(15) : LPRINT CHR$(27);"N"
10
INPUT "Input column width in inches=",CW : LPRINT TAB(10);"Column width =";CW;"inches"
INPUT "Input column depth in inches=",CT : LPRINT TAB(10);"Column depth =";CT;"inches"
INPUT "Input concrete strength f'c in ksi=",FC : LPRINT TAB(10);"Concrete strength f'c
      =";FC;"ksi"
15
INPUT " Type SLW if concrete is sand lt.wt else type NW for normal wt =",T$
IF T$="SLW" THEN GA=.85 : GOTO 20
IF T$="NW" THEN GA=1! : GOTO 20
PRINT "TRY AGAIN " : GOTO 15
20
IF T$="SLW" THEN LPRINT TAB(10);"Semi-light wt concrete" ELSE LPRINT TAB(10);"Normal
      weight concrete"
INPUT "Input corbel steel strength fy in ksi =",FY : LPRINT TAB(10);"Corbel steel strength fy
      =";FY;" ksi"
INPUT "Input bearing pad compressive strength in ksi =",FBP: LPRINT TAB(10);"Bearing pad
      strength = ";FBP;" ksi"
INPUT "Input beam width in inches =",BW: LPRINT TAB(10);"Beam width =";BW;" INCHES"
INPUT " Input Vu in kips=",VU : LPRINT TAB(10);"Vu= ";VU;" kips"
INPUT "Input Nu if known otherwise type 0.0 =",NU
IF NU < .2 * VU THEN NU = .2*VU : LPRINT TAB(10);"Nu = ";NU;"kips"
INPUT "Input your optimum a/d ratio else input 0.5=",R : LPRINT TAB(10);"a/d ratio =";R
B=BW : IF B>CW THEN B=CW
LP=VU/(FBP*B) : LP=INT(LP)+1:A=2/3*LP+1:LC=LP+2.5 : A=INT(A)+1
FY=.2*FC : IF FY > .8 THEN FY=.8
D=VU/(FY*.85*GA*CW) : IF D<A/R THEN D=A/R
IF D>INT(D) THEN D=INT(D)+1
30
H=D+1.25
S1=(VU*A+NU*(H-D))/(.85*FY*D)+NU/(.85*FY)
S2=2/3*VU/(.85*1.4*GA*FY)+NU/(.85*FY)
SA=S1 : IF SA<S2 THEN SA=S2
SA= INT (SA*100)/100
AM=.04*FC/FY*CW*D : AN=4/3*SA : IF AM>AN THEN AM=AN
AM=INT(AM*100)/100
IF AM>SA THEN PRINT " AS MIN. GOVERNS " : SA=AM
IF SA<2*.44 THEN SA$="2-#6" : GOTO 50
IF SA<2*.6 THEN SA$="2-#7" : GOTO 50
IF SA<2*.79 THEN SA$="2-#8" : GOTO 50
IF SA<2*1 THEN SA$="2-#9)":GOTO 50
IF SA< 3*.44 THEN SA$="2#6" : GOTO 50
IF SA<3*.6 THEN SA$="3-#7" : GOTO 50
IF SA<3*.79 THEN SA$="3-#8" : GOTO 50
IF SA<3*1 THEN SA$="3-#9" : GOTO 50
IF SA<3*1.27 THEN SA$="3-#10" : GOTO 50
LPRINT TAB(10);" YOUR D IS TOO SMALL MUST INCREASE BY 2 INCHES "
D=D+2 : GOTO 30
50
AH=.5*(SA-NU/(.85*FY)) : AH= INT(AH*100)/100
IF AH <4*.11 THEN AH$="2-#3 HOOPS " :GOTO 60
IF AH < 6*.11 THEN AH$="3-#3 HOOPS" : GOTO 60
IF AH< 8 * .11 THEN AH$="4-#3 HOOPS" : GOTO 60
IF AH,10*.11 THEN AH$="5-#3 HOOPS": GOTO 60
LPRINT TAB(10);" YOUR AH IS TOO LARGE, INCREASE D BY 2 INCHES " : D=D+2 : GOTO 30
IF LC> INT (LC) THEN LC=INT(LC) + 1
```

**Figure 36.27** Computer program listing: corbel design.

```
60
LP=INT(LP*10)/10 : A= INT(A*10)/10
LPRINT CHR$(27);"n": LPRINT CHR$(14): LPRINT TAB(6);" DESIGN SOLUTION "
LPRINT CHR$(15) : LPRINT CHR$(27);"N"
LPRINT TAB(10);"Corbel height = ";H;" INCHES
LPRINT TAB(10);"effective depth d = ";D;" INCHES
LPRINT TAB(10) ; "Ratio A/D =";INT (A/D*100)/100
LPRINT TAB(10);"Beam width = ";BW;"Column width=";CW;"Corbel width=";B;"inches"
LPRINT TAB(10);"Bearing pad width =";B;" Bearing pad length = ";LP;" inches"
LPRINT TAB(10);"Vu= ";VU;" Nu=";NU;" kips, a= ";A
LPRINT TAB(10);"As= ";SA;" sq.in., use ";SA$
LPRINT TAB(10);"Ah= ";AH;" sq.in. , use ";AH$
LPRINT TAB(10);"Ah min = ";AM;" sq. in."
LPRINT
INPUT " If you wish to increase d by 2 inches type Y else type N",A$
IF A$ = "Y" THEN D=D+2 : GOTO 30
INPUT " If you wish to run program again typt Y else type N ",B$
IF B$ = "Y" THEN GOTO 10
LPRINT
LPRINT TAB(10) ;"END OF RUN"
END
```

**Figure 36.27** (*Continued*)

# Index

## ABOUT THE AUTHORS

DAVID A. SHEPPARD is president of his own structural engineering firm in Sonora, California. He has previously held various management positions with several major precast concrete manufacturing firms. He was executive director of the Prestressed Concrete Manufacturers Association of California from 1977 to 1982.

WILLIAM R. PHILLIPS is professor of architectural engineering at California Polytechnic State University at San Luis Obispo.